D0215727

Digital Signal Processing

A Practical Approach

ELECTRONIC SYSTEMS ENGINEERING SERIES

Consulting editors

E L Dagless
University of Bristol

J O'Reilly
University College North Wales, Bangor

OTHER TITLES IN THE SERIES

Advanced Microprocessor Architectures *L Ciminiera and A Valenzano*

Optical Pattern Recognition Using Holographic Techniques *N Collings*

Modern Logic Design *D Green*

Data Communications, Computer Networks and Open Systems (3rd Edn) *F Halsall*

Multivariable Feedback Design *J M Maciejowski*

Microwave Components and Systems *K F Sander*

Tolerance Design of Electronic Circuits *R Spence and R Soin*

Computer Architecture and Design *A J van de Goor*

Digital Systems Design with Programmable Logic *M Bolton*

Introduction to Robotics *P J McKerrow*

MAP and TOP Communications: Standards and Applications *A Valenzano, C Demartini and L Ciminiera*

Integrated Broadband Networks *R Händel and M N Huber*

The Board Designer's Guide to Testable Logic Circuits *C Maunder*

Electronics: A Systems Approach *N Storey*

Electric Circuit Analysis: Principles and Applications *K F Sander*

Telecommunication Networks and Services *J van Duuren, P Kastelein and F C Schoute*

Digital Signal Processing

A Practical Approach

Emmanuel C. Ifeachor
University of Plymouth

Barrie W. Jervis
Sheffield Hallam University

ADDISON-WESLEY PUBLISHING COMPANY

Wokingham, England • Reading, Massachusetts • Menlo Park, California • New York
Don Mills, Ontario • Amsterdam • Bonn • Sydney • Singapore
Tokyo • Madrid • San Juan • Milan • Paris • Mexico City • Seoul • Taipei

The programs in this book have been included for their instructional value. They have been tested with care but are not guaranteed for any particular purpose. The publisher does not offer any warranties or representations, nor does it accept any liabilities with respect to the programs.

Many of the designations used by manufacturers and sellers to distinguish their products are claimed as trademarks. Addison-Wesley has made every attempt to supply trademark information about manufacturers and their products mentioned in this book. A list of the trademark designations and their owners appears on p. xxiv.

Cover designed by Chris Eley and
printed by The Riverside Printing Co. (Reading) Ltd.
Typeset by Keytec Typesetting Ltd., Bridport, Dorset.
Printed in Great Britain by William Clowes, Beccles, Suffolk.

First printed 1993. Reprinted 1993.

British Library Cataloguing in Publication Data
A catalogue record for this book is available from The British Library.

Library of Congress Cataloging in Publication Data
Ifeachor, Emmanuel C.
Digital signal processing : a practical approach / Emmanuel C.
Ifeachor, Barrie W. Jervis.
p. cm. – (Electronic systems engineering series)
Includes bibliographical references and index.
ISBN 0 201 54413 X
1. Signal processing–Digital techniques. 2. Digital filters
(Mathematics) 3. Adaptive signal processing. I. Jervis, Barrie W.
II. Title. III. Series.
TK5102.I33 1993
621.382′2–dc20

93-15001
CIP

Preface

Purpose of this book

This book was born out of our experience in teaching practically oriented courses in digital signal processing (DSP) to undergraduate students at the University of Plymouth and the Sheffield Hallam University, and to application engineers in industry for many years. It appeared to us that many of the available textbooks were either too elementary or too theoretical to be of practical use for undergraduates or application engineers in industry. As most readers will know from experience, the gap between learning the fundamentals in any subject and actually applying them is quite wide. We therefore decided to write this book which we believe undergraduates will understand and appreciate and which will equip them to undertake practical digital signal processing assignments and projects. We also believe that higher degree students and practising engineers and scientists will find this text most useful.

Our own research work over the last two decades in applied DSP has also inspired the contents, by identifying practical issues for discussion and presentation to bridge the gap between theoretical concepts and practical implementation, and by suggesting application examples, case studies, and problems.

The current great interest and developments in DSP both in industry and academia are likely to continue for the foreseeable future. The availability of numerous digital signal processors recognizes the commercial potential of DSP. Its major attraction lies in the ability to achieve guaranteed accuracy and perfect reproducibility, and in its inherent flexibility compared with analogue signal processing. In industry, many engineers lack the necessary knowledge and expertise in DSP to utilize the immense potential of the very powerful digital signal processors now available off the shelf. This book provides insight and practical guidance to enable engineers to design and develop practical DSP systems using these devices.

In academia, DSP is generally regarded as one of the more mathematical topics in the electrical engineering curriculum, and based on our experiences of teaching we have reduced the mathematical content to what we consider useful, essential, and interesting; we have also emphasized points of difficulty. Our experiences indicate that students learn best if they are aware of the practical relevance of a subject, and while more theoretical texts are essential for completeness and reference as the student matures in the subject, we believe in producing graduates equipped also with practical knowledge and skills. This book was written with these considerations in mind.

The book is not a comprehensive text on DSP, but it covers most aspects of the subject found in undergraduate electrical, electronic or communication engineering degree courses. A number of DSP techniques which are of particular relevance to industry are also covered and in a few years, we believe, these will find their way into undergraduate curricula. These include techniques such as adaptive filtering and multirate processing.

The emphasis throughout the book is on the practical aspects of DSP. C language programs are provided to enable readers to explore the concepts presented in the book, to design and analyse their own DSP systems, and to gain a deeper understanding of DSP.

Commercial DSP software is available and plays a key role in the design and analysis of DSP systems. However, with most commercial software, it is difficult actually to find out or follow how a given operation is performed. The programs given in this book are useful in verifying results obtained manually. For example, by inserting break points the user can check the intermediate results of computation and follow the way the computation is made. Having acquired a sound knowledge of the principles, commercial software packages with good graphics support and user-friendly interfaces may then be used in most designs.

Main features of the book

- Provides an understanding of the fundamentals, implementation, and applications of DSP techniques from a practical point of view.
- Clear and easy to read, with mathematical contents reduced to that which is necessary for comprehension.
- DSP techniques and concepts are illustrated with practically oriented worked examples.
- Provides practical guidance to enable readers to design and develop actual DSP systems. Complete design examples and practical implementation details are given, including assembly language programs for the TMS320C10 and C25 processors.
- Provides C language implementation of many DSP algorithms and functions including programs for
  - digital FIR and IIR filter design,
  - finite wordlength effect analysis of user-designed IIR filters,
  - converting from cascade to parallel realization structures,
  - correlation computation,
  - discrete and fast Fourier transform algorithms,
  - inverse z-transformation,

  - frequency response estimation, and
  - multirate processing systems design.

- PC-based C programs are available on a computer disk to encourage readers to participate more actively in the learning process (see the section 'How to obtain the program disk for this book' in this preface for details).
- Contains many real-world application examples.
- Contains many end-of-chapter problems.
- Use of realistic examples to illustrate important concepts and to reinforce the knowledge gained.

The intended audience

The book is aimed at engineering, science and computer science students, and application engineers and scientists in industry who wish to gain a working knowledge of DSP. In particular, final year students studying for a degree in electronics, electrical or communication engineering will find the book valuable for both taught courses as well as their project work, as increasingly a greater proportion of student project work involves aspects of DSP. Postgraduates studying for a master's degree or PhD in the above subjects will also find the book useful.

Undergraduate students will find the fundamental topics very attractive and, we believe, the book will be a valuable source of information both throughout their course as well as when they go into industry.

Large commercial or government organizations who undertake their own internal DSP short courses could base them on the book. We believe the book will serve as a good teaching text as well as a valuable self-learning text for undergraduate, graduate and application engineers.

Contents and organization

Chapter 1 contains an overview of DSP and its applications to make the reader aware of the meaning of DSP and its importance. The chapter presents, from a practical point of view using real-world examples, many fundamental topics which form the cornerstone of DSP, such as sampling and quantization of signals and their implications in real-time DSP. Discrete-time signals and systems are introduced in this chapter, and discussed further in Chapter 3.

Discrete transforms, particularly the discrete and fast Fourier transforms (FFT), provide important mathematical tools in DSP as well as relating the time and frequency domains. They are introduced and described in Chapter 2 with a discussion of some applications to put them in context. The derivation of the discrete Fourier transform (DFT) from the Fourier transform and the exponential Fourier series provides a logical justification for the DFT which does not require coverage of the discrete Fourier series which would unnecessarily increase the length of the book (and the amount of work for the student!). The discussion has also been restricted to the description and implementation of the transforms. In particular, the topic of windowing has not been included in this chapter but is more appropriately discussed in detail in Chapter 10 on spectrum analysis.

In Chapter 3 the basics of discrete-time signals and systems are discussed. Important aspects of the z-transform, an invaluable tool for representing and analysing discrete-time signals and systems, are discussed. Many applications of the z-transform are highlighted, for example its use in the design, analysis and computation of the frequency response of discrete-time signals and systems. As in the rest of the book, the concepts as well as applications of the z-transform are illustrated with fully worked examples.

Correlation and convolution are fundamental and closely related topics in DSP and are covered in depth in Chapter 4. The authors consider an awareness of all the contents of this chapter to be essential for DSP, but after a preliminary scanning of the contents the reader may well be advised to build up his or her detailed knowledge by progressing through the chapter in stages. The contents might well be spread over several years of an undergraduate course.

Chapters 5, 6 and 7 include detailed practical discussions of digital filter design, one of the most important topics in DSP, being at the core of most DSP systems. Digital filter design is a vast topic and those new to it can find this somewhat overwhelming. Chapter 5 provides a general framework for filter design. A simple but general step-by-step guide for designing digital filters is given.

Techniques for designing FIR (finite impulse response) filters from specifications through to filter implementations are discussed in Chapter 6. Several fully worked examples are given throughout the chapter to consolidate the important concepts. A complete filter design example is included to show how all the stages of filter design fit together.

IIR (infinite impulse response) filter design is discussed in detail in Chapter 7, based on the simple step-by-step guide. The effect of quantization on filter performance in its various forms is discussed. In particular, ADC (analogue-to-digital) noise, coefficient quantization errors and their effects on frequency response and stability, arithmetic roundoff errors and overflow and procedures to minimize the effects of these errors are discussed.

Multirate processing techniques allow data to be processed at more than one sampling rate and have made possible such novel applications as single-bit ADCs and DACs (digital-to-analogue converters), and oversampled digital filtering, which are exploited in a number of modern digital systems, including for example the familiar compact disc player. In Chapter 8, the basic concepts of multirate processing are explained, illustrated with fully worked examples and by the design of actual multirate systems.

In Chapter 9, key aspects of adaptive filters are described, based on the LMS (least-mean-squares) and RLS (recursive least-square) algorithms which are two of the most widely used algorithms in adaptive signal processing. The treatment is practical with only the essential theory included in the main text.

In Chapter 10, the important topic of spectrum estimation and analysis used to describe and study signals in the frequency domain is described with the emphasis being on the traditional nonparametric approach. Readers who are particularly interested in spectral analysis should study both Chapters 10 and 2 as Chapter 10 draws on explanations and worked examples given in Chapter 2. Those who master the contents of these chapters will be well placed to become competent in the analysis of signals in the frequency domain. However, it is worth noting that while the chapter introduces nonparametric methods in depth, the more modern parametric methods are expected to become more important in the future. These methods have not been covered in detail and the reader is referred to the given references.

In the last decade, tremendous progress has been made in DSP hardware, and this has led to the wide availability of low cost digital signal processors. For a successful application of DSP using these processors, it is necessary to appreciate the underlying concepts of DSP hardware and software. Chapter 11 discusses the key issues underlying general- and special-purpose processors for DSP, the impact of DSP algorithms on the hardware and software architectures of these processors, and the architectural requirements for efficient execution of DSP functions. We have used the Texas Instruments devices, TMS320C10, TMS320C25 and TMS320C30, as well as other well-known processors, to illustrate specific points, where possible.

In Chapter 12, we describe the hardware development environment used to implement some of the DSP algorithms described in previous chapters. Two low cost target boards based on the Texas Instruments TMS320C10 and TMS320C25 processors are described. A number of applications of DSP are described in the form of case studies. The presentation draws on many concepts discussed in earlier chapters.

How to use the book

A useful approach for undergraduate teaching will be to cover the materials in Chapter 1, to provide the understanding of fundamental topics such as the sampling theorem and discrete-time signals and systems, and to establish the benefits and applications of DSP. Then discrete transforms should be introduced, starting with the DFT and FFT (Chapter 2), and the z-transform (Chapter 3). Aspects of Chapters 10 and 4 may be used to illustrate the application of the DFT and FFT. After an introduction to correlation processing using a selection of materials from Chapter 4, a detailed treatment of digital filters should be undertaken.

In our experience students learn more when they are given realistic assignments to carry out. To this end we would encourage substantial assignments on, for example, filter design, the inverse z-transform, the DFT and FFT. Laboratory work should also be designed to demonstrate and reinforce the techniques taught. It is important that students actually participate as well as attend lectures.

For postgraduate students the approach could be the same but the pace will be more brisk, and the more specialist topics of multirate processing and adaptive filters will also be included.

How to obtain the program disk for this book

The C-language programs and TMS320C10/25 assembly language programs described in this book are available on a disk for the IBM PC (or compatibles), as 3 1/2″ high density (1.44 Mbyte). The C-language programs are available in both executable form and as source codes. A C compiler is required to run the source codes, but not to run the executable codes. The programs are written in standard ANSI C under Borland Turbo C version 2.0. The TMS320C25 codes in the disk require the Texas Instrument SWDS (software development system) package to run. The TMS320C10 codes run on the target board described in Chapter 12; they will require minor modifications to run on other systems.

The cost per disk is £27.50 or $50 (including sales tax). This price also includes illustrative examples of how to use the C-language programs.

To obtain the program disk, please send your order together with a cheque drawn on a UK bank, bankers draft or proof of direct bank transfer to:

Digital Signal Processing: A Practical Approach
PEP Research & Consultancy Limited
Charles Cross Centre
Constantine Street
Plymouth
Devon PL4 8DE
UK

Payments should be in favour of 'PEP Research & Consultancy Ltd'.

Bank account details are:

Midland Bank plc
City Centre Branch
4 Old Town Street
Plymouth
Devon PL1 1DD
UK

Sort code: 40-36-22

Account number:
Sterling (£) 41537105
US dollars ($) 35454528

A valid VAT invoice will be issued on receipt of payment.

Defective disks will be replaced if returned within a reasonable time, but refunds are unfortunately not possible.

Acknowledgements

We are fortunate to have received many useful comments and suggestions from many of our present and past students which have improved the technical content and clarity of the book. We are grateful to all of them, but especially to Eddie Riddington, Robin Clark, Ian Scholey, François Amand, Nichola Gater, Robert Ruse, and Andrew Paulley.

The authors would like to thank Mr Mike Fraser, until recently a technical member of staff of the University of Plymouth, and formerly a Chief Engineer with Rank Toshiba, Plymouth. His considerable experience and valuable comments have been most useful. We would also like to thank him for developing and constructing the TMS320C10 target board from our initial design, and for developing the environment in which many programs were implemented and tested. Paul Smithson is to be thanked for doing a very good job on the TMS320C25 target board. He developed and constructed the hardware for the system from our initial design, with contributions from Mike Fraser. We acknowledge the comments and assistance from many other colleagues, especially Mr Peter Van Et Velt for deriving the mathematical formulae in Appendices 7B and 7C.

The practical nature of the book made it difficult to keep to deadlines. Each chapter took much longer to write than we had imagined or planned for. We thank the acquisition editor, Tim Pitts, for his patience and encouragement.

Finally, the authors are especially grateful to their families for their tolerance, patience and support throughout this very time-consuming project.

Emmanuel Ifeachor
Barrie Jervis
April 1993

Contents

Preface *v*

Chapter 1
Introduction **1**

1.1 Digital signal processing and its benefits 1
1.2 Application areas 3
1.3 Key DSP operations 4
1.3.1 Convolution 6
1.3.2 Correlation 6
1.3.3 Digital filtering 8
1.3.4 Discrete transformation 8
1.3.5 Modulation 10
1.4 Overview of real-time signal processing 13
1.4.1 Typical real-time DSP systems 13
1.4.2 Analogue-to-digital conversion process 14
1.4.3 Digital-to-analogue conversion process: signal recovery 29
1.4.4 Digital signal processors 33
1.4.5 Constraints of real-time signal processing with analogue input/output signals 34
1.5 Application examples 35
1.5.1 Speech synthesis and recognition 35
1.5.2 Adaptive telephone echo cancellation 37
1.5.3 The compact disc digital audio system 39
1.6 Summary 43
Problems 43
References 46
Bibliography 46

Chapter 2
Discrete transforms **47**

2.1 Introduction 47
2.1.1 Fourier series 49
2.1.2 The Fourier transform 52

2.2 DFT and its inverse 55
2.3 Properties of the DFT 62
2.4 Computational complexity of the DFT 64
2.5 The decimation-in-time fast Fourier transform algorithm 65
2.5.1 The butterfly 70
2.5.2 Algorithmic development 72
2.5.3 Computational advantages of the FFT 76
2.6 Inverse fast Fourier transform 76
2.7 Implementation of the FFT 77
2.7.1 The decimation-in-frequency FFT 78
2.7.2 Comparison of DIT and DIF algorithms 78
2.7.3 Modifications for increased speed 78
2.8 Other discrete transforms 79
2.8.1 Discrete cosine transform 79
2.8.2 Walsh transform 80
2.8.3 Hadamard transform 84
2.9 Worked examples 86
Problems 90
References 92
Appendices 93
2A C language program for direct DFT computation 93
2B C program for radix-2 decimation-in-time FFT 99
References for appendices 102

Chapter 3
The *z*-transform and its applications in signal processing **103**

3.1 Discrete-time signals and systems 104
3.2 The *z*-transform 105
3.3 The inverse *z*-tranform 109
3.3.1 Power series method 111
3.3.2 Partial fraction expansion method 114
3.3.3 Residue method 121
3.3.4 Comparison of the inverse *z*-transform methods 127
3.4 Properties of the *z*-transform 127
3.5 Some applications of the *z*-transform in signal processing 130
3.5.1 Pole–zero description of discrete-time systems 130
3.5.2 Frequency response estimation 134
3.5.3 Geometric evaluation of frequency response 134
3.5.4 Direct computer evaluation of frequency response 138
3.5.5 Frequency response estimation via the FFT 139
3.5.6 Frequency units used in discrete-time systems 139
3.5.7 Stability considerations 142
3.5.8 Difference equations 143

3.5.9 Impulse response estimation 144
3.5.10 Applications to digital filter design 147
3.5.11 Realization structures for digital filters 147
3.6 Summary 152
Problems 152
References 156
Bibliography 157
Appendices 157
3A Recursive algorithm for the inverse z-transform 157
3B C program for evaluating the inverse z-transform and for cascade-to-parallel structure conversion 159
3C C program for estimating frequency response 180
Reference for appendices 182

Chapter 4

Correlation and convolution **183**

4.1 Introduction 184
4.2 Correlation description 184
4.2.1 Cross- and autocorrelation 191
4.2.2 Applications of correlation 199
4.2.3 Fast correlation 207
4.3 Convolution description 213
4.3.1 Properties of convolution 221
4.3.2 Circular convolution 222
4.3.3 Fast linear convolution 222
4.3.4 Computational advantages of fast linear convolution 223
4.3.5 Convolution and correlation by sectioning 224
4.3.6 Overlap–add method 227
4.3.7 Overlap–save method 232
4.3.8 Computational advantages of fast convolution by sectioning 235
4.3.9 The relationship between convolution and correlation 235
4.4 Implementation of correlation and convolution 236
4.5 Application examples 237
4.5.1 Correlation 237
4.5.2 Convolution 242
4.6 Summary 246
Problems 246
References 249
Appendix 250
4A C language program for computing auto- and cross-correlation 250

Chapter 5
A framework for digital filter design **251**

5.1 Introduction to digital filters 252
5.2 Types of digital filters: FIR and IIR filters 254
5.3 Choosing between FIR and IIR filters 255
5.4 Filter design steps 258
5.4.1 Specification of the filter requirements 259
5.4.2 Coefficient calculation 261
5.4.3 Representation of a filter by a suitable structure (realization) 262
5.4.4 Analysis of finite wordlength effects 266
5.4.5 Implementation of a filter 268
5.5 Illustrative examples 269
5.6 Summary 274
Problems 274
Reference 276
Bibliography 276

Chapter 6
Finite impulse response (FIR) filter design **278**

6.1 Introduction 279
6.1.1 Summary of key characteristic features of FIR filters 279
6.1.2 Linear phase response and its implications 280
6.1.3 Types of linear phase FIR filters 283
6.2 FIR filter design 285
6.3 FIR filter specifications 285
6.4 FIR coefficient calculation methods 288
6.5 Window method 288
6.5.1 Some common window functions 291
6.5.2 Summary of the window method of calculating FIR filter coefficients 295
6.5.3 Advantages and disadvantages of the window method 303
6.6 The optimal method 303
6.6.1 Basic concepts 304
6.6.2 Parameters required to use the optimal program 307
6.6.3 Relationships for estimating filter length, N 308
6.6.4 Summary of procedure for calculating filter coefficients by the optimal method 309
6.6.5 Illustrative examples 310
6.7 Frequency sampling method 317
6.7.1 Nonrecursive frequency sampling filters 317
6.7.2 Recursive frequency sampling filters 326

6.7.3 Frequency sampling filters with simple coefficients 328
6.7.4 Summary of the frequency sampling method 335
6.7.5 Comparison of the window, optimum and frequency sampling methods 336
6.7.6 Special filters and transformations for FIR filters 339
6.7.7 Other FIR coefficient calculation methods 342

6.8 Realization structures for FIR filters 344
6.8.1 Transversal structure 344
6.8.2 Linear phase structure 345
6.8.3 Other structures 346
6.8.4 Choosing between structures 347

6.9 Finite wordlength effects in FIR digital filters 348
6.9.1 Coefficient quantization errors 350
6.9.2 Roundoff errors 356
6.9.3 Overflow errors 357

6.10 FIR implementation techniques 358
6.11 Design example 359
6.12 Summary 361
6.13 Application examples of FIR filters 363
Problems 363
References 368
Bibliography 369
Appendix 370
6A C programs for FIR filter design 370

Chapter 7
Design of infinite impulse response (IIR) digital filters 374

7.1 Introduction: summary of the basic features of IIR filters 375
7.2 Design stages for digital IIR filters 376
7.3 Stage 1: performance specification 377
7.4 Stage 2: calculation of IIR filter coefficients 379
7.4.1 Pole–zero placement method 379
7.4.2 Converting analogue filters into equivalent digital filters 383
7.4.3 Impulse invariant method 383
7.4.4 Summary of the impulse invariant method of obtaining IIR coefficients 386
7.4.5 Remarks on the impulse invariant method 387
7.4.6 Bilinear *z*-transform (BZT) method 388
7.4.7 Summary of the procedure for calculating digital filter coefficients by the BZT method 390
7.4.8 Comments on the bilinear transformation method 392

7.4.9 Use of classical analogue filters to design IIR digital filters 394
7.4.10 Designing highpass, bandpass and bandstop filters 398
7.4.11 Calculating IIR filter coefficients: method 1 399
7.4.12 Calculating IIR filter coefficients: method 2 407
7.4.13 Illustrative examples using method 2 411
7.4.14 Using an IIR filter design program 415
7.4.15 Choosing between the coefficient calculation methods 416

7.5 Stage 3: realization structures for IIR digital filters 416
7.5.1 Practical building blocks for IIR filters 418
7.5.2 Cascade and parallel realization structures for higher IIR filters 419

7.6 Stage 4: analysis of finite wordlength effects 425
7.6.1 ADC quantization noise 426
7.6.2 Coefficient quantization errors 427
7.6.3 Coefficient wordlength to maintain stability 427
7.6.4 Coefficient wordlength for desired frequency response 429
7.6.5 A detailed example – coefficient wordlength requirements for stability and frequency response 430
7.6.6 Addition overflow errors 432
7.6.7 Principles of scaling 433
7.6.8 Scaling in cascade realization 436
7.6.9 Scaling in parallel realization 438
7.6.10 Output overflow detection and prevention 440
7.6.11 Product roundoff errors 440
7.6.12 Effects of roundoff errors on the signal-to-noise ratio 442
7.6.13 Illustrative example: assessment of the effect of roundoff errors on signal to noise ratio 445
7.6.14 Roundoff noise in cascade and parallel realizations 446
7.6.15 Effect of product roundoff noise in modern DSP systems 451
7.6.16 Roundoff noise reduction schemes 452
7.6.17 Limit cycles due to product roundoff errors 458
7.6.18 Other nonlinear phenomena 461

7.7 Stage 5: implementation of the filter 461

7.8 A detailed design example of an IIR digital filter 462

7.9 Summary 467

7.10 Application examples 468
7.10.1 Digital audio 468
7.10.2 Digital control 469

7.10.3 Digital frequency oscillators 470
7.10.4 Telecommunication 471
7.10.5 Digital touch-tone generation and receiving 471
7.10.6 Clock recovery for data communication 472
Problems 476
References 480
Bibliography 481
Appendices 483
7A C programs for IIR digital filter design 483
7B Evaluation of complex square roots using real arithmetic 487
7C L_2 scaling factor equations 489

Chapter 8
Multirate digital signal processing **491**

8.1 Introduction 492
8.1.1 Some current uses of multirate processing in industry 492
8.2 Concepts of multirate signal processing 493
8.2.1 Sampling rate: decimation by integer factors 494
8.2.2 Sampling rate increase: interpolation by integer factors 495
8.2.3 Sampling rate conversion by non-integer factors 498
8.2.4 Multistage approach to sampling rate conversion 501
8.3 Design of practical sampling rate converters 502
8.3.1 Filter specification 502
8.3.2 Filter requirements for individual stages 503
8.3.3 Determining the number of stages and decimation factors 505
8.3.4 Illustrative design example 506
8.4 Software implementation of sampling rate converters– decimators 508
8.4.1 Program for multistage decimation 510
8.4.2 Test example for the decimation program 512
8.5 Software implementation of interpolators 514
8.5.1 Program for multistage interpolation 516
8.5.2 Test example 518
8.6 Application examples 520
8.6.1 High quality analogue-to-digital conversion for digital audio 520
8.6.2 Efficient digital-to-analogue conversion in compact hi-fi systems 521
8.6.3 Application in the acquisition of high quality data 523

8.6.4 Efficient implementation of narrowband digital filters 527
8.6.5 High resolution narrowband spectral analysis 531
8.7 Summary 533
Problems 533
References 534
Bibliography 535
Appendix 536
8A C programs for multirate processing and systems design 536

Chapter 9
Adaptive digital filters **541**

9.1 When to use adaptive filters and where they have been used 542
9.2 Concepts of adaptive filtering 543
9.2.1 Adaptive filters as a noise canceller 543
9.2.2 Other configurations of the adaptive filter 544
9.2.3 Main components of the adaptive filter 544
9.2.4 Adaptive algorithms 544
9.3 Basic Wiener filter theory 547
9.4 The basic LMS adaptive algorithm 550
9.4.1 Implementation of the basic LMS algorithm 551
9.4.2 Practical limitations of the basic LMS algorithm 553
9.4.3 Other LMS-based algorithms 556
9.5 Recursive least squares algorithm 557
9.5.1 Recursive least squares algorithm 558
9.5.2 Limitations of the recursive least squares algorithm 559
9.5.3 Factorization algorithms 560
9.6 Application example 1 – adaptive filtering of ocular artefacts from the human EEG 561
9.6.1 The physiological problem 561
9.6.2 Artefact processing algorithm 562
9.6.3 Real-time implementation 563
9.7 Application example 2 – adaptive telephone echo cancellation 563
9.8 Other applications 565
Problems 569
References 569
Bibliography 570
Appendix 571
9A C language programs for adaptive filtering 571

Chapter 10
Spectrum estimation and analysis **577**

10.1 Introduction 578
10.2 Principles of spectrum estimation 580
10.3 Traditional methods 583
10.3.1 Pitfalls 583
10.3.2 Windowing 586
10.3.3 The periodogram method and periodogram properties 597
10.3.4 Modified periodogram methods 600
10.3.5 The Blackman–Tukey method 601
10.3.6 The fast correlation method 602
10.3.7 Comparison of the power spectral density estimation methods 603
10.4 Modern parametric estimation methods 603
10.5 Comparison of estimation methods 604
10.6 Application examples 604
10.6.1 Use of spectral analysis by a DFT for differentiating between brain diseases 604
10.6.2 Spectral analysis of EEGs using autoregressive modelling 608
10.7 Summary 608
10.8 Worked example 609
Problems 610
References 612

Chapter 11
General- and special-purpose hardware for DSP **614**

11.1 Introduction 615
11.2 Computer architectures for signal processing 615
11.2.1 Harvard architecture 617
11.2.2 Pipelining 618
11.2.3 Hardware multiplier–accumulator 625
11.2.4 Special instructions 626
11.2.5 Replication 627
11.2.6 On-chip memory/cache 628
11.3 General-purpose digital signal processors 628
11.3.1 Texas Instruments TMS320 family 628
11.3.2 Motorola DSP56633 family 633
11.3.3 Analog Devices ADSP2100 family 634
11.4 Implementation of DSP algorithms on general-purpose digital signal processors 636
11.4.1 FIR digital filtering 636

11.4.2 IIR digital filtering 642
11.4.3 FFT processing 650
11.4.4 Multirate processing 657
11.4.5 Adaptive filtering 660
11.5 Special-purpose DSP hardware 662
11.5.1 Hardware digital filters 663
11.5.2 Hardware FFT processors 665
11.6 Summary 668
Problems 668
References 671
Bibliography 671
Appendix 672
11A TMS320 assembly language programs for real-time signal processing and a C language program for constant geometry radix-2 FFT 672

Chapter 12
Applications and case studies **679**

12.1 TMS320C10 target board for real-time DSP 680
12.1.1 Background and system specifications 680
12.1.2 System description 681
12.1.3 System use 684
12.2 TMS320C25 target board for real-time DSP 685
12.2.1 Background and system specifications 685
12.2.2 System description 686
12.3 TMS320C25 Software Development System (SWDS) 687
12.3.1 SWDS software 688
12.3.2 SWDS hardware 688
12.3.3 Using SWDS with the analogue interface board 689
12.4 FFT spectrum analyser 689
12.4.1 Features of the analyser 690
12.4.2 Spectrum estimation in the analyser 690
12.4.3 Analyser hardware 692
12.4.4 Analyser software 693
12.4.5 Using the analyser 696
12.5 Detection of foetal heartbeats during labour 697
12.5.1 The foetal electrocardiogram 697
12.5.2 Foetal ECG signal pre-processing 700
12.5.3 QRS template 701
12.5.4 QRS detection methods 702
12.5.5 Performance measure for QRS detection 704
12.5.6 Results 705
12.6 Real-time adaptive removal of ocular artefacts from human EEGs 706
12.6.1 Introduction 706

12.6.2 On-line removal algorithms used in the OAR system 710
12.6.3 Hardware for the on-line ocular artefact removal system 714
12.6.4 Software for the on-line ocular artefact removal system 717
12.6.5 System testing and experimental results 719
12.6.6 Discussion 722
12.6.7 Conclusions 722
12.7 Fixed- and floating point implementation of DSP systems 723
12.7.1 Introduction 723
12.7.2 Fixed-point number system 725
12.7.3 Floating point number system 729
12.8 Equalization of digital audio signals 733
12.9 Adaptive ocular artefact filter 736
12.9.1 Software floating point arithmetic routines 737
12.9.2 Floating point data format 737
12.9.3 Floating point arithmetic routines 738
12.10 Summary 743
Problems 745
References 745
Bibliography 747
Appendices 748
12A The modified UD factorization algorithm 748
12B Programs for the semiparametric equalizer and floating point arithmetic routines 748

Index 757

1

Introduction

1.1	Digital signal processing and its benefits	1
1.2	Application areas	3
1.3	Key DSP operations	4
1.4	Overview of real-time signal processing	13
1.5	Application examples	35
1.6	Summary	43
	References	46
	Bibliography	46

The aims of this chapter are to explain the meaning and benefits of digital signal processing (DSP), to make the reader aware of the wide range of DSP applications, and to introduce key DSP operations with emphasis on their implementational simplicity. The requirements of real-time DSP systems are also described using practical examples to illustrate the concepts presented. The specific practical application examples presented are drawn from areas to which most readers can relate.

1.1 Digital signal processing and its benefits

By a signal we mean any variable that carries or contains some kind of information that can, for example, be conveyed, displayed or manipulated. Examples of the types of signals of particular interest are

- speech, which we encounter for example in telephony, radio and everyday life,
- biomedical signals, such as the electroencephalogram (brain signals),

- sound and music, such as reproduced by the compact disc player,
- video and image, which most people watch on the television, and
- radar signals, which are used to determine the range and bearing of distant targets.

Digital signal processing is concerned with the digital representation of signals and the use of digital processors to analyse, modify, or extract information from signals. Most signals in nature are analogue in form, often meaning that they vary continuously with time, and represent the variations of physical quantities such as sound waves. The signals used in most popular forms of DSP are derived from analogue signals which have been sampled at regular intervals and converted into a digital form.

The specific reason for processing a digital signal may be, for example, to remove interference or noise from the signal, to obtain the spectrum of the data, or to transform the signal into a more suitable form. DSP is now used in many areas where analogue methods were previously used and in entirely new applications which were difficult or impossible with analogue methods. The attraction of DSP comes from key advantages such as the following.

- *Guaranteed accuracy*. Accuracy is only determined by the number of bits used.
- *Perfect reproducibility*. Identical performance from unit to unit is obtained since there are no variations due to component tolerances. For example, using DSP techniques, a digital recording can be copied or reproduced several times over without any degradation in the signal quality.
- No drift in performance with temperature or age.
- Advantage is always taken of the tremendous advances in semiconductor technology to achieve greater reliability, smaller size, lower cost, low power consumption, and higher speed. For example, in the last two or three years it has become possible to produce high speed low power ICs (integrated circuits) using CMOS technology. As a result newer DSP chips are predominantly CMOS devices instead of bipolar, while NMOS devices are being upgraded to CMOS.
- *Greater flexibility*. DSP systems can be programmed and reprogrammed to perform a variety of functions, without modifying the hardware. This is perhaps one of the most important features of DSP.
- *Superior performance*. DSP can be used to perform functions not possible with analogue signal processing. For example, linear phase response can be achieved, and complex adaptive filtering algorithms can be implemented using DSP techniques.
- In some cases information may already be in a digital form and DSP offers the only viable option.

DSP is not without disadvantages. However, the significance of these disadvantages is being continually diminished by new technology.

- *Speed and cost*. DSP designs can be expensive especially when large bandwidth signals are involved. At the present, fast ADCs/DACs (analogue-to-digital converters/digital-to-analogue converters) either are too expensive or do not have sufficient resolution for wide bandwidth DSP applications. Currently, only specialized ICs can be used to process signals in the megahertz range and these are quite expensive. Furthermore, most DSP devices are still not fast enough and can only process signals of moderate bandwidths. Bandwidths in the 100 MHz range are still processed only by analogue methods. Nevertheless, DSP devices are becoming faster and faster.
- *Design time*. Unless you are knowledgeable in DSP techniques and have the necessary resources (software packages and so on), DSP designs can be time consuming and in some cases almost impossible. The acute shortage of suitable engineers in this area is widely recognized. However, the situation is changing as many new graduates now possess some knowledge of digital techniques and commercial companies are beginning to exploit the advantages of DSP in their products.
- *Finite wordlength problems*. In real-time situations, economic considerations often mean that DSP algorithms are implemented using only a limited number of bits. In some DSP systems, if an insufficient number of bits is used to represent variables serious degradation in system performance may result.

1.2 Application areas

DSP is one of the fastest growing fields in modern electronics being used in any area where information is handled in a digital form or controlled by a digital processor. Application areas include the following:

- Image processing
  - pattern recognition
  - robotic vision
  - image enhancement
  - facsimile
  - satellite weather map
  - animation
- Instrumentation/control
  - spectrum analysis
  - position and rate control
  - noise reduction
  - data compression

- Speech/audio
  - speech recognition
  - speech synthesis
  - text to speech
  - digital audio
  - equalization
- Military
  - secure communication
  - radar processing
  - sonar processing
  - missile guidance
- Telecommunications
  - echo cancellation
  - adaptive equalization
  - ADPCM transcoders
  - spread spectrum
  - video conferencing
  - data communication
- Biomedical
  - patient monitoring
  - scanners
  - EEG brain mappers
  - ECG analysis
  - X-ray storage/enhancement

A look at the list, which is by no means complete, will confirm the importance of DSP. A testimony to the recognition of the importance of DSP is the continual introduction of powerful DSP devices by semiconductor manufacturers. However, there are insufficient engineers with adequate knowledge in this area. An objective of this book is to provide an understanding of DSP techniques and their implementation, to enable the reader to gain a working knowledge of this important subject.

1.3 Key DSP operations

Several DSP algorithms exist and many more are being invented or discovered. However, all these algorithms, including the most complex, require similar basic operations. It is instructive to examine some of these operations at the outset so as to appreciate the implementational simplicity of DSP. The basic DSP operations are convolution, correlation, filtering, transformations, and modulation. Table 1.1 summarizes these operations and a brief description of each is given below. An important point to note in the table is that all the basic

Table 1.1 Summary of key DSP operations.

(1) *Convolution*. Given two finite length sequences, $x(k)$ and $h(k)$, of lengths N_1 and N_2, respectively, their linear convolution is

$$y(n) = h(n) \circledast x(n) = \sum_{k=-\infty}^{\infty} h(k)x(n-k) = \sum_{k=0}^{M-1} h(k)x(n-k), \; n = 0, 1, \ldots, M-1 \quad \textbf{(1.1)}$$

where $M = N_1 + N_2 - 1$.

(2) *Correlation*.

(a) Given two N-length sequences, $x(k)$ and $y(k)$, with zero means an estimate of their cross-correlation is given by

$$\rho_{xy}(n) = \frac{r_{xy}(n)}{[r_{xx}(0)r_{yy}(0)]^{1/2}} \qquad n = 0, \pm 1, \pm 2, \ldots \quad \textbf{(1.2)}$$

where $r_{xy}(n)$ is an estimate of the cross-covariance and defined as

$$r_{xy}(n) = \begin{cases} \dfrac{1}{N} \displaystyle\sum_{k=0}^{N-n-1} x(k)y(k+n) & n = 0, 1, 2, \ldots \\ \dfrac{1}{N} \displaystyle\sum_{k=0}^{N+n-1} x(k-n)y(k) & n = 0, -1, -2, \ldots \end{cases}$$

$$r_{xx}(0) = \frac{1}{N}\sum_{k=0}^{N-1} [x(k)]^2, \; r_{yy}(0) = \frac{1}{N}\sum_{k=0}^{N-1} [y(k)]^2$$

(b) An estimate of the autocorrelation, $\rho_{xx}(n)$, of an N-length sequence, $x(k)$, with zero mean is given by

$$\rho_{xx}(n) = \frac{r_{xx}(n)}{r_{xx}(0)} \qquad n = 0, \pm 1, \pm 2, \ldots \quad \textbf{(1.3)}$$

where $r_{xx}(n)$ is an estimate of the autocovariance and defined as

$$r_{xx}(n) = \frac{1}{N}\sum_{k=0}^{N-n-1} x(k)x(k+n) \qquad n = 0, 1, 2, \ldots$$

(3) *Filtering*. The equation for finite impulse response (FIR) filtering is

$$y(n) = \sum_{k=0}^{N-1} h(k)x(n-k) \quad \textbf{(1.4)}$$

where $x(k)$ and $y(k)$ are the input and output of the filter, respectively, and $h(k)$, $k = 0, 1, \ldots, N-1$, are the filter coefficients.

(4) *Discrete transform*.

$$X(n) = \sum_{k=0}^{N-1} x(k)W^{kn} \qquad W = \exp(-\mathrm{j}2\pi/N) \quad \textbf{(1.5)}$$

DSP operations require only simple arithmetic operations of multiply, add/subtract, and shifts to carry out. Notice also the similarity between most of the operations.

1.3.1 Convolution

Convolution is one of the most frequently used operations in DSP. For example, it is the basic operation in digital filtering. Given two finite and causal sequences, $x(n)$ and $h(n)$, of lengths N_1 and N_2, respectively, their convolution is defined as

$$y(n) = h(n) \circledast x(n) = \sum_{k=-\infty}^{\infty} h(k)x(n-k) = \sum_{k=0}^{\infty} h(k)x(n-k), \quad n = 0, 1, \ldots, (M-1)$$

where the symbol $\circledast$ is used to denote convolution and $M = N_1 + N_2 - 1$. As we shall see in later chapters, DSP device manufacturers have developed signal processors that perform efficiently the multiply–accumulate operations involved in convolution. An example of the linear convolution of the two sequences depicted in Figures 1.1(a) and 1.1(b) is given in Figure 1.1(c). In this example, $h(n)$, $n = 0, 1, 2, \ldots$, can be viewed as the impulse response of a digital system, and $y(n)$ the system's response to the input sequence, $x(n)$. The numerical values for the convolution, that is $y(n)$, were obtained by direct evaluation of Equation 1.1. For example, $y(1)$ is obtained as follows:

$$\begin{aligned} y(1) &= h(0)x(1) + h(1)x(0) + h(2)x(-1) + \ldots + h(12)x(-11) \\ &= 0 \times 1 + (-0.02) \times 1 + 0 \times 0 + \ldots + 0 \times 0 = -0.02 \end{aligned}$$

The significance of convolution is more apparent when it is observed in the frequency domain, and use is made of the fact that convolution in the time domain is equivalent to multiplication in the frequency domain. A more detailed discussion of convolution including its properties and graphical interpretation is given in Chapter 4.

1.3.2 Correlation

There are two forms of correlations: auto- and cross-correlations.

(1) The cross-correlation function (CCF) is a measure of the similarities or shared properties between two signals. Applications of CCFs include cross-spectral analysis, detection/recovery of signals buried in noise, for example the detection of radar return signals, pattern matching, delay measurements. CCF is defined in Equation 1.2 in Table 1.1.

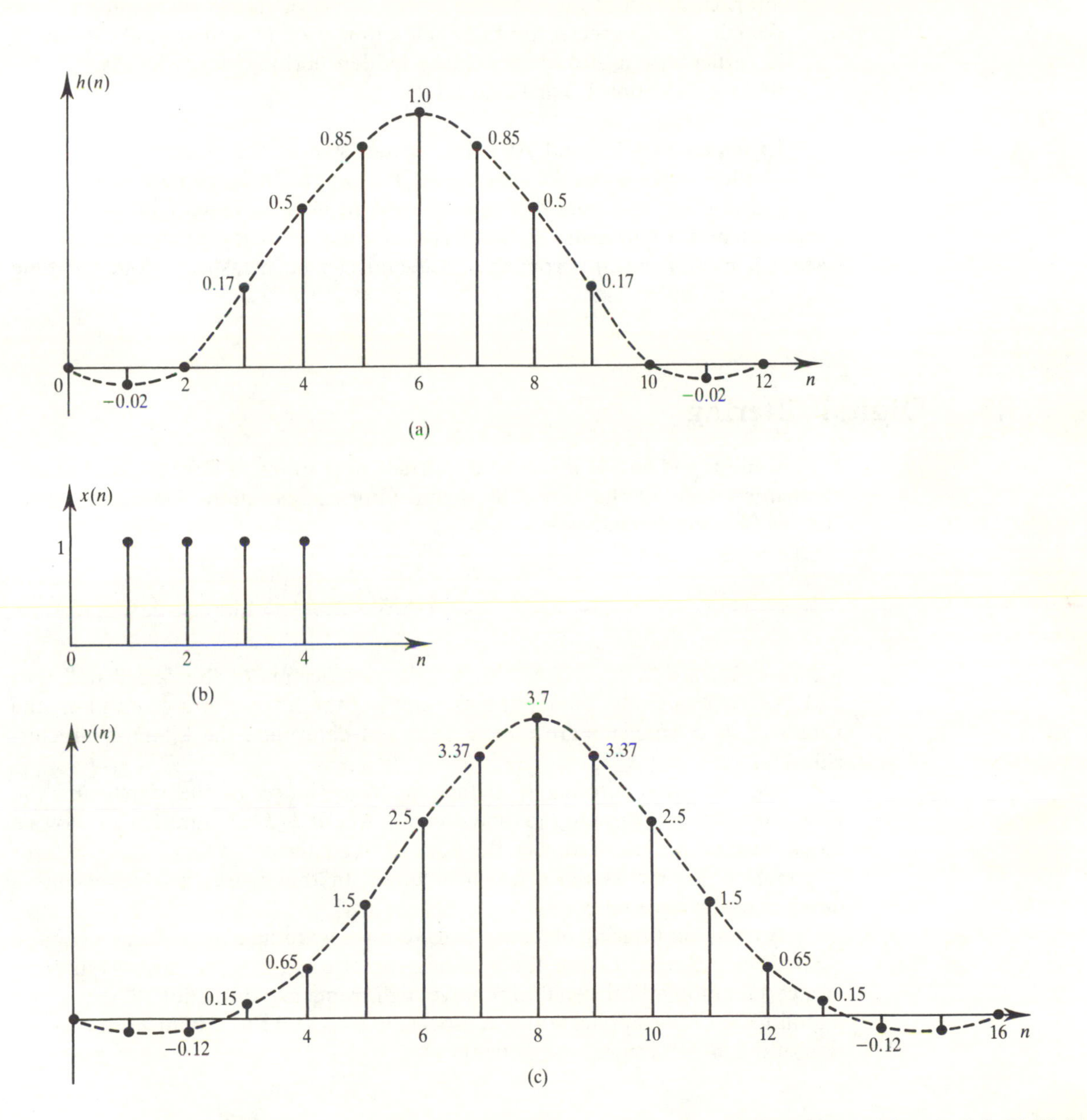

Figure 1.1 An example of the convolution of two sequences. $y(n)$ is the convolution of $h(n)$ and $x(n)$. If $h(n)$ is considered the impulse response of a system, then $y(n)$ is the system's output in response to the input $x(n)$. The values of $y(n)$ above were obtained directly from Equation 1.1.

(2) The autocorrelation function (ACF) involves only one signal and provides information about the structure of the signal or its behaviour in the time domain. It is a special form of CCF and is used in similar applications. It is particularly useful in identifying hidden periodicities. The ACF is defined in Equation 1.3 in Table 1.1.

Examples of CCF and ACF for certain signals are given in Figures 1.2 and 1.3. Notice, for example, that the ACF of the noise-corrupted signal shows clearly that there is a periodic signal buried in noise (Figure 1.2). Figure 1.3 illustrates how to measure delays. The amount of delay introduced by the system is clearly evident from the CCF and can be measured from the time origin to the large peak.

1.3.3 Digital filtering

Digital fitering is one of the most important operations in DSP as will become clear in subsequent chapters. The digital filtering operations for an important class of filters is defined as

$$y(n) = \sum_{k=0}^{N-1} h(k)x(n-k)$$

where $h(k)$, $k = 0, 1, \ldots, N-1$, are the coefficients of the filter, and $x(n)$ and $y(n)$, respectively, the input and output of the filter. For a given filter, the values of its coefficients are unique to it and determine the filter's characteristics.

We note that filtering is in fact the convolution of the signal and the filter's impulse response in the time domain, that is $h(k)$. Figure 1.4(a) shows a block diagram representation of the filter defined above. In this form, the filter is popularly known as the transversal filter. In the figure, z^{-1} represents a delay of one sample time.

A common filtering objective is to remove or reduce noise from a wanted signal. For example, Figure 1.4(b) shows the effects of digital lowpass filtering of a certain biomedical signal to remove high frequency distortion. The use of a digital filter in this application was especially important to minimize the distortion of the in-band signal components.

1.3.4 Discrete transformation

Discrete transforms allow the representation of discrete-time signals in the frequency domain or the conversion between time and frequency domain representations. The spectrum of a signal is obtained by decomposing it into its

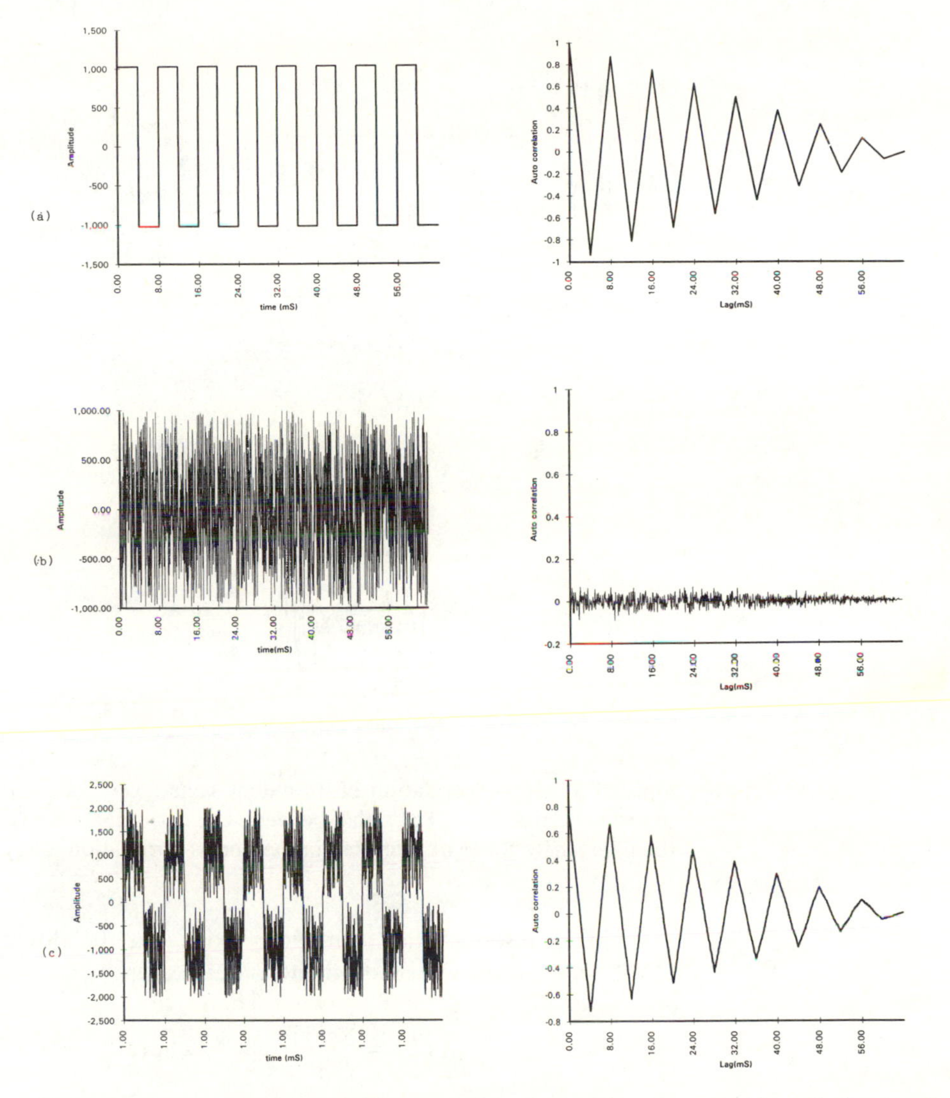

Figure 1.2 Autocorrelations of (a) a periodic signal, (b) noise and (c) periodic signal plus noise. Note that in (c) the periodic nature of the signal buried in noise is still evident, illustrating why autocorrelation is used in detecting hidden periodicity.

constituent frequency components using a discrete transform. A knowledge of such a spectrum is invaluable in, for example, determining the bandwidth required to transmit the signal. Conversion between time and frequency domains is necessary in many DSP applications. For example, it allows for a more efficient implementation of DSP algorithms, such as those for digital filtering, convolution and correlation.

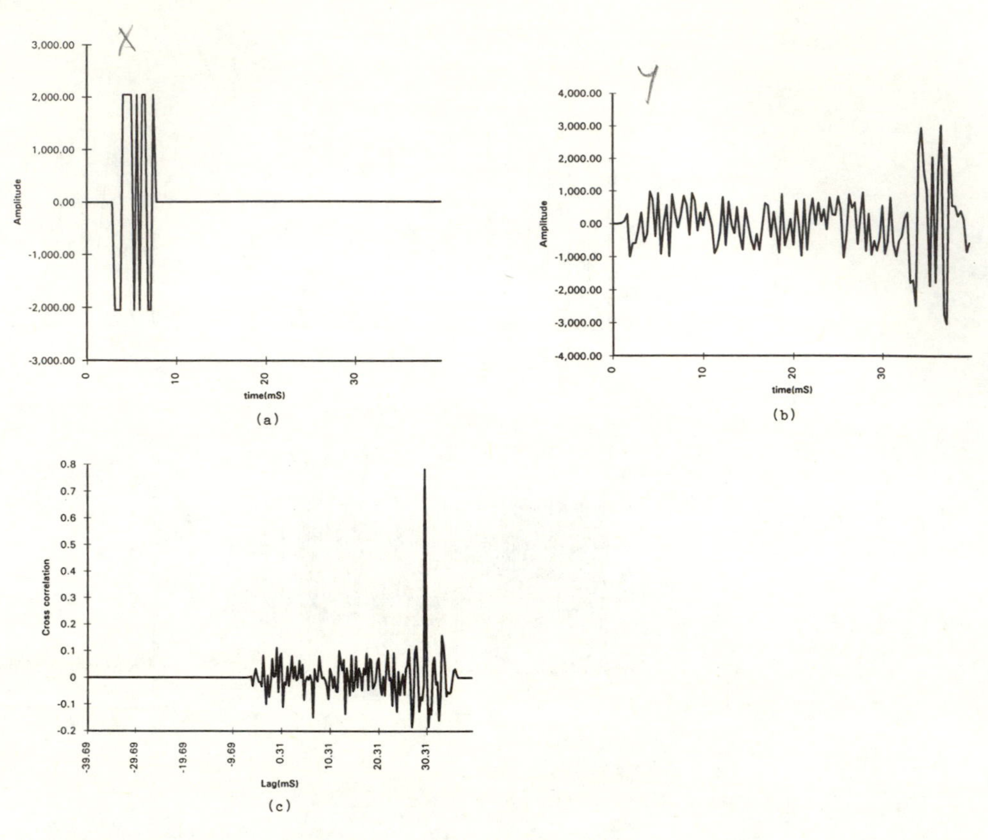

Figure 1.3 Cross-correlation of a random signal, $x(t)$, and a delayed noisy version of the same signal, $y(t)$. The delay between the two signals is the time from the origin to the time where the peak occurred in their cross-correlation in (c).

Many discrete transformations exist, but the discrete Fourier transform (DFT) is the most widely used and is defined as

$$X(k) = \sum_{n=0}^{N-1} x(n)W^{nk}, \text{ where } W = e^{-j2\pi/N}$$

An example of the use of the DFT is given in Figure 1.5. Here, the impulse response of a filter, $h(n)$, $n = 0, 1, \ldots, N-1$, is transformed to give the frequency response of the filter using the DFT. Details of the DFT and its applications are given in Chapters 2, 3 and 10.

1.3.5 Modulation

Digital signals are rarely transmitted over long distances or stored in large quantities in their raw form. The signals are normally modulated to match their frequency characteristics to those of the transmission and/or storage media to

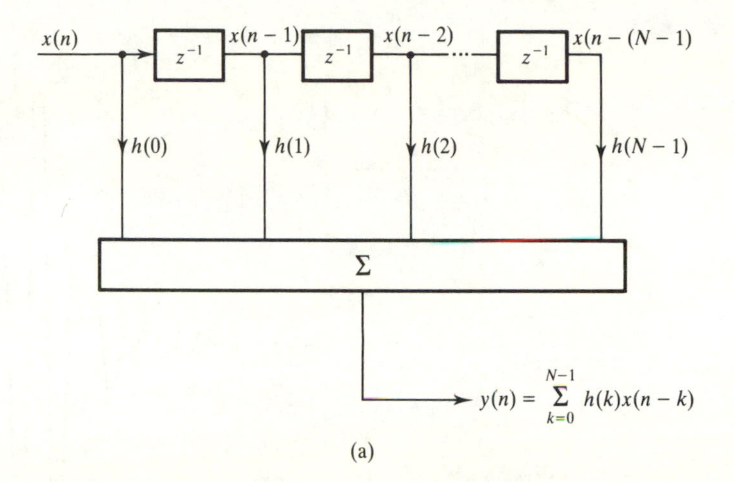

(a)

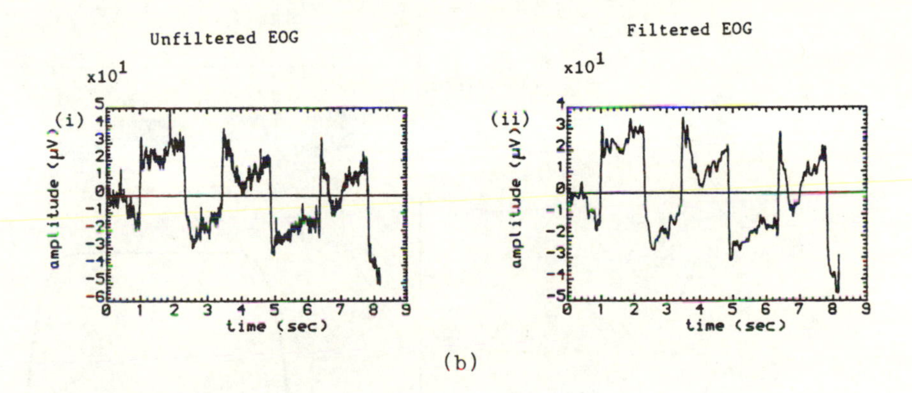

(b)

Figure 1.4 (a) Block diagram representation of the transversal filter. $h(k)$, $k = 0, 1, \ldots, N - 1$, are the filter coefficients, and each box containing Z^{-1} represents a delay of one sampling period. (b) Digital lowpass filtering of a biomedical signal to remove noise.

minimize signal distortion, to utilize the available bandwidth efficiently, or to ensure that the signals have some desirable properties. Perhaps the two application areas where modulation is extensively employed are telecommunications and digital audio engineering.

The process of modulation often involves varying a property of a high frequency signal, known as the carrier, in sympathy with the signal we wish to transmit or store, called the modulating signal. The three most commonly used digital modulation schemes for transmitting digital data over a bandpass channel (for example a microwave link) are amplitude shift keying (ASK), phase

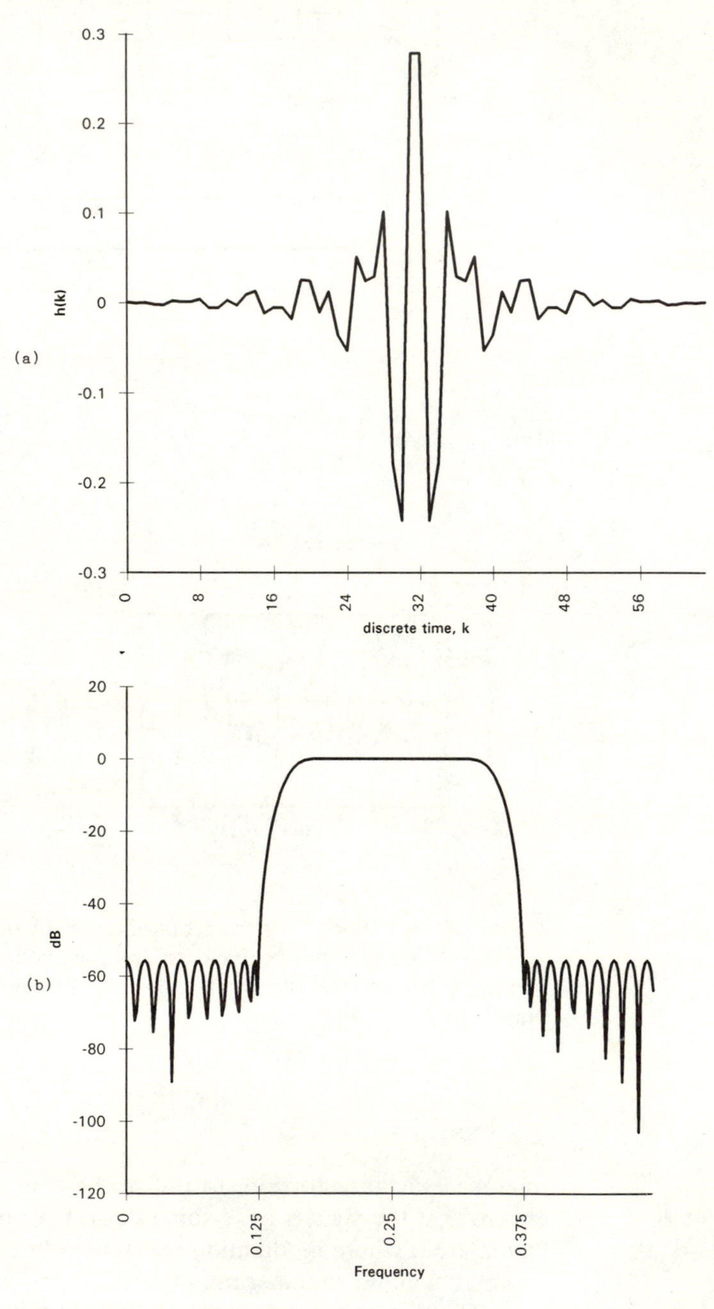

Figure 1.5 Time and frequency domain representations of a digital filter: (a) impulse response; (b) filter spectrum. The filter spectrum was obtained from the discrete transform of $h(n)$, illustrating one of the many uses of the DFT.

shift keying (PSK), and frequency shift keying (FSK). When digital data is transmitted over an all-digital network, a scheme known as pulse code modulation (PCM) is commonly used (see, for example, Bellamy, 1982). Several other modulation schemes have been developed for digital audio details of which can be found in Watkinson (1987).

1.4 Overview of real-time signal processing

1.4.1 Typical real-time DSP systems

The block diagram of a typical DSP system operating in real time is depicted in Figure 1.6. The analogue input filter is used to bandlimit the analogue input signal prior to digitization to reduce aliasing (see later). The ADC converts the analogue input signal into a digital form. For wide bandwidth signals or when a slow ADC is used, it is necessary to precede the ADC with a sample and hold circuit, although newer ADCs now have built-in sample and hold circuits. After digital processing in the processor, the DAC converts the processed signal back into analogue form. The output filter smooths out the outputs of the DAC and removes unwanted high frequency components.

The heart of the system in Figure 1.6 is the digital processor which may be based on a general purpose microprocessor such as the Motorola MC68000, a digital signal processor chip such as the Texas Instruments TMS320C25, or some other piece of hardware. The digital processor may implement one of several DSP algorithms, for example digital filtering, mapping the input, $x(n)$, into the output, $y(n)$.

Signal processing using a digital processor implies that the input signal must be in a digital form before it can be processed. In some real-time applications the data may already be in a digital form or does not need to be converted to an analogue signal. For example, after processing, the signal may be stored in a computer memory for later use or it may be displayed graphically on a display unit. In other applications it may be required to generate signals digitally. Examples of this are in speech synthesis, digital frequency synthesis and pseudorandom binary sequence generators. Much of the discussion in this

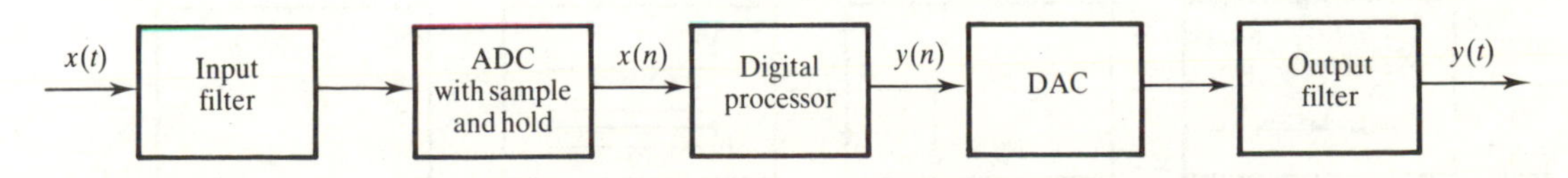

Figure 1.6 Block diagram of a simplified, generalized real-time digital signal processing system. In some applications, the input filter and the ADC or the DAC and the output filter will not be necessary.

book assumes that the signal is in a digital form or has been adequately digitized as described in the next section.

1.4.2 Analogue-to-digital conversion process

As was mentioned above, before any DSP algorithm can be performed, the signal must be in a digital form. Most signals in nature are in analogue form, necessitating an analogue-to-digital conversion process, which involves the following steps.

- The (bandlimited) signal is first sampled, converting the analogue signal into a discrete-time continuous amplitude signal.
- The amplitude of each signal sample is quantized into one of 2^B levels, where B is the number of bits used to represent a sample in the ADC.
- The discrete amplitude levels are represented or encoded into distinct binary words each of length B bits.

The process is depicted in Figure 1.7. Three distinct types of signals can be identified in the figure.

- *The analogue input signal*. This signal is continuous in both time and amplitude.
- *The sampled signal*. This signal is continuous in amplitude but defined only at discrete points in time. Thus the signal is zero except at time $t = nT$ (the sampling instants).
- *The digital signal, $x(n)$ ($n = 0, 1, \ldots$)*. This signal exists only at discrete points in time and at each time point can only have one of 2^B values (discrete-time discrete-value signal). This is the type of signal that is of concern to us in this book.

Note that the discrete-time (that is, sampled) signal and the digital signal can each be represented as a sequence of numbers, $x(nT)$, or simply $x(n)$ ($n = 0, 1, 2, \ldots$). Let us now look more closely at the steps in digitizing a signal.

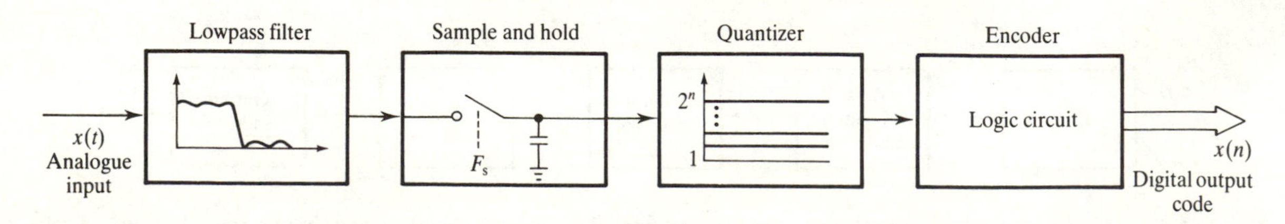

Figure 1.7 A pictorial representation of the analogue-to-digital conversion process.

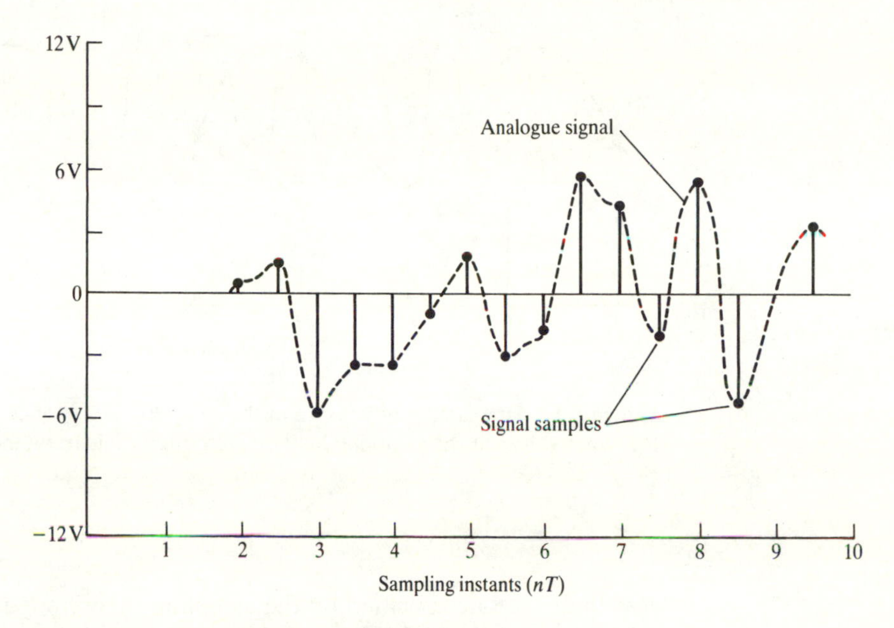

Figure 1.8 An example of a sampled signal (ideal sampling). The values of the signal samples are equal to those of the original analogue signal at the sampling instants.

1.4.2.1 Sampling

Sampling is the acquisition of a continuous (for example analogue) signal at discrete time intervals and is a fundamental concept in real-time signal processing. An example of a sampled analogue signal is shown in Figure 1.8. Note that after sampling, in this ideal case, the analogue signal is now represented only at discrete times, with the values of the samples equal to those of the original analogue signal at the discrete times.

In this chapter, we shall give an intuitive presentation of the sampling theorem, which specifies the rate at which an analogue signal should be sampled to ensure that all the relevant information contained in the signal is captured or retained by sampling.

The sampling theorem

If the highest frequency component in a signal is f_{max}, then the signal should be sampled at the rate of at least $2f_{max}$ for the samples to describe the signal completely:

$$F_s \geqslant 2f_{max} \tag{1.6}$$

where F_s is the sampling frequency or rate. Thus, if the maximum frequency component in an analogue signal is 4 kHz, then to preserve or capture all the information in the signal it should be sampled at 8 kHz or more. Sampling at

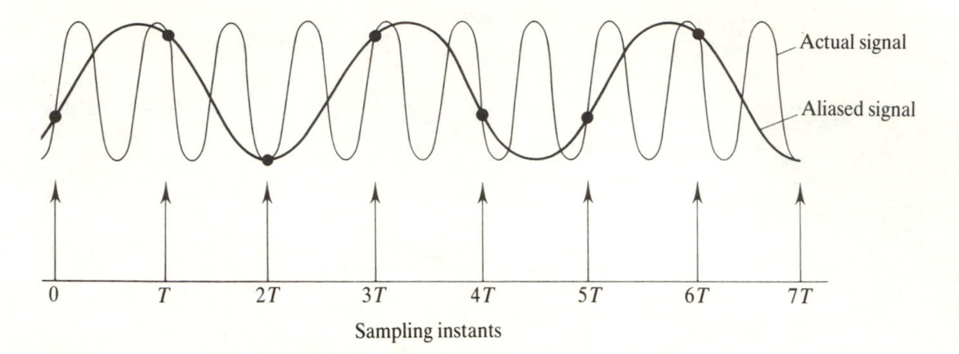

Figure 1.9 An example of aliasing in the time domain. Notice that the two signals have the same values at the sampling instants, although their frequencies are different.

less than the rate specified by the sampling theorem leads to a folding over or 'aliasing' of 'image' frequencies into the desired frequency band so that the original signal cannot be recovered if we were to convert the sampled data back to analogue. An important point to remember is that a signal often has significant energy outside of the highest frequency of interest and/or contains noise, which invariably has a wide bandwidth. For example, in telephony the highest frequency of interest is about 3.4 kHz but speech signals may extend beyond 10 kHz. Thus, the sampling theorem will be violated if we do not remove the signal or noise outside of the band of interest. In practice, this is achieved by first passing the signal through an analogue anti-aliasing filter.

Aliasing and spectra of sampled signals

Suppose we sampled a time domain signal at intervals of T (seconds) (that is, a sampling frequency of $1/T$ (hertz)). It is seen, Figure 1.9, that another frequency component with the same set of samples as the original signal exists. Thus, the frequency component can be mistaken for the lower frequency component and this is what aliasing is about. In practice, it is more instructive, from the point of view of analysing the effects or finding the solution to the problem of aliasing, to examine aliasing in the frequency domain.

Figure 1.10 shows the sampling process, which can be regarded as the multiplication of the analogue signal $x(t)$ by a sampling function, $p(t)$. $p(t)$ consists of pulses of unit amplitudes, width dt (which is infinitesimally small) and period T. The spectra of $x(t)$, $p(t)$, and their product are shown in Figure 1.10. Note that $X'(f)$ is the convolution of $X(f)$ and $P(f)$ – multiplication in the time domain is equivalent to convolution in the frequency domain.

The following points should be noted for the sampled signal in Figure 1.10(d).

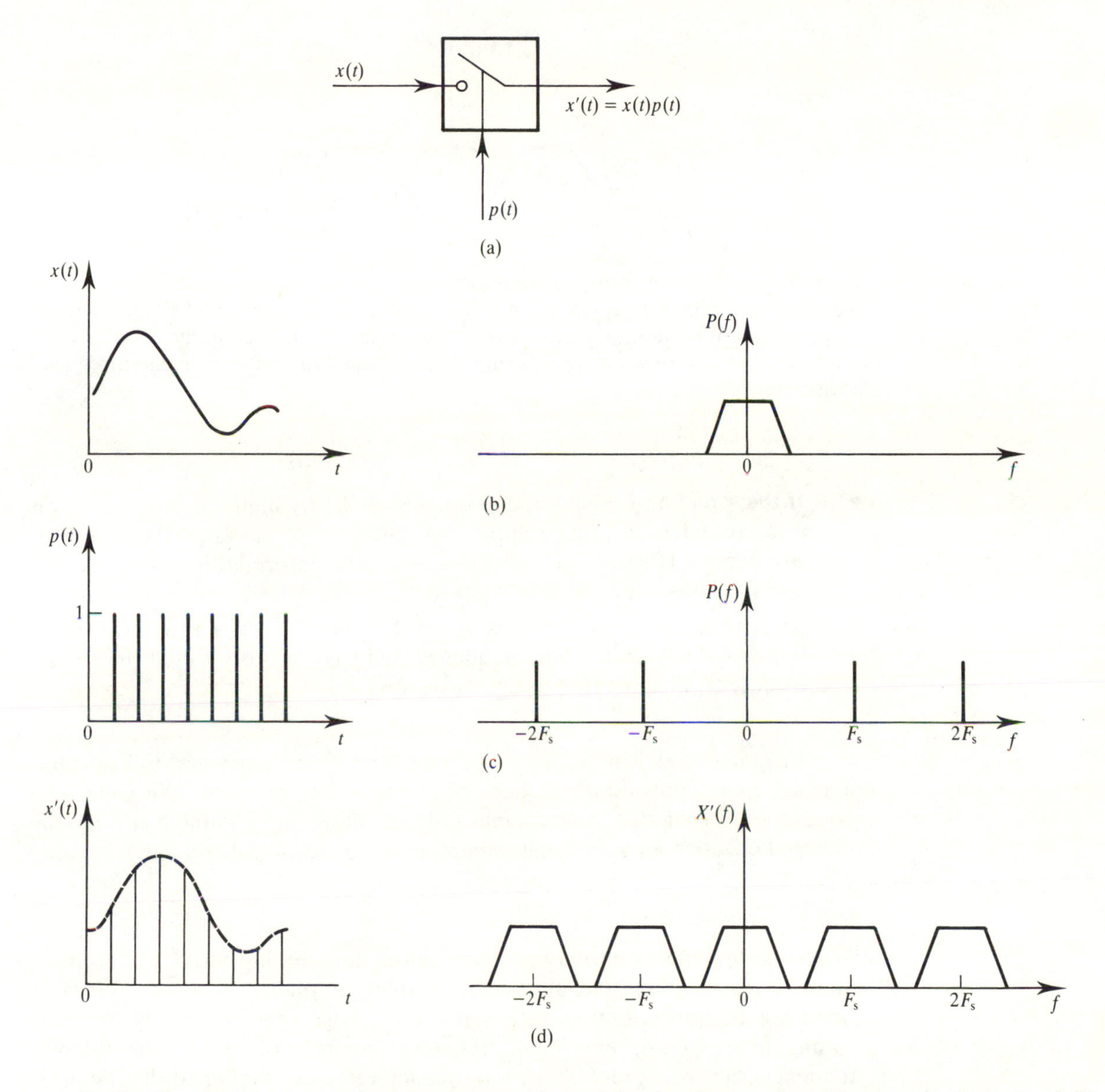

Figure 1.10 Time and frequency domain representations of the sampling process. The spectra of the signal (b) before and (d) after sampling should be compared. (d) Note the changes in the sampled signal and in particular that the spectrum of the sampled signal repeats at multiples of the sampling frequency, F_s.

- The spectrum is the same as the original analogue spectrum, but repeats at multiples of the sampling frequency, F_s. The higher order components which are centred on the multiples of F_s are referred to as image frequencies.

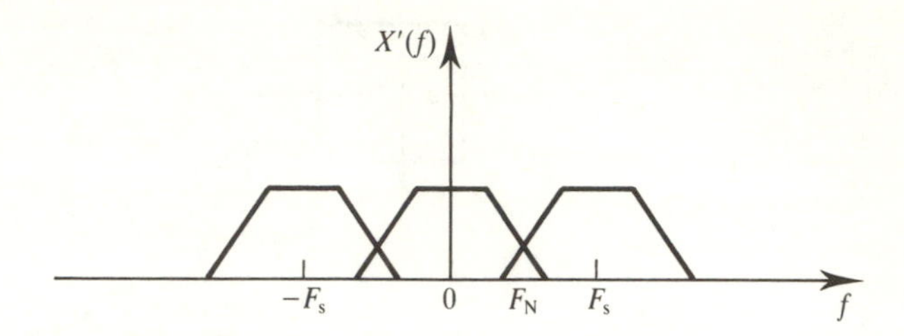

Figure 1.11 Spectrum of an undersampled signal, showing aliasing (foldover region). Signals in the foldover region are not recoverable. F_N is equal to half the sampling frequency and it is often called the Nyquist frequency. To recover all the components of a signal we must sample at a rate greater than (or equal to) twice the highest frequency component.

- If the sampling frequency, F_s, is not sufficiently high the image frequencies centred on F_s, for example, will fold over or alias into the base band frequencies (Figure 1.11). In this case, the information of the desired signal is indistinguishable from its image in the foldover region.
- The overlap or aliasing occurs about the point F_N, that is half the sampling frequency point. This frequency point is variously called the folding frequency, Nyquist frequency, and so on.

In practice, aliasing is always present because of noise and the existence of signal energy outside of the band of interest. The problem then is deciding the level of aliasing that is acceptable and then designing a suitable anti-aliasing filter and choosing an appropriate sampling frequency to achieve this.

Anti-aliasing filtering

To reduce the effects of aliasing sharp cutoff anti-aliasing filters are normally used to bandlimit the signal and/or the sampling frequency is increased so as to widen the separation between the signal and image spectra. Ideally, the anti-aliasing filter should remove all frequency components above the foldover frequency, that is it should have a frequency response similar to that depicted in Figure 1.12(a). A more practical response is given in Figure 1.12(b), where f_c and f_s are the cutoff and stopband frequencies, respectively. We note from Figures 1.12(b) and 1.12(c) that the practical response introduces an amplitude distortion into the signal as it is not flat in the passband. Also, the signal components greater than f_s will be attenuated by A_{min}, but those between f_c and f_s, the transition width, will have their amplitudes reduced monotonically.

The anti-aliasing filter should provide sufficient attenuation at frequencies above the Nyquist frequency. Because of the nonideal response of practical filters, the effective Nyquist frequency is taken as f_s (the stopband edge frequency). In specifying the anti-aliasing filter it is useful to take the ADC

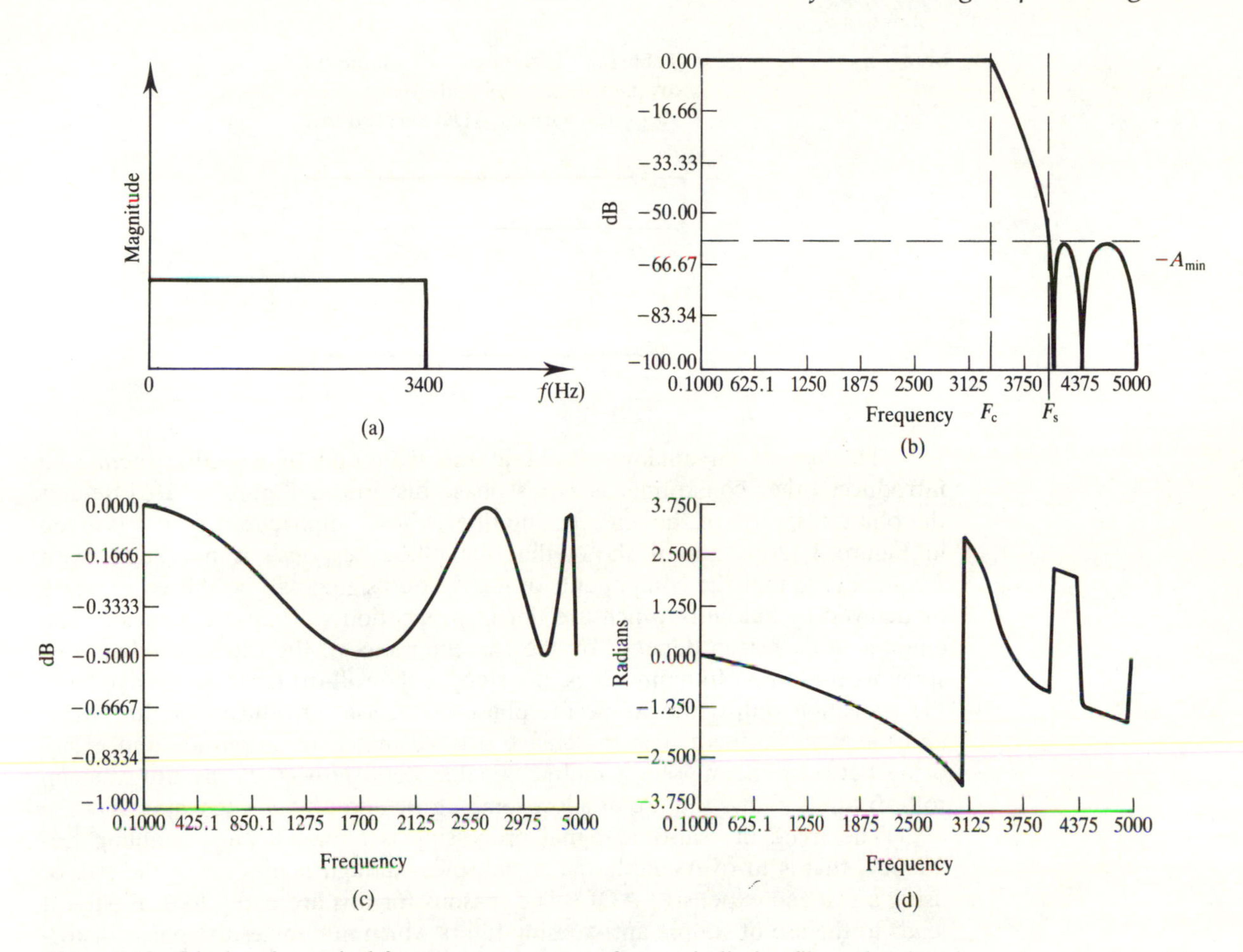

Figure 1.12 Ideal and practical frequency responses of an anti-aliasing filter, showing the errors introduced by the practical responses: (a) ideal response; (b) practical amplitude response; (c) practical passband amplitude response; (d) practical phase response. Compare, for example, the flat passband response in (a) and the practical passband response in (c): the ripples in the practical response would introduce an amplitude distortion of the in-band signal components.

resolution requirements into account. Thus a filter would be designed to attenuate the frequencies above the Nyquist frequency to a level not detectable by the ADC, for example to less than the quantization noise level (see later). Thus, for a system using a B-bit linear ADC, the minimum stopband attenuation of the filter would typically be

$$A_{\min} = 20 \log(\sqrt{1.5} \times 2^{B+1}) \tag{1.7}$$

where B is the number of bits in the ADC (see Example 1.2 for further details). Table 1.2 gives the values for $A_{\min}$ for various values of B.

Table 1.2 Estimates of minimum lowpass filter stopband attenuation, A_{min}, for various ADC resolutions, n.

n	A_{min} (dB)
8	56
10	68
12	80
16	104

The use of an analogue filter at the front-end of a DSP system also introduces other constraints, such as phase distortion. Figure 1.12(d) depicts the phase response of the anti-aliasing filter whose amplitude response is given in Figure 1.12(c), and it shows that the phase response is not linear with frequency, so that the components of the desired signal will be shifted in phase or delayed by amounts which are not in proportion to their frequencies. The amount of distortion depends on the characteristics of the filter including how steep its roll-off is. In many cases, the steeper the roll-off (that is, the narrower the transition width) the worse the phase distortion introduced by the filters, and the more difficult it is to achieve a good match in amplitude and group delay between channels in a multichannel system. However, the use of steep roll-off filters allows the use of a low sample rate and a slower, cheaper ADC.

The trend in real-time signal processing is to use a high sampling frequency, that is to oversample the signal, even though it may be at the risk of using a fast and expensive ADC. The reasons for this are many-fold. Firstly, it leads to the use of simple anti-aliasing filters which minimizes the phase distortion and, for a multichannel system, to lower costs. Secondly, oversampling combined with additional digital signal processing leads to an improved signal-to-noise ratio (see Chapter 8). For a DSP system with an analogue front-end to be usable in different applications the filter cutoff frequency needs to be variable. Programmable analogue filters are coming into use, for example MF10, but their performance is not completely satisfactory and, for multichannel systems, may be expensive. Oversampling of the analogue signal permits the use of digital sample rate conversion techniques (see Chapter 10) to achieve easily the requirements of variable cutoff frequency.

Illustrative examples

There are a number of ways of specifying acceptable aliasing error. For example, we can specify the acceptable aliasing error for a given filter and then determine the sampling frequency necessary to achieve this. Alternatively, for a given sampling frequency, we can work out the minimum stopband attenuation to give a specified aliasing error, taking the resolution of the ADC into account. The following examples illustrate both methods.

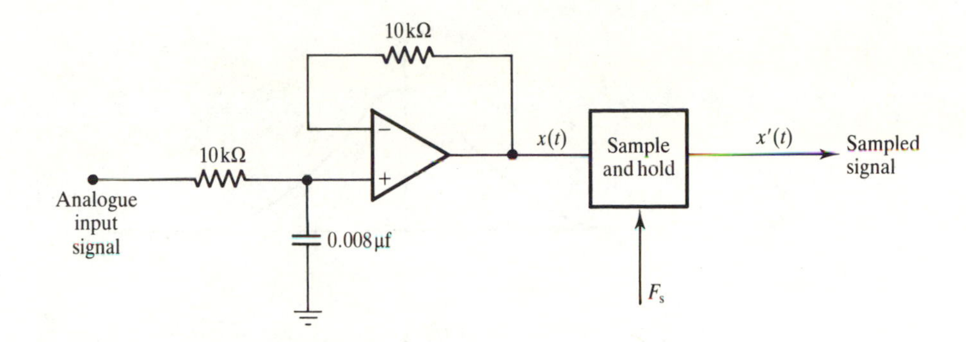

Figure 1.13 Front-end of a simple data acquisition system. The simple active filter is used to bandlimit the signal before it is sampled at the rate of F_s.

Example 1.1

Figure 1.13 depicts the front-end of a simple data acquisition system. Determine the minimum sampling frequency, F_S, to give an aliasing error of less than 2% of the signal level in the passband.

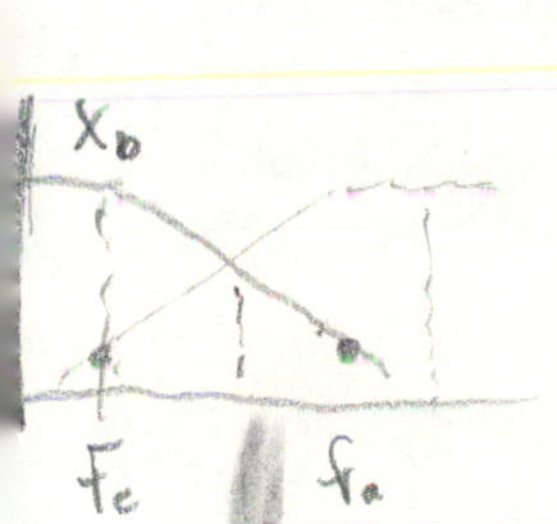

Solution

The amplitude response of the active filter is given by:

$$|H(f)| = \frac{1}{[1 + (f/f_c)^2]^{1/2}} \qquad \text{where } f_c = 1/2\pi RC = 2 \text{ kHz}$$

The spectrum of the bandlimited input signal and the sampled signal are depicted in Figure 1.14, where we have assumed a wideband analogue input.

We note from the figure that the spectrum of the sampled signal repeats at multiples of the sampling frequency. The foldover of the image frequencies into the desired frequency band (0 to 2 kHz) is aliasing.

At 2 kHz, the signal level, $X_b = 0.7071$, so that

$$\text{desired aliasing level} < 0.7071 \times 2/100 = 0.01414$$

Thus,

$$0.01414 < \frac{1}{[1 + (f_a/2)^2]^{1/2}}$$

where f_a is the aliasing frequency. Solving for f_a, we have $f_a < 141.4$ kHz. Thus,

$$F_s(\text{min}) > f_c + f_a = 2 \text{ kHz} + 141.4 \text{ kHz} = 143.4 \text{ kHz}$$

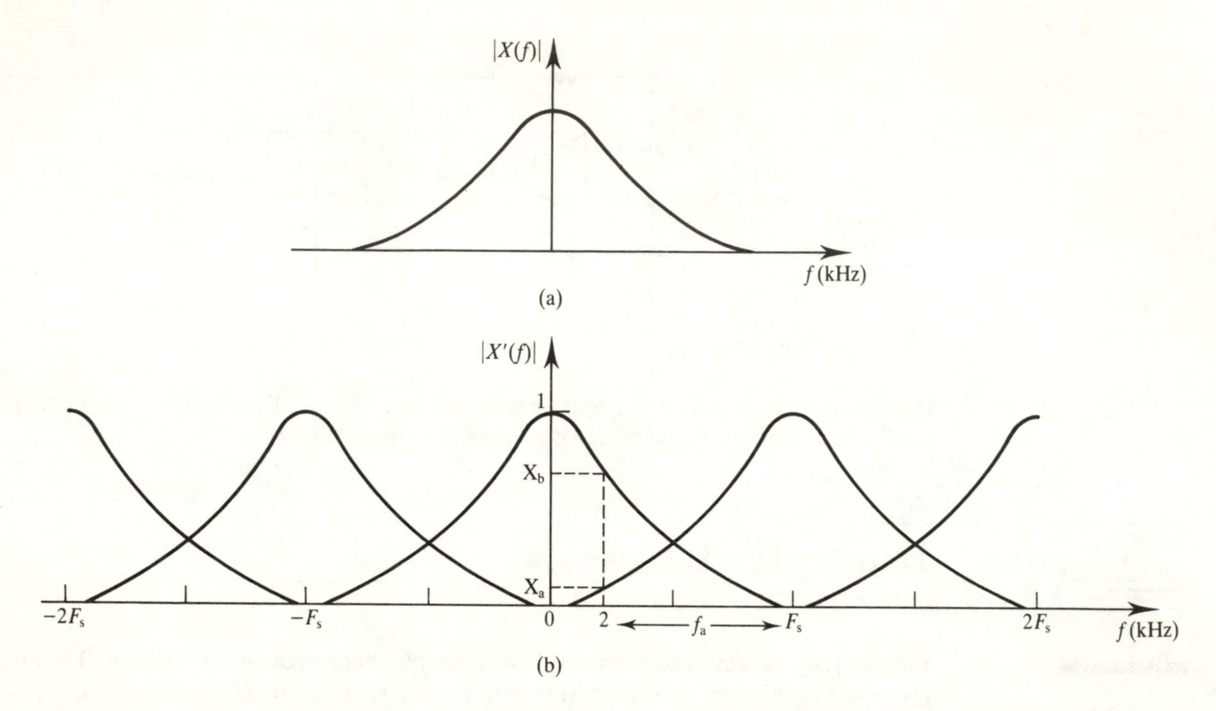

Figure 1.14 Spectrum of signal (a) at the output of the analogue filter and (b) after sampling (Example 1.1).

To meet the specification and to account for the effects of the image frequencies centred on $2F_s$, $3F_s$ and so on (ignored above) then $F_s(\text{min}) > 143.4$ kHz. Let $F_s(\text{min}) = 150$ kHz.

Example 1.2

Figure 1.15 depicts a real-time DSP system. Assuming that the band of interest extends from 0 to 4 kHz and that a 12-bit ADC is used, estimate

(1) the minimum stopband attenuation, A_{min}, for the anti-aliasing filter,

(2) minimum sampling frequency, F_s, and

(3) the level of the aliasing error relative to signal level in the passband for the estimated A_{min} and F_s.

Sketch and label the spectrum of the signal at the output of the analogue filter, assuming a wideband signal at the input, and that of the signal after sampling.

Solution

The anti-aliasing filter should attenuate the levels of frequencies in the stopband to less than the rms quantization noise level for the ADC, so that they are not detectable by the ADC.

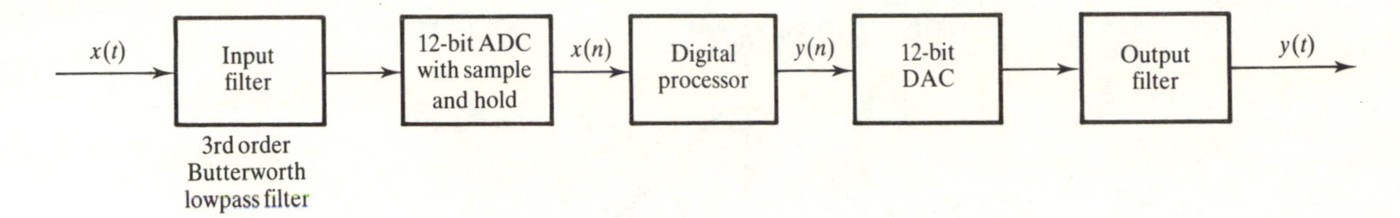

Figure 1.15 Real-time digital signal processing system.

The rms quantization noise level is $(\alpha^2/12)^{1/2} = \alpha/2\sqrt{3}$, where the quantum level $\alpha = V_{fs}/(2^B - 1) \approx V_{fs}/2^B$ and V_{fs} is the full-scale input. Thus, assuming a full-scale input,

$$\frac{\text{maximum passband signal level}}{\text{stopband signal level}} = (\alpha 2^B/\sqrt{2})/(\alpha/2\sqrt{3}) = \sqrt{1.5} \times 2^{B+1}$$

(1) Thus, the minimum stopband attenuation, A_{min}, is given by

$$A_{min} = 20 \log(\sqrt{1.5} \times 2^{B+1})|_{B=12} = 80 \text{ dB}$$

(2) The signal spectrum, before and after sampling (ignoring the effects of higher order image frequencies), is given in Figure 1.16. Choosing the folding frequency, $F_s/2$, as the effective stopband frequency, then for a third-order Butterworth filter we have $A_{min} = 20 \log[1 + (f/f_c)^6]^{1/2}$. With $A_{min} = 83$ dB, $f_c = 4$ kHz, we have $f > 86.2 \text{ kHz} = F_s/2$. Thus $F_s > 172.4$ kHz. Let $F_s = 173$ kHz.

(3) The aliasing level at 4 kHz is

$$\frac{1}{\{1 + [(194 - 4)/4]^6\}^{1/2}} = 9.33 \times 10^{-6}$$

Aliasing level relative to signal level at 4 kHz is $(9.33 \times 10^{-6}) \times 100/0.7071 = 0.0013\%$.

We note from Figure 1.16 that the portion of the signal spectrum between 0 and f_c, that is the band of interest, will have aliasing levels that are less than the rms quantization noise level. A lower sampling frequency will result if we take 86.2 kHz as the aliasing frequency at 4 kHz (as in Example 1.1). In this case, $F_s = 86.2 + 4 = 90.2$ kHz and the aliasing level at 4 kHz is equal to the quantization noise level.

Other practical issues associated with sampling: accuracy and bandwidth limitations

In practical systems, instantaneous sampling shown in Figure 1.10(d) is not possible; instead, the sampling function has a finite width. This leads to a

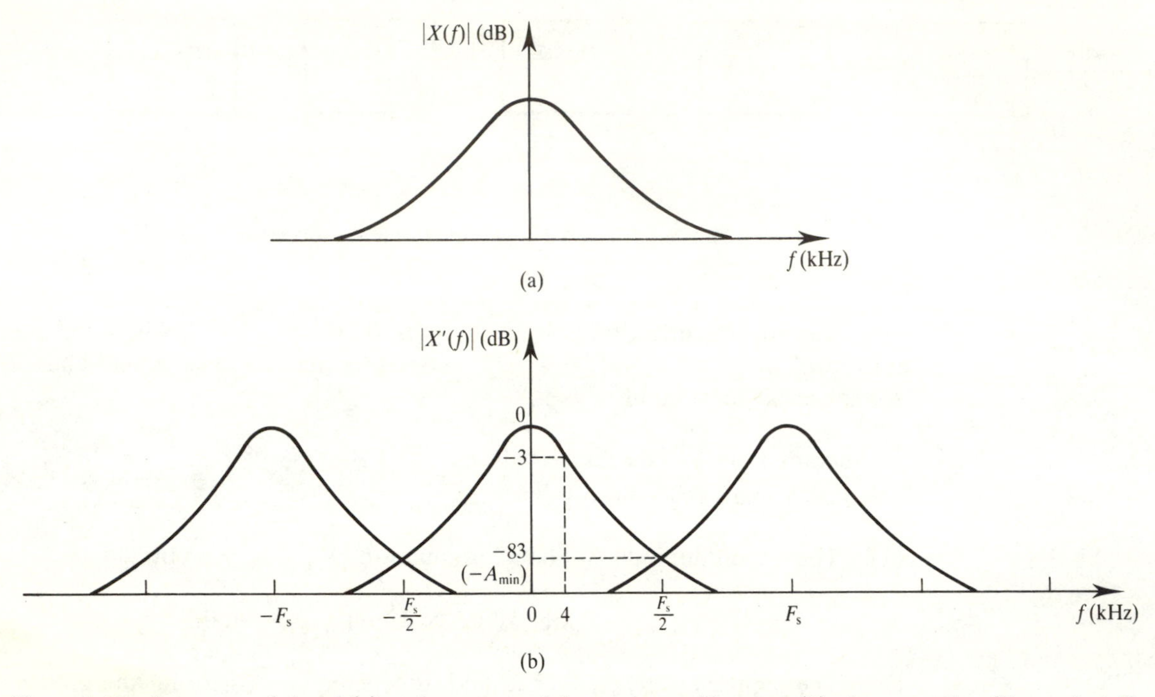

Figure 1.16 Spectrum of signal (a) at the output of the analogue filter and (b) after sampling (Example 1.2).

problem termed the aperture effect, to indicate that the signal is measured over a finite time interval instead of instantaneously. Non-zero aperture time limits the accuracy and maximum signal frequency that can be digitized because the signal may be changing while it is being sampled. A measure of the effects of aperture can be obtained if we assume that the input voltage can only change during the aperture interval by a maximum of 1/2 LSB (least significant bit) (say). Thus, for a sine wave input, the maximum frequency that can be digitized to 1/2 LSB accuracy for a system using a B-bit ADC is given by

$$f_{\max} = \frac{1}{\pi 2^{B+1} \tau} \tag{1.8}$$

where τ is the aperture time (see Example 1.3 for the proof).

Example 1.3

A real-time DSP system uses a 12-bit ADC with a conversion time of 35 μs and no sample and hold. What is the highest frequency that can be digitized to

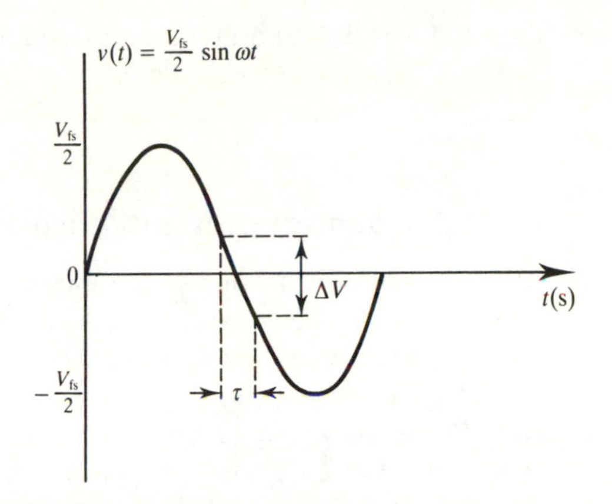

Figure 1.17 Sine wave signal for Example 1.3.

within 1/2 LSB accuracy, assuming a binary system with uniform quantization? Comment on the result.

Solution

Consider a sine wave signal with a peak amplitude equal to half the full-scale range of the ADC, $V_{fs}/2$ (Figure 1.17). In the figure, τ is the aperture time and Δv is the change in $v(t)$ during τ. The point of greatest change is at $t = 0$ and the ADC must handle this to measure the signal with desired accuracy. At this point

$$\left.\frac{dv(t)}{dt}\right|_{t=0} = (V_{fs}/2)w \cos wt = \pi f V_{fs} \quad (\text{V s}^{-1}) = \frac{\Delta v}{\tau}$$

For 1/2 LSB accuracy, $\Delta v = \alpha/2$, where $\alpha = (V_{fs}/2^B)$. Thus $\Delta v/\tau = \pi f V_{fs}$. Substituting for V_{fs} and Δv, and simplifying,

$$f_{max} = \frac{1}{\pi 2^{B+1}\tau}$$

For the DSP system, $B = 12$ and $\tau = 35\ \mu$s. Thus $f_{max} = 1.11$ Hz.

An ADC which can only convert a maximum frequency of 1.11 Hz is clearly of little use. In practice, the ADC is often preceded by a sample and hold which freezes the signal sample during conversion, enabling signals in the kilohertz range to be accurately digitized. For example, if the ADC above is preceded by a sample and hold with an aperture time of 25 ns, and an acquisition time of 2 μs, then the maximum frequency that can be converted becomes

$$2f_{max} \leqslant F_s = 1/(35 + 2 + 0.025) \times 10^{-6} \text{ kHz, that is } f_{max} = 13.5 \text{ kHz}$$

Thus the signal with a maximum frequency of 13.5 kHz would be sampled at a rate of 27 kHz, or at intervals of $(35 + 2 + 0.025)\ \mu s = 37.025\ \mu s$.

1.4.2.2 Quantization and encoding

Before conversion to digital the analogue sample is assigned one of 2^B values (Figure 1.18). This process, termed quantization, introduces an error which cannot be removed. The level of the error is a function of the number of bits of the ADC, being approximately equal to one-half of an LSB (assuming rounding). For example, a 12-bit ADC with an input voltage range of ± 10 V will have an LSB of $20/2^{12}$ mV, that is 4.9 mV and a quantization error of 2.45 mV.

For an ADC with B binary digits the number of quantization levels is 2^B and the interval between the levels, that is the quantization step size, q, is given by

$$q = V_{fs}/(2^B - 1) \approx V_{fs}/2^B \qquad \textbf{(1.9a)}$$

where V_{fs} is the full-scale range of the ADC with bipolar signal inputs. The maximum quantization error, for the case where the values are rounded up or down, is $\pm q/2$. For a sine wave input of amplitude A (such that the peak-to-peak amplitude of the signal just fills the ADC input range), the quantization step size becomes

$$q = 2A/2^B \qquad \textbf{(1.9b)}$$

The quantization error for each sample, e, is normally assumed to be random and uniformly distributed in the interval $\pm q/2$ with zero mean. In this case, the quantization noise power, or variance, is given by

$$\begin{aligned} \sigma_e^2 &= \int_{-q/2}^{q/2} e^2 P(e)\, de \\ &= \frac{1}{q} \int_{-q/2}^{q/2} e^2\, de \\ &= \frac{q^2}{12} \end{aligned} \qquad \textbf{(1.10)}$$

For the sine wave input, the average signal power is $A^2/2$. The signal-to-quantization noise power ratio (SQNR), in decibels, is

$$\begin{aligned} \text{SQNR} &= 10 \log\left(\frac{A^2/2}{q^2/12}\right) = 10 \log\left(\frac{3 \times 2^{2B}}{2}\right) \\ &= 6.02B + 1.76 \text{ dB} \end{aligned} \qquad \textbf{(1.11)}$$

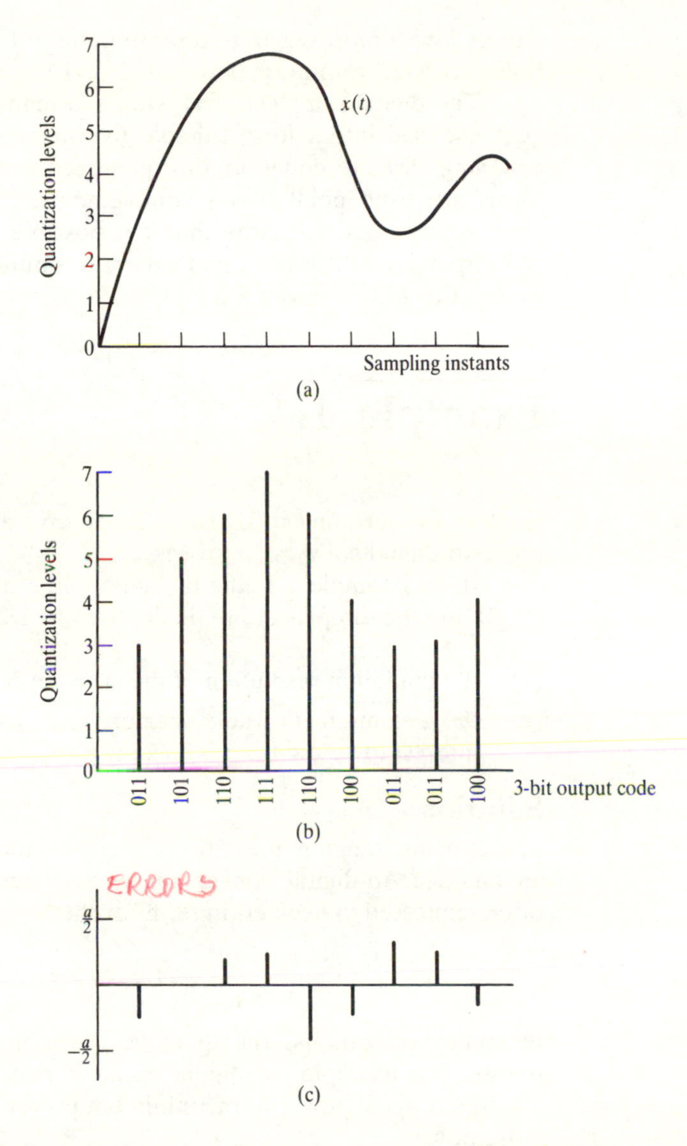

Figure 1.18 Quantization of analogue signal samples (Quantization errors in (c) are obtained by subtracting the signal samples in (a) from the quantized samples (3-bit quantizer) in (b)).

This is a theoretical maximum. In practice, when real-world input signals are used, achievable SQNR is less than this value. However, the SQNR increases with the number of bits, B. Practical factors such as speed, inherent signal-to-noise ratio (SNR) of the analogue signal, and costs limit the number of bits used. For example, it is unnecessary to use a converter that yields an accuracy better than the SNR of the analogue signal to be converted, since this will merely give a more accurate representation of the noise. In many DSP applications, an ADC resolution between 12 and 16 bits is adequate.

The digital samples, $x(n)$, which in many cases are in a binary form, are next encoded into a form suitable for further manipulation. Encoding means assigning discrete codes to the quantized samples. In DSP, the commonest forms are fixed point (two's complement), floating point and block floating point representations. Note that it is possible to perform the three operations of sampling, quantization, and encoding simultaneously. This is the case when we use the ADC without a sample and hold.

Example 1.4

Explain the meaning of dynamic range and aperture time in relation to analogue-to-digital conversion process.

If, in Example 1.1, the dynamic range of the ADC is to be greater than 70 dB and the samples are to be digitized to 1/2 LSB accuracy, determine

(1) the minimum resolution of the ADC in bits, and

(2) the maximum allowable aperture time, assuming the highest frequency of interest to be digitized is 20 kHz.

Solution

The dynamic range is the ratio of the maximum to minimum signal levels that an analogue-to-digital conversion system can handle. The dynamic range is often expressed in decibels in terms of the number of bits in the converter:

$$D = 20 \log_{10} 2^B \tag{1.12}$$

In some applications, the dynamic range is defined in terms of the signal power. For example, in digital audio it may be defined as the ratio of the maximum signal power to the minimum power which can be discerned from the noise power.

For the ADC when used alone, the aperture time is essentially the conversion time of the ADC and refers to the period of time over which the analogue input must remain stable so that accurate conversion may be made. In relation to the sample and hold, it is the time required to achieve a hold following the hold command.

(1) Using the expression for D, we have

$$70 = 20 \log_{10} 2^B$$

from which $B = 11.62$. Let $B = 12$ bits (the nearest integer).

(2) The minimum aperture τ is given by

$$\tau = 1/2^{B+1}\pi f_{\max} = 1/(2^{13} \times \pi \times 20 \times 10^3)\ \text{s} = 1.94\ \text{ns}$$

This small aperture time calls for the use of a sample and hold ahead of the ADC.

1.4.3 Digital-to-analogue conversion process: signal recovery

The digital-to-analogue conversion process is employed to convert the digital signal into an analogue form after it has been digitally processed, transmitted or stored. The reason for such a conversion may be, for example, to generate an audio signal to drive a loudspeaker (as in the compact disc system) or to sound an alarm. The most commonly used arrangement is shown in Figure 1.19, and can be seen to consist of two main components: the DAC (digital-to-analogue converter) and a lowpass filter sometimes called a reconstruction, smoothing or anti-image filter.

1.4.3.1 The DAC

The basic DAC accepts parallel digital data and produces an analogue output signal which is related to the digital code at its input. A register is used to buffer the DAC's input to ensure that its output remains the same until the DAC is fed the next digital input. The register may be external to the DAC or it may form part of the DAC chip, as in Figure 1.19. In some applications,

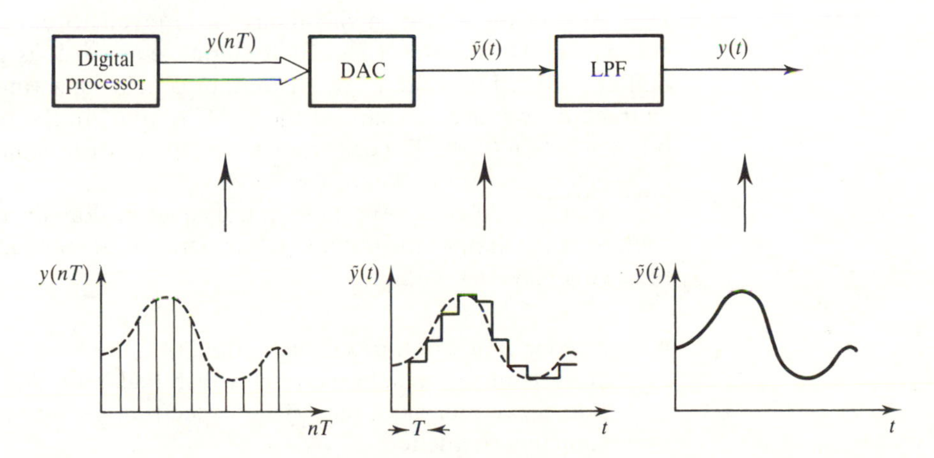

Figure 1.19 Digital-to-analogue conversion process used to recover the analogue signal after digital processing. Note that the inputs to the DAC are a series of impulses, while the output of the DAC has a staircase shape as each impulse is held for a time T (s).

Input

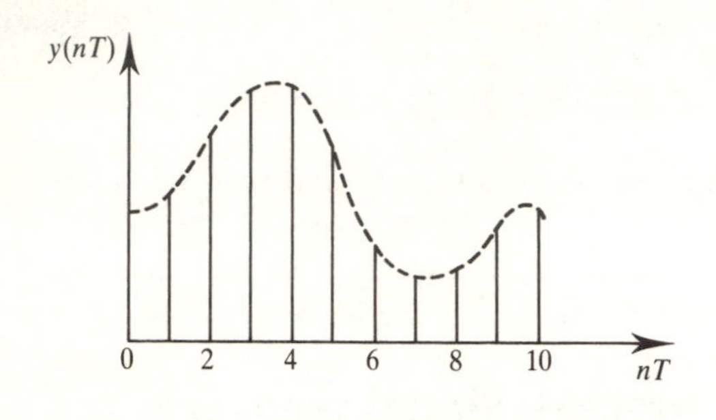

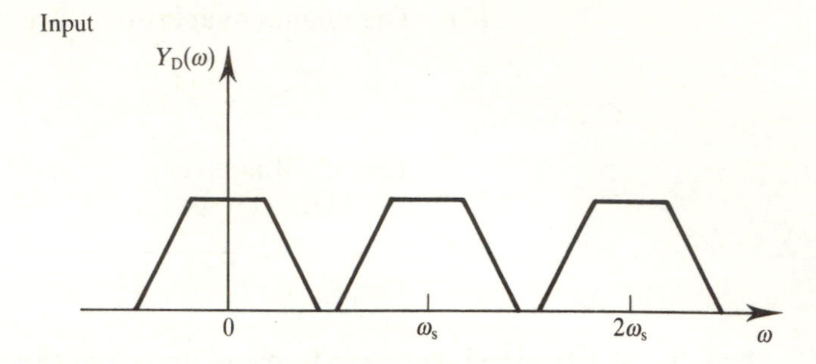

Output

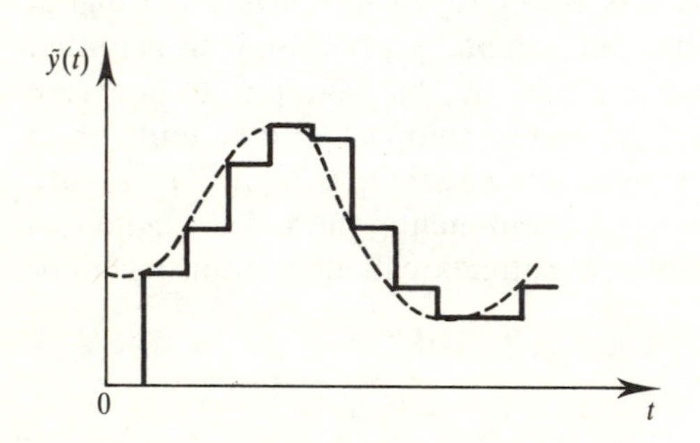

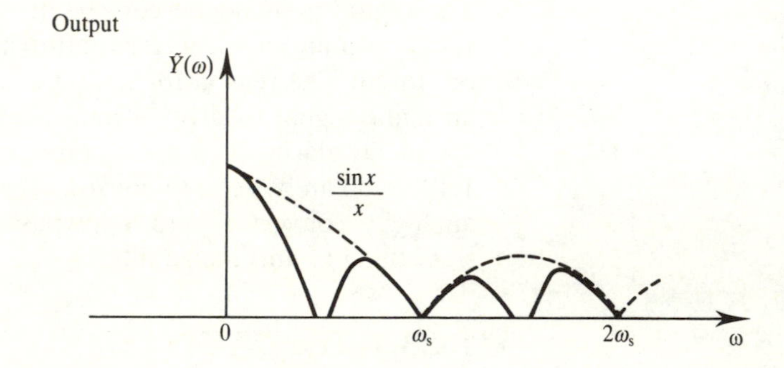

Figure 1.20 Time and frequency domain representation of the DAC input and output signals. Notice that the $\sin x/x$ effect manifests itself as a lowpass filter.

additional circuitry may be required to prevent false digital code from generating transient spikes at the DAC output.

The DAC shown in the figure is referred to as a zero-order hold. By comparing its output, $\tilde{y}(t)$, and its input, $y(nT)$, it is evident that for each digital code fed into the DAC, its output is held for a time T. The result is the characteristic staircase shape at the DAC output. In the frequency domain, the holding action of the DAC introduces a type of distortion known as the $\sin x/x$ or aperture distortion, where $x = wT/2$.

Figure 1.20 shows the time and frequency domain representations of the signals at the input and output of the zero-order hold DAC, from which the following may be noted.

- The input and output signals of the DAC are both wideband signals. Each consists of the signal spectrum (which had been digitized) plus an infinite number of images of the original spectrum centred at the multiples of the sampling frequency.
- The amplitude of the output signal spectrum is multiplied by the $\sin x/x$ function, which acts like a lowpass filter, with the image frequencies heavily attenuated.

The $\sin x/x$ effect is due to the holding action of the DAC and, in signal recovery, introduces an amplitude distortion. The average error due to the effect at a given frequency may be expressed as a percentage deviation from unity:

$$(1 - \sin x/x) \times 100\% \qquad \textbf{(1.13)}$$

For the zero-order hold, the function $\sin x/x$ falls to about 4 dB at half the sampling frequency ($F_s/2$) giving an average error of about 36.4%. Aperture error can be eliminated by equalization. In practice this can be achieved by first applying the signal, before converting it to analogue, through a digital filter whose amplitude–frequency response has an $x/\sin x$ shape.

In some applications, the digital processor may be used to insert or interpolate between the actual sample points applied to the DAC. This helps to smooth out the analogue signal and gives a substantially better result than the simple zero-order hold. Another approach that is becoming popular where high quality audio signals are required is to perform the digital-to-analogue conversion at a much higher rate than specified by the sampling theorem (see, for example, Goedhart *et al.*, 1982) using multirate techniques. This has the advantage of improved signal-to-noise performance and simplifies the anti-image filter. Further details are given in Section 1.5.3 and Chapter 8.

1.4.3.2 Anti-imaging filtering

The output of the DAC contains unwanted high frequency or image components centred at multiples of the sampling (that is, the update) frequency as well as the desired frequency components. Depending on the application, the h.f. components may cause undesirable side effects. For example, in the compact disc player, although the image frequencies are not audible they could overload the player amplifier and cause intermodulation products with the desired baseband frequency components. The result is an unacceptable degradation in the audio signal quality.

The role of the output (that is, anti-imaging) filter is to smooth out the steps in the DAC output thereby removing the unwanted h.f. components. The roll-off requirements for the output filter depend on the effect the analogue signal will have on subsequent analogue stages. In general, the requirements of the anti-imaging filter are similar to those of the anti-aliasing filter.

Example 1.5

Figure 1.21(a) depicts the set-up used to recover an analogue signal, after it has been digitally processed in a certain real-time digital audio system. The analogue signal has a baseband that extends from dc to 20 kHz and the digital-to-analogue converter is updated at a rate of 176.4 kHz. The image frequencies

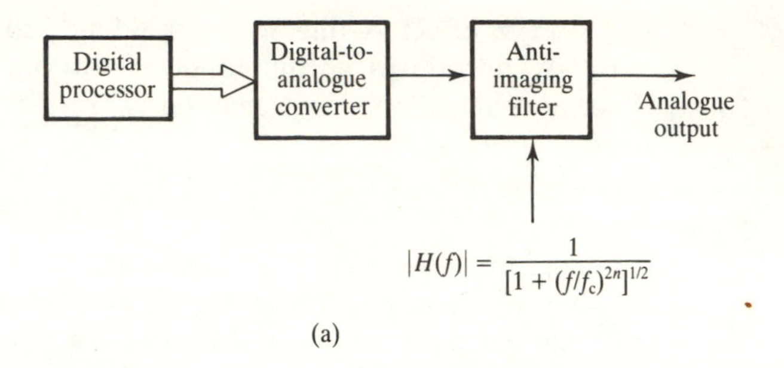

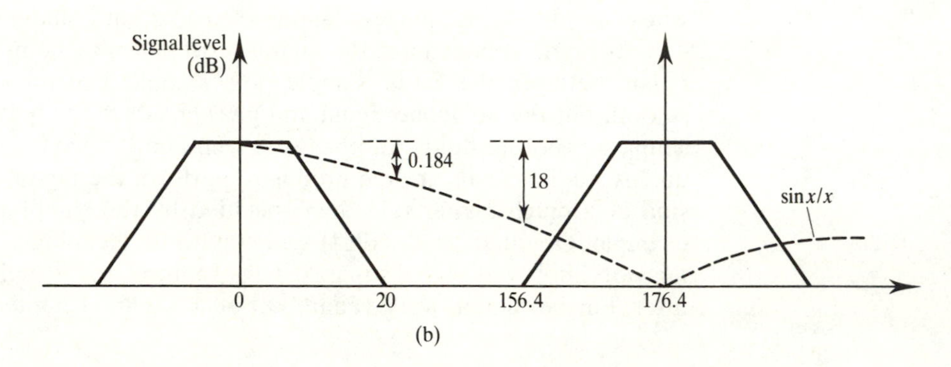

Figure 1.21 (a) Back-end of a real-time DSP system; (b) spectrum at the DAC output.

are to be suppressed by at least 50 dB and the signal components of interest are to be altered by a maximum of 0.5 dB. Determine the minimum values for the order and cutoff frequency for the anti-image filter, assuming that it has a Butterworth characteristic. State any reasonable assumptions made.

Solution

Assuming a zero-order hold, the spectrum at the DAC output is the product of the signal spectrum and the $\sin x/x$ response; see Figure 1.21(b). Attenuation of the signal due to the $\sin x/x$ spectrum at the two critical frequencies, 20 kHz and 156.4 kHz (the image frequencies closest to the baseband), is as follows:

$$\text{at 20 kHz:} \qquad \frac{\sin x}{x} = 0.9789 \text{ (with } x = wT/2) = -0.184 \text{ dB}$$

$$\text{at 156.4 kHz:} \qquad \frac{\sin x}{x} = 0.125 = -18 \text{ dB}$$

Thus, in the passband the output filter should not have more than $0.5 - 0.184 = 0.316$ dB deviation. In the stopband an additional attenuation of at least

$50 - 18 = 32$ dB is required. Thus

$$20 \log [1 + (20/f_c)^{2n}]^{1/2} \leqslant 0.316 \text{ dB}$$
$$20 \log [1 + (156.4/f_c)^{2n}]^{1/2} \geqslant 32 \text{ dB}$$

Solving the simultaneous equations for n gives $n = 2.4 \simeq 3$ (integer) and $f_c = 30.76$ kHz.

1.4.4 Digital signal processors

DSP systems are characterized by real-time operation, with emphasis on high throughput rate, and the use of algorithms requiring intensive arithmetic operations, notably multiplication and addition or multiply–accumulate. These lead to a heavy flow of data through the processor.

The architectures of standard microprocessors are unsuited to the DSP characteristics and this has led to the development of a new kind of processor whose architecture and instruction set are tailored to DSP operations. The new processors or DSP chips have features that include the following.

- Built-in hardware multiplier(s) to allow fast multiplications. Newer DSP chips incorporate single-cycle multiply–accumulate instructions and some have several multipliers working in parallel.
- Separate busses/memories for program and data – the well-known Harvard architecture, which permits an overlap of instruction fetch and execution.
- Cycle-saving instructions for branching or looping. For example, the following instructions for Texas Instruments' TMS320C25 reduce significantly the number of cycles and program size for a digital filter:

```
RPTK  N     ;Repeat the next instruction N times
MACD        ;Move data into memory, multiply and accumulate with delay
```

- Very fast raw speed. For example, the TMS320C25 uses a 40 MHz clock and has a cycle time of 100 ns.
- Use of pipelining which reduces instruction time and increases speed.

Newer DSP chips are faster and more versatile. Some now have floating point arithmetic capabilities, and incorporate features found in standard microprocessors such as serial line, extended memory space, timers, multilevel interrupts. A more detailed discussion of DSP chips and how to design with them is presented in Chapters 11 and 12.

1.4.5 Constraints of real-time signal processing with analogue input/output signals

The main constraints and errors introduced by the analogue-to-digital and digital-to-analogue conversion processes in real-time DSP have already been discussed. We outline here these constraints and possible solutions.

- The use of a finite number of bits to represent data introduces an intrinsic error, the quantization error, which is propagated into subsequent signal processing. Two ways of dealing with this error are to increase the resolution of the ADC and to oversample the signal followed by a further DSP to improve the SNR (see Section 1.4 and Chapter 8 for details).
- High resolution ADC and DAC are in general slow (except for the very expensive converters). Typically, an ADC takes a few microseconds to convert an analogue sample, and a DAC takes a significant fraction of a microsecond to settle. Thus these delays impose a limit on the maximum sampling frequency achievable. In fact, with current technology, the ADCs/DACs are now a major bottleneck in most real-time DSP applications.
- The ADCs/DACs are subject to a variety of other errors which include temperature effects and nonlinearities. Thus a good real-time DSP system with analogue inputs should have good quality analogue input/output sections.
- The output of the sample and hold is wideband (because of the image frequencies) and will increase the noise at the ADC input.
- Aliasing error from signal energy outside of the band of interest is always present. To reduce aliasing to an acceptable level, bandlimit the signal before sampling, and oversample if possible.
- The use of a zero-order DAC introduces the $\sin x/x$ effect which progressively reduces the high frequency components of a signal. This can be compensated for by using a digital filter with an $x/\sin x$ response.
- Errors are introduced by the anti-aliasing filters. Typically, these are amplitude and phase errors. The amplitude responses of these filters are not flat in the band of interest. Analogue filters with reasonably good amplitude response have invariably poor phase response, which means that the harmonic relationships between signal components are distorted. In multichannel systems, the problem is compounded because the distortions introduced by the analogue signal conditioners are different for each channel and may need to be compensated for.
- Sample and hold errors include acquisition time, aperture uncertainty, droop errors during the conversion interval, and feedthrough in the hold mode.
- The trend in modern DSP systems, especially digital audio systems such as the compact disc player, is to use single-bit ADCs and DACs. These

newer devices exploit the advantages of multirate techniques (see Chapter 8 for details).

1.5 Application examples

1.5.1 Speech synthesis and recognition

1.5.1.1 Speech synthesis

Synthetic speech has in the past been perceived as sounding mechanical. However, advances in semiconductor technology and DSP have made it economically possible to obtain a speech quality that is indistinguishable from human speech.

The Speak & Spell (Frantz and Wiggins, 1982) is an example of a successful commercial product with speech output which many readers may be familiar with. It is an electronic learning aid for children and uses the LPC (linear predictive coding) technique, where the actual human speech to be reproduced later is modelled as the response of a time-varying digital filter to a periodic or random excitation signal (Figure 1.22). The periodic excitation is used for voiced sounds (for example vowels), and represents the air flow through the vocal cords as they vibrate. The random excitation is used for unvoiced sounds (for example S, SH), and represents the noise created by forcing air past constrictions in the vocal tract. The filter models the behaviour of the vocal tract. Human speech contains a lot of redundant information. The LPC retains only the relevant information necessary to preserve the characteristics of speech such as intonation, accent, and dialect, allowing minutes of high quality sound to be held in a moderately sized memory.

In the Speak & Spell, use is made of the TMS5100 speech synthesizer

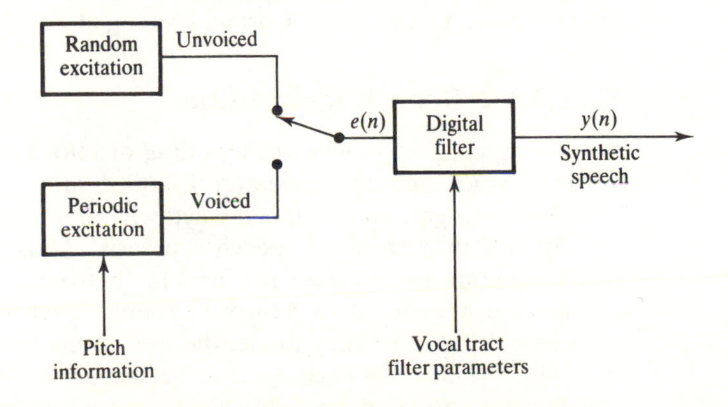

Figure 1.22 Linear predictive coding of speech.

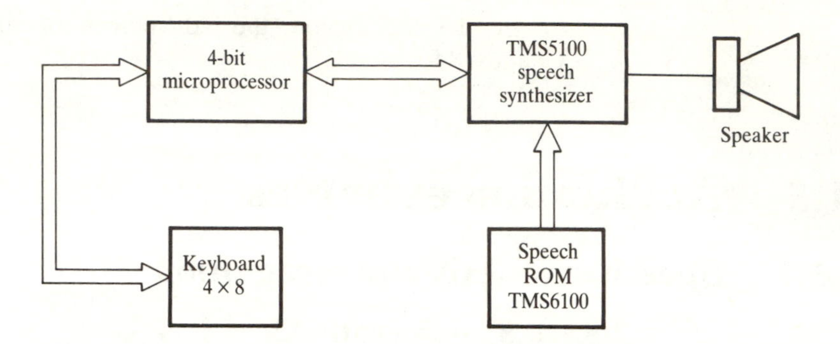

Figure 1.23 A configuration for the Speak & Spell learning aid.

chip which incorporates all the components of an LPC model (a digital filter and excitation sources) as well as a decoder and an 8-bit DAC. The synthesizer chip operates in conjunction with a 4-bit microprocessor and two 128-kbit ROMs which together hold a vocabulary of about 300 words and phrases (Figure 1.23). Speech information is stored in the ROMs in frames (representing 25 ms of speech) where each frame is characterized by a set of 10 or 12 LPC parameters. The frame parameters are fed to the synthesizer once every 25 ms and used to update the digital filter coefficients and to select the excitation source and its energy level. The output of the digital filter is converted to analogue and applied to the loudspeaker to produce the required sound with a specific pitch, amplitude, and harmonic content. To obtain a smooth transition in speech spectrum, once every 3 ms, the synthesizer updates the LPC parameters by interpolating between the previous and the current frame parameters.

In one mode of operation, the child is asked to spell a word. The child enters the word, one letter at a time via the keyboard. If the spelling is correct, on pressing 'enter', the Speak & Spell responds 'That is right' or 'Correct'. If it is wrong it says 'Wrong, try again'. If the next attempt is wrong, it chides, 'That is incorrect', and adds, 'Correct spelling of . . . is . . .'.

1.5.1.2 Speech recognition

Voice recognition involves inputting of information into a computer using human voice and the computer listening and recognizing the human speech. Voice recognition is still being actively researched as problems posed are more difficult than those of speech synthesis. Thus, successful commercial speech recognition systems are few and far between. The more successful ones are speaker-dependent single-word systems. Such systems operate in one of two modes. In the training mode, the user trains the system to recognize his or her voice by speaking each word to be recognized into a microphone. The system digitizes and creates a template of each word and stores this in its memory. In the recognition mode, each spoken word is again digitized and its template

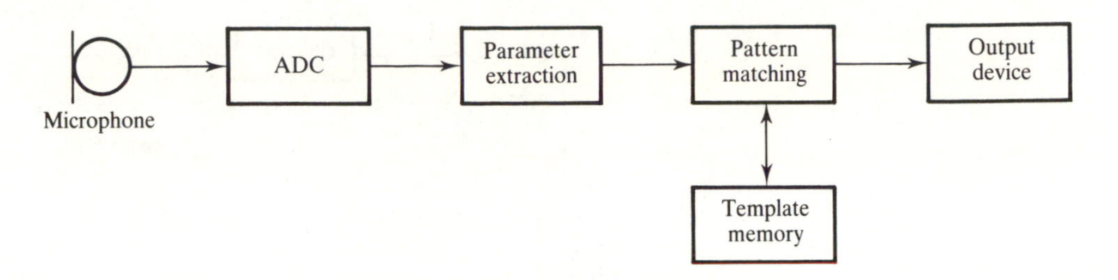

Figure 1.24 Block diagram of a speech recognition system.

compared with the templates in memory. When a match occurs the word has been recognized and the system informs the user or takes some action. The performance of such systems is affected by speakers not pausing long enough after each word, background noise, and how clearly and carefully the word is spoken. The two important DSP operations in a recognizer are parameter extraction, where distinct patterns are obtained from the spoken word and used to create a template, and pattern matching, where the templates are compared with those stored in memory; see Figure 1.24.

For most people, voice is the most natural form of communication, being faster than writing or typing. Thus, in the office environment, voice systems now exist which allow application programs to be driven by voice commands instead of by keyboard entries. Systems which will allow the usual office documents, such as letters and memos, to be created and sent by voice are envisaged. Word recognizers are being incorporated into consumer products, such as voice-operated telephone dialling systems, and are used in voice-activated domestic appliances for disabled people with limited movement. This increases their independence by enabling them to perform simple tasks such as turning on/off lights, radio or TV.

There are of course numerous potential applications of voice recognition. However, it appears that future advances in this area will rely significantly on artificial intelligence (AI) techniques because of the need for machines to understand as well as recognize speech.

1.5.2 Adaptive telephone echo cancellation

Echoes arise primarily in communication systems when signals encounter a mismatch in impedance. Figure 1.25 shows a simplified long-distance telephone circuit. The hybrid circuit at the exchange converts the two-wire circuit from the subscriber premises to a four-wire circuit, and provides separate paths for each direction of transmission. This is largely for economic reasons, for example to allow the multiplexing or simultaneous transmission of many calls.

Ideally, the speech signal originating from customer A travels along the upper transmission path to the hybrid on the right and from there to customer B, while that from B travels along the lower transmission path to A. The

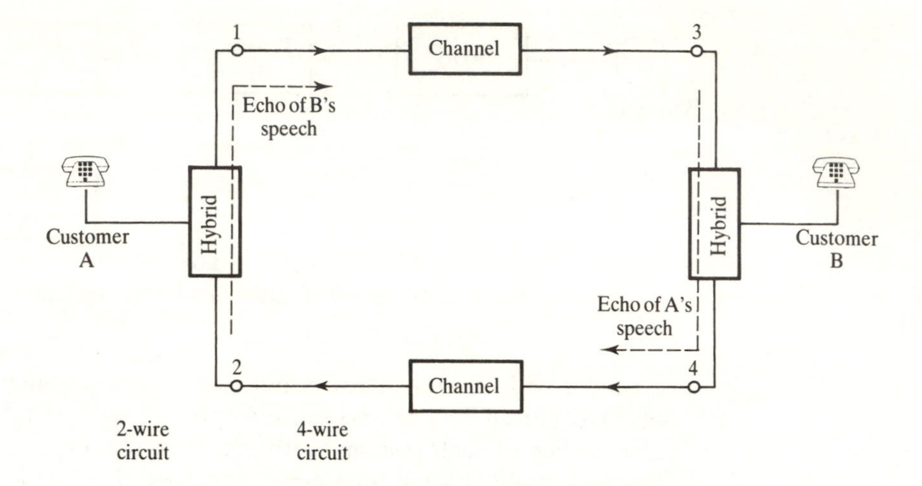

Figure 1.25 Simplified long-distance telephone circuit.

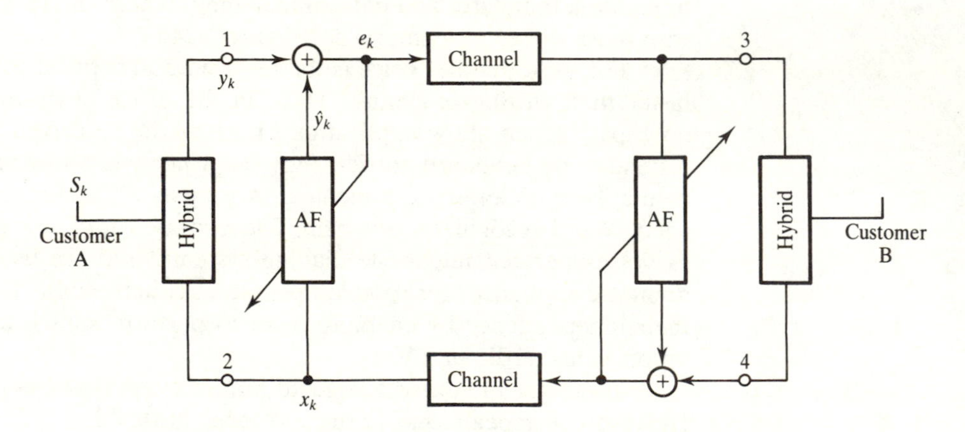

Figure 1.26 Echo cancellation in long-distance voice telephony.

hybrid network at each end should ensure that the speech signal from the distant customer is coupled into its two-wire port and none to its output port. However, because of impedance mismatches, the hybrid network allows some of the incoming signals to leak into the output path and return to the talker as an echo. When the telephone call is made over a long distance (for example using geostationary satellites) the echo may be delayed by as much as 540 ms and represents an impairment that can be annoying to users. The impairment increases with distance. To overcome this problem, echo cancellers are installed in the network in pairs, as illustrated in Figure 1.26 (Duttweiller, 1978).

At each end of the communication system (Figure 1.26), the incoming signal, x_k, is applied both to the hybrid and the adaptive filter (AF). The cancellation is achieved by making an estimate of the echo and subtracting it

Table 1.3 Comparison of features of the LP record and the compact disc (CD).

Features	*LP record*	*Compact disc*
Frequency response	30 Hz to 20 kHz (±3 dB)	20 Hz to 20 kHz (+0.5 to −1 dB)
Dynamic range	70 dB (at 1 kHz)	>90 dB
Signal-to-noise ratio	60 dB	>90 dB
Harmonic distortion	1–2%	0.004%
Separation between stereo channels	25–30 dB	>90 dB
Wow and flutter	0.03%	Not detectable
Effect of dust, scratches and fingermarks	Causes noise	Leads to correctable or concealable errors
Durability	Hf response degrades with playing	Semipermanent
Stylus life	500–600 h	Semipermanent
Playing time	40–45 min (both sides)	50–75 min (extendable)

from the return signal, y_k. The estimate of the echo is given by

$$\hat{y}_k = \sum_{i=0}^{N-1} w_{k+1}(i)x_{k-i}$$

where x_k are the samples of the incoming signal from the far-end speaker, and $w_k(i)$, $i = 0, 1, \ldots, N-1$, is the estimate of the impulse response of the echo path at the discrete time k.

1.5.3 The compact disc digital audio system

Most readers are familiar with the undesirable sounds that accompany music reproduced from LP (long play) records when there is damage, a scratch, dirt or fingermarks on the records. The compact disc (CD) system is a state-of-the-art system which overcomes the disadvantages of the LP record. Table 1.3 compares the important features of the LP and the CD (Bloom, 1985).

In the CD, the information is recorded in digital form as a spiral track that consists of a succession of pits (Figure 1.27) (Carasso *et al.*, 1982). Each bit recorded on the CD occupies an area of only 1 μm^2, that is 10^6 bits per square millimetre, leading to very high density of information on the CD.

A simplified block diagram of the audio signal processing in the CD, during recording, is depicted in Figure 1.28. The analogue audio signal in each of the stereo channels is sampled at 44.1 kHz and digitized. Each sample is represented as a 16-bit code, representing a dynamic range of 90 dB. Thus at each sampling instant 32 bits are obtained, 16 bits each from the left and right

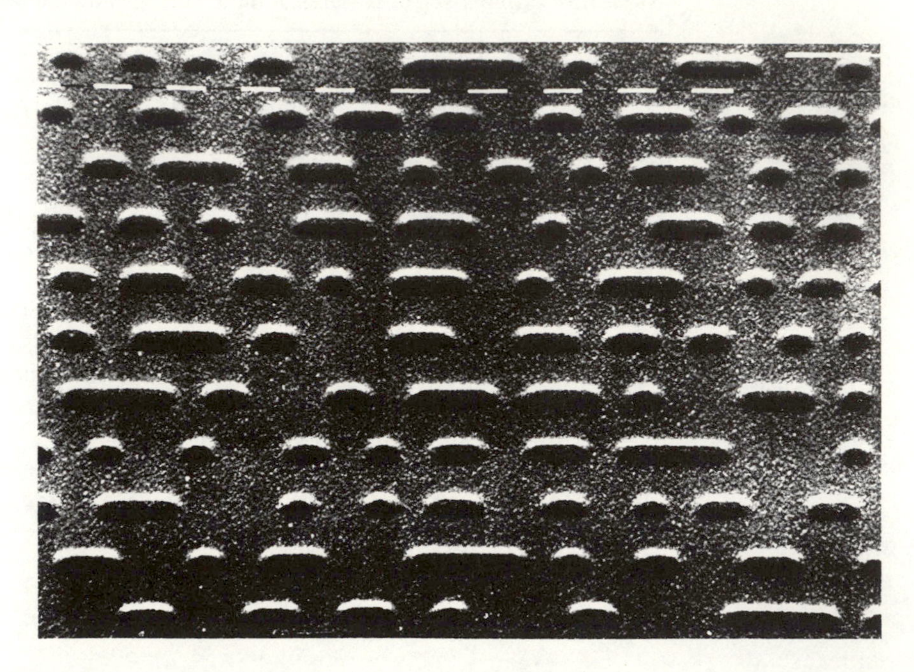

Figure 1.27 Laser-cut pits on a compact disc. Each pit is 0.5 μm wide, 0.8–3.5 μm in length, and has a depth of 0.11 μm. The distance between tracks is 1.6 μm. (Reproduced from *Philips Technical Review,* **40**(6), 1982.)

audio channels. The digital samples are encoded using a two-level Reed–Solomon coding scheme to enable errors to be detected and corrected or concealed during reproduction of the audio signal. Additional bits are added for control and display information for the listener. The resulting data bit streams are then modulated to translate them into a form more suitable for disc storage. The EFM (eight-to-fourteen modulation) scheme translates each byte in the data stream into a 14-bit code. The resulting channel bit stream, after further processing, is used to control a laser beam, causing the digital information to be recorded onto a light-sensitive layer on a rotating glass disc. A photographic developing process is then used to produce a pattern of pits on the master disc from which users' compact discs are subsequently produced.

In the CD player, during reproduction, the tracks on the disc are optically scanned at a constant velocity of 1.2 $\mathrm{m\,s^{-1}}$ while the disc rotates at a speed of between 8 $\mathrm{rev\,s^{-1}}$ and about 3.5 $\mathrm{rev\,s^{-1}}$ to pick up the recorded information (Figure 1.29). The digital signal from the disc is first demodulated, and any errors in the data are detected and, if possible, corrected. The errors may be due to manufacturing defects, damage, fingermarks or dust on the disc. If the errors are not correctable, they are concealed either by replacing the sample in error with a new one obtained by interpolating between adjacent correct samples or, if more than one sample is in error, by zeroing them (muting).

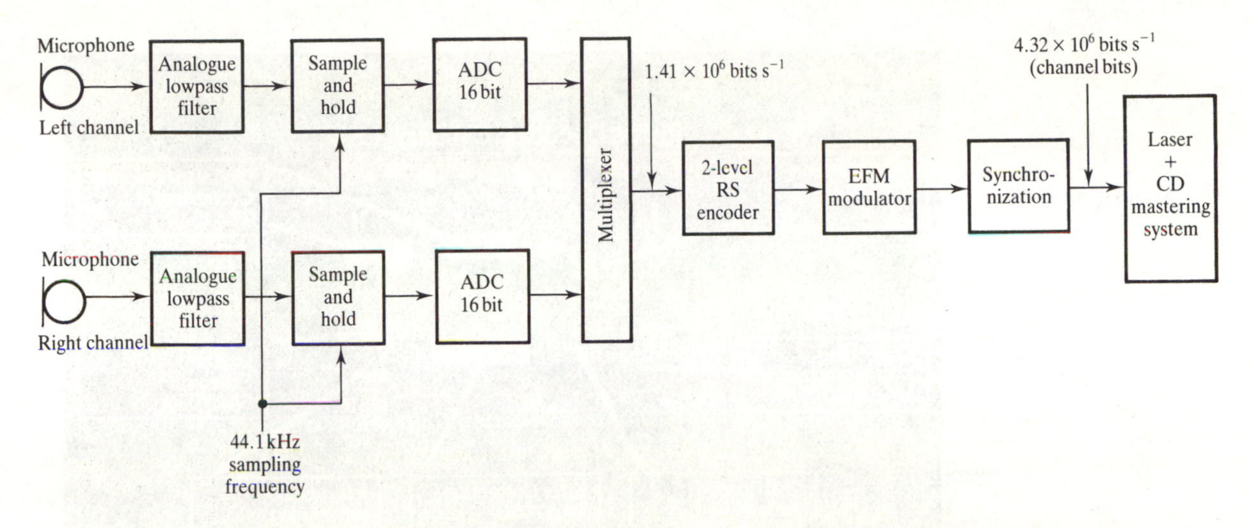

Figure 1.28 Simplified block diagram of the audio signal processing and recording in the compact disc system.

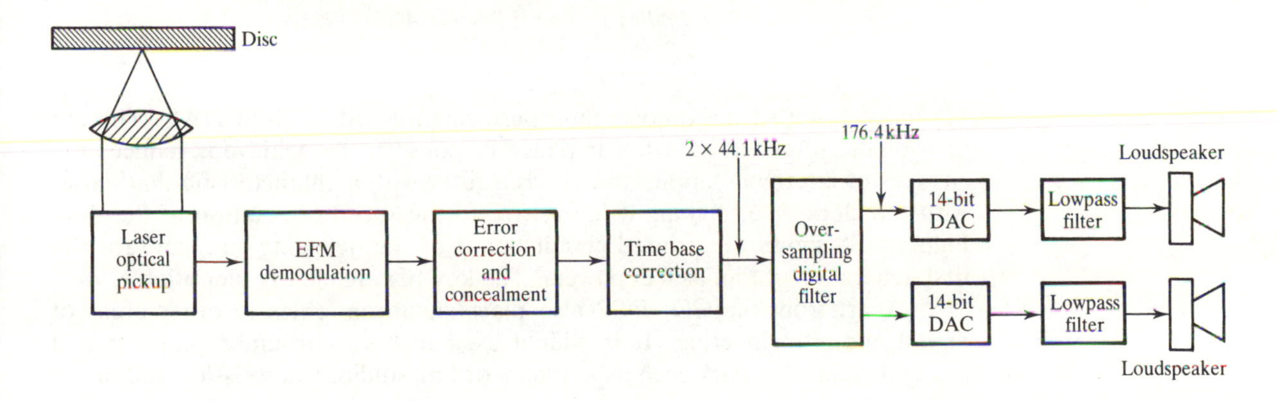

Figure 1.29 Audio signal reproduction in the compact disc system.

After error correction and/or concealment, the resulting data is a series of 16-bit words, each representing an audio signal sample. These samples could be applied directly to a 16-bit DAC and then analogue lowpass filtered. However, this would require analogue filters with very tight specifications. In particular, the levels of frequencies above 20 kHz should be reduced by at least 50 dB relative to the maximum audio signal, and the filter should have a linear phase characteristic in the audio band to avoid impairments to the sound waveform. To avoid this, the digital signals are processed further by passing them through a digital filter operating at four times the audio sampling rate of 44.1 kHz. The effect of raising the sampling frequency is to make the output of the DAC smoother, simplifying the analogue filtering requirements. It also helps to

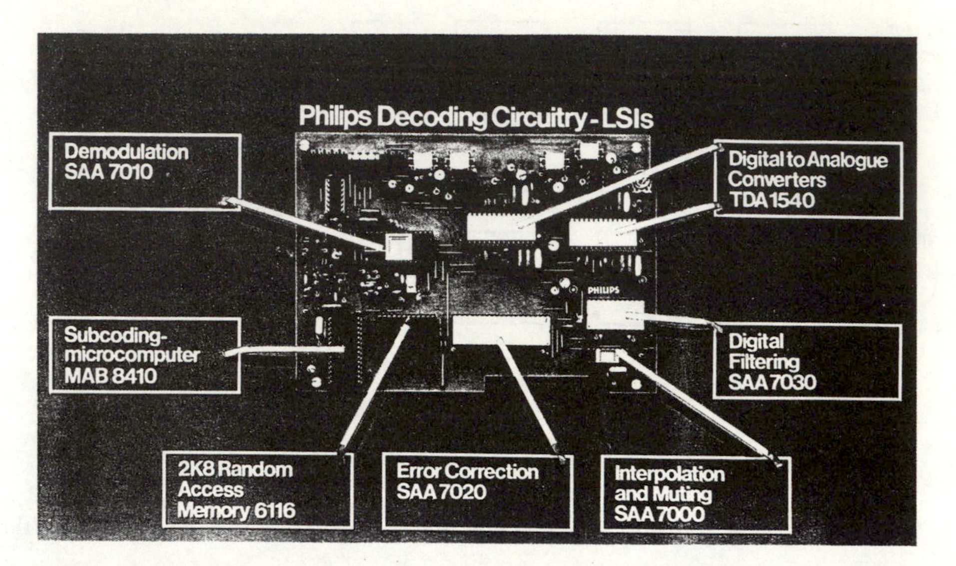

Figure 1.30 The printed circuit board for the decoding circuitry of the Philips CD player. (Reproduced from *Philips Technical Review,* **40**(6), 1982.)

achieve a 16-bit signal-to-noise ratio performance with a 14-bit DAC. The use of a digital filter allows a linear phase response to be achieved, reduces the chances of intermodulation, and yields a filter with a characteristic that varies with the clock rate, making it insensitive to the speed of rotation of the disc. Figure 1.30 shows the printed circuit board of the decoding circuitry, for the first generation of Philips CD players. The key ICs are clearly identifiable.

Apart from the CD, DSP also plays significant roles in other areas of digital audio engineering. It is widely used in both consumer products and professional audio work such as in the recording studio, transmission and distribution of TV programs by the broadcasting authorities, film and music industries. Specific uses of DSP in digital audio work, some of which we have already mentioned, include the following.

- Use of advanced DSP techniques in encoding, detecting, and correcting or concealing errors caused by dropouts, and in eliminating wow and flutter during replay, ensuring that the limitations imposed by the recording medium (magnetic or optical) no longer dictate the quality attainable in recording and reproduction. Thus the output of a recorder will sound the same for different brands of tapes with similar error rates.
- Enhancement of the listening environment and enrichment of the sound. For example, simple digital filter structures have been used to create echoes, natural reverberation, and chorus effects.
- Synthesis of sounds that are close imitations of musical instruments and of those no other instruments can produce.

- In the creation and use of sound effects, for example gunshots, footsteps, applause, car sounds, punches, in TV commercials, cartoons and motion pictures to enhance the illusion of reality or to add credibility to a scene.
- Enhancement of archival recordings or forensic recordings.

1.6 Summary

In this chapter, the meaning of DSP has been explained, the areas of application have been discussed, and key DSP operations have been identified. A generalized view of a real-time DSP system is one which consists of an analogue-to-digital conversion section, a digital processor and a digital-to-analogue conversion section. The time required to convert between analogue and digital in such a system limits the maximum signal bandwidth that can be handled, and the devices used in the conversion processes may introduce significant errors or signal degradation. Most of the errors can be minimized by a careful choice of devices (ADCs, DACs, and so on) and the system parameters (sampling frequency and so on). For example, aliasing can be reduced by sampling at a sufficiently high frequency and employing adequate bandlimiting filters.

Specific application examples discussed have shown that DSP is already making a significant impact in consumer and professional electronics.

Problems

1.1 State, with justifications, two major advantages and two major disadvantages of DSP compared with analogue signal processing system design.

1.2 A signal has the spectrum depicted in Figure 1.31. Determine the minimum sampling frequency to avoid aliasing. Assume the signal is sampled at a rate of 16 kHz, and sketch the spectrum of the sampled signal in the interval ±16 kHz. Indicate the relevant frequencies including the foldover frequency in your sketch.

1.3 Explain why sampling theorem considerations alone are not sufficient for establishing the actual sampling frequencies used in practical DSP systems.

1.4 Explain clearly the role of the anti-aliasing filter and the anti-imaging filter in real-time DSP systems. Why are the requirements of the two filters often the same in DSP systems?

1.5 The requirements for the analogue input section of a certain real-time DSP system are

frequency band of interest	0–4 kHz
maximum permissible passband ripple	$\leqslant$ 0.5 dB
stopband attenuation	$\geqslant$ 50 dB

Determine the minimum order of an anti-aliasing filter with Butterworth characteristics and a suitable sampling frequency to satisfy the requirements.

1.6 A real-time DSP system uses a linear 16-bit ADC in the bipolar mode, with an input range of ±5 V. What is the maximum quantization

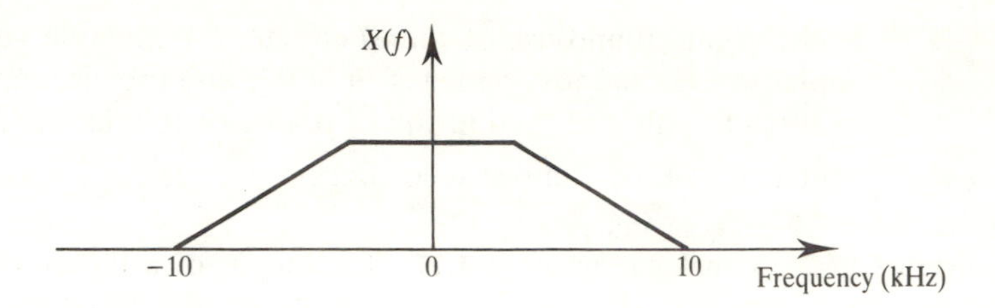

Figure 1.31

error? Calculate the theoretical maximum SQNR, in decibels, for the system.

1.7 A DSP system is preceded by a sample and hold with an aperture time of 10 ns and acquisition time of 1 μs followed by an 8-bit ADC. Determine the maximum ADC conversion time to support a sampling frequency of 100 kHz.

1.8 The analogue input to a real-time DSP system is digitized with a 16-bit ADC in the bipolar mode. The peak-to-peak amplitude of the input signal is in the range ±10 V and lies in the band 0–10 kHz. Estimate the minimum

(1) stopband attenuation, $A_{\min}$, for the anti-aliasing filter, and

(2) sampling frequency, F_s, to keep the aliasing error in the passband to just below the quantization noise level (assume a sixth-order Butterworth filter is used for the anti-aliasing filtering).

1.9 An analogue signal with a uniform power spectral density is bandlimited by a filter with the following amplitude response:

$$|H(f)| = \frac{1}{[1 + (f/f_c)^6]^{1/2}}$$

where $f_c = 3.4$ kHz. The signal is digitized using a linear 8-bit ADC. Determine the minimum sampling frequency so that the maximum aliasing error is less than the quantization error level in the passband.

1.10 A sinusoidal signal with peak-to-peak amplitude of 5 V is digitized with a 16-bit ADC. Assuming linear quantization, determine

(1) the quantization step size, and

(2) the rms signal-to-quantization noise ratio.

State any assumptions made.

1.11 The analogue input to a DSP system is digitized at a rate of 100 kHz with uniform quantization. Assuming a sine wave input with a peak-to-peak amplitude of ±5 V find the minimum number of bits for the ADC to achieve a SQNR of at least 90 dB. State any reasonable assumptions.

1.12 Show that the signal-to-quantization noise ratio of a linear ADC is given by

$$\text{SQNR} = 6.02B + 4.77 - 20\log(A/\sigma_x) \quad \text{(dB)}$$

where B is the number of ADC bits, $\pm A$ is the input range of the ADC, and σ_x is the rms value of the input signal. Determine the SQNR if the resolution of the ADC is 16 bits and the input is

(1) a sine wave signal, and

(2) a signal with an rms value of $A/4$.

State any assumptions made.

1.13 The analogue input signal to a B-bit ADC has an rms value of σ_x (V). The input range of the ADC is adjusted to the range $\pm 3\sigma_x$ (V). Find an expression for the SNQR, in decibels, for the converter. State any reasonable assumptions made.

1.14 The analogue input signal to a real-time DSP system is bandlimited to 30 Hz with an analogue filter having a third-order Butterworth characteristic before it is digitized. If the aliasing error due to sampling is to be less than 1% of the signal level in the passband, determine the minimum sampling frequency, F_s, required for the system.

If the signal, after digitization and processing, was converted back to analogue what will be the average error introduced by the aperture effect at 30 Hz? Assume that the input signal was digitized using an ideal sampler and ADC, but was recovered using a zero-order hold DAC. A common sampling frequency of 128 Hz at the input and output may be assumed.

1.15 The block diagram of a real-time DSP system with analogue output is depicted in Figure 1.32(a), and Figure 1.32(b) shows the baseband spectrum of the signal applied to the DAC. Sketch the spectrum of the signal at the output of the DAC in the interval 0–$2F_s$, where F_s is the sampling frequency. Determine the amplitudes of the signal components in your sketch. Assume a sampling frequency of 15 kHz.

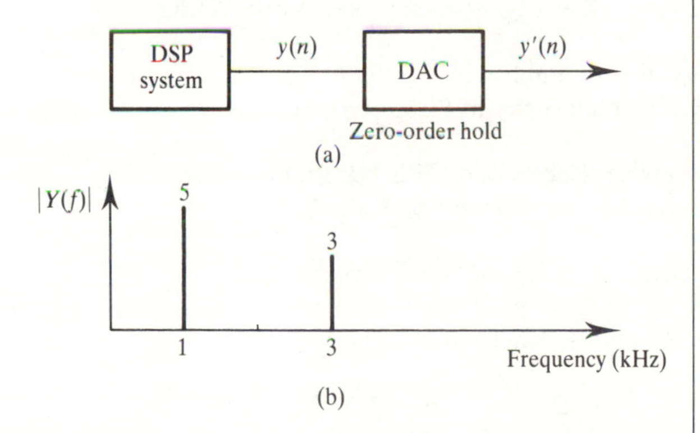

Figure 1.32 (a) Real-time DSP system with analogue output; (b) spectrum of signal applied to the DAC.

1.16 A real-time DSP system uses a 16-bit processor, a 12-bit ADC with a conversion time of 15 μs, and a 12-bit DAC with a settling time of 500 ns. If the required DSP operation is the convolution summation given by

$$y(n) = \sum_{k=0}^{N-1} h(k)x(n-k)$$

where the variables have the usual meanings and the computation must be performed between samples, estimate the real-time capability of the system, stating any assumptions made.

1.17 The output of a digital-to-analogue converter in response to a digital sequence is given by

$$y(t) = \sum_{n} y(n)h(t-nT)$$

where $h(t)$ is the impulse response of the DAC and $1/T$ is the rate at which data is fed to the DAC. Assume that the DAC is a zero-order hold and $h(t)$ a square pulse of duration T (s).

Sketch the output of the DAC in response to the input sequence, $y(n)$, shown in Figure 1.33. Show that the spectral shaping effect of the DAC on the signal spectrum can be compensated for by a digital filter with a spectrum of the form

$$|H(w)| = \frac{wT}{2\sin(wT/2)}$$

1.18 Critically examine the main constraints and errors introduced by analogue/digital conversion processes in real-time digital signal processing, suggesting how each constraint or error may be reduced.

1.19 Describe, with the aid of a block diagram, the audio signal reproduction process in the compact disc player. State and justify four advantages of using DSP techniques in this application.

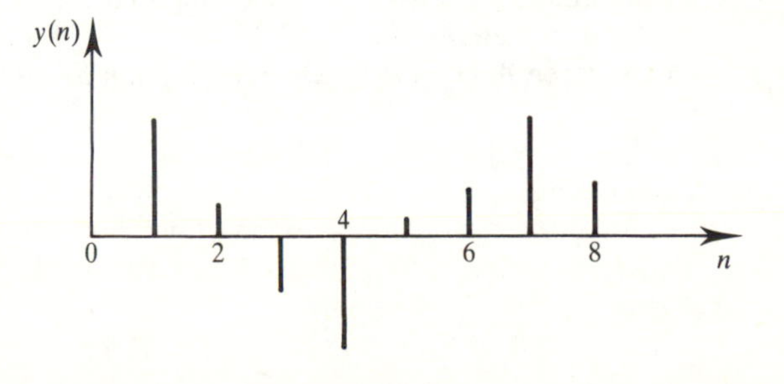

Figure 1.33

References

Bellamy J.C. (1982). *Digital Telephony*. New York: Wiley

Bloom P.J. (1985). High-quality digital audio audio in the entertainment industry: an overview of achievements and challenges. *IEEE ASSP Magazine*, October, 2–25

Carasso M.G., Peek J.B.H. and Sinjou J.P. (1982). The compact disc digital audio system. *Philips Technical Rev.*, **40**(6), 151–6

Duttweiller D.L. (1978). A twelve-channel digital echo canceler. *IEEE Trans. Communications*, **26**, 647–53

Frantz G.A. and Wiggins R.H. (1982). Design case history: Speak and Spell learns to talk. *IEEE Spectrum*, February, 45–9

Goedhart D., Van de Plassche R.J. and Stikvoort E.F. (1982). Digital-to-analog conversion in playing compact disc. *Philips Technical Rev.*, **40**(6), 174–9

Watkinson (1987). The art of digital audio

Bibliography

Berkhout P.J. and Eggermont L.D.J. (1985). Digital audio systems. *IEEE ASSP Magazine*, October, 45–67

Blesser B.A. (1978). Digitization of audio: A comprehensive examination of theory, implementation, and current practice. *J. Audio Eng. Soc.*, **26**(10), 739–71

Blesser B., Locanthi B. and Stockham Th.G., Jr (Eds) (1982). *Digital Audio*. New York: Audio Engineering Society

Garret P.H. (1981). *Analog I/O Design*. Reston VA: Reston Publishing Co. Inc

Oliver B.M., Pierce J.R. and Shannon C.E. (1948). The philosophy of PCM. *Proc IRE*, Nov., 1324–31

Oppenheim and Schaffer (1975). *Digital Signal Processing*. Englewood Cliffs NJ: Prentice-Hall

Papamichalis P. (1987). *Practical Approaches to Speech Coding*. Englewood Cliffs NJ: Prentice-Hall

Rabiner L.R. and Gold B. (1975). *Theory and Applications of Digital Signal Processing*. Englewood Cliffs NJ: Prentice-Hall

Sheingold (Ed.) (1986). *Analog-digital Conversion Handbook*. Englewood Cliffs NJ: Prentice-Hall

Steer Jr, R.W. (1989). Antialiasing filters reduce errors in A/D converters. *EDN*, March, 171–86

Tiefenthaler C. (1987). Oversampling to increase signal to noise ratio of ADCs. *Electronic Product Design*, March, 59–62

Van Doren A.H. (1982). *Data Acquisition Systems*. Reston VA: Reston Publishing Co. Inc

2

Discrete transforms

2.1	Introduction	47
2.2	DFT and its inverse	55
2.3	Properties of the DFT	62
2.4	Computational complexity of the DFT	64
2.5	The decimation-in-time fast Fourier transform algorithm	65
2.6	Inverse fast Fourier transform	76
2.7	Implementation of the FFT	77
2.8	Other discrete transforms	79
2.9	Worked examples	86
	References	92
	Appendices	93

This chapter includes introductory material on the usefulness of discrete transforms in digital signal processing, derivation of the widely used fast Fourier transform and its inverse, and a brief discussion of other transforms, especially the discrete cosine transform and the Walsh transform.

2.1 Introduction

The transformation of discrete data between the time and frequency domains is described in this chapter. Voltage versus time representations become magnitude versus frequency and phase versus frequency representations, and vice versa. The two domains provide complementary information about the same data. Thus it may sometimes be more meaningful in an application to inspect the magnitude versus frequency plot for changes in the voltage amplitude at a particular frequency than to observe the voltage waveform in order, for example, to obtain an early indication of wear in a machine by fast Fourier transforming the output data. Another example might involve the use of a discrete Fourier transform analyser and oscilloscope for checking the output of a modulator in a communication system to ensure correct functioning, in which case the test signals should produce amplitude components at certain known frequencies. Attention in these two cases of spectral analysis to a selected and restricted set of frequencies illustrates that transformation may produce the advantage of data reduction in which unimportant data is ignored, thus resulting in increased ease of interpretation. Discrete transforms, particularly the discrete cosine transform, are used in the data compression of speech and video signals to allow transmission with reduced bandwidth. They are also used in image processing to obtain a reduced set of features for pattern recognition purposes. The transforms are also useful as a mathematical tool for accelerating calculations in other signal processing applications such as correlation as used in, for example, sonars for range detection, or in convolution or deconvolution to determine the interrelationships between a system and its inputs and outputs. For these computations the transformation from the frequency to the time domain is as important as the converse. The entire subject is very mathematical, but it may be true to say that the use of discrete transforms in many applications is now standard so expert knowledge of the mathematics and theorems will rarely be required of the applications engineer. The spectral analysis of waveforms is the exception. Each problem here has to be treated on its own merits and it is important to understand the topic properly to avoid the numerous pitfalls associated with the need to acquire sufficient discrete regular data samples, and to avoid aliasing, picket fencing and spectral leakage. These topics will be discussed in detail in Chapter 10.

Of the available transforms, the discrete Fourier transform (DFT) and the algorithm for its fast computation, the fast Fourier transform (FFT), are the best known and probably the most important. Reasons for this are that they permit adequate representation in the frequency domain for all but the shortest of data lengths (<1 s), that the truncated Fourier frequency components give a more faithful representation of the data than any other exponential series, that the individual components are sinusoidal and are not distorted during transmission through linear systems, thereby constituting good test signals, and that the FFT can be computed so rapidly. Another reason is that Fourier analysis has been in existence since its publication in 1822 by Fourier and has therefore achieved a high degree of familiarity, respectability and development, as well as a wide range of applications.

Electrical and electronic engineering students are initially taught to analyse the electrical behaviour of circuits using the Laplace transform. This is because the Fourier transform can deal with neither non-zero initial conditions nor step inputs. When they progress to the study of the frequency response and stability of discrete systems such as finite impulse response filters use of the z-transform is required. Thus, the applications of the Fourier transform lie primarily in fast signal processing computations using the FFT and in spectral analysis. However, the three transforms are related. The Laplace transform may be regarded as the more general since the other two may be derived from it. Thus, the Laplace variable is $s = \sigma + j\omega$ while the Fourier transform variable is $s = j\omega$ and the z-transform variable is given by $z = e^{sT}$, where T is the time between the discrete samples. Finally, the Fourier transform and z-transform variables are related by $z = e^{j\omega T}$ (see Chapter 3).

2.1.1 Fourier series

Any periodic waveform, $f(t)$, can be represented as the sum of an infinite number of sinusoidal and cosinusoidal terms, together with a constant term, this representation being the Fourier series given by

$$f(t) = a_0 + \sum_{n=1}^{\infty} a_n \cos(n\omega t) + \sum_{n=1}^{\infty} b_n \sin(n\omega t) \tag{2.1}$$

where t is an independent variable which often represents time but could, for example, represent distance or any other quantity, $f(t)$ is often a varying voltage versus time waveform, but could be any other waveform, $\omega = 2\pi/T_p$ is known as the first harmonic, or fundamental, angular frequency, related to the fundamental frequency, f, by $\omega = 2\pi f$, T_p is the repetition period of the waveform,

$$a_0 = \frac{1}{T_p}\int_{-T_p/2}^{T_p/2} f(t)\,dt$$

is a constant equal to the time average of $f(t)$ taken over one period which might represent, for example, a dc voltage level,

$$a_n = \frac{2}{T_p}\int_{-T_p/2}^{T_p/2} f(t)\cos(n\omega t)\,dt$$

and

$$b_n = \frac{2}{T_p}\int_{-T_p/2}^{T_p/2} f(t)\sin(n\omega t)\,dt$$

The frequencies $n\omega$ are known as the nth harmonics of ω. The infinite series 2.1 therefore includes cosinusoidal and sinusoidal frequency-dependent terms of different amplitudes a_n and b_n at the positive harmonic frequencies $n\omega$. This series may be written more compactly by using exponential notation and has the advantage in that form of being more easily manipulated mathematically. The series then becomes

$$f(t) = \sum_{n=-\infty}^{\infty} d_n \mathrm{e}^{\mathrm{j}n\omega t} \tag{2.2}$$

in which

$$d_n = \frac{1}{T_\mathrm{p}} \int_{-T_\mathrm{p}/2}^{T_\mathrm{p}/2} f(t)\mathrm{e}^{-\mathrm{j}n\omega t}\,\mathrm{d}t \tag{2.3}$$

is complex and $|d_n|$ has the unit of volts.

The summation includes negative values of n, so that half of the series consists of negative frequencies, $-n\omega$. These lack physical significance, being purely mathematical, but, as a result, the magnitudes $|d_n|$ of the complex amplitudes d_n are now halved numerically. This represents an equal sharing of the amplitudes between corresponding negative and positive frequencies. Thus the correct amplitude at frequency $n\omega$ is found by doubling the calculated value. The complex and trigonometric forms are related through

$$d_n = (a_n^2 + b_n^2)^{1/2} \tag{2.4}$$

and

$$\phi_n = -\tan^{-1}(b_n/a_n) \tag{2.5}$$

where ϕ_n is the phase angle of the nth harmonic component, also given by the arctangent of the ratio of the imaginary to the real parts of d_n. Thus each harmonic component of the waveform is characterized by both its phase angle and its amplitude.

Example 2.1

As an example consider the periodic unipolar pulse waveform shown in Figure 2.1(a). Deliberate choice of the time origin to be offset from the centre and edge of a pulse is intended to allow illustration of the phase feature of the Fourier series. Substituting appropriate values into Equation 2.3 gives

$$d_n = \frac{1}{T_\mathrm{p}} \int_{-(\tau - x\tau)}^{x\tau} A\mathrm{e}^{-\mathrm{j}n\omega t}\,\mathrm{d}t \tag{2.6}$$

$$= \frac{A}{T_\mathrm{p}} \left[\frac{\mathrm{e}^{-\mathrm{j}n\omega t}}{-\mathrm{j}n\omega}\right]_{-(\tau - x\tau)}^{x\tau}$$

$$= \frac{A}{n\omega T_\mathrm{p}} \frac{\mathrm{e}^{-\mathrm{j}n\omega x\tau} - \mathrm{e}^{\mathrm{j}n\omega(\tau - x\tau)}}{-\mathrm{j}}$$

$$= \frac{A}{n\omega T_p} e^{-jn\omega x\tau}\left[\frac{e^{jn\omega\tau}-1}{j}\right]$$

$$= \frac{2A}{n\omega T_p} e^{-jn\omega x\tau}\left[\frac{e^{jn\omega\tau/2}-e^{-jn\omega\tau/2}}{2j}\right]e^{jn\omega\tau/2}$$

$$= \frac{2A}{n\omega T_p} e^{jn\omega(\tau/2-x\tau)}\sin\left(\frac{n\omega\tau}{2}\right)$$

$$= \frac{2A}{n\omega T_p}\frac{n\omega\tau}{2} e^{jn\omega(\tau/2-x\tau)}\frac{\sin(n\omega\tau/2)}{n\omega\tau/2}$$

$$= \frac{A\tau}{T_p}\mathrm{Sa}\left(\frac{n\omega\tau}{2}\right)e^{jn\omega(0.5-x)\tau} \qquad \textbf{(2.7)}$$

where

$$\mathrm{Sa}\left(\frac{n\omega\tau}{2}\right) = \frac{\sin(n\omega\tau/2)}{n\omega\tau/2}$$

is known as the sampling function of the argument $n\omega\tau/2$. The modulus of d_n is

$$|d_n| = \frac{A\tau}{T_p}\left|\mathrm{Sa}\left(\frac{n\omega\tau}{2}\right)\right|$$

and is plotted in Figure 2.1(b). $n\omega(0.5-x)\tau$ represents the phase angle, ϕ_n, in radians, associated with the nth harmonic component. In order to be able to plot this phase angle against harmonic number, n, a specific case is considered. Set $x = 0$, that is locate the time origin at the lagging edge of a pulse, and let $\tau = T/5$, when

$$\phi_n = \frac{n\omega\tau}{2} = n\frac{2\pi}{T_p}\frac{\tau}{2} = n\frac{2\pi}{T_p}\frac{T_p}{5}\frac{1}{2} = \frac{n}{5}\pi$$

ϕ_n is plotted in Figure 2.1(c), where the convention $-180° \leqslant \phi_n \leqslant 180°$ has been adopted. Selection of a different time origin would result in a different phase spectrum, ϕ_n versus n. The analysis is usually simplified by location of the time origin at a position of symmetry, for example at the midpoint of a pulse in a periodic pulse train. For the chosen case the amplitude spectrum (Figure 2.1(b)) is seen to be an even function ($|d_n| = |d_{-n}|$) while the phase spectrum (Figure 2.1(c)) is seen to be an odd function ($\phi_{-n} = -\phi_n$). The phase angles, ϕ_n, ϕ_{-n}, give the relative phase angles of the harmonic components to one another. At time, t, the absolute phase angles are $\{n\omega(0.5-x)\tau + n\omega t\}$, from Equation 2.2.

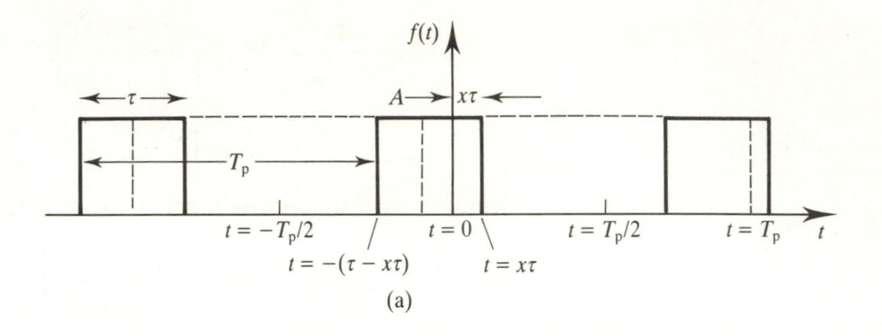

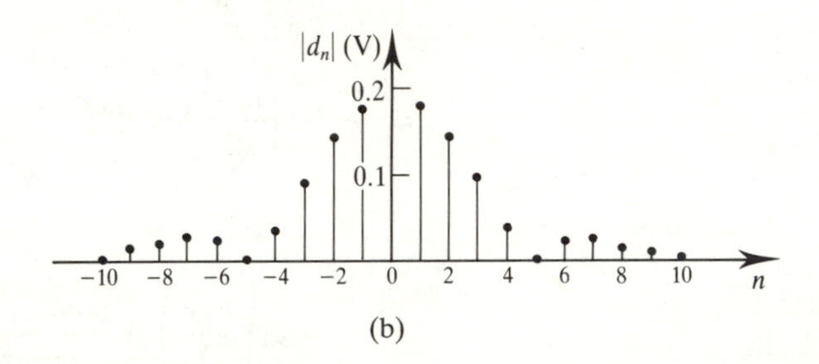

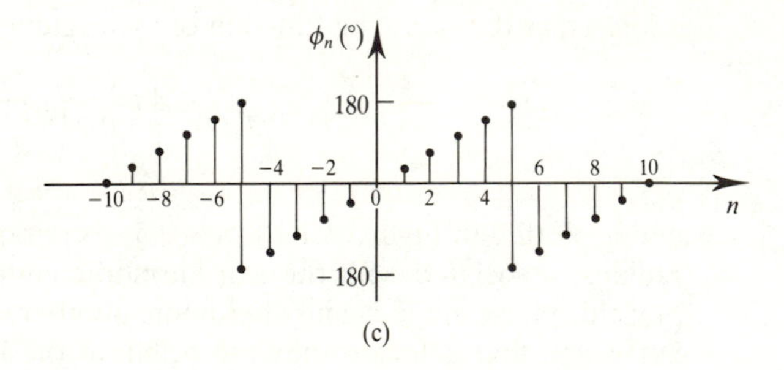

Figure 2.1 (a) Waveform, $f(t)$, (b) amplitude spectrum, $|d_A|$, and (c) phase spectrum, ϕ_n.

2.1.2 The Fourier transform

The Fourier series approach has to be modified when the waveform is not periodic. A most important example of this is the case of a single rectangular pulse such as might be obtained from the periodic waveform of Figure 2.1(a) by increasing the period T_p to be infinite. As T_p is increased the spacing between the harmonic components, $1/T_\mathrm{p} = \omega/2\pi$, decreases to $\mathrm{d}\omega/2\pi$, eventually becoming zero. This corresponds to a change from the discrete frequency variable $n\omega$ to the continuous variable ω, and the amplitude and phase spectra become continuous. Thus $d_n \rightarrow d(\omega)$ as $T_\mathrm{p} \rightarrow \infty$. With these modifications,

Equation 2.3 becomes

$$d(\omega) = \frac{d\omega}{2\pi} \int_{-\infty}^{\infty} f(t)e^{-j\omega t}\,dt \tag{2.8}$$

It is conventional to normalize this formula by dividing by $d\omega/2\pi$ to obtain

$$\frac{d(\omega)}{d\omega/2\pi} = F(j\omega) = \int_{-\infty}^{\infty} f(t)e^{-j\omega t}\,dt \tag{2.9}$$

$F(j\omega)$ is complex and is known as the Fourier integral or, more commonly, the Fourier transform. If we put

$$F(j\omega) = \mathrm{Re}\,(j\omega) + j\,\mathrm{Im}\,(j\omega) = |F(j\omega)|e^{j\phi(\omega)} \tag{2.10}$$

then

$$|F(j\omega)| = [\mathrm{Re}^2\,(j\omega) + \mathrm{Im}^2\,(j\omega)]^{1/2} \tag{2.11}$$

and has the units of volts per hertz rather than volts. $|F(j\omega)|$ is therefore an amplitude density called the amplitude spectral density. The associated phase angle, $\phi(\omega)$, is

$$\phi(\omega) = \tan^{-1}\,[\mathrm{Im}\,(j\omega)/\mathrm{Re}\,(j\omega)] \tag{2.12}$$

$|F(j\omega)|^2$ has the units of $V^2\,Hz^{-2}$. Since normalized electrical power, that is the power dissipated by a 1 Ω resistor, has the units of V^2 which are equivalent to $J\,s^{-1}$ or J Hz (J denotes joules, the unit of energy) then $V^2\,Hz^{-2}$ is J Hz $\times\,Hz^{-2} = J\,Hz^{-1}$. Hence $|F(j\omega)|^2$ has the unit equivalent to energy Hz^{-1}, that is $|F(j\omega)|^2$ is an energy spectral density. The area under the $|F(j\omega)|$ versus f plot between frequencies $f_0 - df$ and $f_0 + df$ gives the mean voltage at frequency f_0, and similarly the area under the corresponding $|F(j\omega)|^2$ versus f plot will yield the mean energy at frequency f_0. It is quite usual in spectral analysis to plot the energy spectral density against frequency.

Example 2.2

Returning to the consideration of a single pulse, let us calculate its amplitude spectral density using Equation 2.9 and Figure 2.1(a). The expression becomes

$$F(j\omega) = \int_{-(\tau - x\tau)}^{x\tau} Ae^{-j\omega t}\,dt \tag{2.13}$$

which differs from Equation 2.6 only by the constant $1/T_p$. It follows that

$$F(j\omega) = A\tau e^{j\omega(1/2-x)\tau}\,\mathrm{Sa}\,(\omega\tau/2) \qquad \textbf{(2.14)}$$

which is a factor T_p larger than d_n corresponding to the fact that $|F(j\omega)|$ has units of volts multiplied by time or $\mathrm{V\,Hz^{-1}}$. Incidentally, the result 2.14 may be obtained more simply than Equation 2.7 was obtained if certain Fourier transform properties are utilized. Thus a pulse of width τ and of unit height centred at $t = 0$ and denoted as $\mathrm{rect}\,(t/\tau)$ has the Fourier transform $\tau\,\mathrm{Sa}\,(\omega\tau/2)$. Since $Af(t)$ has the transform $AF[f(t)]$ where F denotes the Fourier transform then the pulse of height A has the transform $A\tau\,\mathrm{Sa}\,(\omega\tau/2)$. In the case of Figure 2.1(a), the pulse is shifted left by $\tau/2 - x\tau$ and so is actually the rectangular pulse, $\mathrm{rect}\,\{[t + (\tau/2 - x\tau)]/\tau\}$. The delay property of the Fourier transform states that $f(t - t_0) = e^{-j\omega t_0}F[f(t)]$ for a pulse shifted right by t_0. Application of this property to $A\tau\,\mathrm{Sa}\,(\omega\tau/2)$ gives the required form of the transform as

$$F(j\omega) = e^{+j\omega(+\tau/2-x\tau)}A\tau\,\mathrm{Sa}\left(\frac{\omega\tau}{2}\right)$$

$$= A\tau e^{j\omega(1/2-x)\tau}\,\mathrm{Sa}\left(\frac{\omega\tau}{2}\right)$$

which is the same result as Equation 2.14.

If the time origin is located at the centre of the pulse, that is $x = \frac{1}{2}$, then the Fourier transform of the pulse is given by

$$F(j\omega) = \frac{\sin\,(\omega\tau/2)}{\omega\tau/2} = A\tau\,\mathrm{Sa}\left(\frac{\omega\tau}{2}\right) \qquad \textbf{(2.15)}$$

and is real. $|F(j\omega)|$ is continuous and is plotted in Figure 2.2(a) for the values $A = 1$ V, $T_p = 10$ s and $\tau = 2$ s. This amplitude spectrum, shaped in proportion to the sampling function, is always associated with rectangular pulses, and also with any waveform of finite duration, τ. The latter may be regarded as an infinite waveform multiplied by $\mathrm{rect}\,[(t \pm t_0)/\tau]$, that is by a unit pulse. Experimentally determined waveforms fall into this category, being of finite duration τ. The sampling function passes through zero whenever $\sin\,(\omega\tau/2) = 0$, that is whenever $\omega\tau/2 = m\pi$ ($m \neq 0$, m integer). Thus amplitude nulls occur at $f = 1/\tau,\ 2/\tau,\ 3/\tau, \ldots$. When $\omega \to 0$, $\sin\,(\omega\tau/2) \to \omega\tau/2$ and $\mathrm{Sa}\,(\omega\tau/2) = \sin\,(\omega\tau/2)/(\omega\tau/2) \to 1$, so $F(j\omega) = A\tau$ at $\omega = 0$, that is at $f = 0$. The energy spectral density of the pulse of amplitude 2 V is plotted in Figure 2.2(b) for comparison with the amplitude spectral density of Figure 2.2(a).

It is possible to transform from the frequency domain to the time domain by using the inverse Fourier transform. In this case

$$f(t) = \frac{1}{2\pi}\int_{-\infty}^{\infty} F(j\omega)e^{j\omega t}\,d\omega = \int_{-\infty}^{\infty} F(j\omega)e^{j\omega t}\,df \qquad \textbf{(2.16)}$$

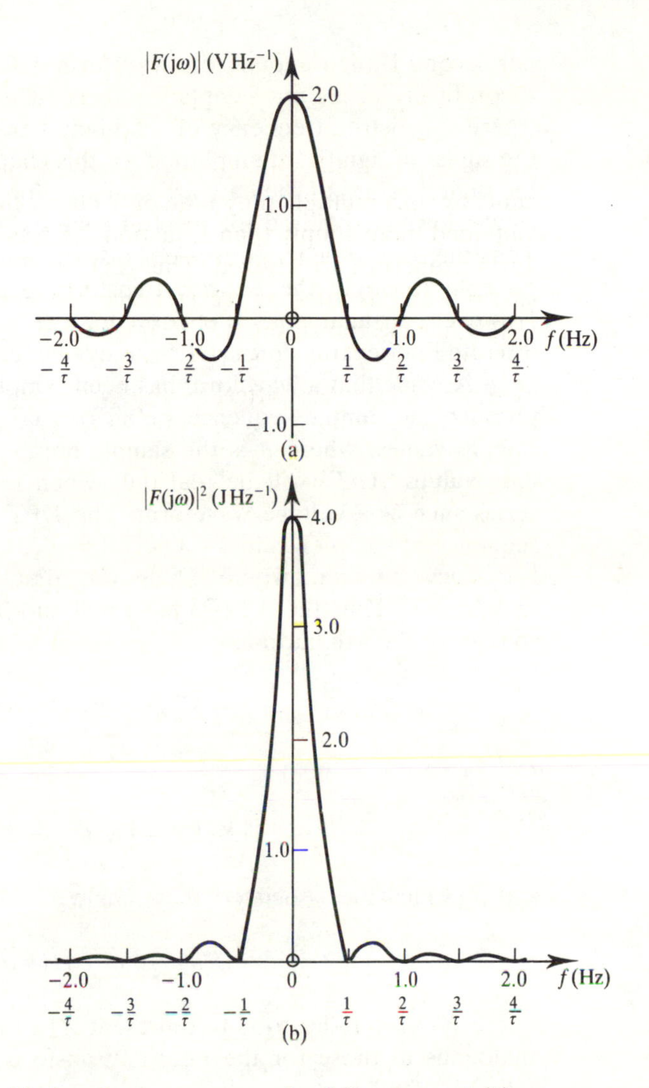

Figure 2.2 (a) Amplitude spectrum of a 2 V pulse and (b) energy spectrum of a 2 V pulse.

2.2 DFT and its inverse

In practice the Fourier components of data are obtained by digital computation rather than by analogue processing. Because the analogue waveform consists of an infinite number of contiguous points, the representation of all their values is a practical impossibility. Thus the analogue values have to be sampled at regular intervals and the sample values are then converted to a digital binary representation. This is achieved using a sample-and-hold circuit followed by an analogue-to-digital (AD) converter. Provided the number of samples recorded per second is high enough the waveform will be adequately represented. The theoretically necessary sampling rate is called the Nyquist rate and is $2f_{max}$ where f_{max} is the frequency of the highest frequency sinusoidal component in the signal of significant amplitude. In this chapter, therefore, it is assumed that the digital values of the data are available for transformation, and aspects such as windowing, required for spectral analysis, will be dealt with in Chapter 10. Thus the data to be transformed is discrete and probably non-periodic. It is not possible to apply the Fourier transform because it is for continuous data. However, an analogous transform for use with discrete data, known as the discrete Fourier transform (DFT), is available.

Assume that a waveform has been sampled at regular time intervals T to produce the sample sequence $\{x(nT)\} = x(0), x(T), \ldots, x[(N-1)T]$ of N sample values, where n is the sample number from $n = 0$ to $n = N - 1$. The data values $x(nT)$ will be real only when representing the values of a time series such as a voltage waveform. The DFT of $x(nT)$ is then defined as the sequence of complex values $\{X(k\Omega)\} = X(0), X(\Omega), \ldots, X[(N-1)\Omega]$ in the frequency domain, where Ω is the first harmonic frequency given by $\Omega = 2\pi/NT$. Thus the $X(k\Omega)$ have real and imaginary components in general so that for the kth harmonic

$$X(k) = R(k) + jI(k) \qquad \textbf{(2.17)}$$

and

$$|X(k)| = [R^2(k) + I^2(k)]^{1/2} \qquad \textbf{(2.18)}$$

and $X(k)$ has the associated phase angle

$$\phi(k) = \tan^{-1}[I(k)/R(k)] \qquad \textbf{(2.19)}$$

where $X(k)$ is understood to represent $X(k\,\Omega)$. These equations are therefore analogous to those for the Fourier transform: compare Equations 2.17–2.19 with 2.10–2.12.

Note that N real data values (in the time domain) transform to N complex DFT values (in the frequency domain). The DFT values, $X(k)$, are given by

$$X(k) = F_D[x(nT)] = \sum_{n=0}^{N-1} x(nT)e^{-jk\Omega nT}, \; k = 0, 1, \ldots, N-1 \qquad \textbf{(2.20)}$$

where F_D denotes the discrete Fourier transformation. In this equation, k represents the harmonic number of the transform component. The equation can be seen to be analogous to the Fourier transform of Equation 2.9 when $f(t) = 0$ for $t < 0$ and $t > (N-1)T$ by putting $x(nT) = f(t)$, $k\Omega = \omega$, and $nT = t$, so the two transforms may be expected to have similar properties. The transforms are not, however, equal. Thus, making these substitutions in Equation 2.9, and putting $\mathrm{d}t = T$, and replacing the integral by a summation gives

$$\sum_{n=0}^{N-1} x(nT)\mathrm{e}^{-\mathrm{j}k\Omega nT}\,T = F(\mathrm{j}\omega) \qquad \textbf{(2.21)}$$

for $0 \leqslant t \leqslant (N-1)T$. Then a comparison of Equation 2.20 with 2.21 reveals that

$$F(\mathrm{j}\omega) = TX(k) \qquad \textbf{(2.22)}$$

showing that the Fourier transform components are related to the DFT components by the sampling interval, and may be obtained by multiplying the DFT components by the sampling interval.

Example 2.3

It is appropriate at this point to illustrate the use of Equation 2.20 by evaluating a simple case. The DFT of the sequence $\{1, 0, 0, 1\}$ will be evaluated. It is worth noting here that if discontinuities in the actual data are present the average value of the data on each side of the discontinuity is taken for computational purposes to represent the value at the discontinuity. This will apply to the first and final data values as well as to other discontinuities. However, when obtaining the spectra of signals it is necessary to carry out a process known as windowing to avoid distortion of the spectra due to discontinuities at the beginning and end of the data. This topic is discussed in Chapter 10. In this section it is assumed that the sequence $\{1, 0, 0, 1\}$ has already been preprocessed. Assume that this data represents four consecutive voltages $x(0) = 1$, $x(T) = 0$, $x(2T) = 0$, $x(3T) = 1$, recorded at time intervals, T. Thus $N = 4$. It is required to find the complex values $X(k)$ for $k = 0$, $k = 1$, $k = 2$ and $k = 3$ (since $N - 1 = 3$). With $k = 0$, Equation 2.20 becomes

$$\begin{aligned} X(0) &= \sum_{n=0}^{3} x(nT)\mathrm{e}^{-\mathrm{j}0} = \sum_{n=0}^{3} x(nT) \\ &= x(0) + x(T) + x(2T) + x(3T) \\ &= 1 + 0 + 0 + 1 = 2 \end{aligned}$$

so $X(0) = 2$ is entirely real, of magnitude 2 and phase angle $\phi(0) = 0$. With $k = 1$, Equation 2.20 becomes

$$X(1) = \sum_{n=0}^{3} x(nT)\mathrm{e}^{-\mathrm{j}\Omega nT}$$

T has not been given, but may be eliminated using $\Omega = 2\pi/NT$, giving

$$X(1) = \sum_{n=0}^{3} x(nT)\mathrm{e}^{-\mathrm{j}\Omega n 2\pi/N\Omega} = \sum_{n=0}^{3} x(nT)\mathrm{e}^{-\mathrm{j}2\pi n/N}$$

$$= 1 + 0 + 0 + 1\mathrm{e}^{-\mathrm{j}2\pi 3/4} = 1 + \mathrm{e}^{-\mathrm{j}3\pi/2}$$

$$= 1 + \cos\left(\frac{3\pi}{2}\right) - \mathrm{j}\sin\left(\frac{3\pi}{2}\right) = 1 + \mathrm{j}$$

Thus $X(1) = 1 + \mathrm{j}$ and is complex with magnitude $\sqrt{2}$ and phase angle $\phi(\Omega) = \tan^{-1} 1 = 45°$. For $k = 2$, Equation 2.20 becomes

$$X(2) = \sum_{n=0}^{3} x(nT)\mathrm{e}^{-\mathrm{j}2\Omega nT} = \sum_{n=0}^{3} x(nT)\mathrm{e}^{-\mathrm{j}2n2\pi/N}$$

$$= \sum_{n=0}^{3} x(nT)\mathrm{e}^{-\mathrm{j}4\pi n/N}$$

$$= 1 + 0 + 0 + 1\mathrm{e}^{-\mathrm{j}4\pi 3/4} = 1 + 0 + 0 + \mathrm{e}^{-\mathrm{j}3\pi} = 1 - 1 = 0$$

Thus $X(2) = 0$, of magnitude zero and phase angle $\phi(2) = 0$. Finally, for $k = 3$, Equation 2.20 becomes

$$X(3) = \sum_{n=0}^{3} x(nT)\mathrm{e}^{-\mathrm{j}3n2\pi/N}$$

$$= 1 + 0 + 0 + \mathrm{e}^{-\mathrm{j}9\pi/2} = 1 - \mathrm{j}$$

Thus $X(3) = 1 - \mathrm{j}$, of magnitude $\sqrt{2}$ and phase angle $\phi(3) = -45°$.

It has therefore been shown that the time series $\{1, 0, 0, 1\}$ has the DFT given by the complex sequence $\{2, 1 + \mathrm{j}, 0, 1 - \mathrm{j}\}$.

It is common practice to represent the DFT by the plots of $|X(k)|$ versus $k\Omega$ and of $\phi(k)$ versus $k\Omega$. This may be done in terms of harmonics of Ω, or in terms of frequency if Ω is known. To find Ω it is necessary to know the value of T, the sampling interval. If it is assumed that the above data sequence had been sampled at 8 kHz then $T = 1/(8 \times 10^3) = 125\ \mu\mathrm{s}$. Then $\Omega = 2\pi/NT = 2\pi/(4 \times 125 \times 10^{-6}) = 12.57$ kHz. Hence $2\Omega = 25.14$ kHz and $3\Omega = 37.71$ kHz. Figure 2.3(a) is a plot of $x(nT)$ versus t, Figure 2.3(b) is a plot of $|X(k)|$ versus $k\Omega$, and Figure 2.3(c) is a plot of $\phi(k)$ versus $k\Omega$. It is noteworthy that the 'amplitude' plot of Figure 2.3(b) is symmetrical about the second harmonic component, that is about harmonic number $N/2$, and that in Figure 2.3(c) the phase angles are an odd function centred round this component. These results are more generally true.

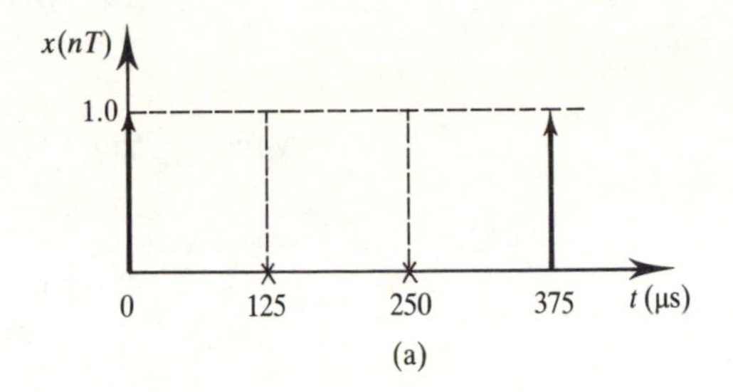

(a)

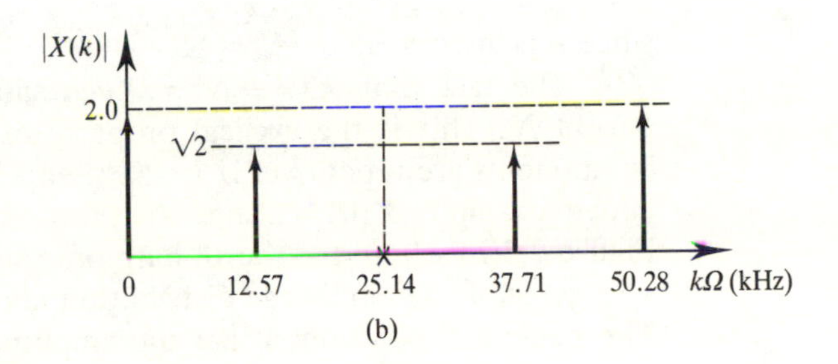

(b)

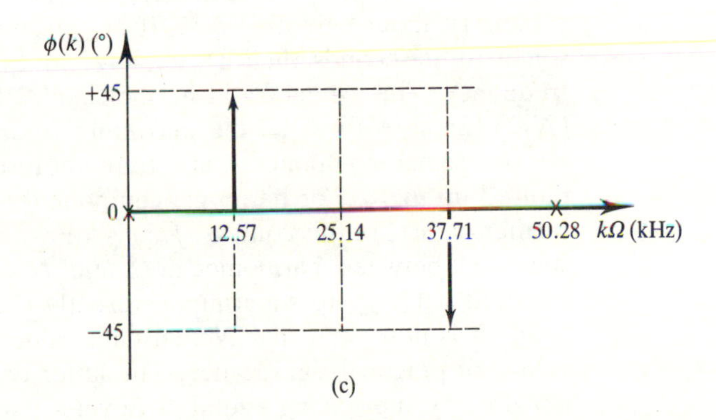

(c)

Figure 2.3 (a) $x(nT)$ versus t, (b) $|X(k)|$ versus k, and (c) $\phi(k)$ versus k.

An important property of the DFT may be deduced if the kth component of the DFT, $X(k)$, is compared with the $(k+N)$th component, $X(k+N)$. Thus

$$X(k) = \sum_{n=0}^{N-1} x(nT)\mathrm{e}^{-\mathrm{j}k\Omega nT}$$

$$= \sum_{n=0}^{N-1} x(nT)\mathrm{e}^{-\mathrm{j}k2\pi n/N}$$

and

$$X(k+N) = \sum_{n=0}^{N-1} x(nT)\mathrm{e}^{-\mathrm{j}k2\pi n/N}\mathrm{e}^{-\mathrm{j}N2\pi n/N}$$

$$= \sum_{n=0}^{N-1} x(nT)\mathrm{e}^{-\mathrm{j}k2\pi n/N}\mathrm{e}^{-\mathrm{j}2\pi n}$$

$$= \sum_{n=0}^{N-1} x(nT)\mathrm{e}^{-\mathrm{j}k2\pi n/N} = X(k)$$

since n is integral so $\mathrm{e}^{-\mathrm{j}2\pi n} = 1$.

The fact that $X(k+N) = X(k)$ shows that the DFT is periodic with period N. This is the cyclical property of the DFT. The values of the DFT components are repetitive. If $k = 0$, then $k + N = N$ and $X(0) = X(N)$. In the above example $X(0) = 2$ and therefore $X(4) = 2$ also. This is illustrated in Figure 2.3(b) where the fourth harmonic amplitude is drawn at 50.28 kHz. The symmetry of the amplitude distribution about the second harmonic is obvious. The general conclusion is that the amplitude spectrum of an N-point DFT is symmetrical about harmonic $N/2$ when both the zero and $(N+1)$th harmonics are included in the plot. Similarly, the phase function being odd exhibits anti-symmetry about harmonic $N/2$. If $2f_{\max}$ samples per second were taken of the signal for t seconds then $2f_{\max}t = N$, so $1/t = 2f_{\max}/N$ is the first harmonic frequency. The symmetry at harmonic $N/2$ therefore occurs at the frequency $(N/2)/(2f_{\max}/N) = f_{\max}$, the maximum frequency present in the signal. Thus all the signal components are fully represented in an amplitude spectrum plotted up to $f_{\max}$ or harmonic component $N/2$, and it is unnecessary to plot further points. In this context, $f_{\max}$ is known as the folding frequency, since the spectrum between harmonics $N/2$ and N may be folded about the axis of symmetry at $f_{\max}$ to superimpose exactly the low frequency half of the spectrum. It is now seen that N real data values transform to $N/2$ complex DFT values of practical significance. The latter consist of $N/2$ real values and $N/2$ imaginary values giving a total of N values derived from the initial N real data values. Finally the values of the Fourier transform components, $F(\mathrm{j}\omega)$, of the data $\{1, 0, 0, 1\}$ in Example 2.3 may be obtained by multiplying the DFT components by $T = 125\ \mu\mathrm{s}$. Therefore $F(0) = 250\ \mu\mathrm{V\,Hz}^{-1}$, $F(12.57\ \mathrm{kHz}) = (125 + \mathrm{j}125)\ \mu\mathrm{V\,Hz}^{-1}$, $F(25.14\ \mathrm{kHz}) = 0\ \mathrm{V\,Hz}^{-1}$, $F(37.71\ \mathrm{kHz}) = (125 - \mathrm{j}125)\ \mu\mathrm{V\,Hz}^{-1}$.

As explained in the introduction to the chapter, it is also necessary to be able to carry out discrete transformation from the frequency to the time domain. This may be achieved using the inverse discrete Fourier transform (IDFT), defined by

$$x(nT) = F_{\mathrm{D}}^{-1}[X(k)] = \frac{1}{N}\sum_{k=0}^{N-1} X(k)\mathrm{e}^{\mathrm{j}k\Omega nT},\ n = 0, 1, \ldots, N-1 \quad \textbf{(2.23)}$$

where F_{D}^{-1} denotes the inverse discrete Fourier transformation.

The analogy with Equation 2.16 for the inverse Fourier transform is obvious. This time it is quite simple to show that the inverse Fourier transform is obtainable from the IDFT by dividing the IDFT by T. The validity of Equation 2.23 can be demonstrated by substituting for $x(nT)$ in Equation 2.20.

Example 2.4

It is useful to illustrate the inverse discrete Fourier transform by using it to derive the time series $\{1, 0, 0, 1\}$ from its DFT components $[2, 1+\mathrm{j}, 0, 1-\mathrm{j}]$.

With $n = 0$,

$$\begin{aligned} x(nT) = x(0) &= \frac{1}{N}\sum_{k=0}^{N-1} X(k) \\ &= \tfrac{1}{4}[X(0) + X(1) + X(2) + X(3)] \\ &= \tfrac{1}{4}[2 + (1+\mathrm{j}) + 0 + (1-\mathrm{j})] = 1 \end{aligned}$$

as expected. With $n = 1$,

$$\begin{aligned} x(nT) = x(T) &= \frac{1}{N}\sum_{k=0}^{N-1} X(k)\mathrm{e}^{\mathrm{j}k\Omega T} \\ &= \frac{1}{N}\sum_{k=0}^{N-1} X(k)\mathrm{e}^{\mathrm{j}k2\pi/N} = \frac{1}{4}\sum_{k=0}^{N-1} X(k)\mathrm{e}^{\mathrm{j}k\pi/2} \\ &= \tfrac{1}{4}[2 + (1+\mathrm{j})\mathrm{e}^{\mathrm{j}\pi/2} + 0 + (1-\mathrm{j})\mathrm{e}^{\mathrm{j}3\pi/2}] \\ &= \tfrac{1}{4}[2 + (1+\mathrm{j})\mathrm{j} + (1-\mathrm{j})(-\mathrm{j})] \\ &= \tfrac{1}{4}(2 + \mathrm{j} - 1 - \mathrm{j} - 1) = 0 \end{aligned}$$

as expected. With $n = 2$,

$$\begin{aligned} x(nT) = x(2T) &= \frac{1}{N}\sum_{k=0}^{N-1} x(k)\mathrm{e}^{\mathrm{j}k\pi} \\ &= \tfrac{1}{4}[2 + (1+\mathrm{j})\mathrm{e}^{\mathrm{j}\pi} + (1-\mathrm{j})\mathrm{e}^{\mathrm{j}3\pi}] = \tfrac{1}{4}[2 - (1+\mathrm{j}) - (1-\mathrm{j})] \\ &= 0 \end{aligned}$$

again, as expected. Finally, with $n = 3$,

$$\begin{aligned} x(nT) = x(3T) &= \frac{1}{N}\sum_{k=0}^{N-1} X(k)\mathrm{e}^{\mathrm{j}k3\pi/2} \\ &= \tfrac{1}{4}[2 + (1+\mathrm{j})\mathrm{e}^{\mathrm{j}3\pi/2} + (1-\mathrm{j})\mathrm{e}^{\mathrm{j}9\pi/2}] \\ &= \tfrac{1}{4}[2 + (1+\mathrm{j})(-\mathrm{j}) + (1-\mathrm{j})\mathrm{j}] = \tfrac{1}{4}(2 - \mathrm{j} + 1 + \mathrm{j} + 1) = 1 \end{aligned}$$

the correct final term of the series.

2.3 Properties of the DFT

The DFT has a number of mathematical properties which can be used to simplify problems or which lead to useful applications. Some of them are listed below. The data sequences $x(nT)$ are written $x(n)$.

(1) *Symmetry*.

$$\text{Re}\,[X(N-k)] = \text{Re}\,X(k) \tag{2.24}$$

(where Re denotes the real part) states the symmetry of the amplitude spectrum, discussed previously, and

$$\text{Im}\,[X(N-k)] = -\text{Im}\,[X(k)] \tag{2.25}$$

(where Im denotes the imaginary part) states the antisymmetrical property of the phase spectrum. This is a property to be aware of when considering component values.

(2) *Even functions*. If $x(n)$ is an even function $x_e(n)$, that is $x_e(n) = x_e(-n)$, then

$$F_D[x_e(n)] = X_e(k) = \sum_{n=0}^{N-1} x_e(n)\cos(k\Omega nT) \tag{2.26}$$

(3) *Odd functions*. If $x(n)$ is an odd function $x_0(n)$, that is $x_0(n) = -x_0(-n)$, then

$$F_D[x_0(n)] = X_0(k) = -\text{j}\sum_{n=0}^{N-1} x_0(n)\sin(k\Omega nT) \tag{2.27}$$

(4) *Parseval's Theorem*. The normalized energy in the signal is given by either of the expressions

$$\sum_{n=0}^{N-1} x^2(n) = \frac{1}{N}\sum_{k=0}^{N-1} |X(k)|^2 \tag{2.28}$$

The right-hand side of Equation 2.28 is the mean square spectral amplitude, while the left-hand side is the sum of the squared magnitudes of the time series.

(5) *Delta function*:

$$F_D[\delta(nT)] = 1 \tag{2.29}$$

(6) The linear cross-correlation of two data sequences or series may be com-

puted using DFTs. The linear cross-correlation of two finite length sequences $x_1(n)$ and $x_2(n)$, each of length N, is defined to be

$$r_{x_1x_2}(\mathrm{j}) = \frac{1}{N} \sum_{n=-\infty}^{\infty} x_1(n)x_2(n + \mathrm{j}), \ -\infty \leqslant \mathrm{j} \leqslant \infty \qquad \textbf{(2.30)}$$

It is also necessary to define the circular correlation of finite length sequences as

$$r_{\mathrm{c}x_1x_2}(\mathrm{j}) = \frac{1}{N} \sum_{n=0}^{N-1} x_1(n)x_2(n + \mathrm{j}), \ \ \mathrm{j} = 0, \ldots, N - 1 \qquad \textbf{(2.31)}$$

because the circular correlation can be evaluated using DFTs. Thus

$$r_{\mathrm{c}x_1x_2}(\mathrm{j}) = F_{\mathrm{D}}^{-1}[X_1^*(k)X_2(k)] \qquad \textbf{(2.32)}$$

Equation 2.32 is known as the correlation theorem. The circular correlation given by Equation 2.32 can be converted into a linear correlation by using augmenting zeros. Now, if the sequences are $x_1(n)$ of length N_1, and $x_2(n)$ of length N_2, their linear correlation will be of length $N_1 + N_2 - 1$. To achieve this $x_1(n)$ is replaced by $x_{1\mathrm{a}}(n)$ which consists of $x_1(n)$ with $N_2 - 1$ zeros added, and $x_2(n)$ is augmented by $N_1 - 1$ zeros to become $x_{2\mathrm{a}}(n)$. The linear cross-correlation of $x_1(n)$ and $x_2(n)$ is then given by

$$r_{x_1x_2}(\mathrm{j}) = F_{\mathrm{D}}^{-1}[X_{1\mathrm{a}}^*(k)X_{2\mathrm{a}}(k)] \qquad \textbf{(2.33)}$$

where

$$X_{1\mathrm{a}}(k) = F_{\mathrm{D}}[x_{1\mathrm{a}}(n)] \text{ and } X_{2\mathrm{a}}(k) = F_{\mathrm{D}}[x_{2\mathrm{a}}(n)]$$

This subject is treated more fully in Chapter 4.

(7) DFTS may also be used in the computation of circular convolutions, and, by using augmenting zeros, in linear convolutions. These may be either time or frequency domain convolutions. The time convolution theorem states that

$$x_3(n) = x_1(n) \circledast x_2(n) = F_{\mathrm{D}}^{-1}[X_1(k)X_2(k)] \qquad \textbf{(2.34)}$$

where $\circledast$ denotes circular convolution, and $x_1(n)$, $x_2(n)$ and $x_3(n)$ are finite sequences of equal length.

In an analogous manner to Equation 2.31, $x_3(n)$ may also be written

$$x_3(n) = \sum_{m=0}^{N-1} x_1(m)x_2(n - m) \qquad \textbf{(2.35)}$$

Furthermore,

$$X_3(k) = X_1(k)X_2(k) \tag{2.36}$$

where $X_3(k) = F_D[x_3(n)]$.

The following equation is the statement of the frequency convolution theorem:

$$\frac{1}{N} X_1(k) \circledast X_2(k) = F_D[x_1(n)x_2(n)] \tag{2.37}$$

where

$$X_1(k) \circledast X_2(k) = \sum_{m=0}^{N-1} X_1(m)X_2(k-m) \tag{2.38}$$

Equation 2.34 gives rise to the statement that convolution in the time domain is equivalent to multiplication in the frequency domain, while Equation 2.37 has led to the observation that convolution in the frequency domain is equivalent to multiplication in the time domain. These statements may provide a means of remembering the relationships. Convolution will also be treated in more detail in Chapter 4.

2.4 Computational complexity of the DFT

A large number of multiplications and additions are required for the calculation of the DFT. For an 8-point DFT the expansion for $X(k)$ becomes (from Equation 2.20)

$$X(k) = \sum_{n=0}^{7} x(n)e^{-jk2\pi n/8}, \; k = 0, \ldots, 7 \tag{2.39}$$

and letting $k2\pi/8 = K$ this expands to

$$\begin{aligned} X(k) = {} & x(0)e^{-jK0} + x(1)e^{-jK1} + x(2)e^{-jK2} + x(3)e^{-jK3} + x(4)e^{-jK4} \\ & + x(5)e^{-jK5} + x(6)e^{-jK6} + x(7)e^{-jK7}, \; k = 0, \ldots, 7 \end{aligned} \tag{2.40}$$

Equation 2.40 contains eight terms on the right-hand side. Each term consists of a multiplication of an exponential term which is always complex by another term which is real or complex (for example, real for a voltage time series). Each of the product terms is added together. There are therefore eight complex multiplications and seven complex additions to be calculated. For an N-point DFT, there will be N and $N-1$ of them respectively. There are also eight harmonic components to be evaluated ($k = 0, \ldots, 7$). This number becomes N for an N-point DFT. Therefore the calculation of the eight-point DFT requires $8^2 = 64$ complex multiplications and $8 \times 7 = 56$ complex additions. For an N-point DFT these become N^2 and $N(N-1)$ respectively. If $N = 1024$, then approximately one million complex multiplications and one million complex additions are required. Clearly some means of reducing these numbers is desirable.

The amount of computation involved may be reduced if we note that there is a considerable amount of built-in redundancy in equations such as Equation 2.4. For example if $k = 1$ and $n = 2$, $\mathrm{e}^{-\mathrm{j}k2\pi n/8} = \mathrm{e}^{-\mathrm{j}\pi/2}$, and if $k = 2$ and $n = 1$, $\mathrm{e}^{-\mathrm{j}k2\pi n/8} = \mathrm{e}^{-\mathrm{j}\pi/2}$ also.

2.5 The decimation-in-time fast Fourier transform algorithm

In this section it will be shown how the computational redundancy inherent in the DFT is used to reduce the number of different calculations necessary, thereby speeding up the computation. For a 1024-point DFT the number of calculations required can be reduced by a factor of 204.8. The algorithms which can achieve this are given the title 'fast Fourier transform', or FFT in short. When applied in the time domain the algorithm is referred to as a decimation-in-time (DIT) FFT. The first DIT algorithm was due to Cooley and Tukey (1965), after whom it is often named. Decimation then refers to the significant reduction in the number of calculations performed on time domain data. It is noteworthy that the computational savings will be seen to increase as $N^2 - (N/2)\log_2 N$.

First the notation will be simplified and some mathematical relationships will be established. Thus Equation 2.20 will be re-written as

$$X_1(k) = \sum_{n=0}^{N-1} x_n \mathrm{e}^{-\mathrm{j}2\pi nk/N}, \; k = 0, \ldots, N-1 \tag{2.41}$$

Also, the factor $\mathrm{e}^{-\mathrm{j}2\pi/N}$ will be written as W_N, thus

$$W_N = \mathrm{e}^{-\mathrm{j}2\pi/N} \tag{2.42}$$

so that Equation 2.41 becomes

$$X_1(k) = \sum_{n=0}^{N-1} x_n W_N^{kn}, \; k = 0, \ldots, N-1 \tag{2.43}$$

It is worthwhile at this point to note some relationships involving W_N. First,

$$W_N^2 = (e^{-j2\pi/N})^2 = e^{-j2\pi 2/N} = e^{-j2\pi/(N/2)} = W_{N/2} \tag{2.44}$$

Second,

$$\begin{aligned} W_N^{(k+N/2)} &= W_N^k W_N^{N/2} = W_N^k e^{-j(2\pi/N)(N/2)} = W_N^k e^{-j\pi} \\ &= -W_N^k \end{aligned} \tag{2.45}$$

Summarizing the useful results concerning W_N for convenience, we have

$$W_N = e^{-j2\pi/N} \tag{2.46a}$$

$$W_N^2 = W_{N/2} \tag{2.46b}$$

$$W_N^{(k+N/2)} = -W_N^k \tag{2.46c}$$

In exploiting the computational redundancy expressed by Equations 2.46 the data sequence is divided into two equal sequences, one of even-numbered data, and one of odd-numbered data. For the sequences to be of equal length, they must all contain an even number of data. If the original sequence consists of an odd number of data, then an augmenting zero should be added to render the number of data even. This allows the DFT, $X_1(k)$, to be written in terms of two DFTs, $X_{11}(k)$ and $X_{12}(k)$, which are the DFTs of the even-valued data and of the odd-valued data respectively (see Table 2.1). Thus the N-point DFT is converted into two DFTs each of $N/2$ points. This process is then repeated until $X_1(k)$ is decomposed into $N/2$ DFTs, each of two points, both of which are initial data. Thus, in practice, the initial data is re-ordered and the $N/2$ two-point DFTs are calculated by taking the data in pairs. These DFT outputs are suitably combined in fours to provide $N/4$ four-point DFTs which are computed and appropriately combined to produce $N/8$ eight-point DFTs which are computed, and so on until the final N-point DFT, $X_1(k)$, is obtained. At each stage common factors which are powers of W_N are incorporated to reduce the number of complex calculations. This procedure is justified as follows.

The suffixes, n, in Equation 2.43 extend from $n=0$ to $n=N-1$, corresponding to the data values $x_0, x_1, x_2, x_3, \ldots, x_{N-1}$. The even-numbered sequence is $x_0, x_2, x_4, \ldots, x_{N-2}$, and the odd-numbered sequence is $x_1, x_3, \ldots, x_{N-1}$. Both sequences contain $N/2$ points. The terms in the even sequence may be designated x_{2n} with $n=0$ to $n=N/2-1$ while those in the odd sequence become x_{2n+1}. Then Equation 2.43 may be re-written

$$\begin{aligned} X_1(k) &= \underbrace{\sum_{n=0}^{N/2-1} x_{2n} W_N^{2nk}}_{\text{even sequence}} + \underbrace{\sum_{n=0}^{N/2-1} x_{2n+1} W_N^{(2n+1)k}}_{\text{odd sequence}} \\ &= \sum_{n=0}^{N/2-1} x_{2n} W_N^{2nk} + W_N^k \sum_{n=0}^{N/2-1} x_{2n+1} W_N^{2nk}, \quad k = 0, \ldots, N-1 \end{aligned} \tag{2.47}$$

Table 2.1 Structure of an 8-point FFT.

Line number	*Line content*									*k ranges*	*N ranges*
1	Data sequence A_0	A_0	$x_0\ x_1\ x_2\ x_3\ x_4\ x_5\ x_6\ x_7$								$0, \ldots, 7$
2	8-point DFT of A_0	$X_1(k) = X_{11}(k) + W_N^k X_{12}(k)$								$0, \ldots, N-1$ $(0, \ldots, 7)$	$0, \ldots, 7$
3	Re-ordered A_0: two sequences, A_1 and A_2	A_1	$x_0\ x_2\ x_4\ x_6$			A_2	$x_1\ x_3\ x_5\ x_7$				$0, \ldots, 3$
4	4-point DFTs of A_1 and A_2	$X_{11}(k) = X_{21}(k) + W_{N/2}^k X_{22}(k)$				$X_{12}(k) = X_{23}(k) + W_{N/2}^k X_{24}(k)$				$0, \ldots, N/2-1$ $(0, \ldots, 3)$	$0, \ldots, 3$
5	Re-ordered sequences A_1 and A_2: four sequences A_3, A_4, A_5, A_6	A_3	$x_0\ x_4$	A_4	$x_2\ x_6$	A_5	$x_1\ x_5$	A_6	$x_3\ x_7$		$0, 1$
6	2-point DFTs of A_3, A_4, A_5, A_6	$X_{21}(k) = x_0 + W_{N/4}^k x_4$		$X_{22}(k) = x_2 + W_{N/4}^k x_6$		$X_{23}(k) = x_1 + W_{N/4}^k x_5$		$X_{24}(k) = x_3 + W_{N/4}^k x_7$		$0, \ldots, N/4-1$ $(0, 1)$	$0, 1$

Using Equation 2.46b gives $W_N^{2nk} = W_{N/2}^{nk}$ so Equation 2.47 becomes

$$X(k) = \sum_{n=0}^{N/2-1} x_{2n} W_{N/2}^{nk} + W_N^k \sum_{n=0}^{n=N/2-1} x_{2n+1} W_{N/2}^{nk}, \quad k = 0, \ldots, N-1 \tag{2.48}$$

Equation 2.48 may be written

$$X_1(k) = X_{11}(k) + W_N^k X_{12}(k), \quad k = 0, \ldots, N-1 \tag{2.49}$$

On comparison of Equation 2.49 with Equation 2.43, it is seen that $X_{11}(k)$ is indeed the DFT of the even sequence, while $X_{12}(k)$ is that of the odd sequence. Therefore, as previously stated, the DFT, $X_1(k)$, can be expressed in terms of two DFTs: $X_{11}(k)$ and $X_{12}(k)$. The factor $W_{N/2}^k$ occurs in both $X_{11}(k)$ and $X_{12}(k)$ and needs calculation once only.

Table 2.1 illustrates the process for an 8-point DFT. Line 1 gives the data while line 2 gives an expression for the DFT of the data in terms of the DFTs of the even and odd sequences, $X_{11}(k)$ and $X_{12}(k)$ respectively. Line 3 shows the re-ordered data from which $X_{11}(k)$ and $X_{12}(k)$ are derived. Line 4 gives the DFTs of the data sequences of line 3 in terms of the DFTs of their even and odd sequences, $X_{21}(k)$, $X_{22}(k)$, $X_{23}(k)$ and $X_{24}(k)$. These sequences are shown in line 5, and are seen to be the ultimate 2-point sequences, whose DFTs are $X_{21}(k)$, $X_{22}(k)$, $X_{23}(k)$ and $X_{24}(k)$ and are expressed in terms of the data in line 6. Thus the single 8-point DFT has been decomposed into four 2-point DFTs, each one of which produces two values, for example $X_{21}(0)$ and $X_{21}(1)$ in the case of $X_{21}(k)$. This process involved two decompositions, and the weights W_N^k were squared at each step. Considering line 6 it is seen that

$$X_{21}(k) = x_0 + W_{N/4}^k x_4 \quad k = 0, \ldots, N/4 - 1, \quad \text{that is } k = 0, 1 \tag{2.50}$$

Thus

$$X_{21}(0) = x_0 + x_4$$

while

$$\begin{aligned} X_{21}(1) &= x_0 + W_{N/4} x_4 \\ &= x_0 + W_2 x_4 = x_0 + \mathrm{e}^{-\mathrm{j}2\pi/2} x_4 = x_0 + \mathrm{e}^{-\mathrm{j}\pi} x_4 = x_0 - x_4 \end{aligned}$$

Similarly,

$$\begin{aligned} X_{22}(0) &= x_2 + x_6, \qquad & X_{22}(1) &= x_2 - x_6 \\ X_{23}(0) &= x_1 + x_5, \qquad & X_{23}(1) &= x_1 - x_5 \\ X_{24}(0) &= x_3 + x_7, \qquad & X_{24}(1) &= x_3 - x_7, \end{aligned}$$

from which we observe that the values with $k = 1$ differ only by a sign from those with $k = 0$. This point is emphasized if $X_{11}(k)$ $(k = 0, 1, 2, 3)$ is considered. Now,

$$X_{11}(k) = X_{21}(k) + W^k_{N/2}X_{22}(k) \tag{2.51}$$

so,

$$X_{11}(0) = X_{21}(0) + W^0_{N/2}X_{22}(0) = X_{21}(0) + X_{22}(0) \tag{2.52}$$

$$\begin{aligned} X_{11}(1) &= X_{21}(1) + W^1_{N/2}X_{22}(1) = X_{21}(1) + e^{-j\pi/2}X_{22}(1) \\ &= X_{21}(1) - jX_{22}(1) \end{aligned} \tag{2.53}$$

$$\begin{aligned} X_{11}(2) &= X_{21}(2) + W^2_{N/2}X_{22}(2) = X_{21}(2) + e^{-j(2\pi/8)2\times 2}X_{22}(2) \\ &= X_{21}(2) + e^{-j\pi}X_{22}(2) = X_{21}(2) - X_{22}(2) \end{aligned} \tag{2.54}$$

Now

$$X_{21}(2) = x_0 + W^2_{N/4}x_4 = x_0 + W^2_2x_4 = x_0 + x_4 = X_{21}(0)$$

and

$$X_{22}(2) = x_2 + W^2_{N/4}x_6 = x_2 + x_6 = X_{22}(0)$$

Hence Equation 2.54 is equivalent to

$$X_{11}(2) = X_{21}(0) - X_{22}(0) \tag{2.55}$$

$$X_{11}(3) = X_{21}(3) + W^3_{N/2}X_{22}(3) \tag{2.56}$$

Now

$$\begin{aligned} X_{21}(3) &= x_0 + W^3_{N/4}x_4 = x_0 + e^{-j(2\pi/2)3}x_4 \\ &= x_0 + e^{-j3\pi}x_4 = x_0 - x_4 = X_{21}(1) \end{aligned}$$

and

$$X_{22}(3) = x_2 - x_6 = X_{22}(1)$$

Hence Equation 2.56 is equivalent to

$$\begin{aligned} X_{11}(3) &= X_{21}(1) + e^{-j(2\pi/4)3}X_{22}(1) = X_{21}(1) + e^{-j3\pi/2}X_{22}(1) \\ &= X_{21}(1) + jX_{22}(1) \end{aligned} \tag{2.57}$$

Drawing these results together gives

$$X_{11}(0) = X_{21}(0) + X_{22}(0) = X_{21}(0) + W_8^0 X_{22}(0) \qquad \textbf{(2.58a)}$$

$$X_{11}(2) = X_{21}(0) - X_{22}(0) = X_{21}(0) - W_8^0 X_{22}(0) \qquad \textbf{(2.58b)}$$

$$X_{11}(1) = X_{21}(1) - \mathrm{j}X_{22}(1) = X_{21}(1) + W_8^2 X_{22}(1) \qquad \textbf{(2.58c)}$$

$$X_{11}(3) = X_{21}(1) + \mathrm{j}X_{22}(1) = X_{21}(1) - W_8^2 X_{22}(1) \qquad \textbf{(2.58d)}$$

Inspection of these equations shows how the DFTs $X_{11}(k)$ are related to the DFTs of their even-numbered and odd-numbered data, and that $X_{11}(0)$ and $X_{11}(2)$ are given by expressions with common terms which differ only in one sign. The same is true of $X_{11}(1)$ and $X_{11}(3)$. Equations such as these are known as recomposition equations because starting from the data pairs and forming $X_{21}(k)$, $X_{22}(k)$, $X_{23}(k)$ and $X_{24}(k)$ allows $X_{11}(k)$ and $X_{12}(k)$ to be found, and hence $X_1(k)$. The number of complex additions and multiplications involved is reduced in this way because (i) the recomposition equations are expressed in terms of powers of the recurring factor W_N, (ii) use is also made of relationships of the type $X_{21}(2) = X_{21}(0)$ and $X_{21}(3) = X_{21}(1)$, and (iii) the presence of only sign differences in the pairs of expressions is exploited. This algorithm is known as the Cooley–Tukey algorithm.

2.5.1 The butterfly

Equations 2.58 may be represented diagrammatically by exploiting the symmetry centred on the sign differences and taking the equations in pairs. Thus from Equations 2.58a and 2.58b the outputs of the recompositions are $X_{11}(0)$ and $X_{11}(2)$ formed from the inputs $X_{21}(0)$ and $X_{22}(0)$. This is illustrated in Figure 2.4(a). The inputs are at the left-hand side of the cross, the outputs to the right. Figure 2.4(b) shows how the outputs $X_{11}(1)$ and $X_{11}(3)$ are obtained diagrammatically. By overlapping Figures 2.4(a) and 2.4(b), a composite diagram is obtained in which the output DFTs are arranged in order of increasing k. This is shown in Figure 2.4(c). The structure of Figure 2.4(a) or 2.4(b) is referred to as a 'butterfly', being reminiscent of the symbolic representation of the insect. The entire 8-point FFT may be depicted in this manner as in Figure 2.5.

Example 2.5

It will now be instructive to obtain the DFT of the sequence $\{1, 0, 0, 1\}$, previously evaluated in Section 2.2, by means of the decimation-in-time FFT algorithm. This is a four-point DFT with $x_0 = 1$, $x_1 = 0$, $x_2 = 0$, $x_3 = 1$, and $X_1(k) = X_{11}(k) + W_N^k X_{12}(k)$, $k = 0, 1, 2, 3$. The re-ordered sequence is x_0, x_2, x_1, x_3.

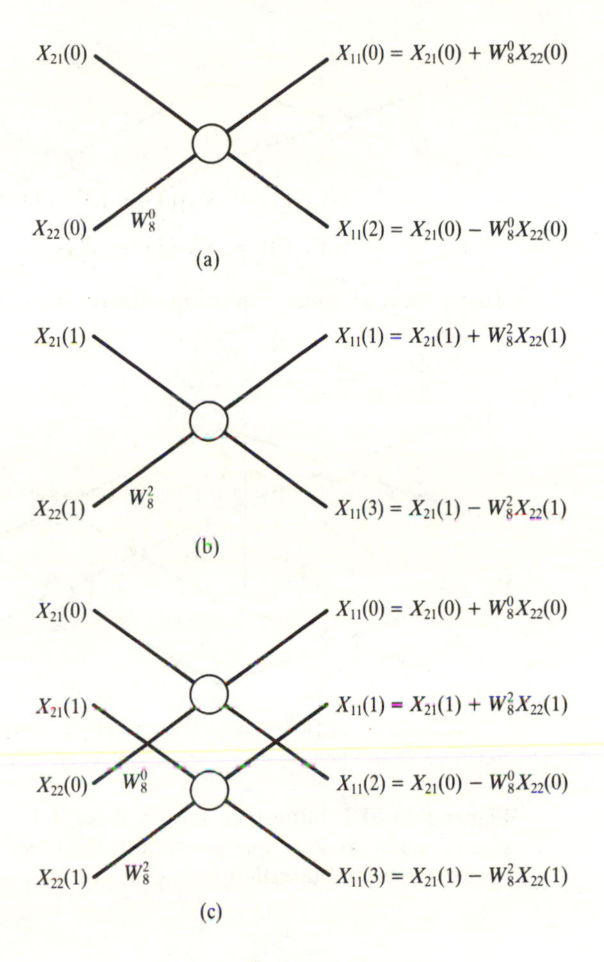

Figure 2.4 FFT butterflies.

We can now utilize the top left-hand corner of Figure 2.5 to work out the DFT. Points x_0, x_4, x_2, x_6 are replaced by x_0, x_2, x_1, x_3, and the required DFT values are $X_{11}(0)$, $X_{11}(1)$, $X_{11}(2)$ and $X_{11}(3)$. Therefore,

$$X_{21}(0) = x_0 + x_2 = 1$$

$$X_{21}(1) = x_0 - x_2 = 1$$

$$X_{22}(0) = x_1 + x_3 = 1$$

$$X_{22}(1) = x_1 - x_3 = -1$$

$$X_{11}(0) = X_{21}(0) + W_8^0 X_{22}(0) = 1 + 1 = 2$$

$$X_{11}(1) = X_{21}(1) + W_8^2 X_{22}(1) = 1 + \mathrm{e}^{-\mathrm{j}\pi/2}(-1) = 1 + \mathrm{j}$$

$$X_{11}(2) = X_{21}(0) - W_8^0 X_{22}(0) = 1 - 1 = 0$$

$$X_{11}(3) = X_{21}(1) - W_8^2 X_{22}(1) = 1 - \mathrm{j}$$

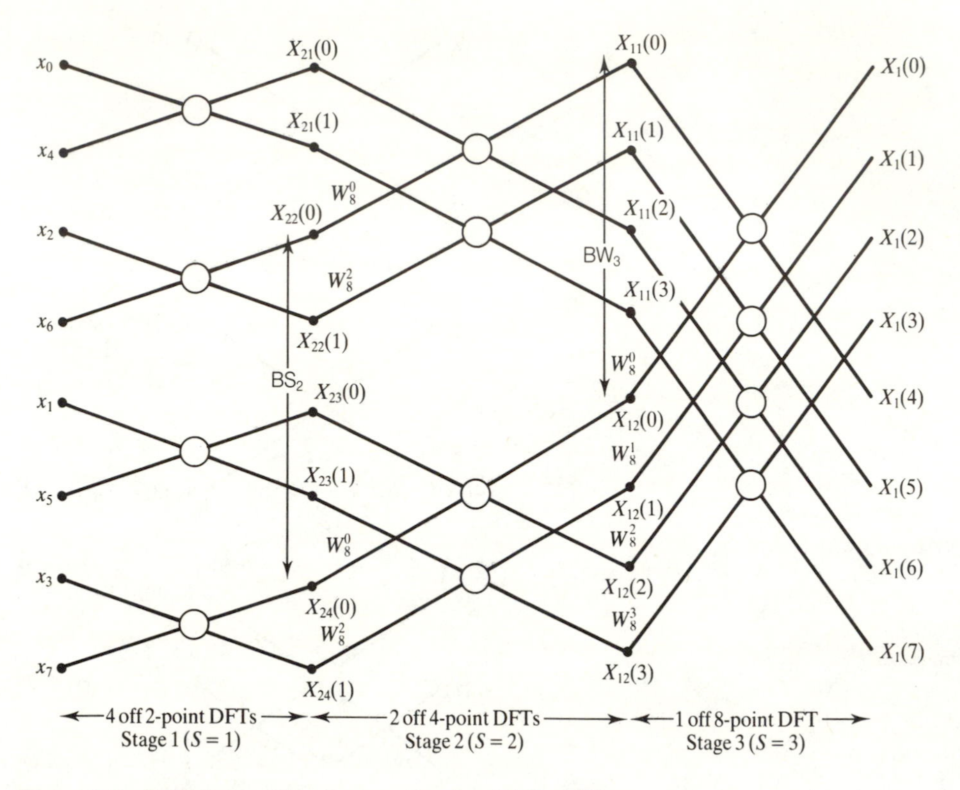

Figure 2.5 FFT butterflies for an 8-point DFT: BW_3, memory separation between points contributing to uppermost butterfly of stage 3; BS_2, memory separation between bottom points of butterflies in stage 2 with the same weighting factor.

These values are the same as those obtained in Section 2.2, but were more readily obtained using the FFT algorithm. This conclusion is general and the computational savings increase as the number of data increases.

2.5.2 Algorithmic development

Inspection of Figure 2.5 shows that in order to execute the FFT the program must re-order the input data and perform the butterfly computations. These will now be considered in turn. (See also Strum and Kirk, 1988.)

2.5.2.1 Re-ordering the input data

While it might at first appear that there is no obvious way to program the re-ordering of the input data, there is. The secret is to think in binary terms. Table 2.2 shows in the first column the required ordering of the data for input

Table 2.2 Sequence re-ordering by bit reversal.

Required sequence for butterfly computation	*Binary addresses of required sequence data*	*Bit-reversed addresses*	*Corresponding sequence = original data sequence*
x_0	000	000	x_0
x_4	100	001	x_1
x_2	010	010	x_2
x_6	110	011	x_3
x_1	001	100	x_4
x_5	101	101	x_5
x_3	011	110	x_6
x_7	111	111	x_7

to the butterfly network as given by Figure 2.5. Each value is assumed to be stored in a binary memory address. These addresses are given in the second column. The third column shows these memory addresses bit reversed. If these bit-reversed addresses are taken to correspond to the binary addresses of the original data sequence, commencing with $x(0)$ at 000, then the corresponding data values are given in the fourth column, which is seen to contain the original data sequence. Thus the addresses of the re-ordered data are seen to be the bit-reversed addresses of the original data sequence. The program is therefore required to convert the data point numbers (0 to $N-1$) to binary, to bit reverse these binary numbers and to convert them back to the denary numbers which are the addresses of the re-ordered data. Conversion to binary may be achieved by repetitively dividing by two when the remainders give the reverse-ordered digits of the corresponding binary number which is therefore the required binary address of the re-ordered sequence. These remainders may be obtained by using the MOD function obtainable in a high level language, for example MOD(K,2) gives the remainder on dividing the denary value K by 2. The integer part of $K/2$ is found by performing integer division. The remaining digits are found by repeating this process until $\log_2 N$ divisions have been made. This is because the data is to consist of $2^m = N$ data points so each address requires m digits and is divisible by two m times, where $m = \log_2 N$. The Ith bit of the new address is the binary coefficient of 2^{m-1-I} so the new address (NADDR) may be found using a DO loop for a particular value of K (K=IADDR) which runs from I=0 to I=m−1 and which takes the remainders (RMNDR) from the successive integers obtained after dividing K=IADDR by two. The pseudo-code required for this is

```
DO FOR I=0 TO m−1
      RMNDR:=MOD(IADDR,2)
      NADDR:=NADDR+RMNDR×2^(m−1−I)
      IADDR:=IADDR/2
END DO
```

This DO loop has to be nested inside another DO loop, the function of which is to extract the original data from a complex array DATA(K), K=0 to N−1 being the datum point number which is also the number of the complex element in the array and corresponds to the initial address of the datum, and to insert the re-ordered data into the array NEWDATA(NADDR). The data in NEWDATA is now in the correct sequence for the butterfly computations. The complete pseudo-code is

```
DO FOR K=0 TO N−1
     NADDR:=0
     IADDR:=K
     DO FOR I=0 TO m−1
          RMNDR:=MOD(IADDR,2)
          NADDR:=NADDR+RMND×2^(m−1−I)
          IADDR:=IADDR/2
     END DO
     NEWDATA(NADDR):=DATA(K)
END DO
```

2.5.2.2 Butterfly computations

Three stages of computation are involved:

(1) calculate weighting factors, $W_N^R = e^{-j(2\pi/N)R}$;

(2) evaluate butterflies in a stage (see Figure 2.5 for stage definitions);

(3) compute over all stages.

An effective procedure is to calculate the weighting factors W_N^R involved in a stage, and for each one to evaluate all the butterflies in the stage with that factor. Having evaluated all the butterflies in that stage, the procedure is repeated for all stages. Thus, referring to stage 2 in Figure 2.5, W_8^0 and the two butterflies involving W_8^0 are calculated. Then W_8^2 and the two butterflies involving it are calculated. This procedure is carried out for each of the three stages in turn starting with stage 1 and the re-ordered data. In order to achieve the FFT computation using a relatively short program, a number of properties of the algorithm are required. Let the butterfly width BWIDTH represent the separation in memory of points which contribute to a butterfly. For the top butterfly of stage 3 in Figure 2.5 this is BW_3. It is a spacing of four points. Consideration of the butterflies in the other stages leads to the conclusion that in general

$$\text{BWIDTH}=2^{S-1} \qquad \textbf{(2.59)}$$

where S is the stage number. Let BSEP, the butterfly separation, be the separation in memory of like points between the nearest butterflies in a stage of those butterflies which have the same weighting factor. For stage 2 in Figure 2.5, BS_2 represents BSEP for butterflies with the weighting factor W_8^0. Inspection of the figure shows that for the Sth stage

$$\text{BSEP}=2^{S} \qquad \textbf{(2.60)}$$

Finally for an N-point FFT the exponents of the weighting factors change by

$$P = N/2^S \quad \textbf{(2.61)}$$

in the Sth stage. This can be seen in Figure 2.5. For example, in stage 2, $S = 2$, $P = 8/2^2 = 2$ and the weighting factors are W_8^0 and W_8^2.

Each butterfly may be calculated as

```
XNEW(TOP)=XOLD(TOP)+W_N^R×XOLD(BOTTOM)
XNEW(BOTTOM)=XOLD(TOP)−W_N^R×XOLD(BOTTOM)        (2.62)
```

but memory space may be conserved by re-writing them as

```
TEMP=W_N^R×X(BOTTOM)
```

where X(BOTTOM) refers to the output of the butterfly in the previous stage,

```
X(BOTTOM)=X(TOP)−TEMP        (2.63a)
```

where X(BOTTOM) now refers to the required butterfly output, X(TOP) refers to the previous value, and

```
X(TOP)=X(TOP)+TEMP        (2.63b)
```

where the new left-hand side value of X(TOP) is the required butterfly output.

With all this knowledge available it is now possible to write the pseudo-code for the FFT as shown below:

```
10   PI:=3.141593
20   DO FOR S=1 TO m                           evaluate for m stages
30        BSEP:=2^S
40        P:=N/BSEP                            (P = N/2^S)
50        BWIDTH:=BSEP/2                       (BWIDTH = 2^(S-1))
60        DO FOR J=0 TO (BSEP−1)               Work out weighting factors
                                               for a particular stage
70             R:=P.J                          calculate power of W_N
80             THETA:=2×PI×R/N                 calculate exponent of e^-j
90             WN:=CMPLX{cos(THETA),−sin(THETA)}
                                               calculate W_N^R
100            DO FOR TOPVAL=J STEP BSEP UNTIL N/2
110                 BOTVAL:=TOPVAL+BSEP
120                 TEMP:=X(BOTVAL)×WN
130                 X(BOTVAL):=X(TOPVAL)−TEMP
140                 X(TOPVAL):=X(TOPVAL)+TEMP
150            END DO
160       END DO
170  END DO
```

Lines 100–150 work out all butterflies in a stage which have the same weighting factors.

2.5.3 Computational advantages of the FFT

The computational advantages of the FFT may be illustrated by considering first the FFT algorithm of Figure 2.5. This figure shows that an N-point FFT contains $N/2$ butterflies per stage and $\log_2 N$ stages, that is it contains a total of $(N/2)\log_2 N$ butterflies. Figure 2.4(a) shows that each butterfly involves one complex multiplication of the form $W_N^R X_{ij}(k)$. Hence the FFT involves $(N/2)\log_2 N$ complex multiplications compared with N^2 in the case of the DFT, as shown in Section 2.4. Thus the computational saving in complex multiplications is $N^2 - (N/2)\log_2 N$. Each butterfly contains two complex additions so the FFT requires $N\log_2 N$ complex additions compared with $N(N-1)$ for the DFT. Thus the saving in complex additions is $N(N-1) - N\log_2 N$. These savings are illustrated in Table 2.3. For a typical 1024-point DFT it is seen that the computation time will be reduced by two orders of magnitude if the FFT algorithm is employed.

2.6 Inverse fast Fourier transform

An FFT algorithm for determining the inverse fast Fourier transform (IFFT) is readily obtainable from the FFT algorithm. Its use lies in transforming spectra into their corresponding waveforms and in checking that the FFT has been correctly computed by using basically the same algorithm to obtain the original data. To see how the IFFT is derived make the following substitutions in Equation 2.20. Sum over the variable λ rather than n, let the variable k become μ, set $\Omega = 2\pi/NT$ so that the exponent of e becomes $-jk(2\pi/N)\lambda$. Again using the notation $x(\lambda T) = x(\lambda)$ and so on, Equation 2.20 then becomes

$$X(\mu) = \sum_{\lambda=0}^{N-1} x(\lambda)e^{-j\mu(2\pi/N)\lambda} \quad \mu = 0, 1, \ldots, N-1 \tag{2.64}$$

Now make similar substitutions in Equation 2.23, that is put $k = \lambda$, and $n = \mu$ so that Equation 2.23 becomes

$$x(\mu) = \frac{1}{N}\sum_{\lambda=0}^{N-1} X(\lambda)e^{j\lambda(2\pi/N)\mu} \quad \mu = 0, 1, \ldots, N-1 \tag{2.65}$$

In the last two equations $X(\mu)$, $X(\lambda)$, $x(\lambda)$ and $x(\mu)$ are all elements of the equidimensional arrays X and x, and so it can be seen that the IFFT, $x(\mu)$, differs from the FFT, $X(\mu)$, only in the factor $1/N$ and the sign of the exponent. Thus with small modifications the FFT may be used to calculate the IFFT. The two transforms may be included in the same algorithm by making the

Table 2.3 Savings in complex multiplications and additions when the FFT is used instead of the DFT.

	DFT		*FFT*		*Ratio of DFT multiplications to FFT multiplications*	*Ratio of DFT additions to FFT additions*
N	*Number of complex multiplications*	*Number of complex additions*	*Number of complex multiplications*	*Number of complex additions*		
2	4	2	1	2	4	1
4	16	12	4	8	4	1.5
8	64	56	12	24	5.3	2.3
16	256	240	32	64	8.0	3.75
32	1024	992	80	160	12.8	6.2
64	4096	4032	192	384	21.3	10.5
128	16384	16256	448	896	36.6	18.1
256	65536	65280	1024	2048	64.0	31.9
512	262144	261632	2304	4608	113.8	56.8
1024	1048576	1047552	5120	10240	204.8	102.3
2048	4194304	4192256	11264	22528	372.4	186.1
4096	16777216	16773120	24576	49152	682.7	341.3
8192	67108864	67100672	53248	106496	1260.3	630.0

following modifications to the preceding pseudo-code:

```
line 5      K:=1 FOR FFT, K:=−1 FOR IFFT
line 80     THETA:=K×2×PI×R/N
line 145    IF K=−1 DO
line 146    X(BOTVAL):=X(BOTVAL)/N
line 147    X(TOPVAL):=X(TOPVAL)/N
line 148    END DO
```

2.7 Implementation of the FFT

Basically it should be clear now that in principle the FFT or IFFT is computed by providing a data array and operating on the data using the FFT or IFFT algorithms (including the bit-reversal algorithms). However, there remain some other considerations. So far the effects of discontinuities which occur at the two ends of the data have been ignored, together with the phenomena known as aliasing and picket fencing. In order to compute good approximations to the true data spectra it is necessary to take these effects into account using the techniques described in Chapter 10. Another aspect lies in the fact that so far in this chapter attention has been confined to the radix-2 decimation-in-time algorithm, but other algorithms, including decimation in frequency (DIF), exist. Some of these points will be discussed in what follows.

2.7.1 The decimation-in-frequency FFT

Section 2.5 described the decimation-in-time fast Fourier transform algorithm which was obtained by repeatedly dividing an initial discrete Fourier transform of the form of Equation 2.43 into two transforms, one consisting of the even terms and one of the odd terms, until the initial transform was reduced to two-point transforms of the initial data. An alternative approach is to separate the initial transform into two transforms, one containing the first half of the data and the other containing the second half of the data. The resulting decimation-in-frequency algorithm was first derived by Gentleman and Sande (1966) and is often known as the Sande–Tukey algorithm, unhappily for Gentleman. Overall there is little to choose between the two algorithms.

2.7.2 Comparison of DIT and DIF algorithms

For the DIF algorithm the order of the input data is unaltered but the output FFT sequence is bit reversed. Both the DIF and DIT algorithms are in-place algorithms. By re-drawing them it is possible to maintain the order of both the inputs and the outputs, but the resulting algorithms are no longer in-place algorithms and extra storage space is required (see Chapter 11). The number of complex multiplications required for both algorithms is the same. Overall, there is little to choose between the two.

2.7.3 Modifications for increased speed

Further increases in computational speed are possible for the DIT algorithm. For example, a radix-4 FFT can be used to reduce the number of complex multiplications by a factor approaching 2. The number of additions is also reduced. Another enhancement in speed may be obtained by removing unnecessary multiplications by the weighting factor W_N(often called the twiddle factor!) which occur when $W_N = \pm 1$, or $\pm$j. This also reduces the number of additions required. Implementation is by including a separate butterfly for the cases for which $W_N = \pm 1$ or $\pm$j. This, for example, would give a radix-2, two-butterfly, in-place, DIT algorithm. Further time saving is obtainable by first calculating the sine and cosine parts of W_N and storing them in a look-up table from which their values are obtained as required. Various other improvements are possible. It is therefore apparent that there exist, not just one FFT, but many. A full appreciation of the subject requires a considerable investment of study time and significant mathematical ability. It is advantageous to be adept with advanced matrix theory, multidimensional index mapping, and number theory. This more comprehensive approach to the subject is beyond the aim of this chapter, which is to explain the concept of discrete transforms in a way that most people involved in digital signal processing might understand. There are a number of other books, for example Burrus and Parks (1985) and

Beauchamp (1987) which present the more specialized approach and these are recommended to the reader. As already stated, there are a number of algorithms from which to choose, and an expert might well write his or her own which is more suitable for the application. However, reducing the numbers of multiplications and additions required does not always result in a proportional increase in speed of computation. Decreasing the multiplications may result in more program code and more additions. If hardware signal processing chips are to be used these will impose their own limitations which may negate the improvements in the algorithm. Since increases in speed are not likely to exceed a factor of 2, the authors are inclined to recommend the basic radix-2 DIT FFT described above for general-purpose use where speed is not that critical. A large number of FORTRAN FFT programs are presented in Burrus and Parks (1985), together with discussion as to their advantages and disadvantages. Programs are also given for hardware implementation. A C language program for the radix-2 DIT FFT is given in Appendix 2A.

2.8 Other discrete transforms

A variety of other transforms are available. The Winograd Fourier transform (Winograd, 1978; see also Burrus and Parks, 1985; Beauchamp, 1987; Rader, 1968; Signal Processing Committee, 1979; McClellan and Rader, 1979) and the prime factor algorithm (Beauchamp, 1987; see also McClellan and Rader, 1979) provide ingenious but complicated methods of increasing the speed of computation of the FFT. The discrete cosine transform is particularly useful in data compression applications (see Section 2.8.1). The Walsh transform (Section 2.8.2) analyses signals into rectangular waveforms rather than sinusoidal ones and is computed more rapidly than the FFT. The Hadamard transform (Section 2.8.3), constructed by re-ordering the Walsh ordered sequence, is even faster to compute. While displaying advantages for some purposes the Walsh and Hadamard transforms suffer from some disadvantages which limit their applicability; see Sections 2.8.2 and 2.8.3. Finally, the Haar transform is particularly useful for edge detection in image processing (Rosenfield and Thurston, 1971) and for similar applications. Beauchamp (1987) provides a good source of starting material for those wishing to know more about the various transforms discussed and their applications.

2.8.1 Discrete cosine transform

In addition to their uses in speeding correlation and convolution computations, and in spectral analysis, transform methods are also used to achieve data compression in, for example, speech and video transmissions and for recording biomedical signals such as EEGs and ECGs. They are also utilized for pattern recognition. In these applications only the more significant transform components are utilized. Hence the number of coding bits required is reduced. This allows faster transmission, the use of narrower bandwidth transmission lines, and easier pattern recognition (because of the data reduction). The three important features of a suitable transform are its compressional efficiency, which relates to concentrating the energy at low frequencies, ease of computation, and minimum mean square error. The ideal transform for achieving these is the Karhunen–Loève transform, but this cannot be represented algorithmically. However, the discrete cosine transform (DCT) has virtually the same properties and does possess an algorithm. It consists essentially of the real part of the DFT. This definition is reasonable since the Fourier series of a real and even function contains only the cosine terms, and in, for example, the case of sampled voltage values the data used is real and can be made symmetrical by doubling the data by adding its mirror image. Thus the DFT is given by (Equation 2.41)

$$X(k) = \sum_{n=0}^{N-1} x_n \mathrm{e}^{-\mathrm{j}2\pi nk/N}, \quad k = 0, 1, \ldots, N-1$$

Defining the DCT $X_c(k)$ as the real part of this gives

$$X_c(k) = \mathrm{Re}\,[X(k)] = \sum_{n=0}^{N-1} x_n \cos\left(\frac{k2\pi n}{N}\right), \quad k = 0, 1, \ldots, N-1 \qquad \textbf{(2.66)}$$

This is one of several forms of the DCT. A more common form is (Beauchamp, 1987; Yip and Ramamohan, 1987; Ahmed and Rao, 1975)

$$X_c(k) = \frac{1}{N} \sum_{n=0}^{N-1} x_n \cos\left(\frac{k2\pi n + k\pi}{2N}\right) = \frac{1}{N} \sum_{n=0}^{N-1} x_n \cos\left[\frac{k\pi(2n+1)}{2N}\right]$$

$$k = 0, 1, \ldots, N-1 \qquad \textbf{(2.67)}$$

and two other forms also exist (Yip and Ramamohan, 1987).

Implementations of the DCT exist based on the FFT as might be expected (Narasinka and Petersen, 1978), and a fast DCT which is six times as fast as these has been developed (Chen *et al.*, 1977). Another version is the C-matrix transform which can be more simply constructed in hardware (Srinivassan and Rao, 1983).

2.8.2 Walsh transform

The transforms discussed so far have been based on cosine and sine functions. Transforms based on pulse-like waveforms which take only values of ±1 are much simpler and faster to compute. They are also more appropriate for the

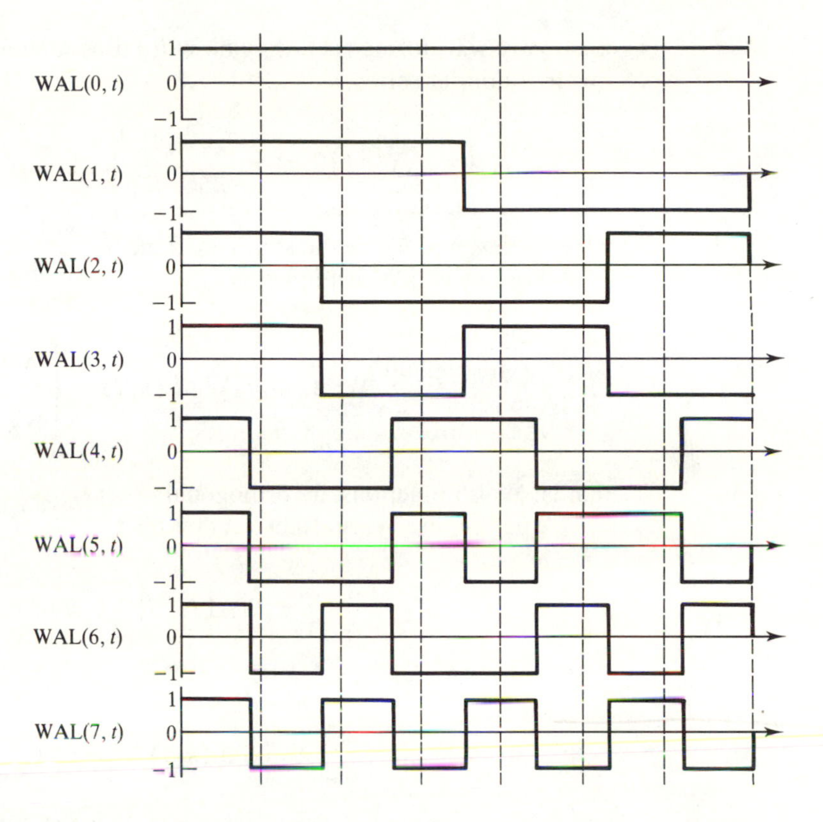

Figure 2.6 Sequency-ordered Walsh functions to $n = 7$ showing sampling times for 8×8 Walsh transform matrix.

representation of waveforms which contain discontinuities, for example in images. Conversely, they are less appropriate for describing continuous waveforms and may not be phase invariant in which case the derived spectrum may be distorted. However, such waveforms are used in image processing (astronomy and spectroscopy), signal coding, and filtering.

Just as the DFT is based on a set of harmonically related cosine and sine waveforms, so is the discrete Walsh transform (DWT) based on a set of harmonically related rectangular waveforms, known as Walsh functions. However, frequency is not defined for rectangular waveforms and so the analogous term sequency is used. Sequency is half the average number of zero crossings per unit time. Figure 2.6 shows the set of Walsh functions up to the order of $N = 8$ drawn in order of increasing sequency. They are said to be sequency, or Walsh, ordered. The Walsh function at time t and of sequency n is designated $\mathrm{WAL}(n, t)$. Inspection of Figure 2.6 shows that there are equal numbers of even and odd Walsh functions, just as there are corresponding cosinusoidal and sinusoidal Fourier series components. The even functions, $\mathrm{WAL}(2k, t)$, are written $\mathrm{CAL}(k, t)$, and the odd functions, $\mathrm{WAL}(2k + 1, t)$, are written $\mathrm{SAL}(k, t)$, where $k = 1, 2, \ldots, N/2 - 1$.

Any waveform, $f(t)$, may be written as a Walsh function series, analogous to a Fourier series, as

$$f(t) = a_0 \,\mathrm{WAL}(0, t) + \sum_{i=1}^{N/2-1} \sum_{j=1}^{N/2-1} [a_i \,\mathrm{SAL}(i, t) + b_j \,\mathrm{CAL}(j, t)] \quad \textbf{(2.68)}$$

where the a_i and b_j are the series coefficients.

For any two Walsh functions,

$$\sum_{t=0}^{N-1} \mathrm{WAL}(m, t)\,\mathrm{WAL}(n, t) = \begin{cases} N \text{ for } n = m \\ 0 \text{ for } n \neq m \end{cases}$$

that is, Walsh functions are orthogonal.

The discrete Walsh transform pair is

$$X_k = \frac{1}{N} \sum_{i=0}^{N-1} x_i \,\mathrm{WAL}(k, i) \quad k = 0, 1, \ldots, N-1 \quad \textbf{(2.69)}$$

and

$$x_i = \sum_{i=0}^{N-1} X_k \,\mathrm{WAL}(k, i) \quad i = 0, 1, \ldots, N-1 \quad \textbf{(2.70)}$$

where we note that, apart from the factor of $1/N$, the inverse transform is the same as the transform, and that $\mathrm{WAL}(k, i) = \pm 1$. The transform pair may, therefore, be calculated by matrix multiplication by digital means as mentioned above. However, the lack of phase invariance means that the DWT is unsuitable for fast correlations or convolutions.

Equation 2.69 shows that the kth DWT component is obtained by multiplying each waveform sample x_i by the Walsh function of sequency k and summing for $k = 0, 1, \ldots, N-1$. This may be expressed for all k DWT components in matrix notation as

$$\mathbf{X}_K = \mathbf{x}_i \mathbf{W}_{ki} \quad \textbf{(2.71)}$$

where $\mathbf{x}_i = [x_0 \;\, x_1 \;\, x_2 \;\, \ldots \;\, x_{N-1}]$, the data sequence,

$$\mathbf{W}_{ki} = \begin{bmatrix} W_{01} & W_{02} & \ldots & W_{0,N-1} \\ W_{11} & & & \\ \vdots & & & \vdots \\ W_{N-1,1} & W_{N-1,2} & \ldots & W_{N-1,N-1} \end{bmatrix}$$

the Walsh transform matrix and $\mathbf{X}_k = [X_0 \;\, X_1 \;\, \ldots \;\, X_{N-1}]$, the $N-1$ components of the DWT. Note that $\mathbf{W}_{ki}$ is an $N \times N$ matrix where N is the number of data points, that is sampled waveform points. Thus, if there are N data points it is necessary to consider the first N sequency-ordered Walsh functions. Each one is sampled N times. The kth row of $\mathbf{W}_{ki}$ corresponds to the N sampled values of the kth sequency component.

Example 2.6

As an example, let us compute the DWT of the data sequence (1, 2, 0, 3). This consists of $N = 4$ data points and so $\mathbf{W}_{ki}$ is a 4×4 matrix obtainable from the first four rows of Figure 2.6 as

$$\mathbf{W}_{ki} = \begin{bmatrix} 1 & 1 & 1 & 1 \\ 1 & 1 & -1 & -1 \\ 1 & -1 & -1 & 1 \\ 1 & -1 & 1 & -1 \end{bmatrix} \tag{2.72}$$

Therefore, from Equation 2.71, $\mathbf{X}_k$ is given by

$$\mathbf{X}_k = \tfrac{1}{4}[1 \quad 2 \quad 0 \quad 3]\begin{bmatrix} 1 & 1 & 1 & 1 \\ 1 & 1 & -1 & -1 \\ 1 & -1 & -1 & 1 \\ 1 & -1 & 1 & -1 \end{bmatrix} = \tfrac{1}{4}[6 \quad 0 \quad 2 \quad -4]$$

so that $X_0 = 1.5$, $X_1 = 0$, $X_2 = 0.5$ and $X_3 = -1$. This is considerably easier to calculate than the corresponding DFT! Needless to say, fast DWTs (FDWTs) exist.

The corresponding spectrum can be calculated with power components given by

$$P(k) = [|\mathrm{CAL}(k, t)|^2 + |\mathrm{SAL}(k, t)|^2]^{1/2}$$

where

$$\begin{aligned} P(0) &= X_c^2(0) \\ P(k) &= X_c^2(k, t) + X_s^2(k, t) \\ P\left(\frac{N}{2}\right) &= X_s^2\left(\frac{N}{2}, t\right) \end{aligned} \tag{2.73}$$

where $k = 1, 2, \ldots, N/2 - 1$, and with phase components

$$\phi(0) = 0, \pi$$

$$\phi(k) = \tan^{-1}\left[\frac{X_s(k)}{X_c(k)}\right], \quad k = 1, 2, \ldots, N/2 - 1 \tag{2.74}$$

and

$$\phi\left(\frac{N}{2}\right) = 2k\pi \pm \pi/2, \quad k = 0, 1, 2, \ldots$$

For the above DWT we have, therefore

$$P(0) = 1.5^2 = 2.25; \ \phi(0) = 0, \ \pi$$

$$P(1) = 0^2 + 0.5^2 = 0.25; \ \phi(1) = \tan^{-1}\left(\frac{0}{0.5}\right) = 0$$

$$P(2) = (-1)^2 = 1; \ \phi(2) = \frac{\pi}{2} + 2k\pi, \quad k = 0, 1, 2$$

2.8.3 Hadamard transform

The Hadamard transform, or Walsh–Hadamard transform, is basically the same as the Walsh transform, but with the Walsh functions and therefore the rows of the transform matrix re-ordered. The resultant Hadamard matrix comprises subarrays of second-order matrices. The Hadamard matrix of order 8 × 8 is given as ${}^{8}\mathbf{H}$ in Figure 2.7, and it can be seen to consist of matrices

$${}^{2}\mathbf{H} = \begin{bmatrix} 1 & 1 \\ 1 & -1 \end{bmatrix} \quad \text{and} \quad -{}^{2}\mathbf{H} = \begin{bmatrix} -1 & -1 \\ -1 & 1 \end{bmatrix}$$

Any Hadamard matrix of order $2N$ may be obtained recursively from ${}^{2}\mathbf{H}$ as

$${}^{2N}\mathbf{H} = \begin{bmatrix} {}^{N}\mathbf{H} & {}^{N}\mathbf{H} \\ {}^{N}\mathbf{H} & -{}^{N}\mathbf{H} \end{bmatrix} \qquad \textbf{(2.75)}$$

$$\begin{array}{cc|cccccccc|} & & \multicolumn{8}{l}{i \rightarrow} \\ & & 0 & 1 & 2 & 3 & 4 & 5 & 6 & 7 \\ k & 0 & 1 & 1 & 1 & 1 & 1 & 1 & 1 & 1 \\ \downarrow & 1 & 1 & -1 & 1 & -1 & 1 & -1 & 1 & -1 \\ & 2 & 1 & 1 & -1 & -1 & 1 & 1 & -1 & -1 \\ & 3 & 1 & -1 & -1 & 1 & 1 & -1 & -1 & 1 \\ & 4 & 1 & 1 & 1 & 1 & -1 & -1 & -1 & -1 \\ & 5 & 1 & -1 & 1 & -1 & -1 & 1 & -1 & 1 \\ & 6 & 1 & 1 & -1 & -1 & -1 & -1 & 1 & 1 \\ & 7 & 1 & -1 & -1 & 1 & -1 & 1 & 1 & -1 \end{array}$$

Figure 2.7 8 × 8 Hadamard transform matrix.

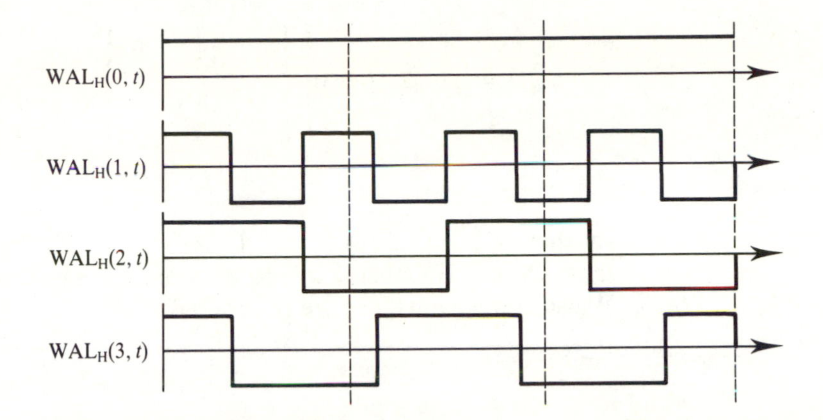

Figure 2.8 Hadamard-ordered Walsh functions to $n = 7$ showing sampling times for 4×4 Hadamard Transform matrix.

The value of this recursive property is that by Hadamard ordering the Walsh functions the resultant fast Walsh–Hadamard transform may be more quickly computed than the DWT. Hadamard-ordered (or naturally ordered) Walsh functions are shown in Figure 2.8. The Hadamard-ordering sequence is obtained from the Walsh-ordered sequence by

(1) expressing the order of the Walsh-ordered function in binary,
(2) bit-reversing the binary values,
(3) converting the binary values to Gray code, and
(4) converting this value to decimal.

Example 2.7

By way of example the discrete Walsh–Hadamard transform of the sequence $(1, 2, 0, 3)$ will now be calculated. The 4×4 Hadamard matrix $\mathbf{H}_{ki}$ is

$$\mathbf{H}_{ki} = \begin{bmatrix} 1 & 1 & 1 & 1 \\ 1 & 1 & -1 & -1 \\ 1 & -1 & 1 & -1 \\ 1 & -1 & -1 & 1 \end{bmatrix} = \begin{bmatrix} 1 & 1 & 1 & 1 \\ 1 & -1 & 1 & -1 \\ 1 & 1 & -1 & -1 \\ 1 & -1 & -1 & 1 \end{bmatrix} \tag{2.76}$$

by the properties of a Hadamard matrix. The DWHT of $(1, 2, 0, 3)$ is, therefore, given by

$$\mathbf{X}_k^{\mathrm{WH}} = \tfrac{1}{4}[1 \quad 2 \quad 0 \quad 3]\begin{bmatrix} 1 & 1 & 1 & 1 \\ 1 & -1 & 1 & -1 \\ 1 & 1 & -1 & -1 \\ 1 & -1 & -1 & 1 \end{bmatrix} = \tfrac{1}{4}[6 \quad -4 \quad 0 \quad 2]$$

so that $X_0^{\mathrm{WH}} = 1.5$, $X_1^{\mathrm{WH}} = -1$, $X_2^{\mathrm{WH}} = 0$, $X_3^{\mathrm{WH}} = 0.5$. The magnitudes of these Walsh–Hadamard components are the same as the previously calculated Walsh components, but are re-ordered.

2.9 Worked examples

Example 2.8

The first four sampled voltage values of a 10 Hz bandwidth signal sampled at 125 Hz were (0.5, 1, 1, 0.5). Demonstrate how the discrete Fourier transform of this sequence may be obtained using the fast Fourier transform, and hence obtain the Fourier transform of the data.

Solution

The flowgraph for the FFT is given in Figure 2.9. We have

$$X(0) = G(0) + W_4^0 H(0) = G(0) + H(0)$$
$$G(0) = x(0) + W_2^0 x(2) = x(0) + x(2)$$
$$H(0) = x(1) + W_2^0 x(3) = x(1) + x(3)$$

Since $W^0 = 1$, substituting values gives

$$X(0) = x(0) + x(2) + x(1) + x(3)$$
$$= 0.5 + 1 + 1 + 0.5 = 3$$
$$X(1) = G(1) + W_4^1 H(1)$$
$$G(1) = x(0) - W_2^0 x(2) = x(0) - x(2)$$
$$H(1) = x(1) - W_2^0 x(3) = x(1) - x(3)$$

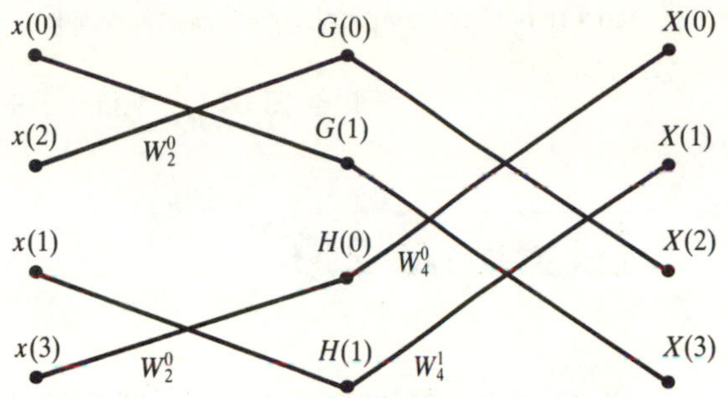

Figure 2.9 Flowgraph of the FFT in Example 2.8.

Now, $W_N = \mathrm{e}^{-\mathrm{j}2\pi/N}$, and therefore $W_N^1 = \mathrm{e}^{-\mathrm{j}2\pi/4} = \mathrm{e}^{-\mathrm{j}\pi/2}$. Substituting gives

$$\begin{aligned}
X(1) &= x(0) - x(2) + \mathrm{e}^{-\mathrm{j}\pi/2}[x(1) - x(3)] \\
&= 0.5 - 1 + \left[\cos\left(\frac{\pi}{2}\right) - \mathrm{j}\sin\left(\frac{\pi}{2}\right)\right](1 - 0.5) \\
&= 0.5 - 1 + (0 - \mathrm{j})0.5 = -0.5 - \mathrm{j}0.5 = -0.5(1 + \mathrm{j}) \\
X(2) &= G(0) - W_4^0 H(0) = G(0) - H(0) \\
&= x(0) + x(2) - [x(1) + x(3)] \\
&= 0.5 + 1 - (1 + 0.5) = 0 \\
X(3) &= G(1) - W_4^1 H(1) \\
&= x(0) - x(2) - \mathrm{e}^{-\mathrm{j}\pi/2}[x(1) - x(3)] \\
&= 0.5 - 1 - \left[\cos\left(\frac{\pi}{2}\right) - \mathrm{j}\sin\left(\frac{\pi}{2}\right)\right](1 - 0.5) \\
&= -0.5 - (-\mathrm{j})0.5 = -0.5 + \mathrm{j}0.5 = 0.5(-1 + \mathrm{j})
\end{aligned}$$

Therefore

$$X(\Omega) = \{3, -0.5(1 + \mathrm{j}), 0, 0.5(-1 + \mathrm{j})\}$$

When T, the sampling interval, is small,

$$\mathrm{FT} = T\,\mathrm{DFT}$$

where FT is the Fourier transform. Here, $T = 1/125\ \mathrm{s} = 0.008\ \mathrm{s}$. The signal period is $1/10\ \mathrm{s} = 0.1\ \mathrm{s}$; therefore

$$\frac{T}{\text{period}} = \frac{0.008}{0.1} = 0.08 \ll 1$$

and it is a good approximation to put FT = T DFT, so that

$$\text{FT} = \{0.024, -0.004(1 + \text{j}), 0, 0.004(-1 + \text{j})\}$$

Example 2.9

In a data compression system the data is first transformed and then the transform values are threshold limited to a threshold magnitude of 0.375. Two transforms are under consideration, the discrete cosine transform, $X_c(k)$, defined by

$$X_c(k) = \frac{1}{N} \sum_{n=0}^{N-1} x_n \cos\left(\frac{k2\pi n}{N}\right), \quad k = 0, 1, \ldots, N - 1$$

and the Walsh transform, X_k, defined by

$$X_k = \frac{1}{N} \sum_{i=0}^{N-1} x_i \,\text{WAL}(k, i), \quad k = 0, 1, \ldots, N - 1$$

Assuming representative results to be given by the data sequence $\{1, 2, 0, 3\}$, determine

(1) which is the more efficient of the two transforms for data compression in this case, and

(2) the percentage data compression achieved.

Inverse transform the compressed data obtained using the Walsh transform and compare with the original data sequence.

Solution

(1) For the DCT with $x_0 = 1, x_1 = 2, x_2 = 0, x_3 = 3$,

$$X_c(0) = \tfrac{1}{4} x_0 \cos 0 + x_1 \cos 0 + x_2 \cos 0 + x_3 \cos 0$$

$$= \tfrac{1}{4}(1 + 2 + 0 + 3) = \tfrac{6}{4} = 1.5$$

$$X_c(1) = \tfrac{1}{4} \sum_{n=0}^{3} x_n \cos\left(\frac{2\pi n}{4}\right) = \tfrac{1}{4} \sum_{n=0}^{3} x_n \cos\left(\frac{n\pi}{2}\right)$$

$$= \tfrac{1}{4}\left[x_0 + x_1 \cos\left(\frac{\pi}{2}\right) + x_2 \cos\left(\frac{2\pi}{2}\right) + x_3 \cos\left(\frac{3\pi}{2}\right)\right]$$

$$= \tfrac{1}{4}[1 + 2 \times 0 + 0 \times (-1) + 3 \times 0] = 0.25$$

$$X_c(2) = \tfrac{1}{4}\left[x_0 \cos\left(\frac{4\pi \times 0}{4}\right) + x_1 \cos\left(\frac{4\pi \times 1}{4}\right) + x_2 \cos\left(\frac{4\pi \times 2}{4}\right) + x_3 \cos\left(\frac{4\pi \times 3}{4}\right)\right]$$

$$= \tfrac{1}{4}(x_0 - x_1 + x_2 - x_3) = \tfrac{1}{4}(1 - 2 + 0 - 3) = -1$$

$$X_c(3) = \tfrac{1}{4}\left[x_0 \cos\left(\frac{6\pi \times 0}{4}\right) + x_1 \cos\left(\frac{6\pi}{4}\right) + x_2 \cos\left(\frac{6\pi \times 2}{4}\right) + x_3 \cos\left(\frac{6\pi \times 3}{4}\right)\right]$$

$$= \tfrac{1}{4}(1 + 2 \times 0 + 0 \times (-1) + 3 \times 0) = 0.25$$

Therefore

$$\text{DCT} = \{1.5, 0.25, -1, 0.25\}$$

The value remaining after thresholding ($|\text{values}| > 0.375$) are 1.5 and -1.

For the Walsh transform, evaluated in Section 2.8.2, $X_k = \{1.5, 0, 0.5, -1\}$, so the values remaining after thresholding in this case are 1.5, 0.5 and -1. Therefore the DCT provides more efficient data compression in this case.

(2) Assuming that the data compression efficiency, η, is given by

$$\eta = \frac{(\text{no. of data in original seq.} - \text{no. of data in transformed seq.}) \times 100\%}{\text{no. of data in original seq.}}$$

then

$$\eta = \frac{4 - 2}{4} \times 100\% = 50\%$$

Finally, the Walsh transform compressed data is $\{1.5, 0, 0.5, -1\}$. The inverse transformation is given by Equation 2.70:

$$x_i = \sum_{i=0}^{N-1} X_k \, \text{WAL}(k, i), \quad i = 0, 1, \ldots, N - 1$$

$$\mathbf{x}_i = [1.5 \quad 0 \quad 0.5 \quad -1]\begin{bmatrix} 1 & 1 & 1 & 1 \\ 1 & 1 & -1 & -1 \\ 1 & -1 & -1 & 1 \\ 1 & -1 & 1 & -1 \end{bmatrix} = [1 \quad 2 \quad 0 \quad 3]$$

This is identical with the initial sequence. This is because $X_1 = 0 < 0.375$ and although it is not transmitted it is represented in the inverse transform as 0, its exact value. Normally the reconstituted sequence will be an approximation to the original one.

Problems

2.1 Determine the Fourier series representation of the periodic waveforms of Figure 2.10.

2.2 Obtain the complex Fourier series representation of the waveforms in Figure 2.10.

2.3 Show that the amplitudes found in Problem 2.1 agree with those found in Problem 2.2.

2.4 Plot the amplitude and phase spectra of the Fourier components of the waveforms in Figure 2.10.

2.5 Calculate the amplitude and energy spectral densities of the voltage waveform $v(t)$ given by

$$v(t) = \begin{cases} \dfrac{A}{\tau} t + A & -\tau \leqslant t \leqslant 0 \\ -\dfrac{A}{\tau} t + A & 0 \leqslant t \leqslant \tau \\ 0 & \tau \leqslant t \leqslant -\tau \end{cases}$$

where $A = 5$ V and $\tau = 20$ ms.

2.6 Calculate the energy spectral density and the phase spectrum of the waveform $v(t)$ given by

$$v = \begin{cases} 2\sin\left[\dfrac{2\pi}{T}\left(t + \dfrac{T}{4}\right)\right] & -\dfrac{T}{4} \leqslant t \leqslant \dfrac{T}{4} \\ 0 & \text{elsewhere} \end{cases}$$

where $T = 0.0167$ s.

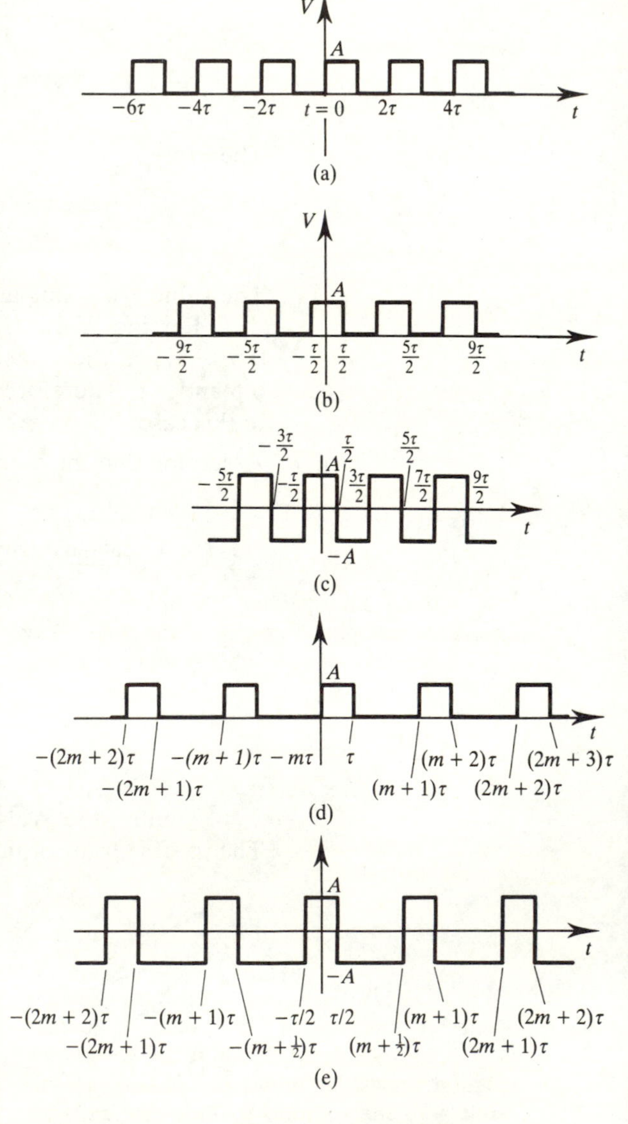

Figure 2.10 Periodic waveforms for Problem 2.1.

2.7 Plot the energy spectral density of the function

$$w(t) = \begin{cases} \sin\left[\frac{2\pi}{T}\left(t + \frac{3T}{4}\right)\right] & -\frac{3T}{4} \leqslant t \leqslant -\frac{T}{2} \\ 1.0 & -\frac{T}{2} \leqslant t \leqslant \frac{T}{2} \\ \sin\left[\frac{2\pi}{T}\left(t - \frac{3T}{4}\right)\right] & \frac{T}{2} \leqslant t \leqslant \frac{3T}{4} \\ 0 & \frac{3T}{4} \leqslant t \leqslant -\frac{3T}{4} \end{cases}$$

where $T = 4$ s.

2.8 Calculate the DFT of the data sequence $\{0, 1, 1, 0\}$ and check the validity of your answer by calculating its IDFT.

2.9 Derive the dimensions of $X(k)$ and of $X^2(k)$. Hence calculate and plot the energy spectrum of the data sequence $\{0, 1, 1, 0\}$ whose DFT was calculated as the solution to Problem 2.8.

2.10 If the sequence $\{0, 1, 1, 0\}$ of Problem 2.8 represented the digitized samples taken from a voltage waveform sampled at 125 Hz, determine the energy spectral density and phase spectrum of the Fourier transform of the data sequence.

2.11 Use the time-shifting property of the DFT and the solution to Problem 2.8 to obtain the amplitude and phase spectra of the time series $\{0, 0, 0, 0, 0, 1, 1, 0\}$ for data sampled at the instants $t = 0, 1, 2, \ldots, 7$ ms.

2.12 Use the results of Problem 2.9 to verify Parseval's theorem for the data $\{0, 1, 1, 0\}$.

2.13 Use the correlation theorem to calculate the circular correlation of the data sequences $\{1, 1, 0, 1\}$ and $\{1, 0, 0, 1\}$. Plot the correlation function against lag number, j.

2.14 Apply the correlation theorem to calculate the linear correlation of the data sequences $\{1, 1, 0, 1\}$ and $\{1, 0, 0, 1\}$. Plot the correlation function against lag numbers and compare the result with the solution to Problem 2.13, explaining any differences.

2.15 Calculate the DFT of the data sequence $\{0, 1, 1, 0\}$ using the decimation-in-time (Cooley–Tukey) FFT algorithm. Check the answer with that of Problem 2.8. Compare the numbers of complex additions and multiplications in the two methods.

2.16 Calculate the IFFT of the answer to Problem 2.15 to verify that the data sequence $\{0, 1, 1, 0\}$ is obtained.

2.17 Calculate the FFT of the data sequence $\{0, 0, 1, 1, 1, 1, 0, 0\}$ and plot the amplitude and phase spectra. Check the answer by calculating its IFFT to obtain the original sequence.

2.18 Write computer programs to compute the FFT and the IFFT. Check the FFT by computing the DFTs of the data sequences $\{0, 1, 1, 0\}$ of Problem 2.8, and $\{1, 1, 0, 1\}$ and $\{1, 0, 0, 1\}$ of Problem 2.13. Check the IFFT by computing the IDFTs of the DFT sequences.

2.19 Use the FFT program to compute the 1024-point DFTs of the waveforms of Problems 2.5 and 2.7. Plot their energy and phase spectra and compare with the plots obtained in the solutions to Problems 2.5 and 2.7.

2.20 Using a 1024-point FFT compute and plot the energy spectrum of the rectangular pulse of amplitude 5 V and width $\tau = 6$ s. Compare the result with that of Problem 2.7.

2.21 (1) Use the convolution theorem (Equation 2.37) to obtain the convolution of the spectra of the two pairs of waveforms:

(a) $v_s = \sin(2\pi \times 100t)$ and the unit height pulse, v_w, centred at $t = 0$ and of width 2 s;

(b) $v_s = \sin(2\pi \times 100t)$

and

$$v_w = \begin{cases} \cos(2\pi \times 0.25t) & \text{for} 1 \leqslant t \leqslant -1 \text{ s} \\ 0 & \text{elsewhere} \end{cases}$$

(2) When the Fourier components of a signal are obtained using the DFT of sampled

data, a sample of the signal of length $(N-1)T$ is in effect used, where N is the number of data and T is the interval between the samples. The signal is said to have been windowed by a data window of length $(N-1)T$. The spectrum computed is then given by the convolution of the signal spectrum by the window spectrum. If in the case of the waveforms of part (1), v_s represents the signal and the v_w are the window data, comment on the relative suitabilities of the two data windows for defining the signal sample.

2.22 Calculate the discrete cosine, discrete Walsh, and discrete Hadamard transforms of the data sequence {0.1, −0.2, 0.3, −0.4, 0.5, 1.5, 2, 1.5, 0.5, −0.4, 0.3, −0.2, 0.1}. Assuming these transforms are being compared with respect to their data compression efficiencies with a preselected threshold value of 0.35, rank them in order of preference.

2.23 The sampled voltages obtained by scanning the intensity distribution of a photographic image are {3.2, 3.6, 3.3, 2.9, 1.7, 1.6, 1.8, 1.5}. Discuss the relative merits of transforming these data by the FFT and the DWT.

2.24 Extend the discussion of Problem 2.23 to include the amount of data compression available (including use of the DCT).

2.25 Draw up a table to show the advantages, disadvantages, and applications of the fast Fourier, discrete Walsh, discrete cosine and discrete Hadamard transforms.

References

Ahmed N. and Rao K.R. (1975). *Orthogonal Transforms for Digital Signal Processing*. Berlin: Springer

Beauchamp K.G. (1987). *Transforms for Engineers. A Guide to Signal Processing*. Oxford: Clarendon

Burrus C.S. and Parks T.W. (1985). *DFT/FFT and Convolution Algorithms. Theory and Implementation*. New York NY: Wiley

Chen W., Smith C.H. and Fialick S.C. (1977). A fast computational algorithm for the discrete cosine transform. *IEEE Trans. Communications*, **25**, 1004–9

Cooley J.W. and Tukey J.W. (1965). An algorithm for the machine calculation of complex Fourier series. *Mathematics Computation*, **19**, 297–301

Gentleman W.M. and Sande G. (1966). Fast Fourier transforms for fun and profit. In *Fall Joint Computing Conf., AFIPS Proc.*, **29**, 563–78

McClellan, J.H. and Rader, C.M. (1979). *Number Theory in Digital Signal Processing*. Englewood Cliffs NJ: Prentice-Hall

Narasinka M.J. and Petersen A.M. (1978). On the computation of the discrete cosine transform. *IEEE Trans. Communications*, **26**, 934–6

Rader C.M. (1968). Discrete Fourier transform when the number of data samples is prime. *IEEE Proc.*, **56**, 1107–8

Rosenfield A. and Thurston M. (1971). Edge and curve detection for visual scene analysis. *IEEE Trans. Computing*, **20**, 562–9

Signal Processing Committee, ed. (1979). *Programs for Digital Signal Processing*. New York NY: IEEE

Srinivassan R. and Rao K.R. (1983). An approximation to the discrete cosine transform. *Signal Processing*, **5**, 81–5

Strum R.D. and Kirk D.E. (1988). *First Principles of Discrete Systems and Digital Signal Processing*. Reading MA: Addison-Wesley

Winograd S. (1978). On computing the discrete Fourier transform. *Mathematics Computation*, **32**, 175–99

Yip P. and Ramamohan (1987). In *Handbook of Digital Signal Processing Engineering Applications* (Elliott D. E., ed.). New York NY: Academic Press

Appendices

2A C language program for direct DFT computation

The C language program given here evaluates, directly, the DFT or the IDFT of a discrete-time sequence, $x(n)$:

$$X(k) = \sum_{n=0}^{N-1} x(n)W^{nk}, \quad k = 0, 1, \ldots, N-1 \qquad \text{DFT} \qquad \textbf{(2A.1a)}$$

$$x(n) = \frac{1}{N}\sum_{k=0}^{N-1} X(k)W^{-nk} \qquad \text{IDFT} \qquad \textbf{(2A.1b)}$$

where $W = e^{j-2\pi/N}$ and N is the sequence length.

The input sequence, $x(n)$, must be in a complex form (real and imaginary). For a real data sequence, the imaginary parts of the data are set to zero. The main function, DFTD.c, is listed in Program 2A.1, and the function that computes the DFT or IDFT in Program 2A.2. Two functions, read_data() and save_data(), are required for reading the input data sequence and for saving the transformed data (Program 2A.3). The input data is held in the input file, coeff.dat, and the output is saved in the file dftout.dat.

Program 2A.1 Main function dftd.c, for derived computation of DFT.

```
/*--------------------------------------------------------------------------*/
/*                                                                          */
/*          Program to compute DFT coefficients directly                    */
/*          3 other functions are used                                      */
/*                                                                          */
/*          E C Ifeachor. July, 1992                                        */
/*                                                                          */
/*--------------------------------------------------------------------------*/
#include    "dsp1.h"      - p.96
#include    "dft.h"       - p.96

main()
{
            extern  long npt;
            extern  int     inv;

            printf("select type of transform\n");
            printf("\n");
            printf("0       for forward DFT\n");
            printf("1       for inverse DFT\n");
```

```
                scanf("%d", &inv);
                read_data();
                dft();
                save_data();
                exit();
}
#include        "dft.c";
#include        "rdata.c";
#include        "sdata.c";
```

Program 2A.2 C language function for direct computation of the DFT of a discrete-time sequence. This function is held in a separate file.

```
/*-----------------------------------------------------------------------*/
/*                                                                       */
/*              Function to compute the DFT of a discrete-time          */
/*              sequence directly                                        */
/*                                                                       */
/*              E C Ifeachor. 31.10.91                                   */
/*                                                                       */
/*-----------------------------------------------------------------------*/
void            dft()
{
                extern int inv;
                extern long npt;
                long    k, n;
                double WN, wk, c, s, XR[size], XI[size];
                extern complex x[size];

                WN=2*pi/npt;
                if(inv==1)
                        WN=-WN;
                for(k=0; k<npt; ++k){
                        XR[k]=0.0; XI[k]=0.0;
                        wk=k*WN;
                        for(n=0; n<npt; ++n){
                                c=cos(n*wk); s=sin(n*wk);
                                XR[k]=XR[k]+x[n+1].real*c+x[n+1].imag*s;
                                XI[k]=XI[k]-x[n+1].real*s+x[n+1].imag*c;
                        }
                        if(inv==1){             /* divide by N for IDFT*/
                                XR[k]=XR[k]/npt;
                                XI[k]=XI[k]/npt;
                        }
                }
                for (k=1; k <= npt; ++k){       /* store transformed data in x*/
                        x[k].real=XR[k-1];
                        x[k].imag=XI[k-1];
                }
}
```

Program 2A.3 Function for reading the data, function for saving transformed data to disk file, header file containing constant structure definitions and header file containing common declarations and variables.

```
/*-----------------------------------------------------------------------------*/
/*                                                                             */
/*        Function to read data, in complex format, for the DFT or FFT         */
/*                                                                             */
/*        E C Ifeachor. Last modification: July, 1992.                         */
/*                                                                             */
/*-----------------------------------------------------------------------------*/
void            read__data()
{
                extern  long    npt;
                int     n;
                extern  complex x[size];

                for(n=0; n < size; ++n){
                        x[n].real=0;
                        x[n].imag=0;
                }
                if((in=fopen("coeff.dat","r"))==NULL){
                        printf("cannot open file coeff.dat\n");
                        exit(1);
                }

                fscanf(in,"%ld",&npt);
                for(n=1; n <=npt; n++){
                        fscanf(in,"%lf %lf",&x[n].real,&x[n].imag);
                }
                fclose(in);
}

void            save__data()                                  /* file name sdata.c */
{
                long            k;
                int             k1;
                extern          long npt;
                extern          complex x[size];

                if((out=fopen("dftout.dat","w"))==NULL){
                                printf("cannot open file dftout.dat \n");
                                exit(1);
                }
                fprintf(out,"k \tXR(k) \t\tXI(k) \n");
                fprintf(out,"\n");
                for(k=1; k <=npt; ++k){
                                k1=k-1;
                                fprintf(out,"%d \t%f \t%f \n", k1, x[k].real, x[k].imag);
                }
                fclose(out);
}
```

```
/* This file contains common definitions and structures
        filename: dsp1.h
*/
#include            <stdio.h>
#include            <math.h>
#include            <dos.h>

#define size        600
#define pi          3.141592654
#define maxbits     30

typedef struct      {
        double      real;
        double      imag;
        double      modulus;
        double      angle;
        }complex;


/*
        filename: dft.h
*/
void            dft();
void            fft();
void            read_data();
void            save_data();
int             inv;
long            npt;
complex         x[size];
FILE            *in, *out, *fopen();
```

Test example 2A.1

Use the direct DFT program to find the DFT coefficients of the following 8-point discrete-time sequence:

$$x(n) = \{4, 2, 1, 4, 6, 3, 5, 2\}$$

The input data file, created with PC edlin (most word processors may be used for this purpose) for the problem, has the following format:

```
8
4 0
2 0
1 0
4 0
6 0
3 0
```

5 0
2 0

The first line specifies the length of the data sequence.

The DFT of the data, using the program, is given below:

k	XR(k)	XI(k)
0	27.000 000	0.000 000
1	−4.121 320	3.292 893
2	4.000 000	1.000 000
3	0.121 320	−4.707 107
4	5.000 000	−0.000 000
5	0.121 320	4.707 107
6	4.000 000	−1.000 000
7	−4.121 320	−3.292 893

Test example 2A.2

Find, using the DFT program, the discrete-time sequence corresponding to the DFT coefficients above. The input data has the following format:

8
27.000 000 0.000 000
−4.121 320 3.292 893
4.000 000 1.000 000
0.121 320 −4.707 107
5.000 000 −0.000 000
0.121 320 4.707 107
4.000 000 −1.000 000
−4.121 320 −3.292 893

The IDFT option was selected in response to the program prompts. The output of the program is the same as the discrete-time sequence in Test example 2A.1.

Test example 2A.3

The third test example uses the complex data sequence (IEEE, 1979, Chapter 1):

$$x(n) = Q^n, \quad n = 0, 1, \ldots, 31$$

where $Q = 0.9 + j0.3$.

The input data sequence, $x(n)$, and its DFT, $X(k)$, using the direct DFT program are listed in Tables 2A.1 and 2A.2 respectively.

Table 2A.1 Complex input data sequence.

n	$x(n)$	
	Real	*Imaginary*
0	0.100000E01	0.
1	0.900000E00	0.300000E00
2	0.720000E00	0.540000E00
3	0.486000E00	0.702000E00
4	0.226800E00	0.777600E00
5	−0.291600E-01	0.767880E00
6	−0.256608E00	0.682344E00
7	−0.435650E00	0.537127E00
8	−0.553224E00	0.352719E00
9	−0.603717E00	0.151480E00
10	−0.588789E00	−0.447828E-01
11	−0.516476E00	−0.216941E00
12	−0.399746E00	−0.350190E00
13	−0.254714E00	−0.435095E00
14	−0.987144E-01	−0.467999E00
15	0.515569E-01	−0.450814E00
16	0.181645E00	−0.390265E00
17	0.280560E00	−0.296745E00
18	0.341528E00	−0.182903E00
19	0.362246E00	−0.621539E-01
20	0.344667E00	0.527352E-01
21	0.294380E00	0.150862E00
22	0.219684E00	0.224090E00
23	0.130488E00	0.267586E00
24	0.371637E-01	0.279974E00
25	−0.50544E-01	0.263125E00
26	−0.124428E00	0.221649E00
27	−0.178480E00	0.162156E00
28	−0.209279E00	0.923965E-01
29	−0.216070E00	0.203732E-01
30	−0.200575E00	−0.464851E-01
31	−0.166572E00	−0.102009E00

Table 2A.2 Transformed output for Test example 2A.3.

0.693972	3.499714
2.792268	8.050456
9.402964	−9.135013
1.866446	−3.833833
1.131822	−2.234158
0.904794	−1.534631
0.799557	−1.139607
0.739607	−0.882315
0.700858	−0.698566
0.673577	−0.558478
0.653112	−0.446244

0.636987	−0.352691
0.623790	−0.272085
0.612613	−0.200642
0.602885	−0.135703
0.594200	−0.075314
0.586276	−0.017948
0.578899	0.037651
0.571898	0.092607
0.565139	0.147983
0.558490	0.204882
0.551858	0.264523
0.545134	0.328363
0.538217	0.398257
0.531000	0.476679
0.523403	0.567133
0.515361	0.674850
0.506928	0.808100
0.498469	0.980906
0.491388	1.219210
0.490730	1.577083
0.517355	2.188832

2B C program for radix-2 decimation-in-time FFT

The FFT program given here is a C language implementation of the radix-2 decimation-in-time FFT (Cooley and Tukey, 1965). The program evaluates the DFT or the IDFT of a discrete-time sequence as defined in Equations 2A.1. The program consists of a main function, dftf.c, and three functions: fft(), read_data(), and save_data(). As in the case of the direct DFT, all the functions are held in separate files and combined during compilation by include statements in the main function. The two functions read_data() and save_data() are used to read the data and to save the transformed data to file. These two files are identical to those used for the direct DFT. The main program, dftf.c, and the function fft() are listed in Programs 2B.1 and 2B.2 respectively.

Applying each of the test data described in Appendix 2A to the FFT program, in exactly the same format, yields identical results to those of the direct DFT. It is left as an exercise to the reader to confirm that this is the case.

Program 2B.1 Main function, dftf.c, for computing DFTs using decimation-in-time FFT.

```
/*-----------------------------------------------------------------------*/
/*                                                                       */
/*        Program to compute DFT coefficients using DIT FFT              */
/*        3 other functions are used                                     */
/*                                                                       */
/*        E C Ifeachor. July, 1992                                       */
/*                                                                       */
/*-----------------------------------------------------------------------*/
#include    "dsp1.h"
#include    "dft.h"
```

```
main()
{
            extern   long npt;
            extern   int        inv;

            printf("select type of transform \n");
            printf("\n");
            printf("0       for forward DFT\n");
            printf("1       for inverse DFT\n");
            scanf("%d", &inv);
            read_data();
            fft();
            save_data()
            exit();
}
#include    "fft.c";
#include    "rdata.c";
#include    "sdata.c";
```

Program 2B.2 C language implementation of radix-2, decimation-in-time FFT algorithm.

```
/*-------------------------------------------------------------------------*/
/*                                                                         */
/*          file name: fft.c                                               */
/*          E C Ifeachor. June, 1992                                       */
/*                                                                         */
/*          Function computes the DFT of a sequence using radix2 FFT       */
/*                                                                         */
/*                                                                         */
/*-------------------------------------------------------------------------*/

void        fft()
{
            int         sign;
            long        m, irem, l, le, le1, k, ip,i,j;
            double      ur, ui, wr, wi, tr, ti, temp;
            extern      long npt;
            extern      int inv;
            extern      complex x[size];

            /* in-place bit reverse shuffling of data */

            j=1;
            for(i=1; i < npt; ++i){
                    if(i<j){
                            tr=x[j].real; ti=x[j].imag;
                            x[j].real=x[i].real;
                            x[j].imag=x[i].imag;
                            x[i].real=tr; x[i].imag=ti;
                            k=npt/2;
                            while(k<j){
                                    j=j-k;
                                    k=k/2;
```

```
                        }
                }
                else{
                        k=npt/2;
                        while(k<j){
                                j=j-k;
                                k=k/2;
                        }
                }
                j=j+k;
}
/* calculate the number of stages: m=log2(npt), and whether FFT or IFFT */
        m=0; irem=npt;
        while(irem>1){
                irem=irem/2;
                m=m+1;
        }
        if(inv==1)
                sign=1;
        else
                sign=-1;

/* perform the FFT computation for each stage */

        for(l=1; l<=m, l++){
                le=pow(2, l);
                le1=le/2;
                ur=1.0; ui=0;
                wr=cos(pi/le1);
                wi=sign*sin(pi/le1);
                for(j=1; j <= le1; ++j){
                        i=j;
                        while(i <= npt){
                                ip=i + le1;
                                tr=x[ip].real*ur-x[ip].imag*ui;
                                ti=x[ip].imag*ur+x[ip].real*ui;
                                x[ip].real=x[i].real-tr;
                                x[ip].imag=x[i].imag-ti;
                                x[i].real=x[i].real+tr;
                                x[i].imag=x[i].imag+ti;
                                i=i+le;
                        }
                        temp=ur*wr-ui*wi;
                        ui=ui*wr+ur*wi;
                        ur=temp;
                }
        }
        /*If inverse fft is desired divide each coefficient by npt */
        if(inv==1){
                for(i=1; i<=npt; ++i){
                        x[i].real=x[i].real/npt;
                        x[i].imag=x[i].imag/npt;
                }
        }
}
```

References for Appendices

Cooley J.W. and Tukey J.W. (1965). An algorithm for the machine calculation of complex Fourier series. *Mathematics Computation*, **19**(90) (April), 297–301

IEEE (1979). *Programs for digital signal processing*. New York NY: IEEE Press

3

The z-transform and its applications in signal processing

3.1	Discrete-time signals and systems	104
3.2	The z-transform	105
3.3	The inverse z-transform	109
3.4	Properties of the z-transform	127
3.5	Some applications of the z-transform in signal processing	130
3.6	Summary	152
	References	156
	Bibliography	157
	Appendices	157

The z-transform is a convenient yet invaluable tool for representing, analysing and designing discrete-time signals and systems. It plays a similar role in discrete-time systems to that which the Laplace transform plays in continuous-time systems.

In this chapter, we present important aspects of the z-transform, especially those that will be used in subsequent chapters, and highlight its applications in discrete-time systems design. The applications include the use of the z-transform to describe discrete-time signals and systems so that we can readily infer their degree of stability and to visualize their frequency responses, the analysis of quantization errors in digital filters and the computation of the frequency response of discrete-time systems. Most of the applications are covered in more detail in subsequent chapters.

As in the rest of the book, the presentation in this chapter is practical in approach. Algorithms and C language programs, where necessary, are provided to enable readers to gain a deeper understanding of the subject. Much of the discussion in this chapter involves linear discrete-time signals and systems, and so we will begin by reviewing very briefly characteristic features of this class of signals and systems.

3.1 Discrete-time signals and systems

A discrete signal has values which are defined only at discrete values of time or some other appropriate variable, for example space. As discussed in Chapter 1, such a signal may be generated by sampling a continuous-time signal at regular time intervals nT, $n = 0, 1, \ldots$, where T is the sampling period. It may also be generated, artificially, via some algorithm in a computer. The amplitude of a discrete-time signal may have discrete values (discrete time, discrete amplitude), or it may be continuous.

By tradition, a discrete-time signal is represented as a sequence of numbers:

$$x(n), \qquad n = 0, 1, \ldots \tag{3.1a}$$

$$x(nT), \qquad n = 0, 1, \ldots \tag{3.1b}$$

$$x_n, \qquad n = 0, 1, \ldots \tag{3.1c}$$

where the symbol, $x(n)$, $x(nT)$ or x_n indicates the value of the signal at the discrete time n (or nT). For convenience we will use the symbol, $x(n)$, to denote both the value of the sequence at the discrete time n and the sequence itself unless we wish to emphasize the difference. The meaning will be clear from the context. It is common practice in DSP to omit the symbol T since the sequence is not always a function of time (it may for example be a function of space). Sometimes T is omitted because the sampling frequency is assumed to be unity (that is, normalized) for convenience.

A discrete-time system is essentially a mathematical algorithm that takes an input sequence, $x(n)$, and produces an output sequence, $y(n)$. Examples of discrete-time systems are digital controllers, digital spectrum analysers, and digital filters. A discrete-time system may be linear or nonlinear, time invariant or time varying. Linear time-invariant (LTI) systems form an important class of systems used in DSP. Examples are digital filters discussed in detail in Chapters 5 to 7.

A discrete-time system is said to be linear if it obeys the principles of superposition. That is, the response of a linear system to two or more inputs is equal to the sum of the responses of the system to each input acting separately in the absence of all the other inputs. For example, if an input, $x_1(n)$, to the system gives rise to the output, $y_1(n)$, and another input $x_2(n)$ produces the output $y_2(n)$, the response of the system to both inputs will be

$$a_1x_1(n) + a_2x_2(n) \rightarrow a_1y_1(n) + a_2y_2(n) \quad (3.2)$$

where a_1 and a_2 are arbitrary constants.

A discrete-time system is said to be time invariant (sometimes referred to as shift invariant) if its output is independent of the time the input is applied. For example, if the input $x(n)$ gives the output $y(n)$, then the input $x(n-k)$ will give the output $y(n-k)$:

$$x(n) \rightarrow y(n) \quad (3.3a)$$

$$x(n-k) \rightarrow y(n-k) \quad (3.3b)$$

that is a delay in the input causes a delay by the same amount in the output signal.

The input–output relationship of an LTI system is given by the convolution sum

$$y(n) = \sum_{k=-\infty}^{\infty} h(k)x(n-k) \quad (3.4)$$

where $h(k)$ is the impulse response of the system. The values of $h(k)$ completely define the discrete-time system in the time domain. An LTI system is stable if its impulse response satisfies the condition

$$\sum_{k=-\infty}^{\infty} |h(k)| < \infty \quad (3.5)$$

This condition is satisfied if $h(k)$ is of finite duration or if $h(k)$ decays towards zero as k increases. Stability considerations are described in more detail in Section 3.5.7.

A causal system is one which produces an output only when there is an input. All physical systems are causal. In general, a causal discrete-time sequence, $x(n)$, or the impulse response, $h(k)$, of a discrete-time system is zero before time 0, that is $x(n)=0$, $n<0$, or $h(k)=0$, $k<0$. Much of the discussion in this book is about practical, that is causal, systems.

3.2 The z-transform

The z-transform of a sequence, $x(n)$, which is valid for all n, is defined as

$$X(z) = \sum_{n=-\infty}^{\infty} x(n)z^{-n} \quad (3.6)$$

where z is a complex variable.

In causal systems, $x(n)$ may be nonzero only in the interval $0 < n < \infty$ and Equation 3.6 reduces to the so-called one-sided z-transform:

$$X(z) = \sum_{n=0}^{\infty} x(n)z^{-n} \tag{3.7}$$

Clearly, the z-transform is a power series with an infinite number of terms and so may not converge for all values of z. The region where the z-transform converges is known as the region of convergence (ROC), and in this region the values of $X(z)$ are finite. Not surprisingly, the region of convergence is determined by the properties of $x(n)$ or equivalently by those of $X(z)$ as illustrated by the following set of examples.

Example 3.1

Find the z-transform and the region of convergence for each of the discrete-time sequences given in Figure 3.1.

(1) The sequence of Figure 3.1(a) is noncausal, since $x(n)$ is not zero for $n < 0$, but it is of a finite duration. The sequence has values $x(-6) = 0$, $x(-5) = 1$, $x(-4) = 3$, $x(-3) = 5$, $x(-2) = 3$, $x(-1) = 1$ and $x(0) = 0$. From Equation 3.6, the z-transform is given by

$$X_1(z) = \sum_{n=-\infty}^{\infty} x(n)z^{-n}$$
$$= z^5 + 3z^4 + 5z^3 + 3z^2 + z$$

It is readily verified that the value of $X(z)$ becomes infinite when $z = \infty$. Thus the ROC is everywhere in the z-plane except at $z = \infty$.

(2) Again, the sequence in Figure 3.1(b) is not causal. It is of a finite duration, and double sided. The values of the sequence are $x(3) = 0$, $x(-2) = 1$, $x(-1) = 3$, $x(0) = 5$, $x(1) = 3$, $x(2) = 1$ and $x(3) = 0$. From Equation 3.6, the z-transform is given by

$$X_2(z) = \sum_{n=-\infty}^{\infty} x(n)z^{-n}$$
$$= z^2 + 3z + 5 + 3z^{-1} + z^{-2}$$

It is evident that the value of $X(z)$ is infinite if $z = 0$ or if $z = \infty$. Therefore the region of convergence is everywhere except at $z = 0$ and $z = \infty$.

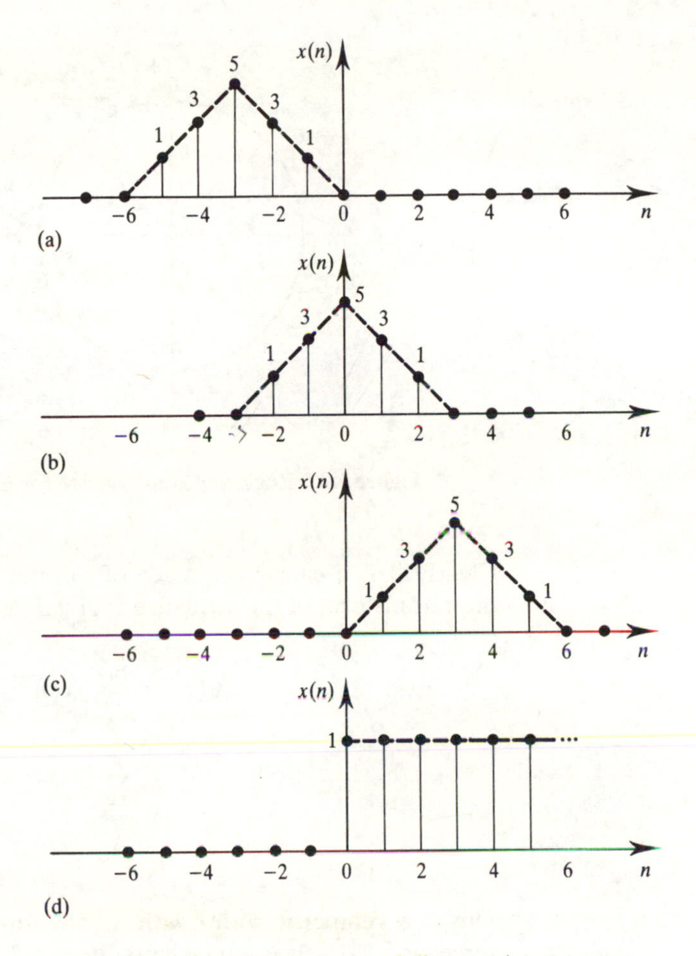

Figure 3.1 Causal and noncausal discrete-time sequences.

(3) Figure 3.1(c) represents a causal, finite duration sequence with values $x(0) = 0$, $x(1) = 1$, $x(2) = 3$, $x(3) = 5$, $x(4) = 3$, $x(5) = 1$ and $x(6) = 0$. The z-transform is given by

$$X_3(z) = \sum_{n=-\infty}^{\infty} x(n)z^{-n}$$

$$= z^{-1} + 3z^{-2} + 5z^{-3} + 3z^{-4} + z^{-5}$$

In this case, $X(z) = \infty$ for $z = 0$. Thus the region of convergence is everywhere except at $z = 0$.

(4) The discrete-time sequence in Figure 3.1(d) may be defined mathematically as

$$x(n) = 1 \qquad 0 \leqslant n \leqslant \infty$$

$$= 0 \qquad n < 0$$

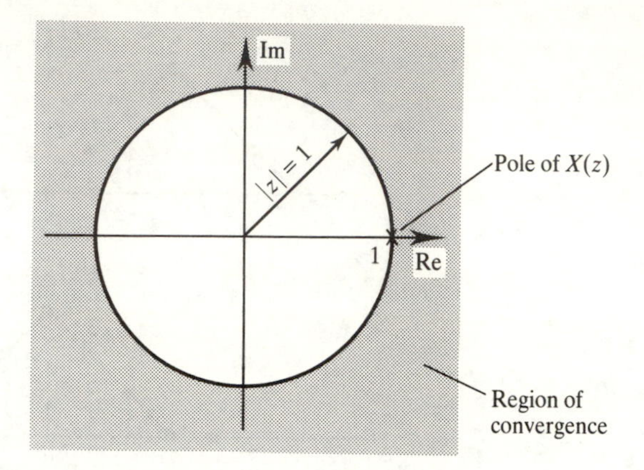

Figure 3.2 Region of convergence for Example 3.1, part (4).

Clearly it is a causal sequence of infinite duration. From Equation 3.6, the z-transform of the sequence is given by

$$X(z) = \sum_{n=-\infty}^{\infty} x(n)z^{-n}$$

$$= \sum_{n=0}^{\infty} z^{-n}$$

$$= 1 + z^{-1} + z^{-2} + \ldots$$

This is a geometric series with a common ratio of z^{-1}. The series converges if $|z^{-1}| < 1$ or equivalently if $|z| > 1$. Thus we may express $X(z)$ in closed form provided that $|z| > 1$:

$$\begin{aligned} X(z) &= 1 + z^{-1} + z^{-2} + \ldots \\ &= 1/(1 - z^{-1}) \\ &= z/(z - 1) \end{aligned} \qquad \textbf{(3.8)}$$

In this case, the z-transform is valid everywhere outside a circle of unit radius whose centre is at the origin. The exterior of the circle is the region of convergence (Figure 3.2). We can readily verify that when $|z| > 1$ $X(z)$ converges whereas when $|z| < 1$ $X(z)$ diverges. For example, if we let $z = 2$ (outside the unit circle) we find that the series on the RHS of Equation 3.8 adds up to 2:

$$X(z) = 1 + 1/2 + (1/2)^2 + (1/2)^3 + \ldots = 2/(2 - 1) = 2$$

as it is clearly a geometric series with a common ration of 1/2 and a first term of 1, giving the sum to infinity of $2/(2-1)=2$. On the other hand if $z=1/2$ (inside the unit circle) the series of Equation 3.8 becomes

$$X(z) = 1 + 1/0.5 + (1/0.5)^2 + (1/0.5)^3 + \ldots$$
$$= 1 + 2 + 4 + 8 + \ldots$$

which is seen to be diverging. In Figure 3.2, the region of convergence (hatched) is seen to be bounded by the circle $|z|=1$, the radius of the pole of $X(z)$. Values of z for which $X(z)=\infty$ are referred to as poles of $X(z)$. Values of z for which $X(z)=0$ are referred to as the zeros of $X(z)$.

From the examples above we may infer that for causal sequences of finite duration the z-transform converges everywhere except at $z=0$. For causal infinite duration sequences the z-transform converges everywhere outside a circle bounded by the radius of the pole with the largest radius. For stable causal systems the ROC always encloses the circle of unit radius which is important for the systems to have a frequency response.

The z-transforms of common sequences are available, in closed form, and are given in the form of tables, such as Table 3.1. Such tables are useful in finding the inverse z-transform.

3.3 The inverse z-transform

The inverse z-transform (IZT) allows us to recover the discrete-time sequence, $x(n)$, given its z-transform. The IZT is particularly useful in DSP work, for example in finding the impulse response of digital filters. Symbolically, the inverse z-transform may be defined as

$$x(n) = Z^{-1}[X(z)] \tag{3.9}$$

where $X(z)$ is the z-transform of $x(n)$ and Z^{-1} is the symbol for the inverse z-transform.

Assuming a causal sequence, the z-transform, $X(z)$, in Equation 3.7 can be expanded into a power series as

$$X(z) = \sum_{n=0}^{\infty} x(n)z^{-n}$$
$$= x(0) + x(1)z^{-1} + x(2)z^{-2} + x(3)z^{-3} + \ldots \tag{3.10}$$

It is seen that the values of $x(n)$ are the coefficients of z^{-n} $(n=0,1,\ldots)$ and so can be obtained directly by inspection. In practice, $X(z)$ is often expressed

Table 3.1 z-transforms of some common sequences.

Entry number	*Discrete-time sequence* $x(n), \quad n \geq 0$	*z-transform* $X(z)$	*Region of convergence of* $X(z)$
1	$k\delta(n)$	k	Everywhere
2	k	$\dfrac{kz}{z-1}$	$\|z\| > 1$
3	kn	$\dfrac{kz}{(z-1)^2}$	$\|z\| > 1$
4	kn^2	$\dfrac{kz(z+1)}{(z-1)^3}$	$\|z\| > 1$
5	$ke^{-\alpha n}$	$\dfrac{kz}{z-e^{-\alpha}}$	$\|z\| > e^{-\alpha}$
6	$kne^{-\alpha n}$	$\dfrac{kze^{-\alpha}}{(z-e^{-\alpha})^2}$	$\|z\| > e^{-\alpha}$
7	$1-e^{-\alpha n}$	$\dfrac{z(1-e^{-\alpha})}{z^2-z(1+e^{-\alpha})+e^{-\alpha}}$	$\|z\| > e^{-\alpha}$
8	$\cos(\alpha n)$	$\dfrac{z(z-\cos\alpha)}{z^2-2z\cos\alpha+1}$	$\|z\| > 1$
9	$\sin(\alpha n)$	$\dfrac{z\sin\alpha}{z^2-2z\cos\alpha+1}$	$\|z\| > 1$
10	$e^{-\alpha n}\sin(\alpha n)$	$\dfrac{ze^{-\alpha}\sin\alpha}{z^2-2e^{-\alpha}z\cos\alpha+e^{-2\alpha}}$	$\|z\| > e^{-\alpha}$
11	$e^{-\alpha n}\cos(\alpha n)$	$\dfrac{ze^{-\alpha}(ze^{\alpha}-\cos\alpha)}{z^2-2ze^{-\alpha}\cos\alpha+e^{-2\alpha}}$	$\|z\| > e^{-\alpha}$
12	$\cosh(\alpha n)$	$\dfrac{z^2-z\cosh\alpha}{z^2-2z\cosh\alpha+1}$	$\|z\| > \cosh\alpha$
13	$\sinh(\alpha n)$	$\dfrac{z\sinh\alpha}{z^2-2z\cosh\alpha+1}$	$\|z\| > \sinh\alpha$
14	$k\alpha^n$	$\dfrac{kz}{z-\alpha}$	$\|z\| > \alpha$
15	$kn\alpha^n$	$\dfrac{k\alpha z}{(z-\alpha)^2}$	$\|z\| > \alpha$
16	$2\|c\|\|p\|^n\cos(n\angle p + \angle c)$	$\dfrac{cz}{z-p}+\dfrac{c^*z}{z-p^*}$	

k and α are constants; c is a complex number.

as a ratio of two polynomials in z^{-1} or equivalently in z:

$$X(z) = \frac{a_0 + a_1z^{-1} + a_2z^{-2} + \ldots + a_Nz^{-N}}{b_0 + b_1z^{-1} + b_2z^{-2} + \ldots + b_Mz^{-M}} \tag{3.11}$$

In this form, the inverse z-transform, $x(n)$, may be obtained using one of several methods including the following three:

(1) power series expansion method;
(2) partial fraction expansion method;
(3) residue method.

Each method has its own merits and demerits. In terms of mathematical rigour, the residue method is perhaps the most elegant. The power series method, however, lends itself most easily to computer implementation.

In the next few sections, we will describe each of the three methods in turn, using numerical examples to illustrate the principles involved. In the appendix, we provide complete C language program listings for evaluating the inverse z-transform for methods (1) and (2).

3.3.1 Power series method

Given the z-transform, $X(z)$, of a causal sequence as in Equation 3.11, it can be expanded into an infinite series in z^{-1} or z by long division (sometimes called synthetic division):

$$\begin{aligned} X(z) &= \frac{a_0 + a_1 z^{-1} + a_2 z^{-2} + \ldots + a_N z^{-N}}{b_0 + b_1 z^{-1} + b_2 z^{-2} + \ldots + b_M z^{-M}} \\ &= x(0) + x(1)z^{-1} + x(2)z^{-2} + x(3)z^{-3} + \ldots \end{aligned} \tag{3.12}$$

In this method, the numerator and denominator of $X(z)$ are first expressed in either descending powers of z or ascending powers of z^{-1} and the quotient is then obtained by long division. We will illustrate the method by an example.

Example 3.2

Given the following z-transform of a causal LTI system, obtain its IZT by expanding it into a power series using long division:

$$X(z) = \frac{1 + 2z^{-1} + z^{-2}}{1 - z^{-1} + 0.3561z^{-2}}$$

Solution

First, we expand $X(z)$ into a power series with the numerator and denominator polynomials in ascending powers of z^{-1} and then perform the usual long division.

$$
\begin{array}{r|l}
 & 1 + 3z^{-1} + 3.6439z^{-2} + 2.5756z^{-3} + \ldots \\
\hline
1 - z^{-1} + 0.3561z^{-2} & 1 + 2z^{-1} + z^{-2} \\
 & \underline{1 - z^{-1} + 0.3561z^{-2}} \\
 & \quad 3z^{-1} + 0.6439z^{-2} \\
 & \quad \underline{3z^{-1} - 3z^{-2} + 1.0683z^{-3}} \\
 & \qquad 3.6439z^{-2} - 1.0683z^{-3} \\
 & \qquad \underline{3.6439z^{-2} - 1.0683z^{-3} + 1.297\,592\,7z^{-4}} \\
 & \qquad\qquad 2.5756z^{-3} - 1.297\,592\,7z^{-4}
\end{array}
$$

Alternatively, we may express the numerator and denominator in positive powers of z, in descending order, and then perform the long division:

$$
\frac{z^2 + 2z + 1}{z^2 - z + 0.3561}
$$

$$
\begin{array}{r|l}
 & 1 + 3z^{-1} + 3.6439z^{-2} + 2.5756z^{-3} + \ldots \\
\hline
z^2 - z + 0.3561 & z^2 + 2z + 1 \\
 & \underline{z^2 - z + 0.3561} \\
 & \quad 3z + 0.6439 \\
 & \quad \underline{3z - 3 + 1.0683z^{-1}} \\
 & \qquad 3.6439 - 3.643\,91z^{-1} + 1.297\,592\,7z^{-2} \\
 & \qquad 2.5756z^{-1} - 1.297\,592\,7z^{-2}
\end{array}
$$

Either way, the z-transform is now expanded into the familiar power series, that is

$$
X(z) = 1 + 3z^{-1} + 3.6439z^{-2} + 2.5756z^{-3} + \ldots
$$

The inverse z-transform can now be written down directly:

$$
x(0) = 1;\ x(1) = 3;\ x(2) = 3.6439;\ x(3) = 2.5756;\ \ldots
$$

The long division approach can be reformulated (see Appendix 3A) so that the values of $x(n)$ are obtained recursively:

$$
\begin{aligned}
x(0) &= a_0/b_0 \\
x(1) &= [a_1 - x(0)b_1]/b_0 \\
x(2) &= [a_2 - x(1)b_1 - x(0)b_2]/b_0 \\
\vdots \quad & \quad \vdots \quad \vdots \quad \vdots \quad \vdots \qquad \vdots
\end{aligned}
$$

$$x(n) = \left[a_n - \sum_{i=1}^{n} x(n-i)b_i\right]/b_0, \quad n = 1, 2, \ldots \qquad \textbf{(3.13a)}$$

where

$$x(0) = a_0/b_0 \qquad \textbf{(3.13b)}$$

We will repeat the previous example to illustrate the recursive approach.

Example 3.3

Find the first four terms of the inverse z-transform, $x(n)$, using the recursive approach. Assume that the z-transform, $X(z)$, is the same as in Example 3.2, that is

$$X(z) = \frac{1 + 2z^{-1} + z^{-2}}{1 - z^{-1} + 0.3561z^{-2}}$$

Solution

Comparing the coefficients of $X(z)$ above with those of the general transform in Equation 3.12 we have

$$a_0 = 1,\ a_1 = 2,\ a_2 = 1,\ b_0 = 1,\ b_1 = -1,\ b_2 = 0.3561;\ N = M = 2$$

From Equations 3.13 we have

$$\begin{aligned}
x(0) &= a_0/b_0 = 1 \\
x(1) &= [a_1 - x(0)b_1]/b_0 = [2 - 1 \times (-1)] = 3 \\
x(2) &= [a_2 - x(1)b_1 - x(0)b_2] = 1 - 3 \times (-1) - 1 \times 0.3561 = 3.6439 \\
x(3) &= [a_3 - x(2)b_1 - x(1)b_2 + x(0)b_3] \\
&= 0 - x(2)b_1 - x(1)b_2 = 0 - 3.6439 \times (-1) - 3 \times 0.3561 = 2.5756
\end{aligned}$$

Thus the first four values of the inverse z-transform are

$$x(0) = 1,\ x(1) = 3,\ x(2) = 3.6439,\ x(3) = 2.5756$$

It is seen that both the recursive and direct, long division methods lead to identical solutions.

The recursion in Equation 3.13 can be readily implemented on a computer as shown in the partial C language code below:

```
x[0]=A[0]/B[0];
for(n=1;n<=npt;++n){
      sum=0;
      k=n;
      if(n>M)
            k=M;
      for(i=1;i<=k;++i){
            sum=sum+x[n-i]*B[i];
      }
      x[n]=(A[n]-sum)/B[0];
}
```

In the code, M is the order of the denominator polynomial and npt is the number of data points for the IZT. The numerator and denominator polynomials are assumed to be in ascending powers of z^{-1}. A complete C language program for evaluating the IZT based on the above code is given in Appendix 3B.

3.3.2 Partial fraction expansion method

In this method, the z-transform is first expanded into a sum of simple partial fractions. The inverse z-transform of each partial fraction is then obtained from tables, such as Table 3.1, and then summed to give the overall inverse z-transform. In many practical cases, the z-transform is given as a ratio of polynomials in z or z^{-1} and has the now familiar form

$$X(z) = \frac{a_0 + a_1 z^{-1} + a_2 z^{-2} + \ldots + a_N z^{-N}}{b_0 + b_1 z^{-1} + b_2 z^{-2} + \ldots + b_M z^{-M}} \tag{3.14}$$

If the poles of $X(z)$ are of first order and $N = M$, then $X(z)$ can be expanded as

$$\begin{aligned} X(z) &= B_0 + \frac{C_1}{1 - p_1 z^{-1}} + \frac{C_2}{1 - p_2 z^{-1}} + \ldots + \frac{C_M}{1 - p_M z^{-1}} \\ &= B_0 + \frac{C_1 z}{z - p_1} + \frac{C_2 z}{z - p_2} + \ldots + \frac{C_M z}{z - p_M} \\ &= B_0 + \sum_{k=1}^{M} \frac{C_k z}{z - p_k} \end{aligned} \tag{3.15}$$

where p_k are the poles of $X(z)$ (assumed distinct), C_k are the partial fraction coefficients and

$$B_0 = a_N/b_N \tag{3.16}$$

The C_k are also known as the residues of $X(z)$; see Section 3.3.3.

If the order of the numerator is less than that of the denominator in Equation 3.14, that is $N < M$, then B_0 will be zero. If $N > M$ then $X(z)$ must be reduced first, to make $N \leq M$, by long division with the numerator and denominator polynomials written in descending powers of z^{-1}. The remainder can then be expressed as in Equation 3.15.

The coefficient, C_k, associated with the pole p_k may be obtained by multiplying both sides of Equation 3.15 by $(z - p_k)/z$ and then letting $z = p_k$:

$$C_k = \frac{X(z)}{z}(z - p_k)\bigg|_{z=p_k} \tag{3.17}$$

If $X(z)$ contains one or more multiple-order poles (that is poles that are coincident) then extra terms are required in Equation 3.15 to take this into account. For example, if $X(z)$ contains an mth-order pole at $z = p_k$ the partial fraction expansion must include terms of the form

$$\sum_{i=1}^{m} \frac{D_i}{(z - p_k)^i} \tag{3.18a}$$

The coefficients, D_i, may be obtained from the relationship

$$D_i = \frac{1}{(m-i)!}\frac{\mathrm{d}^{m-i}}{\mathrm{d}z^{m-i}}[(z - p_k)^m X(z)]_{z=p_k} \tag{3.18b}$$

Evaluation of inverse z-transforms by the partial fraction expansion method is best illustrated by examples.

Example 3.4

$X(z)$ contains simple, first-order poles Find the inverse z-transform of the following:

$$X(z) = \frac{z^{-1}}{1 - 0.25z^{-1} - 0.375z^{-2}}$$

Solution

For simplicity, we first express the z-transform in positive powers of z by multiplying the numerator and denominator by z^2 (the highest power of z):

$$X(z) = \frac{z}{z^2 - 0.25z - 0.375} = \frac{z}{(z - 0.75)(z + 0.5)}$$

$X(z)$ contains first-order poles at $z = 0.75$ and at $z = -0.5$ (that is, only one pole occurs at each pole position). Since the order of the numerator is less than the order of the denominator ($N < M$), the partial fraction expansion has the form

$$X(z) = \frac{z}{(z - 0.75)(z + 0.5)} = \frac{C_1 z}{z - 0.75} + \frac{C_2 z}{z + 0.5} \tag{3.19}$$

To make it easier to find the values of the C_k we divide both sides by z:

$$\frac{X(z)}{z} = \frac{z}{z(z - 0.75)(z + 0.5)} = \frac{C_1}{z - 0.75} + \frac{C_2}{z + 0.5} \tag{3.20}$$

To obtain C_1, we simply multiply both sides of Equation 3.20 by $z - 0.75$ and let $z = 0.75$:

$$\frac{(z - 0.75)X(z)}{z} = \frac{\cancel{(z - 0.75)}}{\cancel{(z - 0.75)}(z + 0.5)} = C_1 + \frac{C_2(z - 0.75)}{z + 0.5}$$

$$C_1 = \left.\frac{1}{z + 0.5}\right|_{z=0.75} = \frac{1}{0.75 + 0.5} = \frac{4}{5}$$

Similarly, C_2 is obtained as

$$C_2 = \left.\frac{(z + 0.5)X(z)}{z}\right|_{z=0.5}$$

$$= \left.\frac{\cancel{(z + 0.5)}}{(z - 0.75)\cancel{(z + 0.5)}}\right|_{z=-0.5} = \frac{1}{-0.5 - 0.75} = -\frac{4}{5}$$

Using the values of C_1 and C_2 in Equation 3.19 we have

$$X(z) = \frac{(4/5)z}{z - 0.75} - \frac{(4/5)z}{z + 0.5} \tag{3.21}$$

From the z-transform table, entry 14 in Table 3.1, the inverse z-transform of each term on the right-hand side of Equation 3.21 is given as

$$Z^{-1}\left[\frac{(4/5)z}{z-0.75}\right] = \frac{4(0.75)^n}{5}$$

$$Z^{-1}\left[\frac{-(4/5)z}{z+0.5}\right] = \frac{-4(-0.5)^n}{5}$$

The desired inverse z-transform, $x(n)$, is the sum of the two inverse z-transforms:

$$x(n) = \frac{4}{5}[(0.75)^n - (-0.5)^n], \quad n > 0$$

Example 3.5

$X(z)$ contains first-order, complex conjugate poles Find the discrete-time signal, $x(n)$, represented by the following z-transform using the partial fraction expansion method:

$$X(z) = \frac{1 + 2z^{-1} + z^{-2}}{1 - z^{-1} + 0.3561z^{-2}}$$

Solution

First, $X(z)$ is expressed in positive powers of z:

$$X(z) = \frac{N(z)}{D(z)} = \frac{z^2 + 2z + 1}{z^2 - z + 0.3561}$$

The poles of $X(z)$ are found by solving the quadratic $D(z) = z^2 - z + 0.3561 = 0$ using the formulae

$$p_1 = \frac{-b + (b^2 - 4ac)^{1/2}}{2a}$$
$$p_2 = \frac{-b - (b^2 - 4ac)^{1/2}}{2a} \quad \textbf{(3.22)}$$

where a and b are the coefficients of z^2 and z, respectively, and c is the constant term. With $a = 1$, $b = 1$, and $c = 0.3561$ the poles are

$$p_1 = \frac{-1 + (1 - 4 \times 0.3561)^{1/2}}{2}$$
$$= 0.5 + 0.3257\mathrm{j} = re^{\mathrm{j}\theta}$$
$$p_2 = p_1^* = 0.5 - 0.3257\mathrm{j} = re^{-\mathrm{j}\theta}$$

where $r = 0.5967$ and $\theta = 33.08°$. Thus we can express $X(z)$ in terms of its poles:

$$X(z) = \frac{z^2 + 2z + 1}{(z - p_1)(z - p_1^*)}$$

Since the numerator and denominator of $X(z)$ are of the same order, the partial fraction expansion has the form

$$\frac{X(z)}{z} = \frac{B_0}{z} + \frac{C_1}{z - p_1} + \frac{C_2}{z - p_1^*} \tag{3.23}$$

From Equation 3.16 $B_0 = 1/0.3561 = 2.8082$. To find C_1, we multiply both sides of Equation 3.23 by $z - p_1$ and then let $z = p_1$:

$$\frac{(z - p_1)X(z)}{z} = \frac{B_0(z - p_1)}{z} + C_1 + \left.\frac{C_2(z - p_1)}{(z - p_2)}\right|_{z=p_1}$$

Thus

$$C_1 = \frac{(z - p_1)X(z)}{z} = \left.\frac{(z - p_1)(z^2 + 2z + 1)}{z(z - p_1)(z - p_2)}\right|_{z=p_1=re^{\mathrm{j}\theta}}$$
$$= \frac{(re^{\mathrm{j}\theta})^2 + 2re^{\mathrm{j}\theta} + 1}{re^{\mathrm{j}\theta}(re^{\mathrm{j}\theta} - re^{-\mathrm{j}\theta})} \tag{3.24}$$

where $r = 0.5967$, $\theta = 33.08°$. After some manipulation and simplification we have

$$C_1 = \frac{2.1439 + 0.977\,19\mathrm{j}}{-0.2122 + 0.3257\mathrm{j}}$$
$$= -0.904\,099\,9 - 5.992\,847\mathrm{j} = 6.060\,66\underline{/-98.58°}$$

Since p_1 and p_2 are complex conjugate pairs then

$$C_2 = C_1^* = -0.904\,099\,9 + 5.992\,847\mathrm{j} = 6.060\,66\underline{/98.58°}$$

Thus the z-transform can be expressed as (from Equation 3.23)

$$X(z) = 2.8082 + \frac{C_1 z}{z - p_1} + \frac{C_2 z}{z - p_1^*} \quad \textbf{(3.25)}$$

where

$$p_1 = 0.5 + 0.3257\text{j} \qquad p_2 = 0.5 - 0.3257\text{j}$$
$$C_1 = -0.9041 - 5.59928\text{j} \qquad C_2 = -0.9041 + 5.59928\text{j}$$

From the z-transform table, entries 1 and 16 in Table 3.1, the inverse z-transform of the terms on the right-hand side of Equation 3.25 is

$$Z^{-1}(2.8082) = 2.8082u(n)$$

$$Z^{-1}\left[\frac{C_1 z}{z - p_1} + \frac{C_2 z}{z - p_1^*}\right] = 2 \times 6.06066(0.5967)^n \cos(33.08n - 98.58°)$$
$$= 12.1213(0.5967)^n \cos(33.08n - 98.58°)$$

Thus the discrete-time signal becomes

$$x(n) = 2.8082u(n) + 12.1213(0.5967)^n \cos(33.08n - 98.58°), \quad n \geqslant 0$$

A useful check for partial fraction results is to compute the values of $x(n)$ for $n = 0, 1, 2$ (say) and then to compare these with values obtained by the power series method. For example, from the expression for $x(n)$ we find that

$$x(0) = 2.8082 - 1.80838 = 1; \; x(1) = 2.99959 = 3; \; x(2) = 3.6436$$

which checks with the results obtained in Example 3.3 using the power series method.

Example 3.6

$X(z)$ contains a second-order pole Find the discrete-time sequence, $x(n)$, with the following z-transform:

$$X(z) = \frac{z^2}{(z - 0.5)(z - 1)^2}$$

$X(z)$ has a first-order pole at $z = 0.5$ and a second-order pole at $z = 1$. In this case, the partial fraction expansion has the form

$$X(z) = \frac{C}{z - 0.5} + \frac{D_1}{z - 1} + \frac{D_2}{(z - 1)^2} \qquad \textbf{(3.26)}$$

To obtain C, we proceed as before and multiply both sides of Equation 3.26 by $z - 0.5$, set $z = 0.5$ and evaluate the expression

$$C = \frac{(z - 0.5)z^2}{z(z - 0.5)(z - 1)^2}\bigg|_{z=0.5}$$
$$= 0.5/(0.5 - 1)^2 = 2$$

To obtain D_1 we use Equation 3.18b, with $i = 1$ and $m = 2$. Thus

$$D_1 = \frac{\mathrm{d}}{\mathrm{d}z}\left[\frac{(z - 1)^2 X(z)}{z}\right]_{z=1} = \frac{(z - 1)^2 z^2}{z(z - 0.5)(z - 1)^2}\bigg|_{z=1}$$
$$= \frac{\mathrm{d}}{\mathrm{d}z}\left(\frac{z}{z - 0.5}\right)_{z=1} = \frac{z - 0.5 - z}{(z - 0.5)^2}\bigg|_{z=1} = -2$$

Similarly, D_2 is obtained from Equation 3.18b by letting $i = 2$ and $m = 2$:

$$D_2 = \frac{(z - 1)^2 X(z)}{z}\bigg|_{z=1} = \frac{(z - 1)^2 z^2}{z(z - 0.5)(z - 1)^2}\bigg|_{z=1}$$
$$= 1/(1 - 0.5) = 2$$

Combining the results, $X(z)$ becomes

$$X(z) = \frac{2z}{z - 0.5} - \frac{2z}{z - 1} + \frac{2z}{(z - 1)^2}$$

The inverse z-transform of each term on the right-hand side is obtained from Table 3.1 and then summed to give $x(n)$:

$$x(n) = 2(0.5)^n - 2 + 2n = 2[(n - 1) + (0.5)^n], \quad n \geqslant 0 \qquad \textbf{(3.27)}$$

The reader can verify that the result is correct by comparing the first few values of $x(n)$ with values calculated with the power series method.

You will agree that the partial fraction expansion method is very tedious, except for simple cases, and mistakes are likely. A C language program is given in the appendix for computing the inverse z-transform, using partial fraction expansions, for $X(z)$ with first-order poles.

3.3.3 Residue method

In this method the IZT is obtained by evaluating the contour integral

$$x(n) = \frac{1}{2\pi j}\oint_C z^{n-1}X(z)\,dz \tag{3.28}$$

where C is the path of integration enclosing all the poles of $X(z)$. For rational polynomials, the contour integral in Equation 3.28 is evaluated using a fundamental result in complex variable theory known as Cauchy's residue theorem (Mathews, 1982):

$$\begin{aligned} x(n) &= \frac{1}{2\pi j}\oint_C z^{n-1}X(z)\,dz \\ &= \text{sum of the residues of } z^{n-1}X(z) \text{ at all the poles inside } C. \end{aligned} \tag{3.29}$$

In the last section, it was stated that the partial fraction coefficients, the C_k, are also referred to as residues of $X(z)$ and a way of obtaining their values given. The key point to remember is that every residue, C_k, is associated with a pole, p_k. In the present method, the residue of $z^{n-1}X(z)$ at the pole p_k (not the residue of $X(z)$) is given by

$$\text{Res}\,[F(z), p_k] = \frac{1}{(m-1)!}\frac{d^{m-1}}{dz^{m-1}}[(z-p_k)F(z)]_{z=p_k} \tag{3.30}$$

where $F(z) = z^{n-1}X(z)$, m is the order of the pole at p_k and $\text{Res}\,[F(z), p_k]$ is the residue of $F(z)$ at $z = p_k$. For a simple (distinct) pole, Equation 3.30 reduces to

$$\begin{aligned} \text{Res}\,[F(z), p_k] &= (z-p_k)F(z) \\ &= (z-p_k)z^{n-1}X(z)\big|_{z=p_k} \end{aligned} \tag{3.31}$$

Example 3.7

Find, using the residue method, the discrete-time signal corresponding to the following z-transform:

$$X(z) = \frac{z}{(z-0.75)(z+0.5)}$$

Assume C is the circle $|z| = 1$.

Solution

This problem is the same as Example 3.4. In factored form, $X(z)$ is given by

$$X(z) = \frac{z}{(z - 0.75)(z + 0.5)}$$

If we let $F(z) = z^{n-1}X(z)$ then

$$F(z) = \frac{z^{n-1}z}{(z - 0.75)(z + 0.5)}$$

$$= \frac{z^n}{(z - 0.75)(z + 0.5)}$$

$F(z)$ has poles at $z = 0.75$ and $z = -0.5$. A sketch of the contour with the positions of the poles indicated by crosses is given in Figure 3.3. Both poles lie inside the contour (unit circle). From Equation 3.29, the inverse z-transform is given by

$$x(n) = \text{Res}\,[F(z), 0.75] + \text{Res}\,[F(z), -0.5]$$

Since the poles are first order, Equation 3.31 will be used. Thus

$$\text{Res}\,[F(z), 0.75] = (z - 0.75)F(z)|_{z=0.75}$$

$$= \left.\frac{\cancel{(z - 0.75)}z^n}{\cancel{(z - 0.75)}(z + 0.5)}\right|_{z=0.75}$$

$$= \frac{(0.75)^n}{0.75 + 0.5}$$

$$= \frac{4}{5}(0.75)^n$$

$$\text{Res}\,[F(z), -0.5] = (z + 0.5)F(z)|_{z=-0.5}$$

$$= \left.\frac{(z + 0.5)z^n}{(z - 0.75)(z + 0.5)}\right|_{z=-0.5}$$

$$= -\frac{4}{5}(-0.5)^n$$

The inverse z-transform is the sum of the residues at $z = 0.75$ and at $z = -0.5$:

$$x(n) = (4/5)[(0.75)^n - (-0.5)^n]$$

which is identical to the result obtained by the partial fraction expansion.

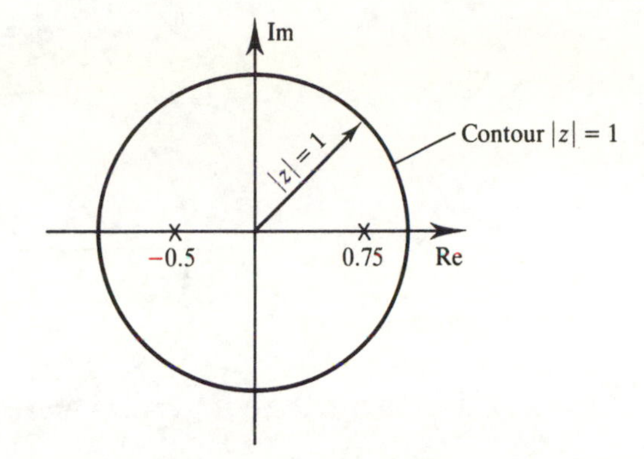

Figure 3.3 A sketch of the contour of integration showing the role of $X(z)$.

Example 3.8

The poles of $X(z)$ are complex conjugate poles Find the inverse z-transform, using the residue method, given the following z-transform:

$$X(z) = \frac{z^2 + 2z + 1}{z^2 - z + 0.3561}$$

Solution

In factored form $X(z)$ is given as:

$$X(z) = \frac{z^2 + 2z + 1}{(z - p_1)(z - p_2)}$$

where $p_1 = 0.5 + 0.3557\text{j}$ and $p_2 = 0.5 - 0.3557\text{j}$, that is $p_2 = p_1^*$. To find the inverse z-transform we evaluate the residues of $F(z)$, where in this case

$$F(z) = z^{n-1}X(z) = \frac{z^{n-1}(z^2 + 2z + 1)}{z^2 - z + 0.3561}$$

$$= \frac{z^n(z^2 + 2z + 1)}{z(z^2 - z + 0.3561)}$$

$F(z)$ has the same poles as $X(z)$, that is at $z = p_1$ and $z = p_2$, plus a pole at $z = 0$ when $n = 0$. Figure 3.4 shows a sketch of the contour with the positions of the poles indicated. All the poles lie inside the contour. The pole at $z = 0$ does not exist for $n > 0$ and so we need to consider the two cases separately.

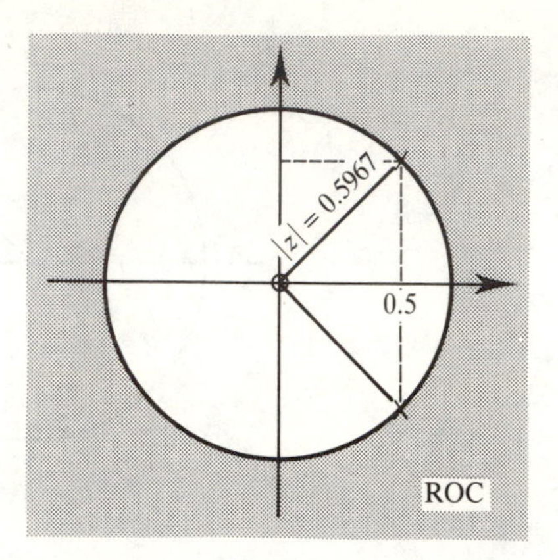

Figure 3.4 Contour for Example 3.8 showing the ROC.

When $n = 0$, $F(z)$ reduces to

$$F(z) = \frac{z^2 + 2z + 1}{z(z^2 - z + 0.3561)}$$

and

$$x(0) = \text{Res}\,[F(z), 0] + \text{Res}\,[F(z), p_1] + \text{Res}\,[F(z), p_2]$$

Therefore

$$\begin{aligned}\text{Res}\,[F(z), 0] &= zF(z)|_{z=0} = \left.\frac{z(z^2 + 2z + 1)}{z(z^2 - z + 0.3561)}\right|_{z=0} \\ &= 1/0.3561 = 2.8082\end{aligned}$$

$$\begin{aligned}\text{Res}\,[F(z), p_1] &= (z - p_1)F(z)|_{z=p_1} \\ &= \frac{(z - p_1)(z^2 + 2z + 1)}{z(z - p_1)(z - p_2)} \\ &= \frac{(re^{j\theta})^2 + 2re^{j\theta} + 1}{re^{j\theta}(re^{j\theta} - re^{-j\theta})}\end{aligned}$$

where $r = 0.5967$ and $\theta = 33.08°$. Noting that this expression is identical to that of Equation 3.24, we can write

$$\text{Res}\,[F(z), p_1] = -0.9041 - 5.9928\text{j}$$

Since p_1 and p_2 are complex conjugate pairs then

$$\text{Res}\,[F(z),\, p_2] = -0.9041 + 5.9928\text{j}$$

Thus

$$\begin{aligned} x(0) &= \text{Res}\,[F(z),\, 0] + \text{Res}\,[F(z),\, p_1] + \text{Res}\,[F(z),\, p_2] \\ &= 2.8082 - 0.9041 - 5.9928\text{j} - 0.9041 + 5.9928\text{j} \\ &= 1 \end{aligned}$$

When $n > 0$, the pole at $z = 0$ vanishes and we have

$$F(z) = \frac{z^n(z^2 + 2z + 1)}{z(z^2 - z + 0.3561)}$$

$$x(n) = \text{Res}\,[F(z),\, p_1] + \text{Res}\,[F(z),\, p_2]$$

$$\begin{aligned} \text{Res}\,[F(z),\, p_1] &= (z - p_1)F(z)\big|_{z=p_1} \\ &= \left.\frac{\cancel{(z - p_1)}z^n(z^2 + 2z + 1)}{z\cancel{(z - p_1)}(z - p_2)}\right|_{z=p_1} \qquad (3.32) \\ &= \frac{(re^{\text{j}\theta})^n[(re^{\text{j}\theta})^2 + 2re^{\text{j}\theta} + 1]}{re^{\text{j}\theta}(re^{\text{j}\theta} - re^{-\text{j}\theta})} \end{aligned}$$

where $r = 0.5967$ and $\theta = 33.08°$. Noting that this expression is similar to Equation 3.24, we can write

$$\begin{aligned} \text{Res}\,[F(z),\, p_1] &= (0.5967e^{\text{j}33.08})^n(6.060\,66e^{-\text{j}98.58}) \\ &= 6.060\,66(0.5967)^n[\cos(33.08n - 98.58) \\ &\quad + \text{j}\sin(33.08n - 98.58)] \end{aligned}$$

Since p_2 and p_1 are complex conjugate pole pairs we can write

$$\begin{aligned} \text{Res}\,[F(z),\, p_2] = 6.060\,66(0.5967)^n[&\cos(33.08n - 98.58) \\ &- \text{j}\sin(33.08n - 98.58)] \end{aligned}$$

Thus

$$\begin{aligned} x(n) &= \text{Res}\,[F(z),\, p_1] + \text{Res}\,[F(z),\, p_2] \\ &= 12.1213(0.5967)^n\cos(33.08n - 98.58°), \quad n > 0 \end{aligned}$$

which checks with the results for the partial fraction expansion.

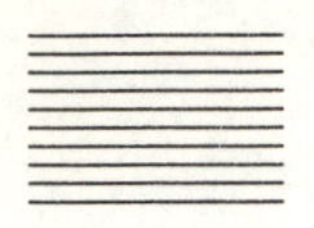

Example 3.9

$X(z)$ contains a second-order pole Find the discrete-time sequence, $x(n)$, with the following z-transform:

$$X(z) = \frac{z^2}{(z - 0.5)(z - 1)^2}$$

Solution

This example is the same as Example 3.6 under partial fraction expansion. According to the residue method the discrete-time sequence is given by

$$x(n) = \sum_{k=1}^{M} \text{Res}\,[F(z), p_k]$$

where

$$F(z) = z^{n-1}X(z) = \frac{z^{n+1}}{(z - 0.5)(z - 1)^2}$$

$F(z)$ has a simple pole at $z = 0.5$ and a second-order pole at $z = 1$; thus $x(n)$ is given by

$$x(n) = \text{Res}\,[F(z), p_1] + \text{Res}\,[F(z), p_2]$$

$$\begin{aligned}\text{Res}\,[F(z), 0.5] &= \frac{\cancel{(z - 0.5)}z^{n+1}}{\cancel{(z - 0.5)}(z - 1)^2} = \left.\frac{z^{n+1}}{(z - 1)^2}\right|_{z=0.5} \\ &= 0.5(0.5)^n/(0.5)^2 = 2(0.5)^2\end{aligned}$$

$$\begin{aligned}\text{Res}\,[F(z), 1] &= \frac{\mathrm{d}}{\mathrm{d}z}\left[\frac{\cancel{(z - 1)^2}z^{n+1}}{(z - 0.5)\cancel{(z - 1)^2}}\right] \\ &= \left.\frac{(z - 0.5)(n + 1)z^n - z^{n+1}}{(z - 0.5)^2}\right|_{z=1} \\ &= [(0.5)(n + 1) - 1]/(0.5)^2 = 2(n - 1)\end{aligned}$$

Combining the results, we have

$$x(n) = 2[(n - 1) + (0.5)^n]$$

which is the same result as for the partial fraction expansion method.

The reader may have noticed that the partial fraction and residue methods are related. Both methods require the evaluation of residues albeit performed in different ways. The partial fraction method requires the evaluation of the residues of $X(z)$, that is the C_k, while the residue method requires the evaluation of the residues of $z^{n-1}X(z)$. When $X(z)$ has first-order poles we have

$$\text{Res}\,[z^{n-1}X(z),\, p_k] = z^n\,\text{Res}\,[X(z),\, p_k] = z^n C_k \qquad \textbf{(3.33)}$$

Thus the C language program for the partial fraction expansion given in the appendix may be exploited to obtain results for the residue method.

3.3.4 Comparison of the inverse z-transform methods

We have discussed in some detail three methods of obtaining the inverse z-transform: the power series, partial fraction expansion and the residue methods. A limitation of the power series method is that it does not lead to a closed form solution (although this can be deduced in simple cases), but it is simple and lends itself to computer implementation. However, because of its recursive nature care should be taken to minimize possible build-up of numerical errors when the number of data points in the inverse z-transform is large, for example by using double precision.

Both the partial fraction expansion and the residue methods lead to closed form solutions. The main disadvantage with both methods is the need to factorize the denominator polynomial, that is finding the poles of $X(z)$. If the order of $X(z)$ is high finding the poles of $X(z)$, if $X(z)$ is not in factored form, is quite a difficult task. This topic is discussed further in Section 3.5.1. Both methods may also involve high-order differentiation if $X(z)$ contains multiple-order poles. Clearly, if closed form solution is required then the partial fraction or residue method is the most appropriate. The partial fraction method is particularly useful in generating the coefficients of parallel structures for digital filters (see Section 3.5.11). The residue method is widely used in the analysis of quantization errors in discrete-time systems. (See Chapter 7.)

3.4 Properties of the z-transform

Some useful properties of the z-transform which have found practical use in DSP are described briefly below. The proofs for some of these properties are given as problems at the end of the chapter.

(1) *Linearity*. If the sequences $x_1(n)$ and $x_2(n)$ have z-transforms $X_1(z)$ and $X_2(z)$, then the z-transform of their linear combination is

$$ax_1(n) + bx_2(n) \rightarrow aX_1(z) + bX_2(z) \tag{3.34}$$

(2) *Delays or shifts*. If the z-transform of a sequence, $x(n)$, is $X(z)$ then the z-transform of the sequence delayed by m samples is $z^{-m}X(z)$. This property is widely used in converting the z transfer function of discrete-time systems into time domain difference equations and vice versa; see Section 3.5.8.

$$x(n) \rightarrow X(z)$$
$$x(n-m) \rightarrow z^{n-m}X(z)$$

(3) *Convolution*. Given a discrete-time LTI system with input, $x(n)$, and impulse response, $h(k)$, the output of the system is given by

$$y(n) = \sum_{k=-\infty}^{\infty} h(k)x(n-k) \tag{3.35a}$$

In terms of the z-transform, the input and output are related as

$$Y(z) = H(z)X(z) \tag{3.35b}$$

where $X(z)$, $H(z)$ and $Y(z)$ are, respectively, the z-transform of $x(n)$, $h(k)$ and $y(n)$. Given $X(z)$ and $H(z)$, the output $y(n)$ can be obtained by inverse z-transforming $Y(z)$.

It is seen that the convolution operation in Equation 3.35a has become a multiplicative process in the z-domain. $H(z)$ is often referred to as the system transfer function.

(4) *Differentiation*. If $X(z)$ is the z-transform of $x(n)$, then the z-transform of $nx(n)$ can be obtained by differentiating $X(z)$:

$$x(n) \rightarrow X(z)$$
$$nx(n) \rightarrow -z\frac{\mathrm{d}X(z)}{\mathrm{d}z} \tag{3.36}$$

This property is useful in obtaining the inverse z-transform when $X(z)$ contains multiple order poles.

(5) *Relationship with the Laplace transform*. Continuous-time systems or signals are normally described using the Laplace transform. If we let $z = \mathrm{e}^{sT}$, where s the complex Laplace variable given by

$$s = d + \mathrm{j}\omega$$

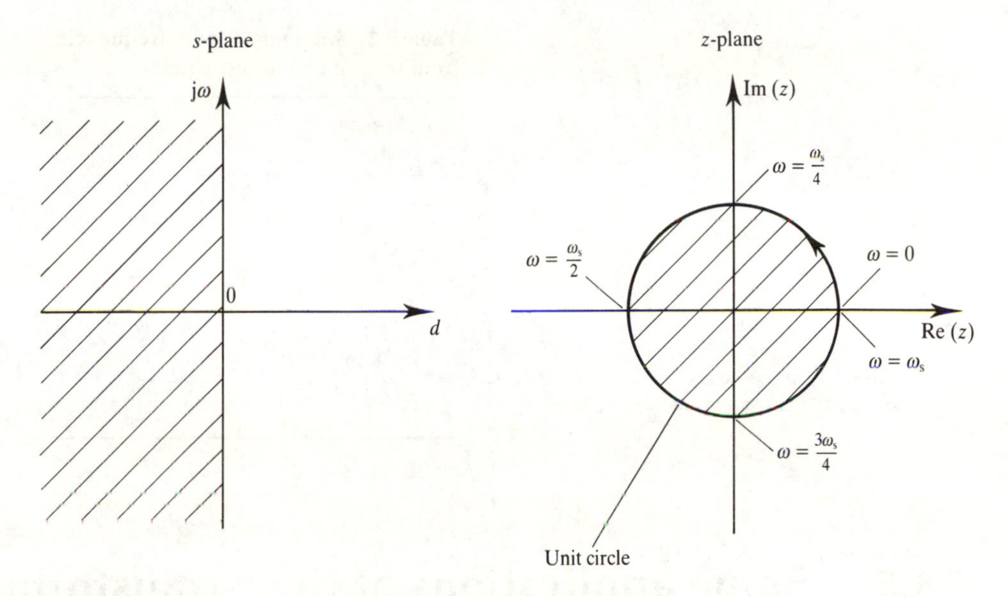

Figure 3.5 Mapping of s-plane to z-plane. The left-hand side of the s-plane maps to the interior of the z-plane, the right-hand side maps to the exterior and the jω-axis maps onto the unit circle.

then

$$z = \mathrm{e}^{(d+\mathrm{j}\omega)T} = \mathrm{e}^{dT}\mathrm{e}^{\mathrm{j}\omega T} \tag{3.37}$$

Thus

$$|z| = \mathrm{e}^{dT} \text{ and } \underline{/z} = \omega T = 2\pi f/F_\mathrm{s} = 2\pi\omega/\omega_\mathrm{s}$$

where ω_s (rad s^{-1}) is the sampling frequency. As ω varies from $-\infty$ to ∞ the s-plane is mapped to the z-plane as shown in Figure 3.5. The entire jω axis in the s-plane is mapped onto the unit circle. The left-hand s-plane is mapped to the inside of the unit circle and the right-hand s-plane maps to the outside of the unit circle.

In terms of frequency response, the jω axis is the most important in the s-plane. In this case, $d = 0$, and frequency points in the s-plane are related to points on the z-plane unit circle by

$$z = \mathrm{e}^{\mathrm{j}\omega T} \tag{3.38}$$

Table 3.2 shows how some specific frequencies are mapped from the s-plane to the z-plane. It is clear that the mapping is not unique since for example the two frequencies $\omega = \omega_\mathrm{s}$ and $\omega = 2\omega_\mathrm{s}$ in the s-plane map to the same point on the unit circle.

Table 3.2 Mapping of frequencies from the s-plane to the z-plane.

s-plane: ω (rad s^{-1})	*z-plane:* ωT (rad)
0	0
$\omega_S/4$	$\pi/2$
$\omega_S/2$	π
$3\omega_S/4$	1.25π
ω_S	2π
$1.25\omega_S$	$\pi/2$
$1.5\omega_S$	π
$1.75\omega_S$	1.25π
$2\omega_S$	2π

3.5 Some applications of the z-transform in signal processing

Applications of the z-transform in DSP are many. Several of these are discussed in more detail in later chapters, especially in Chapter 7. The next few sections are intended to highlight some of these applications and to establish some fundamental issues common to them.

3.5.1 Pole–zero description of discrete-time systems

In most practical discrete-time systems the z-transform, that is the system transfer function $H(z)$, can be expressed in terms of its poles and zeros. Consider for example, the following z-transform representing a general, Nth-order discrete-time filter (where $N = M$):

$$H(z) = \frac{N(z)}{D(z)} \tag{3.39}$$

where

$$N(z) = a_0 z^N + a_1 z^{N-1} + a_2 z^{N-2} + \ldots + a_N$$

$$D(z) = b_0 z^N + b_1 z^{N-1} + b_2 z^{N-2} + \ldots + b_N$$

The a_k and b_k are the coefficients of the filter.

If $H(z)$ has poles at $z = p_1, p_2, \ldots, p_N$ and zeros at $z = z_1, z_2, \ldots, z_N$,

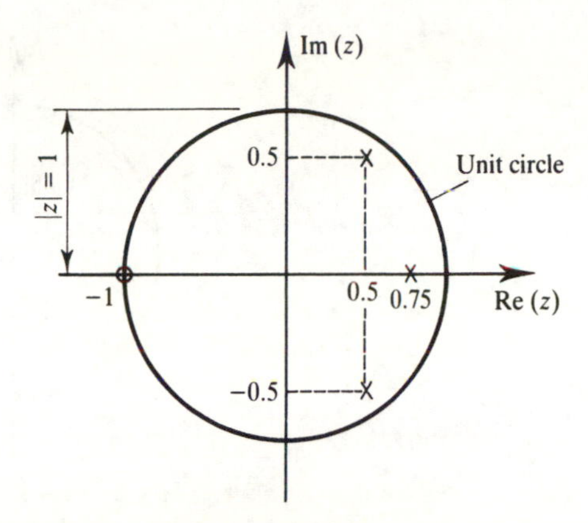

Figure 3.6 Description of a z-transform in the form of a pole–zero diagram: X, pole; o, zero.

then $H(z)$ can be factored and represented as

$$H(z) = \frac{K(z - z_1)(z - z_2) \ldots (z - z_N)}{(z - p_1)(z - p_2) \ldots (z - z_N)} \tag{3.40}$$

where z_i is the ith zero, p_i is the ith pole and K is the gain factor. You will recall that the poles of a z-transform such as $H(z)$ are the values of z for which $H(z)$ becomes infinity. The values of z for which $H(z)$ becomes zero are referred to as zeros. The poles and zeros of $H(z)$ may be real or complex. When they are complex, they occur in conjugate pairs, to ensure that the coefficients, a_k and b_k, are real. It should be clear from Equation 3.40 that if the locations of the poles and zeros of $H(z)$ are known, then $H(z)$ itself can be readily reconstructed to within a constant.

The information contained in the z-transform can be conveniently displayed as a pole–zero diagram; see for example Figure 3.6. In the diagram, X marks the position of a pole and o denotes the position of a zero. For this example, the poles are located at $z = 0.5 \pm 0.5\text{j}$, and at $z = 0.75$. A single zero is at $z = -1$. An important feature of the pole–zero diagram is the unit circle, that is the circle defined by $|z| = 1$; see Figure 3.6. As will become clear, the unit circle plays an important role in the analysis and design of discrete-time systems.

The pole–zero diagram provides an insight into the properties of a given discrete-time system. For example, from the locations of the poles and zeros we can infer the frequency response of the system as well as its degree of stability. For a stable system, all the poles must lie inside the unit circle (or be coincident with zeros on the unit circle).

Often, the z-transform is not available in factored form but as a ratio of polynomials such as Equation 3.39. In these cases, describing the z-transform,

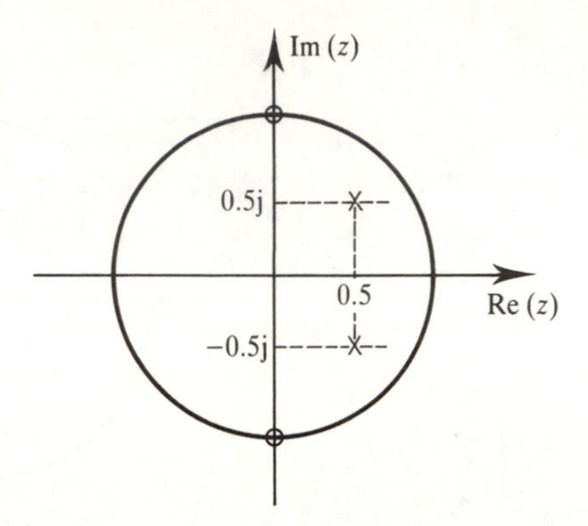

Figure 3.7 Pole–zero diagram for Example 3.10, part (2).

$H(z)$, in terms of its poles and zeros will require finding the roots of the denominator polynomial, $D(z)$, and those of the numerator polynomial, $N(z)$.

For a second-order polynomial which has the form $ax^2 + bx + c$, the roots are given by

$$\frac{-b \pm (b^2 - 4ac)^{1/2}}{2a} \tag{3.41}$$

For higher-order polynomials, finding the roots of $N(z)$ or $D(z)$ is a difficult task. In practice, this is often achieved using numerical methods involving, for example, Newton's and/or Baistow's algorithms (see for example, Atkinson and Harley (1983)). The need to find the poles and zeros often arises in connection with the design of discrete filters and in stability analysis. Fortunately, in the case of discrete-time filter design the poles and zeros are automatically generated by the filter design software, obviating the need to find the roots directly.

Example 3.10

(1) Express the following transfer function in terms of its poles and zeros and sketch the pole–zero diagram:

$$H(z) = \frac{1 - z^{-1} - 2z^{-2}}{1 - 1.75z^{-1} + 1.25z^{-2} - 0.375z^{-3}}$$

(2) Determine the transfer function, $H(z)$, of a discrete-time filter with the pole–zero diagram shown in Figure 3.7.

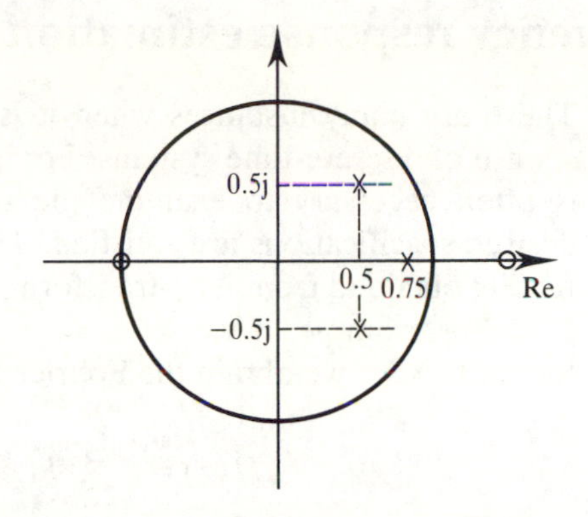

Figure 3.8 Pole–zero diagram for Example 3.10.

Solution

(1) First, we express $H(z)$ in positive powers of z and then factorize it so that its poles and zeros can be determined. If we multiply the denominator and numerator by z^2, the highest power of z, we obtain

$$H(z) = \frac{z^2 - z - 2}{z^3 - 1.75z^2 + 1.25z - 0.375}$$

Factoring, we have

$$H(z) = \frac{(z - 2)(z + 1)}{(z - 0.5 + j0.5)(z - 0.5 - j0.5)(z - 0.75)}$$

Thus, the pole locations are at $z = 0.5 \pm 0.5j$ and at $z = 0.75$. The zeros are at $z = 2$ and $z = -1$. The pole–zero diagram is depicted in Figure 3.8.

(2) From the pole–zero diagram, the zeros of the transfer function are at $z = \pm j$ and the poles are at $z = 0.5 \pm 0.5j$. The transfer function can be written down directly:

$$H(z) = \frac{K(z - j)(z + j)}{(z - 0.5 - 0.5j)(z - 0.5 + 0.5j)}$$

$$= \frac{K(z^2 + 1)}{z^2 - z + 0.5}$$

$$= \frac{K(1 + z^{-2})}{1 - z^{-2} - 0.5z^{-2}}$$

3.5.2 Frequency response estimation

There are many instances when it is necessary to evaluate the frequency response of discrete-time systems. For example, in the design of discrete filters, it is often necessary to examine the spectrum of the filter to ensure that the desired specifications are satisfied. The frequency response of a system can be readily obtained from its z-transform.

For example, if we set $z = e^{j\omega T}$, that is evaluate the z-transform around the unit circle, we obtain the Fourier transform of the system:

$$H(z) = \sum_{n=-\infty}^{\infty} h(n)z^{-n}\Bigg|_{z=e^{j\omega T}} \tag{3.42a}$$

$$= H(e^{j\omega T}) = \sum_{n=-\infty}^{\infty} h(n)e^{-jn\omega T} \tag{3.42b}$$

$H(e^{j\omega T})$ is referred to as the frequency response of the system. We have used the symbol T to emphasize the dependence of the frequency response of discrete-time systems on the sampling frequency. In general, $H(e^{j\omega T})$ is complex. Its modulus gives the magnitude response and its phase the phase response of the system.

The frequency response may be obtained from the z-transform using several methods. We will describe three methods.

3.5.3 Geometric evaluation of frequency response

This is a simple but useful method of obtaining a rough idea of what the frequency response of a discrete-time system would look like, based on its pole–zero diagram. Recall that the z-transform of an LTI system may be expressed in terms of its poles and zeros:

$$H(z) = \frac{K(z - z_1)(z - z_2)\ldots(z - z_N)}{(z - p_1)(z - p_2)\ldots(z - p_N)} = \frac{\prod_{i=1}^{N} K(z - z_i)}{\prod_{i=1}^{N}(z - p_i)} \tag{3.43}$$

where we have assumed, for simplicity, that the orders of the numerator and denominator are equal. The frequency response is obtained by making the substitution $z = e^{j\omega T}$ in Equation 3.43 and evaluating $H(e^{j\omega T})$ in the interval $(0 \leq \omega \leq \omega_S/2)$.

$$H(e^{j\omega T}) = \frac{\prod_{i=1}^{N} K(e^{j\omega T} - z_i)}{\prod_{i=1}^{N}(e^{j\omega T} - p_i)} \tag{3.44}$$

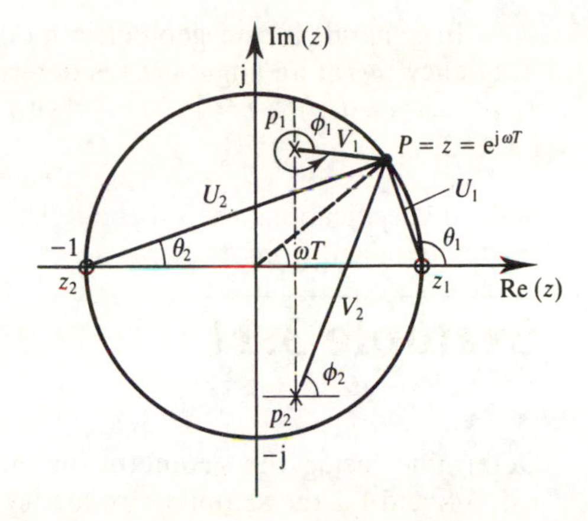

Figure 3.9 Geometric evaluation of the frequency response from the pole–zero diagram.

A geometric interpretation of Equation 3.44 for a z-transform with only two zeros and two poles is shown in Figure 3.9. In this case, the frequency response is given by

$$H(e^{j\omega T}) = \frac{K(e^{j\omega T} - z_1)(e^{j\omega T} - z_2)}{(e^{j\omega T} - p_1)(e^{j\omega T} - p_2)}$$

$$= \frac{KU_1\underline{/\theta_1}U_2\underline{/\theta_2}}{V_1\underline{/\phi_1}V_2\underline{/\phi_2}} \tag{3.45}$$

where U_1 and U_2 represent the distances from the zeros to the point $z = e^{j\omega T}$, and V_1 and V_2 the distances of the poles to the same point as shown in Figure 3.9. Thus the magnitude and phase responses for the system, from Equation 3.45, are

$$|H(e^{j\omega T})| = \frac{U_1U_2}{V_1V_2}, \qquad K = 1$$

$$\underline{/H(e^{j\omega T})} = \theta_1 + \theta_2 - (\phi_1 + \phi_2)$$

The complete frequency response is obtained by evaluating $H(e^{j\omega T})$ as the point P moves from $z = 0$ to $z = -1$. It is evident that, as the point P moves closer to the pole p_1, the length of the vector V_1 decreases and so the magnitude response increases. On the other hand, as the point moves closer to the zero z_1, the zero vector U_1 decreases and so the magnitude response, $|H(e^{j\omega T})|$, decreases. Thus at the pole the magnitude response exhibits a peak whereas, at the zero, the magnitude response falls to zero.

In general, in the geometric method, the frequency response at a given frequency, ω (at an angle ωT) is determined by the ratio of the product of the zero vectors, $U_i\underline{/\theta_i}$, $i = 1, 2, \ldots$, with the product of the pole vectors, $V_i\underline{/\phi_i}$, $i = 1, 2, \ldots$.

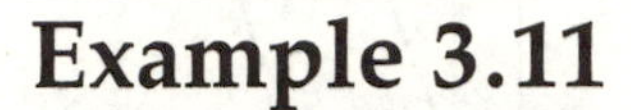

Example 3.11

Determine, using the geometric method, the frequency response at dc, 1/8, 1/4, 3/8 and 1/2 the sampling frequency, of the causal discrete-time system with the following z-transform:

$$H(z) = \frac{z+1}{z-0.7071}$$

Sketch the amplitude frequency response in the interval $0 \leqslant \omega \leqslant \omega_S$, where ω_S (rad s^{-1}) is the sampling frequency.

Solution

In this example, $H(z)$ has a single pole and a single zero, as shown in the pole–zero diagram of Figure 3.10(a). From Equation 3.44, the response at ω is given by

$$H(e^{j\omega T}) = \frac{U\underline{/\theta}}{V\underline{/\phi}} = \frac{e^{j\omega T}+1}{e^{j\omega T}-0.7071} = \frac{1+\cos(\omega T)+j\sin(\omega T)}{\cos(\omega T)-0.7071+j\sin(\omega T)} \qquad \textbf{(3.46)}$$

At dc, $\omega T = 0$ and the zero and pole vectors to the point $z = 0$ are $2\underline{/0^\circ}$ and $0.2929\underline{/0^\circ}$. Thus the frequency response is given by

$$H(e^{j\omega T}) = 2/0.2929 = 6.828\underline{/0^\circ}$$

At $\omega = \omega_s/8$, $\omega T = \omega_s/8F_s = \pi/4$. The pole and zero vectors in this case are shown in Figure 3.10(b). Rather than actually measuring the lengths and angles of the vectors we will use the explicit expression on the right-hand side of Equation 3.46. Thus

$$H(e^{j\omega T}) = \frac{1+\cos(\pi/4)+j\sin(\pi/4)}{\cos(\pi/4)-0.7071+j\sin(\pi/4)}$$

$$= \frac{1.8477\underline{/22.5^\circ}}{0.7071\underline{/90^\circ}} = 2.6131\underline{/-67.5^\circ}$$

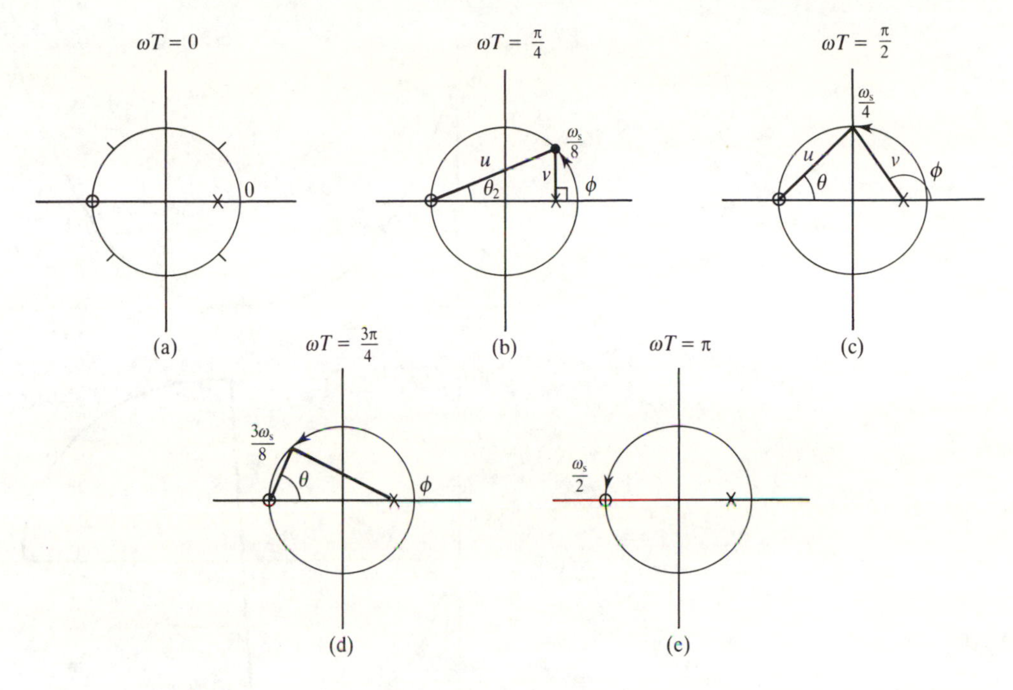

Figure 3.10 Frequency response estimation using geometric method and the pole–zero diagram.

The responses at the remaining frequencies, obtained in a similar way, are summarized below, and the vectors are given in Figures 3.10(c)–3.10(e).

ω (rad s^{-1})	ωT (rad)	$\lvert H(e^{j\omega T})\rvert$	$\underline{/H(e^{j\omega T})}$ (degrees)
0	0	6.828	0
$\omega_s/8$	$\pi/4$	2.6131	−67.5
$\omega_s/4$	$\pi/2$	1.1547	−80.26
$3\omega_s/8$	$3\pi/4$	0.4840	−85.93
$\omega_s/2$	π	0	0

A sketch of the magnitude and phase responses is shown in Figure 3.11. An important point to note is that the magnitude response, $|H(e^{j\omega T})|$, is symmetrical about half the sampling frequency (Nyquist frequency), and the phase response antisymmetrical about the same frequency. This is always the case when the coefficients, a_k and b_k, of a discrete-time system are real. Further, the frequency response of such systems is periodic with a period of ω_s (the sampling frequency), a behaviour that is consistent with the sampling theorem.

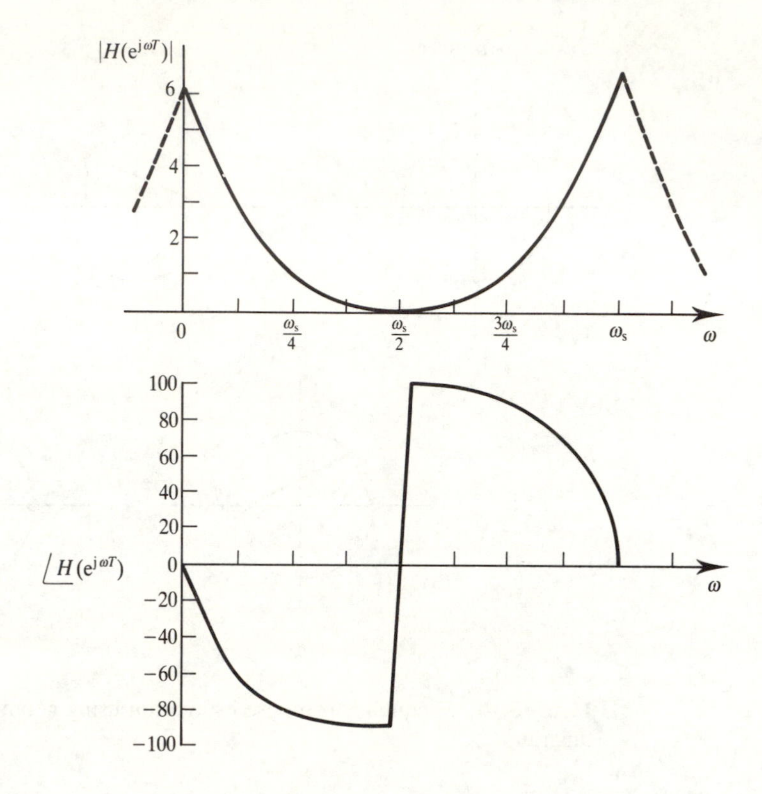

Figure 3.11 A sketch of the frequency response of the discrete-time system of Example 3.11.

3.5.4 Direct computer evaluation of frequency response

Geometric evaluation of the frequency response gives one a feel for the frequency response, but it is clearly very tedious if the precise response is required at many frequencies. Although the process can be automated, the difficulty of finding the locations of the poles and zeros limits its usefulness. If the complete frequency response is required, it is common practice to make the substitution $z = e^{j\omega T}$ directly into the transfer function and then to evaluate the resulting expression:

$$
\begin{aligned}
H(e^{j\omega T}) \\
&= \left.\frac{a_0 + a_1 z^{-1} + \ldots + a_N z^{-N}}{b_0 + b_1 z^{-1} + \ldots + b_N z^{-M}}\right|_{z=e^{j\omega T}} \qquad (3.47) \\
&= \frac{a_0 + a_1 e^{-j\omega T} + \ldots + a_N e^{-jN\omega T}}{b_0 + b_1 e^{-j\omega T} + \ldots + b_M e^{-jM\omega T}}
\end{aligned}
$$

$$= \frac{a_0 + a_1[\cos(\omega T) - \mathrm{j}\sin(\omega T)] + \ldots + a_N[\cos(N\omega T) - \mathrm{j}\sin(N\omega T)]}{b_0 + b_1[\cos(\omega T) - \mathrm{j}\sin(\omega T)] + \ldots + b_M[\cos(M\omega T) - \mathrm{j}\sin(M\omega T)]} \quad \textbf{(3.48)}$$

A C language implementation of Equation 3.48 is discussed in Appendix 3C. The program evaluates $H(\mathrm{e}^{\mathrm{j}\omega T})$ in the interval $(0 \leqslant \omega \leqslant \omega_S/2)$.

3.5.5 Frequency response estimation via FFT

The FFT may also be used to evaluate the frequency response of discrete-time systems. A way of doing this, for IIR systems, is first to obtain the impulse response of the system using, for example, the power series method, and then to compute the FFT of the impulse response. This follows directly from Equation 3.42(b) which shows that the frequency response of a discrete-time system is simply the Fourier transform of its impulse response. To obtain a smooth frequency response, it is important to take a sufficient number of impulse response values and/or to zero-pad the impulse response values before the FFT is taken. A C language implementation is discussed in the appendix.

An alternative technique is first to zero-pad the numerator and denominator coefficients, for example

$$\{b(n)\} = \{b_0, b_1, b_2, \ldots, b_M, 0, 0, \ldots, 0\} \quad \textbf{(3.49)}$$

$$\{a(n)\} = \{a_0, a_1, a_2, \ldots, a_N, 0, 0, \ldots, 0\}$$

and then to obtain the FFTs of $\{b(n)\}$ and $\{a(n)\}$, $A(k)$ and $B(k)$, respectively. The ratio of the two FFTs gives the frequency response:

$$H(\mathrm{e}^{\mathrm{j}\omega_k T}) = A(k)/B(k), \quad k = 0, 1, \ldots, N/2 \quad \textbf{(3.50)}$$

3.5.6 Frequency units used in discrete-time systems

Continuous-time systems or signals are normally described using the Laplace transform. Thus the frequency response of a continuous-time system is traditionally evaluated by letting $s = \mathrm{j}\omega$ in the system transfer function, $H(s)$, where s is the complex Laplace variable. In DSP we deal with discrete-time systems and signals. In this case, the frequency response is found by letting $z = \mathrm{e}^{\mathrm{j}\omega T}$ and then evaluating the z-transfer function, $H(z)$, in the interval $0 \leqslant \omega \leqslant \omega_S/2$. The key point in discrete-time systems is the dependence of the useful frequency range on the sampling frequency, ω_S.

Table 3.3 shows how ωT and z change as ω varies from 0 to ω_S. It can be inferred that as the angle ωT goes from 0 to 2π the value of z varies from 1 through j and back to 1. This information is depicted in Figure 3.12. The figure also makes it evident that the frequency response of a discrete-time system is

Table 3.3 Units of frequency use in discrete-time systems and their relationships to points on the unit circle.

f (Hz)	ω (rad s^{-1})	ωT (rad)	$z = e^{j\omega T}$
0	0	0	1
$\frac{F_S}{8}$	$\frac{\omega_S}{8}$	$\frac{\pi}{4}$	$\frac{\sqrt{2}}{2} + \frac{\sqrt{2}}{2}j$
$\frac{F_S}{4}$	$\frac{\omega_S}{4}$	$\frac{\pi}{2}$	j
$\frac{3F_S}{8}$	$\frac{3\omega_S}{8}$	$\frac{3\pi}{4}$	$-\frac{\sqrt{2}}{2} + \frac{\sqrt{2}}{2}j$
$\frac{F_S}{2}$	$\frac{\omega_S}{2}$	π	-1
$\frac{5F_S}{8}$	$\frac{5\omega_S}{8}$	$\frac{5\pi}{4}$	$-\frac{\sqrt{2}}{2} - \frac{\sqrt{2}}{2}j$
$\frac{2F_S}{4}$	$\frac{3\omega_S}{4}$	$\frac{3\pi}{2}$	$-j$
$\frac{7F_S}{8}$	$\frac{7\omega_S}{4}$	$\frac{7\pi}{4}$	$\frac{\sqrt{2}}{2} - \frac{\sqrt{2}}{2}j$
F_S	ω_S	2π	1

$F_S = 1/T$ is the sampling frequency in Hz; T is the sampling period; $\omega_S = 2\pi/T$ is the sampling frequency in rad s^{-1}.

cyclic: as we go round the circle one or more revolutions the values of z simply repeat.

Two frequency units are normally used to describe the frequency response of discrete-time systems, namely, ω (rad s^{-1}) and f (Hz). When the frequency unit is rad s^{-1} the frequency response goes from $\omega = 0$ to $\omega = \omega_S/2$ or equivalently from $\omega = 0$ to $\omega = \pi/T$ (since $\omega_S = 2\pi F_S = 2\pi/T$). When the standard frequency unit of hertz is used the frequency range is from 0 to $F_S/2$ or equivalently 0 to $1/2T$. Both frequency units may also be expressed in normalized form, that is $T = 1$ or equivalently $F_S = 1$. Table 3.3 shows the relationship between the two frequency units. Thus the frequency ranges of interest may be expressed in one of the following, equivalent, six ways:

$$\left.\begin{array}{ll} 0 \leq \omega \leq \omega_S/2 & \text{rad s}^{-1} \\ 0 \leq \omega \leq \pi/T & \text{rad s}^{-1} \\ 0 \leq \omega \leq \pi & \text{(normalized)} \end{array}\right\} \tag{3.51}$$

$$\left.\begin{array}{ll} 0 \leq f \leq F_S/2 & \text{Hz} \\ 0 \leq f \leq 1/2T & \text{Hz} \\ 0 \leq f \leq 1/2 & \text{(normalized)} \end{array}\right\} \tag{3.52}$$

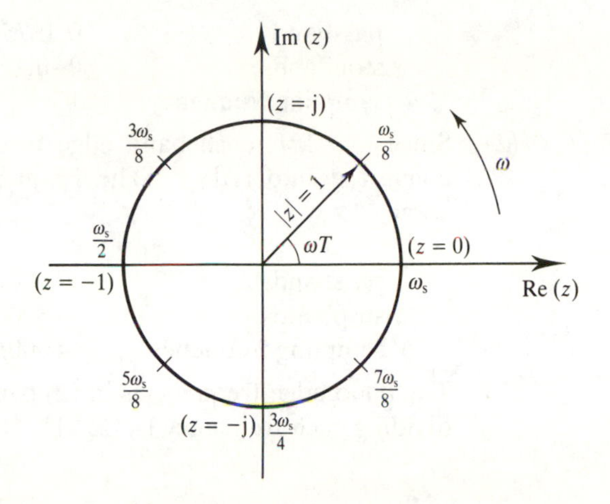

Figure 3.12 *Z*-plane unit circle showing critical frequency points.

The unit of Hz is more appealing (and less confusing) when we are examining frequency response plots or specifying discrete-time systems. However, when evaluating the numerous mathematical formulas in DSP, the unit of rad s^{-1} is more convenient.

Example 3.12

Given the frequency response specification for a bandpass discrete-time filter in Hz as

passband	6–10 kHz
stopbands	0–4 kHz and 12–16 kHz
sampling frequency	32 kHz

(1) express the specifications in normalized frequency, f,

(2) convert the specification from standard units of Hz to rad s^{-1}, and

(3) convert the specifications from the units of rad s^{-1} in part (2) to normalized frequency, ω.

Solution

(1) The band edge frequencies, which are in units of Hz, can be expressed in normalized form by simply dividing each frequency by the sampling frequency. Thus, the specification in normalized form becomes

passband	0.1875–0.3125
stopbands	0–0.125 and 0.375–0.5
sampling frequency	1

(2) Since $\omega = 2\pi f$, each band edge frequency is simply multiplied by 2π to convert it into rad s^{-1}. The frequency response specifications now become

passband	$12\,000\pi$–$20\,000\pi$ rad s^{-1}
stopbands	0–8000π and $24\,000\pi$–$32\,000\pi$ rad s^{-1}
sampling frequency	$64\,000\pi$ rad s^{-1}

(3) The band edge frequencies in (2) can be expressed in normalized form by dividing each frequency by 32 kHz (the sampling frequency), for example

$$12\,000\pi \rightarrow \frac{12\,000\pi}{32\,000} = \frac{3\pi}{8}$$

Thus the specifications become

passband	$3\pi/8$–$3\pi/8$
stopbands	0–$\pi/4$ and $3\pi/4$–π
sampling frequency	2π

3.5.7 Stability considerations

Stability analysis is often carried out as part of the design of discrete-time systems. A useful stability criterion for LTI systems is that all bounded inputs produce bounded outputs. This is the so-called BIBO (bounded input, bounded output) condition. An LTI system is said to be BIBO stable if and only if it satisfies the criterion

$$\sum_{k=0}^{\infty} |h(k)| < \infty \tag{3.53}$$

where $h(k)$ is the impulse response of the system. It is obvious that if the impulse response is of finite length the condition above is satisfied since the sum of the impulse response coefficients will be finite. Thus, stability considerations apply only to systems with impulse response of infinite duration.

For the output to be bounded all the poles must lie inside the unit circle. When a pole lies outside the unit circle the system is unstable. In practice, a system with a pole on the unit circle is also regarded as unstable or potentially unstable, since a minor disturbance or error will invariably push the system into instability. An exception is when a pole is coincident with a zero on the unit

	$h(n)$			
n	α=0.5	α=0.99	α=1	α=1.5
0	0.00000e+00	0.00000e+00	0.00000e+00	0.00000e+00
1	1.00000e+01	1.00000e+01	1.00000e+01	1.00000e+01
2	5.00000e+00	9.90000e+00	1.00000e+01	1.50000e+01
3	2.50000e+00	9.80100e+00	1.00000e+01	2.25000e+01
4	1.25000e+00	9.70299e+00	1.00000e+01	3.37500e+01
5	6.25000e-01	9.60596e+00	1.00000e+01	5.06250e+01
6	3.12500e-01	9.50990e+00	1.00000e+01	7.59375e+01
7	1.56250e-01	9.41480e+00	1.00000e+01	1.13906e+02
8	7.81250e-02	9.32065e+00	1.00000e+01	1.70859e+02
9	3.90625e-02	9.22745e+00	1.00000e+01	2.56289e+02

(a)

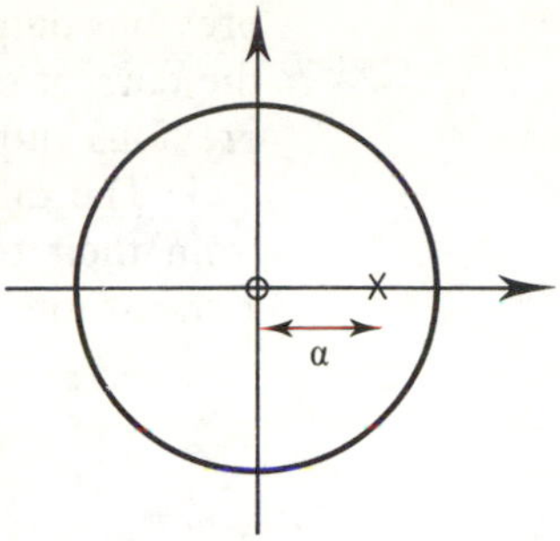

(b)

Figure 3.13 An illustration of the behaviour of the impulse response of a system for various degrees of stability: (a) impulse response; (b) z-plane pole–zero diagram. The system z-transform is $10z^{-1}/(1-\alpha z^{-1})$. (i) For $\alpha = 0.5$ system is stable, (ii) for $\alpha = 0.99$ it is marginally stable, (iii) for $\alpha = 1$ it is potentially unstable, and (iv) for $\alpha = 1.5$ it is unstable. Notice, for example, that for $\alpha = 0.5$ the impulse response values decay rapidly as n increases, while, for $\alpha = 1.5$, the impulse response values increase rapidly.

circle so that its effects are nullified. For an unstable system, the impulse response will increase indefinitely with time.

In principle, testing for stability is simple: find the pole positions of the z-transform. If any pole is on or outside the unit circle (unless it is coincident with a zero on the unit circle) then the system is unstable. In practice, finding the positions of the poles may not be an easy task.

A simple test that may be used when the system z-transform, $H(z)$, is not available in factored form is to obtain and plot a sufficient number of values of the impulse response, by finding the inverse z-transform. If the impulse response increases indefinitely with time or fails to decay fast enough then the system is either unstable or marginally stable. Figure 3.13 shows examples of the behaviour of the impulse response of a simple discrete-time system for various degrees of stability. Other more sophisticated tests for stability are available in more advanced texts on z-transforms (for example Jury, 1964; Proakis and Manolakis, 1992). Further considerations of stability for second-order systems are given in Chapter 7.

3.5.8 Difference equations

The difference equation specifies the actual operations that must be performed by the discrete-time system on the input data, in the time domain, in order to generate the desired output. The difference equation, for most practical cases of interest, may be written as

$$y(n) = \sum_{k=0}^{N} a_k x(n-k) - \sum_{k=1}^{M} b_k y(n-k) \tag{3.54}$$

where $x(n)$ is the input sample, $y(n)$ is the output sample, $y(n-k)$ is the previous output and a_k, b_k are system coefficients. Equation 3.54 indicates that the current output, $y(n)$, is obtained from present and past input samples and previous outputs, $y(n-k)$.

The difference equations for discrete-time systems are readily obtained from their transfer functions and vice versa using the delay property of the z-transform:

$$a_k x(n) \leftrightarrow a_k X(z)$$

$$a_k x(n-k) \leftrightarrow a_k z^{-k} X(z)$$

Thus, Equation 3.54 may be written as

$$Y(z) = \sum_{k=0}^{N} a_k z^{-k} X(z) - \sum_{k=0}^{M} b_k z^{-k} Y(z) \tag{3.55}$$

Simplifying, we obtain the z-domain transfer function of the discrete system, $H(z)$:

$$H(z) = \frac{Y(z)}{X(z)} = \sum_{k=0}^{N} a_k z^{-k} \Bigg/ \left(1 + \sum_{k=0}^{M} b_k z^{-k}\right) \tag{3.56}$$

If all the denominator coefficients, b_k, are zero Equations 3.54 and 3.55 reduce to

$$y(n) = \sum_{k=0}^{N} a_k x(n-k) \tag{3.57a}$$

$$H(z) = \frac{Y(z)}{X(z)} = \sum_{k=0}^{N} a_k z^{-k} \tag{3.57b}$$

The output of the system, $y(n)$, now depends only on the present and past input samples but not on previous outputs as in Equation 3.54. The coefficients, the a_k, in this case represent the impulse response of the system and are normally denoted by the symbol $h(k)$. This class of LTI system is referred to as finite impulse response (FIR) systems, since the length of $h(k)$ is clearly finite.

Systems characterized by Equations 3.54 and 3.56 where at least one of the denominator coefficients is nonzero are referred to as infinite impulse response (IIR) systems. In IIR systems at least one of the poles will be nonzero, but FIR systems normally contain no poles.

3.5.9 Impulse response estimation

In the design of discrete-time systems the need often arises to obtain values of impulse responses. For example, in FIR system design the impulse response is required to implement the system, and in IIR system design the values are required for stability analysis. The impulse response may also be used to evaluate the frequency response of the system.

The impulse response of a discrete-time system may be defined as the inverse z-transform of the system's transfer function, $H(z)$:

$$h(k) = Z^{-1}[H(z)], \quad k = 0, 1, \ldots$$

If the z-transform, $H(z)$, is available as a power series, that is

$$\begin{aligned} H(z) &= \sum_{n=0}^{\infty} h(n)z^{-n} \\ &= h(0) + h(1)z^{-1} + h(2)z^{-1} + \ldots \end{aligned} \tag{3.58}$$

the coefficients of the z-transform give directly the impulse response. For IIR systems, $H(z)$ is often expressed as a ratio of polynomials such as Equation 3.47. In this case, the IZT methods described in Section 3.3 can be used to obtain the impulse response of the system. The C language program given in the appendix may be used for this purpose.

The impulse response may also be viewed as the response of a discrete-time system to a unit impulse, $u(n)$, which has a value of 1 at $n = 0$ and a value of 0 at all other values of n. This view arises from the fact that if we make the input to the system equal to the unit impulse, $x(n) = u(n)$, the output of the system is in fact equal to $h(n)$, the system's impulse response:

$$\begin{aligned} y(n) &= \sum_{k=0}^{\infty} h(k)x(n-k) = \sum_{k=0}^{\infty} h(k)u(n-k) \\ &= h(0)u(n) + h(1)u(n-1) + h(2)u(n-2) + \ldots \\ &= h(n), \quad n = 0, 1, \ldots \end{aligned} \tag{3.59}$$

This provides a simple alternative method of computing $h(n)$ (indeed it provides another method of obtaining the inverse z-transform) as illustrated by the following example.

Example 3.13

Find the impulse response of the discrete-time filter characterized by the following z-transfer function (1) by using the power series method and (2) by applying a unit impulse to the system

$$H(z) = \frac{1 - z^{-1}}{1 + 0.5z^{-1}}$$

Solution

(1) Using the power series method, the values of the impulse response are obtained as

$$
\begin{array}{r|l}
 & 1 - 1.5z^{-1} + 0.75z^{-2} - 0.325z^{-3} \ldots \\
\hline
1 + 0.5z^{-1} & 1 - z^{-1} \\
 & \underline{1 + 0.5z^{-1}} \\
 & \quad -1.5z^{-1} \\
 & \quad \underline{-1.5 - 0.75z^{-2}} \\
 & \qquad\qquad -0.325z^{-3}
\end{array}
$$

From the quotients the impulse response values are

$$h(0) = 1, h(1) = -1.5, h(2) = 0.75, h(3) = -0.325$$

The impulse response values can of course be obtained with the aid of the C language program for the power series method given in the appendix.

(2) First, we need to obtain the difference equation of the filter from the transfer function:

$$H(z) = \frac{Y(z)}{X(z)} = \frac{1 - z^{-1}}{1 + 0.5z^{-1}}$$

Cross-multiplying and using the delay property of the z-transform we obtain the difference equation:

$$Y(z) + 0.5Y(z)z^{-1} = X(z) - X(z)z^{-1}$$
$$y(n) + 0.5y(n-1) = x(n) - x(n-1)$$

Simplifying we have

$$y(n) = x(n) - x(n-1) - 0.5y(n-1)$$

The impulse response of the filter can now be obtained by letting $x(n) = u(n)$ where

$$
\begin{aligned}
u(n) &= 1, \quad n = 0 \\
&= 0, \quad n \neq 0
\end{aligned}
$$

and assuming an initial condition of $y(-1) = 0$:

$$y(0) = 1$$
$$y(1) = x(1) - x(0) - 0.5y(0) = 0 - 1 - 0.5 = -1.5$$
$$y(2) = x(2) - x(1) - 0.5y(1) = -0.5 \times -1.5 = 0.75$$
$$y(3) = x(3) - x(2) - 0.5y(2) = -0.5 \times 0.75 = -0.325$$
$$\vdots$$

From this the impulse response values are

$$h(0) = 1,\ h(1) = -1.5,\ h(2) = 0.75,\ h(3) = -0.325$$

It is seen that both methods lead to identical results.

3.5.10 Applications in digital filter design

One of the most important applications of the z-transform in DSP is in the design and analysis of errors in digital filters, especially IIR filters. It is used extensively to determine the coefficients of digital filters and to analyse the effects of various quantization errors on digital filter performance. For example, it is well known that quantization errors are inherent in discrete-time systems when they are implemented in hardware or software owing to finite register lengths of practical processors. The z-transform provides a convenient means of analysing the effects of such errors on system performance. In particular, errors due to rounding or truncating the result of the multiplication operations indicated in the difference equations are often analysed with the aid of the z-transform. Noise analysis in discrete-time filters is discussed in more detail in Chapter 7.

Another important application of the z-transform in discrete filter design is in the representation of digital filter structures. We will discuss this in more detail here because it requires the use of the partial fraction expansion program mentioned earlier.

3.5.11 Realization structures for digital filters

Discrete-time filters are often represented in the form of block or signal flow diagrams. The diagrams are a convenient way of representing the difference equations or equivalently the transfer functions. Consider for example, a simple discrete filter with the following difference equation:

$$y(n) = x(n-1) - b_1y(n-1) + b_2y(n-2) + b_3y(n-3) \qquad \textbf{(3.60)}$$

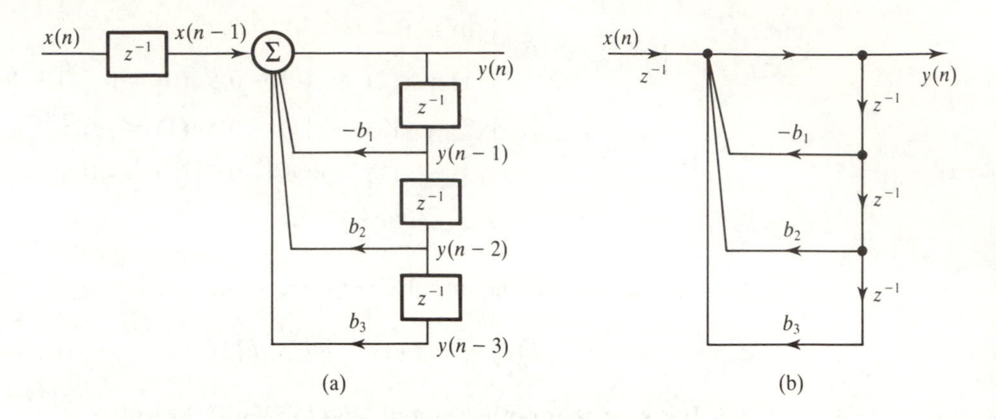

Figure 3.14 Realization diagrams for a difference equation: (a) in block diagram form; (b) signal flow diagram.

The block diagram representation of this equation is shown in Figure 3.14(a). In the figure, the symbol z^{-1} represents a delay of 1 unit of time as may be deduced from the signals at various nodes, the arrows represent multipliers and the constants next to them the multiplication factors. The relationship between the difference equation and the block diagram should be evident. A signal flow diagram representation of the same difference equation is shown in Figure 3.14(b). It is common to refer to the block or flow diagram as a realization diagram.

When the order of $H(z)$ is high, the discrete-time filter is rarely realized directly as Figure 3.14 implies because large errors will result if a small number of bits is used to represent the coefficients and to execute the difference equations (see Chapter 7). The common practice is to decompose the transer function into a cascade or parallel combination of second- and/or first-order z-transforms. For cascade realization, the transfer function, $H(z)$, is factored as

$$H(z) = H_1(z)H_2(z)\ldots H_K(z) = \prod_{i=1}^{K} H_i(z) \tag{3.61}$$

where $H_i(z)$ is either a second- or a first-order section:

$$H_i(z) = \frac{a_0 + a_{1i}z^{-1} + a_{2i}z^{-2}}{1 + b_{1i}z^{-1} + a_{2i}z^{-2}} \qquad \text{second order}$$

$$H_i(z) = \frac{a_0 + a_{1i}z^{-1}}{1 + b_{1i}z^{-1}} \qquad \text{first order}$$

and K is the integer part of $(M + 1)/2$. The overall z-transform is the product of the individual z-transforms: see Figure 3.15.

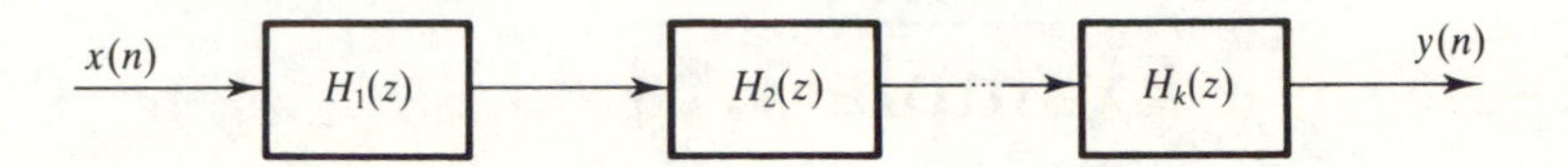

Figure 3.15 General structure for cascade realization.

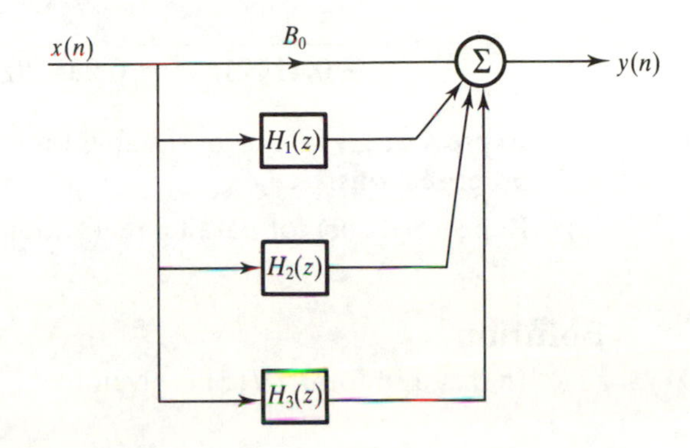

Figure 3.16 General structure for parallel realization.

For parallel realization, the transfer function is decomposed, using partial fractions, to give

$$H(z) = B_0 + \sum_{i=1}^{K} H_i(z) \tag{3.62}$$

where, as above, $H_i(z)$ is either a second-order or a first-order section but of the form

$$H_i(z) = \frac{a_{0i} + a_{1i}z^{-1}}{1 + b_{1i}z^{-1} + b_{2i}z^{-2}} \qquad \text{second order}$$

$$H_i(z) = \frac{a_0}{1 + b_{1i}z^{-1}} \qquad \text{first order}$$

where K is the integer part of $(M + 1)/2$ and

$$B_0 = a_N/b_M$$

Figure 3.16 shows the general structure for the parallel realization.

In the design of digital filters, software packages are normally used to obtain the coefficients, a_{ki} and b_{ki}, above. Unfortunately, most software packages produce coefficients only for the cascade structure. The coefficients for the parallel structure can be obtained from those of the cascade structure by using a partial fraction expansion. We will illustrate this by an example.

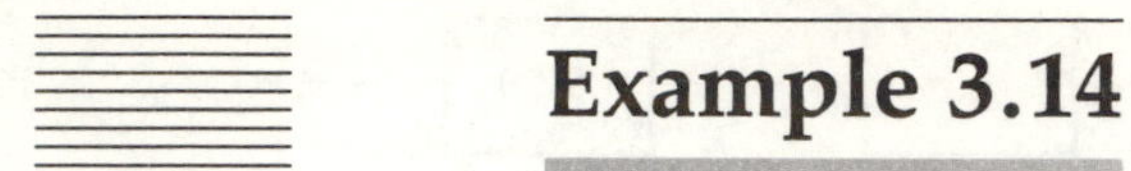

Example 3.14

A discrete-time system is characterized by the following z-transfer function:

$$H(z) = \frac{1 - 2z^{-2} + z^{-4}}{1 - 0.41421z^{-1} + 0.08579z^{-2} + 0.292895z^{-3} + 0.5z^{-4}}$$

(1) Express $H(z)$ in a form suitable for cascade realization using two second-order sections.

(2) Repeat part (1) for parallel realization.

Solution

(1) In factored form, $H(z)$ is given by

$$H(z) = H_1(z)H_2(z)$$

where

$$H_1(z) = \frac{1 - 2z^{-1} + z^{-2}}{1 - 1.41421z^{-1} + z^{-2}} \tag{3.63a}$$

$$H_2(z) = \frac{1 + 2z^{-1} + z^{-2}}{1 + z^{-1} + 0.5z^{-2}} \tag{3.63b}$$

(2) To express $H(z)$ in a form suitable for parallel realization, we first expand it using partial fraction methods. Thus

$$H(z) = B_0 + \frac{C_1}{z - p_1} + \frac{C_2}{z - p_2} + \frac{C_3}{z - p_3} + \frac{C_4}{z - p_4} \tag{3.64}$$

Using the program for the PFE given in Appendix 3B the poles, p_1 to p_4, and coefficients, B_0 and C_1 to C_4, are obtained as

$p_1 = 0.7071 + 0.7071\mathrm{j} = \mathrm{e}^{\mathrm{j}0.785}$; $\quad p_2 = p_1^*$

$p_3 = -0.5 + 0.5\mathrm{j} = 0.7071\mathrm{e}^{\mathrm{j}2.35619}$; $\quad p_4 = p_3^*$

$B_0 = 2$

$C_1 = 0.114383 + 0.666669\mathrm{j} = 0.6764104\angle 1.400877$; $\quad C_2 = C_1^*$

$C_2 = -0.61438276 - 0.58088079\mathrm{j} = 0.8455108976\angle 3.898969$; $\quad C_4 = C_3^*$

where the angles are in radians. Having found the poles and coefficients, C_k and B_0, the partial fractions in Equation 3.64 must be combined such that $H(z)$ is a sum of second-order sections of the form

$$H(z) = B_0 + \sum_{i=1}^{2} H_i(z) \tag{3.65a}$$

where

$$H_i(z) = \frac{a_0 + a_{1i}z^{-1}}{1 + b_{1i}z^{-1} + b_{2i}z^{-2}} \tag{3.65b}$$

To ensure that the coefficients a_{ki} and b_{ki} in Equation 3.65b are real, the partial fractions in Equation 3.64 involving C_1 and C_2 must be combined since they are complex conjugate pairs. Similarly the fractions with coefficients C_3 and C_4 must also be combined. Combining the fractions involving C_1 and C_2 we have

$$\frac{C_1 z}{z - p_1} + \frac{C_2 z}{z - p_2} = \frac{(C_1 + C_2)z^2 - (C_1 p_2 + C_2 p_1)z}{z^2 - (p_1 + p_2)z + p_1 p_2} \tag{3.66}$$

$$= \frac{C_1 + C_2 - (C_1 p_2 + C_2 p_1)z^{-1}}{1 - (p_1 + p_2)z^{-1} + p_1 p_2 z^{-2}} \tag{3.67}$$

Comparing Equations 3.65b and 3.67, with $i = 1$ in Equation 3.65b, we find that

$$\begin{aligned} a_{01} &= C_1 + C_2, & a_{11} &= -(C_1 p_2 + C_2 p_1) \\ b_{11} &= -(p_1 + p_2), & b_{21} &= p_1 p_2 \end{aligned} \tag{3.68}$$

If we make use of the fact that $p_2 = p_1^*$, $C_2 = C_1^*$, and substitute the values of p_1 and C_1, then

$$\begin{aligned} a_{01} &= C_1 + C_1^* = 2 \times 0.114\,383 = 0.2288 \\ a_{11} &= -(C_1 p_1^* + C_1^* p_1) \\ &= -(|C_1|e^{j\theta_1}|p_1|e^{-j\phi_1} + |C_1|e^{-j\theta_1}|p_1|e^{j\phi_1}) \\ &= -|C_1||p_1|[e^{j(\theta_1-\phi_1)} + e^{j(\theta_1-\phi_1)}] \\ &= -2|C_1||p_1|\cos(\theta_1 - \phi_1) \\ &= -2 \times 0.676\,4104 \times 1\cos(1.400\,877 - 0.785\,400\,68) = -1.1046 \end{aligned} \tag{3.69}$$

(where $\theta_1 = \angle C_1$, $\phi_1 = \angle p_1$). Thus we have

$$H_1(z) = \frac{0.2288 - 1.1046z^{-1}}{1 - 1.4142z^{-1} + z^{-2}} \tag{3.70}$$

where the values of the denominator coefficients have been taken straight from Equation 3.63. Similarly, from the partial fractions involving C_3 and C_4 we have

$$H_2(z) = \frac{-1.2288 - 0.0335z^{-1}}{1 + z^{-1} + 0.5z^{-2}} \tag{3.71}$$

Combining the results,

$$H(z) = 2 + \frac{0.2288 - 1.1046z^{-1}}{1 - 1.4142z^{-1} + z^{-2}} + \frac{-1.2288 - 0.0335z^{-1}}{1 + z^{-1} + 0.5z^{-2}}$$

Although the above process is straightforward, it is very tedious and mistakes are likely, especially if the partial fraction coefficients are worked out by hand. We give in Appendix 3B a C language routine that may be used to obtain the coefficients of the parallel structure, given the transfer function in cascade form. The program is really a simple extension of the partial fraction expansion routine in the same appendix. In Chapter 7, we will discuss in some detail applications of cascade and parallel realization structures.

3.6 Summary

A knowledge of the z-transform is very important in DSP work, as it is an invaluable tool for representing, analysing and designing discrete-time systems.

We have shown how to evaluate the z-transform of discrete-time sequences and how to recover the sequences from their z-transforms. Several C language programs are provided to enable readers gain a practical understanding of the concepts and applications of the z-transform in signal processing. Try to use them whenever possible.

Problems

3.1 Find the z-transform of each of the following discrete-time sequences:

(1) $x(n) = \sin(n\omega T), \quad n = 0, 1, \ldots$

(2) $x(n) = a^n, \quad n > 0$
$\quad = 0, \quad n < 0$

(3) $x(n) = 1, \quad 0 \leqslant n \leqslant N - 1$
$\quad = 0, \quad \text{elsewhere}$

3.2 An exponential sequence is defined as

$$x(n) = e^{-an}, \quad n \geqslant 0$$

Find its z-transform, including the constraint on z for the z-transform to converge, for each of the following cases:

(1) k is real;

(2) k is complex.

3.3 Given the causal sequences, $x(n)$ and $nx(n)$, with z-transforms $X(z)$ and $X'(z)$, show that

$$X'(z) = -z \frac{\mathrm{d}X(z)}{\mathrm{d}z}$$

3.4 The z-transform of a discrete-time sequence is given by

$$X(z) = \sum_{n=0}^{\infty} x(n) z^{-n}$$

Starting from the above equation show, stating any assumptions made, that the inverse z-transform is given by

$$x(n) = \frac{1}{2\pi \mathrm{j}} \oint z^{n-1} X(z)\,\mathrm{d}z, \quad n > 0$$

Discuss, briefly, the role of the residue theorem in evaluating the above integral.

3.5 (1) Find, using the power series method, the first five values of the causal discrete-time sequence corresponding to each of the following z-transforms:

(a) $X(z) = \dfrac{z - 1}{(z - 0.7071)^2}$

(b) $X(z) = \dfrac{1}{(z - 0.5)(z + 0.9)^3}$

(c) $X(z) = \dfrac{z^4 - 1}{z^4 - 1}$

(d) $X(z) = \dfrac{z^3 - z^2 + z - 1}{(z + 0.9)^3}$

(2) Repeat part (1) using the partial fraction expansion method.

(3) Repeat part (1) using the residue method.

3.6 (1) Given a z-transform, $X(z)$, in the form

$$X(z) = \frac{N(z)}{D(z)}$$

where $N(z)$ and $D(z)$ are polynomials, and assuming that $X(z)$ has a pole at $z = p_k$, show that

$$\mathrm{Res}[X(z), p_k] = \frac{N(p_k)}{D'(p_k)}$$

where

$$D'(p_k) = \frac{\mathrm{d}D(z)}{\mathrm{d}z}$$

(2) Use this result to find the inverse z-transform of the following:

$$X(z) = \frac{1}{z^4 - 1}$$

3.7 Given the following transfer function for a stable, causal system, find its impulse response, $h(n)$, in closed form using the residue method:

$$H(z) = \frac{1}{1 + b_1 z^{-1} + b_2 z^{-2}}$$

Assume the poles are distinct and complex.

3.8 The partial fraction expansion of an Nth-order discrete-time system is given by

$$X(z) = \frac{N(z)}{D(z)} = B_0 + \sum_{k=1}^{N} \frac{C_k z}{z - p_k z^k}$$

where

$$N(z) = a_0 + a_1 z^{-1} + a_2 z^{-2} + \ldots + a_N z^{-N}$$

$$D(z) = b_0 + b_1 z^{-1} + b_2 z^{-2} + \ldots + b_M z^{-M}$$

p_k are the poles of $X(z)$ (assumed distinct) and C_k are the partial fraction coefficients. Find a general expression for C_k, $k = 1, 2, \ldots$, in terms of the poles, the p_k, and $N(z)$. Assuming that $N = 3$ show, by long division, that B_0 is given by

$$B_0 = a_3/b_3$$

3.9 Given the difference equation

$$y(n) + B_1 y(n-1) + B_2 y(n-2) = A, \quad n > 0$$

where A, B_1 and B_2 are arbitrary constants, find an expression for the z-transform, $Y(z)$. Use an appropriate inverse z-transform technique to obtain a closed form expression for $y(n)$.

3.10 A second-order discrete-time system is characterized by the following z-transfer function:

$$H(z) = \frac{1}{(z - 0.9)^2}, \quad |z| > 0.9$$

Obtain the corresponding discrete-time sequence, $h(n)$, using the residue method.

3.11 For the discrete-time system shown in Figure 3.17, obtain the difference equation relating the output, $y(n)$, and the input, $x(n)$. Derive its transfer function, $H(z)$.

3.12 The transfer function of a discrete-time system has poles at $z = 0.5$, $z = 0.1 \pm j0.2$ and a third-order zero at $z = -1$ and $z = 1$.

(1) Sketch the pole–zero diagram for the system.

(2) Derive the system transfer function, $H(z)$, from the pole–zero diagram.

(3) Develop the difference equation.

(4) Draw the realization diagram in signal flowgraph form.

3.13 The signal flow diagram of a discrete-time system is shown in Figure 3.18. Obtain the two-step difference equation relating the output, $y(n)$, and input, $x(n)$. Derive the transfer function, $H(z)$, from the difference equation.

3.14 The frequency response specification for a bandpass discrete-time filter in normalized form is as follows

passband	0.4π–0.6π
stopbands	0–0.3π and 0.7π–π
sampling interval	$T = 100\ \mu s$

(1) Express the specifications in rad s^{-1} (denormalized).

(2) Convert the specification from rad s^{-1} to standard units of Hz.

(3) Convert the specifications from the units of hertz in part (2) to normalized form.

(4) Sketch the frequency responses for each of the three cases above in the interval from 0 to the sampling frequency.

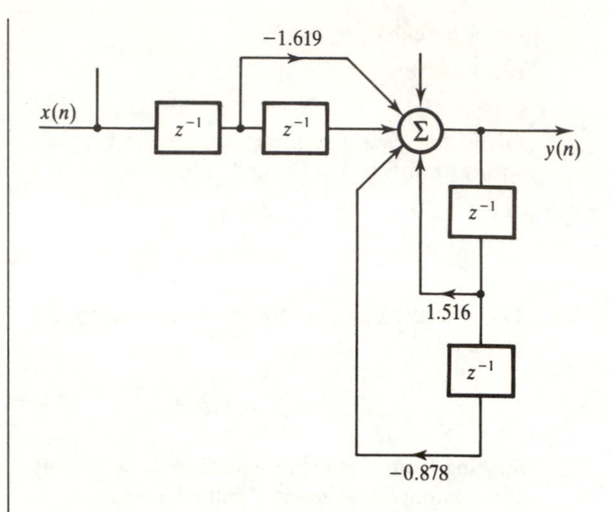

Figure 3.17 Block diagram for the discrete-time system of Problem 3.11.

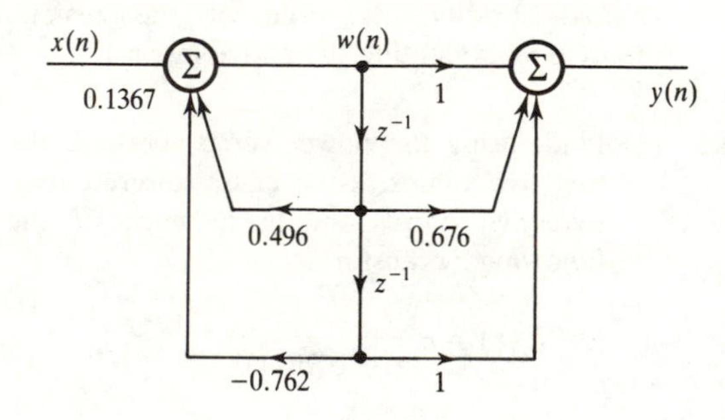

Figure 3.18 The signal flow diagram of the discrete-time system for Problem 3.13.

3.15 An LTI system is characterized by the following z-transform:

$$\frac{1 + z^{-2}}{1 + 0.81z^{-2}}$$

Determine the frequency response at dc, 1/4 and 1/2 the sampling frequency. Sketch the frequency response in the interval $0 \leqslant \omega \leqslant \omega_S$, where ω_S is the sampling frequency in rad s^{-1}.

3.16 A requirement exists for a simple lowpass discrete-time filter with the following specifications:

cutoff frequency	1 kHz
sampling frequency	10 kHz

Specify and sketch a suitable pole–zero diagram for the filter.

Obtain the transfer function of the filter from the pole–zero diagram. Determine the amplitude and phase response at 1 kHz, 2.5 kHz and 5 kHz. Sketch the amplitude–frequency response.

3.17 The transfer function of a certain system is defined as

$$H(z) = \frac{(1-1.094\,621z^{-1}-z^{2})(1-0.350\,754z^{-1}+z^{-2})}{(1-1.340\,228z^{-1}+0.796\,831z^{-2})(1-0.5z^{-1}-0.5z^{-2})}$$

(1) Find the poles and zeros, and sketch the pole–zero diagram.

(2) State, with justification, whether the system is stable or not.

The following are computer-based problems:

3.18 The z-transform of a discrete-time system is given by

$$H(z) = \sum_{k=0}^{8} a_k z^{-k} \Bigg/ \sum_{k=0}^{8} b_k z^{-k}$$

where

$a_0 = 2.740\,584\times10^{-2}$ $\quad b_0 = 1$

$a_1 = 2.825\,341\times10^{-3}$ $\quad b_1 = 2.233\,030\times10^{-1}$

$a_2 = -2.932\,353\times10^{-2}$ $\quad b_2 = 2.353\,762$

$a_3 = 3.563\,199\times10^{-4}$ $\quad b_3 = 4.369\,285\times10^{-1}$

$a_4 = 4.924\,136\times10^{-2}$ $\quad b_4 = 2.712\,411$

$a_5 = 3.563\,226\times10^{-4}$ $\quad b_5 = 3.571\,619\times10^{-1}$

$a_6 = -2.932\,353\times10^{-2}$ $\quad b_6 = 1.593\,957$

$a_7 = 2.825\,337\times10^{-3}$ $\quad b_7 = 1.141\,820\times10^{-1}$

$a_8 = 2.740\,582\times10^{-2}$ $\quad b_8 = 4.143\,201\times10^{-1}$

Obtain and plot the impulse response of the system, with the aid of the inverse z-transform program (power series method) given in the appendix. From your plot, state whether the system is stable, marginally stable or unstable.

3.19 The transfer function of a third-order IIR system, in factored form, is given by

$$H(z) = \frac{N_1(z)N_2(z)}{D_1(z)D_2(z)}$$

where

$$N_1(z) = 1 - 0.971\,426z^{-1} + z^{-2}$$

$$N_2(z) = 1 + z^{-1}$$

$$D_1(z) = 1 - 0.935\,751z^{-1} + 0.726\,879z^{-2}$$

$$D_2(z) = 1 + 0.183\,11z^{-1}$$

(1) Draw a realization diagram for the system, for a cascade structure, using a second-order and one first-order section.

(2) Express $H(z)$ as the sum of partial fractions,

$$H(z) = B_0 + \sum_{k=1}^{3} \frac{C_k}{z - p_k}$$

and use the partial fraction expansion program in Appendix 3C to compute the coefficients, B_0 and C_k.

(3) Combine the partial fractions such that the system can be realized in parallel using a second-order and one first-order section.

(4) Draw a realization diagram for the parallel structure using the result from (3).

3.20 A digital notch filter is characterized by the following z-transform:

$$\frac{z^2 + 1}{z^2 + r^2}$$

Estimate the frequency response, using the FFT approach and a sampling frequency of 1 kHz, for each of the following cases: (i) $r = 0.8$, (ii) $r = 0.95$, (iii) $r = 1$. Explain your results.

3.21 A lowpass discrete-time filter has the following transfer function:

$$H(z) = \frac{a_0 + a_1z^{-1} + a_2z^{-2} + \ldots + a_4z^{-4}}{b_0 + b_1z^{-1} + b_2z^{-2} + \ldots + b_4z^{-4}}$$

where

$a_0 = 0.193\,441$ $\quad b_0 = 1$

$a_1 = 0.378\,331$ $\quad b_1 = -2.516\,884$

$a_2 = 0.524\,14$ $\quad b_2 = 1.054\,118$

$a_3 = 0.378\,331$ $\quad b_3 = -0.240\,603$

$a_4 = 0.193\,441$ $\quad b_4 = 0.198\,5861$

Estimate the frequency response of the filter using

(1) the program for direct frequency response evaluation discussed in Appendix 3C, and

(2) the power series method and FFT.

Compare the two results and comment on any differences.

3.22 The coefficients of a simple bandpass FIR system are listed in Table 3.4. Assuming a sampling frequency of 10 kHz, compute the magnitude–frequency response of the system using the program discussed in Appendix 3C.

3.23 Given the z-transform

$$H(z) = \frac{1 + 3z^{-1} + z^{-2} + z^{-3}}{1 + (1-k)z^{-1} + (k + 0.3561)z^{-2} + 0.3561k}$$

use the computer program for the power series expansion method in Appendix 3B to compute a sufficient number of values of the system's impulse response for each of the following cases:

(1) $k = -1$;

(2) $k = 1$;

(3) $k = 2$;

(4) $k = 0.9$.

For each case, plot the impulse response and from this state whether the system is stable, marginally stable or unstable.

Table 3.4 Coefficients of the FIR bandpass filter for Problem 3.22.

H (1) =	−0.67299600E−02 =	H(35)
H (2) =	0.16799420E−01 =	H(34)
H (3) =	0.17195700E−01 =	H(33)
H (4) =	−0.27849080E−01 =	H(32)
H (5) =	−0.17486810E−01 =	H(31)
H (6) =	0.13515580E−01 =	H(30)
H (7) =	0.45570510E−02 =	H(29)
H (8) =	0.33293060E−01 =	H(28)
H (9) =	0.95162150E−02 =	H(27)
H(10) =	−0.68548560E−01 =	H(26)
H(11) =	−0.68992230E−02 =	H(25)
H(12) =	0.23802370E−01 =	H(24)
H(13) =	−0.11597510E−01 =	H(23)
H(14) =	0.12073780E+00 =	H(22)
H(15) =	0.23806900E−01 =	H(21)
H(16) =	−0.29095690E+00 =	H(20)
H(17) =	−0.12362380E−01 =	H(19)
H(18) =	0.36717700E+00 =	H(18)

3.24 Develop a C language program for checking the result of partial fraction expansion. The program should accept as inputs the values of the poles p_k $(k = 1, 2, \ldots, M)$ and the associated coefficients C_k $(k = 1, 2, \ldots, M)$, and produce as outputs the coefficients of the polynomials $A(z)$ and $B(z)$, where

$$X(z) = \frac{A(z)}{B(z)}$$

and

$$A(z) = a_0 + a_1z^{-1} + a_2z^{-2} + \ldots + a_Nz^{-N}$$

$$B(z) = b_0 + b_1z^{-1} + b_2z^{-2} + \ldots + b_Mz^{-M}$$

Extend your program to check the results of converting cascade structures to parallel.

References

Atkinson L.V. and Harley P.J. (1983). *An Introduction to Numerical Methods with Pascal*, Chapter 3. Wokingham: Addison-Wesley

Jury E.I. (1964). *Theory and Applications of the z-transform Method*. New York NY: Wiley

Mathews J.H. (1982). *Basic Complex Variables for Mathematics and Engineering*. Boston MA: Allyn and Bacon

Proakis J.G. and Manolakis D.G. (1992). *Digital Signal Processing* 2nd edn. New York NY: Macmillan

Bibliography

Ahmed N. and Natarajan T. (1983). *Discrete-time Signals and Systems*. Reston VA: Reston Publishing Co. Inc

Churchhill R.V., Brown J.W. and Verhey R.F. (1976). *Complex Variables and Applications*. New York NY: McGraw-Hill

Jong M.T. (1982). *Methods of Discrete Signal and Systems Analysis*. New York NY: McGraw-Hill

Oppenheim A.V. and Schafer R.W. (1975). *Digital Signal Processing*. Englewood Cliffs NJ: Prentice-Hall

Rabiner L.R. and Gold B. (1975). *Theory and Application of Digital Signal Processing*. Englewood Cliffs NJ: Prentice-Hall

Ragazzini J.R. and Zadeh L.A. (1952). Analysis of sampled data systems. *Trans. AIEE*, **71**(II), 225–34

Steiglitz K. (1974). *An Introduction to Discrete Systems*. New York NY: Wiley

Strum R.D. and Kirk D.E. (1988). *First Principles of Discrete Systems and Digital Signal Processing*. Reading MA: Addison-Wesley

Appendices

3A Recursive algorithm for the inverse z-transform

It was stated in the main text that the long division method can be recast in a recursive form. Specifically, we want to show here that given a z-transform, $X(z)$, such that

$$X(z) = \frac{a_0 + a_1 z^{-1} + a_2 z^{-2}}{b_0 + b_1 z^{-1} + b_2 z^{-2}}$$

the inverse z-transform, $x(n)$, can be obtained as (Jury, 1964)

$$x(n) = \frac{1}{b_0}\left[a_n - \sum_{i=1}^{n} x(n-i)b_i\right], \quad n = 1, 2, \ldots$$

$$x(0) = \frac{a_0}{b_0}$$

The result can be generalized. Using long division, we can express $X(z)$ as a power series as follows:

$$\frac{a_0}{b_0} + \left[\left(a_1 - \frac{a_0}{b_0}b_1\right)\Big/b_0\right]z^{-1} + \frac{1}{b_0}\left[\left(a_2 - \frac{a_0}{b_0}b_2\right) - \frac{b_1}{b_0}\left(a_1 - \frac{a_0}{b_0}b_1\right)\right]z^{-2}$$

$$b_0 + b_1z^{-1} + b_2z^{-2}\ \overline{\big)\ a_0 + a_1z^{-1} + a_2z^{-2}}$$

$$a_0 + \left(\frac{a_0}{b_0}b_1\right)z^{-1} + \left(\frac{a_0}{b_0}b_2\right)z^{-2}$$

$$\overline{\left(a_1 - \frac{a_0}{b_0}b_1\right)z^{-1} + \left(a_2 - \frac{a_0}{b_0}b_2\right)z^{-2}}$$

$$\left(a_1 - \frac{a_0}{b_0}b_1\right)z^{-1} + \frac{b_1}{b_0}\left(a_1 - \frac{a_0}{b_0}b_1\right)z^{-2} + \frac{b_2}{b_0}\left(a_1 - \frac{a_0}{b_0}b_1\right)z^{-3}$$

$$\overline{\left[\left(a_2 - \frac{a_0}{b_0}b_2\right) - \frac{b_1}{b_0}\left(a_1 - \frac{a_0}{b_0}b_1\right)\right]z^{-2} - \frac{b_2}{b_0}\left(a_1 - \frac{a_0}{b_0}b_1\right)z^{-3}}$$

$$\left[\left(a_2 - \frac{a_0}{b_0}b_2\right) - \frac{b_1}{b_0}\left(a_1 - \frac{a_0}{b_0}b_1\right)\right]z^{-2} + \frac{b_1}{b_0}\left[\left(a_2 - \frac{a_0}{b_0}b_2\right)\right.$$

$$\left. - \frac{b_1}{b_0}\left(a_1 - \frac{a_0}{b_0}b_1\right)\right]z^{-3} + \frac{b_2}{b_0}\left[\left(a_2 - \frac{a_0}{b_0}b_2\right) - \frac{b_1}{b_0}\left(a_1 - \frac{a_0}{b_0}b_1\right)\right]z^{-4}$$

$$\overline{\left\{\left[-\frac{b_2}{b_0}\left(a_1 - \frac{a_0}{b_0}b_1\right)\right] - \frac{b_1}{b_0}\left[\left(a_2 - \frac{a_0}{b_0}b_2\right)\right.\right.}$$

$$\left.\left. - \frac{b_1}{b_0}\left(a_1 - \frac{a_0}{b_0}b_1\right)\right]\right\}z^{-3} - \frac{b_2}{b_0}\left[\left(a_2 - \frac{a_0}{b_0}b_2\right)\right.$$

$$\left. - \frac{b_1}{b_0}\left(a_1 - \frac{a_0}{b_0}b_1\right)\right]z^{-4}$$

$$\vdots$$

The quotient of the long division gives the coefficients of the power series:

$$X(z) = \frac{a_0}{b_0} + \left[\left(a_1 - \frac{a_0}{b_0}b_1\right)\Big/b_0\right]z^{-1} + \frac{1}{b_0}\left[\left(a_2 - \frac{a_0}{b_0}b_2\right)\right.$$

$$\left. - \frac{b_1}{b_0}\left(a_1 - \frac{a_0}{b_0}b_1\right)\right]z^{-2}$$

$$+ \frac{1}{b_0}\left\{\left[-\frac{b_2}{b_0}\left(a_1 - \frac{a_0}{b_0}b_1\right)\right] - \frac{b_1}{b_0}\left[\left(a_2 - \frac{a_0}{b_0}b_2\right)\right.\right.$$

$$\left.\left. - \frac{b_1}{b_0}\left(a_1 - \frac{a_0}{b_0}b_1\right)\right]\right\}z^{-3} + \ldots$$

From the definition of the z-transform for causal systems, $X(z)$ is given by

$$X(z) = \sum_{n=0}^{\infty} x(n)z^{-n} = x(0) + x(1)z^{-1} + x(2)z^{-2} + \ldots$$

Thus, we can write

$$x(0) = \frac{a_0}{b_0}$$

$$x(1) = \frac{1}{b_0}\left(a_1 - \frac{a_0}{b_0}b_1\right) = \frac{1}{b_0}[a_1 - x(0)b_1]$$

$$x(2) = \frac{1}{b_0}\left[\left(a_2 - \frac{a_0}{b_0}b_2\right) - \frac{b_1}{b_0}\left(a_1 - \frac{a_0}{b_0}b_1\right)\right]$$

$$= \frac{1}{b_0}[a_2 - x(0)b_2 - b_1x(1)]$$

$$x(3) = \frac{1}{b_0}\left[\left\{-\frac{b_2}{b_0}\left(a_1 - \frac{a_0}{b_0}b_1\right)\right\} - \frac{b_1}{b_0}\left\{\left(a_2 - \frac{a_0}{b_0}b_2\right) - \frac{b_1}{b_0}\left(a_1 - \frac{a_0}{b_0}b_1\right)\right\}\right]$$

$$= \frac{1}{b_0}[-b_2x(1) - b_1x(2)]$$

In general, we can write

$$x(n) = \frac{1}{b_0}\left[a_n - \sum_{i=1}^{n} x(n-i)b_i\right], \quad n = 1, 2, \ldots$$

$$x(0) = a_0/b_0$$

3B C program for evaluating the inverse z-transform and for cascade-to-parallel structure conversion

The program described here computes the inverse z-transform using the power series or the partial fraction expansion method. The program can also be used to convert a discrete-time system transfer function, $H(z)$, from cascade to parallel structure. First, we will summarize the principles on which the program is based and how to use it for each of the above purposes. The program, together with other programs in the book, is available on disk (see Preface).

The program is written in Borland Turbo C, version 2.01, and runs on IBM PC or compatible. It is fairly large and so for efficiency is organized into two program modules, izt.c and dsplib.c, held in separate files, which can be compiled separately and then linked.

The file izt.c contains the main function where the program execution starts. This file is listed in Program 3B.1. The file ltilib.c is essentially a collection of commonly used functions necessary for various discrete-time LTI operations. This file is also used by other programs in the book. The file is listed in Program 3B.2. Three header files (dsp1.h, dsp21.h, dsp21ex.h) define the global variables, structures and functions for the two files. These are listed in Program 3B.3.

Program 3B.1

```
/* ............................................................................ */
/*                                                                              */
/*  Program name: izt.c                                                         */
/*                                                                              */
```

```
/* program for:                                                                 */
/* (1) computing the inverse z-transform via the power series or               */
/*     partial fraction expansion method                                        */
/*                                                                              */
/* (2) converting a transfer function, H(z), in cascade form to an             */
/*     equivalent transfer function in parallel, via partial fraction          */
/*     expansion. The basic building block is the second order biquad          */
/*                                                                              */
/*        input file:          cascade.dat                                      */
/*        output file:         xdata.dat                                        */
/*                                                                              */
/*        Manny Ifeachor, June, 1992.                                           */
/* .......................................................................... */

#include          "dsp1.h"       /* common structures + include files          */
#include          "dsp21.h"      /* global function + variable declarations    */

/* .......................................................................... */
main()
{
extern double     A[size], B[size], Ni[size], Di[size];
extern int        M, N, N1, M1, nstage, iopt, iir;
extern long       npt;
extern FILE       *in, *out, *fopen();

      iir=0;
      read_coeffs();                          /* go read the system coeffs */
      poly_product1(Di,B,nstage);             /* form B(z) */
      poly_product1(Ni,A,nstage);             /* form A(z) */
      M1=M; N1=N;
      if(B[M]==0){                            /* system of odd order */
            M1=M-1;
            N1=M1;
      }
      printf("select desired operation\n");
      printf("0     for power series method of IZT\n");
      printf("1     for partial fraction coeffs estimation\n");
      printf("2     for cascade to parallel conversion\n");
      scanf("%d",&iopt);
      switch(iopt){
            case 0:
                  printf("enter number of data points required\n");
                  scanf("%ld",&npt);
                  power_series();
                  izt_output();
                  break;
            case 1:
                  pfraction();
                  print_pfcoeffs();
                  break;
            case 2:
                  pfraction();                /* compute PF coeffs */
                  cascade_parallel();         /* compute parallel coeffs for H(z)*/
                  print_pfcoeffs();           /* print PF coeffs */
                  printpar();                 /* print parallel coeffs for H(z) */
```

```
                break;
            default:      break;
        }
        exit(0);
}

/* .................................................................................................... */

/* function to evaluate the partial fraction coefficients                                               */

void    pfraction()
{
        int i;
        complex nz[10], dz[10];
        extern double A[size], B[size];
        extern complex pk[10], ck[10];
        extern double B0;
        extern int M, N1, M1, nstage;

        B0=A[M1]/B[M1];                          /* compute constant coeff */

            polar__pole();                       /* find positions of poles p1, p2, p3 */
            for(i=1;i<=M1;++i){
                nz[i]=poly__polar(A,pk[i],M1);
                                                 /* evaluate N(z) at p1 */
                dz[i]=Dz(pk[i],i);               /* evaluate D'(z) at p1 */
                ck[i]=cdiv1(nz[i],dz[i]);        /* obtain c1=N(z)/D'(z), z=p1 */
            }
}
/* .................................................................................................... */

/* function to evaluate the denominator polynomial D'(z) */

complex Dz(complex px, int npfc)
{
int i, j;
complex dk[10], dz;
extern complex p[20];

pole__array(npfc);
                                      /* evaluate Dk(z)=z(z-p[2]...(z-p[M]), z=px */
        if(M1==2){
            dz=Dzz(px,px,p[1]);
        }
        if(M1>2){
            j=1;
            dk[1]=Dzz(px,px,p[1])];
            for(i=2;i<(M1-1);++i){
                dk[i]=Dzz(dk[i-1],px,p[i]);
                j=i;
            }
            dz=Dzz(dk[j],px,p[j+1]);
        }
        return(dz);
}
```

```
/* ......................................................................... */

/* function to evaluate factors of the form z(z-p) */

complex Dzz(complex pa, complex pb, complex pc)
{
        complex t1, t2, dz;
        t1=cmul1(pa,pb);
        t2=cmul1(pa,pc);
        dz=cadd1(t1,t2,0);
        return(dz);
}

/* ......................................................................... */

/* function to re-order the poles */

void    pole_array(npfc)
{
        int i;

        switch(npfc){
        case 1:                                         /* pole array for c1*/
                for(i=1;i<=9;++i){
                        p[i]=pk[i+1];
                }
                break;
        case 2:                                         /* pole array for c2 */
                p[1]=pk[1];
                for(i=2;i<=9;++i){
                        p[i]=pk[i+1];
                }
                break;
        case 3:                                         /* pole array for c3 */
                p[1]=pk[1];
                p[2]=pk[2];
                for(i=3;i<=9;++i){
                        p[i]=pk[i+1];
                }
                break;
        case 4:
                p[1]=pk[1];
                p[2]=pk[2];
                p[3]=pk[3];
                for(i=4;i<=9;++i){
                        p[i]=pk[i+1];
                }
                break;
        case 5:
                for(i=1;i<=4;++i){
                        p[i]=pk[i];
                }
                for(i=5;i<=9;++i){
                        p[i]=pk[i+1];
                }
```

```
            break;
        case 6:
            for(i=1;i<=5;++i){
                p[i]=pk[i];
            }
            for(i=6;1<=9;++i){
                p[i]=pk[i+1];
            }
            break;
        case 7:
            for(i=1;i<=6;++i){
                p[i]=pk[i];
            }
            for(i=7;i<=9;++i){
                p[i]=pk[i+1];
            }
            break;
        case 8:
            for(i=1;i<=7;++i){
                p[i]=pk[i];
            }
            p[8]=pk[9];
            p[9]=pk[10];
            break;
        case 9:
            for(i=1;i<=8;++i){
                p[i]=pk[i];
            }
            p[9]=pk[10];
            break;
        case 10:
            for(i=1;i<=9;++i){
                p[i]=pk[i];
            }

            break;
        }
}

/* ........................................................................................................ */

/* function to print the partial fraction coefficients */

void    print_pfcoeffs()
{
int     k;
long    i;
extern  int M, M1;
extern  complex ck[10], pk[10];
extern  double B0;

        printf("poles of the z-transform\n");
        printf("\n");
        printf("pk  \treal  \t\timag  \t\tmag  \t\tphase\n");
        for(k=1;k<=M1;++k){
            pk[k].angle=pk[k].angle*180/pi;
```

```
			i=(long)(pk[k].angle/360)]
			pk[k].angle=pk[k].angle-i*360;
			if(pk[k].angle<-180)
				pk[k].angle=pk[k].angle+360;
			if(pk[k].angle>180)
				pk[k].angle=pk[k].angle-360;
			printf("%d \t%f \t%f \t%f \t%f\n",k,pk[k].real,pk[k].imag,pk[k].modulus,pk[k].angle);
		}
		printf("\n");
		printf("\n");
		printf("partial fraction coeffs \n");
		printf("\n");
		printf("B0=%f \n",B0);
		printf("\n");
		printf("Ck  \treal  \t\timag  \t\tmag  \t\tphase\n");
		for(k=1; k<=M1; ++k){
			ck[k].angle=ck[k].angle*180/pi;
			i=(long)(ck[k].angle/360);
			ck[k].angle=ck[k].angle-i*360;
			if(ck[k].angle<-180)
				ck[k].angle=ck[k].angle+360;
			if(ck[k].angle>180)
				ck[k].angle=ck[k].angle-360;
			printf("%d \t%f \t%f \t%f \t%f\n",k,ck[k].real,ck[k].imag,ck[k].modulus,ck[k].angle);
		}
		printf("press enter to continue \n");
		getch();
}
/* ........................................................................................................ */

/* Function to compute coefficients for parallel realization */

void	cascade_parallel()
{
	int i;
	extern double ak[size], bk[size];
	extern complex pk[10], ck[10];
	extern double B0;
	extern int M;

		for(i=0; i<M; ++i){
			ncoeff(i);				/* compute coeffs for 2nd order sections */
		}
}
/* ........................................................................................................ */

void printpar()
{
	int		i, j;
	extern		double ak[size];
	extern		int M;

	printf("stage	\tNi(z)\n");
	printf("\n");
	for(i=0; i<=M/2; ++i){
```

```
        j=2*i;
        printf("%d  \t%f  \t%f\n",i,ak[j],ak[j+1]);
    }
    j=0;
    printf("\n");
    printf("\n");
    printf("stage  \tDi(z)\n");
    printf("\n");
    for(i=0;i<=M/2;++i){
        printf("%d  \t%f  \t%f  \t%f\n",i,Di[j],Di[j+1],Di[j+2]);
        j=j+3;
    }

    printf("press enter to continue \n");
    getch();

}
/* ........................................................................................ */

/* function to compute numerator coeffs of second order sections from partial
   fraction coeffs
*/

void    ncoeff(int i)
{
        int             j;
        complex         temp1, temp2, temp3;
        extern          double ak[size];
        extern          complex pk[10], ck[10];

        j=2*i;
        temp1=cadd1(ck[j+1], ck[j+2],1);
        ak[j]=temp1.real;
        temp1=cmul1(ck[j+1],pk[j+2]);
        temp2=cmul1(ck[j+2],pk[j+1]);
        temp3=cadd1(temp1,temp2,1)];
        ak[j+1]=-temp3.real;
}

/* ........................................................................................ */
void    izt_output()
{
        long            i;
        extern          long npt;
        extern          double h[size];

        if((out=fopen("xdata.dat","w"))==NULL){
                printf("cannot open file xdata.dat\n");
                exit(1);
        }
        for(i=0;i<npt;++i){
                fprintf(out,"%15e\n",h[i]);
        }
        fclose(out);
}
```

Program 3B.2

```
/* ........................................................................................ */
/*                                                                                          */
/* program name: ltilib.c                                                                   */
/*                                                                                          */
/* This is a library of commonly used functions for various programs                        */
/* in this and other chapters                                                               */
/*                                                                                          */
/* Manny Ifeachor, June, 1992.                                                              */
/*                                                                                          */
/* ........................................................................................ */


/* ........................................................................................ */

#include    "dsp1.h"
#include    "dsp21ex.h"

/* function to read coefficient values of IIR or FIR systems                                */

void  read_coeffs()
{
        int             i, j;
        extern          int M, N, nstage, iir;
        extern          double A[size], B[size], Ni[size], Di[size];

        if((in=fopen("coeff.dat","r"))==NULL){
                printf("cannot open file coeff.dat\n");
                exit(1);
        }

        zero_arrays();                                          /* initialize all global arrays */

        if(iir==1){
                printf("specify  type of system \n");
                printf("0        for IIR system \n");
                printf("1        for FIR system \n");
                scanf("%d",&iir);
        }
        switch(iir){

        case 0:                                 /* read IIR filter coefficients for H(z) in cascade */
                fscanf(in,"%d",&nstage);
                j=0;
                for(i=0;i<nstage; ++i){
                        fscanf(in,"%lf  %lf  %lf",&Di[j],&Di[j+1],&Di[j+2]);
                        fscanf(in,"%lf  %lf  %lf",&Ni[j],&Ni[j+1],&Ni[j+2]);
                        j=j+3;
                }
                M=nstage*2; N=M;
                break;
        case 1:                                                 /* read FIR coefficients */
                fscanf(in,"%d",&N);
                for(i=0;i<N; ++i){
```

```
                fscanf(in,"%lf",&A[i]);
            }
            B[0]=1.0; M=N;
            break;
        default:
            printf("invalid option selected \n");
            printf("press enter to continue \n");
            getch();
            break;
    }
    fclose(in);
}
/* ............................................................................ */

/* function performs complex addition */

complex cadd1(complex pa, complex pb, int nadd)
{
    long k;
    complex np;
    np.real=0; np.imag=0;
    if(nadd==0){
        np.real=pa.real-pb.real;
        np.imag=pa.imag-pb.imag;
    }
    if(nadd==1){
        np.real=pa.real+pb.real;
        np.imag=pa.imag+pb.imag;
    }
    np.modulus=sqrt(np.real*np.real+np.imag*np.imag);
    np=fixnp(np);
    return(np);
}
/* ............................................................................ */

/* Function to adjust angles of complex numbers to their correct values       */

complex fixnp(complex pa)
{
    long k;
    complex np;

    np=pa;
    if(fabs(pa.real)<0.0000000008){
        np.real=0;
    }
    if(fabs(pa.imag)<0.0000000008){
        np.imag=0;
    }
    if(np.real==0 && np.imag==0){
        np.angle=0;
        return(np);
    }
    if(np.real==0 && np.imag!=0){
        if(np.image>0)
```

```
                np.angle=pi/2;
            if(np.imag<0)
                np.angle=1.5*pi;
            return(np);
        }
        if(np.real!=0 && np.imag==0){
            if(np.real>0)
                np.angle=0;
            if(np.real<0)
                np.angle=pi;
            np.modulus=fabs(pa.real);
            return(np);
        }
        np.angle=atan(np.imag/np.real);
        if(np.real<0)
            np.angle=np.angle+pi;
        k=(long)(np.angle/2*pi);
        np.angle=np.angle - k*2*pi;
        np.modulus=sqrt(pa.real*pa.real+pa.imag*pa.imag);
        return(np);
}
/* ............................................................................................ */

/* function to multiply two complex numbers */
complex cmul1(complex pa, complex pb)
{
        complex np;
        np.real=0; np.imag=0;
        np.modulus=pa.modulus*pb.modulus;
        np.real=np.modulus*cos(pa.angle+pb.angle);
        np.imag=np.modulus*sin(pa.angle+pb.angle);
        np=fixnp(np);
        return(np);
}

/* ............................................................................................ */

/* function to divide two complex numbers */
complex cdiv1(complex pa, complex pb)
{
        complex np;
        np.real=0; np.imag=0;
        np.modulus=pa.modulus/pb.modulus;
        np.real=np.modulus*cos(pa.angle-pb.angle);
        np.imag=np.modulus*sin(pa.angle-pb.angle);
        np=fixnp(np);
        return(np);
}

/* ............................................................................................ */

/* Routine to compute and express the pole of a second order section in polar and rectangular
   forms. The denominator polynomial has the form

                        D(z)=1+b1z^-1+b2z^-2.
```

```
*/
complex polar__pole()
{
        int             i, k, j=0;
        complex         px, ptemp;
        double          temp;
        extern          int nstage;
        extern          double Di[size];
        extern          complex pk[10];

        for(i=0; i<nstage; ++i){
                temp=Di[1+j]*Di[1+j]-4*Di[2+j];
                if(temp>=0){                                    /* poles are real */
                        px.imag=0; ptemp.imag=0;
                        if(Di[2+j]==0){                         /* simple poles */
                                px.real=-Di[1+j];
                                ptemp.real=0;
                        }
                        else{
                                px.real=(-Di[1+j]+sqrt(temp))/2;
                                ptemp.real=(-Di[1+j]-sqrt(temp))/2;
                        }
                        px.modulus=fabs(px.real);
                        ptemp.modulus=fabs(ptemp.real);
                        px.angle=0; ptemp.angle=0;
                        if(px.real<0)
                                px.angle=acos(-1);
                        if(ptemp.real<0)
                                ptemp.angle=acos(-1);
                }
                else{                                   /* complex conjugate poles */
                        px.modulus=sqrt(fabs(Di[2+j]));
                        px.angle=acos(-Di[1+j]/(2*px.modulus));
                        px.real=px.modulus*cos(px.angle);
                        px.imag=px.modulus*sin(px.angle);
                        ptemp.modulus=px.modulus;
                        ptemp.angle=-px.angle;
                        ptemp.real=px.real;
                        ptemp.imag=-px.imag;
                }
                k=2*i+1;
                pk[k]=px;
                pk[k+1]=ptemp;
                j=j+3;
        }
}
/* ..........................................................................................
Routine to evaluate a given polynomial, of order n, at z=pi. The result is returned as a complex
number in polar and rectangular forms
*/

complex poly__polar(double P2[], complex pj, int n)
{
        complex         np;
        int             i, k;
```

```
        i=0; np.real=0; np.imag=0;

        for(i=0; i<n; ++i){
                k=n-i;
                np.real=np.real+P2[i]*pow(pj.modulus,k)*cos(k*pj.angle);
                np.imag=np.imag+P2[i]*pow(pj.modulus,k)*sin(k*pj.angle);
        }
        np.real=np.real+P2[n];
        np.modulus=sqrt(np.real*np.real+np.imag*np.imag);
        np.angle=atan(np.imag/np.real);
        if(np.real<0)
                np.angle=np.angle+pi;
        return(np);
}

/* ..............................................................................
Function to compute the absolute magnitude of a complex number
*/

double cabs1(complex z)
{
        double temp;
        temp=z.real*z.real+z.imag*z.imag;
        temp=sqrt(temp);
        return(temp);
}

/* .............................................................................. */
/* Function to produce a polynomial in direct form */

void    poly_product1(double PI[], double AB[], int nsect)
{

        double    C[size], D1[size], D2[size];
        int       i, NN;
                if(nsect<2){
                        for(i=0; i<3; ++i){
                                AB[i]=P1[i];
                        }

                return;
                }
                D1[0]=P1[0];      /* retrieve F1(z) */
                D1[1]=P1[1];
                D1[2]=P1[2];

                D2[0]=P1[3];      /* retrieve F2(z) */
                D2[1]=P1[4];
                D2[2]=P1[5];
                NN=2;
                poly_product(C,D1,D2,NN);                  /* compute F1(z)F2(z) */


        if(nsect>2){
                D1[0]=C[0];       /* retrieve F1(z)F2(z) */
```

```
        D1[1]=C[1];
        D1[2]=C[2];
        D1[3]=C[3];
        D1[4]=C[4];
        D2[0]=P1[6];      /* retrieve F3(z) */
        D2[1]=P1[7];
        D2[2]=P1[8];
        D2[3]=0.0;
        D2[4]=0.0;
        NN=4;
        poly_product(C,D1,D2,NN);                 /* compute the product F1(z)F2(z)F3(z) */

}
if(nsect>3){

        D1[0]=C[0];       /* retrieve F1(z)F2(z)F3(z) */
        D1[1]=C[1];
        D1[2]=C[2];
        D1[3]=C[3];
        D1[4]=C[4];
        D1[5]=C[5];
        D1[6]=C[6];

        D2[0]=P1[9];      /* retrieve F4(z) */
        D2[1]=P1[10];
        D2[2]=P1[11];
        D2[3]=0.0;
        D2[4]=0.0;
        D2[5]=0.0;
        D2[6]=0.0;

        NN=6;
        poly_product(C,D1,D2,NN);                 /* compute [F1(z)F2(z)][F2(z)F4(z)] */

}
if(nsect>4){

        D1[0]=C[0];       /* retrieve F1(z)F2(z)F3(z)F4(z) */
        D1[1]=C[1];
        D1[2]=C[2];
        D1[3]=C[3];
        D1[4]=C[4];
        D1[5]=C[5];
        D1[6]=C[6];
        D1[7]=C[7];
        D1[8]=C[8];

        D2[0]=P1[9];      /* retrieve F5(z) */
        D2[1]=P1[10];
        D2[2]=P1[11];
        D2[3]=0.0;
        D2[4]=0.0;
        D2[5]=0.0;
        D2[6]=0.0;
        D2[7]=0.0;
```

```
                D2[8]=0.0;

                NN=8;
                poly_product(C,D1,D2,NN);          /* compute [F1(z)F2(z)][F2(z)F4(z)F5(z)] */

                }
                        /* save results */

                for(i=0;i<(2*nsect+1);++i){
                        AB[i]=C[i];
                }
}
/* ..........................................................................
Function to compute the product of two polynomials
*/
void    poly_product(double C[], double D1[], double D2[], int NN)
{
        int         i, j, k, j1;
        double      sum1=0.0, sum2=0.0;

        for(i=0;i<size;++i){    /* initialize the result array */
                C[i]=0.0;
        }
        for(i=0;i<(NN+1);++i){

                j=NN-i;
                j1=i+NN;
                for(k=0;k<(j+1);++k){
                        sum1=sum1+D1[k]*D2[j-k];
                        sum2=sum2+D1[i+k]*D2[NN-k];
                }
                C[j]=sum1;
                C[j1]=sum2;
                sum1=0.0;sum2=0.0;
        }
}
/* ........................................................................ */
/* Function to zero global arrays used as inputs */

void     zero_arrays()
{

        long        i;
        extern      double A[size], B[size], Ni[size], Di[size];
        extern      double ak[size], bk[size], h[size];

        for(i=0;i<size;++i){
                A[i]=0.0;
                B[i]=0.0;
                Ni[i]=0;
                Di[i]=0;
                ak[i]=0;
                bk[i]=0;
                h[i]=0;
```

```
        }
}
/* ..................................................................................... */
/* Function to compute the izt using the power series method */

void    power_series()
{
        int         i, k;
        long        n;
        double      sum1, sum2, sum3=0;
        double      temp;
        extern      double A[size], B[size], h[size];
        extern      float l1n, l2n;
        extern      long npt;
        extern      int M;

        /* compute h[n] recursively */

        h[0]=A[0]/B[0];
        sum2=h[0];
        for(n=1;n<npt;++n){
                sum1=0.0;
                k=n;
                if(n>M)
                        k=M;
                for(i=1;i<=k;++i){
                        sum1=sum1+h[n-i]*B[i];
                }
                h[n]=(A[n]-sum1)/B[0];
                temp=signx(h[n])*h[n];
                sum2=sum2+temp;
                sum3=sum3+temp*temp;
        }
        l1n=sum2;
        l2n=sqrt(sum3);
}
/* ..................................................................................... */
int     signx(double x)
{
        int temp;
        if(x<0)
                temp=-1;
        else
                temp=1;
        return(temp);
}
```

Program 3B.3

```
/*
This file contains common definitions and structures
        filename: dsp1.h
*/
```

```
#include        <stdio.h>
#include        <math.h>
#include        <dos.h>

#define         size 600
#define         pi 3.141592654
#define         maxbits 30

typedef struct  {
        double real;
        double imag;
        double modulus;
        double angle;
        }complex;

/*
                function and variable declarations

                filename: dsp21.h
*/
complex         poly_polar(double P2[], complex, int);
void            pfraction();
complex         polar_pole();
complex         cmul1(complex, complex);
complex         cdiv1(complex, complex);
complex         cadd1(complex, complex, int);
complex         Dz(complex, int);
complex         Dzz(complex, complex, complex);
complex         fixnp(complex);
void            pole_array(int);
void            poly_product(double C[], double D1[], double D2[], int);
void            poly_product1(double P1[], double AB[], int);
void            read_coeffs();
void            print_pfcoeffs();
void            printpar();
void            cascade_parallel();
void            zero_arrays();
double          cabs1(complex);
void            izt_output();
void            ncoeff(int);
double          A[size], B[size], Ni[size], Di[size];
double          ak[size], bk[size, h[size];
void            power_series();
int             signx(double);
int             M, N, N1, M1, nstage, iopt, iir;
long            npt;
complex         pk[10], ck[10], p[20];
float           l1n, l2n, fmax;
double          B0;
FILE            *in, *out, *fopen();

/*
                function declarations
                filename: dsp21ex.h
*/
```

```
complex     poly_polar(double P2[], complex, int);
complex     polar_pole();
complex     cmul1(complex, complex);
complex     cdiv1(complex, complex);
complex     cadd1(complex, complex, int);
complex     fixnp(complex);
void        poly_product(double C[], double D1[], double D2[], int);
void        poly_product1(double P1[], double AB[], int);
void        read_coeffs();
double      cabs1(complex);
void        zero_arrays();
void        power_series();
int         signx(double);
```

3B.1 Power series method

The inverse z-transform, $x(n)$, is computed recursively by the function power_series() (see Program 3B.2) based on the following equation:

$$x(n) = \left[a_n - \sum_{i=1}^{n} x(n-i)b_i\right] \bigg/ b_0, \quad n = 1, 2, \ldots \tag{3B.1a}$$

where

$$x(0) = a_0/b_0 \tag{3B.1b}$$

To use the program to calculate the inverse z-transform, via the power series method, the z-transform may be specified in either direct or cascade form:

$$X(z) = \frac{a_0 + a_1 z^{-1} + a_2 z^{-2} + \ldots + a_N z^{-N}}{b_0 + b_1 z^{-1} + b_2 z^{-2} + \ldots + b_M z^{-N}} \qquad \text{direct form} \tag{3B.2a}$$

$$X(z) = \prod_{k=i}^{K} X_i(z) \qquad \text{cascade} \tag{3B.2b}$$

where $H_i(z)$ is a second-order section given by

$$X_i(z) = \frac{a_{0i} + a_{1i} z^{-1} + a_{2i} z^{-2}}{1 + b_{1i} z^{-1} + b_{2i} z^{-2}} \tag{3B.3}$$

An input data file, called coeff.dat, must be created. The file contains the number of stages, K (for the direct form, $K = 1$), and the denominator and numerator coefficients of the z-transform. The use of the input data file is convenient as it removes the task of typing coefficients and the possibility of mistakes. Further, it makes it easier to utilize the results of one investigation as input to another program. The following examples illustrate the use of the program to calculate inverse z-transform via the power series method.

Example 3B.1

Find the first five values of the inverse z-transform of the discrete-time system characterized by the following z-transform using the power series method:

$$X(z) = \frac{0.183\,301\,5 + 0.341\,956\,1z^{-1} + 0.341\,956\,1z^{-2} + 0.183\,301\,5z^{-3}}{1 - 0.352\,518\,2z^{-1} + 0.419\,402\,3z^{-2} - 0.016\,369z^{-2}}$$

Clearly, $X(z)$ is in direct form. The input data file, created with edlin that comes with all PCs, has the form:

```
1
1  -0.3525182  0.4194023  -0.016369
0.1833015  0.3419561  0.3419561  0.1833015
```

The output of the program is summarized below:

$$h(0) = -0.016369;\ h(1) = 0.177531;\ h(2) = 0.411404;\ h(3) = 0.0705705;$$
$$h(4) = -0.1476666$$

Example 3B.2

Obtain the first five values of the inverse z-transform of the following system using the power series method:

$$X(z) = \frac{N_1(z)N_2(z)N_3(z)}{D_1(z)D_2(z)D_3(z)}$$

where

$$N_1(z) = 1 - 1.122\,346z^{-1} + z^{-2}$$
$$N_2(z) = 1 - 0.437\,833z^{-1} + z^{-2}$$
$$N_3(z) = 1 + z^{-1}$$
$$D_1(z) = 1 - 1.433\,509z^{-1} + 0.858\,110z^{-2}$$
$$D_2(z) = 1 - 1.293\,601z^{-1} + 0.556\,929z^{-2}$$
$$D_3(z) = 1 - 0.612\,159z^{-1}$$

Clearly, the transfer function consists of three stages: two second-order stages and one first-order stage. The first-order stage is entered as a second-order stage with a zero coefficient for the z^{-2} term. The input data file is given below.

```
3                               /*number of stages; maximum 5*/
1  -1.433509   0.858110         /*coefficients of D1(z)*/
1  -1.122346   1                /*coefficients of N1(z)*/
1  -1.293601   0.556929         /*coefficients of D2(z)*/
1  -0.437833   1                /*coefficients of N2(z)*/
```

```
1  −0.6121593  0          /*coefficients of D3(z)*/
1  1  0                   /*coefficients of N3(z)*/
```

The comments on the right-hand side are not part of the file; they are merely for explanation purposes. The output of the program is summarized below:

$$x(0) = 1;\ x(1) = 2.779\,09;\ x(2) = 5.2725$$

$$x(3) = 8.7218;\ x(4) = 11.7438;\ x(5) = 13.4723$$

3B.2 Partial fraction expansion

Given an Nth-order z-transform, with distinct poles, that is

$$X(z) = \frac{N(z)}{D(z)} = \frac{a_0z^N + a_1z^{N-1} + \ldots + a_{N-1}z + a_N}{b_0z^N + b_1z^{N-1} + \ldots + b_{N-1}z + b_N}$$

then $X(z)$ can be expanded into partial fractions as

$$\frac{N(z)}{zD(z)} = \frac{N(z)}{z(z-p_1)(z-p_2)(z-p_3)\ldots(z-p_N)} \tag{3B.4}$$

$$= \frac{B_0}{z} + \sum_{k=1}^{M} \frac{C_k}{z-p_k}$$

where

$$N(z) = a_0z^N + a_1z^{N-1} + \ldots + a_{N-1}z + a_N$$

$$D(z) = b_0z^N + b_1z^{N-1} + \ldots + b_{N-1}z + b_N$$

the p_k are the poles of $X(z)$ (assumed first order) and the C_k are the partial fraction coefficients. The constant, B_0, is given by

$$B_0 = a_N/b_N \tag{3B.5}$$

The partial fraction coefficient, C_k, associated with p_k is obtained by multiplying both sides of Equation 3B.4 by $z - p_k$ and then letting $z = p_k$:

$$C_k = \frac{N(z)(z-p_k)}{zD(z)} = \left.\frac{N(z)}{zD_k(z)}\right|_{z=pk} \tag{3B.6}$$

where

$$D_k(z) = \prod_{\substack{i=1\\ i\neq k}}^{M} (z-p_i)$$

For example, to find C_1, we multiply both sides of Equation 3B.4 by $z - p_1$ and then let $z = p_1$:

$$C_1 = \frac{N(z)(z-p_1)}{zD(z)} = \left.\frac{N(z)\cancel{(z-p_1)}}{z\cancel{(z-p_1)}(z-p_2)(z-p_3)\ldots(z-p_N)}\right|_{z=p1} = \left.\frac{N(z)}{zD_1(z)}\right|_{z=p1}$$

where

$$D_1(z) = (z - p_2)(z - p_3) \ldots (z - p_N)$$

With the poles expressed in polar coordinates, that is $p_k = r_k e^{j\theta k}$, the coefficients are given by

$$C_k = \frac{N(r_k e^{j\theta k})}{r_k e^{j\theta k} D_k(e^{j\theta k})} \quad k = 1, \ldots, N \tag{3B.7}$$

The partial fraction expansion function (see Program 3B.1) first finds the positions of the poles, p_k, $k = 1, 2, \ldots, N$, and then evaluates Equation 3B.7 for each pole.

When the values of B_0 and C_k have been obtained, the z-transform can be written as

$$X(z) = B_0 + \sum_{k=1}^{N} \frac{C_1 z}{z - p_k} \tag{3B.8}$$

For causal sequences, the inverse z-transform is the sum of the inverse z-transform of each term in Equation 3B.8:

$$x(n) = B_0 u(n) + C_1(p_1)^n + C_2(p_2)^n + \ldots + C_N(p_N)^n \tag{3B.9}$$

To use the program to calculate the partial fraction coefficients, the z-transform must be expressed in a cascade form using second-order factors. An example will make this clear.

Example 3B.3

Find, using the partial fraction expansion method, the inverse z-transform of the fifth-order transfer function given in Example 3B.2.

The partial fraction expansion for the transfer function has the form

$$X(z) = B_0 + \sum_{k=1}^{5} \frac{C_k z}{z - p_k} \tag{3B.10}$$

The input data file for the example, coeff.data, is identical to that of Example 3B.2. The output of the program is summarized below:

poles of the z-transform

pk	real	imag	mag	phase
1	0.716754	0.586833	0.926342	39.308436
2	0.716754	−0.586833	0.926342	−39.308436
3	0.646801	0.372261	0.746277	29.922232
4	0.646801	−0.372261	0.746277	−29.922232
5	0.612159	0.000000	0.612159	0.000000

partial fraction coeffs

BO=−3.418163

Ck	real	imag	mag	phase
1	1.611473	5.209672	5.453212	72.811944
2	1.611473	−5.209672	5.453212	−72.811943

3	−19.580860	−9.681908	21.843751	−153.689550
4	−19.580861	9.681908	21.843751	153.689551
5	40.356939	0.000000	40.356939	0.000000

From these values, the inverse z-transform is given by

$$x(n) = B_0 u(n) + \sum_{k=1}^{5} C_k (p_k)^n, \quad n \geqslant 0$$

Table 3.1 may also be used to find the inverse z-transform, $x(n)$, in a form that combines the terms with complex conjugate poles. This is left as an exercise for the reader.

3B.3 Cascade-to-parallel structure conversion

The program can also be used for converting a z-transform from cascade to parallel, based on the principles described in Example 3.14.

Example 3B.4

The transfer function of a fourth-order discrete-time system, in cascade form, is given by

$$H(z) = \frac{N_1(z)N_2(z)}{D_1(z)D_2(z)}$$

where

$$D_1(z) = 1 + 0.052\,921z^{-1} + 0.831\,73z^{-2}$$
$$N_1(z) = 1 + 0.481\,199z^{-1} + z^{-2}$$
$$D_2(z) = 1 - 0.304\,609z^{-1} + 0.238\,865z^{-2}$$
$$N_2(z) = 1 + 1.474\,597z^{-1} + z^{-2}$$

Use the program to convert the transfer function from cascade into parallel structure.

The input data file has the following form:

```
2
1  0.05292   0.83173
1  0.481199   1
1  -0.304609   0.238865
1  1.474597   1
```

This gives the following output from the program:

```
selected desired operation
0         for power series method of IZT
1         for partial fraction coeffs estimation
2         for cascade to parallel conversion
2
```

```
poles of the z-transform
pk      real          imag          mag          phase
1       -0.026460     0.911413      0.911797     91.662967
2       -0.026460     -0.911413     0.911797     -91.662967
3       0.152305      0.464401      0.488738     71.842631
4       0.152305      -0.464401     0.488738     -71.842631

partial fraction coeffs

B0=5.035604

Ck      real          imag          mag          phase
1       -0.257338     0.421333      0.493705     121.415410
2       -0.257338     -0.421333     0.493705     -121.415409
3       -1.760464     -3.766287     4.157421     -115.052650
4       -1.760464     3.766287      4.157421     115.052650
press enter to continue
stage   Ni(z)

0       -0.514677     -0.781635
1       -3.520927     4.034388
2       0.000000      -0.000000

stage   Di(z)

0       1.000000      0.052921      0.831373
1       1.000000      -0.304609     0.238865
2       0.000000      0.000000      0.000000
```

3C C program for estimating frequency response

The program computes the frequency response using either the direct estimation method or via the FFT as described in Section 3.5.5. The z-transform of the system whose frequency response is to be estimated must be in either direct or cascade form. An example will make this clear.

Example 3C.1

Obtain the frequency response of the discrete-time system whose transfer function is given by

$$H(z) = \frac{1 - 1.6180z^{-1} + z^{-2}}{1 - 1.5161z^{-1} + 0.878z^{-2}}$$

using

(1) the direct estimation method and

(2) the FFT approach.

Assuming a sampling frequency of 500 Hz and a resolution of < 1 Hz.

Solution

To satisfy the desired resolution, the number of frequency points to use with the program, npt, is 512 for the FFT approach (500/512 = 0.98 Hz) and 256 for the direct estimation. With the input data file

```
1
1  −1.5161  0.878
1  −1.618   1
```

the frequency response was obtained for each method. In either case, the response is held in ASCII format in three files as follows:

magn.dat	contains the magnitude response in decibels
phase.dat	contains the phase response in radians
fresp.dat	contains the frequency response in rectangular form

The first 10 values of the magnitude and phase responses are listed in Table 3C.1 and the magnitude and phase response for the direct method are depicted in Figures 3C.1(a) and 3C.1(b) respectively.

The program for frequency response consists of five functions held in separate files as follows:

freqres1.c	main function
fixdata.c	computes the magnitude and phase angles
freqd.c	direct frequency response estimation
fft.c	radix-2 decimation in time FFT algorithm
ltilib.c	a collection of common DSP functions. This is identical to ltilib.c described in Section 3B except that it does not require a .h file.

The functions are not listed here because of lack of space, but are available on the PC disk for the book (see the Preface for details).

Table 3C.1 The first 10 values of the magnitude and phase responses for Example 3C.1, using direct estimation or FFT approach.

	Direct estimation		FFT estimation	
k	*Magnitude* (dB)	*Phase* (rad)	*Magnitude* (dB)	*Phase* (rad)
0	0.469 496	0.000 000	0.469 496	0.000 000
1	0.469 39	−0.004 155	0.469 391	−0.004 138
2	0.469 073	−0.008 318	0.469 076	−0.008 286
3	0.468 541	−0.012 500	0.468 549	−0.012 451
4	0.467 791	−0.016 710	0.467 805	−0.016 644
5	0.466 817	−0.020 956	0.466 839	−0.020 873
6	0.465 612	−0.025 249	0.465 643	−0.025 148
7	0.464 165	−0.029 599	0.464 208	−0.029 479
8	0.462 466	−0.034 016	0.462 523	−0.033 876
9	0.460 501	−0.038 511	0.460 574	−0.038 351

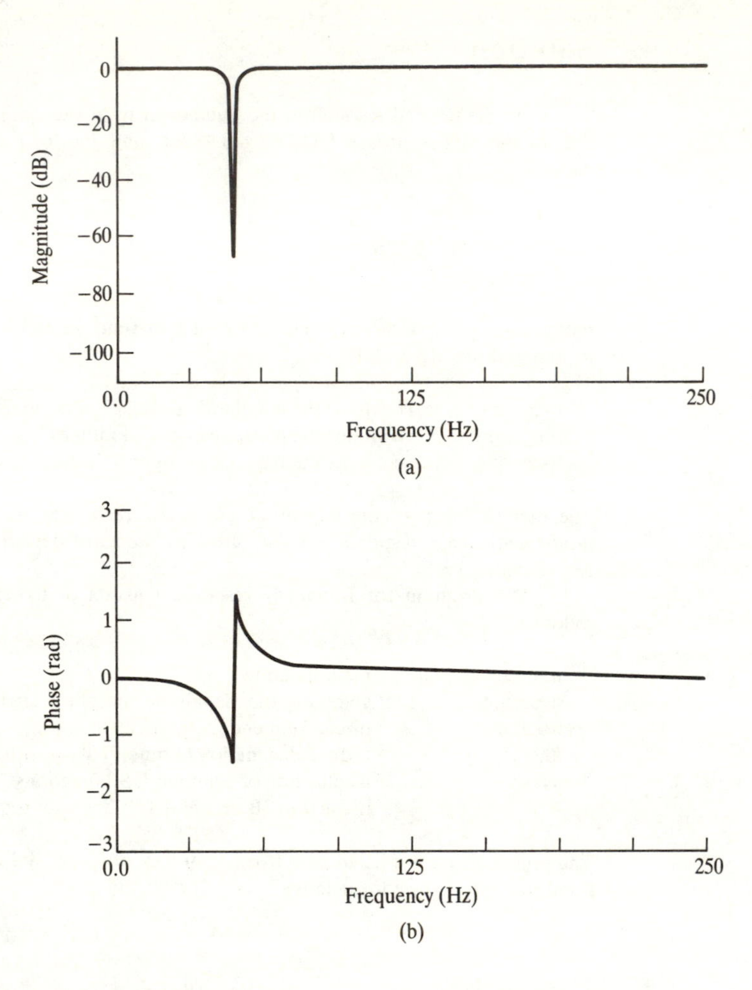

Figure 3C.1 (a) Magnitude–frequency response of the IIR system for Example 3C.1 using direct estimation. (b) Phase response of the IIR system for Example 3C.1 using direct estimation.

Reference for Appendices

Jury E.I. (1964). *Theory and Applications of the z-transform Method.* New York NY: Wiley

4

Correlation and convolution

4.1	Introduction	184
4.2	Correlation description	184
4.3	Convolution description	213
4.4	Implementation of correlation and convolution	236
4.5	Application examples	237
4.6	Summary	246
	References	249
	Appendix	250

The nature of the correlation process is first described in this chapter followed by an explanation using worked examples of the calculation of cross- and autocorrelations. The attenuating effects of correlation on the noise content of signals is described, as are a number of applications of correlation. The technique of fast correlation utilizing the FFT is then explained. The topic of convolution is covered in a similar manner to correlation. The treatment includes circular and linear convolution, fast linear convolution, and the sectioning methods (overlap–add, overlap–save) needed to handle large amounts of input data. The relationship between correlation and convolution is established. The chapter finishes with a section on implementation and some worked application examples.

4.1 Introduction

It is frequently necessary to be able to quantify the degree of interdependence of one process upon another, or to establish the similarity between one set of data and another. In other words, the correlation between the processes or data is sought. Correlation can be defined mathematically and can be quantified. The process of correlation occupies a significant place in signal processing. Applications are found in image processing for robotic vision or remote sensing by satellite in which data from different images is compared, in radar and sonar systems for range and position finding in which transmitted and reflected waveforms are compared, in the detection and identification of signals in noise, in control engineering for observing the effect of inputs on outputs, in the identification of binary codewords in pulse code modulation systems using correlation detectors, as an integral part of the ordinary least squares estimation technique, in the computation of the average power in waveforms, and in many other fields, such as, for example, climatology. Correlation is also an integral part of the process of convolution. The convolution process is essentially the correlation of two data sequences in which one of the sequences has been reversed. This means that the same algorithms may be used to compute correlations and convolutions simply by reversing one of the sequences. The process of convolution gives the output from a system which filters the input. The spectrum of a recorded signal consists of the convolution of the spectrum of the signal with the spectrum of its window function.

4.2 Correlation description

Consider how two data sequences, each consisting of simultaneously sampled values taken from the two corresponding waveforms, might be compared. If the two waveforms varied similarly point for point, then a measure of their correlation might be obtained by taking the sum of the products of the corresponding pairs of points. This proposal becomes more convincing when the case of two independent and random data sequences is considered. In this case the sum of the products will tend towards a vanishingly small random number as the number of pairs of points is increased. This is because all numbers, positive and negative, are equally likely to occur so that the product pairs tend to be self-cancelling on summation. By contrast, the existence of a finite sum will indicate a degree of correlation. A negative sum will indicate negative correlation, that is an increase in one variable is associated with a decrease in the other variable. The cross-correlation $r_{12}(n)$ between two data sequences $x_1(n)$ and $x_2(n)$ each containing N data might therefore be written as

$$r_{12} = \sum_{n=0}^{N-1} x_1(n)x_2(n)$$

This definition of cross-correlation, however, produces a result which depends on the number of sampling points taken. This is corrected for by normalizing the result to the number of points by dividing by N. Alternatively this may be regarded as averaging the sum of products. Thus, an improved definition is

$$r_{12} = \frac{1}{N} \sum_{n=0}^{N-1} x_1(n) x_2(n)$$

Example 4.1

The calculation of r_{12} is illustrated in the following example, in which the point numbers in the data sequences are the n, and the sequences are x_1 and x_2.

n	1	2	3	4	5	6	7	8	9
x_1	4	2	−1	3	−2	−6	−5	4	5
x_2	−4	1	3	7	4	−2	−8	−2	1

$$\begin{aligned} r_{12} &= \frac{1}{9}(4 \times -4 + 2 \times 1 + -1 \times 3 + 3 \times 7 + -2 \times 4 + -6 \times -2 + \\ &\quad -5 \times -8 + 4 \times -2 + 5 \times 1) \\ &= 5 \end{aligned}$$

However, this definition needs modification to be useful. In some cases it may indicate zero correlation although the two waveforms are 100% correlated. This may occur for example when the two waveforms are out of phase, which will often be the case. The situation is illustrated by the waveforms of Figure 4.1. From this figure it is seen that each pair product in the correlation is zero, and hence the correlation is zero, because one of either x_1 or x_2 is always zero. However, the waveforms are clearly highly correlated, although they are out of phase. The phase difference could, for example, occur because x_1 is the reference signal while x_2 is the delayed output from a circuit. To overcome such phase differences it is necessary to shift, or lag, one of the waveforms with respect to the other. Typically x_2 is shifted to the left to align the waveforms prior to correlation. As illustrated in Figure 4.2 this is equivalent to changing $x_2(n)$ to $x_2(n + j)$, where j represents the amount of lag which is the number of sampling points by which x_2 has been shifted to the left. An alternative, but equivalent, procedure is to shift x_1 to the right. The formula for the cross-

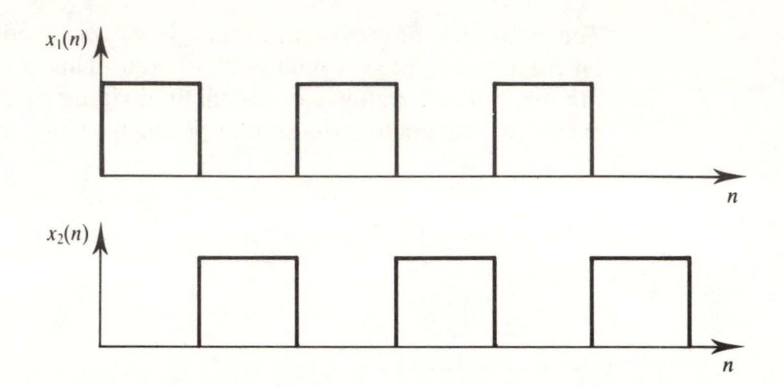

Figure 4.1 Out-of-phase 100% correlated waveforms with zero correlation at lag zero.

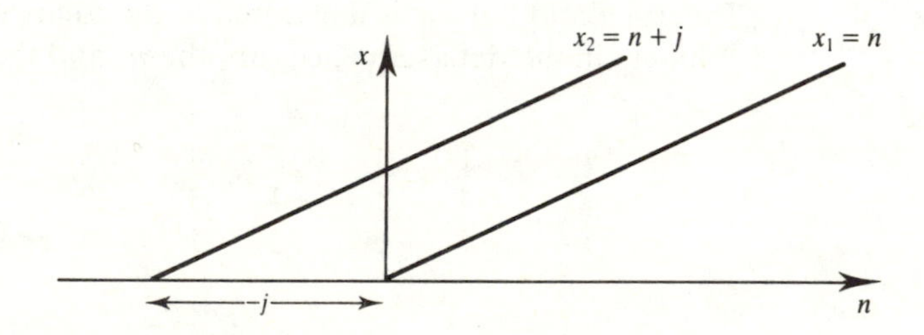

Figure 4.2 Waveform $x_2 = x_1 + j$ shifted j lags to the left of waveform x_1.

correlation thus becomes

$$r_{12}(j) = \frac{1}{N}\sum_{n=0}^{N-1} x_1(n)x_2(n+j)$$

$$= r_{21}(-j) = \frac{1}{N}\sum_{n=0}^{N-1} x_2(n)x_1(n-j) \qquad \textbf{(4.1)}$$

In practice when two waveforms are correlated their phase relationship will probably not be known and so the correlation will be computed for a number of different lags in order to establish the largest value of the correlation which is then taken to be the correct value.

Example 4.2

Consider the cross-correlation of the above two sequences $x_1(n)$ and $x_2(n)$ at a lag of $j = 3$, that is consider $r_{12}(3)$. The two sequences become

n	1	2	3	4	5	6	7	8	9
x_1	4	2	−1	3	−2	−6	−5	4	5
x_2	7	4	−2	−8	−2	1			

so

$$r_{12}(3) = \frac{1}{9}(4 \times 7 + 2 \times 4 + -1 \times -2 + 3 \times -8 + -2 \times -2 + -6 \times -1)$$

$$= 2.667$$

Of course, it is also possible to consider correlation in the continuous time domain, and some analogue signal correlation is implemented this way. In the continuous domain $n \rightarrow t$ and $j \rightarrow \tau$ and

$$r_{12}(\tau) = \lim_{T \to \infty} \frac{1}{T} \int_{-T/2}^{T/2} x_1(t) x_2(t + \tau)\, dt \qquad \textbf{(4.2)}$$

However, if $x_1(t)$ and $x_2(t)$ are periodic with period T_0 Equation 4.2 simplifies to

$$r_{12}(\tau) = \frac{1}{T_0} \int_{-T_0/2}^{T_0/2} x_1(t) x_2(t + \tau)\, dt \qquad \textbf{(4.3)}$$

If the waveforms are finite energy waveforms, for example nonperiodic pulse-type waveforms then the average evaluated over the time T as $T \rightarrow \infty$ is not taken because then $1/T \rightarrow 0$ and $r_{12}(\tau)$ is always vanishingly small. For this case Equation 4.4 is used in principle:

$$r_{12}(\tau) = \int_{-\infty}^{\infty} x_1(t) x_2(t + \tau)\, dt \qquad \textbf{(4.4)}$$

In practice, a finite record length will be processed and so Equation 4.5 or 4.1 will be applied:

$$r_{12}(\tau) = \frac{1}{T} \int_{0}^{T} x_1(t) x_2(t + \tau)\, dt \qquad \textbf{(4.5)}$$

There is another difficulty associated with cross-correlating finite lengths of data. This can be seen in the above example in which $r_{12}(3) = 2.667$ was determined. As x_2 is shifted to the left the waveforms no longer overlap and data at the ends of the sequences no longer form pair products. This is known as the end effect. In the example the number of pairs has dropped from nine to six for a lag of three. The result is a linear decrease in $r_{12}(j)$ as j increases,

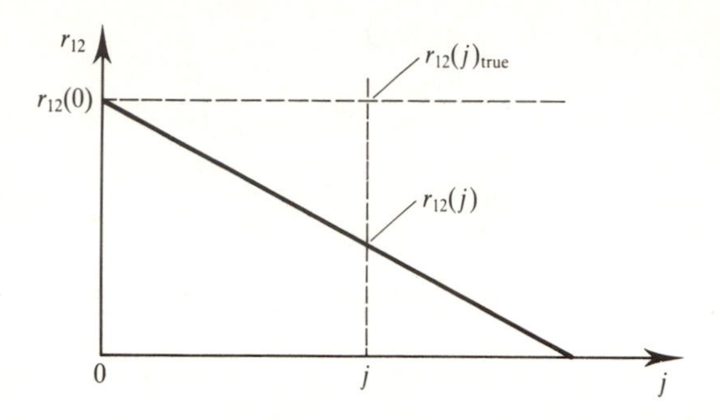

Figure 4.3 The effect of the end-effect on the cross-correlation $r_{12}(j)$.

leading to debatable values of $r_{12}(j)$. One possible solution is to make one of the sequences twice as long as the required length for correlation. This could be achieved by recording more data, or, if one of the sequences were periodic, by repeating the sequence (taking care to match the two ends). Another possibility is to add a correction to all computed values. Figure 4.3 shows how $r_{12}(j)$ decreases with j purely as a result of the end effect, that is actual variations in $r_{12}(j)$ are not included. At $j=0$, $r_{12}(j)=r_{12}(0)$, which can be computed. At $j=N$, $r_{12}(N)=0$, because the waveforms no longer overlap. In between, at some lag j, the true value of $r_{12}(j)$ is $r_{12}(j)_{true}$ while the actual value caused by the end effect is $r_{12}(j)$. Then, from the figure

$$\frac{r_{12}(j)_{true} - r_{12}(j)}{j} = \frac{r_{12}(0)}{N}$$

whence

$$r_{12}(j)_{true} = r_{12}(j) + \frac{j}{N} r_{12}(0) \tag{4.6}$$

Computed values of the cross-correlation are therefore easily corrected for end effects by adding $jr_{12}(0)/N$ to the values of $r_{12}(j)$.

The cross-correlation values computed according to the above formulae depend on the absolute values of the data. It is often necessary to measure cross-correlations according to the fixed scale between -1 and $+1$. This can be achieved by normalizing the values by an amount depending on the energy content of the data. For example, consider the two pairs of waveforms $x_1(n)$, $x_2(n)$, and $x_3(n)$, $x_4(n)$. The data values are given in the table below:

n	0	1	2	3	4	5	6	7	8
$x_1(n)$	0	3	5	5	5	2	0.5	0.25	0
$x_2(n)$	1	1	1	1	1	0	0	0	1
$x_3(n)$	0	9	15	15	15	6	1.5	0.75	0
$x_4(n)$	2	2	2	2	2	0	0	0	2

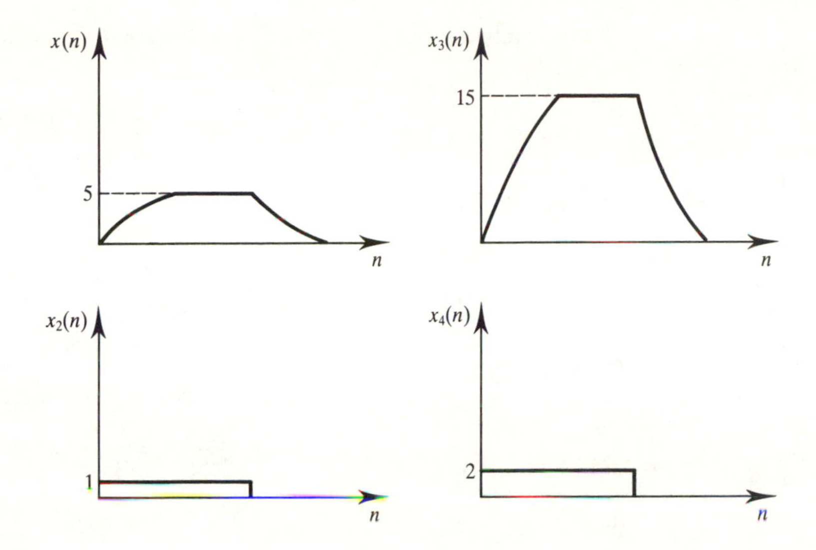

Figure 4.4 Pairs of waveforms $\{x_1(n), x_2(n)\}$, $\{x_3(n), x_4(n)\}$ of different magnitudes but equal cross-correlations.

As may be seen from Figure 4.4, waveforms $x_1(n)$ and $x_3(n)$ are alike, differing only in magnitude. The same is true of the pair $x_2(n)$ and $x_4(n)$. The correlation between $x_1(n)$ and $x_2(n)$ is therefore the same as that between $x_3(n)$ and $x_4(n)$. However, the cross-correlations $r_{12}(1)$ and $r_{34}(1)$ are 1.47 and 8.83 respectively. They are different because they depend on the absolute values of the data. This situation can be rectified by normalizing the cross-correlation $r_{12}(j)$ by the factor

$$\left[\frac{1}{N}\sum_{n=0}^{N-1} x_1^2(n) \times \frac{1}{N}\sum_{n=0}^{N-1} x_2^2(n)\right]^{1/2} = \frac{1}{N}\left[\sum_{n=0}^{N-1} x_1^2(n) \sum_{n=0}^{N-1} x_2^2(n)\right]^{1/2} \tag{4.7}$$

and similarly for $r_{34}(j)$. The normalized expression for $r_{12}(j)$ then becomes

$$\rho_{12}(j) = \frac{r_{12}(j)}{\dfrac{1}{N}\left[\displaystyle\sum_{n=0}^{N-1} x_1^2(n) \sum_{n=0}^{N-1} x_2^2(n)\right]^{1/2}} \tag{4.8}$$

$\rho_{12}(j)$ is known as the cross-correlation coefficient. Its value always lies between -1 and $+1$. $+1$ means 100% correlation in the same sense, -1 means 100% correlation in the opposing sense, for example signals in antiphase. A value of 0 signifies zero correlation. This means the signals are completely independent. This would be the case, for example, if one of the waveforms were completely random. Small values of $\rho_{12}(j)$ indicate very low correlation.

The normalizing factor for $r_{12}(j)$ in the above illustration is

$$\frac{1}{N}\left[\sum_{n=0}^{N-1} x_1^2(n) \sum_{n=0}^{N-1} x_2^2(n)\right]^{1/2} = \frac{1}{9}(88.31 \times 6)^{1/2} = 2.56$$

and for $r_{34}(j)$ it is

$$\frac{1}{N}\left[\sum_{n=0}^{N-1} x_3^2(n) \sum_{n=0}^{N-1} x_4^2(n)\right]^{1/2} = \frac{1}{9}(794.8 \times 24)^{1/2} = 15.34$$

Therefore

$$\rho_{12}(1) = \frac{r_{12}(1)}{2.56} = \frac{1.47}{2.56} = 0.57$$

and

$$\rho_{34}(1) = \frac{r_{34}(1)}{15.34} = \frac{8.83}{15.34} = 0.57$$

Now $\rho_{12}(1) = \rho_{34}(1)$ which demonstrates that this normalization process indeed allows a comparison of cross-correlations independently of the absolute data values.

A special case occurs when $x_1(n) = x_2(n)$. The waveform is then cross-correlated with itself. This process is known as autocorrelation. The autocorrelation of a waveform is given by

$$r_{11}(j) = \frac{1}{N} \sum_{n=0}^{N-1} x_1(n)x_1(n + j)$$

The autocorrelation function has one very useful property in that

$$r_{11}(0) = \frac{1}{N} \sum_{n=0}^{N-1} x_1^2(n) = S$$

where S is the normalized energy of the waveform. This provides a method for calculating the energy of a signal. If the waveform is completely random, for example corresponding to that of white, gaussian noise in an electrical system, then the autocorrelation will have its peak value at zero lag and will reduce to a random fluctuation of small magnitude about zero for lags greater than about unity (see Figure 4.5). This constitutes a test for random waveforms. This topic will be more fully covered in Section 4.2.1. It is also true that

$$r_{11}(0) \geqslant r_{11}(j)$$

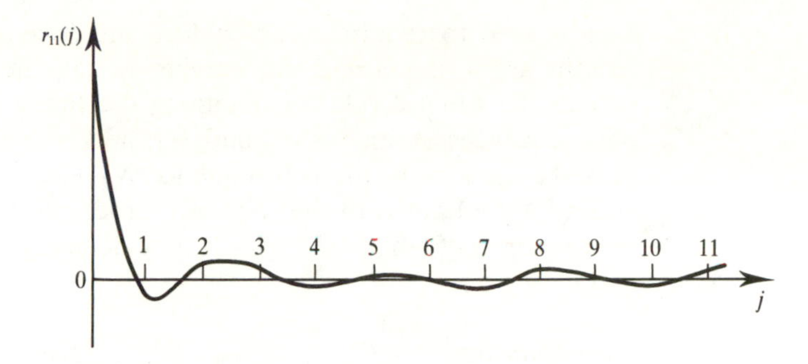

Figure 4.5 Autocorrelation function of a random waveform.

4.2.1 Cross- and autocorrelation

Care has to be exercised when cross-correlating two unequal length sequences when they are periodic. This is because the result of the correlation will be cyclic with the period of the shorter sequence. This result does not represent the full periodicity of the longer sequence and is, therefore, incorrect. This may be demonstrated by cross-correlating the sequences $a = \{4, 3, 1, 6\}$ and $b = \{5, 2, 3\}$ to obtain $r_{ab}(j)$. The sequence b is placed below sequence a, and b is shifted left by one lag on each of the subsequent rows, with the value of the cross-correlation appearing in the final column on the right.

Sequence				*Lag*	$r_{ab}(j)$	
4	3	1	6			
3	5	2	3	0	47	
5	2	3	5	1	59	
2	3	5	2	2	34	
3	5	2	3	3	47	$r_{ab}(j)$ repeats
5	2	3	5	4	59	
etc.						

The result shows that $r_{ab}(j)$ is cyclic, repeating every third lag, that is $r_{ab}(j)$ has the same period as that of the shorter sequence, b. This procedure is known as cyclic correlation. To obtain the correct value in which each value in a is multiplied by each value in b, all the elements in b have to be shifted in turn below each value in a as shown below:

```
            4 3 1 6
                  5 2 3
                5 2 3
              5 2 3
            5 2 3
          5 2 3
        5 2 3
      5 2 3
```

This is seen to require 6 lags before the b sequence repeats. The sequence lengths are 4 and 3 and the number of lags necessary is $4+3-1=6$. This reveals the general rule for obtaining the linear cross-correlation of two periodic sequences of lengths N_1 and N_2: add augmenting zeros to each sequence to make the lengths of each sequence N_1+N_2-1. This may be expressed as adding N_2-1 zeros to the sequence of length N_1 and adding N_1-1 zeros to the sequence of length N_2. This is now demonstrated for the given sequences a and b:

Sequence						*Lag*	$r_{ab}(j)$	
4	3	1	6	0	0			
5	2	3	0	0	0	0	29	
2	3	0	0	0	5	1	17	
3	0	0	0	5	2	2	12	
0	0	0	5	2	3	3	30	
0	0	5	2	3	0	4	17	
0	5	2	3	0	0	5	35	
5	2	3	0	0	0	6	29	$r_{ab}(j)$ repeats
etc.								

Thus, the required linear cross-correlation of a and b is

$$r_{ab}(j) = \{29, 17, 12, 30, 17, 35\}$$

So far, the instances of cross-correlation taken have all assumed digitized data, but cross-correlation may also be performed analytically when analytical expressions can be written for the waveforms, including when this requires sectioning of the waveforms. In practice the analytical procedure has its equivalent in the use of analogue circuits to effect the cross-correlation. An example of analytical cross-correlation follows.

Example 4.3

Obtain the cross-correlation $r_{12}(-\tau)$ between the waveforms $v_1(t)$ and $v_2(t)$ of Figure 4.6.

It is easy to express the waveforms analytically by dividing them into straight-line sections. It is only necessary to do this over one period, T, of the waveforms because $r_{12}(-\tau)$ will be periodic in τ with period T. Therefore, for $0 \leqslant t \leqslant T$, $v_1(t) = t/T$, and for $0 \leqslant t \leqslant T/2$, $v_2(t) = 1.0$, while for $T/2 \leqslant t \leqslant T$, $v_2(t) = -1.0$. The requirement is to obtain an expression for $r_{12}(-\tau)$, that is $v_2(t)$, the rectangular waveform, is to be shifted right wrt $v_1(t)$. For $0 \leqslant \tau \leqslant T$, the situation is described by Figure 4.7 which shows that $v_1(t)$ has to be

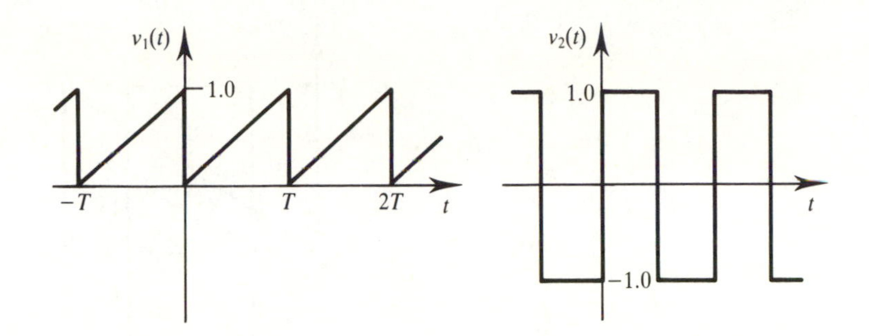

Figure 4.6 The waveforms $v_1(t)$ and $v_2(t)$ for cross-correlation example.

multiplied by three consecutive sections of $v_2(t)$ in which $v_2(t)$ has the consecutive values -1, 1, -1. For $T/2 \leqslant \tau \leqslant T$, Figure 4.8 applies in which the consecutive values of the set of $v_2(t)$ have changed to 1, -1, $+1$. This means there are two parts to the solution which must match at $\tau = T/2$.

Referring to Figure 4.7, the cross-correlation is split into the three sections with boundaries at $t = \tau$, $t = \tau + T/2$, and $t = T$. Hence

$$
\begin{aligned}
r_{12}(-\tau) &= \frac{1}{T}\int_0^T v_1(t)v_2(t-\tau)\,\mathrm{d}t \\
&= \frac{1}{T}\int_0^{\tau} \frac{t}{T}(-1)\,\mathrm{d}t + \frac{1}{T}\int_{\tau}^{\tau+T/2} \frac{t}{T}(1)\,\mathrm{d}t + \frac{1}{T}\int_{\tau+T/2}^{T} \frac{t}{T}(-1)\,\mathrm{d}t \\
&= \frac{-1}{T^2}\left[\frac{t^2}{2}\right]_0^{\tau} + \frac{1}{T^2}\left[\frac{t^2}{2}\right]_{\tau}^{\tau+T/2} - \frac{1}{T^2}\left[\frac{t^2}{2}\right]_{\tau+T/2}^{T}
\end{aligned}
$$

$$
r_{12}(-\tau) = -\frac{1}{4} + \frac{\tau}{T} \quad \text{for } 0 \leqslant \tau \leqslant \frac{T}{2} \tag{4.9}
$$

For $T/2 \leqslant \tau \leqslant T$, and referring to Figure 4.8, it is seen that

$$
r_{12}(-\tau) = \frac{1}{T}\int_0^{\tau-T/2} \frac{t}{T}(1)\,\mathrm{d}t + \frac{1}{T}\int_{\tau-T/2}^{\tau} \frac{t}{T}(-1)\,\mathrm{d}t + \frac{1}{T}\int_{\tau}^{T} \frac{t}{T}(1)\,\mathrm{d}t
$$

$$
r_{12}(-\tau) = \frac{3}{4} - \frac{\tau}{T} \quad \text{for } \frac{T}{2} \leqslant \tau \leqslant T \tag{4.10}
$$

Substituting $\tau = T/2$ into Equations 4.9 and 4.10 gives $r_{12}(-\tau) = 1/4$ in both cases, confirming that the two functions match correctly. Figure 4.9 shows a plot of $r_{12}(-\tau)$ versus τ for $0 \leqslant \tau \leqslant T$.

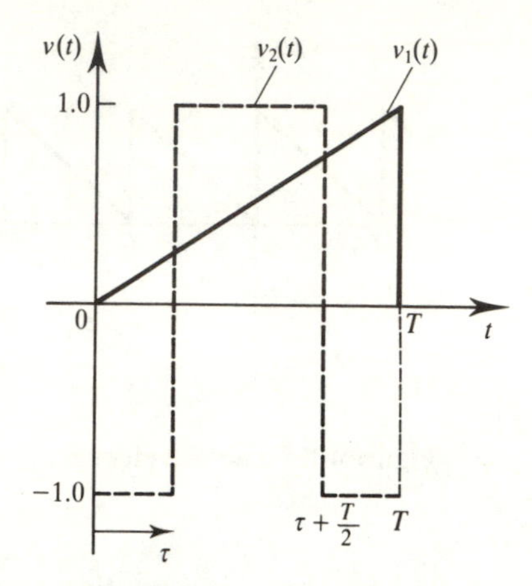

Figure 4.7 Sections of $v_2(t)$ for $\theta \leqslant \tau \leqslant T$.

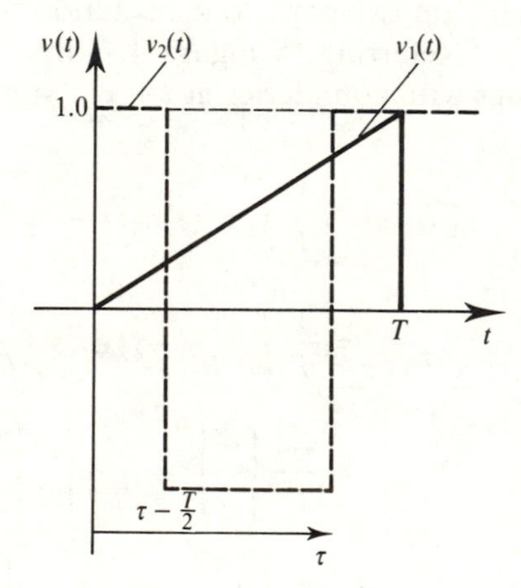

Figure 4.8 Sections of $v_2(t)$ for $T/2 \leqslant \tau \leqslant T$.

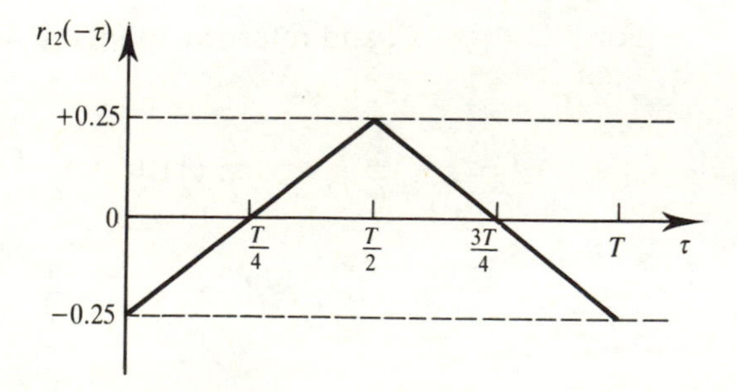

Figure 4.9 $r_{12}(-\tau)$ as a function of τ.

It is of interest to give some consideration to the consequences of using finite lengths of data in the calculation of the correlation. In other words, what is the effect of using Equation 4.5, in which T is finite, instead of Equation 4.2?

This question can be answered by considering just one sinusoidal Fourier harmonic component of the signal. Equation 4.2 will give the correct autocorrelation, in which $T \gg T_p$, where T_p is the period of the sinusoid. Thus

$$r_{11}(\tau) = \lim_{T\to\infty} \frac{1}{2T} \int_{-T}^{T} A \sin(\omega t) A \sin(\omega t + \tau)\, dt$$

$$= \lim_{T\to\infty} \frac{A^2}{2} \left[\cos(\omega\tau) - \frac{\cos(\omega T)}{2\omega T} \sin(\omega\tau)\right] \qquad \textbf{(4.11)}$$

Inspection of this equation shows that the second term in the bracket $\to 0$ when $T \to \infty$, so when $T \neq \infty$ it represents an error. The $\cos(\omega T)$ term represents periodic error effects, while the term $1/2\omega T$ gives the trend in the error. Thus, as far as the correlation length, T, is concerned, the errors are greater the shorter the sequence, and are also largest for the lower frequency components of the waveform. The errors are also periodic in τ.

The $\cos(\omega T)$ term gives least errors when $\omega T = [(2n+1)/2]\pi$. Since $\omega = 2\pi/T_p$ and large values of T are sought, this corresponds to

$$T \geqslant (2n+1)\frac{T_p}{4} \qquad \textbf{(4.12)}$$

The $\sin(\omega\tau)$ term is least when $\omega\tau = m\pi$, where m is integer. Hence,

$$\tau = \frac{m}{2} T_p \qquad \textbf{(4.13)}$$

It is now necessary to make some reasonable assumptions. Assume the condition for large T is satisfied by $n \geqslant 10$. Then $T \geqslant nT_p/2$, or

$$T \geqslant 5T_p \qquad \textbf{(4.14)}$$

From Equation 4.13, the largest allowable value of τ for the lowest frequency component ($m = 1$) satisfies

$$\tau < T_p \qquad \textbf{(4.15)}$$

Combining Equations 4.14 and 4.15,

$$\tau \leqslant T/5$$

This means that when correlating waveforms the errors due to finite data

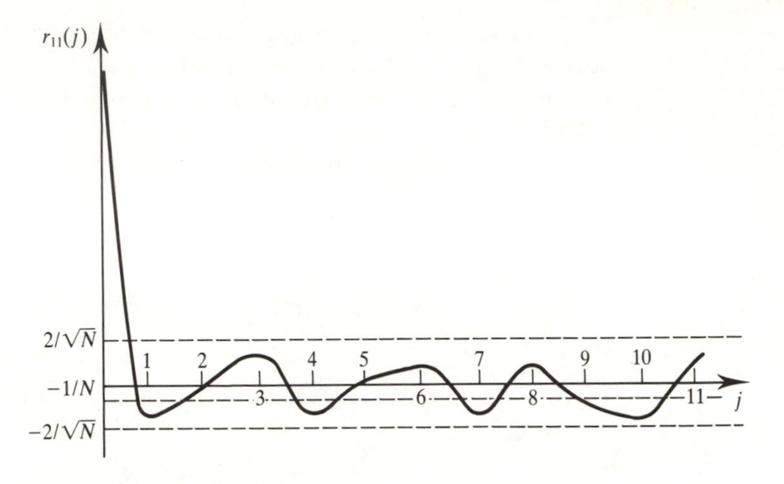

Figure 4.10 The autocorrelation coefficient of a random waveform.

lengths may be minimized by

(1) ensuring that $T \geqslant 5T_{\text{p}}$, where T_{p} is the lowest frequency component of interest, and
(2) overlapping the data by no more than 20% of their length.

Thus, for example, if telephone speech signals with a bandwidth of 300 Hz to 3.4 kHz and sampled at 40 kHz are to be correlated, $T_{\text{p}} = 1/300 = 3.3 \times 10^{-3}$ s. The least acceptable data length would be $5 \times 3.3 \times 10^{-3}$ s $= 16.7$ ms and the largest correlation shift would be 3.33 ms, or 133 data points.

Figure 4.10 shows the plot of $\rho_{11}(j)$, the autocorrelation coefficient of a purely random waveform, for example white noise. The expected value of $r_{11}(j)$ can be shown to be $E[r_{11}(j)] \approx -1/N$ (Chatfield, 1980), where N is the number of data points, and its variance is $\text{var}[r_{11}(j)] \approx 1/N$. The expected value of $-1/N$ is shown on the figure as are the 95% confidence limits of $-1/N$ and $\pm 2/N^{1/2}$. Values of $r_{11}(j)$ which fall outside these confidence limits may be significant, that is they may indicate that the waveform was not truly random. However, it should be noted that as many as one point in twenty may lie outside these limits even when the waveform is completely random. For a random waveform $r_{11}(j)$ should fall to within the 95% confidence limits within one or two lags. Experience and more sophistication is required to be sure that a waveform is random. For example, data pre-whitening may be advisable (Jenkins and Watts, 1968).

The autocorrelation function of a periodic waveform is itself a periodic waveform. This is easily proved as follows. The periodic waveform $x(t)$ of period, T, satisfies

$$x(t) = x(t + n\text{T})$$

so,

$$r_{11}(\tau) = \lim_{T\to\infty} \frac{1}{T} \int_{-T/2}^{T/2} x(t)x(t+\tau)\,\mathrm{d}t$$

$$= \lim_{T\to\infty} \frac{1}{T} \int_{-T/2}^{T/2} x(t)x(t+\tau+nT)\,\mathrm{d}t$$

$$r_{11}(\tau) = r_{11}(\tau + nT) \qquad \textbf{(4.16)}$$

Thus $r_{11}(\tau)$ is seen to be periodic in τ with period, T. This is a useful property because it enables the detection of periodic signals in noise for small signal-to-noise ratios. Autocorrelating the waveform tends to reduce the noise while at the same time developing the periodic autocorrelation function of the signal. Once detected, further processing can be applied to determine its shape if this is required.

Equation 4.11 showed that the autocorrelation function of $A\sin(\omega t)$ is $(A^2/2)\cos(\omega\tau)$. In this case, as in others, the amplitude of the autocorrelation function is related simply to that of the signal, and may be used to estimate the signal amplitude. Another common example is that of the rectangular wave of amplitude, A, which the reader could show has a triangular autocorrelation function of amplitude, A^2. Finally, it should be noted that autocorrelation functions are not unique. This means that a number of different waveforms may share the same autocorrelation function. Hence the shapes of waveforms should not be deduced from the detected autocorrelation functions.

Consider now the case in which the waveform, $v(t)$, is partially random. This represents the case of a noisy signal which may be written as the sum of a signal term, $s(t)$, and a noise term, $q(t)$. Thus

$$v(t) = s(t) + q(t) \qquad \textbf{(4.17)}$$

$s(t)$ and $q(t)$ are assumed to be uncorrelated. The sampled autocorrelation function of $v(t)$ is $r_{vv}(j)$ given by

$$r_{vv}(j) = \frac{1}{N} \sum_{n=0}^{N-1} [s(n) + q(n)][s(n+j) + q(n+j)] \qquad \textbf{(4.18)}$$

$$= \frac{1}{N} \sum_{n=0}^{N-1} s(n)s(n+j) + \frac{1}{N} \sum_{n=0}^{N-1} s(n)q(n+j) + \frac{1}{N} \sum_{n=0}^{N-1} q(n)s(n+j)$$

$$+ \frac{1}{N} \sum_{n=0}^{N-1} q(n)q(n+j) \qquad \textbf{(4.19)}$$

$$= r_{ss}(j) + E[s(n)q(n+j)] + E[q(n)s(n+j)] + E[q(n)q(n+j)]$$

$$= r_{ss}(j) + E[s(n)]E[q(n+j)] + E[q(n)]E[s(n+j)]$$

$$+ E[q(n)]E[q(n+j)]$$

$$= r_{ss}(j) + \overline{s(n)}\,\overline{q(n)} + \overline{q(n)}\,\overline{s(n)} + \overline{q(n)}^2$$

$$= r_{ss}(j) + 2\bar{s}\bar{q} + \bar{q}^2 \qquad \textbf{(4.20)}$$

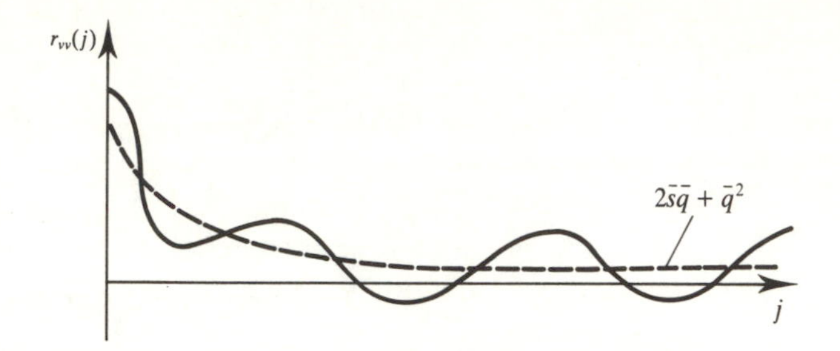

Figure 4.11 The autocorrelation function of a noisy signal.

Now, $\bar{q} \to 0$ for large N, for which

$$r_{vv}(j) \to r_{ss}(j) \tag{4.21}$$

For smaller N, the cross-correlation terms in Equation 4.19 and the autocorrelation of the noise tend towards zero with increasing lag, j.

Thus it is seen that the autocorrelation function of a partially random, or noisy, waveform consists of the autocorrelation function of the signal component superimposed on a noisy decaying function which depends on both the random and signal components and which decays towards the value $2\bar{s}\bar{q} + \bar{q}^2$. Thus the plot of $r_{vv}(j)$ against j will display the periodicity of $s(t)$ provided $|r_{ss}(j)| > |(2\bar{s}\bar{q} + \bar{q}^2)|$: see Figure 4.11. This offers a method for identifying the period of a signal in noise (see Section 4.2.2).

Example 4.4

Derive the cross-correlation function of two noisy waveforms.

Let the two waveforms be $\{s_1(t) + q_1(t)\}$ and $\{s_2(t) + q_2(t)\}$. Their sampled cross-correlation, $r_{12}(j)$ is given by

$$r_{12}(j) = \frac{1}{N}\sum_{n=0}^{N-1}[\{s_1(n) + q_1(n)\}\{s_2(n+j) + q_2(n+j)\}] \tag{4.22}$$

$$= \frac{1}{N}\sum_{n=0}^{N-1}[s_1(n)s_2(n+j) + s_1(n)q_2(n+j) + q_1(n)s_2(n+j)$$
$$+ q_1(n)q_2(n+j)]$$

$$= \frac{1}{N}\sum_{n=0}^{N-1} s_1(n)s_2(n+j) + \frac{1}{N}\sum_{n=0}^{N-1} s_1(n)q_2(n+j) + \frac{1}{N}\sum_{n=0}^{N-1} q_1(n)s_2(n+j)$$

$$+\frac{1}{N}\sum_{n=0}^{N-1} q_1(n)q_2(n+j)$$

$$= r_{s_1s_2}(j) + r_{s_1q_2}(j) + r_{q_1s_2}(j) + r_{q_1q_2}(j) \quad \textbf{(4.23)}$$

As in the previous case of autocorrelation the last three terms on the right-hand side of Equation 4.23 decay towards zero with increasing lag j. For large N, Equation 4.23 becomes

$$r_{12}(j) = r_{s_1s_2}(j) + \overline{s_1}\,\overline{q_2} + \overline{q_1}\,\overline{s_2} + \overline{q_1}\,\overline{q_2} \quad \textbf{(4.24)}$$

Thus as j increases $r_{12}(j) \rightarrow r_{s_1s_2}(j)$, the cross-correlation function of the two signals.

The above analyses illustrate that the cross- and autocorrelation processes emphasize signal properties by reducing the noise content.

4.2.2 Applications of correlation

It can be shown that

$$F[r_{11}(\tau)] = G_E(f) \quad \textbf{(4.25)}$$

where $G_E(f)$ is the energy spectral density of the waveform, that is the energy spectral density and the autocorrelation function constitute a Fourier Transform pair.

It can further be shown that

$$r_{11}(0) = E \quad \textbf{(4.26)}$$

where E is the total energy of the waveform.

Example 4.5

Obtain a relationship between the zero-lag correlation functions of two different waveforms and their total energy content.

Let the waveforms be $v_1(n)$ and $v_2(n)$, and let their summation be $V(n)= v_1(n) + v_2(n)$. The zero-lag autocorrelation function of $V(n)$ is

$$r_{vv}(0) = E_V = \frac{1}{N}\sum_{n=0}^{N-1} V^2(n) = \frac{1}{N}\sum_{n=0}^{N-1} [v_1(n) + v_2(n)]^2$$

where E_V is the energy of the waveform $V(n)$.

$$E_V = \frac{1}{N}\sum_{n=0}^{N-1}[v_1^2(n) + v_2^2(n) + 2v_1(n)v_2(n)]$$

$$= \frac{1}{N}\sum_{n=0}^{N-1} v_1^2(n) + \frac{1}{N}\sum_{n=0}^{N-1} v_2^2(n) + \frac{1}{N}\sum_{n=0}^{N-1} v_1(n)v_2(n)$$

so

$$E_V = r_{v_1}(0) + r_{v_2}(0) + 2r_{v_1v_2}(0) \qquad \textbf{(4.27)}$$

Equation 4.27 is the first form of the required result. Alternatively it may be written as

$$E_V = E_{v_1} + E_{v_2} + 2r_{v_1v_2}(0) \qquad \textbf{(4.28)}$$

Thus the energy of $V(n)$ equals the sum of the energies of its components plus $2r_{v_1v_2}(n)$, where $r_{v_1v_2}(n)$ is the zero-lag cross-correlation function of $v_1(n)$ and $v_2(n)$. If $v_1(n)$ and $v_2(n)$ are uncorrelated then the total energy is just the sum of the component energies.

If the signals $v_1(n)$ and $v_2(n)$ are noisy such that $v_1(n) = v_1'(n) + q_1(n)$ and $v_2(n) = v_2'(n) + q_2(n)$ then it is easy to show that

$$E_V = E_{v_1'} + E_{v_2'} + E_{q_1} + E_{q_2} + r_{v_1'v_2'}(0) \qquad \textbf{(4.29)}$$

The use of cross-correlation to detect and estimate periodic signals in noise will now be considered. The first proposal is that a signal buried in noise can be estimated by cross-correlating it with an adjustable 'template' signal. The template is adjusted by trial and error, guided by any foreknowledge, until the cross-correlation function has been maximized. This template is then the estimate of the signal. This proposal can be justified by referring to Equation 4.22 and assuming that for the template $q_2(n) = 0$. Equation 4.23 then becomes

$$r_{12}(j) = r_{s_1s_2}(j) + r_{q_1s_2}(j) \qquad \textbf{(4.30)}$$

$$= r_{s_1s_2}(j) + \bar{q}_1\bar{s}_2 \qquad \textbf{(4.31)}$$

Then, because $\bar{q}_1 \rightarrow 0$ as N increases,

$$r_{12}(j) \rightarrow r_{s_1s_2}(j) \qquad \textbf{(4.32)}$$

Clearly $r_{s_1s_2}(j)$ will be a maximum when $s_2(n) = s_1(n)$ when $r_{s_1s_2}(j)$ is the autocorrelation function of $s_1(n)$. Thus, changing the shape of the template $s_2(n)$ to maximize the cross-correlation function provides $s_2(n)$ as the estimate of $s_1(n)$.

The template method of signal estimation is convenient sometimes for example when the shape of the signal is known approximately as for certain biomedically evoked potentials, but there is a more scientific approach that may be preferred. In this method the period of the signal is first estimated by autocorrelating the noisy waveform, and then the noisy waveform is cross-correlated with a periodic impulse train of period equal to that of the signal. The resulting cross-correlation function is the signal estimate.

Let the signal of period N_p points ($N_p < N$) be $s(n)$ and let the noise be $q(n)$ so that the noisy waveform is $S(n) = s(n) + q(n)$. Let $\delta(n - kN_p)$ be the periodic impulse train used for the cross-correlation. Then

$$r_s(j) = \frac{1}{N}\sum_{n=0}^{N-1}[s(n) + q(n)]\delta(n - kN_p + j), \quad k = 0, 1, 2, \ldots \tag{4.33}$$

For $j = 0$,

$$r_{12}(0) = \frac{1}{N}[s(0) + q(0) + s(N_p) + q(N_p) + s(2N_p) + q(2N_p) + \ldots + s(N) + q(N)] \tag{4.34}$$

Now, because of the periodicity of the signal $s(n + kN_p) = s(n)$ and so Equation 4.34 becomes

$$r_{12}(0) = \frac{1}{N}[Ns(0) + q(0) + q(N_p) + q(2N_p) + \ldots + q(N)]$$

or

$$r_{12}(0) = s(0) + \frac{1}{N}\sum_{k=0}^{N/N_p} q(kN_p) \tag{4.35}$$

As $N \to \infty$, $1/N\sum_{k=0}^{N/N_p} q(kN_p) \to 0$, and therefore $r_{12}(0) \to s(0)$. Similarly, for other values of j,

$$r(j) = \frac{1}{N}\sum_{n=0}^{N-1}[s(n) + q(n)]\delta[(n + j) - kN_p], \quad k = 0, 1, 2, \ldots$$

which also results in cancellation of the noise and yields values of $s(n)$ for $n = 1, 2, \ldots$. Hence, from Equation 4.33

$$r_s(j) = s(0), s(1), \ldots, s(N - 1), \quad j = 0, 1, 2, \ldots$$

which is the required signal. Thus, a signal lost in a noisy waveform may be estimated by

(1) autocorrelating the waveform to find the period of the signal, and
(2) cross-correlating the waveform with a periodic impulse train of the same period as the signal.

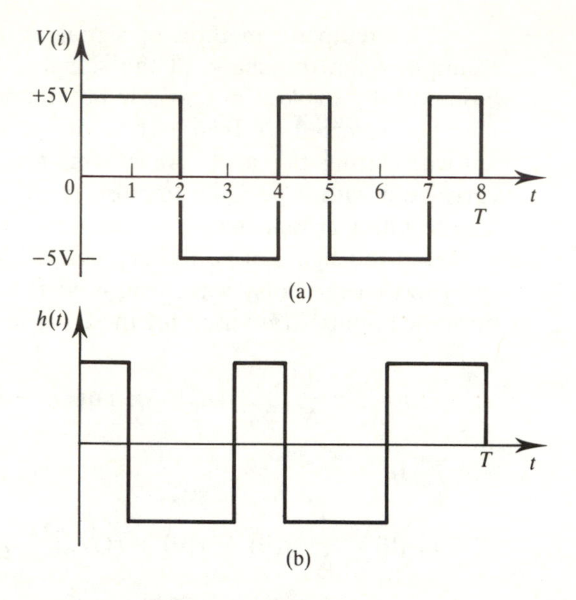

Figure 4.12 (a) A signal, which is an 8-bit PCM codeword. (b) The impulse response of the corresponding matched filter.

Another application of correlation is in the correlation detection implementation of the matched filter. A matched filter is a filter which maximizes the S/N ratio at its output. A matched filter has the impulse response, $h(t)$, given by (Stremler, 1982)

$$h(t) = cs_{\mathrm{i}}(T - t) \tag{4.36}$$

where c is an arbitrary constant, $s_{\mathrm{i}}(t)$ is the input signal (noiseless) given by

$$\begin{aligned} s_{\mathrm{i}}(t) &= s_{\mathrm{i}}(t) && \text{for } 0 \leqslant t \leqslant T \\ &= 0 && \text{for } T < t < 0 \end{aligned}$$

and T is the time at which the filter output is sampled. The impulse response is seen to be obtained by reversing the signal in time and then advancing it T (s) along the time axis. As an example, Figure 4.12(a) shows a signal, which is actually an 8-bit PCM codeword, and Figure 4.12(b) shows the matched filter impulse response which will maximize the detection of this signal.

Matched filter detection will now be shown to be equivalent to correlation. The filter output, $y(t)$, is first expressed in terms of the convolution of its input, $s(t)$, with its impulse response (see Section 4.3 for convolution):

$$y(t) = \int_{-\infty}^{\infty} s(\tau)h(t - \tau)\,\mathrm{d}\tau \tag{4.37}$$

where

$$s(t) = s_1(t) + q(t) \tag{4.38}$$

τ is the lag, and $q(t)$ denotes the noise component as usual. Substituting Equation 4.38 into 4.37 gives

$$y(t) = \int_{-\infty}^{\infty} [s_1(\tau) + q(\tau)]h(t - \tau)\,d\tau$$

$$= \int_{-\infty}^{\infty} s_1(\tau)h(t - \tau)\,d\tau + \int_{-\infty}^{\infty} q(\tau)h(t - \tau)\,d\tau$$

The second term on the right-hand side tends to zero because $q(\tau)$ is random and is uncorrelated with $h(t - \tau)$. Therefore

$$y(t) \approx \int_{-\infty}^{\infty} s_1(\tau)h(t - \tau)\,d\tau \tag{4.39}$$

Now, from Equation 4.36,

$$h(t - \tau) = cs_1(T - t + \tau) \tag{4.40}$$

Combining Equations 4.39 and 4.40 gives

$$y(t) \approx \int_{-\infty}^{\infty} s_1(\tau)cs_1(T - t + \tau)\,d\tau \tag{4.41}$$

If this output is sampled at time $t = T$,

$$y(T) \approx \int_{-\infty}^{\infty} s_1(\tau)cs_1(\tau)\,d\tau$$

$$\approx \int_{-\infty}^{\infty} s_1^2(\tau)\,d\tau = \int_{-\infty}^{\infty} s_1^2(t)\,dt = r_{11}(0) \tag{4.42}$$

if $c = 1$.

Thus $y(T)$ is the autocorrelation at lag zero of $s_1(t)$, and may be obtained by cross-correlating the noisy input with a locally produced noise-free signal. This constitutes the correlation detector. Figure 4.13 shows the schematic circuit of a correlation detector. For example, a PCM codeword detector would contain a correlation detector for each codeword as indicated in Figure 4.14.

In a digital m-bit codeword detector the codewords are stored and multiplied with m of the incoming bits. A peak value will occur whenever (i) the m incoming bits correspond exactly to the m-bit codeword or (ii) the m incoming bits correspond by chance with the m-bit codeword. The second outcome is

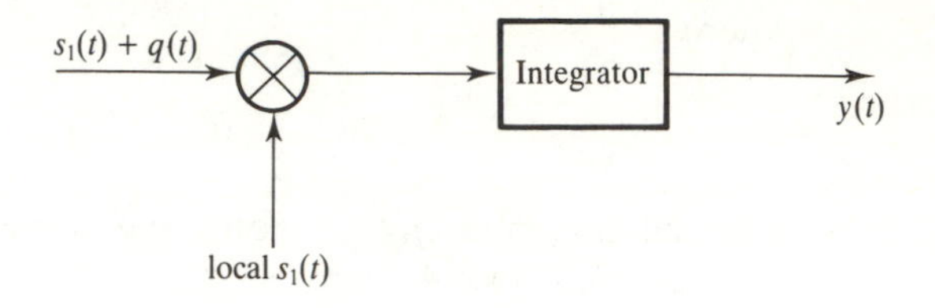

Figure 4.13 A schematic correlation detector.

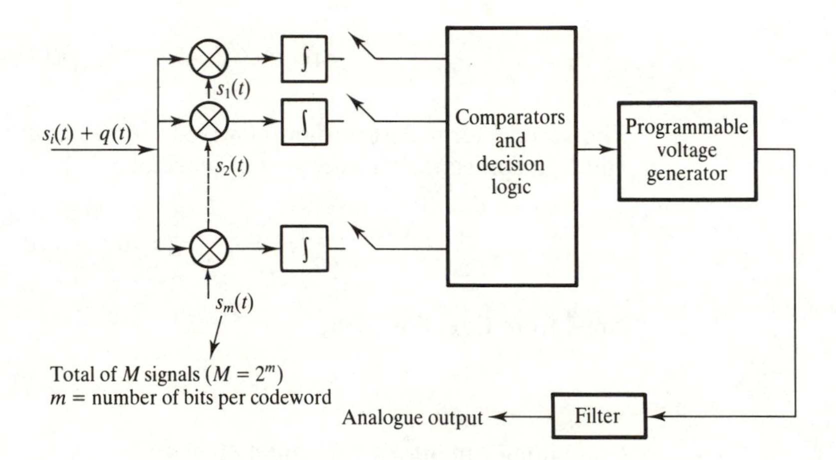

Figure 4.14 A PCM codeword detector based on the correlation detector.

highly undesirable. It might occur if two adjacent codewords happened to contain a bit sequence identical to the required m-bit codeword, or if a codeword had become corrupted. This possibility makes it necessary to arrange for word synchronization as well as bit synchronization in the correlation receiver.

This leads to the requirement for synchronization codewords which have properties which include the following:

(1) the correlation is small for sampling times $t \neq T$:

(2) the correlation is large for samples taken at $t = T$.

A codeword with these properties will be one which has a large autocorrelation at zero lag, but a small autocorrelation at other lags. Thus detection of a large cross-correlation at the receiver will indicate alignment of the incoming codeword with the stored one. This synchronizes the receiver. Now, random waveforms have this autocorrelation property, and may be implemented in a digital receiver by a pseudonoise (PN) sequence, easily generated from a tapped shift register. A three-stage PN sequence generator is shown in Figure

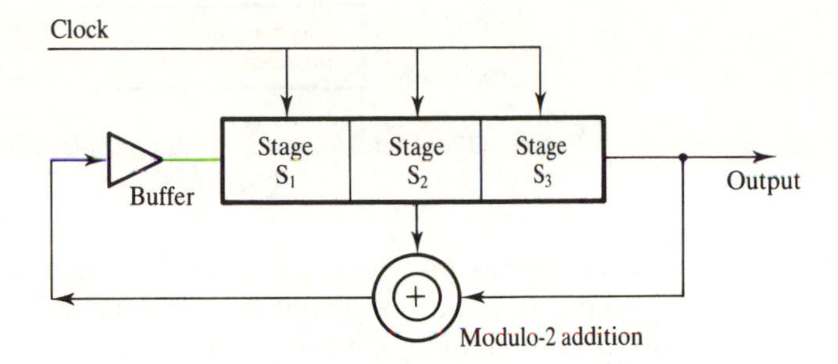

Figure 4.15 A three-stage pseudonoise sequence generator.

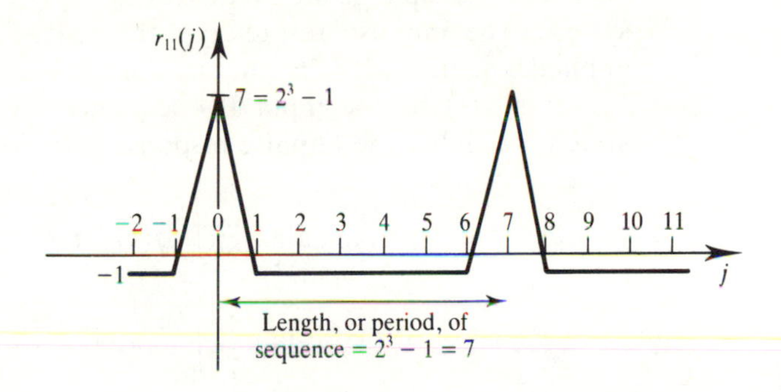

Figure 4.16 The autocorrelation function of a bipolar waveform three-stage pseudonoise generator.

4.15 as an example. The output sequence produced is 1, 1, 1, 0, 0, 1, 0 which then repeats. The autocorrelation function generated when the sequence is represented as a bipolar waveform is shown in Figure 4.16.

Some of the properties of the PN sequences are as follows.

(1) An m-bit codeword produces a sequence of length $2^m - 1$.
(2) The peak values are $2^m - 1$.
(3) The autocorrelation function is equal to -1 other than at the peaks.
(4) The output sequence contains 2^{m-1} 1s and $2^{m-1} - 1$ 0s.
(5) Their power density spectrum is uniform so they may be used as white noise sources.

The last of these properties offers another application for a PN sequence as a source of white noise.

A further application of correlation and of the PN sequence lies in the determination of the impulse response of electrical systems. There can be difficulties in impulse testing systems. For example, in the presence of noise,

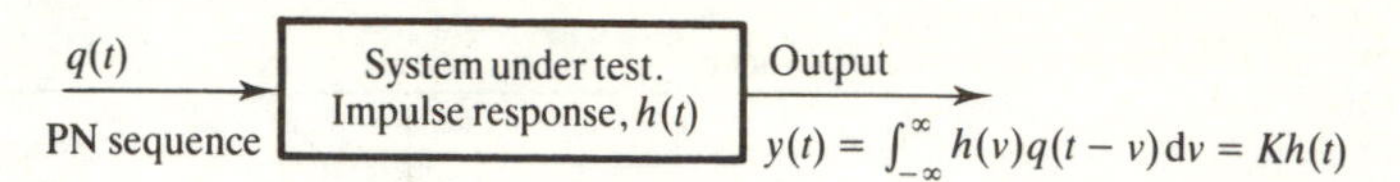

Figure 4.17 Determination of the impulse response of an electrical system.

small impulses may be masked by the noise while larger impulses may cause system overload. It is also difficult to maintain a uniform energy spectral density over the bandwidth using a single impulse. The PN sequence, however, has a uniform energy spectrum, as explained above. Also, if the measurement time is a multiple of the sequence length the variance in the measurement will be zero. This leads to short measurement times and high accuracy.

The principle of the method is to apply a PN sequence to the input of the system. The impulse response is then given by the cross-correlation of the applied sequence with the output. This may be proved as follows.

Let $q(t)$ be the input PN sequence, and let $y(t)$ be the output of the system which has the impulse response $h(t)$. Then,

$$r_{qy}(\tau) = \lim_{T\to\infty} \frac{1}{T}\int_0^T q(t)y(t+\tau)\,dt \tag{4.43}$$

$$= \lim_{T\to\infty} \frac{1}{T}\int_0^T q(t)\,dt \int_{-\infty}^{\infty} h(v)q(t-v+\tau)\,dv \tag{4.44}$$

since $y(t)$ is given as the convolution of the input with the impulse response

$$y(t) = \int_{-\infty}^{\infty} h(v)q(t-v)\,dv \tag{4.45}$$

Changing the order of integration in Equation 4.44 gives

$$r_{qy}(\tau) = \int_{-\infty}^{\infty} h(v)\,dv \lim_{T\to\infty} \frac{1}{T}\int_0^T q(t)q(t-v+\tau)\,dt \tag{4.46}$$

$$= \int_{-\infty}^{\infty} h(v)r_{qq}(\tau - v)\,dv \tag{4.47}$$

Now, $r_{qq}(\tau - v)$ approximates to a delta function because it is the autocorrelation function of the PN sequence. Hence Equation 4.47 may be expressed as

$$r_{qy}(t) = K\int_{-\infty}^{\infty} h(v)\delta(\tau - v)\,dv = Kh(t) \tag{4.48}$$

where K is the area of the impulse function and is equal to the rms value of the noise (Beauchamp, 1973). Figure 4.17 illustrates the arrangement. The method is prone to some errors and precautions should be taken to avoid these.

4.2.3 Fast correlation

The correlation computation may be speeded up by exploiting the correlation theorem, usually stated as

$$r_{12}(j) = F_D^{-1}[X_1^*(k)X_2(k)] \tag{4.49}$$

but which is correctly written as

$$r_{12}(j) = \frac{1}{N} F_D^{-1}[X_1^*(k)X_2(k)] \tag{4.50}$$

where F_D^{-1} denotes the inverse discrete Fourier transform. This approach requires computation of two discrete Fourier transforms (DFTs) and one inverse DFT, each of which is most easily executed using an FFT algorithm (see Chapter 2). If the number of terms in the sequences is sufficiently large, it is quicker to use this FFT method than to calculate the cross-correlation directly.

Proof of the correlation theorem

Let $x_1(l)$, $x_2(r)$, and $x_3(n)$ be periodic sequences of length N, and let their DFTs be $X_1(k)$, $X_2(k)$, and $X_3(k)$ respectively. Furthermore, let

$$X_3(k) = X_1^*(k)X_2(k) \tag{4.51}$$

Now,

$$X_1^*(k) = \sum_{l=0}^{N-1} x_1(l)e^{j(2\pi/N)lk} \tag{4.52}$$

and

$$X_2(k) = \sum_{r=0}^{N-1} x_2(r)e^{j(2\pi/N)(-rk)} \tag{4.53}$$

Substituting Equations 4.52 and 4.53 into 4.51 gives

$$X_3(k) = \sum_{l=0}^{N-1} x_1(l)e^{j(2\pi/N)lk} \sum_{r=0}^{N-1} x_2(r)e^{j(2\pi/N)(-rk)} \tag{4.54}$$

$$= \sum_{l=0}^{N-1} \sum_{r=0}^{N-1} x_1(l)x_2(r)e^{j(2\pi/N)(lk-rk)} \tag{4.55}$$

Now,

$$x_3(n) = \frac{1}{N} \sum_{k=0}^{N-1} X_3(k)e^{j(2\pi/N)nk} \tag{4.56}$$

So, substituting Equation 4.55 into 4.56 yields

$$x_3(n) = \frac{1}{N}\sum_{k=0}^{N-1}\sum_{l=0}^{N-1}\sum_{r=0}^{N-1} x_1(l)x_2(r)e^{j(2\pi/N)(lk-rk+nk)}$$

$$= \frac{1}{N}\sum_{l=0}^{N-1} x_1(l)\sum_{r=0}^{N-1} x_2(r)\left[\sum_{k=0}^{N-1} e^{j(2\pi/N)(l-r+n)k}\right] \tag{4.57}$$

When $r = n + 1$, the term in square brackets equals N. When $r \neq n + 1$ it may be treated as a geometric series of the form

$$\sum ax^n$$

which sums over N terms to

$$\frac{a(1 - x^N)}{(1 - x)}$$

In this case the sum becomes

$$\frac{1[1 - e^{j(2\pi/N)(l-r+n)N}]}{1 - e^{j(2\pi/N)(l-r+n)}} \tag{4.58}$$

The exponent in the numerator is always an integral multiple of 2π, and so the exponential term is unity. Hence the summation equates to zero when $r \neq n + 1$. Equation 4.57 can therefore be written

$$x_3(n) = \frac{1}{N}\sum_{l=0}^{N-1} x_1(l)\sum_{r=0}^{N-1} x_2(r)N\delta(l - r + n) \tag{4.59}$$

in which $\delta(l - r + n) = 1$ when $r = n + 1$ and $\delta(l - r + n) = 0$ when $r \neq n + 1$. Simplifying and putting $r = n + 1$ gives

$$x_3(n) = \sum_{l=0}^{N-1} x_1(l)x_2(l + n) \tag{4.60}$$

or

$$\frac{1}{N}x_3(n) = \frac{1}{N}\sum_{l=0}^{N-1} x_1(l)x_2(l + n) \tag{4.61}$$

The right-hand side of this equation is equivalent to the cross-correlation of $x_1(n)$ and $x_2(n)$ and is seen to be equal to $(1/N)x_3(n)$. From Equation 4.56

$$x_3(n) = F_D^{-1}[X_3(k)] \tag{4.62}$$

Hence, by combining Equations 4.61, 4.62 and 4.50,

$$\frac{1}{N} F_D^{-1}[X_3(k)] = r_{12}(n) = \frac{1}{N} F_D^{-1}[X_1^*(k)X_2(k)] \quad \textbf{(4.63)}$$

Finally, replacing n by j gives

$$r_{12}(j) = \frac{1}{N} F_D^{-1}[X_1^*(k)X_2(k)] \quad \textbf{(4.64)}$$

Example 4.6

Work out the cross-correlation of the two sequences $x_1(n)$ and $x_2(n)$ below by applying the correlation theorem:

$$x_1(n) = \{1, 0, 0, 1\}$$
$$x_2(n) = \{0.5, 1, 1, 0.5\}$$

First use the correlation theorem, Equation 4.64. $X_1(k)$ was found in Section 2.5 to be

$$X_1(k) = 2, 1 + \text{j}, 0, 1 - \text{j}$$

and so

$$X_1^*(k) = 2, 1 - \text{j}, 0, 1 + \text{j}$$

$X_2(k)$ is most easily obtained using the FFT algorithm given in Section 2.5. Thus, with $x_0 = 0.5$, $x_2 = 1$, $x_1 = 1$, and $x_3 = 0.5$,

$$X_{21}(0) = x_0 + x_2 = 1.5$$
$$X_{21}(1) = x_0 - x_2 = -0.5$$
$$X_{22}(0) = x_1 + x_3 = 1.5$$
$$X_{22}(1) = x_1 - x_3 = 0.5$$
$$X_{11}(0) = X_{21}(0) + X_{22}(0) = 3$$
$$X_{11}(1) = X_{21}(1) + (-\text{j})X_{22}(1) = -0.5 - \text{j}0.5$$
$$X_{11}(2) = X_{21}(0) - X_{22}(0) = 0$$
$$X_{11}(3) = X_{21}(1) - (-\text{j})X_{22}(1) = -0.5 + \text{j}0.5$$

Bringing the values of the FFTs together gives

$$X_1^*(k) = 2, 1 - \text{j}, 0, 1 + \text{j}$$
$$X_2(k) = 3, -0.5 - \text{j}0.5, 0, -0.5 + \text{j}0.5$$

So,

$$X_1^*(1)X_2(1) = 2 \times 3 = 6$$
$$X_1^*(2)X_2(2) = (1 - \text{j})(-0.5 - \text{j}0.5) = -1$$
$$X_1^*(3)X_2(3) = 0 \times 0 = 0$$
$$X_1^*(4)X_2(4) = 0.5(1 + \text{j})(-1 + \text{j}) = -1$$

Thus

$$[X_1^*(k)X_2(k)] = 6, -1, 0, -1$$

It is now necessary to take the inverse DFT (IDFT) of this. As explained in Section 2.6, the IDFT is obtainable by changing the signs of the exponents (in the weighting factors W_N) in the above FFT algorithm and dividing the result by N. Hence, and without changing the notation in the algorithm,

$$X_{21}(0) = x_0 + x_2 = 6$$
$$X_{21}(1) = x_0 - x_2 = 6$$
$$X_{22}(0) = x_1 + x_3 = -2$$
$$X_{22}(1) = x_1 - x_3 = 0$$
$$X_{11}(0) = X_{21}(0) + X_{22}(0) = 4$$
$$X_{11}(1) = X_{21}(1) + \text{j}X_{22}(1) = 6$$
$$X_{11}(2) = X_{21}(0) - X_{22}(0) = 8$$
$$X_{11}(3) = X_{21}(1) - \text{j}X_{22}(1) = 6$$

The components of $F_D^{-1}[X_1^*(k)X_2(k)]$ are obtained by dividing the values of $X_{11}(0)$, $X_{11}(1)$, $X_{11}(2)$, and $X_{11}(3)$ by $N = 4$. Thus

$$F_D^{-1}[X_1^*(k)X_2(k)] = 1, 1.5, 2, 1.5$$

Now, from Equation 4.64,

$$r_{12}(j) = \frac{1}{4} F_D^{-1}[X_1^*(k)X_2(k)] = \{0.25, 0.375, 0.5, 0.375\} \qquad \textbf{(4.65)}$$

This correlation will be circular because all the data is periodic with period N.

The cross-correlation $r_{12}(j)$ may be worked out directly to be

$$r_{12}(0) = (1 \times 0.5 + 0 + 0 + 1 \times 0.5)/4 = 0.25$$

$$r_{12}(1) = (1 \times 1 + 0 + 0 + 1 \times 0.5)/4 = 0.375$$

$$r_{12}(2) = (1 \times 1 + 0 + 0 + 1 \times 1)/4 = 0.5$$

$$r_{12}(3) = (1 \times 0.5 + 0 + 0 + 1 \times 1)/4 = 0.375$$

The next value, $r_{12}(4)$, is 0.25, the same as $r_{12}(0)$, and the sequence repeats periodically. This is circular correlation, as discussed in Section 4.2.1, and this result agrees with that derived above using the correlation theorem. The correlation theorem can be used to obtain the linear correlation by adding augmenting zeros to the two sequences as explained in Section 4.2.1. Thus, if the sequence lengths are N_1 for $x_1(n)$ and N_2 for $x_2(n)$, then $N_2 - 1$ zeros are added to $x_1(n)$ and $N_1 - 1$ zeros are added to $x_2(n)$. The cross-correlation is then computed using these two augmented sequences. This method of evaluating cross-correlations by using the correlation theorem and FFTs is known as fast correlation.

Cross-correlation calculations may also be speeded up by implementing them recursively, and this will be illustrated for the case of zero lag. The cross-correlation at zero lag of two sampled waveforms $x_1(n)$ and $x_2(n)$ is

$$r_{12}(0) = \frac{1}{N} \sum_{n=0}^{N-1} x_1(n)x_2(n) \tag{4.66}$$

This involves the computation of N products, $N - 1$ sums and one division. This may occupy excessive time in an on-line application in which pairs of new data values are arriving at the sampling rate. The calculation has to be repeated when the next data pair is available. The new calculation will differ from the previous one only in that the product of the new data has to be added to the sum of product pairs, and the first product has to be subtracted. Thus, for each cross-correlation:

$$\begin{aligned} \text{new value} = \text{previous value} &+ \frac{1}{N}(\text{product of two new data}) \\ &- \frac{1}{N}(\text{product of first two data}) \end{aligned} \tag{4.67}$$

This is the basis of the recursive algorithm. Each cross-correlation now only requires one multiplication, one subtraction, one addition, and one division, provided that the data pair products are saved. For an N-point correlation, the recursive approach gives correct values after the first $N - 1$ points have been computed.

In many applications it is required to set the means of the data to zero, for example to remove dc levels from electrical waveforms. This requires calculation of the average value of the waveforms and then subtraction of it from all the sampled values. This mean value calculation may also be done recursively since for each new data pair:

$$\text{new average} = \text{previous average} + \frac{1}{N}(\text{new datum} - \text{first datum}) \quad \textbf{(4.68)}$$

It is also possible to combine the mean level subtraction and cross-correlation calculations in one recursive algorithm. Consider

$$\bar{x}_1(k) = \frac{1}{N}\sum_{n=0}^{N-1} x_1(n) \quad \textbf{(4.69)}$$

and

$$\bar{x}_2(k) = \frac{1}{N}\sum_{n=0}^{N-1} x_2(n) \quad \textbf{(4.70)}$$

The value of the cross-correlation function of the kth set of N points is

$$r_{12}(k) = \frac{1}{N}\sum_{n=0}^{N-1} x_1(n)x_2(n) \quad \textbf{(4.71)}$$

When the means have been removed the cross-correlation function value becomes $r_{12}^0(k)$ where

$$r_{12}^0(k) = \frac{1}{N}\sum_{n=0}^{N-1} [x_1(n) - \bar{x}_1(k)][x_2(n) - \bar{x}_2(k)] \quad \textbf{(4.72)}$$

which, on expansion, simplifies to

$$r_{12}^0(k) = r_{12}(k) - \bar{x}_1(k)\bar{x}_2(k) \quad \textbf{(4.73)}$$

Combining Equations 4.67 and 4.70 gives

$$r_{12}(k) = r_{12}(k-1) + \frac{1}{N}[x_1(k)x_2(k) - x_1(k-N)x_2(k-N)] \quad \textbf{(4.74)}$$

From Equation 4.68,

$$\bar{x}_1(k) = \bar{x}_1(k-1) + \frac{1}{N}[x_1(k) - x_1(k-N)] \quad \textbf{(4.75)}$$

and

$$\bar{x}_2(k) = \bar{x}_2(k-1) + \frac{1}{N}[x_2(k) - x_2(k-N)] \qquad \textbf{(4.76)}$$

Equations 4.73–4.76 constitute the recursive algorithm which combines subtraction of the mean from the data with the calculation of the cross-correlation. Each calculation requires only three multiplications, four subtractions, three additions, and four divisions. Practitioners should note that care has to be exercised in the choice of N when the mean values of the data are varying, or inaccurate results may be obtained.

4.3 Convolution description

The term convolution describes, amongst other things, how the input to a system interacts with the system to produce the output. Generally the system output will be a delayed and attenuated or amplified version of the input. It is particularly useful to consider the output from the system owing to an impulse input. This is because any input may be represented as a sequence of impulses of different strengths. The output of the system owing to the impulse input will not be a corresponding impulse, but will vary with time, passing through a maximum, as shown in Figure 4.18. This figure shows that at sampling instant m the output owing to the unit impulse applied at sampling instant 0 is $h(m)$. The characteristic is known as the impulse response $h(m)$ of the system.

Consider now the application of a sequence of impulses $x(m)$ to the system, applied at the sampling instants m. Referring to Figure 4.19, the output at instant 0 is $y(0)$ given by

$$y(0) = h(0)x(0)$$

At the sampling instant $m = 1$ the output will be given by $h(0)x(1)$, the effect of the current input $x(1)$, plus the delayed effect $h(1)x(0)$ of the input applied at sampling instant $m = 0$. Thus

$$y(1) = h(1)x(0) + h(0)x(1)$$

Similarly, subsequent outputs will be given by

$$\begin{aligned} y(2) &= h(2)x(0) + h(1)x(1) + h(0)x(2) \\ y(3) &= h(3)x(0) + h(2)x(1) + h(1)x(2) + h(0)x(3) \\ &\vdots \\ y(n) &= h(n)x(0) + h(n-1)x(1) + \ldots + h(0)x(n) \end{aligned} \qquad \textbf{(4.77)}$$

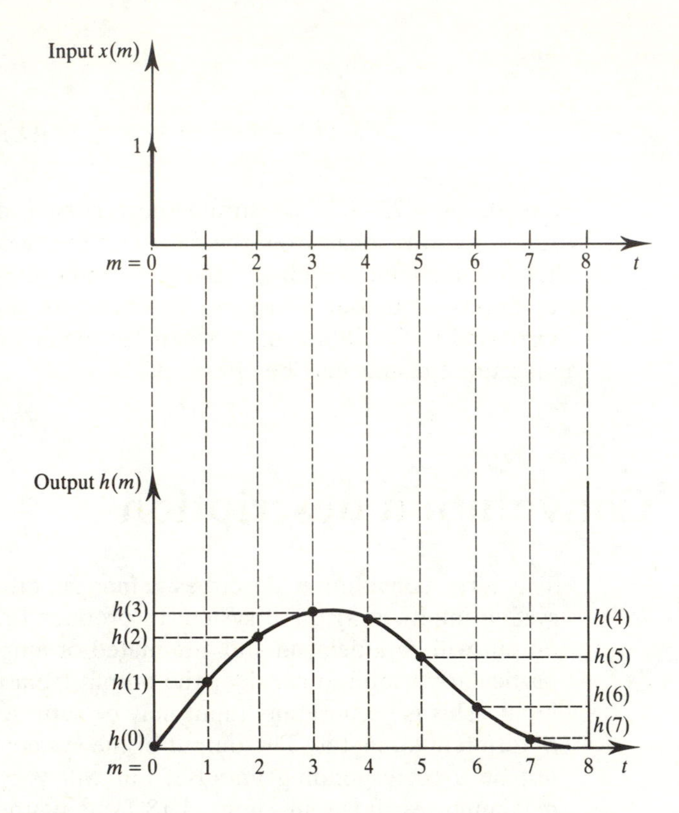

Figure 4.18 Impulse input and corresponding impulse response of a system.

The output may only be written in this manner as a linear sum of the effects of previous inputs if the system is a linear one. Equation 4.77 describes the output of a first-order linear system.

Inspection of the above expressions reveals that the output is obtained by multiplying the input sequence by the corresponding points of the time-reversed impulse response function. Alternatively, since Equation 4.77 may equally well be written as

$$y(n) = h(0)x(n) + h(1)x(n-1) + \ldots + h(n)x(0) \tag{4.78}$$

the output may be regarded as the product of the corresponding pairs of points in the impulse response function by the time-reversed input sequence. Thus the convolution sum is equivalent to the cross-correlation of the one sequence by the time-reversed second sequence.

Equations 4.77 and 4.78 may be written compactly as

$$y(n) = \sum_{m=0}^{n} h(n-m)x(m) \tag{4.79}$$

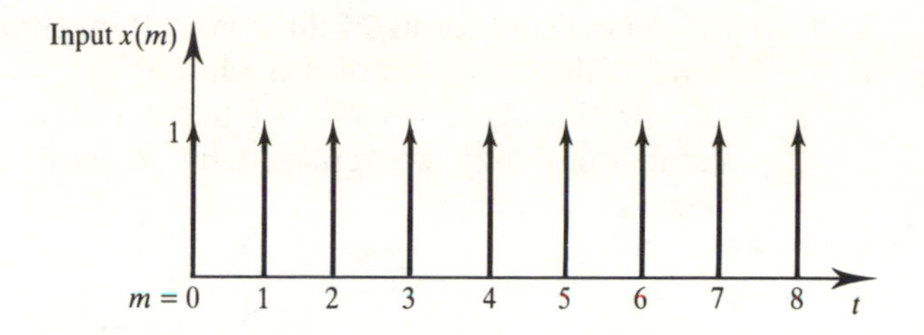

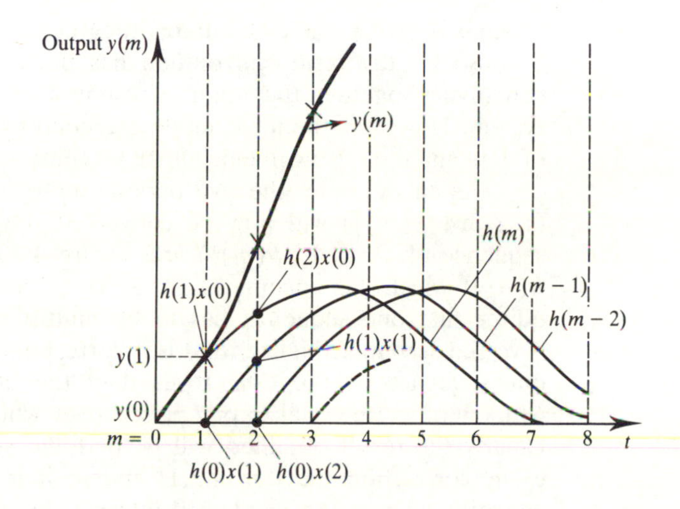

Figure 4.19 Applied impulse sequence and system response derived from the individual impulse responses.

and

$$y(n) = \sum_{m=0}^{n} h(m)x(n-m) \tag{4.80}$$

These latter two functions are referred to as the convolution sums of the inputs by the impulse response function, and the output is said to be given by the convolution of the input by the impulse response of the system.

Equations 4.79 and 4.80 may be extended to waveforms of infinite duration by writing them as

$$y(n) = \sum_{m=-\infty}^{\infty} x(m)h(n-m) = x(n) \circledast h(n) \tag{4.81}$$

and

$$y(n) = \sum_{m=-\infty}^{\infty} h(m)x(n-m) = h(n) \circledast x(n) \tag{4.82}$$

the generalized forms of the convolution sum. In these equations the symbol ⊛ signifies the operation of convolution.

If the input consists of a continuous sequence of impulses the above summations may be replaced by integrals, so, for example, Equation 4.81 becomes

$$y(t) = \int_{-\infty}^{\infty} x(\lambda)h(t - \lambda)\,\mathrm{d}\lambda \tag{4.83}$$

which is known as the convolution integral.

So far the term convolution has been taken to describe the result of convolving together the impulse response of a system and the input to the system. However, the idea can be extended to the convolution of any two sets of data and the term will henceforth be considered in this more general sense.

As an example, the two periodic time sequences ($4, 3, 2, 1\,\{h(m)\}$) and ($1, 2, 3, 4\,\{x(m)\}$) will now be convolved. Figure 4.20(a) shows the periodic sequence (4, 3, 2, 1 $\{h(m)\}$) and Figure 4.20(b) the time-reversed sequence $h(-m)$ which has become (1, 2, 3, 4). (Recall that the convolution sum requires that one sequence has to be multiplied point by point by the time-reversed second sequence, that is it corresponds to the cross-correlation of the one sequence by the time reversal of the other.) The figure also shows a window of width equal to one period over which the convolution is evaluated. Clearly the result obtained will be periodic as for the corresponding case of cyclic correlation (Section 4.2.1) and so it is only necessary to evaluate the convolution over the windowed interval. Figure 4.20(f) shows the second sequence (1, 2, 3, 4 $\{x(m)\}$) for reference.

Now, when $n = 0$, Equation 4.79 becomes

$$y(0) = \sum_{m=0}^{n} h(-m)x(m)$$

and is obtained by cross-correlating the windowed data in Figures 4.20(b) and 4.20(f):

$$y(0) = 4 \times 1 + 1 \times 2 + 2 \times 3 + 3 \times 4 = 24$$

When $n = 1$ Equation 4.79 becomes

$$y(1) = \sum_{m=0}^{n} h(1 - m)x(m)$$

and is obtained by cross-correlating the windowed data in Figures 4.20(c) and 4.20(f), giving

$$y(1) = 3 \times 1 + 4 \times 2 + 1 \times 3 + 2 \times 4 = 22$$

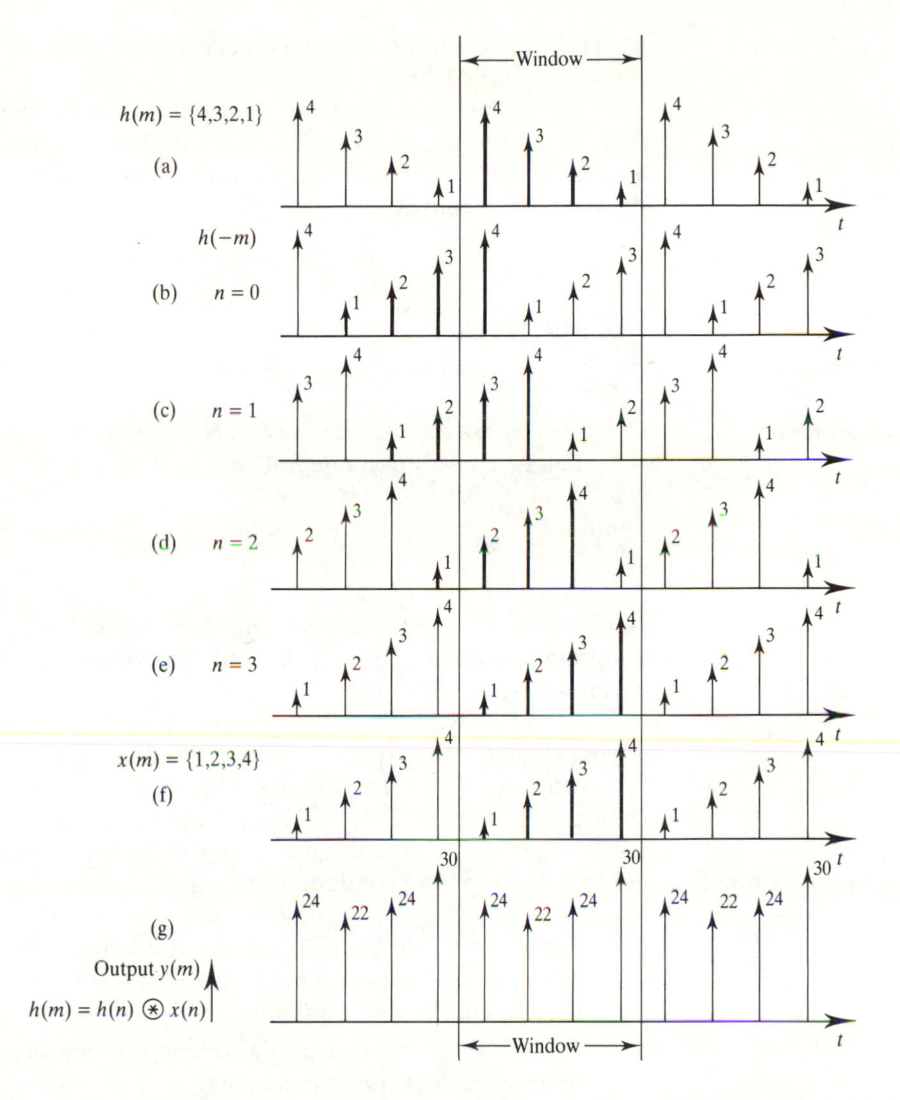

Figure 4.20 The convolution $y(m)$ of $h(n)$ and $x(n)$. (a) The periodic sequence h(m). (b) The time-reversed sequence $h(-m)$. (c) to (e) Versions of $h(-m)$ shifted right by incremental lags. (f) The sequence $x(m)$. (g) The output sequence $y(m) = h(n) \circledast x(n)$.

Similarly, it is found that

$$y(2) = 2 \times 1 + 3 \times 2 + 4 \times 3 + 1 \times 4 = 24$$

and

$$y(3) = 1 \times 1 + 2 \times 2 + 3 \times 3 + 4 \times 4 = 30$$

whereafter the output sequence repeats cyclically. This output sequence is shown in Figure 4.20(g).

When the waveforms are well defined mathematically the convolution may be performed analytically. By considering a similar example, and also illustrating the steps involved graphically, it is possible to obtain a better understanding of the convolution process.

Example 4.7

Convolve the waveforms $x(t)$ and $h(t)$ of Figure 4.21(a) analytically.

Let the convolution integral be

$$y(t) = x(t) \circledast h(t) = \int_{-\infty}^{\infty} x(\tau)h(t-\tau)\,\mathrm{d}\tau \qquad \textbf{(4.84)}$$

Equation 4.84 corresponds to Equation 4.83 in which the variable λ has been replaced by τ to indicate that time lags are being applied. The convolution integral depends on the variable τ and so Figure 4.21(a) has to be replaced by Figure 4.21(b).

It is now necessary to time reverse $h(\tau)$ as shown in Figure 4.21(c). $h(-\tau)$ is next shifted with respect to $x(\tau)$ in the direction of positive τ. The resulting waveform $h(t-\tau)$ then overlaps $x(\tau)$ in five separate geometrical stages as shown in Figures 4.21(d), 4.21(e), 4.21(g), 4.21(h), and 4.21(i). For each of these stages there is a corresponding convolution integral. Thus $x(t) \circledast h(t)$ exists as five separate contiguous regions.

- *Stage 1* $t < 0$ and $h(t-\tau)$ does not overlap $x(\tau)$ (Figure 4.21(d)). As the functions do not overlap $x(\tau)h(t-\tau) = 0$ for all t and there is no contribution to the convolution integral.
- *Stage 2* $0 < t \leqslant 2$ and partial overlap occurs between $h(t-\tau)$ and $x(\tau)$ (Figure 4.21(e)). Over this range

$$y(t) = \int_{\tau=0}^{\tau=t} x(\tau)h(t-\tau)\,\mathrm{d}\tau = \int_{\tau=0}^{\tau=t} (3)\times(2)\,\mathrm{d}\tau$$

$$y(t) = 6[\tau]_0^t = 6t, \quad 0 < t \leqslant 2 \qquad \textbf{(4.85)}$$

This geometrical stage terminates when $t = 2$, as shown in Figure 4.21(f).

- *Stage 3* $2 \leqslant t \leqslant 3$ and there is complete overlap of $h(t-\tau)$ and $x(\tau)$ (Figure 4.21(g)). Over this range of t,

$$y(t) = \int_{\tau=t-2}^{t} (3)\times(2)\,\mathrm{d}\tau = 6[\tau]_{t-2}^{t}$$

$$y(t) = 6(t - t + 2) = 12, \quad 2 \leqslant t \leqslant 3 \qquad \textbf{(4.86)}$$

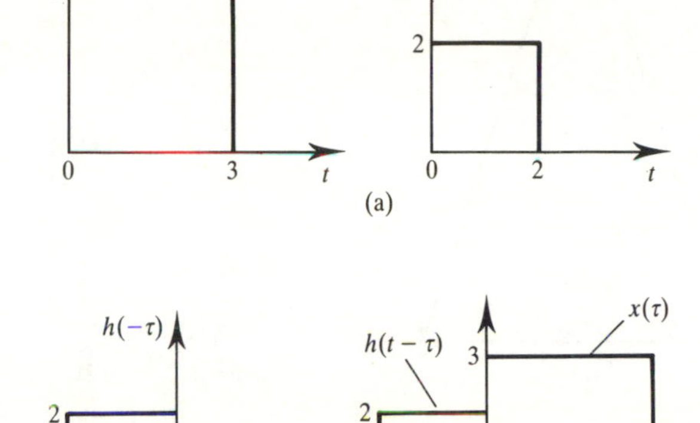

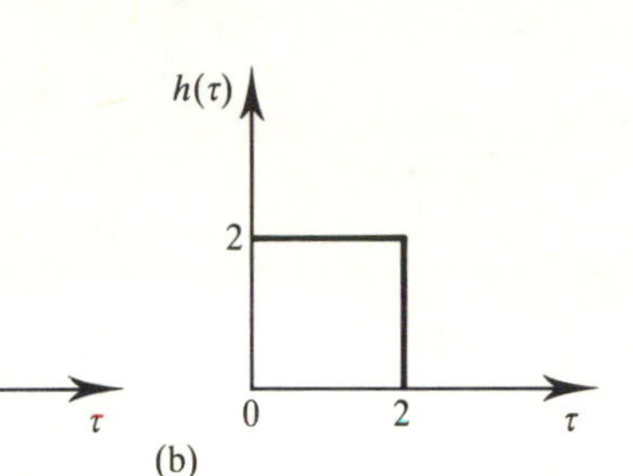

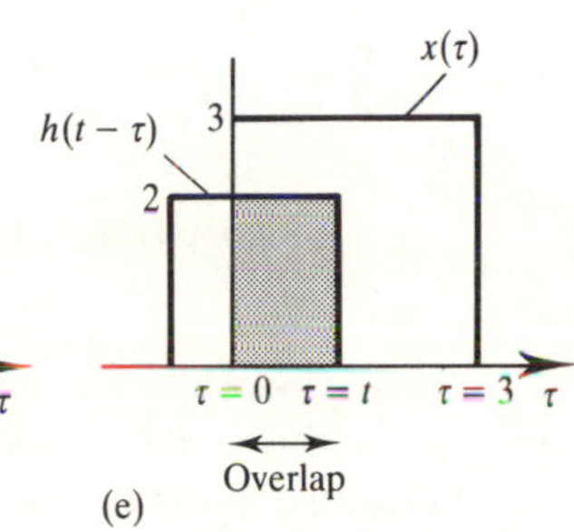

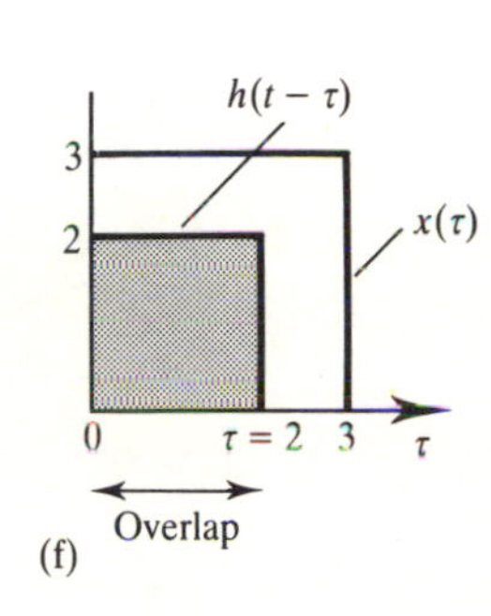

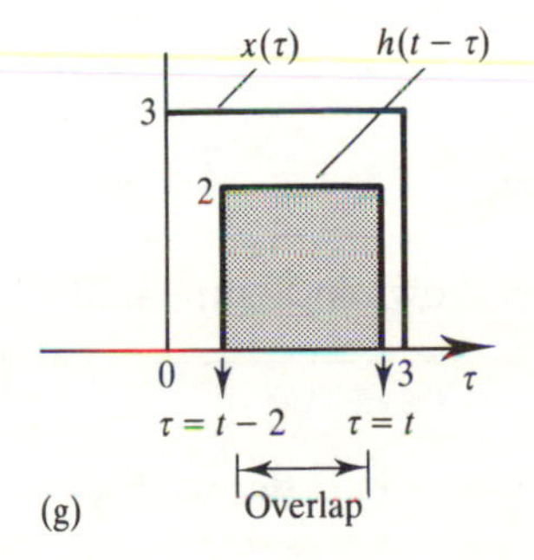

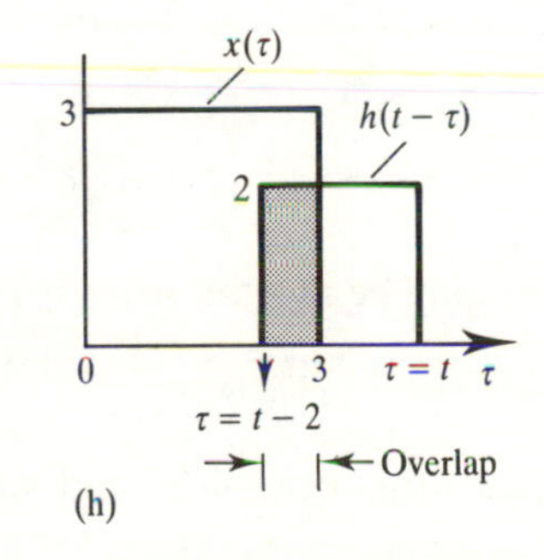

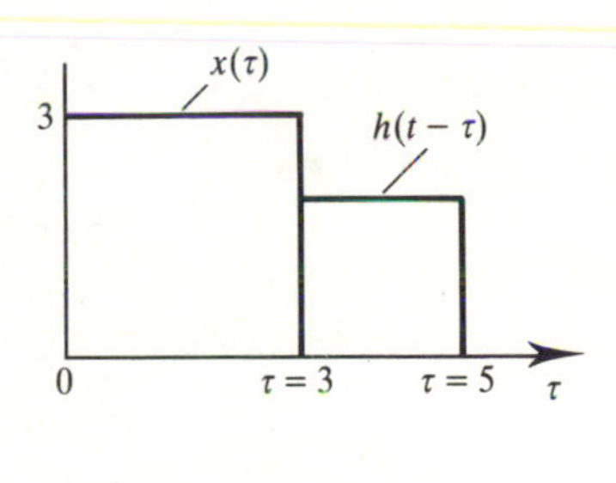

Figure 4.21 (a) The waveforms $x(t)$ and $h(t)$ which are to be convolved analytically. (b) $x(\tau)$ and $h(\tau)$ vs τ. (c) $h(-\tau)$ vs τ. (d) $h(t-\tau)$ and $x(\tau)$ vs τ. $t<0$, $h(t-\tau)$ does not overlap $x(\tau)$. (e) $h(t-\tau)$ and $x(\tau)$ vs τ. $0<t\leqslant 2$. First partial overlap occurs between $h(t-\tau)$ and $x(\tau)$. (f) $h(t-\tau)$ and $x(\tau)$ vs (τ). $t=2$. End of first partial overlap. (g) $h(t-\tau)$ and $x(\tau)$ vs (τ). $2\leqslant t\leqslant 3$. Complete overlap between $h(t-\tau)$ and $x(\tau)$ occurs. (h) $h(t-\tau)$ and $x(\tau)$ vs (τ). $3\leqslant t\leqslant 5$. Second partial overlap occurs between $h(t-\tau)$ and $x(\tau)$. (i) $h(t-\tau)$ and $x(\tau)$ vs τ. $t>5$. $h(t-\tau)$ does not overlap $x(\tau)$.

- *Stage 4* $3\leqslant t\leqslant 5$. This is another type of overlap region shown in Figure 4.21(h):

$$y(t)=\int_{\tau=t-2}^{\tau=3}(3)\times(2)\,\mathrm{d}\tau=6[\tau]_{t-2}^{3}=6(5-t)=30-6t \qquad \textbf{(4.87)}$$

- *Stage 5* $t>5$. As seen in Figure 4.21(i) this is a second region of no overlap, and so there is no contribution to the convolution integral.

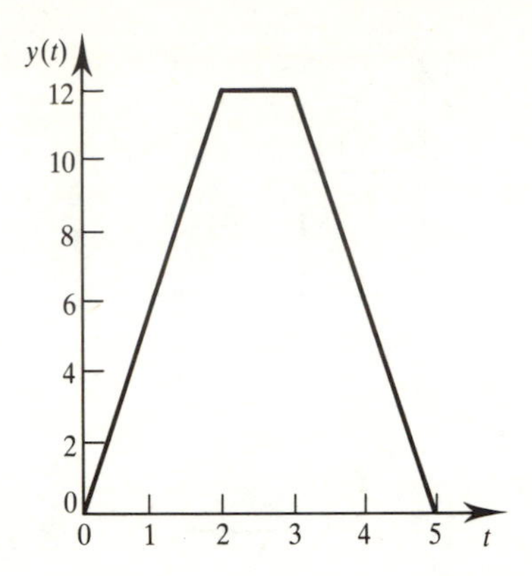

Figure 4.22 The convolution $y(t) = x(t) \circledast h(t)$ vs t.

Stages 2 to 4 thus each make a contribution to the convolution integral, with the convolution integral having a different expression for each of the three regions corresponding to the three stages, as summarized below:

$$0 < t \leqslant 2 \qquad y(t) = 6t$$

$$2 \leqslant t \leqslant 3 \qquad y(t) = 12$$

$$3 \leqslant t \leqslant 5 \qquad y(t) = 30 - 6t$$

From these expressions $y(t)$ may be plotted against t as shown in Figure 4.22.

It is now convenient to repeat Equations 4.81 and 4.83 here in order to discuss them.

$$y(n) = \sum_{m=-\infty}^{\infty} x(m)h(n - m) = x(n) \circledast h(n) \tag{4.81}$$

and

$$y(t) = \int_{-\infty}^{\infty} x(\lambda)h(t - \lambda)\,\mathrm{d}\lambda \tag{4.83}$$

Inspection of these equations serves as a reminder that the convolution was carried out in time. This is known as convolution in the time domain. It is known that in the frequency domain the output component of a system at frequency, f, is $Y(f)$, given by

$$Y(f) = H(f)X(f) \tag{4.88}$$

where $H(f)$ is the frequency response function of the system at frequency, f, and $X(f)$ is the Fourier transform of the input $x(t)$. It can also be demonstrated that $H(f)$ is the Fourier transform of $h(t)$. The inverse Fourier transform of Equation 4.88 is

$$F^{-1}[Y(f)] = y(t) = F^{-1}[H(f)X(f)] \quad \textbf{(4.89)}$$

Bringing together Equations 4.83 and 4.89 shows that

$$y(t) = \int_{-\infty}^{\infty} x(\lambda)h(t-\lambda)\,\mathrm{d}\lambda = x(y) \circledast h(t) = F^{-1}[H(f)X(f)] \quad \textbf{(4.90)}$$

Thus, it is seen that the convolution of two waveforms in the time domain is equivalent to the inverse Fourier transform of the product of the Fourier transforms of the two waveforms. This useful fact is often stated in the abbreviated form that convolution in the time domain is equivalent to multiplication in the frequency domain.

The dual of this relationship exists, that is convolution in the frequency domain is equivalent to multiplication in the time domain. Thus it can be shown (McGillem and Cooper, 1974) that

$$\begin{aligned} X(\omega) &= \frac{1}{2\pi}\int_{-\infty}^{\infty} X(\omega-u)H(u)\,\mathrm{d}u = X(f) \circledast H(f) \\ &= F[y(t)] = F[x(t)h(t)] \end{aligned} \quad \textbf{(4.91)}$$

Thus, the Fourier transform of the product of two time sequences corresponds to the convolution of the Fourier transforms of the two sequences. This result is of practical use in explaining one of the effects of windowing data prior to spectral analysis (see Chapter 10). In this procedure the digitized data sequence is multiplied point by point by another sequence which consists of the sampled points of a window function. This is known as windowing and is carried out to reduce the errors on computing the energy spectrum of the data. The discrete Fourier transform of the windowed data is then computed and from this the energy spectrum is calculated. The purpose is to obtain the energy spectrum of the data sequence, but, from the above, what is actually obtained is the spectrum of the data sequence convolved with the spectrum of the window sequence.

4.3.1 Properties of convolution

(1) *Commutative law*

$$x_1(t) \circledast x_2(t) = x_2(t) \circledast x_1(t) \quad \textbf{(4.92)}$$

Note that this is identical to

$$\int_{-\infty}^{\infty} x_1(\tau)x_2(t-\tau)\,d\tau = \int_{-\infty}^{\infty} x_2(\tau)x_1(t-\tau)\,d\tau$$

(2) *Distributive law*

$$x_1(t) \circledast [x_2(t) + x_3(t)] = x_1(t) \circledast x_2(t) + x_1(t) \circledast x_3(t) \tag{4.93}$$

(3) *Associative law*

$$x_1(t) \circledast [x_2(t) \circledast x_3(t)] = [x_1(t) \circledast x_2(t)] \circledast x_3(t) \tag{4.94}$$

These properties may be proved either by manipulating the integrations involved or by considering the convolutions in terms of cross-correlating the one sequence by the time-reversed second sequence.

4.3.2 Circular convolution

Section 4.2.1 illustrated that the result of correlating two unequal length periodic sequences was a cyclic sequence of period equal to that of the shorter sequence, which was, therefore, an incorrect result. Because convolution is equivalent to the cross-correlation of one sequence by the reverse of a second sequence, the same will be true of convolution. Therefore, as with correlation, in convolution it is necessary that the two sequences be of the same length. Thus if the sequence lengths are N_1 and N_2, then $N_2 - 1$ augmenting zeros must be added to the sequence of length N_1, and $N_1 - 1$ augmenting zeros must be added to the sequence of length N_2. Both sequences will now be of identical length $N_1 + N_2 - 1$, and the correct linear convolution will be obtained, subject to taking other precautions as described for correlation.

4.3.3 Fast linear convolution

In Section 4.2.3 it was shown that correlation computations could be speeded up using the correlation theorem. A similar theorem, the convolution theorem, exists in the case of convolution. Thus, in discrete terminology and for the time domain

$$x_1(l) \circledast x_2(r) = F_D^{-1}[X_1(k)X_2(k)] \tag{4.95}$$

Equation 4.95 is the convolution theorem, in which F_D^{-1} denotes the inverse discrete Fourier transform, $X_1(k)$ is the discrete Fourier transform of $x_1(l)$, and $X_2(k)$ is the discrete Fourier transform of $x_2(r)$. As in Section 4.2.3, $x_1(l)$ and $x_2(r)$ are periodic sequences of length, N.

Proof of convolution theorem

The proof of this theorem is almost identical to that of the correlation theorem as given in Section 4.2.3. In convolution one of the data sequences is reversed, and so, instead of Equation 4.52, its conjugate is employed, that is

$$X_1(k) = \sum_{l=0}^{N-1} x_1(l) e^{j(2\pi/N)(-lk)} \quad \textbf{(4.96)}$$

while Equation 4.53 is used again:

$$X_2(k) = \sum_{r=0}^{N-1} x_2(r) e^{j(2\pi/N)(-rk)} \quad \textbf{(4.97)}$$

Then, once more defining $x_3(n)$ as a periodic sequence of length N with DFT $X_3(k)$, $X_3(k)$ is written as

$$X_3(k) = X_1(k) X_2(k) \quad \textbf{(4.98)}$$

The procedure of Section 4.2.3 is then followed, leading to the required result

$$x_1(l) \circledast x_2(r) = F_D^{-1}[X_1(k) X_2(k)] \quad \textbf{(4.99)}$$

for time domain convolution. For convolution in the frequency domain the analogous equation below applies:

$$\frac{1}{N}[X_1(k) \circledast X_2(k)] = F_D[x_1(l) x_2(r)] \quad \textbf{(4.100)}$$

The last two equations represent periodic, or circular, convolutions which may be converted to linear by adding augmenting zeros as described in Section 4.3.2.

4.3.4 Computational advantages of fast linear convolution

The method of fast linear convolution only offers the advantage of greater computational speed over the direct approach if the number of values to be convolved is sufficiently large. The number of multiplications required to perform the convolution by the direct and fast methods is compared here as a measure of their relative computational efficiencies.

The necessary computations for the direct method were given in Equations 4.77. It can be seen from these equations that to obtain the linear convolution of the two N-point sequences $h(n-m)$ and $x(m)$ it is necessary to multiply each value of $h(n-m)$ by each value of $x(m)$. Thus N values of $h(n-m)$ are each to be multiplied by N values of $x(m)$, making $N \times N = N^2$ multiplications in all.

Table 4.1 The numbers of real multiplications necessary for the convolution of two N-point sequences.

N	*Direct method*	*Fast convolution*	*Number ratio, fast: direct*
8	64	448	7
16	256	1 088	4.25
32	1 024	2 560	2.5
64	4 096	5 888	1.4375
128	16 384	13 312	0.8125
256	65 536	29 696	0.4531
512	262 144	65 536	0.250
1024	1 048 576	143 360	0.1367
2048	4 194 304	311 296	0.0742

Now consider linear convolution of the same two N-point sequences by the fast method according to Equation 4.99. The addition of the necessary augmenting zeros means each sequence is of length $2N-1$ points. Assume $2N-1 \approx 2N$, for example $N \geqslant 8$, and that in order to use a radix-2 FFT N is given by an integer power of 2, that is $N=2^d$ where d is an integer. The number of complex multiplications for an N-point FFT was shown to be $(N/2)\log_2 N$ (Section 2.5.3), so for the $2N$-point FFT $(2N/2)\log_2 2N$ or $N\log_2 2N$ complex multiplications are necessary. Equation 4.99 requires two DFTs and one inverse DFT to be computed. The inverse will be computed using the modified DFT (Section 2.6). The computation therefore requires computation of three $2N$-point FFTs involving $3N\log_2 2N$ complex multiplications. Further, for each of the $2N$ values of Equation 4.99 it is necessary to evaluate the complex multiplications $X_1(k)X_2(k)$, thus increasing the number of complex multiplications necessary to $3N\log_2 2N + 2N$. Now, each complex multiplication of the form $(A+\mathrm{j}B)(C+\mathrm{j}D)$ requires four real multiplications: AC, AD, BC, and BD. Hence $12N\log_2 2N + 8N$ real multiplications are necessary.

It is therefore concluded that the direct method requires N^2 real multiplications while the method of fast convolution requires $12N\log_2 2N + 8N$ of them. Table 4.1 compares the numbers of real multiplications required for different values of N. The table demonstrates that fast convolution is faster than the direct method for sequences containing in excess of 128 data points, being approximately ten times faster for sequences containing 1024 points. The same conclusion also holds for direct and fast correlation.

4.3.5 Convolution and correlation by sectioning

So far it has been assumed that the two functions to be convolved (or correlated) are of finite duration. This, however, may not be the case. For example the input data may be considered to be of infinite duration, either because it is

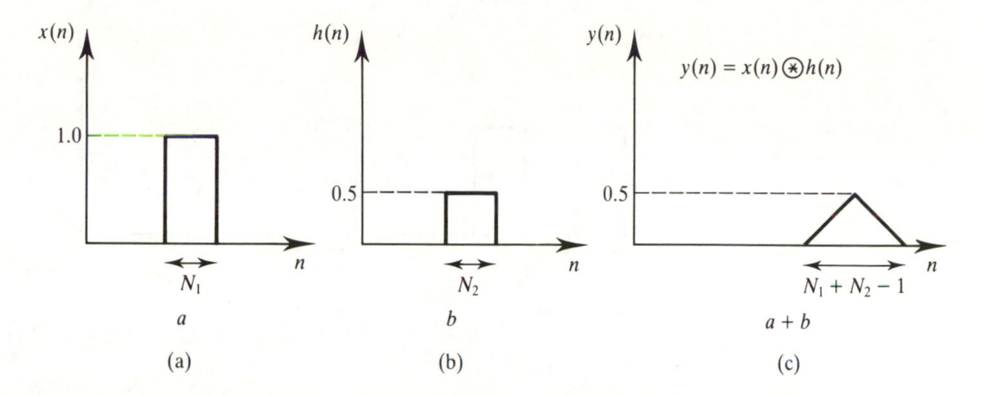

Figure 4.23 The convolution, $y(n) = x(n) \circledast h(n)$, of the two waveforms $x(n)$ and $h(n)$ which do not commence at the origin.

in fact continuous, or more likely because the available memory is not large enough to store all of it. In these cases it is necessary to perform the convolution (or correlation) in stages by dividing the input data into separate sections, performing the calculation for each of the input sections and then combining the results. The two methods used for this are known as the overlap–add and the overlap–save methods and will be described below but, first, these methods will be introduced by considering how to make the computations more efficient when the two functions do not start at the time origin.

Figure 4.23 shows two sampled waveforms $x(n)$ and $h(n)$ and their convolution $x(n) \circledast h(n) = y(n)$. $x(n)$ and $h(n)$ commence respectively at sample points a and b, so that if a and b are large compared with the number of data, N_1 and N_2 in $x(n)$ and $h(n)$ respectively, then a considerable number of calculations involving zero data will be performed. The number of such calculations can be reduced by shifting the waveforms to commence at the origin as in Figure 4.24. Augmenting zeros must then be added to each waveform so that they both contain the same number of points $N = N_1 + N_2 - 1$ so that their periodic convolution corresponds to the linear convolution of the two waveforms. The convolution is performed by applying the convolution theorem, Equation 4.95, and an FFT algorithm. The correct result is then obtained by displacing the resulting convolution along the n-axis to commence at $n = a + b$ (Figure 4.24(d)). It is assumed in this figure than $N = 2^d$ where d is integral so that a radix-2 FFT may be used.

Figure 4.25 shows the analogous case for the correlation of $x(n)$ and $h(n)$, $r_{xh}(n)$. When these waveforms are transposed to the origin, augmenting zeros added so that $N = 2^d \geqslant N_1 + N_2 - 1$, and the correlation carried out using the correlation theorem, Equation 4.64, the resulting waveform is as shown in Figure 4.26. This is not a periodic version of Figure 4.25(c), although it has the correct basic waveshape. The desired periodic result may be obtained by commencing $x(n)$ at the point $n = N - N_1 + 1$ while continuing to commence $h(n)$ at $n = 0$, as in Figure 4.27 which shows that this produces the

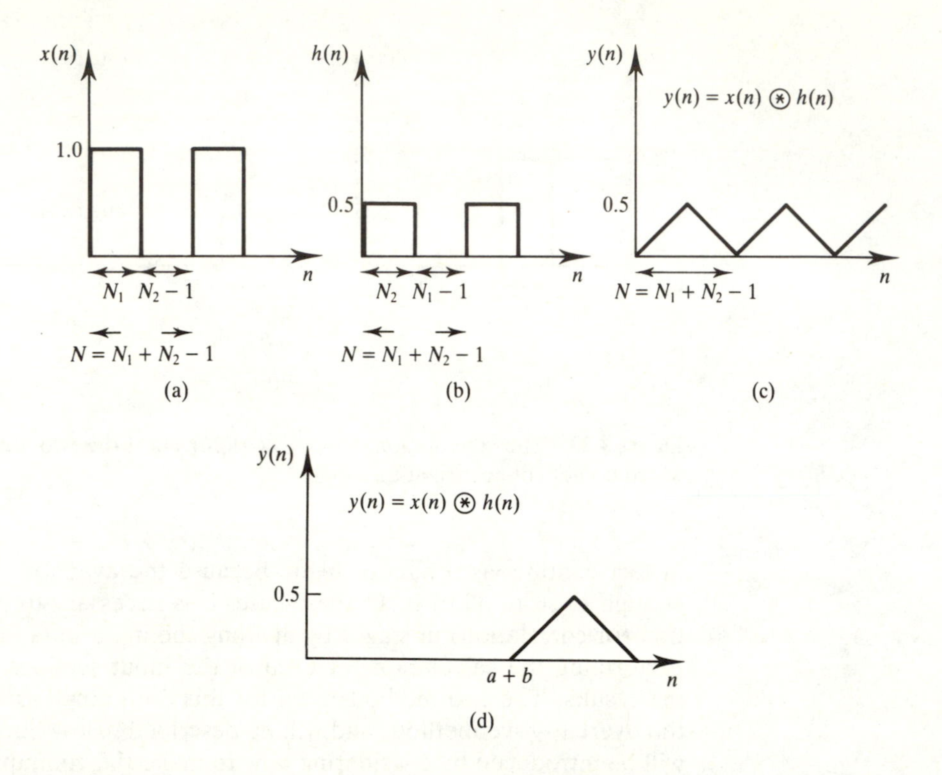

Figure 4.24 The convolution of the two waveforms of Figure 4.23 obtained by translating $x(n)$ and $h(n)$ to commence at the origin. (a) Addition of $N_2 - 1$ augmenting zeros to $x(n)$. (b) Addition of $N_1 - 1$ augmenting zeros to $h(n)$. (c) The convolution $y(n) = x(n) \circledast h(n)$. (d) The correct linear convolution obtained by displacing $y(n)$ along the n-axis to commence at $n = a + b$.

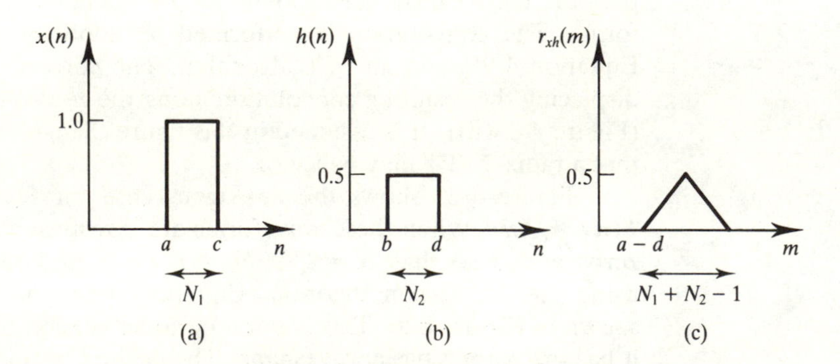

Figure 4.25 The cross-correlation, $r(m)$, of the two waveforms $x(n)$ and $h(n)$ which do not commence at the origin. (a) $x(n)$. (b) $h(n)$. (c) $r_{xh}(m)$.

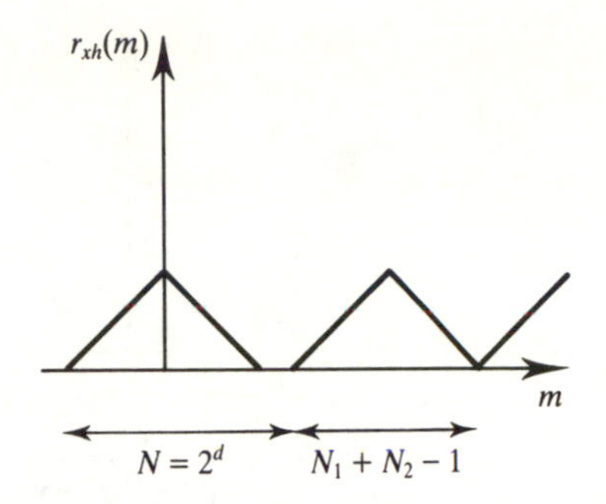

Figure 4.26 Incorrect cross-correlation obtained when $x(n)$ and $h(n)$ are translated to the origin for the cross-correlation.

required periodic correlation function, (Figure 4.27(c)). The result has to be shifted by $a - d - N + N_1 + N_2$ data points to the right to commence at the correct value of $a - d$ (compare with Figure 4.27(c)).

It is now possible to extend these considerations to the case of an infinite sequence $x(n)$ to be convolved with the finite one $h(n)$.

4.3.6 Overlap–add method

Assume $x(n)$ is divided up into equal length sections of N_1 data points. Now assume that these are periodic and are convolved with the N_2 $h(n)$ data augmented with $N_1 - N_2$ zeros so that both sequences are periodic and of length N_1. The result of this convolution will be incorrect because to obtain a correct result each sequence has to be of length $N = N_1 + N_2 - 1$. However, each section of $x(n)$ is of length N_1 (and cannot be increased). The problem may be overcome by considering sections of $x(n)$ of length N and replacing the last $N_2 - 1$ data by zeros which augment the first $N - N_2 + 1 = N_1$ data (Figure 4.28). In this way a sequence of N_1 data of $x(n)$ with $N_2 - 1$ augmenting zeros is convolved with N_2 data of $h(n)$ with $N_1 - 1$ augmenting zeros. Both sequences contain $N = N_1 + N_2 - 1$ data and are correctly convolved (Figure 4.29). The same procedure is carried out on the remaining sequences of $x(n)$ of length N. Because the final $N_2 - 1$ data of the $x(n)$ sections have been replaced by zeros, the resulting convolution functions are erroneous for the first and last $N_2 - 1$ points of each convolution, but these points sum to give the correct convolution when each convolved waveform is translated to its proper origin $(a + b)$ and the final $N_2 - 1$ points of the convolution derived from one section overlap those of the next. Figure 4.29 illustrates the process. Thus enough zeros are first added to eliminate end effects and then the convolution results are overlapped and added together exactly where the zeros were added to the sequence N_1. This is why it is called the overlap–add method.

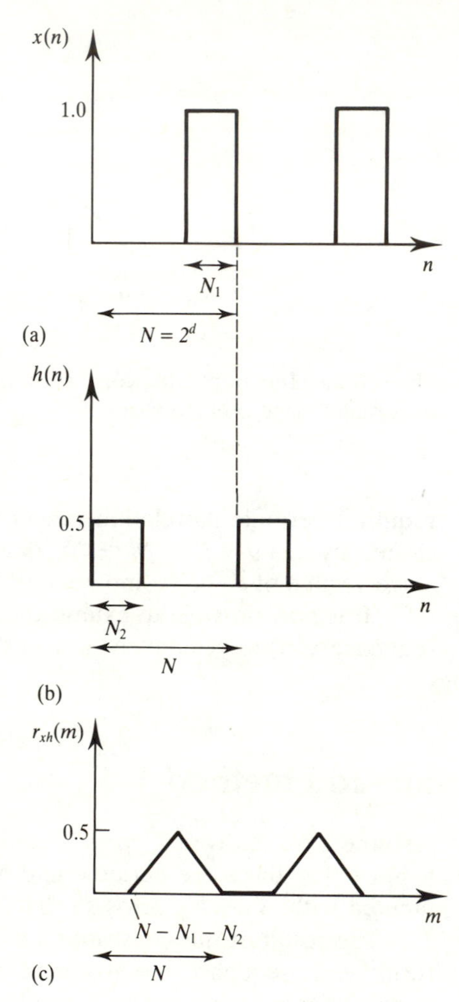

Figure 4.27 Method for obtaining the correct cross-correlation of the two sequences $x(n)$ and $h(n)$. (a) $x(n)$ commencing at $N - N_1 + 1$. (b) $h(n)$ commencing at the origin. (c) The resultant correct periodic cross-correlation coefficient, $r_{xh}(m)$.

Example 4.8

Use the overlap–add procedure to convolve the two sequences $h(n) = \{1, 0, 1\}$ and $x(n) = \{1, 3, 2, -3, 0, 2, -1, 0, -2, 3, -2, 1, \ldots\}$.

Solution

Let $x(n)$ be divided up into sections of length $N_1 = 6$ so that N, the number of points in the DFT, is $N_1 + N_2 - 1 = 6 + 3 - 1 = 8 = 2^d$ where $d = 3$, thus satisfying the requirements for linear convolution and for the use of a radix-2 FFT.

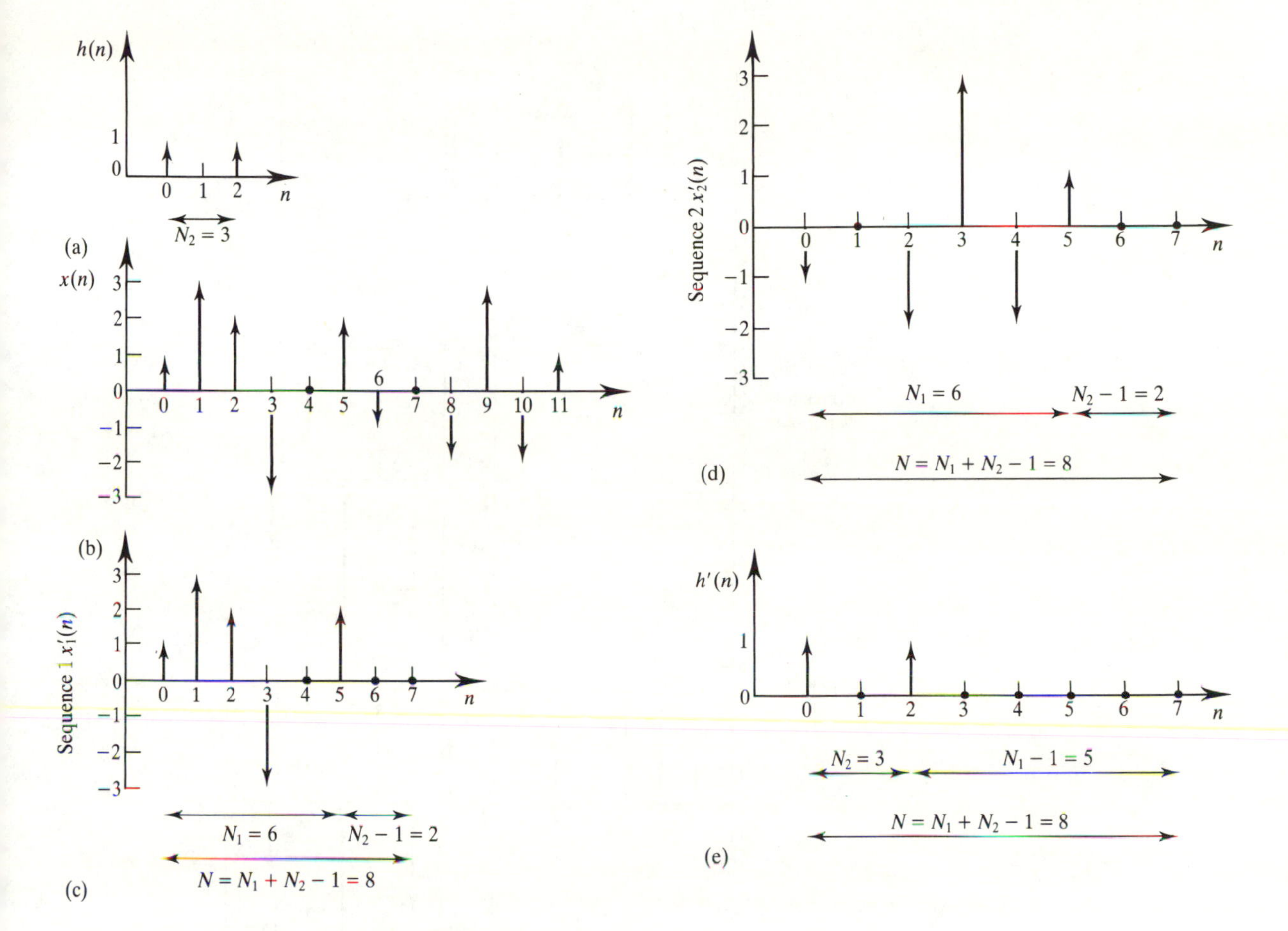

Figure 4.28 The overlap–add method of convolution.

Adding zeros to $h(n)$ gives the augmented sequence $h'(n)$:

$$h'(n) = \{1, 0, 1, 0, 0, 0, 0, 0\}$$

The first two augmented sequences of $x(n)$ are

$$x_1'(n) = \{1, 3, 2, -3, 0, 2, 0, 0\}$$

and

$$x_2'(n) = \{-1, 0, -2, 3, -2, 1, 0, 0\}$$

The terms in the convolution sum $x_1'(n) \circledast h'(n)$ are

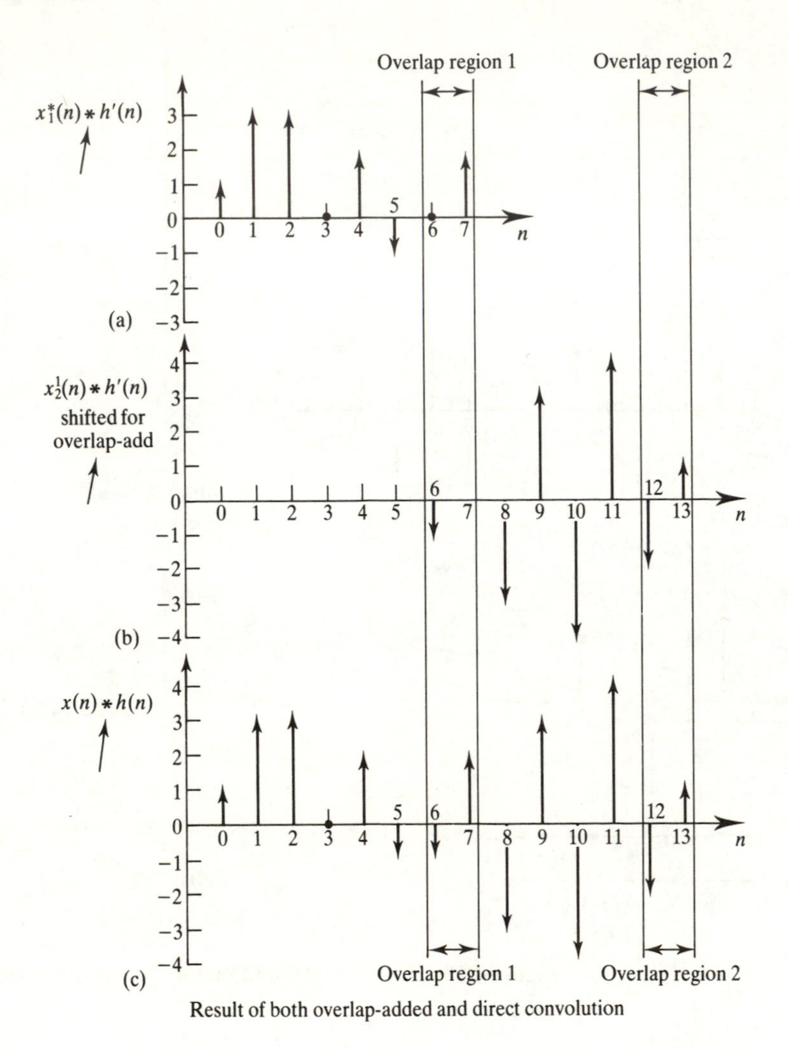

Figure 4.29 Equivalence of overlap–add method of convolution and of direct convolution.

$$y_{10} = h'_0x'_{10} = 1$$

$$y_{11} = h'_0x'_{11} + h'_1x'_{10} = 3 + 0 = 3$$

$$y_{12} = h'_0x'_{12} + h'_1x'_{11} + h'_2x'_{10} = 2 + 0 + 1 = 3$$

$$y_{13} = h'_0x'_{13} + h'_1x'_{12} + h'_2x'_{11} = -3 + 0 + 3 = 0$$

$$y_{14} = h'_0x'_{14} + h'_1x'_{13} + h'_2x'_{12} = 0 + 0 + 2 = 2$$

$$y_{15} = h'_0x'_{15} + h'_1x'_{14} + h'_2x'_{13} = 2 + 0 - 3 = -1$$

$$y_{16} = h'_0x'_{16} + h'_1x'_{15} + h'_2x'_{14} = 0 + 0 + 0 = 0$$

$$y_{17} = h'_0x'_{17} + h'_1x'_{16} + h'_2x'_{15} = 0 + 0 + 2 = 2$$

The terms in the convolution sum $x'_2(n) \circledast h'(n)$ are

$$y_{20} = h'_0 x'_{20} = -1$$

$$y_{21} = h'_0 x'_{21} + h'_1 x'_{20} = 0 + 0 = 0$$

$$y_{22} = h'_0 x'_{22} + h'_1 x'_{21} + h'_2 x'_{20} = -2 + 0 - 1 = -3$$

$$y_{23} = h'_0 x'_{23} + h'_1 x'_{22} + h'_2 x'_{21} = 3 + 0 - 2 = 1$$

$$y_{24} = h'_0 x'_{24} + h'_1 x'_{23} + h'_2 x'_{22} = -2 + 0 - 2 = -4$$

$$y_{25} = h'_0 x'_{25} + h'_1 x'_{24} + h'_2 x'_{23} = 1 + 0 + 3 = 4$$

$$y_{26} = h'_0 x'_{26} + h'_1 x'_{25} + h'_2 x'_{24} = 0 + 0 - 2 = -2$$

$$y_{27} = h'_0 x'_{27} + h'_1 x'_{26} + h'_2 x'_{25} = -2 + 0 + 1 = -1$$

The above two convolution sums are shown in Figures 4.29(a) and 4.29(b) respectively. If the first $N_2 - 1 = 2$ data of x'_2 are overlapped with the last $N_2 - 1$ data of x'_1 and the convolution sums added then the first 12 data of the resulting convolution waveform given by this overlap–add method are as shown in Figure 4.29(c).

The above result can be demonstrated to be identical to that obtained by carrying out the convolution directly as follows. The original sequence $x(n)$ contains 12 data and $h(n)$ contains 3 data. To obtain the linear convolution of the two, augmenting zeros must be added to both sequences so that they both contain $12 + 3 - 1 = 14$ data. Thus the sequences become

$$h'(n) = \{1, 0, 1, 0, 0, 0, 0, 0, 0, 0, 0, 0, 0, 0\}$$

and

$$x'(n) = \{1, 3, 2, -3, 0, 2, -1, 0, -2, 3, -2, 1, 0, 0\}$$

The first nine terms in the convolution sum are

$$y_0 = h'_0 x'_0 = 1$$

$$y_1 = h'_0 x'_1 + h'_1 x'_0 = 3$$

$$y_2 = h'_0 x'_2 + h'_1 x'_1 + h'_2 x'_0 = 2 + 0 + 1 = 3$$

$$y_3 = h'_0 x'_3 + h'_1 x'_2 + h'_2 x'_1 = -3 + 0 + 3 = 0$$

$$y_4 = h'_0 x'_4 + h'_1 x'_3 + h'_2 x'_2 = 0 + 0 + 2 = 2$$

$$y_5 = h'_0 x'_5 + h'_1 x'_4 + h'_2 x'_3 = 2 + 0 - 3 = -1$$

$$y_6 = h'_0 x'_6 + h'_1 x'_5 + h'_2 x'_4 = -1 + 0 + 0 = -1$$

$$y_7 = h'_0 x'_7 + h'_1 x'_6 + h'_2 x'_5 = 0 + 0 + 2 = 2$$

$$y_8 = h'_0 x'_8 + h'_1 x'_7 + h'_2 x'_6 = -2 + 0 - 1 = -3$$

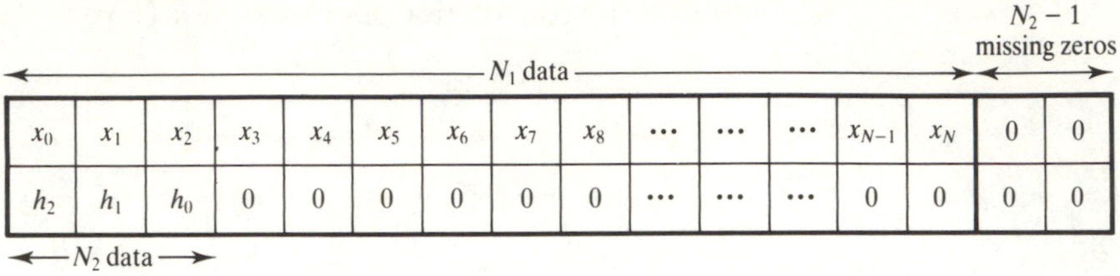

$N_2 - 1$ missing zeros

N_1 data

x_0	x_1	x_2	x_3	x_4	x_5	x_6	x_7	x_8	...	...	...	x_{N-1}	x_N	0	0
h_2	h_1	h_0	0	0	0	0	0	0	...	...	...	0	0	0	0

N_2 data

Figure 4.30 $x(n)$ and $h(n)$ with $N_2 - 1$ zeros added to $h(n)$.

The terms of this convolution sum are indeed identical to those obtained by the overlap–add method plotted in Figure 4.29(c).

The overlap–add procedure for fast convolution (or correlation) by sectioning is therefore as follows.

(1) Select the number of $x(n)$ data, N_1, to be of the order of the number of $h(n)$ data, N_2, with $N_1 > N_2$, and also select the number of DFT points to be $N = 2^d$ where d is an integer and $N \gg N_1 + N_2 - 1$. Satisfy these conditions by adding augmenting zeros to the data sequences as necessary.

(2) Shift the augmented sections of $x(n)$ data to the origin.

(3) For each section of augmented $x(n)$ data, $x'(n)$, perform the fast convolution $x'(n) \circledast h'(n)$, that is compute $X(k)H(k)$ and then its inverse transform.

(4) Sequentially overlap the resulting convolutions by their final and first $N_2 - 1$ values and then add them.

4.3.7 Overlap–save method

Consider again the convolution $x(n) \circledast h(n)$ illustrated in Figure 4.30 in which $N_2 - 1$ zeros have been added to $h(n)$ so that both sequences are of length N_1. The linear convolution sum of these may be obtained by successively sliding $h(n)$ to the right by one datum, cross-multiplying corresponding terms, and summing. However, because neither sequence is of length $N_1 + N_2 - 1$ the result will not be $x(n) \circledast h(n)$. There are in fact $N_2 - 1$ zeros missing from $x(n)$ of length N_1. This means that the first $N_2 - 1$ terms of the convolution sum will be incorrect and should be discarded. Therefore if the data $x(n)$ is divided up into contiguous sections of length N_1 the first $N_2 - 1$ values of each of the convolution sums must be discarded. The convolution of $x(n) \circledast h(n)$ will therefore contain a periodic sequence of gaps of length $N_2 - 1$. These gaps may be correctly filled by overlapping the final $N_2 - 1$ data of each $x(n)$ sequence of length N_1 with the first $N_2 - 1$ data of the following sequence, and then discarding these first $N_2 - 1$ data.

Example 4.9

Use the overlap–save procedure to convolve the same two sequences convolved in Section 4.3.6, that is

$$h(n) = \{1, 0, 1\}$$

and

$$x(n) = \{1, 3, 2, -3, 0, 2, -1, 0, -2, 3, -2, 1\}$$

Solution

Since $h(n)$ has $N_2 = 3$ the amount of overlap is $N_2 - 1 = 2$. The overlapping of the sections is as shown in Figure 4.31. The convolutions are evaluated for each section as below.

For section 1,

$$y_{10} = h_0 x_{10} = 1$$

$$y_{11} = h_0 x_{11} + h_1 x_{10} = 3 + 0 = 3$$

$$y_{12} = h_0 x_{12} + h_1 x_{11} + h_2 x_{10} = 2 + 0 + 1 = 3$$

$$y_{13} = h_1 x_{13} + h_1 x_{12} + h_2 x_{11} + h_3 x_{10} = -3 + 0 + 3 + 0 = 0$$

Thus

$$y_1 = \{1, 3, 3, 0\}$$

$h(n)$	1	0	1									
$x(n)$	1	3	2	−3	0	2	−1	0	−2	3	−2	1
Section 1	1	3	2	−3								
Section 2			2	−3	0	2						
Section 3					0	2	−1	0				
Section 4							−1	0	−2	3		
Section 5									−2	3	−2	1

Figure 4.31 Overlapping of sections for the overlap–save method of convolution.

Table 4.2 Results for Example 4.9.

Section 1	y_0	~~1 3~~	3 0				
Section 2	y_1		~~2 3~~	2 −1			
Section 3	y_2			~~0 2~~	−1 2		
Section 4	y_3				~~−1 0~~	−3 3	
Section 5	y_4					~~−2 3~~	−4 4
$x(n) \circledast h(n)$		~~1 3~~	3 0	2 −1	−1 2	−3 3	−4 4

For the remaining sections remember $h_1 = h_3 = 0$. We obtain, for section 2,

$$y_{20} = h_0 x_{20} = 2$$
$$y_{21} = h_0 x_{21} = -3$$
$$y_{22} = h_0 x_{22} + h_2 x_{20} = 2 + 0 = 2$$
$$y_{23} = h_0 x_{23} + h_2 x_{21} = 2 - 3 = -1$$

$$y_2 = \{2, -3, 2, -1\}$$

Similarly, for section 3,

$$y_3 = \{0, 2, -1, 2\}$$

For section 4,

$$y_4 = \{-1, 0, -3, 3\}$$

and finally, for section 5,

$$y_5 = \{-2, 3, -4, 4\}$$

These results are illustrated in Table 4.2 which shows that the first $N_2 - 1$ results of each sequence are discarded. Apart from the first $N_2 - 1$ points the last row in the table corresponds to the correct convolution.

The overlap–save procedure is therefore as follows.

(1) Select the number of $x(n)$ data, $N_1 = 2^d$, to be convolved with $h(n)$ and add $N_2 - 1$ zeros to $h(n)$ so that both sequences are of length N_1.
(2) Locate both sequences at the origin.
(3) For each sequence compute the corresponding values of $X(k)$ and $H(k)$ using an FFT.
(4) Compute $X(k)H(k)$ and its inverse, which is the convolution of each sequence with $h(n)$.

Table 4.3 Ratio $R_m(S)/R_m(N)$, number of real multiplications for sectioning method: number of real multiplications by straightforward method for fast convolution.

N	N^1	N_1	N/N_1	N_2	$R_m(S)/R_m(N)$	*Comment*
1020	8	6	170	3	0.54	Best result with N^1 short, $N_1 \approx N^1$
1024	256	254	4	3	0.83	$N_1 \approx N^1$
1020	128	102	10	3	0.93	
1020	256	204	5	3	1.04	

(5) Adjust each of the convolutions to overlap the preceding one by $N_2 - 1$ data.

(6) Discard the first $N_2 - 1$ data of each convolution and read out the remaining values which correspond to the correct convolution.

4.3.8 Computational advantages of fast convolution by sectioning

It was shown in Section 4.3.5 that unnecessary computational effort may be avoided by first setting every section of waveform to be convolved at the origin and it is assumed here that this has been done. It is further assumed that the computational requirements of the overlap–add and overlap–save methods are similar so that it is only necessary to consider the overlap–add method. It is assumed that the sequence $x(n)$ of length N is divided up into N/N_1 sections each of length N_1, that the sequence $h(n)$ is of length N_2, and that the lengths of the sequences for linear convolution are $N^1 = 2^d \geqslant N_1 + N_2 - 1$. It has also been shown, Section 4.3.4, that $12N^1 \log_2 2N^1 + 8N^1$ real multiplications are required to perform the fast convolution of two N^1-point sequences. Thus, in order to carry out the fast convolution of the N-point sequence $x(n)$ by the overlap–add method $(N/N_1)(12N^1 \log_2 2N^1 + 8N^1) = R_m(S)$ real multiplications will be required. This shows that the sequence lengths to be convolved, N^1, should be short while the lengths, N_1, of the $x(n)$ data sections should approach N^1. Ideally $N^1 = 2^d = N_1 + N_2 - 1$. The number of real multiplications for the original N-point sequence is $12N \log_2 2N + 8N = R_m(N)$. Table 4.3 shows that for the example of Section 4.3.6 the ratio $R_m(S)/R_m(N) \leqslant 1$ typically, with savings in computational time of the order of 50% being possible.

4.3.9 The relationship between convolution and correlation

In convolution the value of the nth output is given by the convolution sum of Equation 4.80:

$$y(n) = \sum_{m=0}^{n} h(m)x(n-m) = h(0)x(n) + h(1)x(n-1) + \ldots + h(n)x(0) \quad \textbf{(4.101)}$$

The value of the cross-correlation function for the waveforms $h(n)$ and $x(n)$ for the jth lag is given by Equation 4.1 modified slightly to be

$$r_{hx}(j) = \frac{1}{N}\sum_{n=0}^{N-1} h(n)x(j+n)$$

$$= \frac{1}{N}[h(0)x(j) + h(1)x(j+1) + \ldots + h(N-1)x(j+N-1)] \quad \textbf{(4.102)}$$

It is easier to compare $y(n)$ and $r_{hx}(j)$ if the case $j = 0$, that is zero lag in the cross-correlation, is considered. Equation 4.102 then becomes

$$r_{hx}(0) = \frac{1}{N}\sum_{n=0}^{N-1} h(n)x(n)$$

$$= \frac{1}{N}[h(0)x(0) + h(1)x(1) + \ldots + h(N-1)x(N-1)] \quad \textbf{(4.103)}$$

Comparing Equations 4.101 and 4.103 reveals that they are of similar form except that the $x(n)$ sequence in the cross-correlation is in the reverse order to that in the convolution. Thus convolution equates to the cross-correlation of the two waveforms in which one of the original sequences has been time reversed, and the normalizing factor $1/N$ has been set to unity. This means that convolutions and correlations may be computed by the same computer programs simply by reversing one of the sequences.

4.4 Implementation of correlation and convolution

In considering the implementation of these operations it should be remembered that the two are intimately related. Two data sequences may be either correlated or convolved simply by reversing the order of one of the data sequences. Furthermore, for longer data sequences the operations can be speeded up using fast Fourier transform methods to achieve fast correlation or fast convolution. Where one of the data sequences is very long the overlap–add or overlap–save technique will be appropriate: see Sections 4.3.6 and 4.3.7 and Brigham (1974), Strum and Kirk (1988) and DeFatta *et al.* (1988).

Convolution or correlation can be achieved using, for example, an FIR filter, which may be implemented by FFTs. Correlation or convolution can also be achieved using a matched filter as shown in Section 4.2.2, Figure 4.13 which illustrates the correlation detector. For digital processing charge-coupled device (CCD) technology may be used to implement transversal filters. These give a linear phase response with data rates in excess of 100 MHz possible for basic delay line configurations (Grant *et al.*, 1989). Analogue processing may be carried out by implementing tapped delay lines using surface acoustic wave (SAW) devices (Grant *et al.*, 1989). These operate over a range from about 2 MHz to 2 GHz. Other implementations include dedicated convolver and correlator chips, general-purpose digital signal processors, standard microprocessors, and transputers. An example of the latter would be a real-time system for the removal of ocular artefacts from all 16 channels of human EEG (Jervis *et al.*, 1990).

The computational time required for fast correlations and convolutions may be further halved as follows (Brigham, 1974). Consider the convolution of $x(n)$ with $h(n)$. When computing $X(k)$ fill the real part of the FFT with the even terms of $x(n)$ and the imaginary part with the odd terms, thereby halving the length of the FFT. The real part of $(1/N)F_D^{-1}[X(k)H(k)]$ then gives the even terms of the desired convolution and the imaginary part gives the odd terms.

Likewise, the convolutions of two data sequences $x_1(n)$ and $x_2(n)$ with $h(n)$ may be computed simultaneously. Fill the real part of an FFT with $x_1(n)$ and the imaginary part with $x_2(n)$ and transform to obtain $X^1(k)$. Then the real part of $(1/N)F_D^{-1}[X^1(k)H(k)]$ is $x_1(n) \circledast h(n)$ and the imaginary part is $x_2(n) \circledast h(n)$.

4.5 Application examples

4.5.1 Correlation

Example 4.10

This simplified example concerns the application of correlation theory to the control of the attitude of a spacecraft to ensure that the solar panel always faces the sun. Attitude errors are represented as multilevel pulses with level separations $a = 0.2$ mV and pulse widths $T_s = 1$ μs. An initial attempt is made to control the attitude error by transmitting a sequence of negative pulses of height a when there is a positive error. The control system is only considered satisfactory if the correlation coefficient between the error and control signals

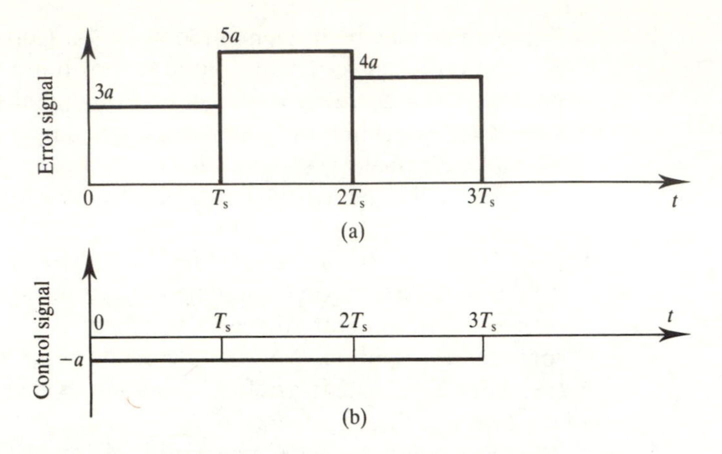

Figure 4.32 Spacecraft attitude control. (a) Error signal. (b) Control signal.

is better than −0.5. Figure 4.32(a) shows three error pulses while Figure 4.32(b) shows the corresponding control signal pulses. For this example it is assumed that these pulses are sufficient, and that it is unnecessary to consider a lag greater than T_s. The problem is to determine whether or not the system may be regarded as satisfactory.

The system can be proven by ascertaining that $|r_{12}(\tau)| > 0.5$ for $0 \leq \tau \leq T_s$. The cross-correlation may be achieved by shifting the control signal to the right while keeping the error signal fixed. This means that $r_{12}(-\tau)$ should be determined.

Now

$$r_{12}(-\tau) = \int_{-\infty}^{\infty} v_1(t)v_2(t-\tau)\,dt$$

where $v_1(t)$ is the error signal and $v_2(t)$ is the control signal.

$$r_{12}(-\tau) = \frac{1}{3T_s}\int_{\tau}^{T_s} 3a(-a)\,dt + \frac{1}{3T_s}\int_{T_s}^{2T_s} 5a(-a)\,dt + \frac{1}{3T_s}\int_{2T_s}^{3T_s} 4a(-a)\,dt$$

$$= \frac{a^2}{3T_s}\{[-3t]_{\tau}^{T_s} + [-5t]_{T_s}^{2T_s} + [-4t]_{2T_s}^{3T_s}\}$$

$$= \frac{a^2}{3T_s}(-3T_s + 3\tau - 10T_s + 5T_s - 12T_s + 8T_s)$$

$$= \frac{a^2}{3T_s}(-12T_s + 3\tau)$$

$r_{12}(\tau)$ is now normalized to confine values to the range $-1 \ll r_{12}(\tau) \ll 1$ by dividing by the normalizing factor

$$\frac{1}{3T_s}\left[\int_{-\infty}^{\infty} v_1^2(t)\,\mathrm{d}t \int_{-\infty}^{\infty} v_2^2(t)\,\mathrm{d}t\right]^{1/2}$$

Now,

$$\begin{aligned}\int_{-\infty}^{\infty} v_1^2(t)\,\mathrm{d}t &= \int_0^{T_s}(3a)^2\,\mathrm{d}t + \int_{T_s}^{2T_s}(5a)^2\,\mathrm{d}t + \int_{2T_s}^{3T_s}(4a)^2\,\mathrm{d}t \\ &= a^2\{[9t]_0^{T_s} + [25t]_{T_s}^{2T_s} + [16t]_{2T_s}^{3T_s}\} \\ &= a^2(9T_s + 25T_s + 16T_s) = 50a^2T_s\end{aligned}$$

Also

$$\int_{-\infty}^{\infty} v_2^2(t)\,\mathrm{d}t = \int_0^{3T_s}(-a)^2\,\mathrm{d}t = a^2[t]_0^{3T_s} = 3a^2T_s$$

Hence the normalizing factor is

$$\frac{1}{3T_s}[(50a^2T_s)(3a^2T_s)]^{1/2} = \frac{1}{3T_s}150^{1/2}a^2T_s$$

and the normalized expression for $r_{12}(-\tau)$ is

$$r_{12}^{\mathrm{N}}(-\tau) = \frac{3\tau - 12T_s}{150^{1/2}T_s} = \frac{3\tau}{12.25T_s} - \frac{12}{12.25}$$

$$r_{12}^{\mathrm{N}}(-\tau) = 0.245\times 10^6\tau - 0.98$$

When $\tau = 0$,

$$r_{12}^{\mathrm{N}}(0) = -0.98$$

When $\tau = 1$ μs (the largest allowed value),

$$r_{12}^{\mathrm{N}}(10^{-6}) = -0.735$$

Thus, over the range considered $|r_{12}^{\mathrm{N}}(-\tau)| > |0.5|$ which satisfies the criterion for good control of the spacecraft's attitude.

Example 4.11

A sonar system is required for the determination of the distance of a sound source. The source is broadband and gaussian with zero mean. The system

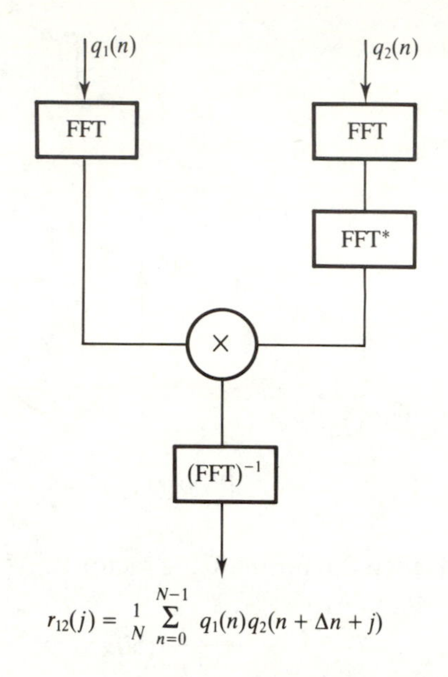

Figure 4.33 Sonar system block diagram.

consists of two sonar transducers separated by a distance, d, and an associated signal processing system. The transducers, T_1 and T_2, receive the broadband noise signals $q_1(t)$ and $q_2(t) = Aq_1(t + \Delta t)$ respectively, Δt being the time lag due to the different path lengths to the source from the two transducers, and A the associated attenuation factor (assume $A = 1$ in this case). The signal processing system computes the correlation function of equal lengths of the two transducer outputs.

Draw a labelled block diagram of a simple system designed to achieve the correlation in the shortest possible time, and explain the principles upon which it is based.

Sketch the transducer output signals and their cross-correlation function, indicating noteworthy features.

If the peak value of the cross-correlation function is 10, and the receiver has a bandwidth of 1–10 Hz, calculate the received energy.

Solution

The system block diagram is shown in Figure 4.33. This system speeds up the correlation calculations in this design by applying the correlation theorem and computing the FFTs involved. This will be faster than a straightforward computation of correlation when the number of data points in the data sequences exceeds about 128. Thus the system computes $r_{12}(\tau)$ which expressed digitally is

$$r_{12}(j) = F_D^{-1}[F_1(k)F_2^*(k)]$$

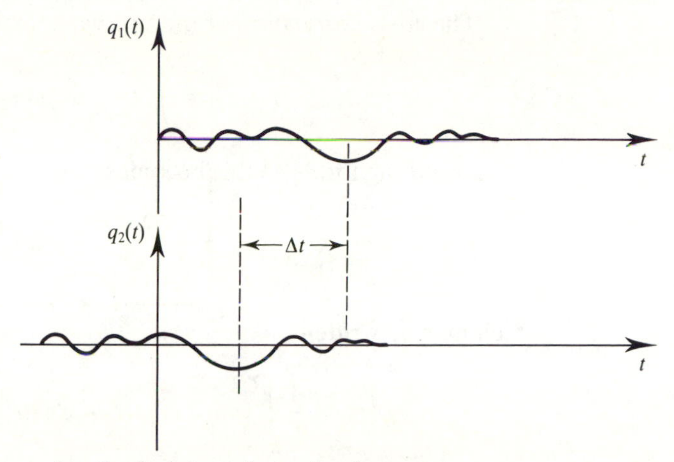

Figure 4.34 Broadband noise signals detected by sonar system.

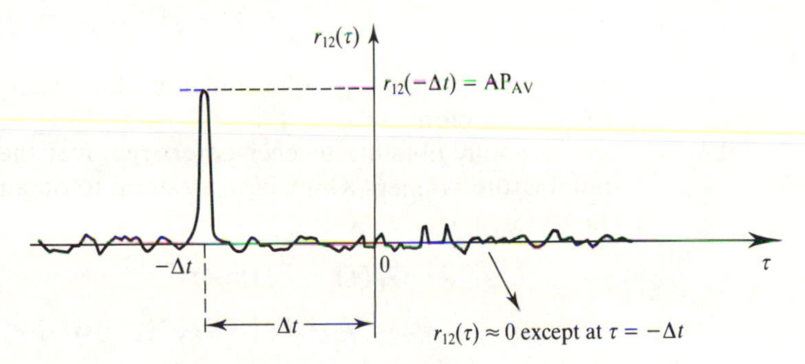

Figure 4.35 Cross-correlation function of signals detected by the sonar system.

The system output, $r_{12}(j)$, is

$$r_{12}(j) = \frac{1}{N} \sum_{n=0}^{N-1} q_1(n) A q_1(n + \Delta n + j)$$

As $q_1(n)$ and $q_2(n)$ are random the system will only produce a significant output when the waveforms are relatively shifted to be in phase. This occurs when $j = -\Delta n$. The output then is

$$\frac{1}{N} \sum_{n=0}^{N-1} q_1^2(n) = P_{AV}, \text{ the average power}$$

Figures 4.34 and 4.35 show the waveforms and their cross-correlation function.

The cross-correlation of the two waveforms is

$$r_{12}(\tau) = \frac{1}{T}\int_{-T/2}^{T/2} q_1(t)q_2(t+\tau)\,\mathrm{d}t$$

On substituting for $q_2(t)$ this becomes

$$r_{12}(\tau) = \frac{1}{T}\int_{-T/2}^{T/2} q_1(t)Aq_1(t+\Delta t+\tau)\,\mathrm{d}t$$

which may be written

$$r_{12}(\tau) = \frac{A}{T}\int_{-T/2}^{T/2} q_1(t)q_1(t+\tau')\,\mathrm{d}t, \quad \text{where } \tau' = \Delta t + \tau$$

The integrand is seen to be equivalent in magnitude to the autocorrelation of $q_1(t)$ at zero lag, and therefore represents the power of this signal, P_{AV}. Hence

$$r_{12}(\tau) = AP_{\mathrm{AV}}\delta(t+\Delta t)$$

where δ represents the delta function. The magnitude of $r_{12}(\tau)$ is seen to be AP_{AV}. Therefore $AP_{\mathrm{AV}} = 10$.

We may obtain the received energy over the required bandwidth by first applying the Wiener–Khintchine theorem to obtain the energy spectral density. The theorem is

$$\begin{aligned} G_{\mathrm{E}}(f) &= F_{\mathrm{D}}[r_{12}(\tau)] \\ &= F_{\mathrm{D}}[AP_{\mathrm{AV}}\delta(t+\Delta t)] = AP_{\mathrm{AV}}\mathrm{e}^{\mathrm{j}\omega\Delta t} \end{aligned}$$

Therefore $|G_{\mathrm{E}}(f)| = AP_{\mathrm{AV}} = 10\,\mathrm{J\,Hz^{-1}}$. The signal bandwidth is $10 - 1$ Hz $=$ 9 Hz. Therefore the received energy is $10 \times 9 = 90$ J.

4.5.2 Convolution

4.5.2.1 FIR and IIR filters

The operation of transversal filters, both FIR and IIR, provides good application examples of convolution (Stremler, 1982; DeFatta *et al.*, 1988). They may be designed to convolve sequences or to perform more general digital filtering, for example two-dimensional filtering as employed in image processing (Grant *et al.*, 1989), for noise reduction, image enhancement, and pattern recognition.

Consider a linear time-independent (LTI) system describable by

$$y(n) = \sum_{k=1}^{N} a_k y(n-k) + \sum_{k=0}^{L} b_k x(n-k) \tag{4.104}$$

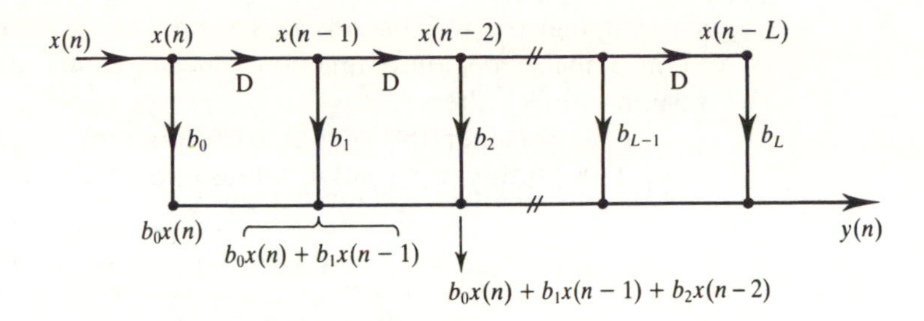

Figure 4.36 Diagrammatic representation of a nonrecursive filter.

in which $y(n)$ represents the output sequence and $x(n)$ the input sequence. The output is seen to depend on the current input and also on the previous inputs and outputs. a_k and b_k are real constants and N is the order of the equation and represents the number of previous outputs that have to be considered.

Because the current output depends on previous outputs the system is recursive. If the system output depends only on previous inputs it is said to be nonrecursive and is describable by

$$y(n) = \sum_{k=0}^{L} b_k x(n-k) \tag{4.105}$$

and this equation is descriptive of a transversal filter (or tapped delay line).

Figure 4.36 shows a diagrammatic description of the system of Equation 4.105. The terms in the summation, which represents the system output, are obtained by summing the delayed and weighted values of the inputs. It will now be shown that these weights correspond to the impulse response of the system. Assume that the input, $x(n)$, is a unit impulse, $\delta(n)$, where

$$x(n) = \delta(n) = \begin{cases} 1, & n = 0, \quad \text{that is } x(0) = 1 \\ 0, & n \neq 0, \quad \text{that is } x(n \neq 0) = 0 \end{cases}$$

The corresponding output is the impulse response, $h(n)$. Substitution of consecutive input values into Equation 4.105 gives

$$y(0) = h(0) = b_0 x(0) + b_1 \times 0 = b_0 \times 1 = b_0$$

$$y(1) = h(1) = b_0 \times 0 + b_1 x(0) + b_2 \times 0 = b_1 \times 1 = b_1$$

$$\vdots$$

$$y(L) = h(L) = b_0 \times 0 + 0 + 0 + \ldots + 0 + b_L \times 1 = b_L$$

Therefore

$$h(n) = \{b_0, b_1, \ldots, b_L\} \tag{4.106}$$

showing that the weights on the system diagram correspond to the coefficients of its impulse response function. Such systems are known as finite impulse response (FIR) filters.

Now consider the output corresponding to the general input sequence $x(n)$. Substituting consecutive values into Equation 4.105 gives

$$\begin{aligned} y(n) &= b_0x(n) + b_1x(n-1) + \ldots + b_nx(0) \\ &\equiv h(0)x(n) + h(1)x(n-1) + \ldots + h(n)x(0) \end{aligned} \qquad \textbf{(4.107)}$$

which can be recognized as the convolution of the input with the output as would be expected. Thus FIR filters may also be regarded as convolvers in which the filter weights correspond to the coefficients of their impulse response.

A different, but similar, relationship applies in the case of infinite impulse response (IIR) filters. Consider the first-order recursive filter described by the equation

$$y(n) = a_1y(n-1) + b_0x(n) \qquad \textbf{(4.108)}$$

It is easily demonstrable that for a unit impulse input

$$y(n) = h(n) = b_0a_1^n \quad n \geqslant 0 \qquad \textbf{(4.109)}$$

For the general input sequence $x(n)$, assuming $y(-1) = 0$,

$$\begin{aligned} y(0) &= b_0x(0) \\ y(1) &= a_1b_0x(0) + b_0x(1) \\ y(2) &= a_1^2b_0x(0) + a_1b_0x(1) + b_0x(2) \\ &\vdots \\ y(n) &= a_1^nb_0x(0) + a_1^{n-1}b_0x(1) + \ldots + a_1b_0x(n-1) + b_0x(1) \end{aligned}$$

Substituting the known values of the weights from Equation 4.109 gives

$$y(n) = h(n)x(0) + h(n-1)x(1) + \ldots + h(0)x(n) \qquad \textbf{(4.110)}$$

Equations 4.109 and 4.110 show that the IIR filter corresponding to the first-order system is a convolver for which the impulse response coefficients are given by $h(n) = b_0a_1^n$.

FIR filters are used in speech processing (Grant *et al.*, 1989) to achieve reduced bandwidth PCM, in sub-band coders, for parametric spectral analysis, and in linear predictive vocoders. FIR filters also find applications in radars, and in spread spectrum communications (Grant *et al.*, 1989).

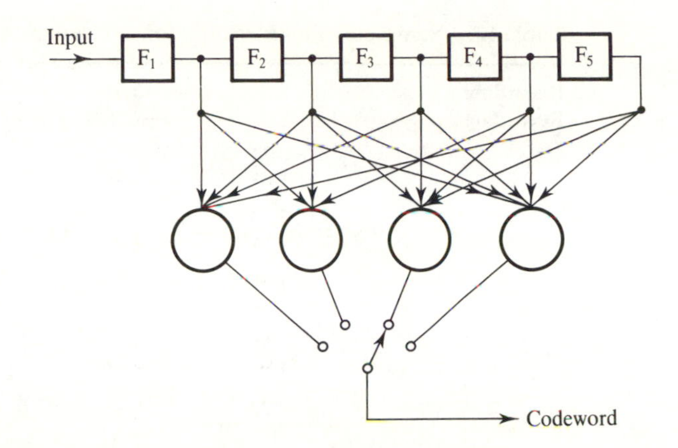

Figure 4.37 Convolution coder.

4.5.2.2 Convolution coding

Convolutional codes allow burst error correction by distributing the parity check digits of the code over a long stream of symbols (Stremler, 1982; Taub and Schilling, 1986). The outputs of the flip–flops of a shift register provide delays and are tapped and appropriately combined using modulo-2 adders. This produces a number of outputs which are read consecutively each clock cycle (Figure 4.37). The system is essentially causal and nonrecursive and produces an output which depends on its previous inputs and convolves new input data with its impulse response.

4.5.2.3 Deconvolution

The input to all systems is convolved with the impulse response of the system and may distort the output. This occurs, for example, in telecommunication systems and may necessitate the design of an equalizer, which is a linear filter which deconvolves the convolved output. Before a deconvolution filter can be designed the impulse response of the system must be measured (system identification). The subject of system identification and deconvolution is quite extensive (Proakis and Manolakis, 1988) and is not discussed here.

4.5.2.4 Speech

There is great interest in analysing and coding speech for purposes such as human–machine interaction and data compression. Use is sometimes made of the fact that the speech waveform can be modelled as the convolution of a train of impulses representing pitch, an excitation pulse, and the impulse response of the vocal tract (Rabiner and Gold, 1975). The resulting triple convolution can easily be converted to a form suitable for processing by an LTI system. The use of FIR filters in speech processing was highlighted in Section 4.5.2.1.

Table 4.4 Sampled voltages (volts) for two separate recordings from the same channel.

Recording 1	6.02	−5.98	7.92	−7.96	−0.78
Recording 2	8.93	−7.20	−0.82	3.23	1.44
	−8.34	9.22	−2.65	−3.7	9.51
	5.43	−9.88	−1.13	0.79	9.83
	5.53	3.50	−3.18	−8.85	8.21
	−8.73	4.64	−8.49	−4.66	−8.84
	1.69	−0.06	6.65	−8.00	−9.21
	5.55	−8.24	−0.37	2.71	4.63
	−0.78	7.27	−5.98	−3.97	9.11
	1.88	−0.92	−5.33	9.01	9.23
	4.23	2.99	−1.85	−5.27	3.81
	−3.7	5.08	−0.72	−5.08	−2.6
	6.62	−2.64	2.08	−5.91	−3.58
	9.67	−8.55	−3.08	4.18	8.11
	−1.65	3.64	−8.19	−3.50	4.84
	0.74	−3.87	−4.09	8.03	6.91
	7.25	2.93	−4.42	−8.21	3.61
	−9.87	−3.62	−8.29	−5.8	−7.04

4.6 Summary

The topics of correlation and convolution and their interrelationship have been thoroughly discussed in this chapter. Normalization procedures and the avoidance of end effects have been discussed for correlation. The effects of correlating noisy signals and the identification of signals in noise by correlation procedures have been described as well as other applications. The techniques of fast correlation and convolution based on the correlation and convolution theorems and using FFTs have been described, and it was shown how to obtain linear convolution. The fast overlap–add and overlap–save methods for obtaining the convolution of a long data sequence were obtained. A further means of speeding up the computations for real data by a factor of 2 by exploiting both the real and the imaginary parts of an FFT has also been described.

Problems

4.1 Two separate recordings of equal length are made of a periodic pulse train being transmitted down a noisy channel. Table 4.4 shows the recorded values of the sampled voltages.

(1) Determine the amount of lag between the two recordings and the period of the waveform.

(2) Derive the periodic waveform.

4.2 Evaluate the cross-correlation functions of recordings 1 and 2 of Table 4.4 with and without

Table 4.5 The sampled voltages from a noisy waveform.

−7.37,	−7.99,	3.31,	−8.59,	−1.68,	3.01,	12.21,	−2.38,	7.46,
−9.84,	1.48,	1.1,	−1.8,	5.48,	8.93,	0,	−9.36,	−10.11,
1.61,	3.36,	−4.86,	6.27					

Table 4.6 Digitized voltage values.

−0.92,	−3.71,	3.11,	−0.24,	4.65,	0.84,	−2.98,	−3.94,	−4.03,	−2.51,	0.17,
3.85,	2.58,	0.38,	4.58,	3.4,	−3.46					

correcting for the end effect. Estimate the errors introduced by the end effect.

4.3 What is the percentage correlation between recordings 1 and 2 of Table 4.4 evaluated at zero lag? Assume percentage correlation is defined as the correlation coefficient, ρ_{12}, multiplied by 100%.

4.4 The sampled voltages from a noisy waveform are given in Table 4.5. Use the technique of cross-correlation with a template waveform to discover the exact shape of the periodic waveform present. Check your conclusion using another method.

4.5 Calculate the autocorrelation function of the periodic waveform of Problem 4.4 (a) numerically and (b) analytically. Compare these solutions with each other and with the autocorrelation function of the noisy waveform. Account for any discrepancies from the anticipated results.

4.6 A voltage waveform is sampled and digitized. The digitized voltage values are given in Table 4.6. Determine whether or not the waveform may be regarded as random. On the assumption that the sampling interval was 1 ms and that a periodic signal component was present with a period of 4 ms, obtain an estimate of the periodic waveform and plot it.

4.7 Compare the signal-to-noise ratios of

(1) the noisy periodic waveform of recording 1 in Table 4.4,

(2) the autocorrelation function of recording 1 in Table 4.4, and

(3) the cross-correlation function of recordings 1 and 2 of Table 4.4.

4.8 Evaluate the theoretical signal-to-noise ratios for

(1) the periodic waveform obtained from the data of recording 1, Table 4.4, by cross-correlation with the appropriate impulse train,

(2) the autocorrelation function of recording 1 in Table 4.4, given that (i) the signal-to-noise ratio $(S/N)_{r0}$ of the autocorrelation function of a noisy sinusoidal signal is

$$(S/N)_{r0} = \frac{N}{1 + 8\Big/\left(\dfrac{S_i}{N_i}\right) + 2\Big/\left(\dfrac{S_i}{N_i}\right)^2}$$

where N is the number of data, S_i is the signal power, and N_i is the noise power, and (ii) the signal-to-noise ratio $(S/N)_\delta$ of the cross-correlation of a noisy-sinusoidal signal with a periodic impulse train of the same period as the signal is

$$(S/N)_\delta = \frac{N}{1 + 1\Big/\left(\dfrac{S_i}{N_i}\right)}.$$

4.9 Compare the answers obtained to Problems 4.7 and 4.8.

4.10 A matched filter has the impulse response function $h(n) = \{1, -1, -1, 1, 1, -1, 1\}$ and is

Table 4.7 Sampled voltages from a noisy bipolar signal.

t (μs)	0	1	2	3	4	5	6
Voltage	0.14	0.48	1.61	2.09	−2.40	0.40	2.35
t (μs)	7	8	9	10	11	12	13
Voltage	−0.59	−1.81	0.32	−0.47	1.81	−1.63	−2.28

used to detect the arrival at a receiver of the corresponding signal transmitted down a noisy channel. Table 4.7 shows the sampled signal values, each of which represents the value of a pulse in a bipolar pulse train of amplitude ±1.5 V and of pulse width 1 μs. Determine the time of arrival of the signal and the value of the matched filter constant.

4.11 Find the impulse response function of a system if its response to the PN sequence {1, 1, −1, 1, −1, −1, 1, −1} is $y(n)$ = {0, 0, 0.5, 1.5, 1.5, 1.5, 1, −1, −1, −1, −1.5, −0.5, −0.5, −0.5}.

4.12 A unit amplitude pulse of width 2 ms is applied to the circuit which has the impulse response shown in Figure 4.38. Determine the output waveform numerically. Sample the waveforms at 0.5 ms intervals.

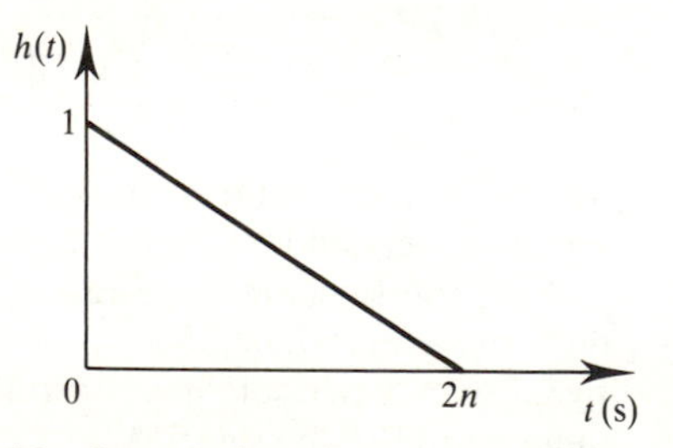

Figure 4.38 System impulse response for Problem 4.12.

4.13 (1) Figure 4.39 shows two functions $x_1(t)$ and $x_2(t)$. Evaluate

(a) their convolution, $x_3(t)$, numerically, taking sampled values at $t = 0, 1, 2, 3, 4, 5$ s, and

(b) $x_3(t)$ analytically.

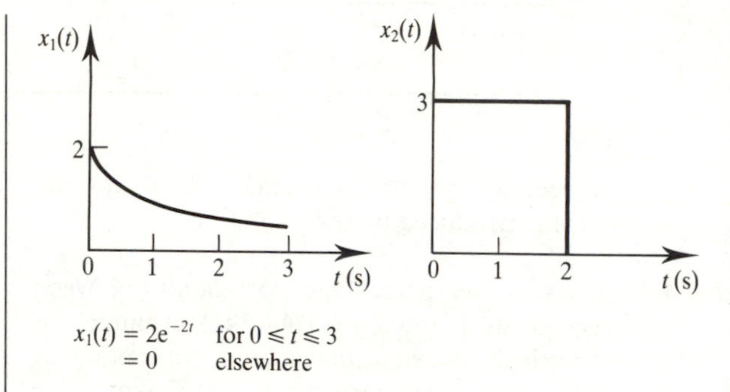

Figure 4.39 The functions $x_1(t)$ and $x_2(t)$ for Problem 4.13.

(2) Sketch the functions $x_3(t)$ and give reasons for any differences between them.

4.14 Determine the shape of the output pulse when a rectangular pulse of magnitude 5 V and width 0.4 μs is applied at the input of a single-stage lowpass RC filter with a cutoff frequency of 6 MHz. Assume that the impulse response of the filter is given by

$$h(t) = \frac{1}{CR} e^{-t/CR} u(t)$$

4.15 A rectangular pulse of height 5 V and width 1.0 μs is applied to the input of a system with the response function, $h(t)$, given by

$$h(t) = 0.1\,[1 - e^{-t/(1.09\times10^{-6})}] \quad 0 \leq t < 10\ \mu s$$
$$= 0 \quad 10\ \mu s < t < 0$$

Determine the system output

(1) analytically, and

(2) by sampling $h(t)$ every 1 μs and representing the pulse by an impulse function located at $t = 0$ s.

Critically compare your results.

4.16 Obtain the cross-correlation function between the two sets of data $\{1.5, 2.0, 1.5, 2.0, 2.5\}$ and $\{0, 0.33, 0.67, 1.0\}$

(1) by direct cross-correlation, and

(2) by applying the correlation theorem.

4.17 Determine the output of an electrical system of impulse response function {0, 0.899, 0.990, 0.991, 1} when the input {0, 2.5, 5.0, 0} (volts) is applied

(1) by direct convolution, and

(2) by applying the convolution theorem.

4.18 Use the overlap–add method to calculate the output of the system with the impulse response function $h(n) = \{0, 0.899, 0.990, 0.999, 1\}$ for the input data given in Table 4.5 (but ignoring the last two data). Assume the data were sampled every 2.5 μs, and divide the input data into five equal length sections. Calculate the phase shift between the output and input to the system. Check your answer using the method of direct convolution.

4.19 Repeat Problem 4.18 using the overlap–add method in which the convolutions are obtained using the convolution theorem. Compare the result with the answer to Problem 4.18.

4.20 Find the output of the system of Problem 4.18 which has the impulse response function $h(n) = \{0, 0.899, 0.990, 0.999, 1\}$ for all but the last two data of Table 4.5 using the overlap–save method. Compare the result with the solution to Problem 4.18.

4.21 Repeat Problem 4.20 applying the convolution theorem to evaluate the convolutions. Compare the result with the answers to Problems 4.18–4.20.

4.22 Consider Problems 4.18–4.21 and your solutions to them and compare the numbers of calculations required for the different methods with each other and with the calculation by direct convolution.

4.23 Write a program to perform convolution by the overlap–add method. Use it to confirm the results of Problem 4.18 and then to investigate the output of various systems for inputs of your choice.

4.24 (1) Write a program to carry out fast correlation and use it to cross-correlate the data recordings 1 and 2 of Table 4.4.

(2) Investigate the cross-correlations and auto-correlations of various waveforms such as square waves, rectangular waves, sine waves, random noise, and waveforms with various signal-to-noise ratios.

(3) Compare the relative abilities of correlation and spectrum estimation methods to detect signals in noise.

References

Beauchamp K.G. (1973). *Signal Processing Using Analog and Digital Techniques*. London: Allen and Unwin

Brigham E.O. (1974). *The Fast Fourier Transform*, Sections 13.3 and 13.4. Englewood Cliffs NJ: Prentice-Hall

Chatfield C. (1980). *The Analysis of Time Series*, p. 62. London: Chapman and Hall

DeFatta D.J., Lucas J.G. and Hodgkiss W.S. (1988). *Digital Signal Processing: A System Design Approach*, Section 6.9, p. 306. New York NY: Wiley

Grant P.M., Cowan C.F.N., Mulgrew B. and Dripps J.H. (1989). *Analogue and Digital Signal Processing and Coding*, Chapters 16, 17, 19 and 20. Chartwell Bratt

Jenkins G.M. and Watts D.G. (1968). *Spectral Analysis and its Applications*. San Francisco CA: Holden-Day

Jervis B.W., Goude A., Thomlinson M., Mir S. and Miller G. (1990). Least squares artefact removal by transputer. In *IEE Colloq. on the Transputer and Signal Processing*, Savoy Place, London, March 5, 1990

McGillem C.D. and Cooper G.R. (1974). *Continuous and Discrete Signal and System Analysis*. New York NY: Holt, Rinehart, and Winston

Proakis J.G. and Manolakis D.G. (1988). *Introduction to Digital Signal Processing*, p. 429. Basingstoke: Macmillan

Rabiner L.R. and Gold B. (1975). *Theory and Application of Digital Signal Processing*, Chapters 12 and 13. Englewood Cliffs NJ: Prentice-Hall

Stremler F.G. (1982). *Introduction to Communication Systems* 2nd edn., Section 3.10 and p. 407. Reading MA: Addison-Wesley

Strum R.D. and Kirk D.E. (1988). *First Principles of Discrete Systems and Digital Signal Processing*, Chapter 3. Reading MA: Addison-Wesley

Taub H. and Schilling D.L. (1986). *Principles of Communication Systems* 2nd edn., p. 562. New York NY: McGraw-Hill

Appendix

4A C language program for computing auto- and cross-correlation

The program, correltn.c, for computing the auto- or cross-correlation values of data sequences is available on the computer disk for the book. Details of how to obtain a copy are given in the preface under 'How to obtain the program disk'. The program is not listed here to limit the size of the book.

5

A framework for digital filter design

5.1	Introduction to digital filters	252
5.2	Types of digital filters: FIR and IIR filters	254
5.3	Choosing between FIR and IIR filters	255
5.4	Filter design steps	258
5.5	Illustrative examples	269
5.6	Summary	274
	Reference	276
	Bibliography	276

The purpose of this chapter is to provide a common framework for digital filter design. A simple step-by-step guide for designing digital filters, from specifications to implementation, is described. The options open to the designer at each step of the design process and factors that influence their choice are highlighted using several illustrative examples. Most DSP texts devote substantial space to the theory of digital filters, especially approximation methods, reflecting the considerable research effort that has gone into finding useful methods of calculating filter coefficients and the significant advances that have been made in filter design. However, such a coverage often overwhelms the inexperienced filter designer and leaves him/her not knowing how actually to go about designing a filter or how it all fits together. Thus the framework provided here, in our experience, is valuable to the designer who actually wants to design digital filters, as opposed to just learning about them from a purely theoretical point of view. This chapter sets the scene for Chapters 6 and 7 in which actual digital filter design is fully covered.

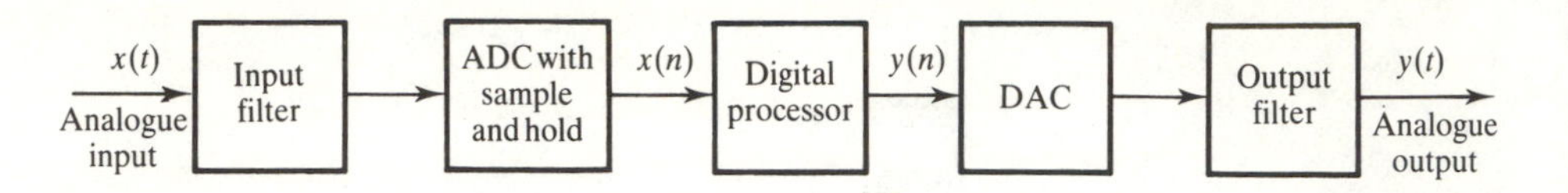

Figure 5.1 A simplified block diagram of a real-time digital filter with analogue input and output signals.

5.1 Introduction to digital filters

A filter is essentially a system or network that selectively changes the wave-shape, amplitude–frequency and/or phase–frequency characteristics of a signal in a desired manner. Common filtering objectives are to improve the quality of a signal (for example, to remove or reduce noise), to extract information from signals or to separate two or more signals previously combined to make, for example, efficient use of an available communication channel.

A digital filter, as we shall see later, is a mathematical algorithm implemented in hardware and/or software that operates on a digital input signal to produce a digital output signal for the purpose of achieving a filtering objective. The term digital filter refers to the specific hardware or software routine that performs the filtering algorithm. Digital filters often operate on digitized analogue signals or just numbers, representing some variable, stored in a computer memory.

A simplified block diagram of a real-time digital filter, with analogue input and output signals, is given in Figure 5.1. The bandlimited analogue signal is sampled periodically and converted into a series of digital samples, $x(n)$, $n = 0, 1, \ldots$. The digital processor implements the filtering operation, mapping the input sequence, $x(n)$, into the output sequence, $y(n)$, in accordance with a computational algorithm for the filter. The DAC converts the digitally filtered output into analogue values which are then analogue filtered to smooth and remove unwanted high frequency components.

Digital filters play very important roles in DSP. Compared with analogue filters they are preferred in a number of applications (for example data compression, biomedical signal processing, speech processing, image processing, data transmission, digital audio, telephone echo cancellation) because of one or more of the following advantages.

- Digital filters can have characteristics which are not possible with analogue filters, such as a truly linear phase response.
- Unlike analogue filters, the performance of digital filters does not vary with environmental changes, for example thermal variations. This eliminates the need to calibrate periodically.

- The frequency response of a digital filter can be automatically adjusted if it is implemented using a programmable processor, which is why they are widely used in adaptive filters.
- Several input signals or channels can be filtered by one digital filter without the need to replicate the hardware.
- Both filtered and unfiltered data can be saved for further use.
- Advantage can be readily taken of the tremendous advancements in VLSI technology to fabricate digital filters and to make them small in size, to consume low power, and to keep the cost down.
- In practice, the precision achievable with analogue filters is restricted; for example, typically a maximum of only about 60 to 70 dB stopband attenuation is possible with active filters designed with off-the-shelf components. With digital filters the precision is limited only by the word length used.
- The performance of digital filters is repeatable from unit to unit.
- Digital filters can be used at very low frequencies, found in many biomedical applications for example, where the use of analogue filters is impractical. Also, digital filters can be made to work over a wide range of frequencies by a mere change to the sampling frequency.

The following are the main disadvantages of digital filters compared with analogue filters:

- *Speed limitation.* The maximum bandwidth of signals that digital filters can handle, in real time, is much lower than for analogue filters. In real-time situations, the analogue–digital–analogue conversion processes introduce a speed constraint on the digital filter performance. The conversion time of the ADC and the settling time of the DAC limit the highest frequency that can be processed. Further, the speed of operation of a digital filter depends on the speed of the digital processor used and on the number of arithmetic operations that must be performed for the filtering algorithm, which increases as the filter response is made tighter.
- *Finite wordlength effects.* Digital filters are subject to ADC noise resulting from quantizing a continuous signal, and to roundoff noise incurred during computation. With higher order recursive filters, the accumulation of roundoff noise could lead to instability.
- *Long design and development times.* The design and development times for digital filters, especially hardware development, can be much longer than for analogue filters. However, once developed the hardware and/or software can be used for other filtering or DSP tasks with little or no modifications (several examples of this are given in subsequent chapters). Good computer-aided design (CAD) support can make the design of digital filters an enjoyable task, but some expertise is required to make full and effective use of such design aids.

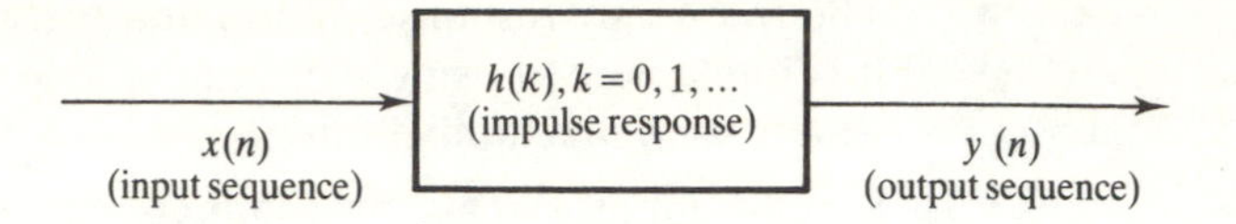

Figure 5.2 A conceptual representation of a digital filter.

5.2 Types of digital filters: FIR and IIR filters

Digital filters are broadly divided into two classes, namely infinite impulse response (IIR) and finite impulse response (FIR) filters. Either type of filter, in its basic form, can be represented by its impulse response sequence, $h(k)$ ($k = 0, 1, \ldots$), as in Figure 5.2. The input and output signals to the filter are related by the convolution sum, which is given in Equations 5.1 for the IIR and in 5.2 for the FIR filter.

$$y(n) = \sum_{k=0}^{\infty} h(k)x(n-k) \tag{5.1}$$

$$y(n) = \sum_{k=0}^{N-1} h(k)x(n-k) \tag{5.2}$$

It is evident from these equations that, for IIR filters, the impulse response is of infinite duration whereas for FIR it is of finite duration, since $h(k)$ for the FIR has only N values. In practice, it is not feasible to compute the output of the IIR filter using Equation 5.1 because the length of its impulse response is too long (infinite in theory). Instead, the IIR filtering equation is expressed in a recursive form:

$$y(n) = \sum_{k=0}^{\infty} h(k)x(n-k) = \sum_{k=0}^{N} a_k x(n-k) - \sum_{k=1}^{M} b_k y(n-k) \tag{5.3}$$

where the a_k and b_k are the coefficients of the filter. Thus, Equations 5.2 and 5.3 are the difference equations for the FIR and IIR filters respectively. These equations, and in particular the values of $h(k)$, for FIR, or a_k and b_k, for IIR, are often very important objectives of most filter design problems. We note that, in Equation 5.3, the current output sample, $y(n)$, is a function of past outputs as well as present and past input samples, that is the IIR is a feedback system of some sort. This should be compared with the FIR equation in which the current output sample, $y(n)$, is a function only of past and present values of the input. Note, however, that when the b_k are set to zero, Equation 5.3 reduces to the FIR equation 5.2.

Alternative representations for the FIR and IIR filters are given in Equations 5.4a and 5.4b respectively. These are the transfer functions for these filters and are very useful in evaluating their frequency responses (see Chapters 3, 6 and 7 for details).

As will become clear in the next few sections, factors that influence the choice of options open to the digital filter designer at each stage of the design process are strongly linked to whether the filter in question is IIR or FIR. Thus, it is very important to appreciate the differences between IIR and FIR, their peculiar characteristics, and more importantly, how to choose between them.

$$H(z) = \sum_{k=0}^{N-1} h(k)z^{-k} \tag{5.4a}$$

$$H(z) = \sum_{k=0}^{N} a_k z^{-k} \Bigg/ \left(1 + \sum_{k=1}^{M} b_k z^{-k}\right) \tag{5.4b}$$

5.3 Choosing between FIR and IIR filters

The choice between FIR and IIR filters depends largely on the relative advantages of the two filter types.

(1) FIR filters can have an exactly linear phase response. The implication of this is that no phase distortion is introduced into the signal by the filter. This is an important requirement in many applications, for example data transmission, biomedicine, digital audio and image processing. The phase responses of IIR filters are nonlinear, especially at the band edges.

(2) FIR filters realized nonrecursively, that is by direct evaluation of Equation 5.2, are always stable. The stability of IIR filters cannot always be guaranteed.

(3) The effects of using a limited number of bits to implement filters such as roundoff noise and coefficient quantization errors are much less severe in FIR than in IIR.

(4) FIR requires more coefficients for sharp cutoff filters than IIR. Thus for a given amplitude response specification, more processing time and storage will be required for FIR implementation. However, one can readily take advantage of the computational speed of the FFT and multirate techniques (see Chapter 8) to improve significantly the efficiency of FIR implementations.

(5) Analogue filters can be readily transformed into equivalent IIR digital filters meeting similar specifications. This is not possible with FIR filters as they have no analogue counterpart. However, with FIR it is easier to synthesize filters of arbitrary frequency responses.

(6) In general, FIR is algebraically more difficult to synthesize, if CAD support is not available.

From the above, a broad guideline on when to use FIR or IIR would be as follows.

- Use IIR when the only important requirements are sharp cutoff filters and high throughput, as IIR filters, especially those using elliptic characteristics, will give fewer coefficients than FIR.
- Use FIR if the number of filter coefficients is not too large and, in particular, if little or no phase distortion is desired. One might also add that newer DSP processors have architectures that are tailored to FIR filtering, and indeed some are designed specifically for FIRs (see Chapter 11).

Example 5.1

The following transfer functions represent two different filters meeting identical amplitude–frequency response specifications:

(1) $$H(z) = \frac{a_0 + a_1 z^{-1} + a_2 z^{-2}}{1 + b_1 z^{-1} + b_2 z^{-2}}$$

where

$$a_0 = 0.498\,181\,9$$
$$a_1 = 0.927\,477\,7$$
$$a_2 = 0.498\,181\,9$$
$$b_1 = -0.674\,487\,8$$
$$b_2 = -0.363\,348\,2$$

(2) $$H(z) = \sum_{k=0}^{11} h(k) z^{-k}$$

where

$$h(0) = 0.546\,032\,80 \times 10^{-2} = h(11)$$
$$h(1) = -0.450\,687\,50 \times 10^{-1} = h(10)$$
$$h(2) = 0.691\,694\,20 \times 10^{-1} = h(9)$$
$$h(3) = -0.553\,843\,70 \times 10^{-1} = h(8)$$
$$h(4) = -0.634\,284\,10 \times 10^{-1} = h(7)$$
$$h(5) = 0.578\,924\,00 \times 10^{0} = h(6)$$

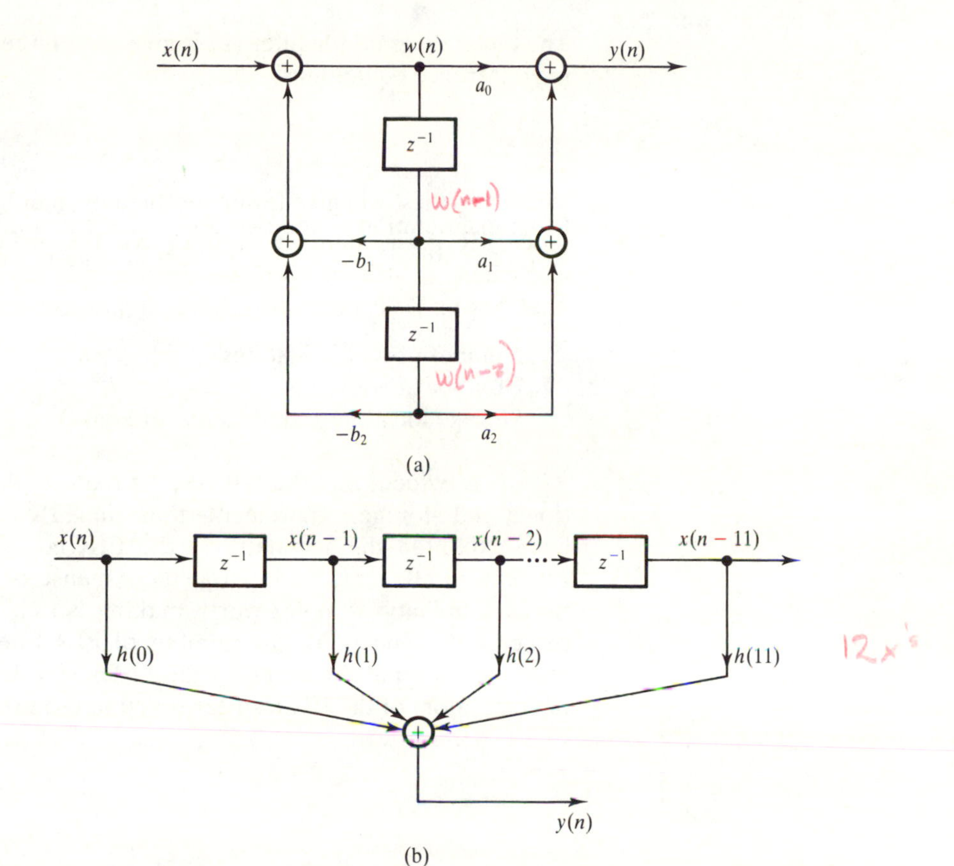

Figure 5.3 (a) Block diagram representation of the IIR filter of Example 5.1. (b) Block diagram representation of the FIR filter of Example 5.1.

For each filter,

(a) state whether it is an FIR or IIR filter,
(b) represent the filtering operation in a block diagram form and write down the difference equation, and
(c) determine and comment on the computational and storage requirements.

Solution

(a) Filters (1) and (2) are IIR and FIR respectively.
(b) The block diagram for filter (1) is given in Figure 5.3(a). The corresponding set of difference equations is:

$$w(n) = x(n) - b_1 w(n-1) - b_2 w(n-2)$$
$$y(n) = a_0 w(n) + a_1 w(n-1) + a_2 w(n-2)$$

The block diagram for filter (2) is given in Figure 5.3(b). The corresponding difference equation is

$$y(n) = \sum_{k=0}^{11} h(k)x(n-k)$$

(c) From examination of the two difference equations the computational and storage requirements for both filters are summarized below:

	FIR	IIR
Number of multiplications	12	5
Number of additions	11	4
Storage locations (coefficients and data)	24	8

It is evident that the IIR filter is more economical in both computational and storage requirements than the FIR filter. However, we could have exploited the symmetry in the FIR coefficients to make the FIR filter more efficient, although at the expense of its obvious implementational simplicity. A point worth making is that, for the same amplitude response specifications, the number of FIR filter coefficients (12 in this example) is typically six times the order (the highest power of z in the denominator) of the IIR transfer function (2 in this case).

5.4 Filter design steps

The design of a digital filter involves five steps:

- specification of the filter requirements;
- calculation of suitable filter coefficients;
- representation of the filter by a suitable structure (realization);
- analysis of the effects of finite wordlength on filter performance;
- implementation of filter in software and/or hardware.

The five steps are not necessarily independent; nor are they always performed in the order given. In fact techniques are now available that combine the second and aspects of the third and fourth steps. However, the approach discussed here gives a simple step-by-step guide that will ensure a successful design. To arrive at an efficient filter, it may be necessary to iterate a few times between the steps, especially if the problem specification is not watertight, as is often the case, or if the designer wants to explore other possible designs. Detailed discussions of these steps now follow.

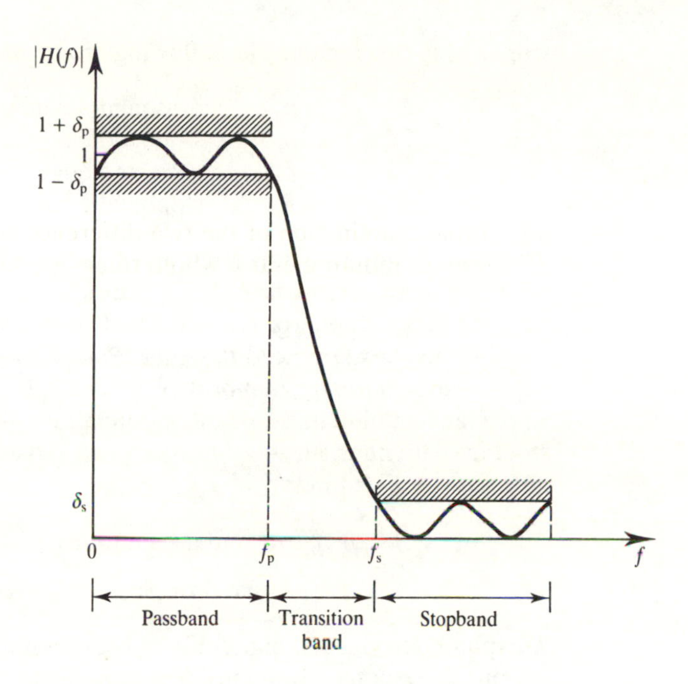

Figure 5.4 Tolerance scheme for a lowpass filter.

5.4.1 Specification of the filter requirements

Requirement specifications include specifying (i) signal characteristics (types of signal source and sink, I/O interface, data rates and width, and highest frequency of interest), (ii) the characteristics of the filter (the desired amplitude and/or phase responses and their tolerances (if any), the speed of operation and modes of filtering (real time or batch), (iii) the manner of implementation (for example, as a high level language routine in a computer or as a DSP processor-based system, choice of signal processor), and (iv) other design constraints (for example, the cost of the filter). The designer may not have enough information to specify the filter completely at the outset, but as many of the filter requirements as possible should be specified to simplify the design process.

Although the above requirements are application dependent it will be helpful to devote some time on some aspects of (ii). The characteristics of digital filters are often specified in the frequency domain. For frequency selective filters, such as lowpass and bandpass filters, the specifications are often in the form of tolerance schemes. Figure 5.4 depicts such a scheme for a lowpass filter. The shaded horizontal lines indicate the tolerance limits. In the passband, the magnitude response has a peak deviation of δ_p and, in the stopband, it has a maximum deviation of δ_s.

The width of the transition band determines how sharp the filter is. The magnitude response decreases monotonically from the passband to the stopband in this region. The following are the key parameters of interest:

δ_p	passband deviation
δ_s	stopband deviation
f_p	passband edge frequency
f_s	stopband edge frequency

The edge frequencies are often given in normalized form, that is as a fraction of the sampling frequency (f/F_s), but specifications using standard frequency units of hertz, kilohertz are valid and sometimes are more meaningful, especially to the inexperienced designer. Passband and stopband deviations may be expressed as ordinary numbers or in decibels when they specify the passband ripple and minimum stopband attenuation respectively. Thus the minimum stopband attenuation, A_s, and the peak passband ripple, A_p, in decibels are given as (for FIR filters)

$$A_s \text{ (stopband attenuation)} = -20\log_{10}\delta_s \quad \textbf{(5.5a)}$$

$$A_p \text{ (passband ripple)} = 20\log_{10}(1+\delta_p) \quad \textbf{(5.5b)}$$

The phase response of digital filters is often not as meticulously specified as the amplitude response. In many cases it is sufficient to indicate that phase distortion is of concern or that linear phase response is desirable. However, in some applications where filters are used to equalize or compensate a system's phase response, for example, or as phase shifters, then the desired phase response will need to be specified.

Example 5.2

An FIR bandpass filter is to be designed to meet the following frequency response specifications:

passband	0.18–0.33	(normalized)
transition width	0.04	(normalized)
stopband deviation	0.001	
passband deviation	0.05	

(1) Sketch the tolerance scheme for the filter.

(2) Express the filter band edge frequencies in the standard unit of kilohertz, assuming a sampling frequency of 10 kHz, and the stopband and passband deviations in decibels.

Solution

(1) The tolerance scheme for the filter is given in Figure 5.5.

(2) The band edge frequencies, at a sampling frequency of 10 kHz, and the

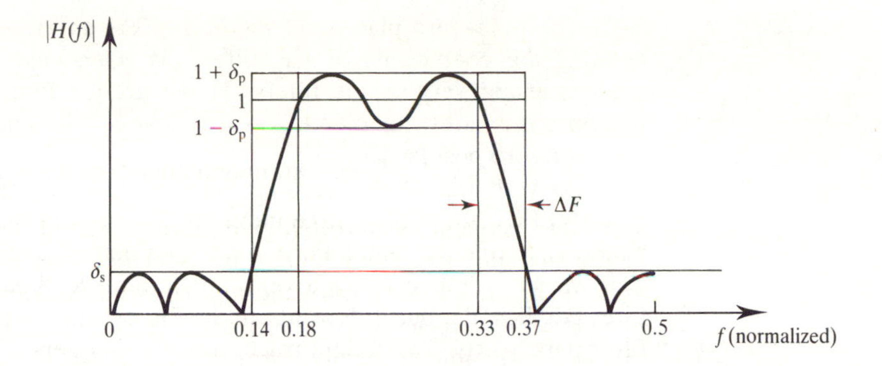

Figure 5.5 Tolerance scheme for the bandpass filter of Example 5.2.

stopband and passband deviations are given below:

passband	1.8–3.3 kHz
stopbands	0–1.4 kHz and 3.7–5 kHz
stopband attenuation	$-20\log_{10}(0.001) = 60$ dB
passband ripple	$20\log_{10}(1 + 0.05) = 0.42$ dB

5.4.2 Coefficient calculation

In this step, we select one of a number of approximation methods and calculate the values of the coefficients, $h(k)$, for FIR, or a_k and b_k, for IIR, such that the filter characteristics given in Section 5.4.1 are satisfied. The method used to calculate the filter coefficients depends on whether the filter is IIR or FIR type.

Calculations of IIR filter coefficients are traditionally based on the transformation of known analogue filter characteristics into equivalent digital filters. The two basic methods used are the impulse invariant and the bilinear transformation methods. With the impulse invariant method, after digitizing the analogue filter, the impulse response of the original analogue filter is preserved, but not its magnitude–frequency response. Because of inherent aliasing, the method is inappropriate for highpass or bandstop filters. The bilinear method, on the other hand, yields very efficient filters and is well suited to the calculation of coefficients of frequency selective filters. It allows the design of digital filters with known classical characteristics such as Butterworth, Chebyshev and elliptic. Digital filters resulting from the bilinear transform method will, in general, preserve the magnitude response characteristics of the analogue filter but not the time domain properties. Efficient computer programs now exist for calculating filter coefficients, using the bilinear method, by merely specifying filter parameters of interest (see Chapter 7). The impulse invariant method is good for simulating analogue systems, but the bilinear method is best for frequency selective IIR filters.

The pole–zero placement method offers an alternative approach to calculating the coefficients of IIR filters. It is an easy way of calculating the coefficients of very simple filters. However, for filters with good amplitude response it is not recommended as it relies on 'trial and error' shuffling of the pole and zero positions.

As with IIR filters there are several methods of calculating the coefficients of FIR filters. The three methods discussed in this book in detail are the window, frequency sampling, and the optimal (Parks–McClellan algorithm). The window method offers a very simple and flexible way of computing FIR filter coefficients, but it does not allow the designer adequate control over the filter parameters. The main attraction of the frequency sampling method is that it allows a recursive realization of FIR filters which can be computationally very efficient. However, it lacks flexibility in specifying or controlling filter parameters. With the availability of an efficient and easy-to-use program, the optimal method is now widely used in industry and, for most applications, will yield the desired FIR filters. Thus, for FIR filters, the optimal method should be the method of first choice unless the particular application dictates otherwise or a CAD facility is unavailable.

In summary, there are several methods of calculating filter coefficients of which the following are the most widely used:

- impulse invariant (IIR);
- bilinear transformation (IIR);
- pole–zero placement (IIR);
- window (FIR);
- frequency sampling (FIR);
- optimal (FIR).

We choose the method that best suits our particular application. Our choice will be influenced by several factors, the most important of which are the critical requirements in the specifications. In general, the crucial choice is really between FIR and IIR. In most cases, if the FIR properties are vital then a good candidate is the optimal method, whereas, if IIR properties are desirable, then the bilinear method will in most cases suffice.

5.4.3 Representation of a filter by a suitable structure (realization)

Realization involves converting a given transfer function, $H(z)$, into a suitable filter structure. Block or flow diagrams are often used to depict filter structures and they show the computational procedure for implementing the digital filter. The structure used depends on whether the filter is an IIR or FIR filter.

For IIR filters, three structures commonly used are the direct form, cascade and parallel forms. The direct form is simply a straightforward representation of the IIR transfer function. In the cascade form, the transfer function of

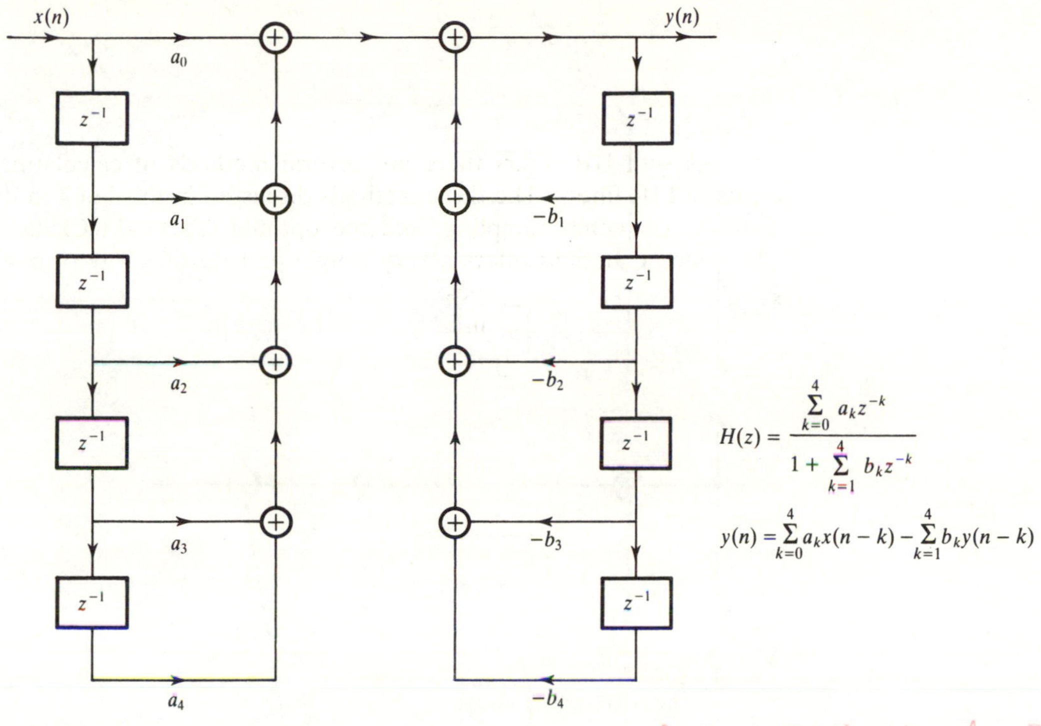

(a)

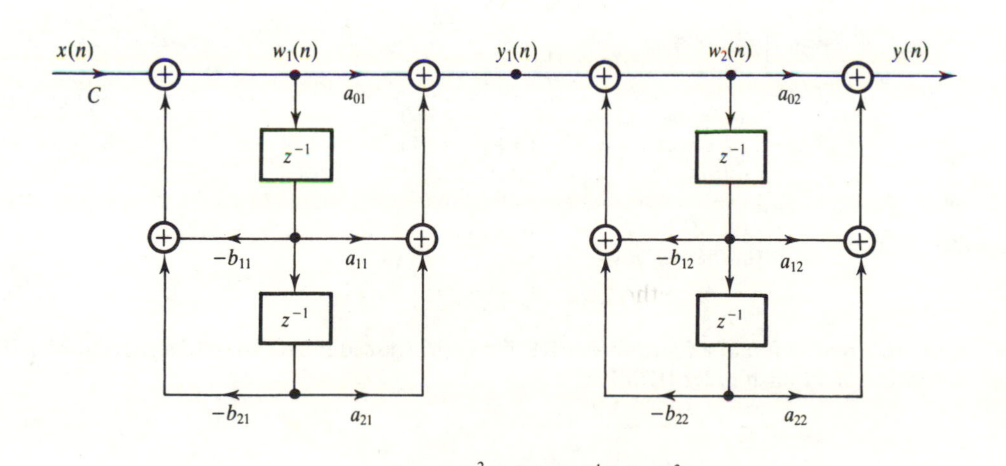

$$H(z) = C \prod_{k=1}^{2} \frac{1 + a_{1k} z^{-1} + b_{2k} z^{-2}}{1 + b_{1k} z^{-1} + b_{2k} z^{-2}}$$

$$w_1(n) = Cx(n) - b_{11} w_1(n) - b_{21} w_1(n-2)$$
$$y_1(n) = a_{01} w_1(n) + a_{11} w_1(n-1) + a_{21} w_1(n-2)$$
$$w_2(n) = y_1(n) - b_{12} w_2(n-1) - b_{22} w_2(n-2)$$
$$y(n) = a_{02} w_2(n) + a_{12} w_2(n-1) + a_{22} w_2(n-2)$$

(b) cascade

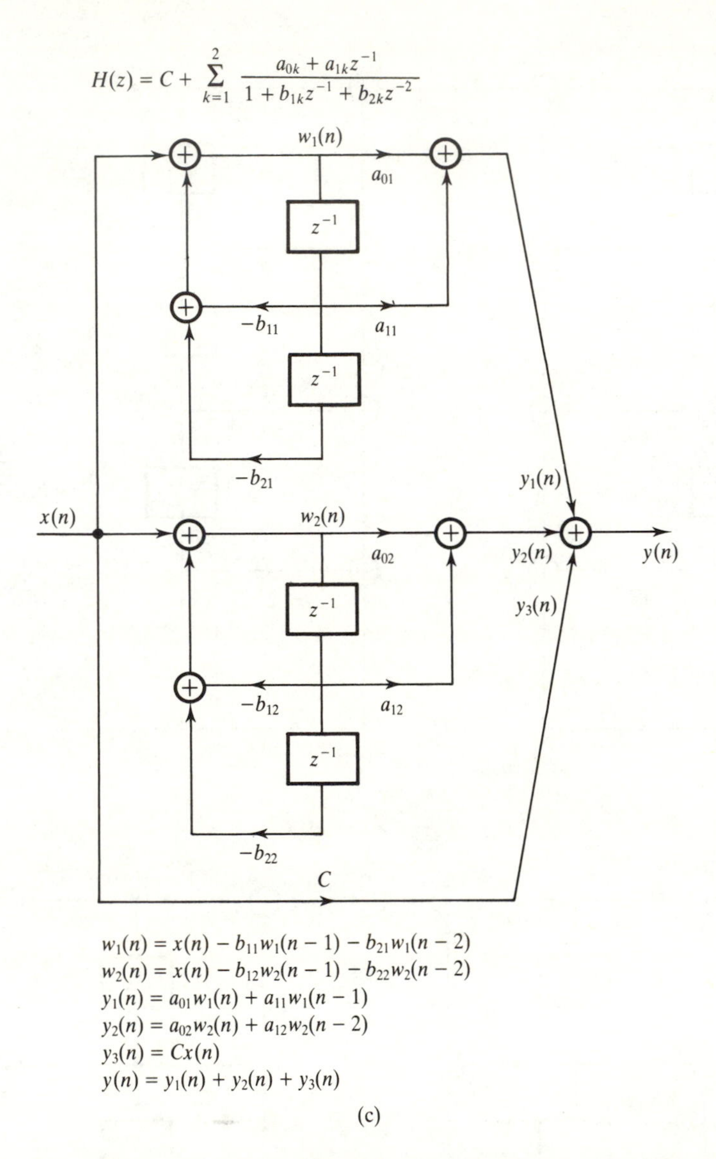

Figure 5.6 (a) Direct realization of a fourth-order IIR filter. (b) Cascade realization of a fourth-order IIR filter. (c) Parallel realization of a fourth-order IIR filter.

the IIR filter, Equation 5.4b, is factored and expressed as the product of second-order sections. In the parallel form, $H(z)$ is expanded, using partial fractions, as the sum of second-order sections. As an illustration and for comparison, Figure 5.6 shows the block diagrams for a fourth-order (that is, $N = 4$) IIR filter represented in the direct, cascade and parallel structures. The corresponding set of transfer functions and the difference equations describing the filter structures are also given in the figure.

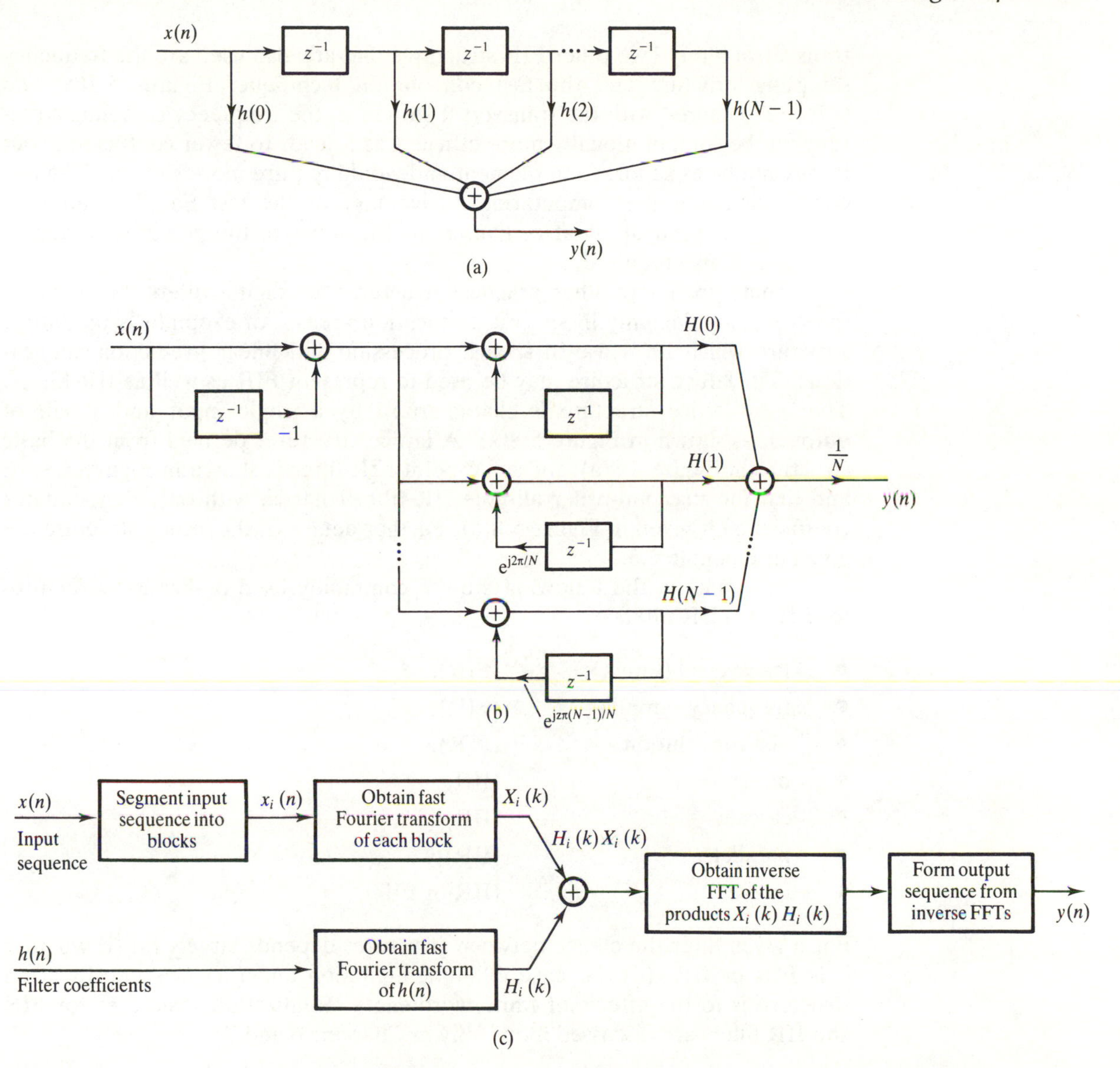

Figure 5.7 Realization structures for FIR filters: (a) transversal filter (or direct); (b) frequency sampling structure; (c) fast convolution.

The parallel and cascade structures are the most widely used for IIR because they lead to simpler filtering algorithms and are far less sensitive to the effects of implementing the filter using a finite number of bits than the direct structure. The direct structure suffers severe coefficient sensitivity problems, especially for high order filters, and should be avoided in these cases.

The most widely used structure for FIR is the direct form, Figure 5.7(a), because it is particularly simple to implement. In this form, the FIR is sometimes called a tapped delay line (because it resembles a tapped delay line) or transversal filter. Two other FIR structures that are also used are the frequency sampling structure and the fast convolution technique, Figures 5.7(b) and 5.7(c). Compared with the transversal structure, the frequency sampling structure can be computationally more efficient as it leads to fewer coefficients, but it may not be as simple to implement and would require more storage. The fast convolution uses the computational advantage of the fast Fourier transform (FFT) and is particularly attractive in situations where the power spectrum of the signal is also required.

There are many other practical structures for digital filters, but most of these are popular only in specific application areas. An example is the lattice structure which finds use in speech processing and linear prediction applications. The lattice structure may be used to represent FIR as well as IIR filters. The basic lattice structure is characterized by a single input and a pair of outputs, as shown in Figure 5.8(a). A lattice structure, derived from the basic structure in Figure 5.8(a), for an N-point FIR filter is shown in Figure 5.8(b), and that for a second-order all-pole IIR filter (that is, with only denominator coefficients) is given in Figure 5.8(c). Further details of the lattice structure are given in Example 5.5.

In summary, the following are the commonly used realization structures for FIR and IIR filters:

- transversal (direct) (FIR);
- frequency sampling (FIR);
- fast convolution (FIR);
- direct form (IIR);
- cascade (IIR);
- parallel (IIR);
- lattice (IIR or FIR).

For a given filter the choice between structures depends largely on (i) whether it is FIR or IIR, (ii) the ease of implementation and (iii) how sensitive the structure is to the effects of finite wordlength. Realization structures for FIR and IIR filters are discussed more fully in Chapters 6 and 7 respectively.

5.4.4 Analysis of finite wordlength effects

The approximation and realization steps assume infinite or very high precision. However, in actual implementation it is often necessary to represent the filter coefficients using limited number of bits, typically 8 to 16 bits, and the arithmetic operations indicated in the difference equation are performed using finite precision arithmetic.

The effects of using a finite number of bits are to degrade the performance of the filter and in some cases to make it unusable. The designer must analyse these effects and choose suitable wordlengths (that is, numbers of bits),

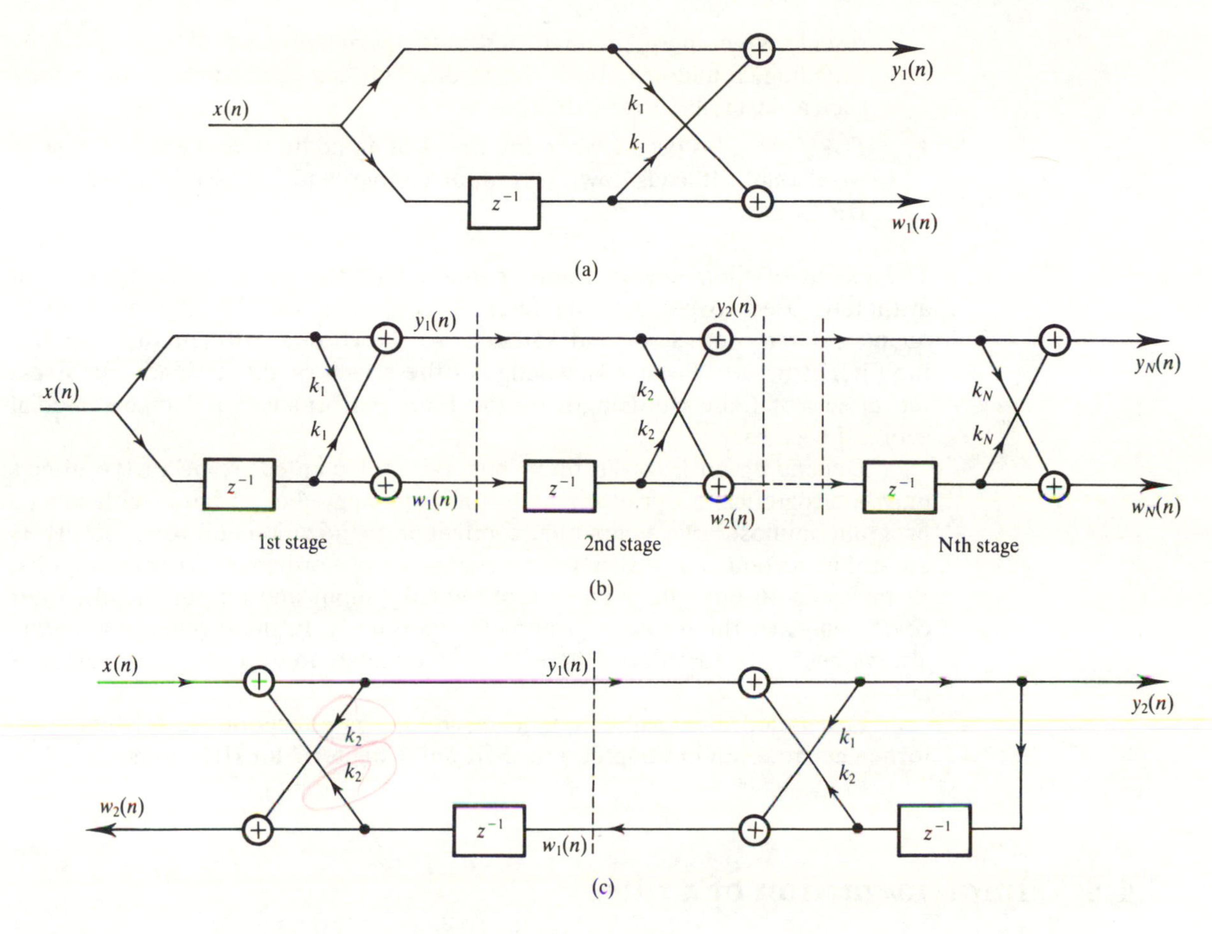

Figure 5.8 (a) The basic lattice structure. (b) An N-stage FIR lattice structure. (c) A 2-stage all-pole IIR lattice structure.

for the filter coefficients, filter variables, that is the input and output samples, and for the arithmetic operations within the filter.

The main sources of performance degradation in digital filters are as follows.

- *Input/output signal quantization*. In particular, the ADC noise due to quantizing of the input signal samples is significant (see Chapter 1 for details).
- *Coefficient quantization*. This leads to deviations in the frequency response of both FIR and IIR filters, and possibly to instability in IIR filters.
- *Arithmetic roundoff errors*. The use of finite precision arithmetic to perform filtering operations yields results that require additional bits to rep-

resent. When these are quantized to the permissible wordlength, often by rounding, roundoff noise is the result. This can cause undesirable effects such as instability in an IIR filter.

- *Overflow*. This occurs when the result of an addition exceeds permissible wordlength. It leads to wrong output samples and to possible instability in IIR filters.

The extent of filter degradation depends on (i) the wordlength and type of arithmetic used to perform the filtering operation, (ii) the method used to quantize filter coefficients and variables to the chosen wordlengths, and (iii) the filter structure. From a knowledge of these factors, the designer can assess the effects of finite wordlength on the filter performance and take remedial action if necessary.

Depending on how the filter is to be implemented, some of the effects may be insignificant. For example, when implemented as a high level language program on most large computers, coefficient quantization and roundoff errors are not important. For real-time processing, finite wordlengths (typically 8 bits, 12 bits, and 16 bits) are used to represent the input and output signals, filter coefficients and the results of arithmetic operations. In these cases, it is nearly always necessary to analyse the effects of quantization on the filter performance.

A detailed discussed of quantization and its effects on digital filter performance are given in Chapter 6 for FIR and Chapter 7 for IIR filters.

5.4.5 Implementation of a filter

Having calculated the filter coefficients, chosen a suitable structure, and verified that the filter degradation, after quantizing the coefficients and filter variables to the selected wordlengths, is acceptable, the difference equation must be implemented as a software routine or in hardware. Whatever the method of implementation, the output of the filter must be computed, for each sample, in accordance with the difference equation (assuming a time domain implementation).

As the examination of any difference equation will show (Equations 5.2 and 5.3), the computation of $y(n)$ (the filter output) involves only multiplications, additions/subtractions, and delays. Thus to implement a filter, we need the following basic building blocks:

- memory (for example ROM) for storing filter coefficients;
- memory (such as RAM) for storing the present and past inputs and outputs, that is $\{x(n), x(n-1), \ldots\}$ and $\{y(n), y(n-1), \ldots\}$;
- hardware or software multiplier(s);
- adder or arithmetic logic unit.

The designer provides these basic blocks and also ensures that they are suitably configured for the application. The manner in which the components are configured depends to a large extent on whether batch (that is, non-real-time) or real-time processing is required. In batch processing, the entire data is already available in some memory device. Such is the case in applications where, for example, experimental data is acquired from elsewhere for later analysis. In this case, the filter is often implemented in a high level language and runs in a general-purpose computer, such as a personal computer or a mainframe computer, where all the basic blocks are already configured. Thus, batch processing may be described as a purely software implementation (although the designer may wish to incorporate additional hardware to increase the speed of processing).

In real-time processing, the filter is required either (i) to operate on the present input sample, $x(n)$, to produce the current output sample, $y(n)$, before the next input sample arrives, that is within the intersample period, or (ii) to operate on an input block of data, using an FFT algorithm for example, to produce an output block of data within a period proportional to the block length. Real-time filtering may require fast and dedicated hardware if the sample rate is very high or if the filter is of a high order. For most audio frequency work, DSP processors such as the DSP56000 (by Motorola) and the TMS320C25 (by Texas Instruments) will be adequate and offer considerable flexibility. These processors have all the basic blocks on board, including built-in hardware multiplier(s). In some applications, standard 8-bit or 16-bit microprocessors such as the Motorola 6800 or 68000 families offer attractive alternative implementations. In addition to the signal processing hardware, the designer must also provide suitable input–output (for example, analogue–digital conversion) interfaces to the digital hardware, depending on the type of data source and sink. Detailed discussions of filter implementations are given in Chapters 6 for FIR and 7 for IIR filters. DSP hardware is covered in Chapters 11 and 12.

5.5 Illustrative examples

Example 5.3

Discuss the five main steps involved in the design of digital filters, using the following design problem to illustrate your answer.

A digital filter is required for real-time physiological noise reduction. The filter should meet the following amplitude response specifications:

passband	0–10 Hz
stopband	20–64 Hz
sampling frequency	128 Hz
maximum passband deviation	< 0.036 dB
stopband attenuation	> 30 dB

Other important requirements are that

(1) minimal distortion of the harmonic relationships between the components of the in-band signals is highly desirable,

(2) the time available for filtering is limited, the filter being part of a larger process, and

(3) the filter will be implemented using the Texas Instruments TMS32010 DSP processor with the analogue input digitized to 12 bits.

Solution

This filter was designed and used in a certain biomedical signal processing project. We shall give here only an outline discussion of the design, postponing detailed discussion to Chapter 6 where FIR filter design methods are fully covered.

(1) *Requirement specification*. As discussed previously, the designer must give the exact role and performance requirements for the filter together with any important constraints. These have already been given for the example.

(2) *Calculation of suitable coefficients*. The requirements of minimal distortion and limited processing time are best achieved with a linear phase FIR filter, with coefficients obtained using the optimal method.

(3) *Selection of filter structure*. The transversal structure will lead to the most efficient implementation using the TMS32010 processor.

(4) *Analysis of finite wordlength effects*. Since the TMS32010 processor will be used, fixed point arithmetic should be used with each coefficient represented by 16 bits (after rounding) for efficiency. FIR filter degradation may result from input signal quantization, coefficient quantization, roundoff and overflow errors. A check should be made to ensure that the wordlengths are sufficiently long. Analysis of finite wordlength effects for this case showed that the input quantization noise and deviation in the frequency response due to coefficient quantization are both insignificant. The use of the TMS32010's 32-bit accumulator to sum the coefficient–data products, rounding only the final sum, would reduce roundoff errors to negligible levels. To avoid overflow, each coefficient should be divided by $\sum_{k=0}^{N-1} |h(k)|$ before quantizing to 16 bits.

(5) *Implementation*. Design and configure the TMS32010-based hardware (if it does not already exist) with the necessary input/output interfaces. Then write a TMS32010 program to handle the I/O protocols and calculate filter output, $y(n) = \sum_{k=0}^{N-1} h(k)x(n-k)$, for each new input, $x(n)$.

Example 5.4

An analogue filter is to be converted into an equivalent digital filter that will operate at a sampling frequency of 256 Hz. The analogue filter has the transfer function:

$$H(s) = \frac{1}{s^3 + 2s^2 + 2s + 1}$$

(1) Obtain suitable coefficients for the digital filter.

(2) Assuming that the digital filter is to be realized using the cascade structure, draw a suitable realization block diagram and develop the difference equations.

(3) Repeat (2) for the parallel structure.

Solution

(1) To preserve the amplitude response of the analogue function, the bilinear method was used to obtain the filter coefficients. The application of the bilinear transformation approach to the analogue transfer function (see Chapter 7 for details) yields the following transfer functions:

$$H(z) = \frac{0.1432(1 + 3z^{-1} + 3z^{-2} + z^{-3})}{1 - 0.1801z^{-1} + 0.3419z^{-2} - 0.0165z^{-3}}$$

(2) For the cascade realization, $H(z)$ is factorized using partial fractions:

$$H(z) = 0.1432\frac{1 + 2z^{-1} + z^{-2}}{1 - 0.1307z^{-1} + 0.3355z^{-2}}\frac{1 + z^{-1}}{1 - 0.0490z^{-1}}$$

The block diagram representation and the corresponding set of difference equations are given in Figure 5.9 and as follows:

$$w_1(n) = 0.1432x(n) + 0.1307w_1(n-1) - 0.3355w_1(n-2)$$

$$y_1(n) = w_1(n) + 2w_1(n-1) + w_1(n-2)$$

$$w_2(n) = y_1(n) + 0.049w_2(n-1)$$

$$y_2(n) = w_2(n) + w_2(n-1)$$

(3) For the parallel realization, $H(z)$ is expressed using partial fractions (see Chapters 3 and 7 for details) as

$$H(z) = \frac{1.2916 - 0.08407z^{-1}}{1 - 0.131z^{-1} + 0.3355z^{-2}} + \frac{7.5268}{1 - 0.049z^{-1}} - 8.6753$$

The parallel realization diagram and its corresponding set of difference

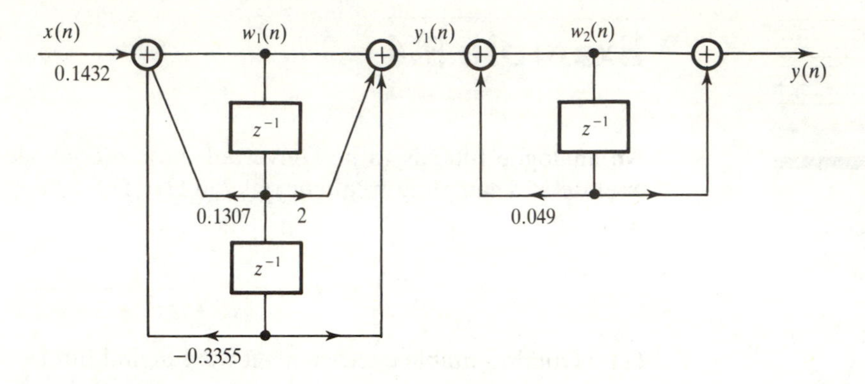

Figure 5.9

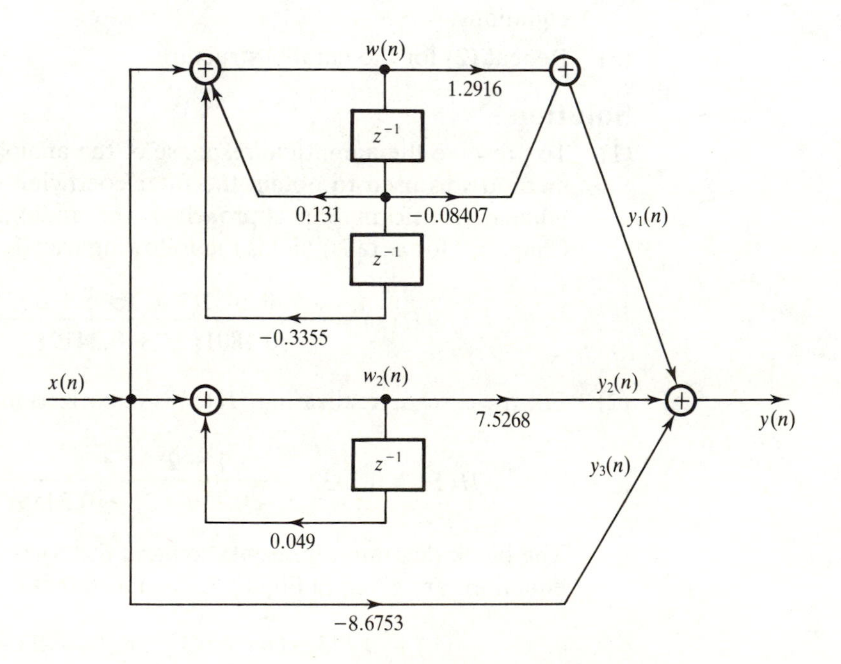

Figure 5.10

equations are given in Figure 5.10 and as follows:

$$w_1(n) = x(n) + 0.131w_1(n-1) - 0.3355w_1(n-2)$$

$$y_1(n) = 1.2916w_1(n) - 0.08407w_1(n-1)$$

$$w_2(n) = x(n) + 0.049w_2(n-1)$$

$$y_2(n) = 7.5268w_2(n)$$

$$y_3(n) = -8.6753x(n)$$

$$y(n) = y_1(n) + y_2(n) + y_3(n)$$

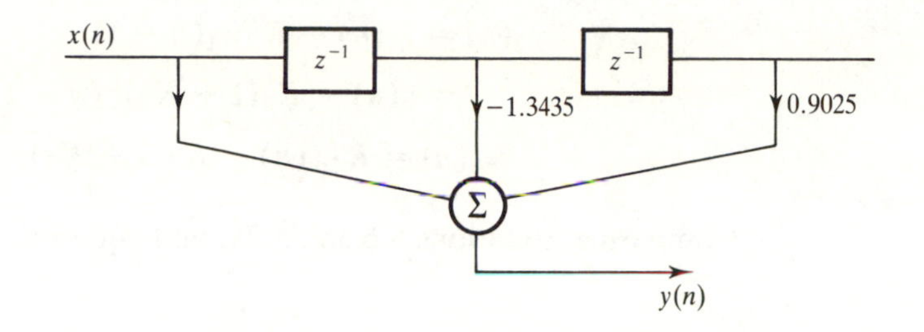

Figure 5.11

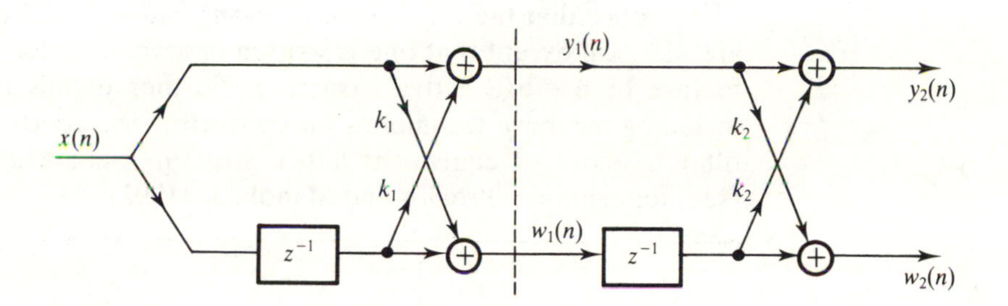

Figure 5.12

Example 5.5

The transfer function for an FIR filter is given by

$$H(z) = 1 - 1.3435z^{-1} + 0.9025z^{-2}$$

Draw the realization block diagram for each of the following cases:

(1) transversal structure;

(2) a two-stage lattice structure.

Calculate the values of the coefficients for the lattice structure.

Solution

(1) From the transfer function, the diagram for the transversal structure is given in Figure 5.11. The input and output of the transversal structure are given by

$$y(n) = x(n) + h(1)x(n-1) + h(2)x(n-2) \tag{5.6}$$

(2) A two-stage lattice structure for the filter is given in Figure 5.12. The outputs of the structure are related to the input as

$$y_2(n) = y_1(n) + K_2 w_1(n-1)$$
$$= x(n) + K_1(1 + K_2)x(n-1) + K_2 x(n-2) \quad \textbf{(5.7a)}$$
$$w_2(n) = K_2 x(n) + K_2(1 + K_2)x(n-1) + x(n-2) \quad \textbf{(5.7b)}$$

Comparing Equations 5.6 and 5.7a, and equating coefficients, we have

$$K_1 = \frac{h(1)}{1 + h(2)} \qquad K_2 = h(2)$$

from which $K_2 = 0.9025$ and $K_1 = -1.3435/(1 + 0.9025) = -0.7062$.

Notice that the coefficients of $y_2(n)$ and $w_2(n)$ (Equations 5.7a and 5.7b) are identical except that one is written in reverse order. This is a characteristic feature of the FIR lattice structure. Further details of the lattice structure including recursive techniques for converting the coefficients of an FIR or IIR filter to those of equivalent lattice structures are available in several books (see, for example, Proakis and Manolakis, 1992).

5.6 Summary

The term digital filter refers to a hardware or software implementation of a mathematical algorithm which accepts, as input, a digital signal, and produces as output another digital signal whose waveshape and/or amplitude and phase responses have been modified in a specified manner. In many applications, the use of digital filters is preferred to analogue filters because they can meet much tighter magnitude and phase specifications, which eliminates the temperature and voltage drifts common with analogue filters.

We have given in this chapter a common framework for designing FIR and IIR filters, from specification to implementation. A simple step-by-step procedure for designing these filters involves five key steps: (i) filter specification, (ii) calculation of suitable filter coefficients, (iii) realization of the filter using a suitable structure, (iv) quantization of filter coefficient and variables to suitable wordlengths and analysis of any resultant errors, and, finally, (v) implementation which is concerned with the hardware or software coding of the filter in a processor which will perform the actual filtering on input data.

Problems

5.1 Assume that the six methods of calculating filter coefficients given in Section 5.4.2 are all available. State and justify which of the methods you should use in each of the following applications:

(1) phase (delay) equalization for a digital communication system;

(2) simulation of analogue systems;
(3) a high throughput noise reduction system requiring a sharp magnitude frequency response filter;
(4) image processing;
(5) high quality digital audio processing;
(6) real-time biomedical signal processing with minimal distortion.

5.2 The following transfer functions represent two different filters meeting identical amplitude frequency response specifications:

$$(1)\quad H(z) = \frac{a_0 + a_1z^{-1} + a_2z^{-2}}{1 + b_1z^{-1} + b_2z^{-2}} \times \frac{a_3 + a_4z^{-1} + a_5z^{-2}}{1 + b_3z^{-1} + b_4z^{-2}}$$

where

$a_0 = 3.136362 \times 10^{-1}$
$a_1 = -5.456657 \times 10^{-2}$
$a_2 = 4.635728 \times 10^{-1}$
$a_3 = -5.456657 \times 10^{-2}$
$a_4 = 3.136362 \times 10^{-1}$
$b_1 = -8.118702 \times 10^{-1}$
$b_2 = 3.339288 \times 10^{-1}$
$b_3 = -2.794577 \times 10^{-1}$
$b_4 = 3.030631 \times 10^{-1}$

$$(2)\quad H(z) = \sum_{K=0}^{22} h_k z^{-k}$$

where

$h_0 = 0.39826480 \times 10^{-1} = h_{22}$
$h_1 = -0.16874380 \times 10^{-1} = h_{21}$
$h_2 = 0.34781130 \times 10^{-1} = h_{20}$
$h_3 = 0.12052890 \times 10^{-1} = h_{19}$
$h_4 = -0.44731860 \times 10^{-1} = h_{18}$
$h_5 = 0.27894610 \times 10^{-1} = h_{17}$
$h_6 = -0.87573360 \times 10^{-1} = h_{16}$
$h_7 = -0.90972060 \times 10^{-1} = h_{15}$
$h_8 = -0.15667550 \times 10^{-1} = h_{14}$
$h_9 = -0.28499560 \times 10^{0} = h_{13}$
$h_{10} = 0.74035030 \times 10^{-1} = h_{12}$
$h_{11} = 0.62349560 \times 10^{0}$

For each filter,

(a) state whether it is an FIR or IIR filter;
(b) represent the filtering operation in a block diagram form and write down the difference equation;
(c) determine and comment on the computational and storage requirements.

5.3 A digital filter is required to remove the mains component from foetal electrocardiogram (ECG) data stored in the memory of a main frame computer. The data was digitized to 12-bit accuracy.

The specification for the filter includes:

attenuation at the mains frequency	>50 dB
passband ripple	<0.05 dB
passband edges	0–0.09and0.11–0.5 (normalized)
sampling frequency	500 Hz

The filter should be implemented as a high level language routine, callable from a main analysis program. Any signal distortion by the filter should be kept to a minimum as important ECG waves are easily distroyed.

Discuss fully the issues involved in the design of such a filter, pointing out the options open to the designer and making recommendations, with justifications.

5.4 A digital filter is required to pre-process raw foetal electrocardiogram (ECG) data, that is the electrical activity of the heart, to make it easier to detect foetal heartbeats in the presence of baseline wander, mains interference, uterine contractions, and movements of the baby or mother. Baseline sways and movement artefacts tend to occupy the frequency range

0–10 Hz while the mains interference is centred around about 50 or 60 Hz (depending on the country). Most of the energy in the ECG necessary for detecting heart beats is thought to lie between about 5 and 50 Hz.

Assume that you have available foetal ECG data, analogue bandpass filtered between 0.05 and 100 Hz and digitized at 500 samples s^{-1} at a resolution of 8 bits.

(1) Assume that an IIR filter is to be used, and develop a set of specifications for the digital filter. Justify your answer.

(2) Repeat (1) for an FIR filter.

Which of the two filter types is best for this application and why?

5.5 A digital filter is to be used to provide the bulk of the anti-aliasing filtering in a certain speech transmission system in preference to an existing active filter. Currently, the analogue input signal to the system is sampled at 8 kHz after bandlimiting by the active filter with the following specifications:

passband	0–3.4 kHz
stopband edge frequency	8 kHz
attenuation in the stopband	30 dB
attenuation at 4 kHz	14 dB
passband ripple	< 0.1 dB

Discuss, with the aid of a block diagram, how the specifications can be met by digital filtering. Specify, with reasons, the type of digital filter that will be used and its characteristics.

5.6 A requirement exists in a communication system to recover the clock frequency from noisy data received from a remote transmitter, using digital filtering techniques, to allow for the data to be extracted reliably. The clock frequency at the remote transmitter is known to be 2.048 MHz. Discuss the characteristics of a suitable digital filter for the task and specify its transfer function.

5.7 The coefficients of a lattice FIR filter are $K_1 = -0.266$ and $K_2 = 0.69$. Draw the realization diagram for the lattice filter. Compute the impulse response coefficients for the filter and draw the diagram for the equivalent transversal structure.

5.8 A second-order IIR digital filter is characterized by the following transfer function:

$$H(z) = \frac{1}{1 - 0.9z^{-1} + 0.81z^{-2}}$$

Draw the realization diagram for each of the following structures:

(1) direct;

(2) lattice.

Find the coefficients of the lattice structure from the given transfer function.

Reference

Proakis J.G. and Manolakis D.G. (1992). *Digital Signal Processing* 2nd edn. New York NY: Macmillan

Bibliography

DeFatta D.J., Lucas J.G. and Hodgkiss W.S. (1988). *Digital Signal Processing*. New York NY: Wiley

Elliott D.F., ed. (1987). *Handbook of Digital Signal Processing*. London: Academic Press

Oppenheim A.V. and Schafer R.W. (1975). *Digital Signal Processing*. Englewood Cliffs NJ: Prentice-Hall

Parks T.W. and Burrus C.S. (1987). *Digital Filter Design*. New York NY: Wiley

Rabiner L.R. and Gold B. (1975). *Theory and Application of Digital Signal Processing*. Englewood Cliffs NJ: Prentice-Hall

Rabiner L.R., Cooley J.W., Helms H.D., Jackson L.B., Kaiser J.F., Rader C.M., Schafer R.W., Steiglitz K. and Weinstein C.J. (1972). Terminology in digital signal processing. *IEEE Trans. Audio Electroacoustics*, **20** (December), 322–37

Taylor F.J. (1983). *Digital Filter Design Handbook*. New York NY: Dekker

6

Finite impulse response (FIR) filter design

6.1	Introduction	279
6.2	FIR filter design	285
6.3	FIR filter specifications	285
6.4	FIR coefficient calculation methods	288
6.5	Window method	288
6.6	The optimal method	303
6.7	Frequency sampling method	317
6.8	Realization structures for FIR filters	344
6.9	Finite wordlength effects in FIR digital filters	348
6.10	FIR implementation techniques	358
6.11	Design example	359
6.12	Summary	361
6.13	Application examples of FIR filters	363
	References	368
	Bibliography	369
	Appendix	370

This chapter is concerned with the design of FIR filters from specifications, through coefficient calculation, to analysis of finite wordlength effects and implementations. Several fully worked examples are given throughout the chapter to illustrate the various design stages and to consolidate the important concepts. A complete filter design is included to show how all the stages fit together and to assist the readers who wish to design their own filters. PC-based C language programs are given in the appendix and on the PC disk for the book which may be used to replicate the results presented here or for designing user-specified filters.

6.1. Introduction

First, we will summarize important characteristics of FIR filters before devoting attention to their design.

6.1.1 Summary of key characteristic features of FIR filters

(1) The basic FIR filter is characterized by the following two equations:

$$y(n) = \sum_{k=0}^{N-1} h(k)x(n-k) \tag{6.1a}$$

$$H(z) = \sum_{k=0}^{N-1} h(k)z^{-k} \tag{6.1b}$$

where $h(k)$, $k = 0, 1, \ldots, N-1$, are the impulse response coefficients of the filter, $H(z)$ is the transfer function of the filter and N is the filter length, that is the number of filter coefficients. Equation 6.1a is the FIR difference equation. It is a time domain equation and describes the FIR filter in its nonrecursive form: the current output sample, $y(n)$, is a function only of past and present values of the input, $x(n)$. When FIR filters are implemented in this form, that is by direct evaluation of Equation 6.1a, they are always stable. Equation 6.1b is the transfer function of the filter. It provides a means of analysing the filter, for example evaluating the frequency response.

(2) FIR filters can have an exactly linear phase response. The implications of this will be discussed in the next section.

(3) FIR filters are very simple to implement. All DSP processors available have architectures that are suited to FIR filtering. Nonrecursive FIR filters suffer less from the effects of finite wordlength than IIR filters. Recursive FIR filters also exist and may offer significant computational advantages (see Section 6.7 for details).

FIR filters should be used whenever we wish to exploit any of the advantages above, in particular the advantage of linear phase. Issues to consider when choosing between FIR and IIR filters are given in Section 5.3.

6.1.2 Linear phase response and its implications

The ability to have an exactly linear phase response is one of the most important properties of FIR filters. For this reason we shall look more closely at this property. When a signal passes through a filter, it is modified in amplitude and/or phase. The nature and extent of the modification of the signal is dependent on the amplitude and phase characteristics of the filter. The phase delay or group delay of the filter provides a useful measure of how the filter modifies the phase characteristics of the signal. If we consider a signal that consists of several frequency components (such as a speech waveform or a modulated signal) the phase delay of the filter is the amount of time delay each frequency component of the signal suffers in going through the filter. The group delay on the other hand is the average time delay the composite signal suffers at each frequency. Mathematically, the phase delay is the negative of the phase angle divided by frequency whereas the group delay is the negative of the derivative of the phase with respect to frequency:

$$T_{\mathrm{p}} = -\,\theta(\omega)/\omega \tag{6.2a}$$

$$T_{\mathrm{g}} = -\,\mathrm{d}\theta(\omega)/\mathrm{d}\omega \tag{6.2b}$$

A filter with a nonlinear phase characteristic will cause a phase distortion in the signal that passes through it. This is because the frequency components in the signal will each be delayed by an amount not proportional to frequency thereby altering their harmonic relationships. Such a distortion is undesirable in many applications, for example music, data transmission, video, and biomedicine, and can be avoided by using filters with linear phase characteristics over the frequency bands of interest.

A filter is said to have a linear phase response if its phase response satisfies one of the following relationships:

$$\theta(\omega) = -\alpha\omega \tag{6.3a}$$

$$\theta(\omega) = \beta - \alpha\omega \tag{6.3b}$$

where α and β are constant. If a filter satisfies the condition given in Equation 6.3a it will have both constant group and constant phase delay responses. It can be shown that for condition 6.3a to be satisfied the impulse response of the filter must have positive symmetry. The phase response in this case is simply a function of the filter length:

$$h(n) = h(N - n - 1), \qquad \begin{cases} n = 0, 1, \ldots, (N-1)/2 & (N \text{ odd}) \\ n = 0, 1, \ldots, (N/2) - 1 & (N \text{ even}) \end{cases}$$

$$\alpha = (N - 1)/2$$

When the condition given in Equation 6.3b only is satisfied the filter will have a constant group delay only. In this case, the impulse response of the filter has negative symmetry:

$$h(n) = -h(N - n - 1)$$
$$\alpha = (N - 1)/2$$
$$\beta = \pi/2$$

Linear phase FIR filters form an important class of FIR filters. They possess a unique set of properties that influence how they are designed and implemented. We will explore some of these by way of an example.

Example 6.1

(1) Discuss briefly the conditions necessary for a realizable digital filter to have a linear phase characteristic, and the advantages of filters with such a characteristic.

(2) An FIR digital filter has impulse response, $h(n)$, defined over the interval $0 \leqslant n \leqslant N-1$. Show that if $N = 7$ and $h(n)$ satisfies the symmetry condition

$$h(n) = h(N - n - 1)$$

the filter has a linear phase characteristic.

(3) Repeat (2) if $N = 8$.

Solution

(1) The necessary and sufficient condition for a filter to have a linear phase response is that its impulse response must be symmetrical (Rabiner and Gold, 1975):

$$h(n) = h(N - 1 - n) \text{ or } h(n) = -h(N - 1 - n)$$

For nonrecursive FIR filters, the storage space for coefficients and the number of arithmetic operations are reduced by nearly a factor of 2. For recursive FIR filters, the coefficients can be made to be simple integers, leading to increased speed of processing. In linear phase filters, all frequency components experience the same amount of delay through the filter, that is no phase distortion.

(2) Using the symmetry condition we find that for $N = 7$:

$$h(0) = h(6);\ h(1) = h(5);\ h(2) = h(4)$$

The frequency response, $H(\omega)$, for the filter is given by

$$\begin{aligned}H(\omega) &= \sum_{n=0}^{6} h(n)\mathrm{e}^{-\mathrm{j}\omega n}\\ &= h(0) + h(1)\mathrm{e}^{-\mathrm{j}\omega} + h(2)\mathrm{e}^{-\mathrm{j}2\omega} + h(3)\mathrm{e}^{-\mathrm{j}3\omega} + h(4)\mathrm{e}^{-\mathrm{j}4\omega}\\ &\quad + h(5)\mathrm{e}^{-\mathrm{j}5\omega} + h(6)\mathrm{e}^{-\mathrm{j}6\omega}\\ &= \mathrm{e}^{-\mathrm{j}3\omega}[h(0)\mathrm{e}^{\mathrm{j}3\omega} + h(1)\mathrm{e}^{\mathrm{j}2\omega} + h(2)\mathrm{e}^{\mathrm{j}\omega} + h(3) + h(4)\mathrm{e}^{-\mathrm{j}\omega}\\ &\quad + h(5)\mathrm{e}^{-\mathrm{j}2\omega} + h(6)\mathrm{e}^{-\mathrm{j}3\omega}]\end{aligned}$$

Using the symmetry condition we can group terms whose coefficients are numerically equal:

$$\begin{aligned}H(\omega) &= \mathrm{e}^{-\mathrm{j}3\omega}[h(0)(\mathrm{e}^{\mathrm{j}3\omega} + \mathrm{e}^{-\mathrm{j}3\omega}) + h(1)(\mathrm{e}^{\mathrm{j}2\omega} + \mathrm{e}^{-\mathrm{j}2\omega})\\ &\quad + h(2)(\mathrm{e}^{\mathrm{j}\omega} + \mathrm{e}^{-\mathrm{j}\omega}) + h(3)]\\ &= \mathrm{e}^{-\mathrm{j}3\omega}[2h(0)\cos(3\omega) + 2h(1)\cos(2\omega) + 2h(2)\cos(\omega) + h(3)]\end{aligned}$$

If we let $a(0) = h(3)$ and $a(n) = 2h(3-n)$, $n = 1, 2, 3$, then $H(\omega)$ can be written in a compact form:

$$H(\omega) = \sum_{n=0}^{3} a(n)\cos(\omega n)\mathrm{e}^{-\mathrm{j}3\omega} = \pm|H(\omega)|\mathrm{e}^{\mathrm{j}\theta(\omega)}$$

where

$$\pm|H(\omega)| = \sum_{n=0}^{3} a(n)\cos(\omega n);\ \theta(\omega) = -3\omega$$

Clearly, the phase response is linear.

(3) In this case, the symmetry conditions lead to

$$h(0) = h(7);\ h(1) = h(6);\ h(2) = h(5);\ h(3) = h(4)$$

Following a similar approach to the above and using the symmetry condition we have

$$\begin{aligned}H(\omega) &= \mathrm{e}^{-\mathrm{j}7\omega/2}[h(0)(\mathrm{e}^{\mathrm{j}7\omega/2} + \mathrm{e}^{-\mathrm{j}7\omega/2}) + h(1)(\mathrm{e}^{\mathrm{j}5\omega/2} + \mathrm{e}^{-\mathrm{j}5\omega/2})\\ &\quad + h(2)(\mathrm{e}^{\mathrm{j}3\omega/2} + \mathrm{e}^{-\mathrm{j}3\omega/2}) + h(3)(\mathrm{e}^{\mathrm{j}\omega/2} + \mathrm{e}^{-\mathrm{j}\omega/2})]\\ &= \mathrm{e}^{-\mathrm{j}7\omega/2}[2h(0)\cos(7\omega/2) + 2h(1)\cos(5\omega/2) + 2h(2)\cos(3\omega/2)\\ &\quad + 2h(3)\cos(\omega/2)]\\ &= \pm|H(\omega)|\mathrm{e}^{\mathrm{j}\theta(\omega)}\end{aligned}$$

Table 6.1 A summary of the key points about the four types of linear phase FIR filters.

Impulse response symmetry	*Number of coefficients N*	*Frequency response* $H(\omega)$	*Type of linear phase*
Positive symmetry, $h(n) = h(N-1-n)$	Odd	$e^{-j\omega(N-1)/2} \sum_{n=0}^{(N-1)/2} a(n)\cos(\omega n)$	1
	Even	$e^{-j\omega(N-1)/2} \sum_{n=1}^{N/2} b(n)\cos[w(n-\frac{1}{2})]$	2
Negative symmetry, $h(n) = -h(N-1-n)$	Odd	$e^{-j[\omega(N-1)/2-\pi/2]} \sum_{n=1}^{(N-1)/2} a(n)\sin(\omega n)$	3
	Even	$e^{-j[\omega(N-1)/2-\pi/2]} \sum_{n=1}^{N/2} b(n)\sin[\omega(n-\frac{1}{2})]$	4

$a(0) = h(N-1)/2]$; $a(n) = 2h[(N-1)/2-n]$
$b(n) = 2h(N/2-n)$

where

$$\pm|H(\omega)| = \sum_{n=1}^{4} b(n)\cos[\omega(n-1/2)];\ \theta(\omega) = -(7/2)\omega$$

$$b(n) = 2h(N/2-n), \quad n = 1, 2, \ldots, N/2$$

The results above can be generalized for FIR filters, see Table 6.1.

6.1.3 Types of linear phase FIR filters

There are exactly four types of linear phase FIR filters, depending on whether N is even or odd and whether $h(n)$ has positive or negative symmetry. Two of the four types of linear phase filters were considered in the example above. Figure 6.1 illustrates how the impulse responses of the four types of linear phase FIR filters differ. Table 6.1 summarizes their key features.

The frequency response of a type 2 filter (positive symmetry and even length) is always zero at $f = 0.5$ (half the sampling frequency, as all frequencies are normalized to the sampling frequency); see Problem 6.1. Thus this type of filter is unsuitable as a highpass filter. Type 3 and 4 (both negative symmetry) each introduce a 90° phase shift. The frequency response is always zero at $f = 0$, making them unsuitable for lowpass filters. In addition, the type 3 response is always zero at $f = 0.5$, making it also unsuitable as a highpass filter. Type 1 is the most versatile of the four. Types 3 and 4 are often used to design

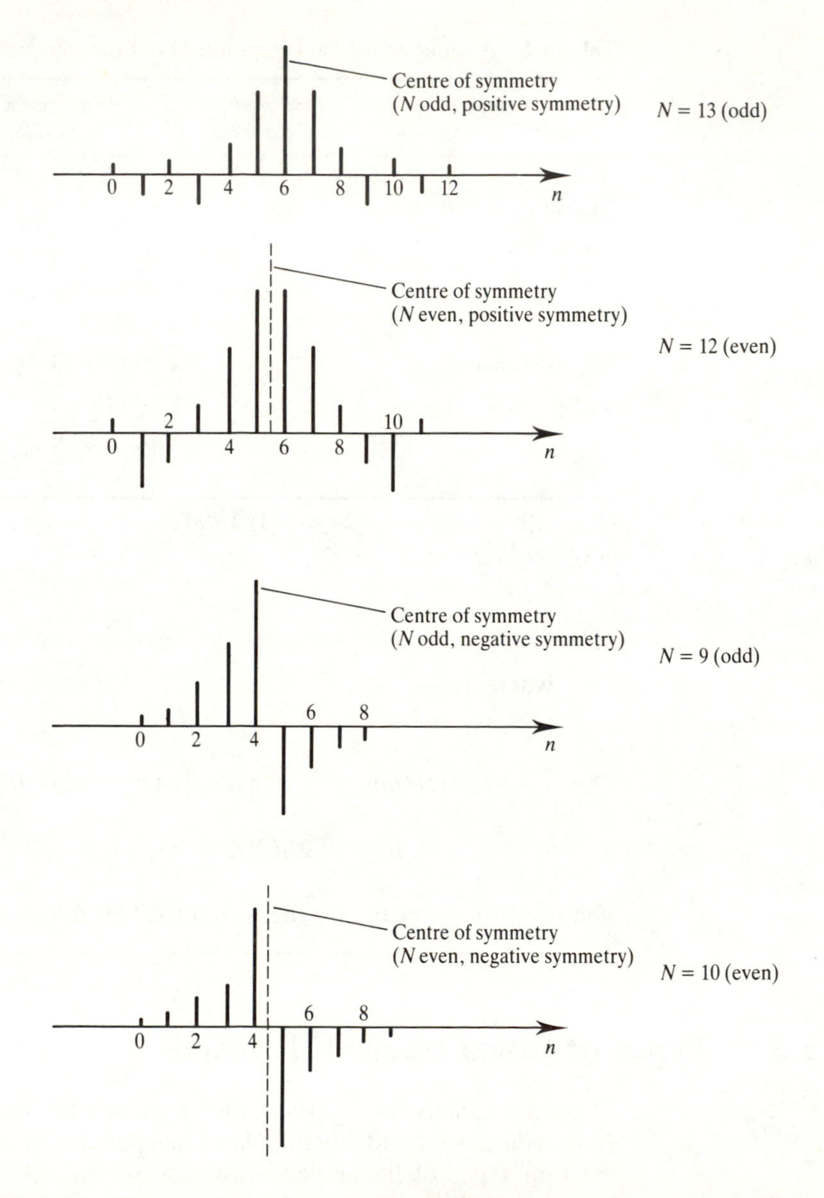

Figure 6.1 A comparison of the impulse of the four types of linear phase filters.

differentiators and Hilbert transformers, because of the 90° phase shift that each can provide.

It is important to note that the phase delay (for type 1 and 2 filters) or group delay (for all four types of filters) is expressible in terms of the number of coefficients of the filter and so can be corrected to give a zero phase or group delay response. For example, for type 1 and 2 filters, the phase delay is given by

$$T_{\mathrm{p}} = \left(\frac{N-1}{2}\right)T \tag{6.4a}$$

and for types 3 and 4 the group delay is given by

$$T_{\mathrm{p}} = \left(\frac{N-1-\pi}{2}\right)T \tag{6.4b}$$

where T is the sampling period.

6.2 FIR filter design

As discussed in Chapter 5, the design of a digital filter involves five steps, viz:

(1) *Filter specification*. This may include stating the type of filter, for example lowpass filter, the desired amplitude and/or phase responses and the tolerances (if any) we are prepared to accept, the sampling frequency, and the wordlength of the input data.

(2) *Coefficient calculation*. At this step, we determine the coefficients of a transfer function, $H(z)$, which will satisfy the specifications given in (1). Our choice of coefficient calculation method will be influenced by several factors, the most important of which are the critical requirements in step (1).

(3) *Realization*. This involves converting the transfer function obtained in (2) into a suitable filter network or structure.

(4) *Analysis of finite wordlength effects*. Here, we analyse the effect of quantizing the filter coefficients and the input data as well as the effect of carrying out the filtering operation using fixed wordlengths on the filter performance.

(5) *Implementation*. This involves producing the software code and/or hardware and performing the actual filtering.

These five interrelated steps are summarized in Figure 6.2. We shall now go through the steps in detail for FIR filters, illustrating with examples as appropriate.

6.3 FIR filter specifications

Filter specifications were discussed in detail in Chapter 5. Here we shall deal with aspects of filter specification relating to the FIR filter. Several examples in this chapter will also illustrate various aspects of filter specification.

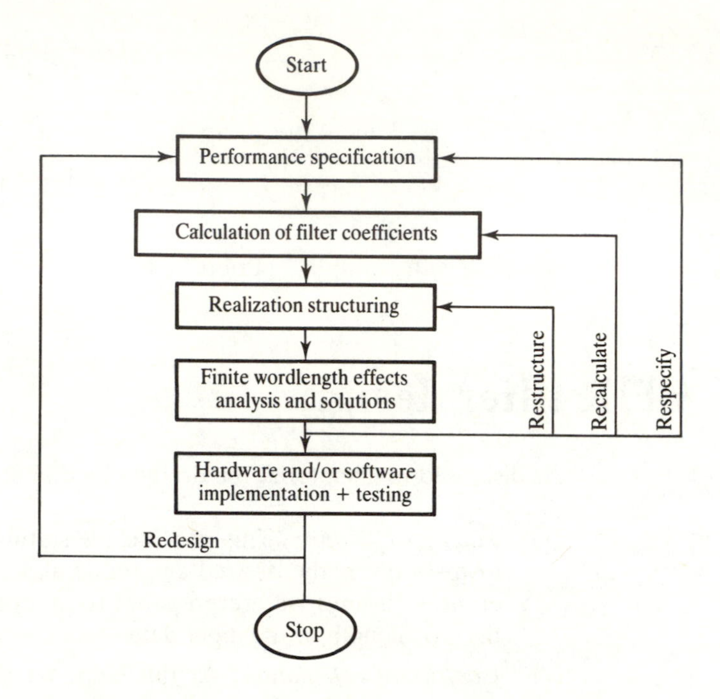

Figure 6.2 Summary of design stages for digital filters.

For the phase response, we need only state whether positive symmetry or negative symmetry is required (assuming linear phase). The amplitude–frequency response of an FIR filter is often specified in the form of a tolerance scheme. Figure 6.3 shows such a scheme for the bandpass filter. A similar scheme can be readily drawn for other frequency selective filters. Referring to the figure, the following parameters are of interest:

δ_p	peak passband deviation (or ripples)
δ_s	stopband deviation
f_p	passband edge frequency
f_s	stopband edge frequency
F_s	sampling frequency

In practice it is often more convenient to express δ_p and δ_s in decibels as indicated in the figure. The difference between f_s and f_p gives the transition width of the filter. Another important parameter is the filter length, N, which defines the number of filter coefficients given. These parameters, in most cases, define completely the frequency response of the FIR filter.

Other specifications that may be of interest include the maximum number of filter coefficients we can accept (this may be forced on us by the particular application, such as the speed at which we wish to operate). We may not have

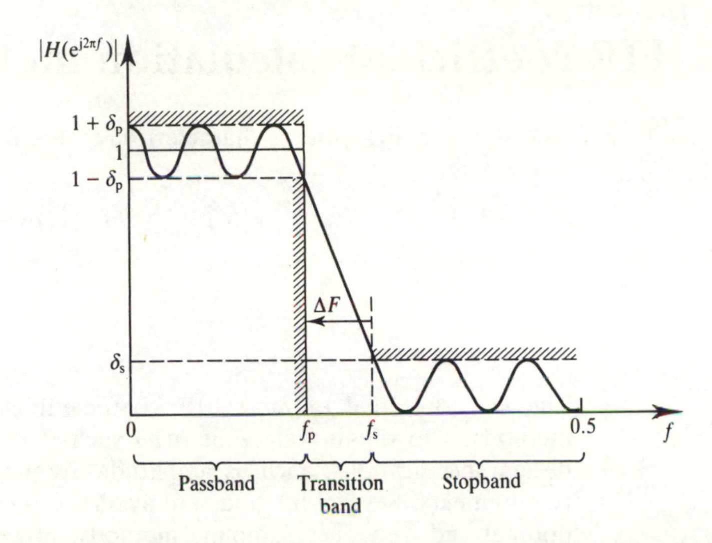

Figure 6.3 Magnitude–frequency response specification for a lowpass filter. The passband and stopband deviations are often expressed in decibels: passband deviation, $20 \log(1 + \delta_p)$ dB; stopband deviation, $-20 \log(\delta_s)$ dB.

any idea what constitutes a good choice for one or more of the parameters above and so may have to deduce them by trial and error.

Example 6.2

Amplitude specification example A lowpass digital filter is required for physiological noise reduction. The filter should meet the following specifications:

passband edge frequency	10 Hz
stopband edge frequency	< 20 Hz
stopband attenuation	> 30 dB
passband ripple	< 0.026 dB
sampling frequency	256 Hz

Important requirements in this application are (i) the filter should introduce as small a distortion as possible to the in-band signals and (ii) the length of filter should be as small as possible and should not exceed 37.

6.4 FIR coefficient calculation methods

Recall that an FIR filter is characterized by the following equations:

$$y(m) = \sum_{n=0}^{N-1} h(n)x(m-n)$$

$$H(z) = \sum_{n=0}^{N-1} h(n)z^{-n}$$

The sole objective of most FIR coefficient calculation (or approximation) methods is to obtain values of $h(n)$ such that the resulting filter meets the design specifications, such as amplitude–frequency response and throughout requirements. Several methods are available for obtaining $h(n)$. The window, optimal and frequency sampling methods, however, are the most commonly used. All three can lead to linear phase FIR filters.

6.5 Window method

In this method, use is made of the fact that the frequency response of a filter, $H_D(\omega)$, and the corresponding impulse response, $h_D(n)$, are related by the inverse Fourier transform:

$$h_D(n) = \frac{1}{2\pi}\int_{-\pi}^{\pi} H_D(\omega)e^{j\omega n}\,d\omega \tag{6.5}$$

The subscript D is used to distinguish between the ideal and practical impulse responses. The need for this distinction will soon become clear. If we know $H_D(\omega)$ we can obtain $h_D(n)$ by evaluating the inverse Fourier transform of Equation 6.5. As an illustration, suppose we wish to design a lowpass filter. We could start with the ideal lowpass response shown in Figure 6.4(a) where ω_c is the cutoff frequency and the frequency scale is normalized: $T = 1$. By letting the response go from $-\omega_c$ to ω_c we simplify the integration operation. Thus the impulse response is given by:

$$h_D(n) = \frac{1}{2\pi}\int_{-\pi}^{\pi} 1 \times e^{j\omega n}\,d\omega = \frac{1}{2\pi}\int_{-\omega_c}^{\omega_c} e^{j\omega n}\,d\omega$$

$$= \frac{2f_c \sin(n\omega_c)}{n\omega_c}, \quad n = 0, -\infty \leqslant n \leqslant \infty \tag{6.6}$$

$$= 2f_c, \qquad n = 0 \text{ (using L'Hôpital's rule)}$$

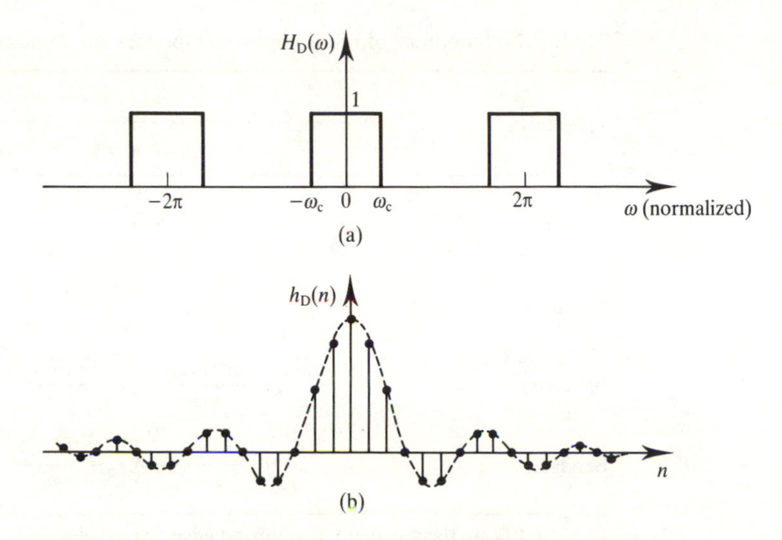

Figure 6.4 (a) Ideal frequency response of a lowpass filter. (b) Impulse response of the ideal lowpass filter.

The impulse responses for the ideal highpass, bandpass and bandstop filters are obtained from the lowpass case of Equation 6.6 and are summarized in Table 6.2. The impulse response for the lowpass filter is plotted in Figure 6.4(b) from which we note that $h_D(n)$ is symmetrical about $n = 0$ (that is $h_D(n) = h_D(-n)$), so that the filter will have a linear (in this case zero) phase response. Several practical problems with this simple approach are apparent. The most important of these is that, although $h_D(n)$ decreases as we move away from $n = 0$, it nevertheless carries on, theoretically, to $n = \pm\infty$. Thus the resulting filter is not an FIR.

An obvious solution is to truncate the ideal impulse response by setting $h_D(n) = 0$ for n greater than M (say). However, this introduces undesirable ripples and overshoots – the so-called Gibb's phenomenon. Figure 6.5 illustrates the effects of discarding coefficients on the filter response. The more coefficients that are retained, the closer the filter spectrum is to the ideal response (Figures 6.5(b) and 6.5(c)). Direct truncation of $h_D(n)$ as described above is equivalent to multiplying the ideal impulse response by a rectangular window of the form

$$\begin{aligned} w(n) &= 1, \quad n = 0, 1, \ldots, M - 1 \\ &= 0, \quad \text{elsewhere} \end{aligned}$$

In the frequency domain this is equivalent to convolving $H_D(\omega)$ and $W(\omega)$, where $W(\omega)$ is the Fourier transform of $w(n)$. As $W(\omega)$ has the classic $(\sin x)/x$ shape, truncation of $h_D(n)$ leads to the overshoots and ripples in the frequency response.

Table 6.2 Summary of ideal impulse responses for standard frequency selective filters.

Filter type	Ideal impulse response, $h_D(n)$	
	$h_D(n),\ n \neq 0$	$h_D(0)$
Lowpass	$2f_c \dfrac{\sin(n\omega_c)}{n\omega_c}$	$2f_c$
Highpass	$-2f_c \dfrac{\sin(n\omega_c)}{n\omega_c}$	$1 - 2f_c$
Bandpass	$2f_2 \dfrac{\sin(n\omega_2)}{n\omega_2} - 2f_1 \dfrac{\sin(n\omega_1)}{n\omega_1}$	$2(f_2 - f_1)$
Bandstop	$2f_1 \dfrac{\sin(n\omega_1)}{n\omega_1} - 2f_2 \dfrac{\sin(n\omega_2)}{n\omega_2}$	$1 - 2(f_2 - f_1)$

f_c, f_1 and f_2 are the passband or stopband edge frequencies; N is the length of filter.

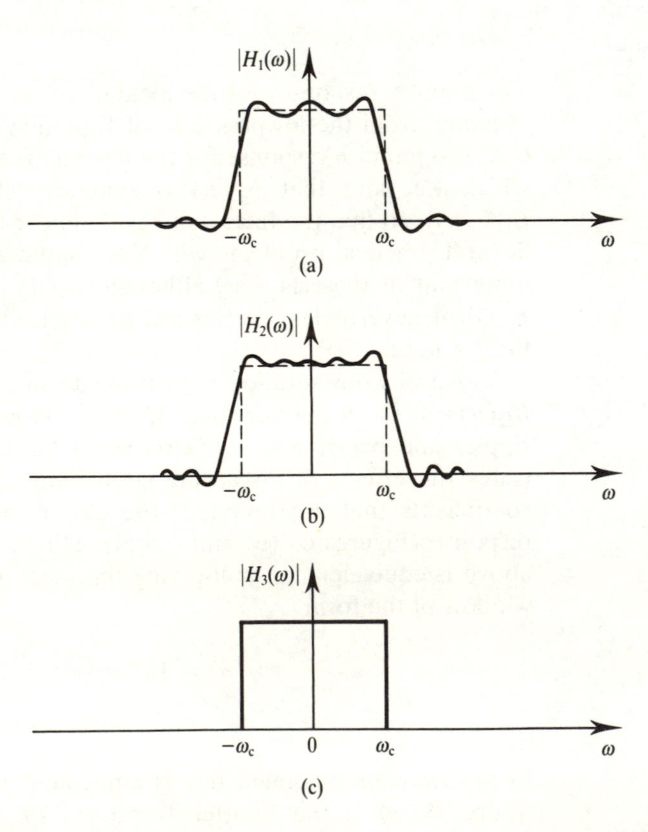

Figure 6.5 Effects on the frequency response of truncating the ideal impulse response to (a) 13 coefficients, (b) 25 coefficients and (c) an infinite number of coefficients.

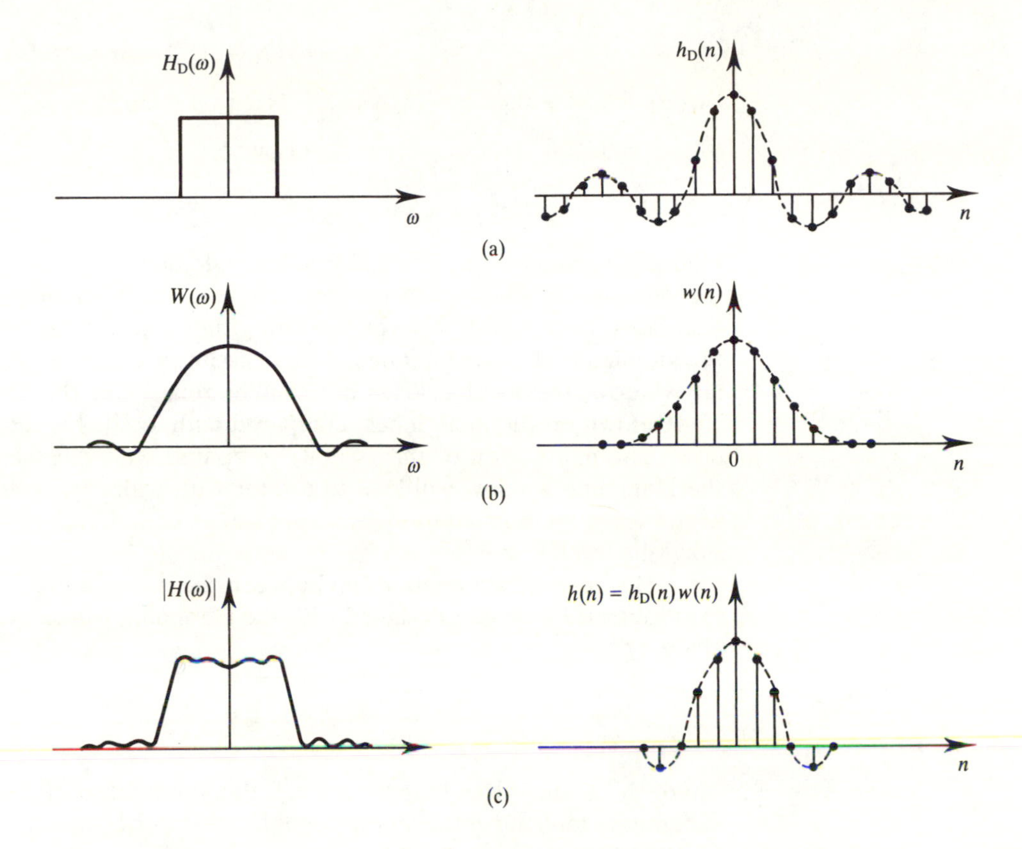

Figure 6.6 An illustration of how the filter coefficients, $h(n)$, are determined by the window method.

A practical approach is to multiply the ideal impulse response, $h_D(n)$, by a suitable window function, $w(n)$, whose duration is finite. This way the resulting impulse response decays smoothly towards zero. The process is illustrated in Figure 6.6. Figure 6.6(a) shows the ideal frequency response and the corresponding ideal impulse response. Figure 6.6(b) shows a finite duration window function and its spectrum. Figure 6.6(c) shows $h(n)$ which is obtained by multiplying $h_D(n)$ by $w(n)$. The corresponding frequency response shows that the ripples and overshoots, characteristic of direct truncation, are much reduced. However, the transition width is wider than for the rectangular case. The transition width of the filter is determined by the width of the main lobe of the window. The side lobes produce ripples in both passband and stopband.

6.5.1 Some common window functions

Several window functions have been proposed. One of the most widely used window functions is the Hamming window which is defined as

$$w(n) = 0.54 + 0.46\cos(2\pi/N) \quad \begin{cases} -(N-1)/2 \leqslant n \leqslant (N-1)/2 & (N \text{ odd}) \\ -N/2 \leqslant n \leqslant N/2 & (N \text{ even}) \end{cases}$$
$$= 0 \quad \text{elsewhere} \tag{6.7}$$

Figure 6.7 compares its time and frequency domain characteristics with those of the rectangular window. In the time domain, the Hamming window function decreases more gently towards zero on either side. In the frequency domain, the amplitude of the main lobe is wider (about twice) than that of the rectangular window, but its side lobes are smaller relative to the main lobe – about 40 dB down on the main lobes, compared with 14 dB for the rectangular window. The implication of this is that, compared with the rectangular window, the Hamming window will lead to a filter with wider transition width because of the wider main lobe but higher stopband attenuation (because of the smaller side lobe levels).

The appropriate relationship between the transition width (from passband to stopband) for a filter designed with the Hamming window and filter length is given by

$$\Delta f = 3.3/N \tag{6.8}$$

where N is the filter length and Δf the normalized transition width. The maximum stopband attenuation possible with the Hamming window is about 53 dB, and the minimum peak passband ripple is about 0.0194 dB.

The most relevant features of some of the most popular window functions are summarized in Table 6.3. We note that the first four window functions have fixed characteristics, such as transition width and stopband attenuation. Thus their use imposes a restriction on the filter designer. We also note that a filter designed by the window method has equal passband and stopband ripples, that is $\delta_p = \delta_s$ (in Figure 6.3). In practice, this restriction may lead to a filter whose passband ripple is unnecessarily small.

The Kaiser window function goes some way in overcoming the above problems by incorporating a ripple control parameter, β, which allows the designer to trade-off the transition width against ripple. The Kaiser window is given by

$$w(n) = I_0\left\{\beta\left[1 - \left(\frac{2n}{N-1}\right)^2\right]^{1/2}\right\}\Big/ I_0(\beta) \qquad -(N-1)/2 \leqslant n \leqslant (N-1)/2$$
$$= 0 \qquad \text{elsewhere} \tag{6.9}$$

where $I_0(x)$ is the zero-order modified Bessel function of the first kind. β controls the way the window function tapers at the edges in the time domain. $I_0(x)$ is normally evaluated using the following power series expansion

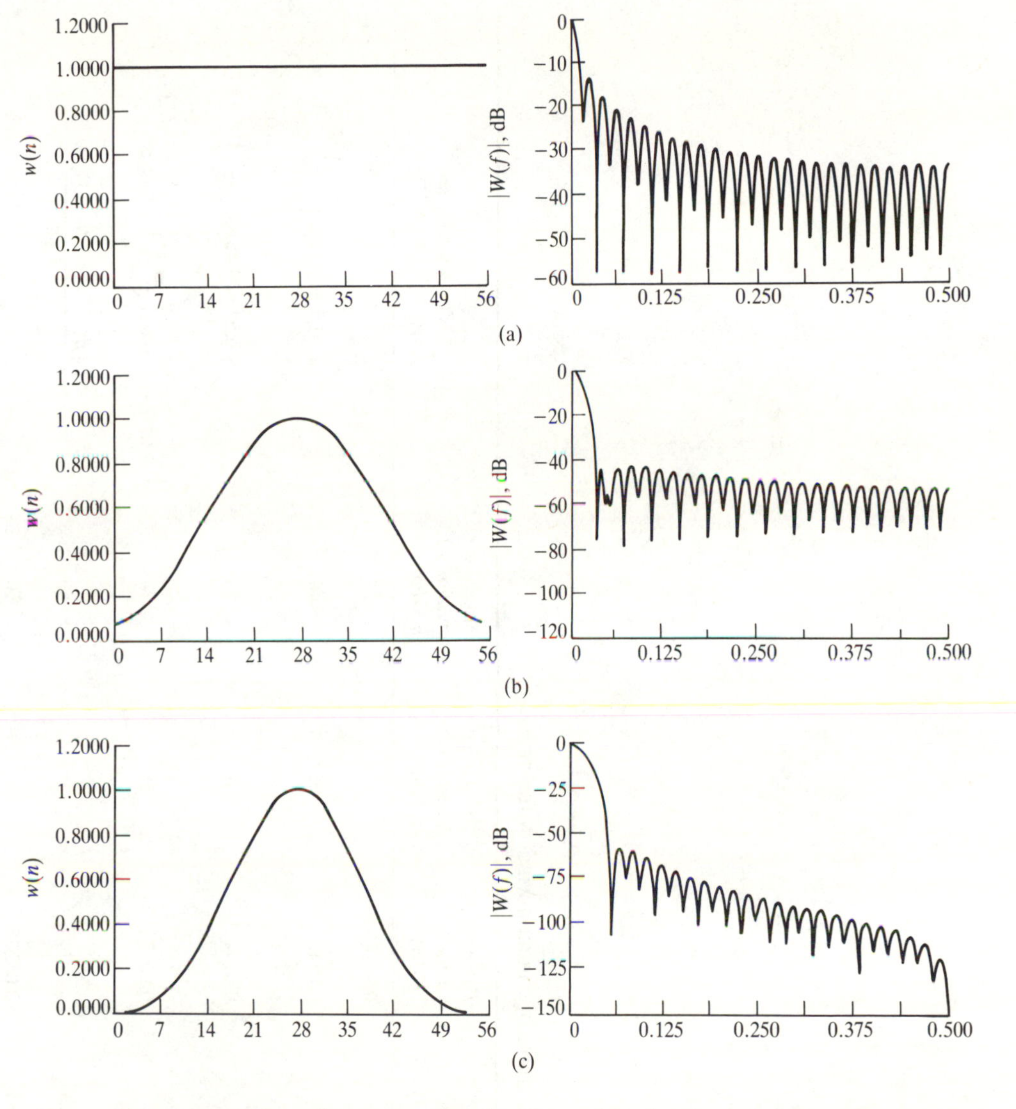

Figure 6.7 Comparison of the time and frequency domain characteristics of common window functions: (a) rectangular; (b) Hamming; (c) Blackman.

(Rabiner and Gold, 1975):

$$I_0(x) = 1 + \sum_{k=1}^{L} \left[\frac{(x/2)^k}{k!} \right]^2$$

where typically L < 25. An algorithm due to Kaiser (Rabiner and Gold, 1975) gives an efficient implementation of this equation (see program window.c on the PC disk for the book).

Table 6.3 Summary of important features of common window functions.

Name of window function	*Transition width* (Hz) (*normalized*)	*Passband ripple* (dB)	*Main lobe relative to side lobe* (dB)	*Stopband attenuation* (dB) (*maximum*)	*Window function* $w(n), \|n\| \leqslant (N-1)/2$
Rectangular	$0.9/N$	0.7416	13	21	1
Hanning	$3.1/N$	0.0546	31	44	$0.5 + 0.5\cos\left(\frac{2\pi n}{N}\right)$
Hamming	$3.3/N$	0.0194	41	53	$0.54 + 0.46\cos\left(\frac{2\pi n}{N}\right)$
Blackman	$5.5/N$	0.0017	57	74	$0.42 + 0.5\cos\left(\frac{2\pi n}{N-1}\right) + 0.08\cos\left(\frac{4\pi n}{N-1}\right)$
	$2.93/N$ ($\beta = 4.54$)	0.0274		50	$\frac{I_0(\beta\{1 - [2n/(N-1)]^2\}^{1/2})}{I_0(\beta)}$
Kaiser	$4.32/N$ ($\beta = 6.76$)	0.002 75		70	
	$5.71/N$ ($\beta = 8.96$)	0.000 275		90	

When $\beta = 0$, the Kaiser window corresponds to the rectangular window, and when it is 5.44, the resulting window is very similar, though not identical, to the Hamming window. The value of β is determined by the stopband attenuation requirements and may be estimated from one of the following empirical relationships:

$$\beta = 0 \qquad \text{if } A \leqslant 21 \text{ dB} \tag{6.10a}$$

$$\beta = 0.5842(A - 21)^{0.4} + 0.07886(A - 21) \qquad \text{if } 21 \text{ dB} < A < 50 \text{ dB} \tag{6.10b}$$

$$\beta = 0.1102(A - 8.7) \qquad \text{if } A \geqslant 50 \text{ dB} \tag{6.10c}$$

where $A = -20 \log_{10}(\delta)$ is the stopband attenuation, $\delta = \min(\delta_p, \delta_s)$, since the passband and stopband ripples are nearly equal, δ_p is the desired passband ripple and δ_s is the desired stopband ripple. The number of filter coefficients, N, is given by

$$N \geqslant \frac{A - 7.95}{14.36\,\Delta f} \tag{6.11}$$

where Δf is the normalized transition width. The values of β and N are used to compute the coefficients for the Kaiser window $w(n)$.

6.5.2 Summary of the window method of calculating FIR filter coefficients

- *Step 1.* Specify the 'ideal' or desired frequency response of filter, $H_D(\omega)$.
- *Step 2.* Obtain the impulse response, $h_D(n)$, of the desired filter by evaluating the inverse Fourier transform (Equation 6.6b). For the standard frequency selective filters the expressions for $h_D(n)$ are summarized in Table 6.2.
- *Step 3.* Select a window function that satisfies the passband or attenuation specifications and then determine the number of filter coefficients using the appropriate relationship between the filter length and the transition width, Δf (expressed as a fraction of the sampling frequency).
- *Step 4.* Obtain values of $w(n)$ for the chosen window function and the values of the actual FIR coefficients, $h(n)$, by multiplying $h_D(n)$ by $w(n)$:

$$h(n) = h_D(n)w(n) \tag{6.12}$$

It is clear that the window method is straightforward and involves a minimal amount of computational effort. Indeed you could obtain the coefficients with your pocket calculators. However, a PC-based program is available on the PC disk for this book for calculating $h(n)$. It should be said that the resulting filter is not optimal, that is in many cases a filter with a smaller number of coefficients can be designed using other methods.

Example 6.3

Obtain the coefficients of an FIR lowpass filter to meet the specifications given below using the window method.

passband edge frequency	1.5 kHz
transition width	0.5 kHz
stopband attenuation	> 50 dB
sampling frequency	8 kHz

Solution

From Table 6.2, we select $h_D(n)$ for lowpass filter which is given by

$$h_D(n) = 2f_c \frac{\sin(n\omega_c)}{n\omega_c} \qquad n = 0$$

$$h_D(n) = 2f_c \qquad n \neq 0$$

Table 6.3 indicates that the Hamming, Blackman or Kaiser window will satisfy the stopband attenuation requirements. We will use the Hamming window for simplicity. Now $\Delta f = 0.5/8 = 0.0625$. From $N = 3.3/\Delta f = 3.3/0.0625 = 52.8$, let $N = 53$. The filter coefficients are obtained from

$$h_D(n)w(n) \qquad -26 \leqslant n \leqslant 26$$

where

$$h_D(n) = \frac{2f_c \sin(n\omega_c)}{n\omega_c} \qquad n = 0$$

$$h_D(n) = 2\text{f}_c \qquad n \neq 0$$

$$w(n) = 0.54 + 0.46\cos(2\pi n/53) \qquad -26 \leqslant n \leqslant 26$$

Because of the smearing effect of the window on the filter response, the cutoff frequency of the resulting filter will be different from that given in the specifications. To account for this, we will use an f_c that is centred on the transition band:

$$f'_c = f_c + \Delta f/2 = (1.5 + 0.25)\text{ kHz} = 1.75\text{ kHz} \equiv 1.75/8 = 0.218\,75$$

Noting that $h(n)$ is symmetrical; we need only compute values for $h(0)$, $h(1)$, . . ., $h(26)$ and then use the symmetry property to obtain the other coefficients.

$$n = 0: \quad h_D(n) = 2f_c = 2 \times 0.218\,75 = 0.4375$$

$$w(0) = 0.54 + 0.46\cos(0) = 1$$

$$h(0) = h_D(0)w(0) = 0.4375$$

$$n = 1: \quad h_D(1) = \frac{2 \times 0.218\,75}{2\pi \times 0.218\,75}\sin(2\pi \times 0.218\,75)$$

$$= \frac{\sin(360 \times 0.218\,65)}{\pi} = 0.312\,19$$

$$w(1) = 0.54 + 0.46\cos(2\pi/53) = 0.54 + 0.46\cos(360/53) = 0.987\,13$$

$$h(1) = h(-1) = h_D(1)w(1) = 0.311\,19$$

$$n = 2: \quad h_D(2) = \frac{2 \times 0.218\,75}{2 \times 2\pi \times 0.218\,75}\sin(2 \times 2\pi \times 0.218\,75)$$

$$= \frac{\sin(157.5°)}{2\pi} = 0.060\,91$$

$$w(2) = 0.54 + 0.46\cos(2\pi \times 2/53)$$

$$= 0.54 + 0.46\cos(720°/53) = 0.987\,13$$

$$h(2) = h(-2) = h_D(2)w(2) = 0.060\,12$$

$$\vdots \quad \vdots \qquad \vdots \qquad \vdots$$

$$n = 26: \quad h_D(26) = \frac{2 \times 0.218\,75}{26 \times 2\pi \times 0.15}\,\frac{\sin(26 \times 2\pi \times 0.218\,75)}{2\pi}$$

$$= 0.011\,31$$

$$w(26) = 0.54 + 0.46\cos(2\pi \times 26/53)$$

$$= 0.54 + 0.46\cos(720°/53) = 0.080\,81$$

$$h(26) = h(-26) = h_D(26)w(26) = 0.000\,913$$

We note that the indices of the filter coefficients run from −26 to 26. To make the filter causal (necessary for implementation) we add 26 to each index so that the indices start at zero. The filter coefficients, with indices adjusted, are listed in Table 6.4. The spectrum of the filter (not shown) indicates that the specifications were satisfied.

Table 6.4 FIR coefficients for Example 6.3 ($N = 53$, Hamming window, $f_c = 1750$ Hz).

h[0] =	−9.1399895e−04	= h[52]
h[1] =	2.1673690e−04	= h[51]
h[2] =	1.3270280e−03	= h[50]
h[3] =	3.2138355e−04	= h[49]
h[4] =	−1.9238177e−03	= h[48]
h[5] =	−1.4683633e−03	= h[47]
h[6] =	2.3627318e−03	= h[46]
h[7] =	3.4846558e−03	= h[45]
h[8] =	−1.9925839e−03	= h[44]
h[9] =	−6.2837232e−03	= h[43]
h[10] =	4.5320247e−09	= h[42]
h[11] =	9.2669460e−03	= h[41]
h[12] =	4.3430586e−03	= h[40]
h[13] =	−1.1271299e−02	= h[39]
h[14] =	−1.1402453e−02	= h[38]
h[15] =	1.0630714e−02	= h[37]
h[16] =	2.0964392e−02	= h[36]
h[17] =	−5.2583216e−03	= h[35]
h[18] =	−3.2156086e−02	= h[34]
h[19] =	−7.5449714e−03	= h[33]
h[20] =	4.3546153e−02	= h[32]
h[21] =	3.2593190e−02	= h[31]
h[22] =	−5.3413653e−02	= h[30]
h[23] =	−8.5682029e−02	= h[29]
h[24] =	6.0122145e−02	= h[28]
h[25] =	3.1118568e−01	= h[27]
h[26] =	4.3750000e−01	= h[26]

Example 6.4

A requirement exists for an FIR digital filter to meet the following specifications:

passband	150–250 Hz
transition width	50 Hz
passband ripple	0.1 dB
stopband attenuation	60 dB
sampling frequency	1 kHz

Obtain the filter coefficients and spectrum using the window method.

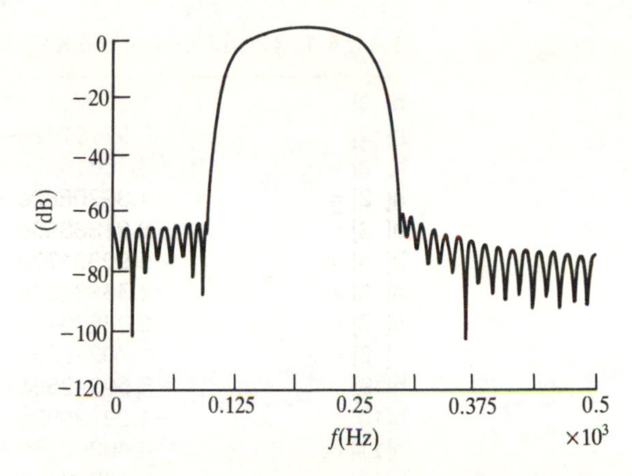

Figure 6.8 Spectrum of the filter (Example 6.4).

Solution

From the specification, the passband and stopband ripples are

$$20\log(1+\delta_p) = 0.1 \text{ dB, giving } \delta_p = 0.0115$$

$$-20\log(\delta_s) = 60 \text{ dB, giving } \delta_s = 0.001$$

Thus

$$\delta = \min(\delta_p, \delta_s) = 0.001$$

The attenuation requirements can only be met by the Kaiser or the Blackman window. For the Kaiser window, the number of filter coefficients is

$$N \geqslant \frac{A - 7.95}{14.36\,\Delta F} = \frac{60 - 7.95}{14.36(50/1000)} = 72.49$$

Let $N = 73$. The ripple parameter is given by

$$\beta = 0.1102(60 - 8.7) = 5.65$$

With $N = 73$, $\beta = 5.65$, the program window.c (see the appendix) is used to compute the values of $w(n)$, the ideal impulse response $h_D(n)$ and the filter coefficients. To account for the smearing effects of the window functions, in computing the ideal impulse response cutoff frequencies of $f_{c1} - \Delta f/2$ and $f_{c2} + \Delta f/2$ were used, that is $f_{c1} = 125$ Hz and 275 Hz respectively. The filter coefficients are given in Table 6.5 and the filter spectrum in Figure 6.8.

For the Blackman window, an estimate of the number of filter coefficients is obtained as

$$N = 5.5/\Delta f = 5.5/(50/1000) \approx 110$$

The filter coefficients for the Blackman window are not given here owing to lack of space. It is evident that the Kaiser window is more efficient than the Blackman window in terms of the number of coefficients required to meet the same specifications. In general, the Kaiser window is more efficient compared to the other windows in this respect.

Table 6.5 Coefficients of the Kaiser filter (Example 6.4).

h[0] =	−1.0627330e−04	= h[72]
h[1] =	−3.9118142e−04	= h[71]
h[2] =	−7.5561629e−05	= h[70]
h[3] =	−1.3695577e−04	= h[69]
h[4] =	−6.8122013e−04	= h[68]
h[5] =	5.0929290e−04	= h[67]
h[6] =	2.3413494e−03	= h[66]
h[7] =	8.0280013e−04	= h[65]
h[8] =	−1.7031635e−04	= h[64]
h[9] =	−5.5034956e−04	= h[63]
h[10] =	−4.9912488e−04	= h[62]
h[11] =	−4.4036355e−03	= h[61]
h[12] =	−2.1639856e−03	= h[60]
h[13] =	6.9094151e−03	= h[59]
h[14] =	6.6067599e−03	= h[58]
h[15] =	−1.6445200e−03	= h[57]
h[16] =	4.5229777e−09	= h[56]
h[17] =	2.1890066e−03	= h[55]
h[18] =	−1.1720511e−02	= h[54]
h[19] =	−1.6377726e−02	= h[53]
h[20] =	6.8804519e−03	= h[52]
h[21] =	1.8882837e−02	= h[51]
h[22] =	2.9068601e−03	= h[50]
h[23] =	4.3925286e−03	= h[49]
h[24] =	1.8839744e−02	= h[48]
h[25] =	−1.2481155e−02	= h[47]
h[26] =	−5.2063428e−02	= h[46]
h[27] =	−1.6557375e−02	= h[45]
h[28] =	3.3298453e−02	= h[44]
h[29] =	1.0439025e−02	= h[43]
h[30] =	9.4320244e−03	= h[42]
h[31] =	8.5673629e−02	= h[41]
h[32] =	4.5314758e−02	= h[40]
h[33] =	−1.6657147e−01	= h[39]
h[34] =	−2.0669512e−01	= h[38]
h[35] =	8.9135544e−02	= h[37]
h[36] =	3.0000000e−01	= h[36]

Example 6.5

Obtain the coefficients of a linear phase FIR filter using the Kaiser window to satisfy the following amplitude response specifications:

stopband attenuation	40 dB
passband ripple	0.01 dB
transition width	500 Hz
sampling frequency	10 kHz
ideal cutoff frequency	1200 Hz

Solution

From the specifications,

$$20\log(1+\delta_p) = 0.01 \text{ dB, giving } \delta_p = 0.001\,15$$
$$-20\log(\delta_s) = 40 \text{ dB, giving } \delta_s = 0.01$$

Since both the passband and stopband ripples are equal (as they cannot be specified independently) in the window method, we use the smaller of the ripples:

$$\delta = \delta_s = \delta_p = 0.001\,15$$

This means that the stopband attenuation is more than actually required, in this case $-20\log(0.001\,15) = 58.8$ dB.

From Equation 6.11, the number of filter coefficients required is

$$N = \frac{A - 7.95}{14.36\,\Delta f} = \frac{58.8 - 7.95}{14.36(500/10000)} \approx 71$$

If the required attenuation specification of 40 dB was used N would have been 45. Thus the need for δ_p to be equal to δ_s in the window method has led to a higher than necessary number of filter coefficients.

The ripple parameter is obtained from Equation 6.10:

$$\beta = 0.5842(58.8 - 21)^{0.4} + 0.078\,86(58.8 - 21) = 5.48$$

The FIR coefficients are obtained from $h(n) = h_D(n)w(n)$ where, from Table 6.2,

$$h_D(n) = 2f_c \frac{\sin(n\omega_c)}{n\omega_c} \qquad n = 0$$
$$h_D(n) = 2f_c \qquad n \neq 0$$

and $w(n)$ is given by Equation 6.9.

As explained before, the cutoff frequency, f_c, used in calculating $h(n)$ is different to that given in the specifications to account for the smearing effect of the window function. We select f_c that is in the middle of the transition band: $f_c' = 1200 + \Delta f/2 = 1450$ Hz.

The following filter parameters are used with the computer program

Table 6.6 Filter coefficients using Kaiser window (Example 6.5).

h[0] =	9.8470163e−05	= h[70]
h[1] =	−1.3972411e−04	= h[69]
h[2] =	−4.5442489e−04	= h[68]
h[3] =	−4.8756977e−04	= h[67]
h[4] =	2.6173965e−05	= h[66]
h[5] =	8.6653647e−04	= h[65]
h[6] =	1.2967984e−03	= h[64]
h[7] =	6.1688894e−04	= h[63]
h[8] =	−1.0445340e−03	= h[62]
h[9] =	−2.4646644e−03	= h[61]
h[10] =	−2.1059775e−03	= h[60]
h[11] =	4.4371801e−04	= h[59]
h[12] =	3.5954580e−03	= h[58]
h[13] =	4.5526695e−03	= h[57]
h[14] =	1.5922295e−03	= h[56]
h[15] =	−3.8904820e−03	= h[55]
h[16] =	−7.6398162e−03	= h[54]
h[17] =	−5.6061945e−03	= h[53]
h[18] =	2.2010888e−03	= h[52]
h[19] =	1.0450148e−02	= h[51]
h[20] =	1.1760002e−02	= h[50]
h[21] =	2.8239875e−03	= h[49]
h[22] =	−1.1380549e−02	= h[48]
h[23] =	−1.9631856e−02	= h[47]
h[24] =	−1.2665935e−02	= h[46]
h[25] =	8.0061777e−03	= h[45]
h[26] =	2.8182781e−02	= h[44]
h[27] =	2.9474031e−02	= h[43]
h[28] =	3.8724896e−03	= h[42]
h[29] =	−3.5942288e−02	= h[41]
h[30] =	−5.9766794e−02	= h[40]
h[31] =	−3.7113570e−02	= h[39]
h[32] =	4.1378026e−02	= h[38]
h[33] =	1.5291289e−01	= h[37]
h[34] =	2.5100632e−01	= h[36]
h[35] =	2.9000000e−01	= h[35]

window.c (see the appendix):

cutoff frequency	1450 Hz
ripple parameter, β	5.48
number of filter coefficients	71
sampling frequency	10 kHz

The resulting filter coefficients are given in Table 6.6 and the filter spectrum in Figure 6.9.

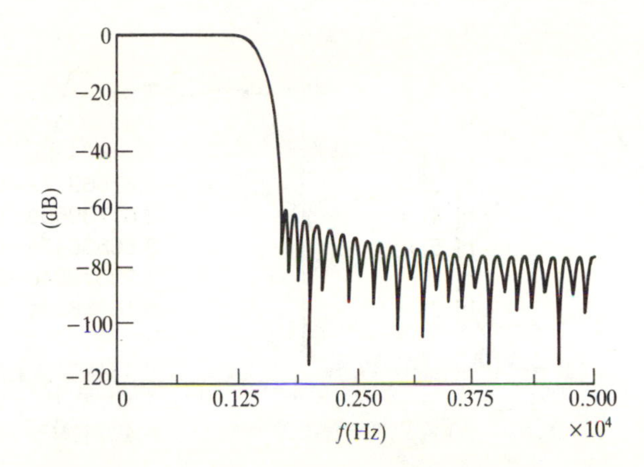

Figure 6.9 Filter spectrum using the Kaiser window (Example 6.5).

6.5.3 Advantages and disadvantages of the window method

- An important advantage of the window method is its simplicity: it is simple to apply and simple to understand. It involves a minimum amount of computational effort, even for the more complicated Kaiser window.
- The major disadvantage is its lack of flexibility. Both the peak passband and stopband ripples are approximately equal, so that the designer may end up with either too small a passband ripple or too large a stopband attenuation.
- Because of the effect of convolution of the spectrum of the window function and the desired response, the passband and stopband edge frequencies cannot be precisely specified.
- For a given window (except the Kaiser) the maximum ripple amplitude in the filter response is fixed regardless of how large we make N. Thus the stopband attenuation for a given window is fixed. Thus, for a given attenuation specification, the filter designer must find a suitable window.
- In some applications, the expression for the desired filter response, $H_D(\omega)$, will be too complicated for $h_D(n)$ to be obtained analytically from Equation 6.5. In these cases $h_D(n)$ may be obtained via the frequency sampling method before the window function is applied (see Section 6.7.1).

6.6 The optimal method

The optimal (in the Chebyshev sense) method of calculating FIR filter coefficients is very powerful, very flexible and, because of the existence of an excel-

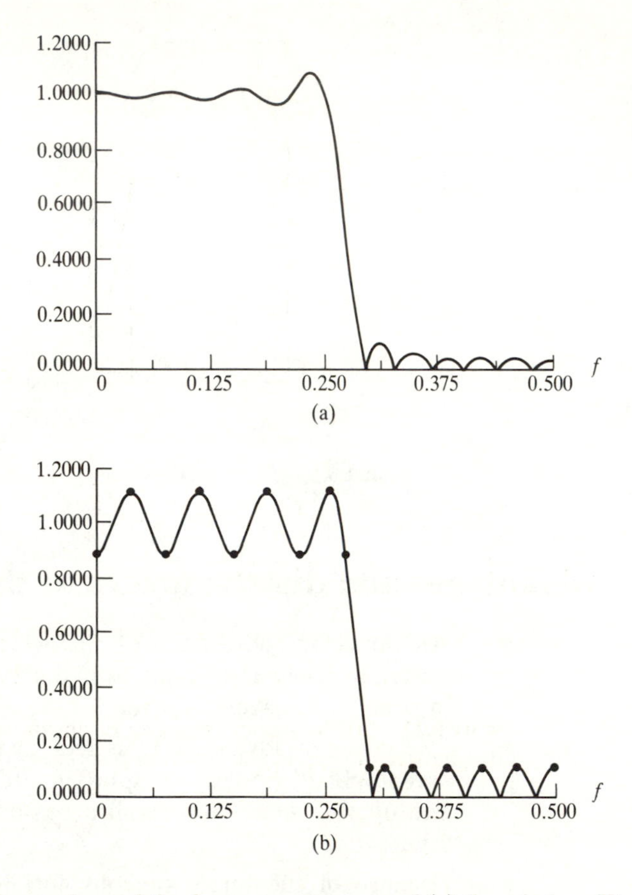

Figure 6.10 Comparison of the frequency response of (a) the window filter and (b) the optimal filter. In (a) the ripples are largest near the band edge; in (b) the ripples have the same peaks (equiripple) in the passband or stopband.

lent design program, very easy to apply. For these reasons and because the method yields excellent filters it has become the method of first choice in many FIR applications. The concept on which the method is based, the design program and its use will be discussed. Several design examples are presented to illustrate the method.

6.6.1 Basic concepts

It is evident in the window method that inherent in the process of calculating suitable filter coefficients is the problem of finding a suitable approximation to a desired or ideal frequency response. The peak ripple of filters designed by the window method occurs near the band edges, and decreases away from the band edges (Figure 6.10(a)). It turns out that if the ripples were distributed

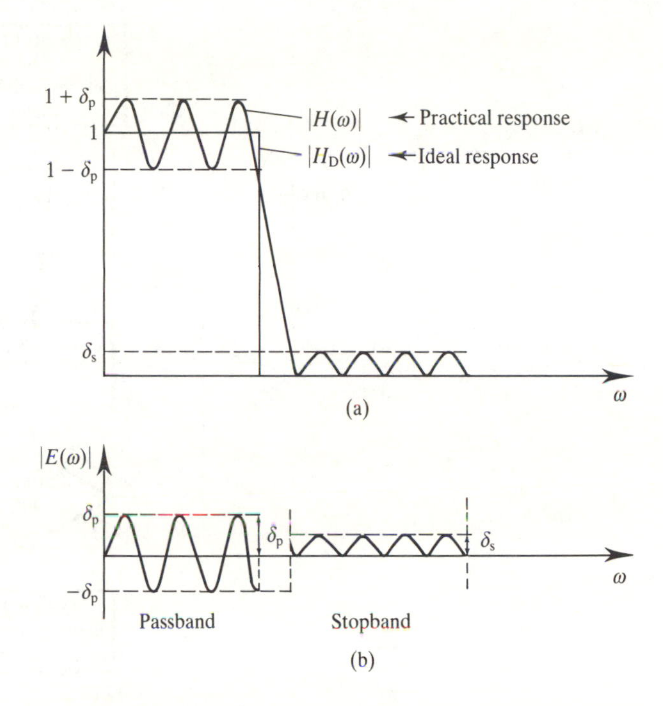

Figure 6.11 (a) Frequency response of an optimal lowpass filter. (b) Response of the error between the ideal and practical responses ($\delta_p = 2\delta_s$).

more evenly over the passband and stopband, for example as in Figure 6.10(b), a better approximation of the desired frequency response can be achieved.

The optimal method is based on the concept of equiripple passband and stopband. Consider the lowpass filter frequency response depicted in Figure 6.11. In the passband, the practical response oscillates between $1 - \delta_p$ and $1 + \delta_p$. In the stopband the filter response lies between 0 and δ_s. The difference between the ideal filter and the practical response can be viewed as an error function:

$$E(\omega) = W(\omega)[H_D(\omega) - H(\omega)] \tag{6.13}$$

where $H_D(\omega)$ is the ideal or desired response and $W(\omega)$ is a weighting function that allows the relative error of approximation between different bands to be defined. In the optimal method, the objective is to determine the filter coefficients, $h(n)$, such that the value of the maximum weighted error, $|E(\omega)|$, is minimized in the passband and stopband. Mathematically, this may be expressed as

$$\min[\max|E(\omega)|]$$

over the passbands and stopbands. It has been established (see for example

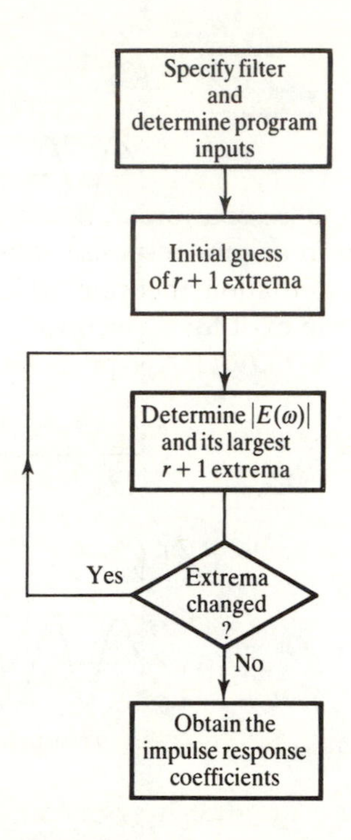

Figure 6.12 Simplified flowchart of the optimal method.

Rabiner and Gold, 1975) that when max $|E(\omega)|$ is minimized the resulting filter response will have equiripple passband and stopband, with the ripple alternating in sign between two equal amplitude levels (Figure 6.10(b)). The minima and maxima are known as extrema. For linear phase lowpass filters for example, there are either $r + 1$ or $r + 2$ extrema, where $r = (N + 1)/2$ (for type 1 filters) or $r = N/2$ (for type 2 filters). The extremal frequencies are indicated as small circles in Figure 6.10(b).

For a given set of filter specifications, the locations of the extremal frequencies, apart from those at band edges (that is at $f = f_p$ and $f = F_s/2$), are not known *a priori*. Thus the main problem in the optimal method is to find the locations of the extremal frequencies. A powerful technique which employs the Remez exchange algorithm to find the extremal frequencies has been developed (Rabiner and Gold, 1975; McClellan *et al.*, 1973; Oppenheim and Schafer, 1975). Knowing the locations of the extremal frequencies, it is a simple matter to work out the actual frequency response and hence the impulse response of the filter. For a given set of specifications (that is passband edge frequencies, N, and the ratio between the passband and stopband ripples) the optimal method involves the following key steps (see Figure 6.12):

- use the Remez exchange algorithm to find the optimum set of extremal frequencies;

- determine the frequency response using the extremal frequencies;
- obtain the impulse response coefficients.

The heart of the optimal method is the first step where an iterative process is used to determine the extremal frequencies of a filter whose amplitude–frequency response satisfies the optimality condition. This step relies on the alternation theorem which specifies the number of extremal frequencies that can exist for a given value of N.

A FORTRAN program that implements the above process is available and is now widely used (for example McClellan *et al*., 1973). An equivalent C language program is given in the PC disk for the book. The program allows the design of a variety of frequency selective filters including lowpass, highpass, bandpass and bandstop as well as differentiators and Hilbert transformers. It is also capable of computing the coefficients of a user-specified arbitrary frequency response. Further details of the optimal method can be found in the references given above.

6.6.2 Parameters required to use the optimal program

To use the design program, the user must provide a set of input parameters describing the filter. These consist of the following parameters:

N — Number of filter coefficients, that is filter length. An estimate of this can be found from the relationships given in the next section.

Jtype — This parameter specifies the type of filter. Three types of filter are possible: Jtype = 1 (multiple passband/stopband filters, including lowpass, highpass, bandpass, and bandstop filters), Jtype = 2 (specifies differentiator), Jtype = 3 (specifies Hilbert transformer).

$W(\omega)$ — The weighting function. This parameter specifies the relative importance of each band. In effect it allows a trade-off between the passband ripple and stopband attenuation. A weight is specified for each band.

Ngrid — This parameter specifies the grid density. This is the number of frequency points at which, during the process of finding the extremal frequencies, the frequency response is checked to see whether the optimality condition has been satisfied (optimum in the sense that the maximum amplitude of the error, $|E(\omega)|$, is minimized in the passband(s) and stopband(s)). The default value for Ngrid is 16. In most designs, an Ngrid value of 16, 32 or 64 is adequate.

Edge — This specifies the band edge frequencies (that is, the lower and upper band edge frequencies for the filter). All the frequencies must be entered in normalized form. The first edge is normally 0 and the last 0.5 (corresponding to half the sampling frequency). A maximum of 10 bands (passbands and/or stopbands) is supported.

The symbols used above are not necessarily those used in the more recent implementations. In fact some implementations now have a more user-friendly interface to the design program than the original implementation.

6.6.3 Relationships for estimating filter length, *N*

In practice, the number of filter coefficients is unknown. Its value may be estimated using the empirical relationships below.

6.6.3.1 Lowpass filter (Herrman *et al.*, 1973)

$$N \simeq \frac{D_\infty(\delta_p, \delta_s)}{\Delta F} - f(\delta_p, \delta_s)\,\Delta F + 1 \qquad \textbf{(6.14)}$$

where ΔF is the width of the transition band normalized to the sampling frequency,

$$\begin{aligned} D_\infty(\delta_p, \delta_s) &= \log_{10}\delta_s\,[a_1(\log_{10}\delta_p)^2 + a_2\log_{10}\delta_p + a_3] \\ &\quad + [a_4(\log_{10}\delta_p)^2 + a_5\log_{10}\delta_p + a_6] \\ f(\delta_p, \delta_s) &= 11.012\,17 + 0.512\,44[\log_{10}\delta_p - \log_{10}\delta_s] \\ a_1 &= 5.309 \times 10^{-3}; \qquad a_2 = 7.114 \times 10^{-2} \\ a_3 &= -4.761 \times 10^{-1}; \qquad a_4 = -2.66 \times 10^{-3} \\ a_5 &= -5.941 \times 10^{-1}; \qquad a_6 = -4.278 \times 10^{-1} \end{aligned}$$

δ_p is the passband ripple or deviation and δ_s is the stopband ripple or deviation.

6.6.3.2 Bandpass filter (Mintzer and Liu, 1979)

$$N \simeq \frac{C_\infty(\delta_p, \delta_s)}{\Delta F} + g(\delta_p, \delta_s)\,\Delta F + 1 \qquad \textbf{(6.15)}$$

where

$$\begin{aligned} C_\infty(\delta_p, \delta_s) &= \log_{10}\delta_s\,[b_1(\log_{10}\delta_p)^2 + b_2\log_{10}\delta_p + b_3] \\ &\quad + [b_4(\log_{10}\delta_p)^2 + b_5\log_{10}\delta_p + b_6] \\ g(\delta_p, \delta_s) &= -14.6\log_{10}\left(\frac{\delta_p}{\delta_s}\right) - 16.9 \\ b_1 &= 0.012\,01; \qquad b_2 = 0.096\,64 \\ b_3 &= -0.513\,25; \qquad b_4 = 0.002\,03 \\ b_5 &= -0.5705; \qquad a_6 = -0.443\,14 \end{aligned}$$

and ΔF is the transition width normalized to the sampling frequency.

A C language program is given in the appendix for computing the value of N using Equation 6.14 or 6.15.

6.6.4 Summary of procedure for calculating filter coefficients by the optimal method

- *Step 1.* Specify the band edge frequencies (that is, passband and stopband frequencies), passband ripple and stopband attenuation (in decibels or ordinary units), and sampling frequency.
- *Step 2.* Normalize each band edge frequency by dividing it by the sampling frequency, and determine the normalized transition width.
- *Step 3.* Use the passband ripple and stopband attenuation, expressed in ordinary units, and the normalized transition width (see note below) to estimate the filter length, N, from Equations 6.14 or 6.15. Typically, the value of N required to meet the specifications would be slightly higher (2 or 3) than the value determined from these equations.
- *Step 4.* Obtain the weights for each band from the ratio of the passband to stopband ripples (or stopband to passband ripples), expressed in ordinary units. It is convenient to express the weights for each band as an integer. For example, a lowpass filter with a passband and stopband ripples of 0.01 and 0.03 (passband ripple and stopband attenuation of 0.09 dB and 30.5 dB respectively) would have a weight of 1 for the passband and 3 for the stopband. Bandpass filter deviations (ripples) of 0.001 in the passband and 0.0105 for each of the stopbands would have weights of 21 for the passband and 2 for each of the stopbands.
- *Step 5.* Input the parameters to the optimal design program to obtain the coefficients: N, band edge frequencies and weights for each band, together with a suitable grid density (typically 16 or 32).
- *Step 6.* Check the passband ripple and stopband attenuation produced by the program.
- *Step 7.* If the specifications are not satisfied, increase the value of N and repeat steps 5 and 6 until they are; then obtain and check the frequency response to ensure that it satisfies the specifications.

It should be noted that the optimal program considers only the passband and stopband during its approximation stage, treating the transition region as a 'don't care' region. To avoid failure or problems with convergence of the algorithm, it is best to set the transition regions equal to the width of the smallest transition region when designing bandpass or multiple-band filters. If unequal transition widths are used, the frequency response should always be checked to ensure that the specification is met. Local maxima and minima may occur in the transition bands, giving unexpected filter characteristics.

6.6.5 Illustrative examples

The following examples illustrate the use of the optimal program.

Example 6.6

A linear phase bandpass filter is required to meet the following specifications:

passband	900–1100 Hz
passband ripple	<0.87 dB
stopband attenuation	>30 dB
sampling frequency	15 kHz
transition frequency	450 Hz

Use the optimal method to obtain suitable coefficients. Plot the filter spectrum.

Solution

From the specifications, the filter has three bands: a lower stopband (0 to 450 Hz), a passband (900 to 1100 Hz), and an upper stopband (1550 to 7500 Hz). To use the optimal design program the band edge frequencies must be normalized, that is expressed as fractions of the sampling frequency:

$$450 = 450/15\,000 = 0.03$$

$$900 = 900/15\,000 = 0.06$$

$$1100 = 1100/15\,000 = 0.0733$$

$$1550 = 1550/15\,000 = 0.1033$$

$$7500 = 7500/15\,000 = 0.5$$

Thus the three normalized bands are (0 to 0.03), (0.06 to 0.0733), (0.1033 to 0.5).

Next, we must choose weights for the bands. The weights are dependent on the passband and stopband deviations. The deviations, in ordinary units, can be obtained from the given passband ripple and stopband attenuation:

$$0.87\text{ dB ripple: } 20\log(1+\delta_p) \quad \rightarrow \delta_p = 0.105\,35$$

$$29.78\text{ dB attenuation: } -20\log(\delta_s) \quad \rightarrow \delta_s = 0.031\,623$$

The ratio of δ_p to δ_s is $3.36 = 10/3$:

$$\frac{\delta_p}{\delta_s} = \frac{10}{3} = \frac{\text{stopband weight}}{\text{passband weight}}$$

Thus we could use weights of 3 for the passband and 10 for the stopband

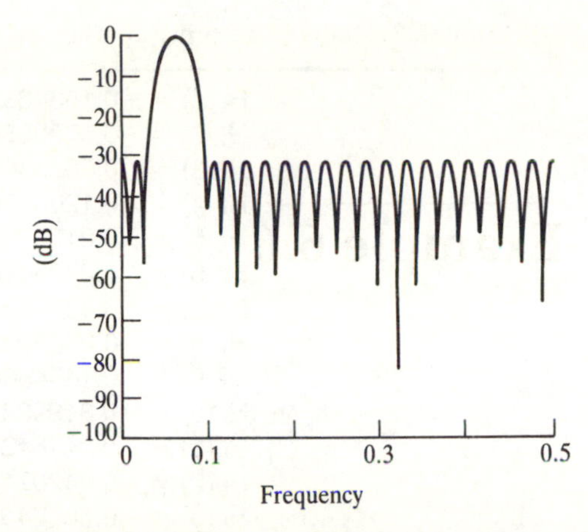

Figure 6.13 Frequency response of filter (normalized frequency scale).

(notice that the weighting is applied in the opposite sense to the ratio of δ_p/δ_s). Equally valid are weights of 1 for the passband and 3.33 for the stopband. A grid density of 32 is used. Using the program which implements the relationships for N, the filter length was found to be 40. We will use $N = 41$.

The input to the optimal program can be summarized:

filter length, N	41
type of filter, Jtype	1
weights, $W(\omega)$	10, 3, 10
Ngrid	32
edge frequencies	0, 0.03, 0.06, 0.0733, 0.1033, 0.5

A printout of the output of the design program is given in Table 6.7 and the frequency spectrum is given in Figure 6.13. A few comments are in order.

- The passband deviation is 3.33 times that of the stopband. This is because the errors in the passband and stopbands were given weights of 3 and 10, respectively. The higher the weight is for a band, the smaller is the resulting ripple or deviation.
- The passband ripples and stopband attenuation (in decibels) are well within the specifications.
- There are 22 extremal frequencies, that is $(N + 3)/2$ maxima and minima in the amplitude response. Notice that the band edge frequencies are also extremal frequencies, as are $f = 0$ and $f = 0.5$ Hz. The band edge frequencies are always extrema frequencies.
- The impulse response is symmetrical about the middle coefficient. The symmetry property is a necessary condition for a linear phase response. Note that, for type 1 filter, the middle coefficient has the maximum value.

For a given design, with the band edges fixed, the designer can adjust the passband ripple and stopband attenuation relative to each other as desired using the weights and N.

Table 6.7 Impulse response coefficients of the optimal filters (Example 6.6).

```
H( 1) = −0.15346380E−01 = H(41)
H( 2) = −0.57805500E−04 = H(40)
H( 3) =  0.50234820E−02 = H(39)
H( 4) =  0.12667060E−01 = H(38)
H( 5) =  0.21082060E−01 = H(37)
H( 6) =  0.27764180E−01 = H(36)
H( 7) =  0.30053620E−01 = H(35)
H( 8) =  0.25869350E−01 = H(34)
H( 9) =  0.14445660E−01 = H(33)
H(10) = −0.31893230E−02 = H(32)
H(11) = −0.24161370E−01 = H(31)
H(12) = −0.44207120E−01 = H(30)
H(13) = −0.58574530E−01 = H(29)
H(14) = −0.63185570E−01 = H(28)
H(15) = −0.55754610E−01 = H(27)
H(16) = −0.36546910E−01 = H(26)
H(17) = −0.85400990E−02 = H(25)
H(18) =  0.23083860E−01 = H(24)
H(19) =  0.52013800E−01 = H(23)
H(20) =  0.72248070E−01 = H(22)
H(21) =  0.79516810E−01 = H(21)
```

	BAND 1	BAND 2	BAND 3
LOWER BAND EDGE	0.000000000	0.060000000	0.103300000
UPPER BAND EDGE	0.030000000	0.073300000	0.500000000
DESIRED VALUE	0.000000000	1.000000000	0.000000000
WEIGHTING	10.000000000	3.000000000	10.000000000
DEVIATION	0.028891690	0.096305620	0.028891690
RIPPLE IN DB	−30.784510000	0.798631800	−30.784510000

EXTREMA FREQUENCIES

0.0000000	0.0208333	0.0300000	0.0600000	0.1033000
0.1122285	0.1308297	0.1538951	0.1777045	0.2015139
0.2260674	0.2506209	0.2759184	0.3004719	0.3257694
0.3503229	0.3756204	0.4001739	0.4254714	0.4500249
0.4753224	0.5000000			

Example 6.7

A digital FIR notch filter satisfying the specifications given below is required:

notch frequency	1.875 kHz
attenuation at notch frequency	> 60 dB

passband edge frequencies	1.575 and 2.175 kHz
passband ripple	< 0.01 dB
sampling frequency	7.5 kHz
number of coefficients	61

Use the optimal method to obtain the filter coefficients of the FIR filter satisfying the specifications.

Solution

The filter has three bands. The normalized frequencies of the three bands and the deviations are

lower passband	0 to 0.21
notch frequency	0.25
upper passband	0.29 to 0.5
passband deviation	0.001 15 (from $20\log_{10}(1 + \delta_p)$)
stopband deviation	0.001 (from $-20\log_{10}(\delta_s)$)

The weights for the bands are 1, 1.1519, 1 (from δ_p/δ_s). The results are summarized in Table 6.8 and Figure 6.14. Notice that for a notch the stopband is effectively a single frequency.

Thus the band edge frequencies entered into the design program are 0, 0.21, 0.25, 0.25, 0.29 and 0.5. By entering the notch frequency twice the stopband is effectively reduced to a single frequency, which is the desired result.

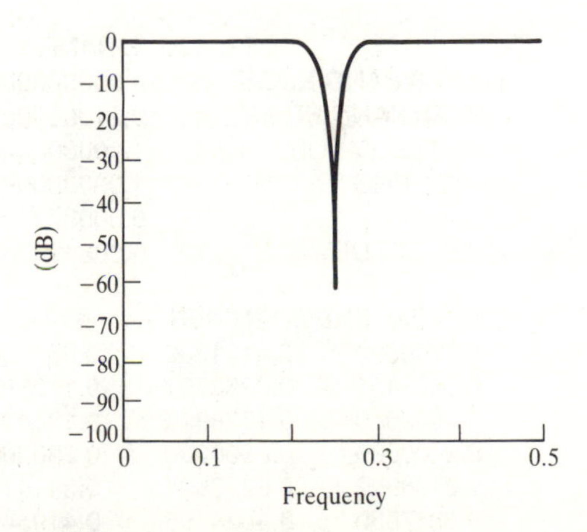

Figure 6.14 Filter response (normalized frequency scale).

Table 6.8 Impulse response coefficients for optimal filter (Example 6.7).

```
H( 1) =  0.12743640E−02 = H(61)
H( 2) =  0.26730640E−05 = H(60)
H( 3) = −0.23681110E−02 = H(59)
H( 4) = −0.17416350E−05 = H(58)
H( 5) =  0.43428480E−02 = H(57)
H( 6) =  0.53579250E−05 = H(56)
H( 7) = −0.71570240E−02 = H(55)
H( 8) = −0.49028620E−05 = H(54)
H( 9) =  0.10897540E−01 = H(53)
H(10) =  0.89629280E−05 = H(52)
H(11) = −0.15605960E−01 = H(51)
H(12) = −0.85508990E−05 = H(50)
H(13) =  0.21226410E−01 = H(49)
H(14) =  0.12250150E−04 = H(48)
H(15) = −0.27630130E−01 = H(47)
H(16) = −0.11091200E−04 = H(46)
H(17) =  0.34579770E−01 = H(45)
H(18) =  0.13800660E−04 = H(44)
H(19) = −0.41774130E−01 = H(43)
H(20) = −0.11560390E−04 = H(42)
H(21) =  0.48832790E−01 = H(41)
H(22) =  0.12787590E−04 = H(40)
H(23) = −0.55359840E−01 = H(39)
H(24) = −0.90065860E−05 = H(38)
H(25) =  0.60944450E−01 = H(37)
H(26) =  0.88997300E−05 = H(36)
H(27) = −0.65232190E−01 = H(35)
H(28) = −0.38167120E−05 = H(34)
H(29) =  0.67925720E−01 = H(33)
H(30) =  0.27041150E−05 = H(32)
H(31) =  0.93115220E+00 = H(31)
```

	BAND 1	BAND 2	BAND 3
LOWER BAND EDGE	0.000000000	0.250000000	0.290000000
UPPER BAND EDGE	0.210000000	0.250000000	0.500000000
DESIRED VALUE	1.000000000	0.000000000	1.000000000
WEIGHTING	1.000000000	1.151900000	1.000000000
DEVIATION	0.000978727	0.000849663	0.000978727
RIPPLE IN DB	0.008496785	−61.414990000	0.008496785

EXTREMA FREQUENCIES

0.0000000	0.0161290	0.0322580	0.0483871	0.0645161
0.0806451	0.0962701	0.1123991	0.1280241	0.1431450
0.1582660	0.1728829	0.1864918	0.1980845	0.2066530
0.2100000	0.2500000	0.2900000	0.2930243	0.3020971
0.3136902	0.3272994	0.3414128	0.3565342	0.3721596
0.3877850	0.4034105	0.4195400	0.4356695	0.4517990
0.4679285	0.4840580			

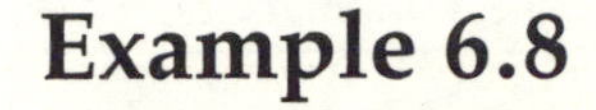

Example 6.8

It is important that the designer appreciates how the parameters interact so that appropriate trade-offs can be made when necessary. This example allows us to examine the effects of the parameters, δ_p, δ_s, W and the various possibilities.

A requirement exists for a linear phase FIR filter for reducing physiological noise (Hamer *et al.*, 1985). The filter is intended to be part of a large time-critical DSP system and so the number of coefficients should be kept as low as possible. The filter characteristics should meet the following specifications:

passband ripple	< 0.026 dB
stopband	> 30 dB
passband edge frequency	10 Hz
stopband edge	< 20 Hz
sampling frequency	128 Hz

Solution

The normalized band edge frequencies, passband and stopband deviations are

passband edge frequency	0.078
stopband edge frequency	< 0.15625
passband deviation	< 0.003
stopband deviation	> 0.0316

Since most of the filter specifications are variable, it is clear that there will be a range of possible solutions. The problem then is one of finding the best solution.

Now using the limiting values above in Equation 6.14, we find that $N > 25.6$ (this represents the smallest possible value of N). For each value of N in the range 25–37, the stopband frequency, f_s, that satisfies the specifications is computed using the following relationship:

$$f_s = f_p + \Delta f$$

where f_s and f_p are the stopband and passband edge frequencies, and Δf the transition width given by ($\Delta f_{max} = 20 - 10\text{ Hz} = 10\text{ Hz}$):

$$\Delta f = \frac{N-1}{2f(\delta_p, \delta_s)}\left[1 + \frac{4f(\delta_p, \delta_s)D_\infty(\delta_p, \delta_s) - 1}{(N-1)^2}\right]^{1/2}$$

Figure 6.15 gives the solution space (above the curve) which is bounded by stopband edge frequency of 20 Hz and $N = 26$ and 37. A value of 27 is chosen

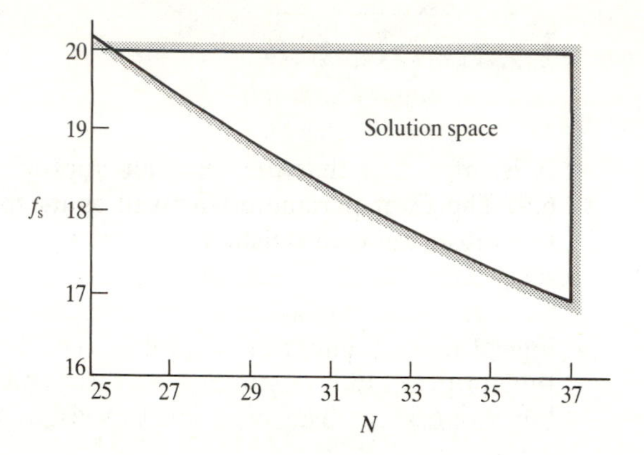

Figure 6.15 Stopband frequency versus filter length showing the range of possible solutions.

Table 6.9 Impulse response coefficients for the optimal filter (Example 6.8).

```
H( 1) = -0.13614960E-01 = H(27)
H( 2) =  0.34793330E-02 = H(26)
H( 3) =  0.11140420E-01 = H(25)
H( 4) =  0.16664540E-01 = H(24)
H( 5) =  0.12807340E-01 = H(23)
H( 6) = -0.33202110E-02 = H(22)
H( 7) = -0.26167170E-01 = H(21)
H( 8) = -0.42047790E-01 = H(20)
H( 9) = -0.34767040E-01 = H(19)
H(10) =  0.55338630E-02 = H(18)
H(11) =  0.75072090E-01 = H(17)
H(12) =  0.15527810E+00 = H(16)
H(13) =  0.21933680E+00 = H(15)
H(14) =  0.24378330E+00 = H(14)
```

	BAND 1	BAND 2
LOWER BAND EDGE	0.000000000	0.152388500
UPPER BAND EDGE	0.078000000	0.500000000
DESIRED VALUE	1.000000000	0.000000000
WEIGHTING	10.500000000	1.000000000
DEVIATION	0.002604177	0.027343860
RIPPLE IN DB	0.022589730	-31.262770000

EXTREMA FREQUENCIES

0.0089286	0.0468750	0.0691964	0.0780000	0.1523885
0.1668974	0.1981473	0.2338614	0.2706916	0.3086379
0.3465842	0.3845305	0.4235928	0.4615391	0.5000000

as a good solution. An odd value of N is preferred to avoid a noninteger number of sample delays through the filter. The following parameters were used: passband (0 to 0.078), stopband (0.1523885 to 0.5, that is 19 Hz to 64 Hz). Weights of 10.5 and 1 were used for the passband and stopband respectively. The resulting filter coefficients and parameters are given in Table 6.9. The filter parameters as well as its spectrum (not given) showed that the specifications were satisfied.

6.7 Frequency sampling method

The frequency sampling method allows us to design nonrecursive FIR filters for both standard frequency selective filters (lowpass, highpass, bandpass filters) and filters with arbitrary frequency response. A unique attraction of the frequency sampling method is that it also allows recursive implementation of FIR filters, leading to computationally efficient filters. With some restrictions, recursive FIR filters whose coefficients are simple integers may be designed, which is attractive when only primitive arithmetic operations are possible, as in systems implemented with standard microprocessors.

6.7.1 Nonrecursive frequency sampling filters

Suppose we wish to obtain the FIR coefficients of the filter whose frequency response is depicted in Figure 6.16(a). We could start by taking N samples of the frequency response at intervals of kF_s/N, $k = 0, 1, \ldots, N-1$. The filter coefficients $h(n)$ can be obtained as the inverse DFT of the frequency samples:

$$h(n) = \frac{1}{N}\sum_{k=0}^{N-1} H(k)e^{j(2\pi/N)nk} \tag{6.16}$$

where $H(k)$, $k = 0, 1, \ldots, N-1$, are samples of the ideal or target frequency response.

It can be shown (see Example 6.9) that for linear phase filters, with positive symmetrical impulse response, we can write (for N even),

$$h(n) = \frac{1}{N}\left[\sum_{k=1}^{N/2-1} 2|H(k)|\cos|[2\pi k(n-\alpha)/N]| + H(0)\right] \tag{6.17}$$

where $\alpha = (N-1)/2$. For N odd, the upper limit in the summation is $(N-1)/2$. The resulting filter will have a frequency response that is exactly the same as the original response at the sampling instants. However, between the sample instants, the response may be significantly different (Figure 6.16(c)). To

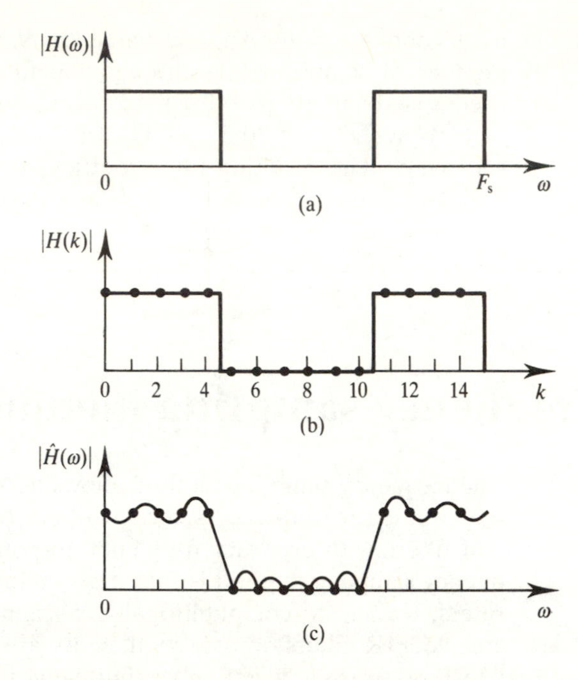

Figure 6.16 Concept of frequency sampling. (a) Frequency response of an ideal lowpass filter. (b) Samples of the ideal lowpass filter. (c) Frequency response of lowpass filter derived from the frequency samples of (b).

obtain a good approximation to the desired frequency response, clearly we must take a sufficient number of frequency samples.

An alternative frequency sampling filter, known as type 2, results if we take frequency samples at intervals of

$$f_k = (k + 1/2)F_s/N, \quad k = 0, 1, \ldots, N-1 \tag{6.18}$$

Figure 6.17 compares the sampling grids for both types of frequency sampling schemes. For a given filter specification, both methods will lead to somewhat different frequency responses. The designer needs to decide which of the two types best suits his/her needs.

Example 6.9

(1) Show that the impulse response coefficients of a linear phase FIR filter with positive symmetry, for N even, can be expressed as

$$h(n) = \frac{1}{N}\left[\sum_{k=1}^{N/2-1} 2|H(k)| \cos|[2\pi k(n-\alpha)/N]| + H(0)\right]$$

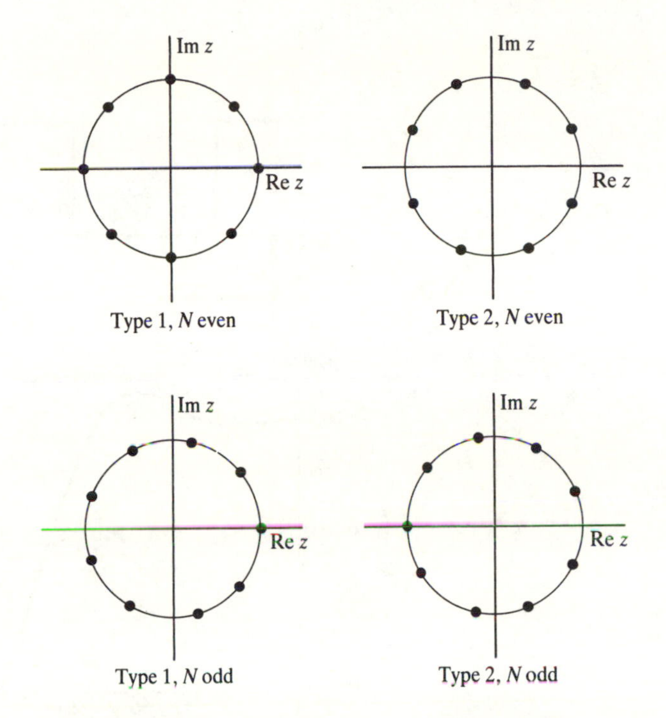

Figure 6.17 The four possible z-plane sampling grids for the two types of frequency sampling filters.

where $\alpha = (N-1)/2$, and $H(k)$ are the samples of the frequency response of the filter taken at intervals of kF_s/N.

(2) A requirement exists for a lowpass FIR filter satisfying the following specifications:

passband	0–5 kHz
sampling frequency	18 kHz
filter length	9

Obtain the filter coefficients using the frequency sampling method.

Solution

(1)
$$h(n) = \frac{1}{N}\sum_{k=0}^{N-1} H(k)e^{j(2\pi/N)nk} \tag{6.19}$$

$$= \frac{1}{N}\sum_{k=0}^{N-1} |H(k)|e^{-j2\pi\alpha k/N}e^{j2\pi kn/N}$$

$$= \frac{1}{N}\sum_{k=0}^{N-1} |H(k)|e^{j2\pi k(n-\alpha)/N}$$

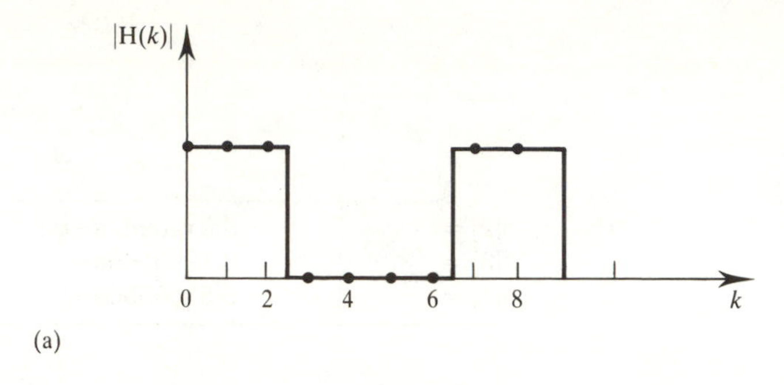

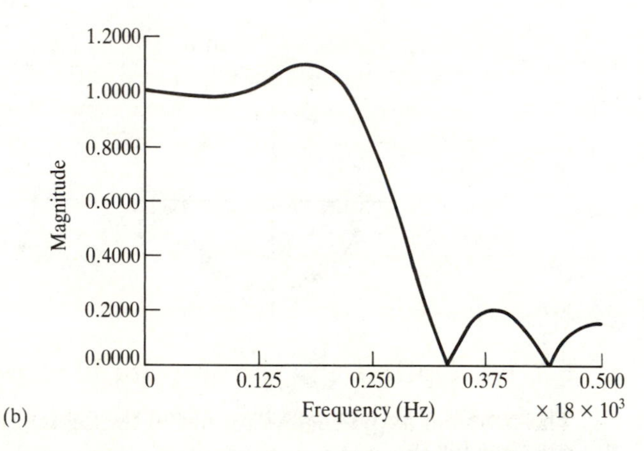

Figure 6.18 (a) Ideal frequency response showing sampling points. (b) Frequency response of frequency sampling filter.

$$= \frac{1}{N} \sum_{k=0}^{N-1} |H(k)| \cos [2\pi k(n - \alpha)/N] + \mathrm{j} \sin [2\pi k(n - \alpha)/N]$$

$$= \frac{1}{N} \sum_{k=0}^{N-1} |H(k)| \cos [2\pi k(n - \alpha)/N] \quad \textbf{(6.20)}$$

since $h(n)$ is entirely real. For the important case of linear phase $h(n)$ will be symmetrical and so we can write

$$h(n) = \frac{1}{N} \left[\sum_{k=1}^{N/2-1} 2|H(k)| \cos [2\pi k(n - \alpha)/N] + H(0) \right] \quad \textbf{(6.21)}$$

For N odd, the upper limit in the summation is $(N - 1)/2$.

(2) The ideal frequency response is depicted in Figure 6.18(a). The frequency samples are taken at intervals of kF_s/N, that is at intervals of $18/9 = 2$ kHz. Thus the frequency samples are given by

$$\begin{aligned} |H(k)| &= 1 \quad k = 0, 1, 2 \\ &\;\;\;\, 0 \quad k = 3, 4 \end{aligned}$$

Table 6.10 Nonrecursive coefficients for the FIR filter of Example 6.9.

h[0] =	7.2522627e−02	= h[8]
h[1] =	−1.1111111e−01	= h[7]
h[2] =	−5.9120987e−02	= h[6]
h[3] =	3.1993169e−01	= h[5]
h[4] =	5.5555556e−01	= h[4]

Using Equation 6.21 with a limit of $(N-1)/2$ and the frequency samples we obtain the impulse response coefficients (see Table 6.10).

The PC disk for this book contains a program to compute the FIR coefficients given the values of the frequency samples. The frequency response for the filter is shown in Figure 6.18(b). It is seen that the filter has a poor amplitude response, caused by the abrupt change from the passband (where $|H(k)| = 1$) to the stopband (where $|H(k)| = 0$).

6.7.1.1 Optimizing the amplitude response

The problem above is akin to that of the rectangular window. We recall that, in the case of the window method, we can trade off wider transition width for improved amplitude response. To improve the amplitude response of the frequency sampling filters, at the expense of wider transition, we can introduce frequency samples in the transition band. Figure 6.19 illustrates a typical specification for a lowpass filter with three transition band frequency samples. For a lowpass filter, the stopband attenuation increases, approximately, by 20 dB for each transition band frequency sample (Rabiner *et al.*, 1970), with a corresponding increase in the transition width:

approximate stopband attenuation	$(25 + 20M)$ dB
approximate transition width	$(M + 1)F_s/N$

where M is the number of transition band frequency samples and N is the filter length.

The values of the transition band frequency samples that will give the optimum stopband attenuation are determined by an optimization process (Rabiner *et al.*, 1970). A useful optimization objective is to find values of the transition band frequency samples, $T_1, T_2, \ldots, T_M$ that minimize the peak stopband ripple (that is, they maximize the stopband attenuation). Mathematically, this may be stated as:

$$\underset{\{T_1, T_2, \ldots, T_M\}}{\text{minimize}} \left[\underset{\{\omega \text{ in the stopband}\}}{\max} |W[H_D(\omega) - H(\omega)]| \right] \tag{6.22}$$

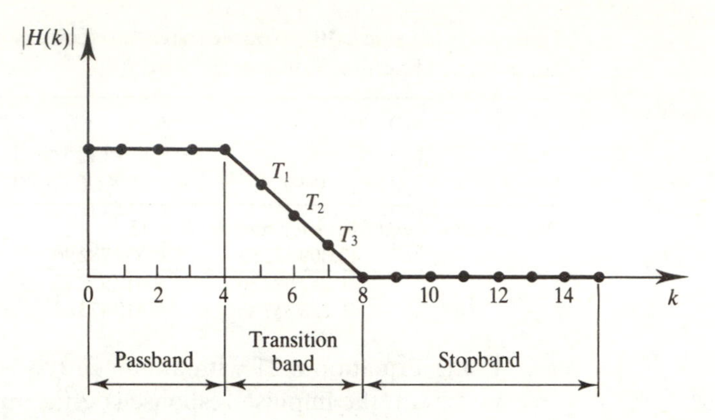

Figure 6.19 Lowpass filter frequency samples including three transition band samples. Note: because of the symmetry in the amplitude response only one half of the filter response is shown.

where $H_D(\omega)$ and $H(\omega)$ are, respectively, the ideal and actual frequency responses of the filter; W is a weighting factor.

Rabiner *et al.* (1970) have provided a table of optimal (in the sense of Equation 6.22) values of transition band frequency samples which is widely used. A sample of the optimal values of transition band frequency samples is given in Table 6.11 for $N = 15$. In the table, the bandwidth refers to the number of frequency samples in the passband of the filter.

In most cases, the values of the transition band frequency samples normally lie in following ranges: for one transition frequency sample,

$$0.250 < T_1 < 0.450$$

for two transition frequency samples,

$$0.040 < T_1 < 0.150$$

$$0.450 < T_2 < 0.650$$

for three transition frequency samples,

$$0.003 < T_1 < 0.035$$

$$0.100 < T_2 < 0.300$$

$$0.550 < T_3 < 0.750$$

The lower values are for filters with wide bandwidth and lead to more stopband attenuation.

Table 6.11 Optimum transition band frequency samples for type 1 lowpass frequency sampling filters for $N = 15$ (adapted from Rabiner *et al.*, 1970).

BW	*Stopband attenuation* (dB)	T_1	T_2	T_3
One transition band frequency sample, $N = 15$				
1	42.309 322 83	0.433 782 96		
2	41.262 992 86	0.417 938 23		
3	41.253 337 86	0.410 473 63		
4	41.949 077 13	0.404 058 84		
5	44.371 245 38	0.392 681 89		
6	56.014 165 88	0.357 665 25		
Two transition band frequency samples, $N = 15$				
1	70.605 405 85	0.095 001 22	0.589 954 18	
2	69.261 681 56	0.103 198 24	0.593 571 18	
3	69.919 734 95	0.100 836 18	0.589 432 70	
4	75.511 722 56	0.084 074 93	0.557 153 12	
5	103.460 783 00	0.051 802 06	0.499 174 24	
Three transition band frequency samples,				
1	94.611 661 91	0.014 550 78	0.184 578 82	0.668 976 13
2	104.998 130 80	0.010 009 77	0.173 607 13	0.659 515 26
3	114.907 193 18	0.008 734 13	0.163 973 10	0.647 112 64
4	157.292 575 84	0.003 787 99	0.123 939 63	0.601 811 54

BW refers to the number of frequency samples in the passband.

Example 6.10

(1) A linear phase 15-point FIR filter is characterized by the following frequency samples:

$$|H(k)| = 1 \quad k = 0, 1, 2, 3$$
$$0 \quad k = 4, 5, 6, 7$$

Assuming a sampling frequency of 2 kHz, obtain its frequency response.

(2) Compare the frequency response of the filter if (a) one transition band frequency sample is used, (b) two transition band frequency samples are used, and (c) three transition band frequency samples are used.

Solution

(1) With the frequency samples as input to the design program fresamp.c (see the appendix), the coefficients of the filter are given in column 2 of Table 6.12. The corresponding frequency response is given in Figure 6.20(a).

(2) For case (a) the value of the transition band frequency sample, from

Table 6.12 Nonrecursive filter coefficients for various transition bands frequency samples.

	No transition samples	*One transition sample*	*Two transition samples*	*Three transition samples*
h[0] =	−4.9815884e−02	−1.3766696e−02	−5.7195305e−03	−4.2282741e−03
h[1] =	4.1202267e−02	−2.3832554e−03	−7.6781827e−03	−7.6031627e−03
h[2] =	6.6666666e−02	3.9729333e−02	2.3920000e−02	1.8793332e−02
h[3] =	−3.6487877e−02	1.2729081e−02	2.5763613e−02	2.8145113e−02
h[4] =	−1.0786893e−01	−9.1220745e−02	−7.3701817e−02	−6.6396840e−02
h[5] =	3.4078020e−02	−1.8619356e−02	−4.4185450e−02	−5.2511978e−02
h[6] =	3.1889241e−01	3.1326097e−01	3.0552137e−01	3.0183514e−01
h[7] =	4.6666667e−01	5.2054133e−01	5.5216000e−01	5.6393334e−01

Because of symmetry, only the first half of the coefficients are listed here.

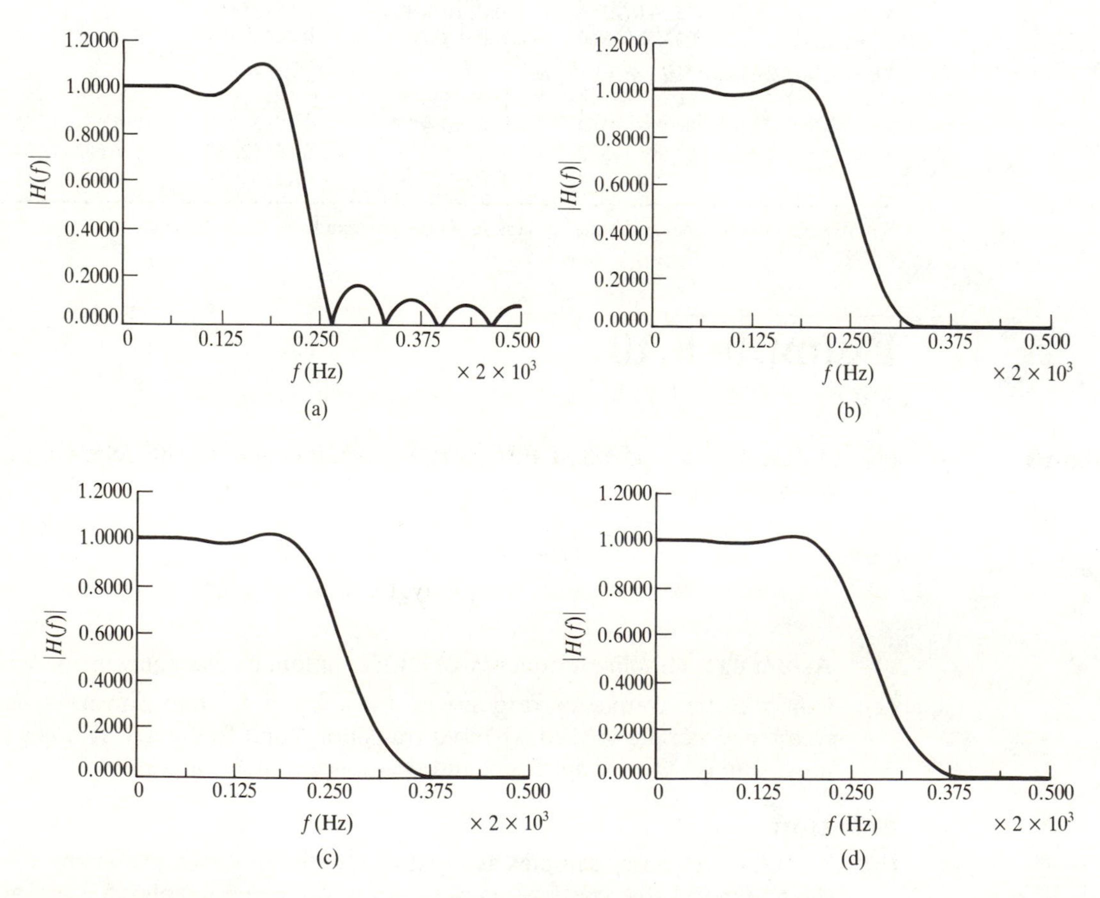

Figure 6.20 Frequency response of frequency sampling filter with (a) no transition band samples; (b) one transition band sample; (c) two transition band frequency samples and (d) three transition band frequency samples.

Table 6.11, is 0.4041. Thus, the frequency samples for the filter are:

$$
\begin{aligned}
|H(k)| &= 1 && k = 0, 1, 2, 3 \\
& \;\; 0.40406 && k = 4 \\
& \;\; 0 && k = 5, 6, 7
\end{aligned}
$$

With these frequency samples as input to the design program, the coefficients of the filter were computed and are summarized in Table 6.12. The corresponding frequency response is given in Figure 6.20(b).

For cases (b) and (c), the frequency samples are defined respectively as

$$
\begin{aligned}
|H(k)| &= 1 && k = 0, 1, 2, 3 \\
& \;\; 0.5571 && k = 4 \\
& \;\; 0.0841 && k = 5 \\
& \;\; 0 && k = 6, 7
\end{aligned}
$$

$$
\begin{aligned}
|H(k)| &= 1 && k = 0, 1, 2, 3 \\
& \;\; 0.6018 && k = 4 \\
& \;\; 0.1239 && k = 5 \\
& \;\; 0.0038 && k = 6 \\
& \;\; 0 && k = 7
\end{aligned}
$$

The coefficients for these cases are summarized in the fourth and fifth columns of Table 6.12. The corresponding frequency responses are given in Figures 6.20(c) and 6.20(d). It is seen that, as the number of transition band frequency samples increases, the amplitude response (in terms of the passband and stopband ripples) improves, but at the expense of increasing transition width or roll-off.

An alternative approach that may be used to improve the amplitude response is to obtain a large number of frequency samples, by sampling at closer intervals, to compute the impulse response using Equation 6.21, and then to apply one of the window functions discussed earlier to reduce the filter to the desired length.

Example 6.11

Obtain the coefficients of the filter given in Example 6.10, using the Blackman window function. In this case let us take 80 frequency samples:

Table 6.13 Impulse response coefficients using Blackman window ($N = 15$, Example 6.11).

h[0] =	−4.5997579e−19	= h[14]
h[1] =	8.8547181e−04	= h[13]
h[2] =	−1.1306679e−03	= h[12]
h[3] =	−1.8677815e−02	= h[11]
h[4] =	−2.3861187e−02	= h[10]
h[5] =	7.3871142e−02	= h[9]
h[6] =	2.7492398e−01	= h[8]
h[7] =	3.8750000e−01	= h[7]

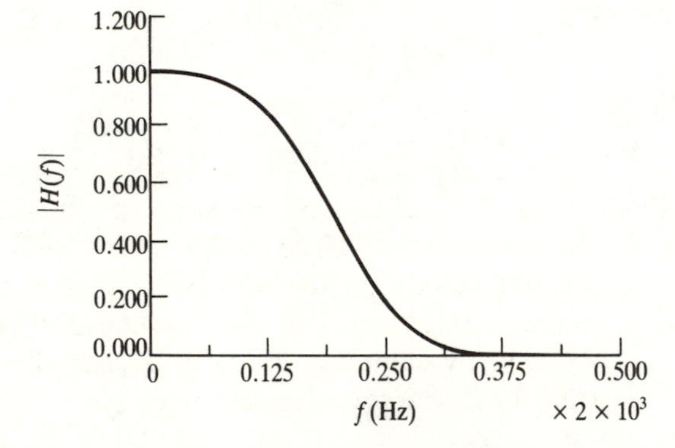

Figure 6.21 An attempt to use a Blackman window to improve the frequency response (Example 6.11).

$$|H(k)| = 1 \quad k = 0, 1, \ldots, 16$$
$$0 \quad k = 17, 12, \ldots, 39$$

Solution

Using the program in the appendix, 80 impulse response coefficients were obtained and a Blackman window of length 15 was then applied to give the coefficients listed in Table 6.13. The corresponding frequency response is given in Figure 6.21.

6.7.2 Recursive frequency sampling filters

Recursive forms of the frequency sampling filter offer significant computational advantages over the nonrecursive forms if a large number of frequency samples are zero valued. It can be shown (see Example 6.12) that the transfer function of an FIR filter, $H(z)$, can be expressed in a recursive form:

$$H(z) = \frac{1 - z^{-N}}{N} \sum_{k=0}^{N-1} \frac{H(k)}{1 - \mathrm{e}^{\mathrm{j}2\pi k/N} z^{-1}} = H_1(z) H_2(z) \tag{6.23}$$

where

$$H_1(z) = \frac{1 - z^{-N}}{N}$$

$$H_2(z) = \sum_{k=0}^{N-1} \frac{H(k)}{1 - \mathrm{e}^{\mathrm{j}2\pi k/N} z^{-1}}$$

Thus we see that, in recursive form, $H(z)$ can be viewed as a cascade of two filters: a comb filter, $H_1(z)$, which has N zeros uniformly distributed around the unit circle, and a sum of N single all-pole filters, $H_2(z)$. The zeros of the comb filter and the poles of the single pole filters are coincident on the unit circle at points $z_k = \mathrm{e}^{2\pi k/N}$. Thus the zeros cancel the poles, making $H(z)$ an FIR as it effectively has no poles.

In practice, finite wordlength effects cause the poles of $H_2(z)$ not to be located exactly on the unit circle so that they are not cancelled by the zeros, making $H(z)$ an IIR and potentially unstable. Stability problems can be avoided by sampling $H(z)$ at a radius, r, slightly less than unity. Thus the transfer function in this case becomes

$$H(z) = \frac{1 - r^N z^{-N}}{N} \sum_{k=0}^{N-1} \frac{H(k)}{1 - r\mathrm{e}^{\mathrm{j}2\pi k/N} z^{-1}} \tag{6.24}$$

In general, the frequency samples, $H(k)$, are complex. Thus a direct implementation of Equation 6.23 or 6.24 would require complex arithmetic. To avoid this complication, we make use of the symmetry inherent in the frequency response of any FIR filter with real impulse response, $h(n)$. For the standard frequency selective linear phase filters (positive symmetrical impulse response), it can be shown (see Example 6.12) that Equation 6.24 can be expressed as

$$H(z) = \frac{1 - r^N z^{-N}}{N} \times \left[\sum_{k=1}^{M} \frac{|H(k)| \{2\cos(2\pi k\alpha/N) - 2r\cos[2\pi k(1+\alpha)/N] z^{-1}\}}{1 - 2r\cos(2\pi k/N) z^{-1} + r^2 z^{-2}} + \frac{H(0)}{1 - z^{-1}} \right] \tag{6.25}$$

where $\alpha = (N-1)/2$. For N odd $M = (N-1)/2$ and for N even $M = N/2 - 1$. The realization diagram for Equation 6.25 is depicted in Figure 6.22.

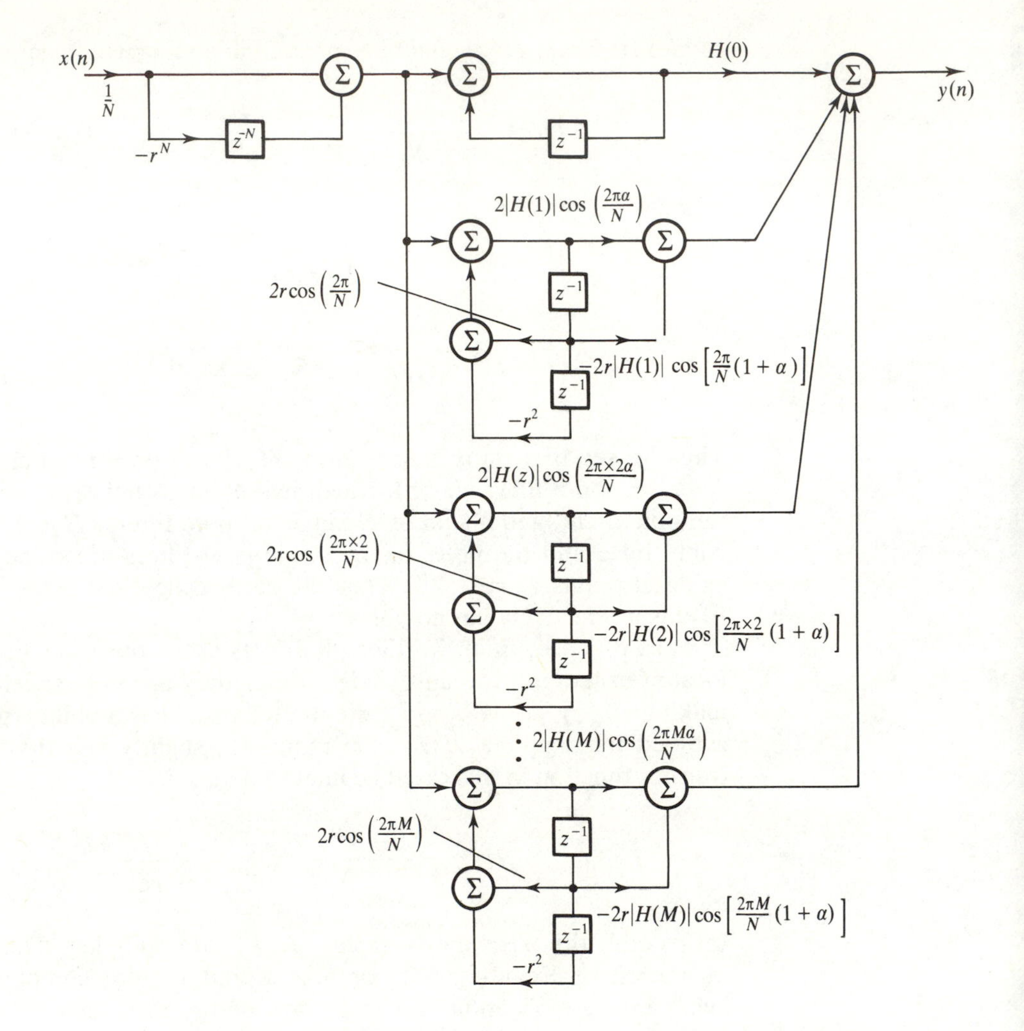

Figure 6.22 Realization diagram for the frequency sampling filter.

6.7.3 Frequency sampling filters with simple coefficients

Recursive implementation of FIR filters greatly reduces the number of arithmetic operations in digital filters. If a filter in addition has coefficients that are simple integers (or powers of 2) its computational efficiency is greatly improved, making it attractive in applications where processors with primitive arithmetic operations, such as ordinary microprocessors, are used. Lynn (1975) has developed a family of frequency sampling filters with small integer coefficients.

However, integer coefficients are only possible if some restrictions are placed on the locations of the poles of the transfer function (Equation 6.25). Equivalently, the passbands of the filters with integer coefficients can only be centred at restricted frequencies. Note that since the coefficients are integers we can place the poles on the unit circle and obtain a perfect cancellation. These filters are a special case of the frequency sampling filters.

Example 6.12

(1) The transfer function of an FIR filter is defined as

$$H(z) = \sum_{n=0}^{N-1} h(n)z^{-n} \tag{6.26a}$$

Starting with the above equation, show that $H(z)$ for a linear phase FIR filter, with positive symmetrical impulse response, can be expressed in the following recursive form:

$$H(z) = \frac{1 - r^N z^{-N}}{N} \times \left[\sum_{k=1}^{M} \frac{|H(k)|\{2\cos(2\pi k\alpha/N) - 2r\cos[2\pi k(1+\alpha)/N]z^{-1}\}}{1 - 2r\cos(2\pi k/N)\, z^{-1} + r^2 z^{-2}} + \frac{H(0)}{1 - z^{-1}} \right]$$

where $\alpha = (N-1)/2$ and $H(k)$ are samples of the frequency response of the filter, taken at intervals of kF_s/N.

(2) A requirement exists for a lowpass filter satisfying the following specifications:

passband	0–4 kHz
sampling frequency	18 kHz
filter length	9

Find the transfer function of the filter, in recursive form, using the frequency sampling method. Assume a radius $r = 1$. Draw the realization diagram and compare the computational complexities with direct form FIR.

Solution

(1) The impulse response of the filter may be defined in terms of the frequency samples:

$$h(n) = \frac{1}{N} \sum_{k=0}^{N-1} H(k)\, r^n e^{j2\pi nk/N} \quad k = 0, 1, \ldots, N-1, r \leqslant 1 \tag{6.26b}$$

Using Equation 6.26b in Equation 6.26a the transfer function, $H(z)$, becomes

$$H(z) = \sum_{n=0}^{N-1} h(n)z^{-n} = \sum_{n=0}^{N-1}\left[\frac{1}{N}\sum_{k=0}^{N-1} H(k)r^n e^{j2\pi nk/N}\right]z^{-n}$$

Interchanging the order of the two summations we have

$$H(z) = \frac{1}{N}\sum_{k=0}^{N-1} H(k)\left\{\sum_{n=0}^{N-1}[re^{j(2\pi k/N)}z^{-1}]^n\right\} \tag{6.27}$$

Now the finite geometric series can be expressed as

$$S_N = \sum_{n=0}^{N-1} \delta^n = \frac{1-\delta^N}{1-\delta} \quad \delta \neq 1$$

In our case, with $\delta = re^{j2\pi k/N}z^{-1}$, we can write

$$\sum_{n=0}^{N-1}[re^{j(2\pi k/N)}z^{-1}]^n = \frac{1-(re^{j2\pi k/N}z^{-1})^N}{1-re^{j2\pi k/N}z^{-1}} = \frac{1-r^N e^{j2\pi k}z^{-N}}{1-re^{j2\pi k/N}z^{-1}}$$

$$= \frac{1-r^N z^{-N}}{1-re^{j2\pi k/N}z^{-1}}$$

since $e^{j2\pi k} = \cos(2\pi k) = 1$, $k = 0, 1, \ldots$. Thus we can write Equation 6.27 as

$$H(z) = \frac{1-r^N z^{-N}}{N}\sum_{k=0}^{N-1}\frac{H(k)}{1-re^{j2\pi k/N}z^{-1}} = H_1(z)H_2(z) \tag{6.28}$$

where

$$H_1(z) = \frac{1-r^N z^{-N}}{N}$$

$$H_2(z) = \sum_{k=0}^{N-1}\frac{H(k)}{1-re^{j2\pi k/N}z^{-1}}$$

Now $H_2(z)$, on expansion, has the form:

$$H_2(z) = \frac{H(0)}{1-rz^{-1}} + \frac{H(1)}{1-re^{j2\pi/N}z^{-1}} + \frac{H(2)}{1-re^{j2\pi 2/N}z^{-1}}$$

$$+ \ldots + \frac{H(N-2)}{1-re^{j2\pi(N-2)/N}z^{-1}} + \frac{\mathrm{H(N-1)}}{1-re^{j2\pi(N-1)/N}z^{-1}}$$

For a filter with real coefficients, the following symmetry conditions hold:

$$H(k) = H^*(N - k),\ e^{j2\pi(N-k)/N} = e^{-j2\pi k/N}$$

Thus we can write $H_2(z)$ as

$$H_2(z) = \frac{H(0)}{1 - rz^{-1}} + \frac{H(1)}{1 - re^{j2\pi/N}z^{-1}} + \frac{H(2)}{1 - re^{j2\pi 2/N}z^{-1}}$$

$$+ \ldots + \frac{H^*(2)}{1 - re^{-j2\pi 2/N}z^{-1}} + \frac{H^*(1)}{1 - re^{-j2\pi/N}z^{-1}}$$

Thus the poles occur in complex conjugate pairs (except the one at $k = 0$, for N odd, and the ones at $k = 0$ and $k = N/2$, for N even). For linear phase filters of even length, $H(N/2) = 0$. Combining the kth single-pole section and its conjugate we have

$$\frac{H(k)}{1 - re^{j2\pi k/N}z^{-1}} + \frac{H^*(k)}{1 - re^{-j2\pi k/N}z^{-1}}$$

$$= \frac{H(k)(1 - re^{-j2\pi k/N}z^{-1}) + H^*(k)(1 - re^{j2\pi k/N}z^{-1})}{(1 - re^{j2\pi k/N}z^{-1})(1 - re^{-j2\pi k/N}z^{-1})} \quad \textbf{(6.29)}$$

The denominator simplifies to

$$(1 - re^{j2\pi k/N}z^{-1})(1 - re^{-j2\pi k/N}z^{-1}) = 1 - 2r\cos(2\pi k/N)z^{-1} + r^2z^{-2} \quad \textbf{(6.30)}$$

For a linear phase filter, with positive symmetrical impulse response, $H(k)$ is given by

$$H(k) = |H(k)|e^{-j2\pi k\alpha/N}$$

where $\alpha = (N - 1)/2$. Thus the numerator may be simplified:

$$|H(k)|e^{-j2\pi k\alpha/N}(1 - re^{-j2\pi k/N}z^{-1}) + |H(k)|e^{j2\pi k\alpha/N}(1 - re^{j2\pi k/N}z^{-1})$$
$$= |H(k)|[e^{-j2\pi k\alpha/N}(1 - re^{-j2\pi k/N}z^{-1}) + e^{j2\pi k\alpha/N}(1 - re^{j2\pi k/N}z^{-1})]$$
$$= |H(k)|(e^{-j2\pi k\alpha/N} - re^{-j2\pi k\alpha/N}e^{-j2\pi k/N}z^{-1} + e^{j2\pi k\alpha/N}$$
$$- re^{j2\pi k\alpha/N}e^{-j2\pi k/N}z^{-1})$$
$$= |H(k)|\{2\cos(2\pi k\alpha/N) - [re^{-j2\pi k(1+\alpha)/N}z^{-1} + re^{j2\pi k(1+\alpha)}z^{-1}]\}$$
$$= |H(k)|\{2\cos(2\pi k\alpha/N) - 2r\cos[2\pi k(1 + \alpha)/N]z^{-1}\} \quad \textbf{(6.31)}$$

Combining Equations 6.30 and 6.31 we can write $H(z)$ as:

$$H(z) = \frac{1 - r^N z^{-N}}{N}\left[\sum_{k=1}^{M} \frac{|H(k)|\{2\cos(2\pi k\alpha/N) - 2r\cos[2\pi k(1+\alpha)/N]z^{-1}\}}{1 - 2r\cos(2\pi k/N)z^{-1} + r^2 z^{-2}} + \frac{H(0)}{1 - rz^{-1}}\right] \tag{6.32a}$$

For N odd $M = (N-1)/2$ and for N even $M = (N/2) - 1$.

(2) With $N = 9$, we will sample the response at intervals of $18/9 = 2$ kHz. Thus the frequency samples are defined as

$$|H(k)| = 1 \quad k = 0, 1, 2$$
$$0 \quad k = 3, 4$$

In this case, $\alpha = (N-1)/2 = (9-1)/2 = 4$ and $r = 1$.

From Equation 6.32a and using the values of the frequency samples above $H(z)$ becomes

$$H(z) = \frac{1 - z^{-9}}{9}\left\{\frac{2|H(1)|[\cos(2\pi 4/9) - \cos(2\pi 5/9)z^{-1}]}{1 - 2\cos(2\pi/9)z^{-1} + z^{-2}} + \frac{2|H(2)|[\cos(2\pi \times 2 \times 4/9) - \cos(2\pi \times 2 \times 5/9)z^{-1}]}{1 - 2z^{-1}\cos(4\pi/9) + z^{-2}} + \frac{1}{1 - z^{-1}}\right\}$$

Now $\cos(8\pi/9) = -0.9397$, $\cos(10\pi/9) = -0.9397$, $\cos(2\pi/9) = 0.7660$, $\cos(16\pi/9) = 0.7660$, $\cos(20\pi/9) = 0.7660$ and $\cos(4\pi/9) = 0.1736$. Substituting these values into the equation above, we have

$$H(z) = \frac{1 - z^{-9}}{9}\left[\frac{2(-0.9397 + 0.9397z^{-1})}{1 - 2 \times 0.7660z^{-1} + z^{-2}} + \frac{2(0.7660 - 0.7660z^{-1})}{1 - 2 \times 0.1736z^{-1} + z^{-2}} + \frac{1}{1 - z^{-1}}\right]$$

$$= \frac{1 - z^{-9}}{9}\left[\frac{-1.8794(1 - z^{-1})}{1 - 1.5320\,z^{-1} + z^{-2}} + \frac{1.5320(1 - z^{-1})}{1 - 0.3472\,z^{-1} + z^{-2}} + \frac{1}{1 - z^{-1}}\right]$$

The realization diagram is given in Figure 6.23. The computational complexities for both direct and frequency sampling filters are summarized below.

	Number of additions	Number of multiplications	Storage
direct	8	9	18
frequency sampling	10	7	25

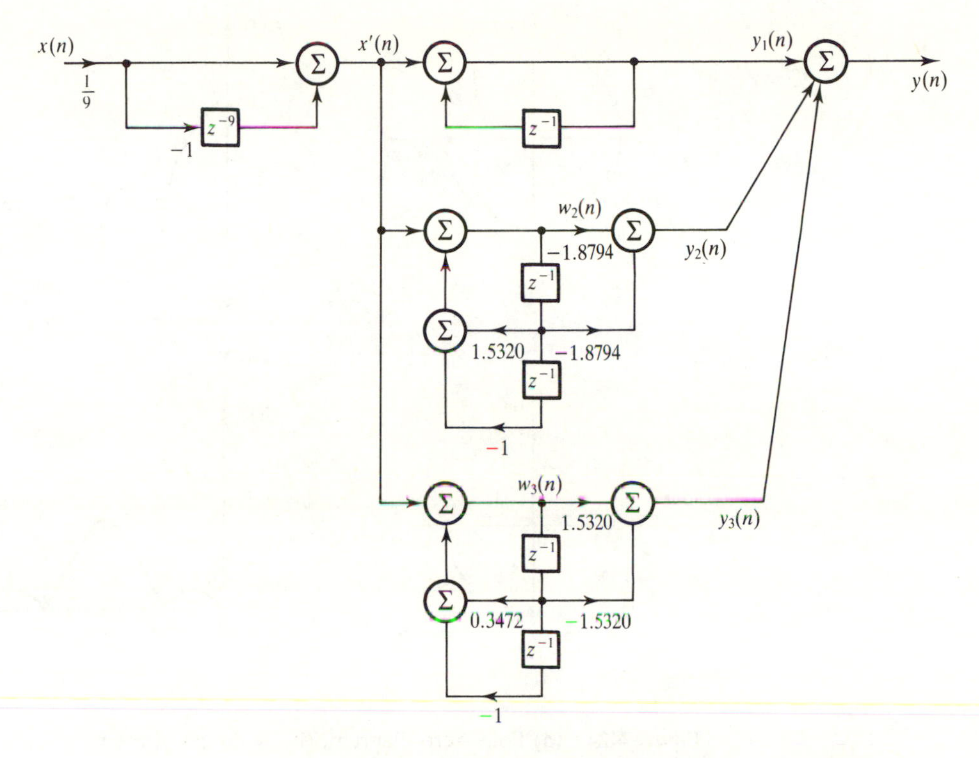

Figure 6.23 Realization diagram for the frequency sampling filter of Example 6.12.

With respect to Figure 6.23, the difference equations are

$$x'(n) = 1/9[x(n) - x(n-9)]$$

$$y_1(n) = x'(n) + y(n-1)$$

$$w_2(n) = 1.5320w_2(n-1) - w_2(n-2) + x'(n)$$

$$y_2(n) = -1.8794w(n) + 1.8794w_2(n-1) \tag{6.32b}$$

$$w_3(n) = 0.3472w_3(n-1) - w_3(n-2) + x'(n)$$

$$y_3(n) = 1.5320w_3(n) - 1.5320w_3(n-1)$$

$$y(n) = y_1(n) + y_2(n) + y_3(n)$$

Example 6.13

Obtain the transfer function and difference equation for

(1) a recursive FIR lowpass filter with simple integer coefficients meeting the

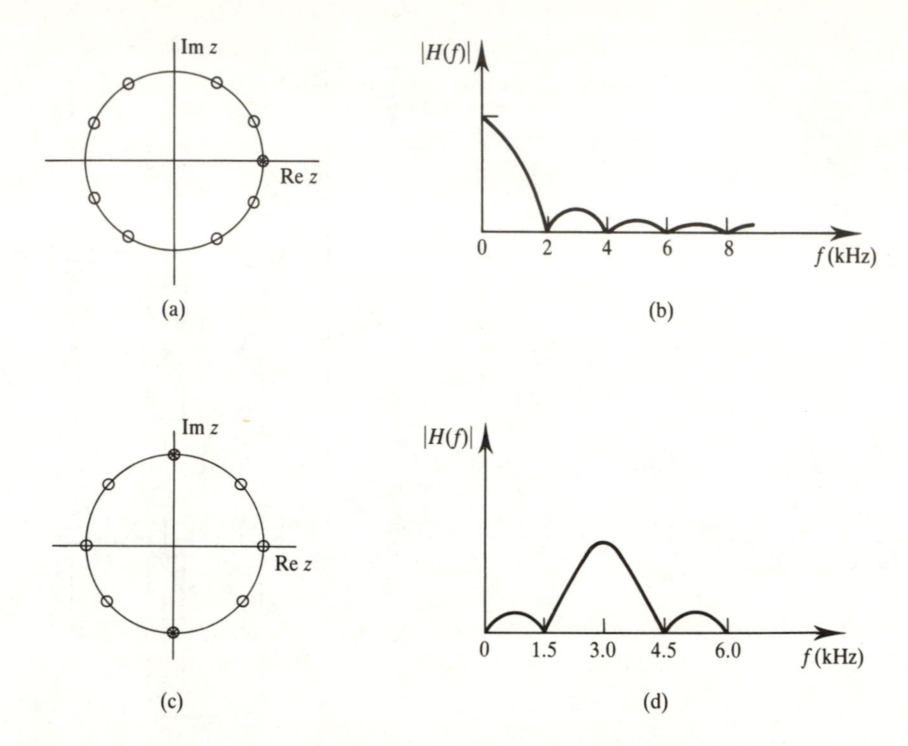

Figure 6.24 (a) Pole–zero diagram. (b) A sketch of the magnitude response of a recursive FIR filter with integer coefficients. (c) Pole–zero diagram for a simple bandpass filter with integer coefficients. (d) The corresponding magnitude response.

following specifications:

centre frequency	0 Hz
sampling frequency	18 kHz

(2) a recursive FIR bandpass filter with simple integer coefficients meeting the following specifications:

centre frequency	3 kHz
sampling frequency	12 kHz

Solution

(1) If we set $N = 9$, the interval between frequency samples is $18/9 = 2$ kHz. The pole–zero diagram for this case is shown in Figure 6.24(a) and the corresponding magnitude response in Figure 6.24(b). From Figure 6.24(a), the transfer function is

$$H(z) = \frac{1 - z^{-9}}{9} \frac{1}{1 - z^{-1}}$$

The corresponding difference equation is

$$y(n) = y(n-1) + 1/9[x(n) - x(n-9)]$$

(2) Since the passband is centred at 3 kHz, we must choose the sampling instants carefully to ensure this is the case. Assuming $N = 8$, the z-plane diagram and the corresponding magnitude response for one possible sampling grid are shown in Figures 6.24(c) and 6.24(d) respectively.

With $N = 8$, the sampling interval is $12/8 = 1.5$ kHz. The transfer function is

$$H(z) = \frac{1 - z^{-8}}{8} \frac{1}{1 + z^{-2}}$$

The corresponding difference equation is

$$y(n) = -y(n-2) + 1/8[x(n) - x(n-8)]$$

It is apparent that the determination of the transfer function for frequency sampling with integer coefficients is a very simple process. However, the amplitude response of such filters is often poor and the designer is restricted as to where the passband can be located. To improve the attenuation and cutoff frequency characteristics of these filters, the transfer function can be raised to an integer value (Lynn, 1973, 1975).

6.7.4 Summary of the frequency sampling method

- *Step 1.* Specify the ideal or desired frequency response, the stopband attenuation and band edge frequencies of the target filter.
- *Step 2.* From the specification select a type 1 frequency sampling filter, where frequency samples are taken at intervals of kF_s/N, or a type 2 frequency sampling filter, where frequency samples are taken at intervals of $(k + 1/2)F_s/N$.
- *Step 3.* Use the specification in step 1 and the design tables (Rabiner *et al.*, 1970) to determine N, the number of frequency samples of the ideal frequency response, M, the number of transition band frequency samples, BW, the number of frequency samples in the passband, and T_i, the values of the transition band frequency samples ($i = 1, 2, \ldots, M$).
- *Step 4.* Use the appropriate equation to calculate the filter coefficients.

6.7.5 Comparison of the window, optimum and frequency sampling methods

The optimum method provides an easy and efficient way of computing FIR filter coefficients. Although the method provides total control of filter specifications, the availability of the optimal filter design software is mandatory. For most applications the optimal method will yield filters with good amplitude response characteristics for reasonable values of N. The method is particularly good for designing Hilbert transformers and differentiators. Other methods will yield larger approximation errors for differentiators and Hilbert transformers than the optimal method.

In the absence of the optimal software or when the passband and stopband ripples are equal, the window method represents a good choice. It is a particularly simple method to apply and conceptually easy to understand. However, the optimal method will often give a more economic solution in terms of the number of filter coefficients. The window method does not allow the designer a precise control of the cutoff frequencies or ripples in the passband and stopband.

The frequency sampling approach is the only method that allows both nonrecursive and recursive implementation of FIR filters, and should be used when such implementations are envisaged as the recursive approach is computationally economical. The special form with integer coefficients should be considered only when primitive arithmetic and programming simplicity are vital (for example assembly language programming in a standard microprocessor), but a check should always be made to see whether its poor amplitude response is acceptable. Filters with arbitrary amplitude–phase response can be readily designed by the frequency sampling method. The frequency sampling method lacks precise control of the location of the band edge frequencies or the passband ripples and relies on the availability of the design table of Rabiner *et al*.

Example 6.14

Two linear phase FIR bandpass filters are required to satisfy the following specifications: for filter 1,

passband	8–12 kHz
stopband ripple	0.001
peak passband ripple	0.001
sampling frequency	44.14 kHz
transition width	3 kHz

for filter 2,

passband	8–12 kHz
stopband ripple	0.001
peak passband ripple	0.01
sampling frequency	44.14 kHz
transition width	3 kHz

Obtain and compare the frequency response for each filter using

(1) the window method,

(2) frequency sampling method, and

(3) the optimal method.

Solution

(1) *Window method.* For filter 1, from the specifications the passband ripple is $20\log(1+0.001) = 0.008\,68$ dB and the stopband attenuation is $-20\log(0.001) = 60$ dB. From Equations 6.10 and 6.11 the parameters for the Kaiser window are

cutoff frequencies	6.5 kHz, 13.5 kHz
ripple parameter, β	5.653
number of filter coefficients	53
sampling frequency	44.14 kHz

For filter 2, the result is the same as for filter 1 since in the window method the passband and stopband ripples are always approximately equal.

The resulting filter spectra are given in Figure 6.25(a).

(2) *Frequency sampling method.* For filter 1, we assume a type 1 sampling filter, and the filter length, N, is chosen as 53, the same as for the window method. From the design tables (Rabiner *et al.*, 1970), we find that we require two transition band frequency samples to achieve the desired stopband attenuation of 60 dB, with $F_s = 44.14$ kHz, $M = 2$, $N = 53$. Sampling of the ideal frequency response, for $N = 53$, gives

$\lvert H(k)\rvert = 0$	$k = 0, 1, \ldots, 7$
0.10689	$k = 8$
0.59253	$k = 9$
1	$k = 10$–14
0.59253	$k = 15$
0.10689	$k = 16$
0	$k = 17$–26

Using the program fresamp.c (see the appendix), the filter was obtained and the corresponding frequency response is depicted in Figure 6.25(b).

Since the stopband attenuation is the same for both filters, filter 2 is the same as filter 1.

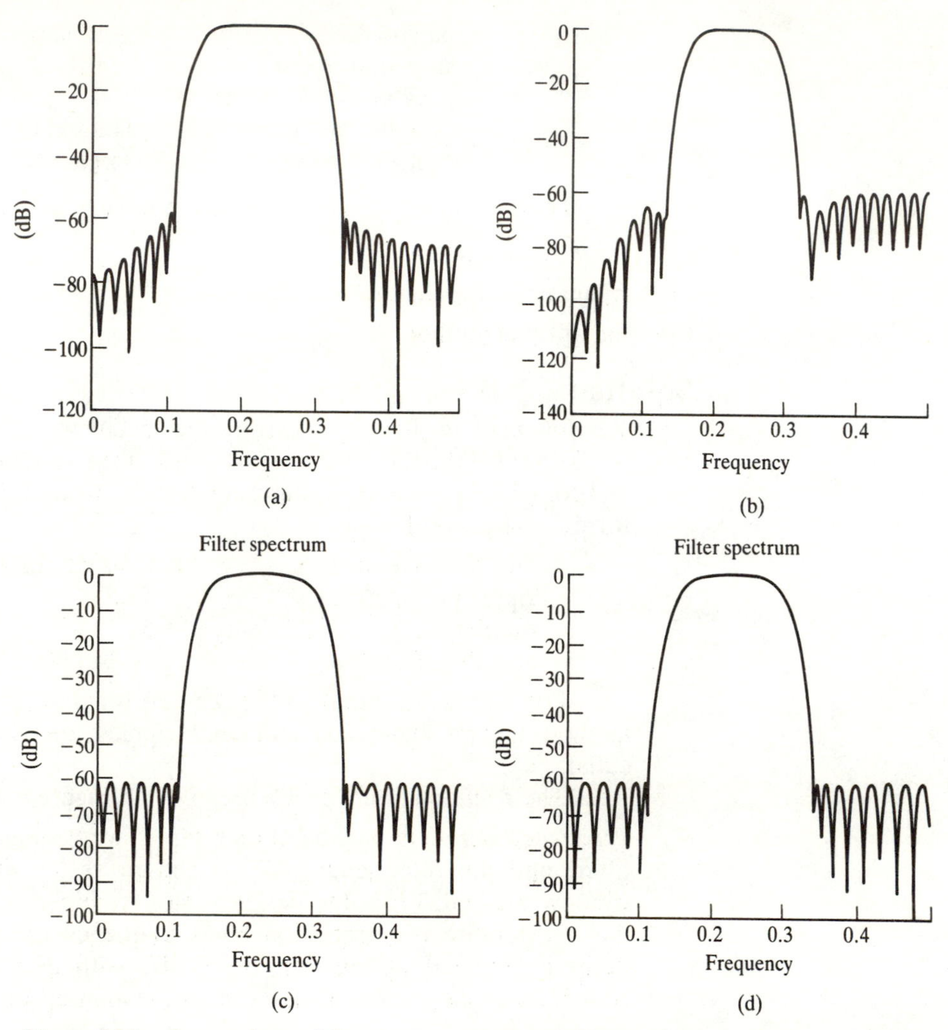

Figure 6.25 Comparison of frequency response of filters using the window, frequency sampling and optimal methods. (a) Filter response using Kaiser window (filter 1 & filter 2). (b) Filter response using frequency sampling method (filter 1 & filter 2). (c) Frequency response of optimal filter 1. (d) Frequency response of optimal filter 2.

(3) *Optimal method.* For filter 1, from the specifications the normalized band edge frequencies are 0, 5/44.14, 8/44.14, 12/44.14, 15/44.14 and 22.07/44.14, that is 0, 0.113 28, 0.181 24, 0.271 86, 0.339 83 and 0.5. Using the program in the appendix, we find $N = 49.6$. Since both the passband and stopband ripples are equal, the weights in the three bands are the same. The input parameters to the optimal design program are

number of filter coefficients	49
band edge frequencies	0, 0.113 28, 0.181 24, 0.271 86, 0.339 83, 0.5
weights	5, 5, 5

The input parameters for filter 2 are

number of filter coefficients	39 (39.45)
band edge frequencies	0, 0.113 28, 0.181 24, 0.271 86, 0.339 83, 0.5
weights	10, 1, 10

The resulting frequency responses for the optimal method are shown in Figures 6.25(c) and 6.25(d).

6.7.6 Special filters and transformations for FIR filters

6.7.6.1 Halfband filters

In some applications, such as multirate processing, a special type of FIR filter known as a halfband filter is required. Causal halfband filters are characterized by an impulse response in which, for N odd, every other coefficient is zero except $h(N-1)/2$:

$$h(2n) = 0 \qquad n = 0, 1, \ldots, (N-1)/2$$
$$0.5 \qquad n = (N-1)/4$$

The passband of a halfband filter normally extends from zero to a quarter of the sampling frequency, i.e half the available bandwidth is passed and the other half rejected.

These filters are readily obtained by the FIR methods described earlier (although with the optimal method some modifications are necessary because of inaccuracies during optimization).

Example 6.15

Obtain the coefficients of an FIR lowpass filter to meet the specifications given below using the window method:

passband edge frequency	2 kHz
transition width	0.5 kHz
stopband attenuation	> 50 dB
sampling frequency	8 kHz

Solution

The filter coefficients are listed in Table 6.14 and the filter spectrum is shown in Figure 6.26(a). It is evident from Table 6.14 that, starting from $h(0)$, every other coefficient is zero (ignoring errors due to numerical inaccuracy in com-

Table 6.14 Coefficients of half-band lowpass filter (Hamming window, $N = 53$, $f_c = 2000$ Hz).

h[0] =	−1.1243421e−09	= h[52]
h[1] =	1.1109516e−03	= h[51]
h[2] =	1.3921496e−09	= h[50]
h[3] =	−1.6473646e−03	= h[49]
h[4] =	−2.0024685e−09	= h[48]
h[5] =	2.6429869e−03	= h[47]
h[6] =	2.9211490e−09	= h[46]
h[7] =	−4.1909615e−03	= h[45]
h[8] =	−4.0967870e−09	= h[44]
h[9] =	6.4068290e−03	= h[43]
h[10] =	5.4636006e−09	= h[42]
h[11] =	−9.4484947e−03	= h[41]
h[12] =	−6.9451110e−09	= h[40]
h[13] =	1.3555871e−02	= h[39]
h[14] =	8.4584215e−09	= h[38]
h[15] =	−1.9134767e−02	= h[37]
h[16] =	−9.9188559e−09	= h[36]
h[17] =	2.6953222e−02	= h[35]
h[18] =	1.1244697e−08	= h[34]
h[19] =	−3.8674295e−02	= h[33]
h[20] =	−1.2361758e−08	= h[32]
h[21] =	5.8666205e−02	= h[31]
h[22] =	1.3207536e−08	= h[30]
h[23] =	−1.0304890e−01	= h[29]
h[24] =	−1.3734705e−08	= h[28]
h[25] =	3.1728215e−01	= h[27]
h[26] =	5.0000000e−01	= h[26]

Note that every other coefficient is zero (numerical errors do not allow these coefficients to be exactly zero).

puting $h(n)$). The implication of this is that during filtering every other input data sample can be ignored (effectively reducing the sampling frequency by a factor of 2).

6.7.6.2 Frequency transformation

In some applications a need may arise to change, in real time, the characteristics of a filter from a lowpass to an equivalent highpass filter. A simple relationship exists between a lowpass and a highpass filter that permits such a change. The coefficients of an FIR highpass filter can be trivially obtained from those of an equivalent lowpass filter by changing the signs of the coefficients as follows:

$$h_{\text{hp}}(n) = (-1)^n h_{\text{lp}}(n)$$

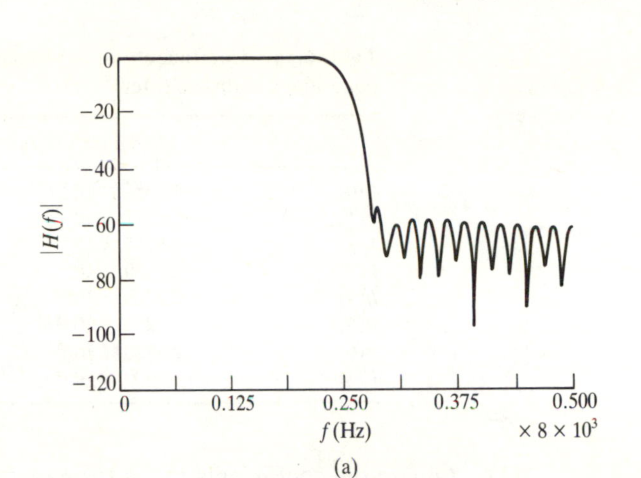

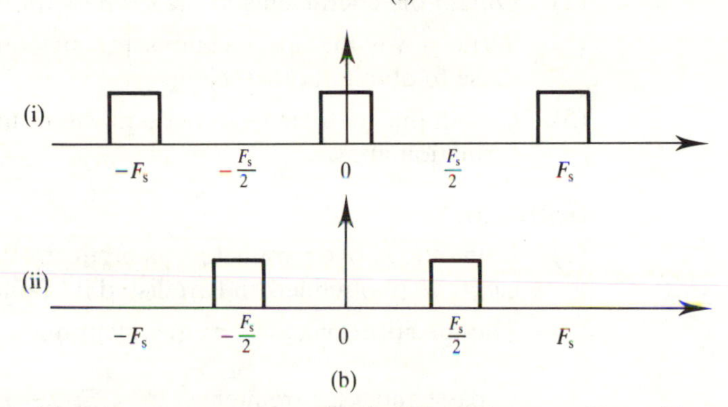

Figure 6.26 (a) Frequency response of a half-band lowpass filter. (b) Frequency response of (i) an ideal lowpass filter and (ii) an equivalent ideal highpass filter.

This relationship is based on the knowledge that the frequency response of a highpass filter is the same as that of a lowpass filter but frequency translated by half the sampling frequency (see Figure 6.26(b)). Thus the frequency response of a highpass filter can be obtained from that of the lowpass by replacing f by $F_s/2 - f$:

$$H_{\text{hp}}(f) = H_{\text{1p}}\left(\frac{F_s}{2} - f\right)$$

Example 6.16

A lowpass filter is characterized by the following:

passband edge frequency	1.5 kHz
sampling frequency	10 kHz
number of coefficients	15

Table 6.15 Coefficients of a lowpass filter and those of an equivalent highpass filter.

	Lowpass	*Highpass*
$h(0)$	1.2654×10^{-3}	1.2654×10^{-3}
$h(1)$	-5.2341×10^{-3}	5.2341×10^{-3}
$h(2)$	-1.9735×10^{-3}	-1.9735×10^{-3}
$h(3)$	-2.3009×10^{-3}	2.3009×10^{-3}
$h(4)$	2.2366×10^{-2}	2.2366×10^{-2}
$h(5)$	1.2833×10^{-1}	-1.2833×10^{-1}
$h(6)$	2.4728×10^{-1}	2.4728×10^{-1}
$h(7)$	3.0000×10^{-1}	-3.0000×10^{-1}

(1) Obtain the coefficients of the lowpass filter using the Hamming window.

(2) Write down the specifications for an equivalent highpass filter and use these to obtain its coefficients.

(3) Obtain the coefficients of the equivalent highpass filter by using the transformation above.

Solution

(1) Using the above parameters as inputs to the program window.c, the coefficients were obtained and are listed in Table 6.15.

(2) The specifications for the equivalent highpass filter are

passband edge frequency	$F_s/2 - f_c = 5000 - 1500\,\text{kHz} = 3500\,\text{kHz}$
sampling frequency	10 kHz
number of coefficients	15

Using these parameters as inputs to the design program window.c, the coefficients for the highpass filter were obtained and are listed in Table 6.15.

(3) Applying the simple transformation above, the coefficients for the highpass filter were obtained and are identical to those obtained in (2).

6.7.7 Other FIR coefficient calculation methods

In some applications, the methods described in this chapter may not be appropriate. For example, in some applications the phase delay introduced by linear phase FIR filters may be unacceptably long (for example, the phase delay for type 1 FIR filters is $(N-1)T/2$ which is large for large N). In a control system, for example, the use of such filters inside a feedback loop could cause instability. In such cases a minimum phase filter may be more appropriate (see Parks and Burrus, 1987).

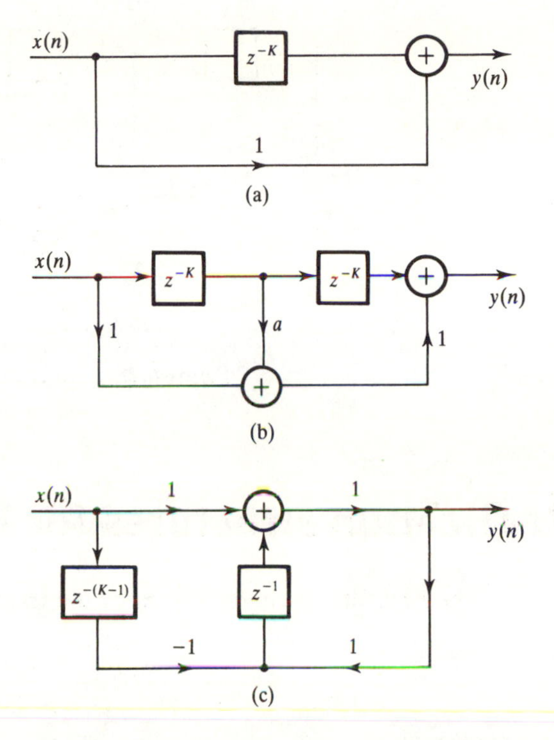

Figure 6.27 Examples of primitive FIR filter sections (a) single adder section ($H(z) = 1 + z^{-k}$); (b) two-adder section, ($H(z) = 1 + az^{-k} + z^{-2k}$; to avoid an explicit multiplication $a = \pm 2^n$, n integer; integer multiplications can be implemented as shifts); (c) recursive running sum lowpass section ($H(z) = (1 - z^{-k})/(1 - z^{-1}) = 1 + z^{-1} + z^{-2} + \dots + z^{-(k-1)}$). A typical filter would consist of 4 to 7 such sections connected in cascade.

The equiripple characteristics of the optimal method can lead to echoes in the impulse response of the filter which may be an undesirable effect. A smooth frequency response characteristic reduces the echoes in the impulse response tails.

In other applications, such as image processing, the number of arithmetic operations when standard FIR filters are used may be too large. Unfortunately, the integer coefficients filters are not suitable for such applications because of their poor amplitude response characteristics. FIR filters which require only very simple arithmetic operations, but whose amplitude responses are comparable with those of standard FIR filters, may be more appropriate (see, for example, Wade *et al.*, 1990).

The basis of the method is to cascade two or more elementary filter sections such as shown in Figure 6.27. Each elementary section involves hardly any multiplications. The main problems with this method include the difficulty of finding an efficient way of selecting the elementary filter sections to cascade and the fact that only filters of low order can be designed efficiently. Genetic algorithms have been employed to tackle the problems (Suckley, 1990).

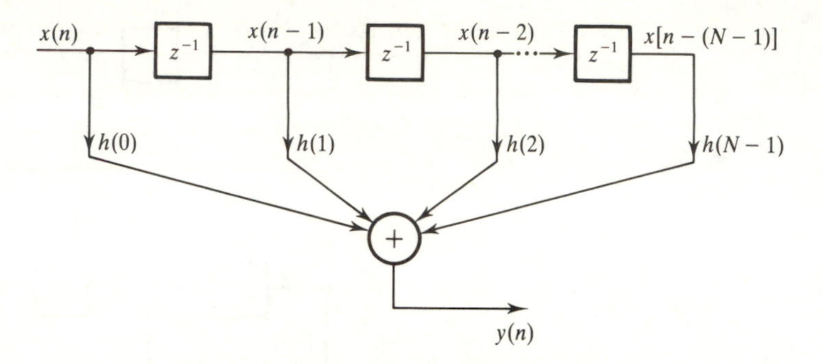

Figure 6.28 Transversal filter structure.

6.8 Realization structures for FIR filters

The FIR filter is characterized by the transfer function, $H(z)$, given by

$$H(z) = \sum_{n=0}^{N-1} h(n)z^{-n}$$

Realization structures are essentially block (or flow) diagram representations of the different theoretically equivalent ways the transfer function can be arranged. In most cases, they consist of an interconnection of multipliers, adders/summers and delay elements. There are many FIR realization structures, but only those that are in common use are presented here.

6.8.1 Transversal structure

The transversal (or tapped delay) structure is depicted in Figure 6.28. The input, $x(n)$, and output of the filter for this structure are related simply by

$$y(n) = \sum_{m=0}^{N-1} h(m)x(n - m) \tag{6.33}$$

In the figure, the symbol z^{-1} represents a delay of one sample or unit of time. Thus $x(n-1)$ is $x(n)$ delayed by one sample. In digital implementations, the boxes labelled z^{-1} could represent shift registers or more commonly memory locations in a RAM. The transversal filter structure is the most popular FIR structure.

The output sample, $y(n)$, is a weighted sum of the present input, $x(n)$, and $N-1$ previous samples of the input, that is $x(n-1)$ to $x(n-N)$. For the transversal structure, the computation of each output sample, $y(n)$, requires

- $N-1$ memory locations to store the $N-1$ input samples,
- N memory locations to store the N coefficients,
- N multiplications, and
- $N-1$ additions.

6.8.2 Linear phase structure

A variation of the transversal structure is the linear phase structure which takes advantage of the symmetry in the impulse response coefficients for linear phase FIR filters to reduce the computational complexity of the filter implementation.

In a linear phase filter, the coefficients are symmetrical, that is $h(n) = \pm h(N-n-1)$. Thus the filter equation can be re-written to take account of this symmetry with a consequent reduction in both the number of multiplications and additions. For type 1 and 2 linear phase filters, the transfer function can be written as

$$H(z) = \sum_{n=0}^{(N-1)/2-1} h(n)[z^{-n} + z^{-(N-1-n)}] + h\left(\frac{N-1}{2}\right)z^{-(N-1)/2} \qquad N \text{ odd} \quad \textbf{(6.34a)}$$

$$H(z) = \sum_{n=0}^{N/2-1} h(n)[z^{-n} + z^{-(N-1-n)}] \qquad N \text{ even} \quad \textbf{(6.34b)}$$

The corresponding difference equations are given by:

$$y(n) = \sum_{k=0}^{(N-1)/2-1} h(k)\{x(n-k) + x[n-(N-1-k)]\} + h[(N-1)/2]x[n-(N-1)/2] \quad \textbf{(6.35a)}$$

$$y(n) = \sum_{k=0}^{(N-1)/2-1} h(k)\{x(n-k) + x[n-(N-1-k)]\} \quad \textbf{(6.35b)}$$

A comparison of Equations 6.33 and 6.35 shows that the linear phase structure is computationally more efficient, requiring approximately half the number of multiplications and additions. However, in most DSP processors Equation 6.33 leads to a more efficient implementation, because the computational advantage in Equation 6.35 is lost in the more complex indexing of data implied.

Example 6.17

A linear phase FIR filter has seven coefficients which are listed below. Draw the realization diagrams for the filter using (a) direct (transversal) and (b) linear phase structures. Compare their computational complexities.

$$
\begin{aligned}
h(0) = h(6) &= -0.032 \\
h(1) = h(5) &= 0.038 \\
h(2) = h(4) &= 0.048 \\
h(3) &= -0.048
\end{aligned}
$$

Solution

The realization diagrams are shown in Figure 6.29.

6.8.3 Other structures

6.8.3.1 Fast convolution

The fast convolution method involves performing the convolution operation of Equation 6.33 in the frequency domain. As was discussed in Chapter 4, convolution in the time domain is equivalent to multiplication in the frequency domain. In simple terms, filtering here is performed by first computing the DFTs of $x(n)$ and $h(n)$ (suitably zero padded), multiplying these together and then obtaining their inverse. The concept is depicted in Figure 6.30. In practice, techniques known as overlap–add and overlap–save are used in real-time filtering. These are discussed in Chapter 4.

6.8.3.2 Frequency sampling structure

In the frequency sampling structure, the filters are characterized by the samples of the desired frequency response, $H(k)$, instead of its impulse response coefficients. This case has already been discussed in detail. For narrowband filters, most of the frequency samples will be zero, and so the resulting frequency sampling filter will require a smaller number of coefficients and hence multiplications and additions than an equivalent transversal structure. A typical realization diagram is given in Figure 6.22.

6.8.3.3 Transpose and cascade structures

The transpose structure is similar to the direct structure, except that the partial sums feed into succeeding stages. This method is more susceptible to roundoff noise than the direct method. In the cascade realization, the transfer function,

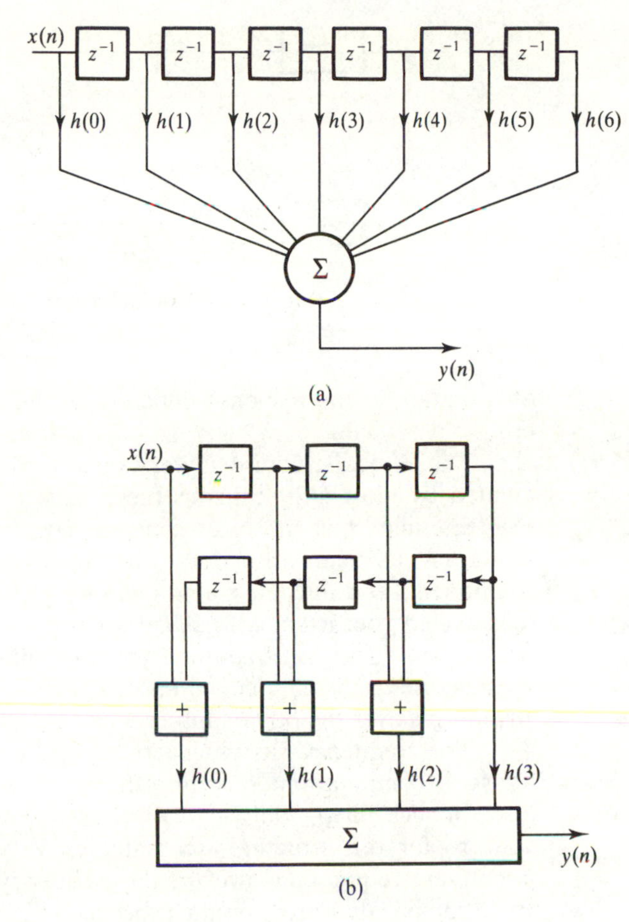

Figure 6.29 (a) Transversal and (b) linear phase structure for Example 6.17.

$H(z)$, is expressed as the product of second-order and first-order sections. The transpose and cascade structures are seldom used for FIR filters in current DSP implementations.

6.8.4 Choosing between structures

The choice between the structures depends on a number of factors and trade-offs which include ease of implementation, that is the implied hardware or software complexity, how difficult it is to obtain the impulse response or transfer function coefficients, and their relative sensitivity to coefficient quantization. In practice, the accuracy to which each coefficient is represented is limited by the wordlength of the processor used. The use of only a finite number of bits to represent each coefficient tends to move the zeros away from the desired locations, resulting in a deviation in the frequency response. The extent

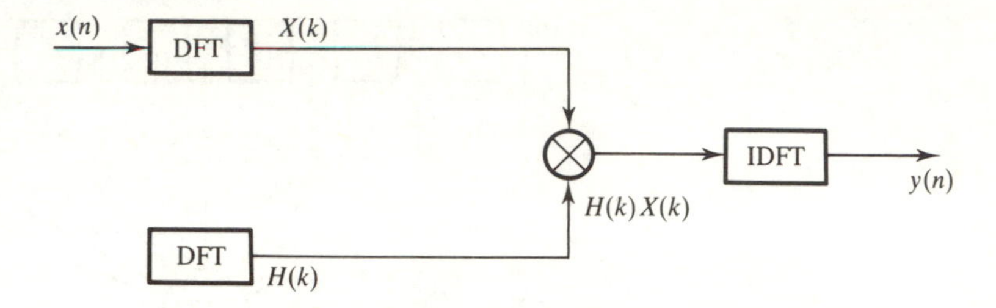

Figure 6.30 An illustration of fast convolution.

of deviation in the response depends on the number of bits and the structure used.

The direct structure is very easy to program and is efficiently implemented by most DSP chips as these have instructions tailored to transversal FIR filtering. It is the most common structure used to realize nonrecursive filters and its main attraction is its simplicity, requiring only a minimum of components and uncomplicated memory accesses for data. The cascade is less sensitive to coefficient errors and quantization noise, but the coefficients require more effort to obtain and the programming is not suited to DSP chips' architectures. The fast convolution structure offers significant computational advantages over the others, but requires the availability of the FFT.

The frequency sampling structure, for narrowband frequency selective filters, is computationally more efficient than an equivalent transversal structure. In such filters, only a relatively small number of frequency samples are nonzero for this structure, so that only very few multiplications per output sample are required. However, the frequency sampling structure may require more complicated programming, because of the more complex indexing inherent in its difference equation (compare for example Equations 6.33 and 6.32b). To avoid stability problems the poles and zeros of the frequency sampling structure should be located slightly inside the unit circle, for example at a radius $r = 0.99$. This structure is the natural choice when recursive implementation of FIR filters is mandatory. The structure is very modular and lends itself to parallel processing.

In general, the tranversal structure should be used unless the specification requirements dictate the use of the frequency sampling structure or there is a need to compute the spectrum of data as well when the fast convolution should be used.

6.9 Finite wordlength effects in FIR digital filters

In practice, FIR digital filters are often implemented using DSP processors (for example the Texas Instruments TMS320C25), algorithmic-specific DSP chips designed for FIR filtering (such as the INMOS A100) or, where high speed is desired, building blocks of multipliers, memory elements, adders and controllers (for example Plessey's PDSP1600 family). In these cases, the number of bits used to represent the input data to the filter and the filter coefficients and in performing arithmetic operations must be small for efficiency and to limit the cost of the digital filter. The problems caused by using a finite number of bits are referred to as finite wordlength effects, and in general lead to a lowering of the performance of the filter.

In this section, we will discuss the effects of finite wordlength on the performance of FIR digital filters and suggest ways of minimizing these effects. The discussion will centre on the direct form FIR structure as it is the most attractive FIR structure in modern signal processing, and rounding will be used being the simplest and most widely used method of quantization.

There are four ways in which finite wordlength affects the performance of FIR digital filters.

(1) *ADC noise*. This is the familiar ADC quantization noise which results when the filter input is derived from analogue signals. ADC noise limits the signal-to-noise ratio (SNR) obtainable. The effects can be reduced by using additional bits, consistent with inherent signal noise (see Chapter 1), and/or by using multirate techniques to enhance the signal to noise ratio (see Chapter 8).

(2) *Coefficient quantization errors*. These result from representing filter coefficients with a limited number of bits. This has the adverse effect of modifying the desired frequency response. In the stopband of a filter, for example, it limits the maximum attenuation possible, thus allowing additional signal transmission. A straightforward solution is to use enough bits to represent filter coefficients. However, optimization techniques allow efficient selection of coefficients to minimize coefficient wordlength.

(3) *Roundoff errors from quantizing results of arithmetic operations*. These can occur, for example, by discarding the lower-order bits before storing the results of a multiplication. This is normally forced on us by the wordlength of the processor used. This error reduces the SNR and may be reduced by rounding after double-length summing of products. The extent of the errors introduced depends on the type of arithmetic used and the filter structure.

(4) *Arithmetic overflow*. This occurs when partial sums or filter output exceeds the permissible wordlength of the system. Essentially, when an overflow occurs, the output sample will be wrong (normally the sign changes). A way to reduce or avoid an overflow is to scale the filter coefficients by dividing each coefficient by a factor such that the filter output sample never exceeds the permissible wordlength. This is clearly at the cost of reduced signal to noise ratio.

In the next few sections we will discuss items (2) to (4).

6.9.1 Coefficient quantization errors

Filter coefficients obtained by any of the approximation methods, for example window or optimal (Remez exchange), are usually very accurate to several places of decimals. To implement the filter, the coefficients must be represented by a fixed number of bits and very often this is determined by the wordlength of the processor used. For example, if we use one of the 16-bit DSP processors to implement the filter then the logical thing to do is to represent each filter coefficient with 16 bits. In doing so, however, we automatically introduce an error which causes the frequency response of the finite wordlength filter to deviate from the desired response. This deviation, in some cases, will mean that the initial specifications are no longer met.

Example 6.18

Determine the effects of quantizing, by rounding, the coefficients of the following filter to 8 bits:

stopband attenuation	> 90 dB
passband ripple	< 0.002 dB
passband edge frequency	3.375 kHz
stopband edge frequency	5.625 kHz
sampling frequency	20 kHz
number of coefficients	45

Solution

Use the design program optimal.c with the following input:

number of filter coefficients	45
band edge frequencies	0, 0.16875, 0.28125, 0.5
weights	1, 7.28

The coefficients of the filter, before and after rounding to 8 bits, are listed in Table 6.16. The corresponding frequency responses are given in Figure 6.31. It is seen that, after quantization, the minimum stopband attenuation is 36 dB, a degradation of more than 58 dB. Clearly, more than 8 bits of resolution is required for the coefficients in this particular example.

The effect of coefficient errors is to cause the frequency response to deviate from the desired response. This deviation in the extreme case will mean that the specifications are no longer met. For a particular filter design problem, a suitable coefficient wordlength can be determined by obtaining the frequency

Table 6.16 Filter coefficients before and after quantization to 8 bits.

$h(n)$	$h_q(n)$
−1.05023e−04	0.00000e+00
−1.25856e−04	0.00000e+00
3.07141e−04	0.00000e+00
6.79484e−04	0.00000e+00
−2.89029e−04	0.00000e+00
−1.77474e−03	0.00000e+00
4.08318e−04	0.00000e+00
3.43482e−03	0.00000e+00
2.66515e−03	0.00000e+00
−5.00314e−03	−7.81250e−03
−7.30591e−03	−7.81250e−03
5.09712e−03	7.81250e−03
1.48422e−02	1.56250e−02
−1.40255e−03	0.00000e+00
−2.49785e−02	−2.34375e−02
−9.39383e−03	−7.81250e−03
3.64568e−02	3.90625e−02
3.28505e−02	3.12500e−02
−4.72008e−02	−4.68750e−02
−8.52427e−02	−8.59375e−02
5.48855e−02	5.46875e−02
3.10921e−01	3.12500e−01
4.42322e−01	4.45212e−01
3.10921e−01	3.12500e−01
5.48855e−02	5.46875e−02
−8.52427e−02	−8.59375e−02
−4.72008e−02	−4.68750e−02
3.28505e−02	3.12500e−02
3.64568e−02	3.90625e−02
−9.39383e−03	−7.81250e−03
−2.49785e−02	−2.34375e−02
−1.40255e−03	0.00000e+00
1.48422e−02	1.56250e−02
5.09712e−03	7.81250e−03
−7.30591e−03	−7.81250e−03
−5.00314e−03	−7.81250e−03
2.66515e−03	0.00000e+00
3.43482e−03	0.00000e+00
4.08318e−04	0.00000e+00
−1.77474e−03	0.00000e+00
−2.89029e−04	0.00000e+00
6.79484e−04	0.00000e+00
3.07141e−04	0.00000e+00
−1.25856e−04	0.00000e+00
−1.05023e−04	0.00000e+00

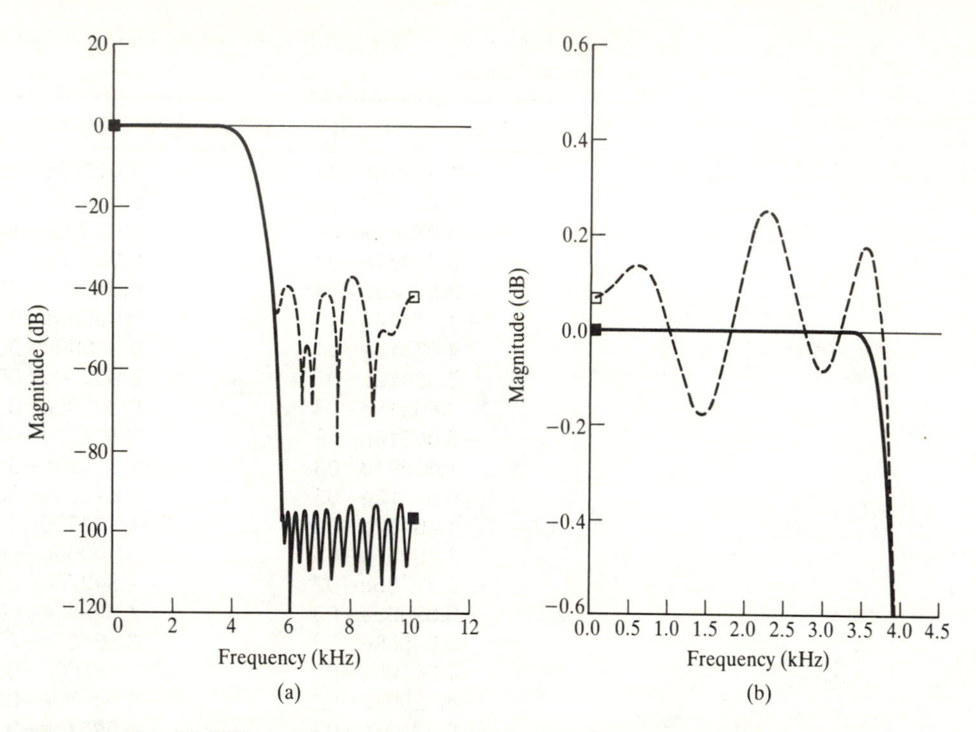

Figure 6.31 (a) Effects of coefficient quantization (Example 6.18). (b) Passband. ■, unquantized; □, quantized.

response for the filter for different coefficient wordlengths. From these, the minimum number of bits required to meet the desired specifications can be determined. However, valuable insight into the design of finite wordlength filters can be gained by analysing the errors introduced by coefficient quantization.

Now, the quantized and unquantized coefficients, $h(n)$ and $h_q(n)$, respectively, are related as

$$h_q(n) = h(n) + e(n), \quad n = 0, 1, \ldots, N-1 \tag{6.36}$$

where $e(n)$ is the error between the quantized and unquantized coefficients. In the frequency domain, Equation 6.36 can be written as

$$H_q(\omega) = H(\omega) + E(\omega) \tag{6.37}$$

where $E(\omega)$, the error in the desired frequency response, is given by

$$E(\omega) = \sum_{m=0}^{N-1} e(m) \exp(-j\omega m)$$

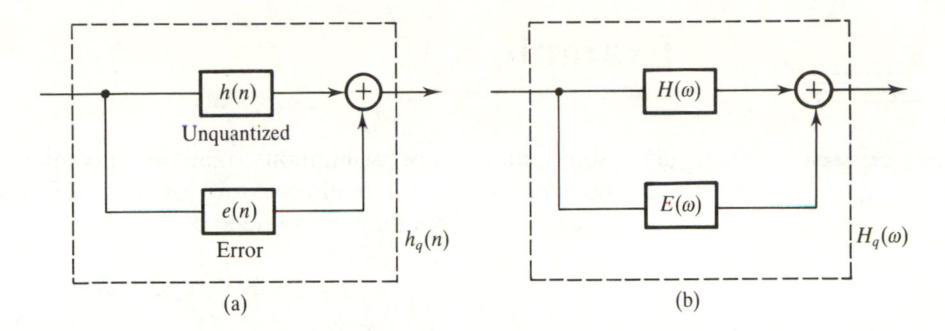

Figure 6.32 Illustration of the effects of coefficient quantizations. (a) Effects on the impulse response coefficients. (b) Effects on the frequency response of the filter.

and $H_q(\omega)$ and $H(\omega)$ are the frequency responses of the filters with quantized and unquantized coefficients, respectively. Figure 6.32(a) and 6.33(b) give diagrammatic representations of Equations 6.36 and 6.37, respectively. We see that physically $e(n)$ can be viewed as the impulse response of another filter in parallel with the impulse response of the desired filter (Rabiner and Gold, 1975). The effect of coefficient error in the frequency domain is represented as a stray transfer function in parallel with that of the very accurate filter. An objective of the designer is to limit the amplitude of $E(\omega)$ so that the frequency response of the actual filter meets the specification.

For frequency selective filters (lowpass, bandpass, bandstop filters), several researchers have developed bounds on the errors in the frequency response. These bounds could serve as useful guides in determining a suitable coefficient wordlength for a given filter. The bounds are useful in estimating coefficient wordlength requirements for adaptive FIR filters as the exact characteristics of these filters are not known *a priori* (see Chapter 9).

For direct form FIR structure, assuming rounding, the following are the most widely used bounds:

$$|E(\omega)| = N2^{-B} \tag{6.38a}$$

$$|E(\omega)| = 2^{-B}(N/3)^{1/2} \tag{6.38b}$$

$$|E(\omega)| = 2^{-B}[(N \log_e N)/3]^{1/2} \tag{6.38c}$$

where B is the number of bits used to represent each coefficient and N is the filter length. Bound 6.38a is an absolute upper bound, derived under worst-case assumptions (see Example 6.19) and so it is overly pessimistic. Bounds 6.38b and 6.38c are statistical bounds and could give a more accurate estimate of the errors in the frequency response and coefficient wordlengths to use. The statistical bounds assume that the quantization errors, $e(n)$, are uniformly distributed and have zero means.

Example 6.19

(1) Show, stating any assumptions, that the maximum stopband attenuation possible, A_{max}, for a direct form lowpass FIR filter, with coefficients quantized by rounding, is bounded by

$$A_{max} \leqslant 20 \log_{10}(2^{-B} N) \tag{6.39}$$

(2) A lowpass FIR filter has the following specifications:

passband deviation	0.05 dB
sampling frequency	10 kHz
passband edge	1.8 kHz
transition width	500 Hz
number of coefficients	65

(a) Estimate the number of bits required to represent each coefficient for the filter to have an attenuation of at least 60 dB in the stopband.

(b) If the coefficient wordlength in (a) is used, estimate the expected increase in passband ripple and the reduction in stopband attenuation in decibels.

(c) Compare the actual stopband attenuation and passband ripple of the filter using the coefficient wordlength obtained in (a).

Solution

(1) Define the response, $E(\omega)$, due to the coefficient quantization error, $e(m)$, as

$$E(\omega) = \sum_{m=0}^{N-1} e(m) \exp(-j\omega m)$$

where N is the filter length. For rounding, the worst-case quantization error is $|e(m)| = 2^{-(B-1)}/2 = 2^{-B}$ where B is the coefficient wordlength (assuming two's complement representation). If we assume the worst-case error for all the coefficients, then we have

$$|E(\omega)| = \sum_{m=0}^{N-1} |e(m)| \exp(-j\omega m) = \sum_{m=0}^{N-1} 2^{-B} \exp(-j\omega m)$$

$$= 2^{-B} \sum_{m=0}^{N-1} \exp(-j\omega m) = 2^{-B} N$$

If $e(m)$ is viewed as the impulse response of another filter in parallel with the desired filter, then the limiting deviation in the passband or stopband is $2^{-B}N$, and so

$$A_{\max} < 20 \log_{10} (2^{-B}N) \text{ dB}$$

Clearly, this bound is overly conservative. Fewer bits than it suggests will often suffice. However, the bound serves as a guide that can be applied simply.

(2) (a) From the bound above, and setting $A_{\max} = 60$ dB, $N = 65$, we find that $B = 15.988$ bits. The required coefficient wordlength is therefore $B = 16$ bits.

(b) After quantization, the worst case peak ripple in the passband, $R_{\max}$, and the stopband attenuation, $A_{\max}$, may be expressed as

$$R_{\max} = 20 \log (1 + \delta_p + |E(\omega)|) = 20 \log (1 + 0.005\,773 + 0.001)$$
$$= 0.0586 \text{ dB}$$

that is an increase of 0.0086 dB, and

$$A_{\max} = -20 \log (\delta_s + |E(\omega)|) = -20 \log (0.001 + 0.001) = 54 \text{ dB}$$

that is a reduction of 6 dB (δ_p and δ_s are the passband and stopband deviations for the unquantized filter).

(c) Use the optimal design program and the following parameters:

number of coefficients	65
band edge frequencies	0, 0.18, 0.23, 0.5
passband–stopband weights	1, 5.773

The filter spectrum before quantization is shown in Figure 6.33. There was no significant difference between the quantized (16-bit wordlength) and unquantized frequency responses. The passband ripple and stopband attenuation after quantization were 0.0227 dB and 64.15 dB compared with 0.0224 dB and 66.96 dB before quantization.

It is clear that the main effects of coefficient quantization are a possible increase in the peak passband ripple and a reduction in the maximum attenuation in the stopband. Practical procedures are available for taking these effects into account when computing the coefficients of filters. Essentially, this involves mapping the unquantized filter specifications into a new set of specifications which are then used to obtain the coefficients. This mapping is such that after coefficient quantization the original specifications would be satisfied.

The resulting filter may not be optimal. This has led to the development of optimization techniques, such as mixed integer programming algorithms, for

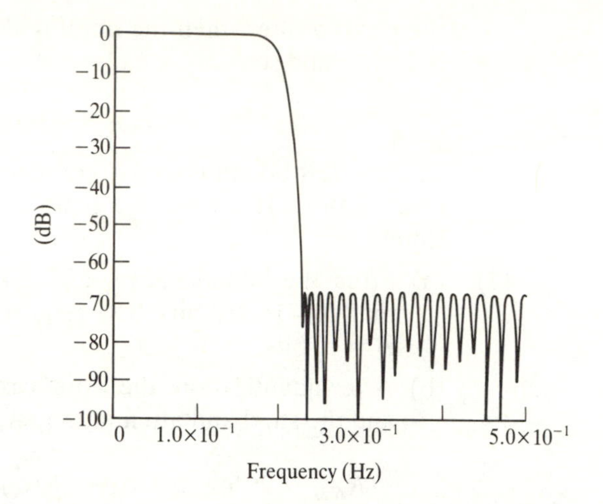

Figure 6.33 Filter spectrum before quantization for Example 6.19.

obtaining the coefficients of finite wordlength FIR filters (for example, Lawrence and Salazar, 1980). The new approaches lead to a significant reduction in coefficient wordlengths compared with straightforward rounding, but finding suitable coefficients often involves high computational overheads for even moderately large N. A pragmatic approach is to use one of the bounds in Equations 6.38 to estimate the number of bits required to represent the coefficients. The required coefficient wordlength will often be 1 to 4 bits above or below this value and can be determined by studying the frequency response corresponding to wordlengths in this range.

6.9.2 Roundoff errors

Recall that the difference equation of the FIR filter is given by

$$y(n) = \sum_{m=0}^{N-1} h(m)x(n-m) \tag{6.40}$$

where each variable is represented by a fixed number of bits. Typically, the input and output samples, $x(n-m)$ and $y(n)$, are each represented by 12 bits and the coefficients by 16 bits in 2's complement format.

It is seen from Equation 6.40 that the output of the filter is obtained as the sum of products of $h(m)$ and $x(n-m)$. After each multiplication, the product contains more bits than either $h(m)$ or $x(n-m)$. For example, if a 12-bit input is multiplied by a 16-bit coefficient the result is 28 bits long and will need to be quantized back to 16 bits (say) before it can be stored in a memory or to 12 bits before it can be output to the DAC (say). This quantization leads to errors whose effects are similar to those of the ADC noise, but could be more severe. The common way to quantize the result of an arithmetic operation is either (a) to truncate the results, that is to retain the most significant higher-order bits and to discard the lower-order bits, or (b) to round the results, that is to choose the higher-order bits closest to the unrounded result. This is achieved by adding half an LSB to the result.

Roundoff errors can be minimized by representing all products exactly, with double-length registers, and then rounding the results after obtaining the final sum, that is after obtaining $y(n)$. This approach introduces a smaller error than the alternative approach of rounding each product separately before summing.

6.9.3 Overflow errors

Overflow occurs when the sum of two numbers, usually two large numbers of the same sign, exceeds the permissible wordlength. Thus in Equation 6.40 overflow could occur when we add the two products: $h(0)x(n)$ and $h(1)x(n-1)$.

Provided that the final output, $y(n)$, is within the permissible wordlength overflow in partial sums are unimportant. This is a desirable property of 2's complement arithmetic. However, if the output, $y(n)$, exceeds the permissible limit then clearly the value of the output sample to the DAC, for example, will be wrong and steps should be taken to prevent this. An approach is to detect and correct for an overflow, but this may be an expensive overhead. Another alternative is to avoid or allow limited overfow by scaling the coefficients and/or the input data. The coefficients may be scaled in one of the following ways:

$$h(m) = \frac{h(m)}{\sum_{k=0}^{N-1} |h(k)|} \qquad \textbf{(6.41a)}$$

$$h(m) = \frac{h(m)}{\left[\sum_{k=0}^{N-1} h^2(k)\right]^{1/2}} \qquad \textbf{(6.41b)}$$

For the method given in Equation 6.41a overflow will never occur, but this form of scaling is often unnecessary because it is based on the worst-case conditions for overflow which are unlikely in practice. It will also introduce more coefficient quantization noise than the method given in Equation 6.41b which allows for occasional overflow.

The input data may be scaled in a similar manner to the coefficients. Often this leads to a better SNR. A third approach is to scale both the input and the output in such a manner that the best possible SNR is achieved. An efficient scaling would use a scale factor for the input that is a power of 2.

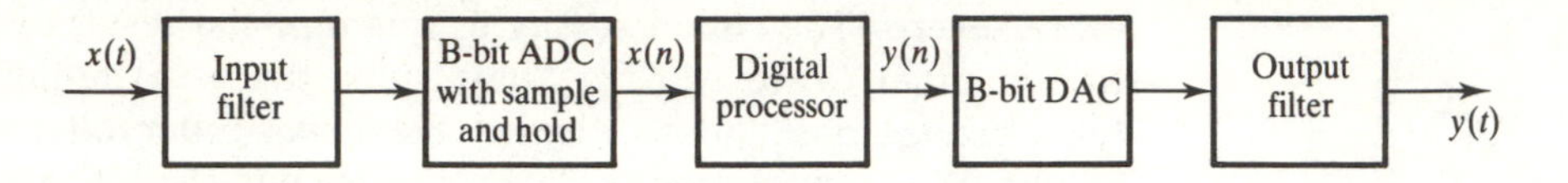

Figure 6.34 A simplified block diagram of a real-time digital filter with analogue input/output signals.

6.10 FIR implementation techniques

The different equation for an FIR digital filter is given by

$$y(n) = \sum_{k=0}^{N-1} h(k)x(n-k) \tag{6.42}$$

The coefficients $h(k)$ will have been obtained at the approximation stage, a suitable structure chosen, and an analysis carried out to verify that the number of bits to be used to represent variables and in carrying out arithmetic operations is adequate. The final stage is to implement the filter, and the key issue here is essentially to produce software code and/or hardware realization of the chosen filter structure. The discussions here will be based on the transversal structure which is characterized by Equation 6.42 as it is the most popular.

As examination of the equation will show, the computation of $y(n)$ involves only multiplications, additions/subtractions, and delays. Thus, to implement a filter, we need the following basic components:

- memory (RAM) to store the present and past input samples, $x(n)$ and $x(n-k)$;
- memory (RAM or ROM) for storing the filter coefficients, the $h(k)$;
- a multiplier (software or hardware);
- adders or arithmetic logic unit (ALU).

These components together with a means of controlling them constitute the digital filter. If the source of the input data is analogue, then we need an ADC as well. Similarly, if the output destination is analogue we need a DAC. Thus the structure of a real-time filter has the form depicted in Figure 6.34. Filter implementation is by tradition divided into two parts: hardware and software. This division, however, is somewhat artificial in modern DSP because there is hardly any truely hardware solution these days as most devices used in filtering are programmable. In this book, we will regard any implementation on large systems, such as mainframe computers and personal computers, as a software implementation. In such cases a high level language would be used to code the filter equation and the operation would be carried out off line. Implementations using DSP devices and special purpose hardware, including standard microprocessors, would be regarded as hardware. In these cases, the filtering equation may be firmware or assembly language code for the particular device.

In most applications, real-time operation is often the main goal. In these cases hardware implementation is the best option. Hardware implementation offers the greatest speed, but is less flexible. Three approaches are now common in hardware implementations: standard microprocessors (such as the Motorola 68000) and DSP processors (such as the Texas Instruments TMS320), building block, and algorithmic specific. In the building block approach, dedicated pieces of hardware are used. In both the algorithm-specific and DSP processors the various devices required for filtering – multipliers, adders, and so on – are implemented in hardware and incorporated in a single IC using VLSI technology. Algorithm-specific processors, however, are already configured to perform FIR filtering. The designer merely supplies the filter coefficients and the necessary glue logic to interface the processor to the outside world. Examples are the Motorola DSP56200 and the INMOS A100. DSP processors have architectures and instruction sets optimized for FIR filtering operations. They are more flexible than algorithm-specific processors, but they are slower.

The design of systems making use of software or hardware approaches is covered in Chapters 11 and 12.

Figure 6.35 depicts a flowchart for the general FIR filtering operations from which we see that, at each sampling instant, we must first shift the data by one place, read and save the latest input sample, $x(n)$, and compute the current output sample using the difference equation.

6.11 Design example

Example 6.20

Design and implement a linear phase bandpass filter meeting the following specifications:

passband	900–1100 Hz
passband ripple	0.87 dB
stopband attenuation	> 30 dB
sampling frequency	15 kHz
number of coefficients	41

The TMS32010 target board (see Chapter 12) is to be used to implement the filter.

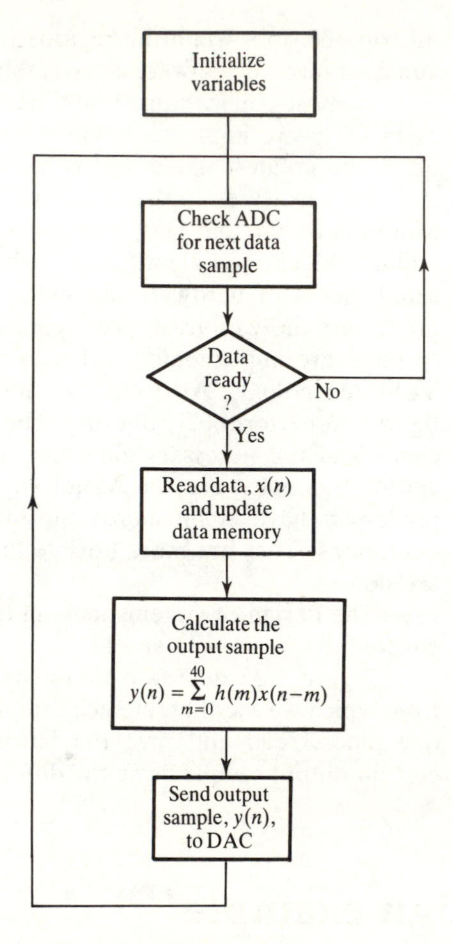

Figure 6.35 A simplified flowchart for a real-time, transversal, FIR filter.

Solution

As discussed earlier, there are five steps involved in the design of an FIR filter.

- *Step 1: specifications*. The specifications are already given.
- *Step 2: coefficient calculation*. We will use the optimal method to calculate the filter coefficients because it would yield the lowest number of filter coefficients (for nonrecursive FIR), and because it is available. We have already computed the coefficients for this filter in a previous example; see Table 6.7. The corresponding frequency response is given in Figure 6.13.
- *Step 3: realization*. The transversal structure is selected (Figure 6.29(a)), as it leads to the most efficient implementation using the TMS32010 processor. The difference equation for this structure is

$$y(n) = \sum_{m=0}^{40} h(m)x(n - m)$$

- *Step 4: quantization and analysis of errors.* Since the TMS32010 is to be used, each coefficient should be quantized to 16 bits for efficient operation. To do this, we multiply each coefficient by 2^{15} and then round up to the nearest integer. For example, the first two coefficients are quantized as follows:

$$h(0) = -0.015\,346\,38 \times 2^{15} = -502.87 \approx 503$$

$$h(1) = -0.000\,057\,805\,5 \times 2^{15} = -1.89 \approx 2$$

The quantized and unquantized coefficients are listed in Table 6.17. The frequency response of the quantized filter should be checked to verify that the specifications are still met, particularly in the stopband. We found that, after quantization to 16 bits, there was little difference between the response of the quantized and unquantized filter.

With the TMS32010, partial sums implied in the difference equation will be carried out in a 32-bit accumulator. A fairly wide product register (32 bits) is used. Thus the effects of roundoff errors for $N = 41$ will be small. In this example, overflow is ignored. If it was a concern, we could overcome it by simply dividing each coefficient obtained in step 2 by a suitable scale factor, SF, for example

$$\text{SF} = \sum_{m=0}^{40} |h(m)|$$

The target board has only an 8-bit ADC. This would restrict the dynamic range of the signal that can be handled to only about 48 dB. In a high quality audio system, for example, the level of quantization noise would have been unacceptable, and in such cases the ADC resolution must be increased.

- *Step 5: Implementation.* A flowchart for the FIR filtering operation is given in Figure 6.35. The flowchart is next translated into a TMS32010 assembly code and stored in the program memory (see Chatper 11 for the development and coding of FIR filtering operations).

6.12 Summary

The design of digital filters can be divided into five interdependent stages: filter specifications, coefficient calculation, realization, analysis of errors and filter implementation.

Table 6.17 Unquantized, $h(m)$, and quantized, $h_q(m)$, coefficients for the design example.

m	*Unquantized coefficients,* $h(m)$	*Quantized coefficients* $h_q(m)$
0	−1.534638e−02	− 503
1	−5.780550e−05	− 2
2	5.023483e−03	165
3	1.266706e−02	415
4	2.108206e−02	691
5	2.776418e−02	910
6	3.005362e−02	985
7	2.586935e−02	848
8	1.444566e−02	473
9	−3.189323e−03	− 105
10	−2.416137e−02	− 792
11	−4.420712e−02	−1449
12	−5.857453e−02	−1919
13	−6.318557e−02	−2070
14	−5.575461e−02	−1827
15	−3.654699e−02	−1198
16	−8.540099e−03	− 280
17	2.308386e−02	756
18	5.201380e−02	1704
19	7.224807e−02	2367
20	7.951681e−02	2606
21	7.224807e−02	2367
22	5.201380e−02	1704
23	2.308386e−02	756
24	−8.540099e−03	− 280
25	−3.654699e−02	−1198
26	−5.575461e−02	−1827
27	−6.318557e−02	−2070
28	−5.857453e−02	−1919
29	−4.420712e−02	−1449
30	−2.416137e−02	− 792
31	−3.189323e−03	− 105
32	1.444566e−02	473
33	2.586935e−02	848
34	3.005362e−02	985
35	2.776418e−02	910
36	2.108206e−02	691
37	1.266706e−02	415
38	5.023482e−03	165
39	−5.780550e−05	− 2
40	−1.534638e−02	− 503

Filter specification is application dependent, but should include a specification of the amplitude and/or phase characteristics.

Coefficient calculation essentially involves finding values of $h(m)$ that will satisfy the desired specifications. The three most common methods of calculating FIR filter coefficients are (1) the window, (2) the frequency sampling, and (3) the optimal methods. The window method is the easiest, but lacks flexibility especially when the passband and stopband ripples are different. The frequency sampling method is well suited to recursive implementation of FIR filters and when filters other than the standard frequency selective filters (lowpass, highpass, bandpass and bandstop) are required. The optimal method is the most powerful and flexible. All three methods were covered in detail in this chapter.

The three most common FIR filter structures are the transversal, which involves a direct convolution using the filter coefficients, the frequency sampling structure, which is directly linked to the frequency sampling method of coefficient calculation, and the fast convolution. The choice between the structures is influenced by the intended application.

The performance of FIR filters of long lengths or high stopband attenuation may be affected by finite wordlength effects. For example, their frequency responses could be altered after coefficient quantization. Thus the characteristics of such filters should be checked to ensure that adequate wordlengths have been allowed, especially when wordlengths of less than about 12 bits are contemplated.

Implementation is normally embarked on when the first four steps are satisfactory and involves software coding or hardware realization of the chosen structure.

6.13 Application examples of FIR filters

There are many areas where FIR filters have been employed, including multirate processing (Crochiere and Rabiner, 1981), noise reduction (Hamer *et al.*, 1985), matched filtering (see Chapter 12), and image processing (Wade *et al.*, 1990).

In multirate processing, for example, FIR filters have been successfully used for efficient digital anti-aliasing and anti-imaging filtering for multirate systems such as high quality data acquisition and the compact disc player (see Chapter 8).

Problems

6.1 The frequency response, $H(\omega)$, of a type 2, linear phase FIR filter may be expressed as (see Table 6.1)

$$H(\omega) = e^{-j\omega(N-1)/2} \sum_{n=1}^{N/2} b(n) \cos[\omega(n - \tfrac{1}{2})]$$

where $b(n)$ is related to the filter coefficients.

Explain why filters with the response above are unsuitable as highpass filters. Use a simple case (such as $N = 4$) to illustrate your answer.

6.2 An FIR filter has an impulse response, $h(n)$, which is defined over the interval $0 \leq n \leq N-1$. Show that if N is even and $h(n)$ satisfies the positive symmetry condition, that is $h(n) = h(N-n-1)$, the filter has a linear phase response. Obtain expressions for the amplitude and phase responses of the filter.

6.3 Show that the impulse response for an ideal bandpass filter (see Table 6.2) is given by

$$h_D(n) = 2f_2 \frac{\sin n\omega_2}{n\omega_2} - 2f_1 \frac{\sin n\omega_1}{n\omega_1} \qquad n \neq 0$$

$$= 2(f_2 - f_1) \qquad n = 0$$

where f_1 is the lower passband frequency and f_2 is the upper passband frequency.

6.4 (1) Obtain the coefficients of an FIR lowpass digital filter to meet the following specifications using the window method:

stopband attenuation	50 dB
passband edge frequency	3.4 kHz
transition width	0.6 kHz
sampling frequency	8 kHz

Include in your answer the type of window used and the reason for your choice.

(2) Assuming that the filter coefficients are stored in contiguous memory locations in a microcomputer, list the values of the coefficients in the order in which they are stored.

(3) Draw and briefly describe a flowchart of the direct software implementation of the filter in real time, and suggest two ways of improving the efficiency of the software implementation.

Note: you may use the information given in Table 6.2 in your design.

6.5 (1) A linear phase FIR filter has an impulse response that satisfies the following symmetry condition:

$$h(n) = h(N - n - 1), \qquad n = 0, 1, \ldots, (N-1)/2$$

where N is the number of filter coefficients. Assuming that N is odd, determine the magnitude and phase responses of the filter and show that the filter has both constant phase and group delays. Comment on the practical significance of a linear phase response in a digital filter.

(2) A linear phase bandpass digital filter is required for feature extraction in a certain signal analyser. The filter is required to meet the following specification:

passband	12–16 kHz
transition width	3 kHz
sampling frequency	96 kHz
passband ripple	0.01 dB
stopband attenuation	80 dB

Assume that the coefficients of the filter are to be calculated using the optimal (Remez exchange) method. Determine the following parameters for the filter:

(a) the number of filter coefficients, N;

(b) suitable weights for the filter bands;

(c) band edge frequencies, in a form suitable for the optimal method.

Explain briefly the roles of the weights and grid frequencies in the optimal method. Suggest a suitable grid density for the above problem.

You may use the information given in Table 6.18.

6.6 (1) Discuss briefly the conditions necessary for a realizable digital filter to have a linear phase characteristic, and the advantages of filters with such a characteristic.

(2) In a certain signal processing application, the input signal, with significant frequency components in the range $0 \leq f \leq 10$ Hz, is contaminated by a 50 Hz mains interference. It is decided to remove the interference using a linear phase digital filter after digitizing the composite signal at a rate of 500 samples s^{-1}. As a first step in the design of the filter, the pole–zero diagram

Table 6.18 Relationship for estimating the length, N, for a bandpass filter.

$$N \approx \frac{C_\infty(\delta_p, \delta_s)}{\Delta F} + g(\delta_p, \delta_s)\,\Delta F + 1$$

where

$$C_\infty(\delta_p, \delta_s) = [\log_{10}\delta_s][b_1(\log_{10}\delta_p)^2 + b_2\log_{10}\delta_p + b_3] + [b_4(\log_{10}\delta_p)^2 + b_5\log_{10}\delta_p + b_6]$$

$$g(\delta_p, \delta_s) = -14.6\log_{10}\left(\frac{\delta_p}{\delta_s}\right) - 16.9$$

$b_1 = 0.01202$;	$b_2 = 0.09664$
$b_3 = -0.51325$	$b_4 = 0.00203$
$b_5 = -0.5705$	$b_6 = -0.44314$

ΔF, transition width normalized to the sampling frequency
δ_p, passband ripple or deviation
δ_s, stopband ripple or deviation

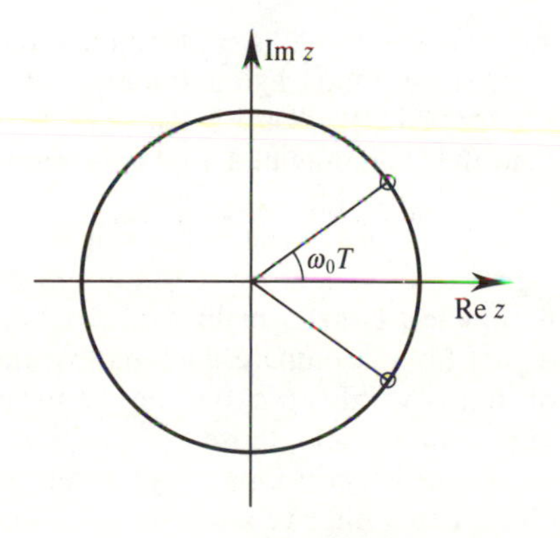

Figure 6.36 Pole–zero diagram for Problem 6.6: $\omega_0 T = \pi/5$ rad.

given in Figure 6.36 was developed. Obtain the transfer function, $H(z)$, of the filter and its difference equation.

(3) The filter obtained in part 2 is to be implemented in a microcomputer with simple arithmetic limited to only additions-/subtractions and shifts. Redesign the filter so that its coefficients are integers. There should be no increase in the number of filter coefficients or sampling rate.

(4) Show that the phase response, $\theta(\omega)$, of the filter of part (3) is given by

$$\theta(\omega) = -\omega T$$

6.7 A highpass FIR filter is characterized by the following impulse response coefficients:

$$\{h(n)\} = \{0.127, -0.026, -0.237, 0.017, 0.434\}$$

Write down the coefficients of an equivalent lowpass filter with the aid of the frequency transformation given in Section 6.7.6.

6.8 An FIR lowpass digital filter is required to meet the following specifications:

stopband attenuation	>40 dB
passband edge frequency	100 Hz
passband ripple	<0.05 dB
transition width	10 Hz
sampling frequency	1024 Hz

(1) Calculate and list the coefficients of the filter, indicating clearly the method you used and why you chose it.

(2) The filter is to be implemented for real-time operation using the fast convolution method. Outline how you would implement the filter with the FT using the overlap–save technique. Indicate clearly parameters such as the number of samples

by which the input sections overlap, the length of sections, the size of the transforms used and how the output samples are extracted from the transforms.

6.9 (1) A requirement exists for a real-time, narrowband, linear phase digital filter for a certain biomedical system. Justify the use of a frequency sampling filter for the system.

Assume that the transfer function of an N-point frequency sampling filter is given by

$$H(z) = \frac{1 - r^N z^{-N}}{N} \times \left(\sum_{k=1}^{M} \frac{|H(k)|[2\cos(2\pi k\alpha/N) - 2r\cos[2\pi k(1+\alpha)/N]}{1 - 2r\cos(2\pi k/N)z^{-1} + r^2 z^{-2}} + \frac{H(0)}{1 - z^{-1}} \right)$$

where the $H(k)$ are the samples of the desired frequency response taken at intervals of F_s/N, $\alpha = (N-1)/2$.

(2) The desired filter is characterized by the following specifications:

passband	48–52 Hz
transition width	2 Hz
sampling frequency	500 Hz
stopband attenuation	> 60 dB

Specify suitable frequency samples, $|H(k)|$. Develop and draw the realization diagram for the filter. How does the filter compare with an equivalent transversal structure in terms of storage and computational requirements?

(3) Comment on the $H(z)$ above and the difficulties that may be encountered in practice with recursive frequency sampling filters, and indicate how these may be overcome.

Explain why $H(z)$ describes a recursive filter and yet its unit impulse response, $h(n)$, is of a finite duration.

6.10 A requirement exists for an N-point FIR filter with the frequency response

$$H(e^{j\omega}) = |H(e^{j\omega})|e^{-j\omega\alpha}$$

where $\alpha = (N-1)/2$. Assume that N samples of $H(e^{j\omega})$ are taken at intervals of $f_k = (k + 1/2)F_s/N$, $k = 0, 1, \ldots, N-1$.

(1) Show that, for N even, the impulse response is given by

$$h(n) = \frac{1}{N}\left\{ \sum_{k=0}^{N/2-1} 2|H(k)|\cos[2\pi(n-\alpha)(k+1/2)/N] \right\}$$

(2) Show that, for N odd, the impulse response is given by

$$h(n) = \frac{1}{N}\left\{ \sum_{k=0}^{(N-3)/2} 2|H(k)|\cos[2\pi(n-\alpha)(k+1/2)/N] + H[(N-1)/2]\cos[\pi(n-\alpha)] \right\}$$

(3) Obtain an expression for the transfer function, $H(z)$, in recursive form for each of parts (1) and (2).

6.11 An analogue signal is contaminated by a 50 Hz component and its harmonics at 100 Hz, 150 Hz, 200 Hz, 250 Hz and 300 Hz. Assume that the contaminated signal is sampled and digitized at 1 kHz.

Find the transfer function of a simple digital filter to remove the interference and its harmonics. Draw a realization diagram for the digital filter. Compare and contrast the effects of finite wordlength on the performance of digital filters and those of component tolerances on the performance of analogue filters. Use a notch filter to illustrate your answer.

6.12 (1) Assess the effects of finite wordlength constraints on the performance of real-time FIR digital filter implementations, and suggest how these may be minimized.

(2) In a certain real-time digital signal processing system, each coefficient of an N-point FIR filter is represented as an n-bit two's complement number. Show that the maximum stopband attenuation, A_{max}, is bounded by

$$A_{max} < 20\log_{10} N2^{-B}$$

where B is the coefficient wordlength and N is the filter length.

State any assumptions made and comment on the bound given above.

(3) The coefficients of a 7-point FIR filter are listed below. Draw a realization diagram for the filter such that a minimum number of multiplications is required for each output computation.

$$
\begin{aligned}
h(0) &= -0.3\\
h(1) &= 0.4\\
h(2) &= 0.2\\
h(3) &= 0.5\\
h(4) &= 0.2\\
h(5) &= 0.4\\
h(6) &= -0.3
\end{aligned}
$$

6.13 A fixed point FIR digital filter implementation uses 2's complement fractional arithmetic with the coefficients represented by 3 bits (including the sign).

(1) Calculate and list all the possible decimal numbers that can be represented. State the largest and smallest representable decimal numbers.

(2) The unquantized coefficients of the FIR filter are listed below. Assume that the coefficients are quantized to 3 bits after truncation (sign included). List the quantized coefficients together with their quantization errors.

n	$h(n)$
0	−0.14975
1	0.256872
2	0.69940
3	0.256872
4	−0.149725

(3) Repeat part (2) if the coefficients are rounded.

6.14 The coefficients of an FIR filter are $\{h(n)\} = \{-1, 0.5, 0.75\}$.

(1) Draw the structure for the filter assuming transversal realization.

(2) Assuming that the coefficients as well as the input data samples are represented by 3 bits (including the sign bit) after truncation, determine and tabulate the values of the quantized coefficients in both binary and decimal.

(3) Show that if the data $\{x(n)\} = \{0.5, -1, -0.5\}$ is applied to the filter the output, $y(n)$, will still be correct despite an overflow in the intermediate result (assume a double-length accumulator).

(4) Show that the input $\{x(n)\} = \{-1, -0.75, 0.5\}$ will lead to wrong output values owing to overflow. How can the overflow be prevented?

6.15 Design a real-time lowpass digital filter for physiological noise reduction. The filter is intended to be part of a larger DSP system and so the number of arithmetic operations in the filter should be kept as low as possible.

The filter should meet the following amplitude specifications:

passband	8–12 Hz
passband ripple	0.1 dB
transition width	2 Hz
stopband attenuation	30 dB
sampling frequency	100 Hz

Other requirements are that

(1) minimal distortion of the harmonic relationships between the components of the in-band signals is highly desirable, and

(2) the filter will be implemented using the TMS320C25 DSP processor with analogue input digitized to 12 bits.

6.16 Design a multiband FIR digital filter to meet the following specifications

band 1	0–0.5 kHz	
	stopband attenuation	49 dB
band 2	1–1.5 kHz	
	passband ripple	0.3 dB
band 3	1.8–2.5 kHz	
	stopband attenuation	38 dB
band 4	3–3.6 kHz	
	passband ripple	0.3 dB
band 5	4.1–5 kHz	
	stopband attenuation	55 dB

The filter is to be implemented using a system with the TMS320C25 processor, a 12-bit ADC and 12-bit DAC at a sampling frequency of 10 kHz.

6.17 Discuss the five main steps involved in the design of digital filters, using the following design problem to illustrate your answer.

A digital filter is required for real-time physiological noise reduction. The filter should meet the following specifications:

passband	0–10 Hz
stopband	20–64 Hz
sampling frequency	256 Hz
maximum passband ripple	0.026 dB
stopband attenuation	30 dB

Other important requirements are that

(1) the filter should have a linear phase response so as to introduce as small a distortion as possible to the in-band signal components,

(2) the time available for filtering is limited, the filter being part of a larger process, and

(3) the filter will be implemented using a TMS32010 processor with the input digitized to 12 bits.

References

Crochiere R.E. and Rabiner L.R. (1981). Interpolation and decimation of digital signals – a tutorial review. *Proc. IEEE,* **69**(3), 300–31

Hamer C.F., Ifeachor E.C. and Jervis B.W. (1985). Digital filtering of physiological signals with minimal distortion. *Medical and Biol. Eng. and Computing*, **23**, 274–8

Herrman O., Rabiner R.L. and Chan D.S.K. (1973). Practical design rules for optimum finite impulse response digital filters. *Bell System Technical J.*, **52**, 769–99

Lawrence V.B. and Salazar A.C. (1980). Finite precision design of linear-phase FIR filters. *Bell System Technical J.* **59**(9), 1575–98

Lynn P.A. (1973). Recursive digital filters with linear phase characteristics. *Computer J.*, **15**, 337

Lynn P.A. (1975). Frequency sampling filters with integer multipliers. In *Introduction to Digital Filtering* (Bogner R. E. and Constantinides A. G., eds.). New York NY: Wiley

McClellan J.H., Parks T.W. and Rabiner L.R. (1973). A computer program for designing optimum FIR linear phase digital filters. *IEEE Trans. Audio Electroacoustics*, **21**, 506–26

Mintzer F. and Liu B. (1979). Practical design rules for optimum FIR bandpass digital filters. *IEEE Trans. Acoustics, Speech Signal Processing*, **27**(2), 204–6

Oppenheim A.V. and Schaffer R.W. (1975). *Digital Signal Processing.* Englewood Cliffs NJ: Prentice-Hall

Parks T.W. and Burrus C.S. (1987). *Digital Filter Design*. New York NY: Wiley

Rabiner L.R. and Gold B. (1975). *Theory and Applications of Digital Signal Processing.* Englewood Cliffs NJ: Prentice-Hall

Rabiner L.R., Gold B. and McGonegal C.A. (1970). An approach to the approximation problem for nonrecursive digital filters. *IEEE Trans. Audio Electroacoustics*, **18**, 83–106

Suckley D. (1990). Genetic algorithm in the design of FIR filters. *IEE Proc. Part G*, **138**(2), 234–8

Wade G., van-Eetvelt P. and Darwen H. (1990). Synthesis of efficient low-order FIR filters from primitive sections. *IEE Proc. Part G*, **137**(5), 367–72

Bibliography

Bateman A. and Yates W. (1988). *Digital Signal Processing Design*. London: Pitman

Chan D.S.K. and Rabiner L.R. (1973). Analysis of quantization errors in the direct form for finite impulse response digital filters. *IEEE Trans. Audio Electroacoustics,* **21**(4), 354–66

Chan D.S.K. and Rabiner L.R. (1973). An algorithm for minimizing roundoff noise in cascade realizations of finite impulse response digital filters. *Bell System Technical J.*, **52**(3), 347–85

DeFatta D.J., Lucas J.G. and Hodgkiss W.S. (1988). *Digital Signal Processing: A System Design Approach*. New York NY: Wiley

Gersho A., Gopinath B. and Odlyzko A.M. (1979). Coefficient inaccuracy in transversal filtering. *Bell System Technical J.*, **58**(10), 2401–2316

Gold B. and Jordan K.L., Jr. (1968). A note on digital filter synthesis. *Proc. IEEE (Lett.)*, **56**, 1717–18

Gold B. and Jordan K.L., Jr. (1969). A direct search procedure for designing finite duration impulse response filters. *IEEE Trans. Audio Electroacoustics*, **17**, 33–6

Gold B. and Rader C.M. (1969). *Digital Processing of Signals.* New York NY: McGraw-Hill

Gore A.E. (1986). Cascadable digital signal processor. *New Electronics*, (October), **19**, 39–41

Heute U. (1977). Comments on "Rabiner L. R. A simplified computational algorithm for implementing FIR digital filters". *IEEE Trans. Acoustics, Speech Signal Processing*, (June), **25**, 266–7

Hillman G.D. (1987). DSP56200: an algorithm-specific digital signal processor peripheral. *Proc. IEEE*, (September), **75**, 1185–91

Knowles J.B. and Olcayto E.M. (1968). Coefficient accuracy and digital filter response. *IEEE Trans. Circuit Theory*, **15**, 31–41

Lin K., Frantz G.A. and Simar R. (1987). The TMS320 family of digital signal processors. *Proc. IEEE*, **75**, 1143–59

Lynn P.A. (1970). Economic linear-phase recursive digital filters. *Electronics Lett*. **6**, 143–5

Lynn P.A. and Fuerst W. (1989). *Introductory Digital Signal Processing with Computer Applications*. New York NY: Wiley

Mintzer F. (1982). On half-band, third-band and *N*th-band FIR filters and their design. *IEEE Trans. Acoustics, Speech Signal Processing,* **30**, 734–8

Mitra S.K. and Sherwood R.J. (1972). Canonic realizations of digital filters using the continued fraction expansion. *IEEE Trans. Audio Electroacoustics*, **20**, 185–94

Proakis J.G. and Manolakis D.G. (1992). *Introduction to Digital Signal Processing*. New York: Macmillan

Rabiner L.R. (1971). Techniques for designing finite-duration impulse response digital filters. *IEEE Trans. Communication Technology*, **19**, 188–95

Rabiner L.R. (1973). Approximate design relationships for lowpass FIR digital filters. *IEEE Trans. Audio Electroacoustics,* **21**, 456–60

Rabiner L.R. (1977). A simplified computational algorithm for implementing FIR digital filters. *IEEE Trans. Acoustics, Speech Signal Processing*, (June), **25**, 259–61

Rabiner L.R. and Schafer R.W. (1971). Recursive and nonrecursive realizations of digital filters designed by frequency sampling techniques. *IEEE Trans. Audio Electroacoustics,* **19**, 200–7

Rabiner L.R. and Schafer R.W. (1972). Correction to "Recursive and nonrecursive realizations of digital filters designed by frequency sampling techniques". *IEEE Trans. Audio Electroacoustics (Corresp.)*, **20**, 104–5

Rabiner L.R., Kaiser J.F. and Schafer R.W. (1974). Some considerations in the design of multiband finite impulse response digital filters. *IEEE Trans. Acoustics, Speech Signal Processing*, **22**(6), 462–72

Rabiner L.R., McClellan J.H. and Parks T.W. (1975). FIR digital filter design techniques using weighted Chebyshev approximation. *Proc. IEEE*, **63**(4), 595–610.

Appendix

6A C programs for FIR filter design

The following C language programs for designing FIR filters are available on the computer disk for this book and may be obtained for a small fee (see the Preface for details):

- fresamp.c, a program for computing filter coefficients via the frequency sampling approach;
- optimal.c, a program for computing filter coefficients via the optimal method;
- window.c, a program for computing filter coefficients via the window method;
- firfilt.c, a program for FIR filtering on data;
- ncoeff.c, a program for estimating the number of filter coefficients for optimal lowpass or bandpass filter.

To limit the size of the book, only the last program, ncoeff.c, is listed here (Program 6A.1). This program is a direct implementation of the equations given in Section 6.6.3. To illustrate the use of the program we will use it to estimate the length of a bandpass filter with the following specifications:

passbands	1800–3300 Hz
stopbands	0–1400, 3700–5000 Hz
sampling frequency	10 kHz
passband ripple	1 dB
stopband attenuation	40 dB

Program 6A.1

```
/* ............................................................................ *
*                                                                              *
*       program for estimating the number of coefficients of                  *
*       optimal FIR lowpass or bandpass filter                                 *
*                                                                              *
*       program name: ncoeff.c                                                 *
*                                                                              *
*       Manny Ifeachor, 17.10.91                                               *
*                                                                              *
* ............................................................................ *
* /
#include        <stdio.h>
#include        <math.h>
#include        <dos.h>

int             filter_spec();
double          lpfcoeff();
double          bpfcoeff();
```

```
float           dp, ds, df;
int             ftype;

main()
{
                double N;
                ftype=filter_spec();                        /* obtain filter specifications */
                switch(ftype){
                        case 1:
                                N=lpfcoeff(); break;
                        case 2:
                                N=bpfcoeff(); break;
                        default:
                                printf("illegal filter type selected \n");
                                break;
}
                printf("Number of coefficients       \t%f\n",N);
                printf("passband ripple in dB        \t%f\n",dp);
                printf("stopband attenuation in dB \t%f\n",ds);
                printf("\n");
                printf("press enter to continue \n");
                getch();
                exit(0);
}
/* ............................................................................................................ */
int             filter_spec()
{
                int     itype;
                printf("program to estimate optimal filter length\n");
                printf("\n");
                printf("select filter type\n");
                printf("1       for optimal lowpass filter\n");
                printf("2       for optimal bandpass filter\n");
                scanf("%d",&itype);
                printf("\n");
                printf("enter passband and stopband deviations in ordinary units\n");
                printf("deviations must be between 0 and 1\n");
                scanf("%f%f",&dp,&ds);
                switch(itype){
                        case 1:
                                printf("enter normalized transition width \n");
                                scanf("%f", &df);
                                break;
                        case 2:
                                printf("enter normalized transition width
                                        – the smaller width\n");
                                scanf("%f", &df);
                                break;
                }
                        return(itype);
}
/* ............................................................................................................ */
double          lpfcoeff()
{
                float           ddp, dds, a1, a2, a3, a4, a5, a6, b1, b2;
```

```
            double      dinf, ff, t1, t2, t3, t4, Nl;

            /*     constants     */
            a1=0.005309; a2=0.07114; a3=-0.4761; a4=-0.00266;
            a5=-0.5941; a6=-0.4278;
            b1=11.01217; b2=0.5124401;

            ddp=log10(dp);
            dds=log10(ds);
            t1=a1*ddp*ddp;
            t2=a2*ddp;
            t3=a4*ddp*ddp;
            t4=a5*ddp;
            dinf=((t1+t2+a3)*dds)+(t3+t4+a6);
            ff=b1+b2*(ddp-dds);
            Nl=((dinf/df)-(ff*df)+1);
            dp=20*log10(1+dp); ds=-20*log10(ds);
            return(Nl);
}
/* ...................................................................... */
double      bpfcoeff()
{
            float       a1, a2, a3, a4, a5, a6, ddp, dds;
            double      t1, t2, t3, t4, cinf, ginf, Nb;

            a1=0.01201, a2=0.09664, a3=-0.51325; a4=0.00203;
            a5=-0.57054; a6=-0.44314;

            ddp=log10(dp);
            dds=log10(ds);
            t1=a1*ddp*ddp;
            t2=a2*ddp;
            t3=a4*ddp*ddp;
            t4=a5*ddp;
            cinf=dds*(t1+t2+a3)+t3+t4+a6;
            ginf=-14.6*log10(dp/ds)-16.9;
            Nb=(cinf/df)+ginf*df+1;
            dp=20*log10(1+dp); ds=-20*log10(ds);
            return(Nb);
}
```

From the specifications, the normalized transition width is 0.04 (450/10 000), the passband deviation is 0.122, from $20\log(1+1)$, and the stopband deviation is 0.01, from $-20\log(40)$. The prompts, responses and output of the program for the above example are given in Table 6A.1. The number of filter coefficients, 31 in this case, is only an estimate. In most practical cases, a higher value of filter length (that is, the number of filter coefficients) than given by the program is necessary to meet the specifications. In the above example, the actual filter length required to meet the specifications was 35. The designer should bear this in mind when using the program.

Table 6A.1 Prompts, responses and output of ncoeff.c.

```
program to estimate optimal filter length

select filter type
1      for optimal lowpass filter
2      for optimal bandpass filter
2

enter passband and stopband deviations in ordinary units deviations
must be between 0 and 1
0.122 0.01
enter normalized transition width – the smaller width
0.04
Number of coefficients            31.261084
passband ripple in dB             0.999857
stopband attenuation in dB        40.000000

press enter to continue
```

7

Design of infinite impulse response (IIR) digital filters

7.1	Introduction: summary of the basic features of IIR filters	375
7.2	Design stages for digital IIR filters	376
7.3	Stage 1: performance specification	377
7.4	Stage 2: calculation of IIR filter coefficients	379
7.5	Stage 3: realization structures for IIR digital filters	416
7.6	Stage 4: analysis of finite wordlength effects	425
7.7	Stage 5: implementation of the filter	461
7.8	A detailed design example of an IIR digital filter	462
7.9	Summary	467
7.10	Application examples	468
	References	480
	Bibliography	481
	Appendices	483

This chapter presents practical design methods for digital infinite impulse response (IIR) filters, including popular methods which permit analogue filters to be converted into equivalent digital filters. A simple but general step-by-step guide for designing digital IIR filters, from specifications to implementation, is described. Several fully worked examples are given to illustrate various aspects of digital IIR filter design, including the analysis of the effects of finite precision arithmetic on filter performance and real-time implementation.

A number of C language programs are provided to enable users to calculate filter coefficients and to carry out finite wordlength analysis. The reader is referred to Chapter 5 for a description of a general framework for filter design, comparison between IIR and FIR, and between digital and analogue filters. In this chapter we will concentrate on IIR filter design and applications.

7.1 Introduction: summary of the basic features of IIR filters

Realizable IIR digital filters are characterized by the following recursive equation:

$$y(n) = \sum_{k=0}^{\infty} h(k)x(n-k) = \sum_{k=0}^{N} a_k x(n-k) - \sum_{k=1}^{M} b_k y(n-k) \tag{7.1}$$

where $h(k)$ is the impulse response of the filter which is theoretically infinite in duration, a_k and b_k are the coefficients of the filter, and $x(n)$ and $y(n)$ are the input and output to the filter. The transfer function for the IIR filter is given by

$$H(z) = \frac{a_0 + a_1 z^{-1} + \ldots + a_N z^{-N}}{1 + b_1 z^{-1} + \ldots + b_M z^{-M}} = \frac{\sum_{k=0}^{N} a_k z^{-k}}{1 + \sum_{k=1}^{M} b_k z^{-k}} \tag{7.2}$$

An important part of the IIR filter design process is to find suitable values for the coefficients a_k and b_k such that some aspect of the filter characteristic, such as frequency response, behaves in a desired manner. Equations 7.1 and 7.2 are the characteristic equations for IIR filters.

Note that, in Equation 7.1, the current output sample, $y(n)$, is a function of past outputs, $y(n-k)$, as well as present and past input samples, $x(n-k)$, that is the IIR filter is a feedback system of some sort. The strength of IIR filters comes from the flexibility the feedback arrangement provides. For example, an IIR filter normally requires fewer coefficients than an FIR filter for the same set of specifications, which is why IIR filters are used when sharp cutoff and high throughput are the important requirements. The price for this is that the IIR filter can become unstable or its performance significantly degraded if adequate care is not taken in its design.

The transfer function of the IIR filter, $H(z)$, given in Equation 7.2 can be factored as

$$H(z) = \frac{K(z - z_1)(z - z_2) \ldots (z - z_N)}{(z - p_1)(z - p_2) \ldots (z - p_M)} \tag{7.3}$$

where $z_1, z_2, \ldots$ are the zeros of $H(z)$, that is those values of z for which $H(z)$ becomes zero, and $p_1, p_2, \ldots$ are the poles of $H(z)$, that is values of z for which $H(z)$ is infinite.

A plot of the poles and zeros of the transfer function is known as the pole–zero diagram and provides a very useful way of representing and analysing the filter in the complex z-plane; see Chapter 2 for details. For the filter to be stable, all its poles must lie inside the unit circle (or be coincident with zeros on the unit circle). There is no restriction on the zero locations.

7.2 Design stages for digital IIR filters

The design of IIR filters can be conveniently broken down into five main stages.

(1) Filter specification, at which stage the designer gives the function of the filter (for example, lowpass) and the desired performance.

(2) Approximation or coefficient calculation, where we select one of a number of methods and calculate the values of the coefficients, a_k and b_k, in the transfer function, $H(z)$, such that the specifications given in stage 1 are satisfied.

(3) Realization, which is simply converting the transfer function into a suitable filter structure. Typical structures for IIR filters are parallel and cascade of second and/or first-order filter sections.

(4) Analysis of errors that would arise from representing the filter coefficients and carrying out the arithmetic operations involved in filtering with only a finite number of bits.

(5) Implementation, which involves building the hardware and/or writing the software codes, and carrying out the actual filtering operation.

These stages are summarized in Figure 7.1. As indicated in the figure, the five stages are not independent and they are not always performed in the order given. In fact, techniques are now available that combine the second, third and fourth steps. However, the approach discussed here will ensure a successful design. To arrive at an efficient filter, it may be necessary to iterate a few times within and/or between the stages.

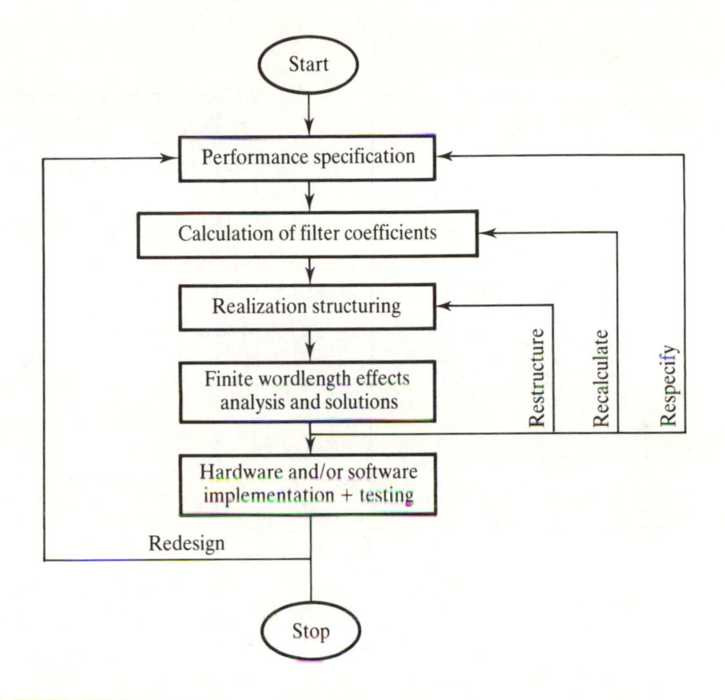

Figure 7.1 Summary of design stages for digital filters.

7.3 Stage 1: performance specification

As with most other engineering problems, the design of digital IIR filters starts with an explicit specification of the performance requirements. These should include (i) signal characteristics (types of signal sources and sinks, I/O interface, data rates and wordlengths, and frequencies of interest), (ii) the frequency response characteristics of the filter (the desired amplitude and/or phase responses and their tolerances (if any), the speed of operation), (iii) the manner of implementation (for example, as a high level language routine in a computer or as a DSP processor-based system, choice of signal processor, modes of filtering (real-time or batch)), and (iv) other design constraints (such as costs and permissible signal degradation through the filter). In general, most of the above requirements are application dependent. The designer may not have enough information to specify the filter completely at the outset, but as many of the filter requirements as possible should be specified to simplify the design process.

For frequency selective filters, such as lowpass and bandpass filters, the frequency response specifications are often in the form of a tolerance scheme.

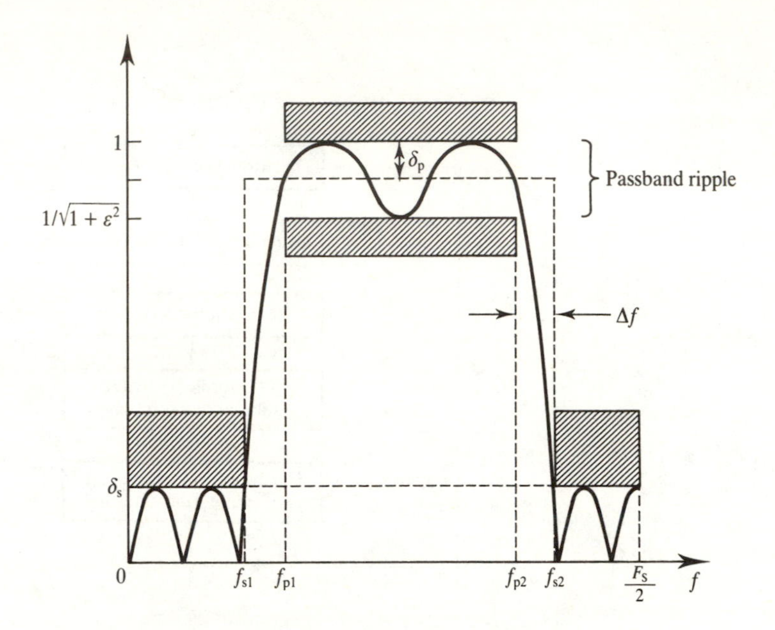

Figure 7.2 Tolerance scheme for an IIR bandpass filter.

Figure 7.2 depicts such a scheme for a bandpass IIR filter. The shaded horizontal lines indicate the tolerance limits. The following parameters are normally used to specify the frequency response.

ε^2	passband ripple parameter
δ_p	passband deviation
δ_s	stopband deviation
f_{p1} and f_{p2}	passband edge frequencies
f_{s1} and f_{s2}	stopband edge frequencies

The band edge frequencies are sometimes given in normalized form, that is a fraction of the sampling frequency (f/F_s), but we shall specify them in standard frequency units of hertz or kilohertz as these are less confusing, especially to the inexperienced designer. Passband and stopband deviations may be expressed as ordinary numbers or in decibels: the passband ripple in decibels is

$$A_p = 10\log_{10}(1+\varepsilon^2) = -20\log_{10}(1-\delta_p) \qquad \textbf{(7.4a)}$$

and the stopband attenuation in decibels is

$$A_s = -20\log_{10}(\delta_s) \qquad \textbf{(7.4b)}$$

As discussed in Chapter 5 and is evident in Figure 7.2, for IIR filters the passband ripple is the difference between the minimum and maximum devi-

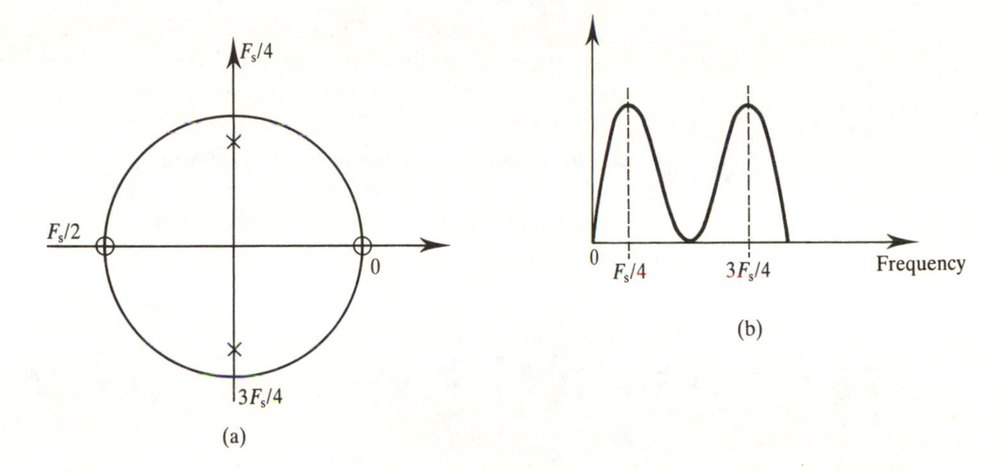

Figure 7.3 (a) Pole–zero diagram of a simple filter, and (b) a sketch of its frequency response.

ation in the passband. For FIR filters, the passband ripple is the difference between the ideal response and the maximum (or minimum) deviation in the passband. Thus, for IIR, when we say passband ripple, what is meant is the peak-to-peak passband ripple.

7.4 Stage 2: calculation of IIR filter coefficients

The task at this stage is to select one of a number of approximation methods and calculate the values of the coefficients, a_k and b_k, in Equation 7.2, such that the filter specifications given in stage 1 are satisfied.

A simple way to obtain the IIR filter coefficients is to place poles and zeros judiciously in the z-plane such that the resulting filter has the desired frequency response. This approach is only useful for very simple filters, for example notch filtering, where the filter parameters (such as passband ripple) need not be specified precisely. A more efficient approach is first to design an analogue filter satisfying the desired specifications and then to convert it into an equivalent digital filter. Most digital IIR filters are designed this way.

7.4.1 Pole–zero placement method

When a zero is placed at a given point on the z-plane, the frequency response will be zero at the corresponding point. A pole on the other hand produces a peak at the corresponding frequency point; see Figure 7.3. Poles that are close to the unit circle give rise to large peaks, whereas zeros close to or on the circle produce troughs or minima. Thus, by strategically placing poles and zeros on the z-plane, we can obtain simple lowpass, or other frequency selective, filters. Lynn and Fuerst (1989) provide a more detailed discussion of filters of this type.

An important point to bear in mind is that for the coefficients of the filter to be real, the poles and zeros must either be real (that is lie on the positive or negative real axes) or occur in complex conjugate pairs. We will illustrate the method with examples.

Example 7.1

Illustrating the simple pole–zero method of calculating filter coefficients

A bandpass digital filter is required to meet the following specifications:

(1) complete signal rejection at dc and 250 Hz;

(2) a narrow passband centred at 125 Hz;

(3) a 3 dB bandwidth of 10 Hz.

Assuming a sampling frequency of 500 Hz, obtain the transfer function of the filter, by suitably placing z-plane poles and zeros, and its difference equations.

Solution

First, we must determine where to place the poles and zeros on the z-plane. Since a complete rejection is required at 0 and 250 Hz, we need to place zeros at corresponding points on the z-plane. These are at angles of 0° and $360° \times 250/500 = 180°$ on the unit circle. To have the passband centred at 125 Hz requires us to place poles at $\pm 360° \times 125/500 = \pm 90°$. To ensure that the coefficients are real, it is necessary to have a complex conjugate pole pair.

The radius, r, of the poles is determined by the desired bandwidth. An approximate relationship between r, for $r > 0.9$, and bandwidth, bw, is given by

$$r \simeq 1 - (\mathrm{bw}/F_s)\pi \tag{7.5}$$

For the problem, $\mathrm{bw} = 10$ Hz and $F_s = 500$ Hz, giving $r = (1 - 10/500)\pi = 0.937$. The pole–zero diagram is given in Figure 7.4(a). From the pole–zero diagram, the transfer function can be written down by inspection:

$$H(z) = \frac{(z-1)(z+1)}{(z - re^{j\pi/2})(z - re^{-j\pi/2})}$$

$$= \frac{z^2 - 1}{z^2 + 0.877\,969} = \frac{1 - z^{-2}}{1 + 0.877\,969z^{-2}}$$

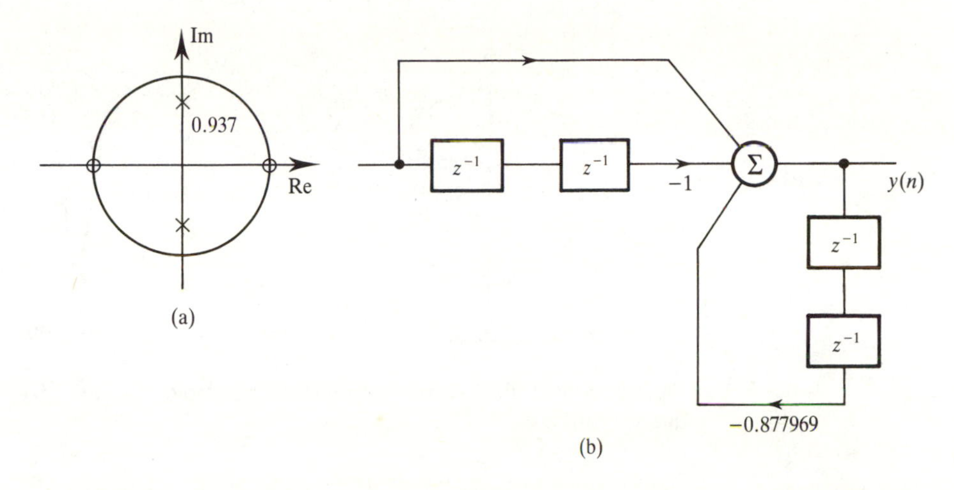

Figure 7.4 (a) Pole–zero diagram. (b) Block diagram representation of filter.

The difference equation (see Chapter 2) is

$$y(n) = -0.877\,969y(n-2) + x(n) - x(n-2)$$

Comparing the transfer function, $H(z)$ with the general IIR equation (Equation 7.2), we find that the filter is a second-order section with the following coefficients.

$$a_0 = 1 \qquad b_1 = 1$$
$$a_1 = 0 \qquad b_2 = 0.877\,969$$
$$a_2 = -1$$

Example 7.2

Using the pole–zero placement method to calculate coefficients of a notch filter Obtain, by the pole–zero placement method, the transfer function and the difference equation of a simple digital notch filter that meets the following specifications:

notch frequency	50 Hz
3 dB width of notch	± 5 Hz
sampling frequency	500 Hz

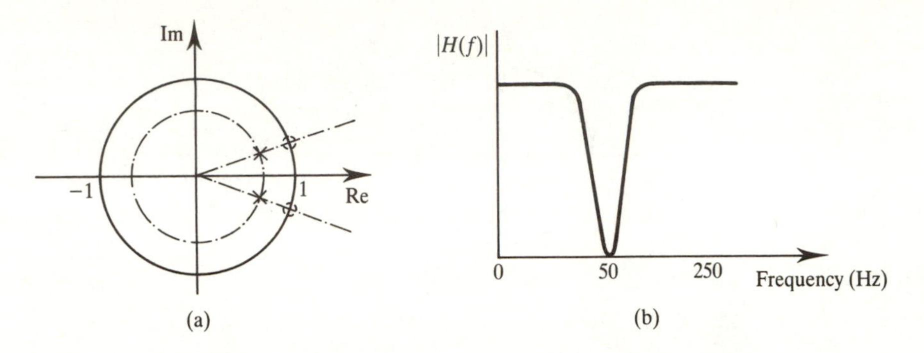

Figure 7.5 (a) Pole–zero diagram for Example 7.2 and (b) the corresponding frequency response.

Solution

- To reject the component at 50 Hz, we place a pair of complex zeros at points on the unit circle corresponding to 50 Hz, that is at angles of $360° \times 50/500 = \pm 36°$.
- To achieve a sharp notch filter and improved amplitude response on either side of the notch frequency, a pair of complex conjugate poles are placed at a radius $r < 1$. The width of the notch is determined by the locations of the poles. The relationship between the bandwidth and the radius of example 7.1 is applicable. Thus the radius of the poles is 0.937.
- The pole–zero diagram is given in Figure 7.5(a). From the figure, the transfer function of the filter is given by

$$H(z) = \frac{[z - \exp(-j36°)][z - \exp(j36°)]}{[z - 0.937\exp(-39.6°)][z - 0.937\exp(39.6°)]}$$

$$= \frac{z^2 - 1.6180z + 1}{z^2 - 1.5161z + 0.87} = \frac{1 - 1.6180z^{-1} + z^{-2}}{1 - 1.5161z^{-1} + 0.8780z^{-2}}$$

The difference equation is

$$y(n) = x(n) - 1.6180x\,(n-1) + x(n-2) + 1.5161y(n-1) - 0.8780y(n-2)$$

Comparing $H(z)$ with Equation 7.2 shows that the coefficients for the notch filter are

$$a_0 = 1 \qquad b_1 = -1.5161$$
$$a_1 = -1.6180 \qquad b_2 = 0.8780$$
$$a_2 = 1$$

7.4.2 Converting analogue filters into equivalent digital filters

This represents the most successful approach of obtaining the coefficients of IIR filters. The rationale behind this method is that there already exists a wealth of information on analogue filters in the literature which can be utilized. We will discuss the two common approaches used to convert analogue filters into equivalent digital filters: the impulse invariant and the bilinear z-transform methods.

7.4.3 Impulse invariant method

In this method, starting with a suitable analogue transfer function, $H(s)$, the impulse response, $h(t)$, is obtained using the Laplace transform. The $h(t)$ so obtained is suitably sampled to produce $h(nT)$, and the desired transfer function, $H(z)$, is then obtained by z-transforming $h(nT)$, where T is the sampling interval. We will illustrate the method by examples.

Example 7.3

Illustrating the impulse invariant method Digitize, using the impulse variant method, the simple analogue filter with the transfer function given by

$$H(s) = \frac{C}{s - p} \tag{7.6}$$

The impulse response, $h(t)$, is given by the inverse Laplace transform:

$$h(t) = L^{-1}[H(s)] = L^{-1}\left(\frac{C}{s - p}\right) = Ce^{pt}$$

where L^{-1} symbolizes the inverse Laplace transform. According to the impulse invariant method, the impulse response of the equivalent digital filter, $h(nT)$, is equal to $h(t)$ at the discrete times $t = nT$, $n = 0, 1, 2, \ldots$, that is

$$h(nT) = h(t)|_{t = nT} = Ce^{pnT}$$

The transfer function of $H(z)$ is obtained by z-transforming $h(nT)$:

$$H(z) = \sum_{n=0}^{\infty} h(nT)z^{-n} = \sum_{n=0}^{\infty} Ce^{pnT}z^{-1}$$

$$= \frac{C}{1 - e^{pT}z^{-1}}$$

Thus, from the result above, we can write

$$\frac{C}{s-p} \to \frac{C}{1-e^{pT}z^{-1}} \tag{7.7}$$

To apply the impulse invariant method to a high-order (for example, Mth-order) IIR filter with simple poles, the transfer function, $H(s)$, is first expanded using partial fractions as the sum of single-pole filters:

$$\begin{aligned} H(s) &= \frac{C_1}{s-p_1} + \frac{C_2}{s-p_2} + \dots + \frac{C_M}{s-p_M} \\ &= \sum_{K=1}^{M} \frac{C_K}{s-p_K} \end{aligned} \tag{7.8}$$

where the p_K are the poles of $H(s)$. Each term on the right-hand side of Equation 7.8 has the same form as Equation 7.6 and so the transformation given in Equation 7.8 is applicable. Thus:

$$\sum_{K=1}^{M} \frac{C_K}{s-p_K} \to \sum_{K=1}^{M} \frac{C_K}{1-e^{p_K T}z^{-1}} \tag{7.9}$$

High-order IIR filters are normally realized as cascades or parallel combinations of standard second-order filter sections. Thus the case when $M = 2$ is of particular interest. In this case the transform of Equation 7.9 becomes

$$\begin{aligned} \frac{C_1}{s-p_1} + \frac{C_2}{s-p_2} &\to \frac{C_1}{1-e^{p_1T}z^{-1}} + \frac{C_2}{1-e^{p_2T}z^{-1}} \\ &= \frac{C_1 + C_2 - (C_1e^{p_2T} + C_2e^{p_1T})z^{-1}}{1-(e^{p_1T}+e^{p_2T})z^{-1} + e^{(p_1+p_2)T}z^{-2}} \end{aligned} \tag{7.10}$$

If the poles, p_1 and p_2, are complex conjugates, then C_1 and C_2 will also be complex conjugates and Equation 7.10 reduces to

$$\begin{aligned} &\frac{C_1}{1-e^{p_1T}z^{-1}} + \frac{C_1^*}{1-e^{p_1^*T}z^{-1}} \\ &\qquad = \frac{2C_r - [C_r\cos(p_iT) + C_i\sin(p_iT)]2e^{p_rT}z^{-1}}{1-2e^{p_rT}\cos(p_iT)z^{-1} + e^{p_rT}z^{-2}} \end{aligned} \tag{7.11}$$

where C_r and C_i are the real and imaginary parts of C_1, P_r and P_i are the real and imaginary parts of P_1, and * symbolizes a complex conjugate.

For most practical impulse invariant IIR filters, the transformations given in Equations 7.7, 7.10 and/or 7.11 are the only transformations required to obtain the coefficients of the transfer function. A C language program for computing the coefficients of impulse invariant filters is given in the appendix. We will illustrate the use of the transformations by an example.

Example 7.4

Applying the impulse invariant method to filter design It is required to design a digital filter to approximate the following normalized analogue transfer function:

$$H(s) = \frac{1}{s^2 + \sqrt{2}s + 1}$$

Using the impulse invariant method obtain the transfer function, $H(z)$, of the digital filter, assuming a 3 dB cutoff frequency of 150 Hz and a sampling frequency of 1.28 kHz.

Solution

Before applying the impulse invariant method, we need to frequency scale the normalized transfer function. This is achieved by replacing s by s/α, where $\alpha = 2\pi \times 150 = 942.4778$, to ensure that the resulting filter has the desired response. Thus

$$H'(s) = H(s)|_{s=s/\alpha} = \frac{\alpha^2}{s^2 + \sqrt{2}\,\alpha s + \alpha^2} = \frac{C_1}{s - p_1} + \frac{C_2}{s - p_2}$$

where

$$p_1 = \frac{-\sqrt{2}\alpha(1 - \mathrm{j})}{2} = -666.4324(1 - \mathrm{j}),\ p_2 = p_1^*$$

$$C_1 = -\frac{\alpha}{\sqrt{2}}\mathrm{j} = -666.4324\mathrm{j};\ C_2 = C_1^*$$

Since the poles are complex conjugates, the transformation in Equation 7.11 is used to obtain the discrete-time transfer function, $H(z)$. For the problem, $C_r = 0$, $C_i = -666.4324$, $P_iT = 0.5207$, $P_rT = -0.5207$, $e^{P_rT} = 0.5941$, $\sin(P_iT) = 0.4974$, $\cos(p_iT) = 0.8675$, and $e^{P_rT} = 0.3530$. Substituting these values into Equation 7.11, we obtain $H(z)$:

$$H(z) = \frac{393.9264z^{-1}}{1 - 1.0308z^{-1} + 0.3530z^{-2}}$$

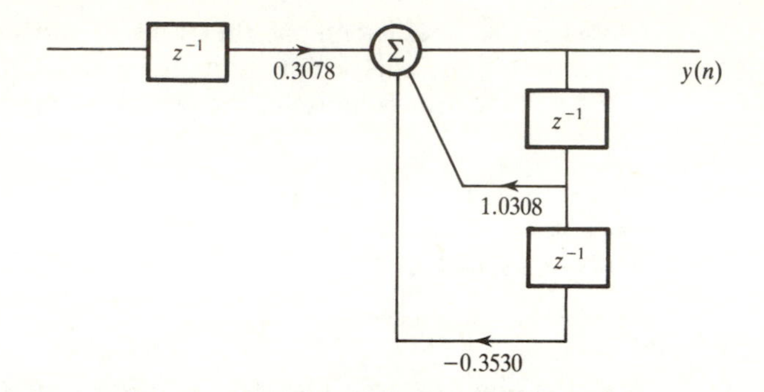

Figure 7.6 A block diagram representation of the filter in Example 7.4.

If we substitute $z = e^{i\omega T}$ in the equation above, the value of $H(z)$ at $\omega = 0$ is 1223, approximately equal to the sampling frequency. Such a large gain is characteristic of impulse invariant filters. In general, the gain of the transfer function obtained by this method is equal to the sampling frequency, that is $1/T$, and results from sampling the impulse response. To keep the gain down and to avoid overflows when the filter is implemented, it is common practice to multiply $H(z)$ by T (or equivalently to divide it by the sampling freqency). Thus, for the problem, the transfer function becomes

$$H(z) = \frac{0.3078z^{-1}}{1 - 1.0308z^{-1} + 0.3530z^{-2}}$$

Thus we have

$$a_0 = 0 \qquad b_1 = -1.0308$$
$$a_1 = 0.3078 \qquad b_2 = 0.3530$$

An alternative method of removing the effect of the sampling frequency on the filter gain is to work with normalized frequencies. Thus in the last example we would use $T = 1$ and $\alpha = 2\pi \times 150/1280 = 0.7363$. Using these values in Equation 7.11 leads directly to the desired transfer function above. An important advantage of working with normalized frequencies is that the numbers involved are much simpler. It also means that the results can be generalized. The filter is represented in the form of a block diagram in Figure 7.6.

7.4.4 Summary of the impulse invariant method of obtaining IIR coefficients

(1) Determine a normalized analogue filter, $H(s)$, that satisfies the specifications for the desired digital filter.

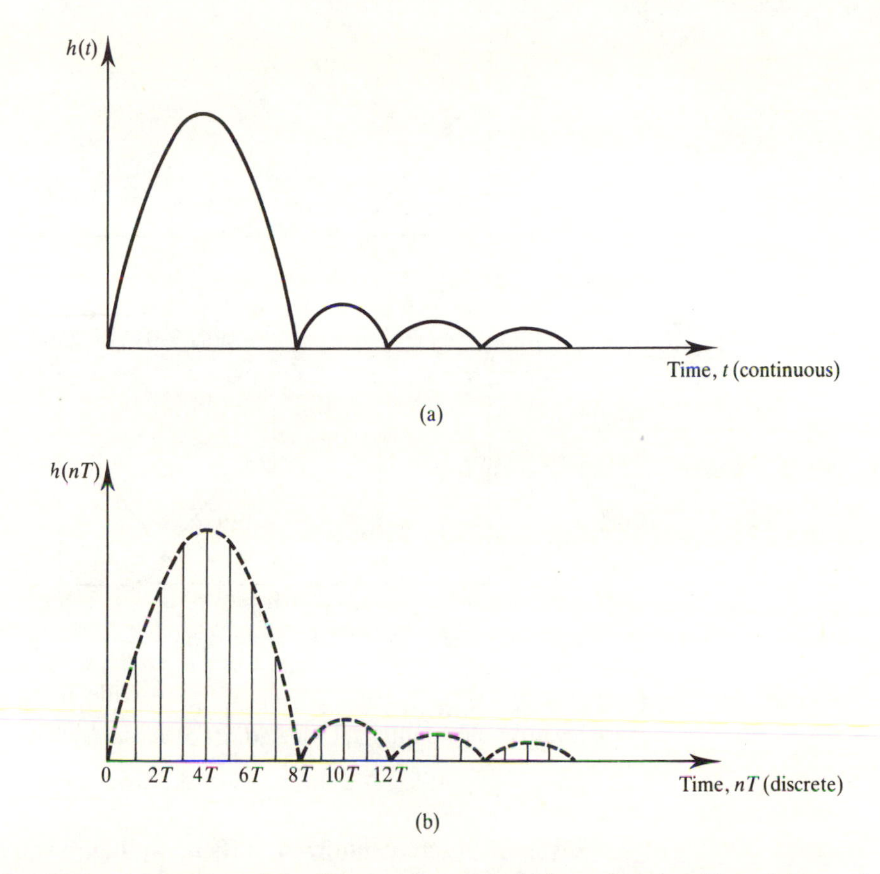

Figure 7.7 Comparison of the impulse response of (a) an analogue filter, $h(t)$, and (b) its digital filter equivalent, $h(nT)$. In the impulse invariant method, the two impulse responses are identical at the sampling instants.

(2) If necessary, expand $H(s)$ using partial fractions to simplify the next step.

(3) Obtain the z-transform of each partial fraction to obtain Equation 7.9.

(4) Obtain $H(z)$ by combining the z-transforms of the partial fractions into second-order terms and possibly one first-order term. If the actual sampling frequency is used then multiply $H(z)$ by T.

7.4.5 Remarks on the impulse invariant method

(1) The impulse response of the discrete filter, $h(nT)$, is identical to that of the analogue filter, $h(t)$, at the discrete time instants $t = nT$, $n = 0, 1, \ldots$; see Figure 7.7 for example. It is for this reason that the method is called the impulse invariant method.

(2) The sampling frequency affects the frequency response of the impulse

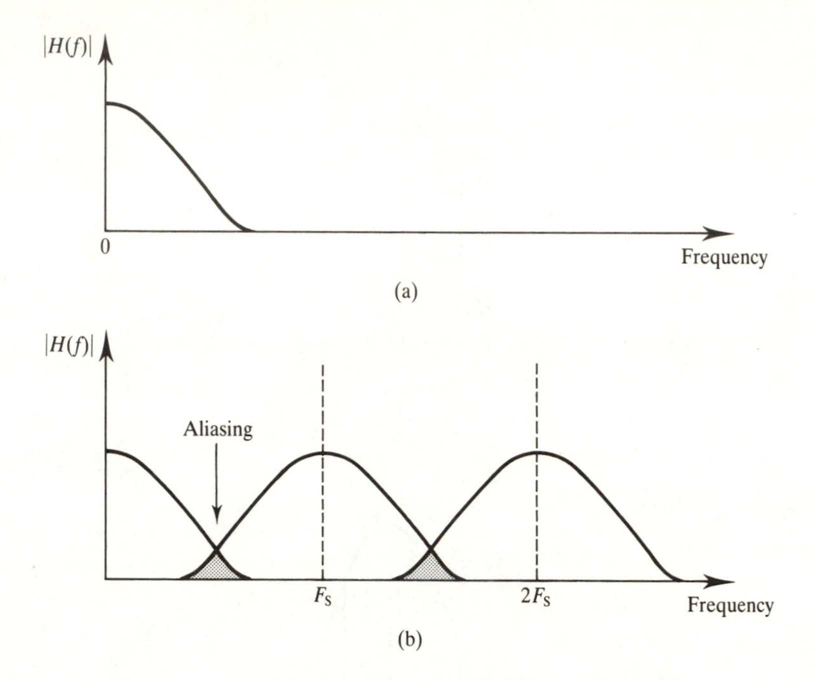

Figure 7.8 (a) Spectrum of an analogue filter and (b) spectrum of an equivalent impulse invariant digital filter showing effects of aliasing.

invariant discrete filter. A sufficiently high sampling frequency is necessary for the frequency response to be close to that of the equivalent analogue filter.

(3) As is the case with sampled data systems, the spectrum of the impulse invariant filter corresponding to $H(z)$ would be the same as that of the original analogue filter, $H(s)$, but repeats at multiples of the sampling frequency as shown in Figure 7.8, leading to aliasing. However, if the roll-off of the original analogue filter is sufficiently steep or if the analogue filter is bandlimited before the impulse invariant method is applied, the aliasing will be low. Low aliasing can also be achieved by making the sampling frequency high. We conclude that the method may be used for very sharp cutoff lowpass filters with little aliasing, provided that the sampling frequency is reasonably high, but it is unsuitable for highpass or bandstop filters unless an anti-aliasing filter is used.

7.4.6 Bilinear z-transform (BZT) method

This is by far the more important method of obtaining IIR filter coefficients. In the BZT method, the basic operation required to convert an analogue filter

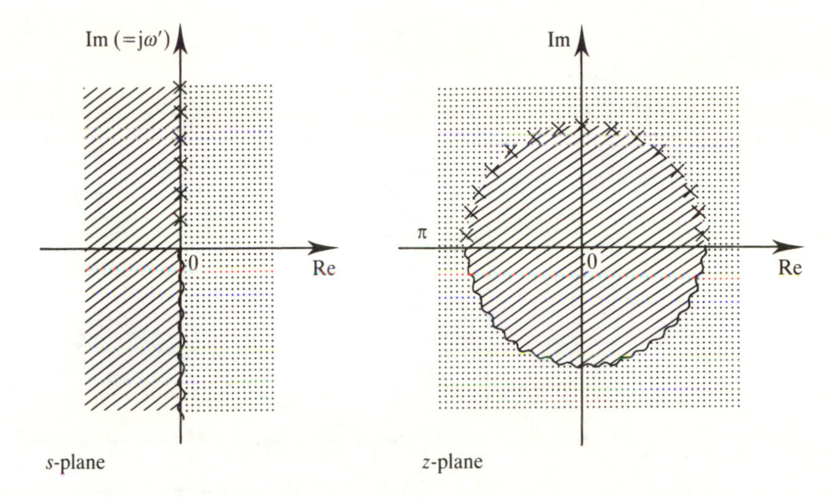

Figure 7.9 An illustration of the s-plane to z-plane mapping using the bilinear z transformation. Note that the positive $j\omega'$ axis in the s-plane (that is, $s = 0$ to $s = j\infty$) maps to the upper half of the unit circle, and the negative $j\omega'$ axis maps to the lower half.

$H(s)$ into an equivalent digital filter is to replace s as follows:

$$s = k\frac{z-1}{z+1}, \quad k = 1 \text{ or } \frac{2}{T} \tag{7.12a}$$

The above transformation maps the analogue transfer function, $H(s)$, from the s-plane into the discrete transfer function, $H(z)$, in the z-plane as shown in Figure 7.9. Notice that in the figure the entire $j\omega$ axis in the s-plane is mapped onto the unit circle, the left-half s-plane is mapped inside the unit circle, and the right-half s-plane is mapped outside the z-plane unit circle. Thus, a stable analogue filter, with poles on the left half of the s-plane, will lead to a digital filter with poles inside the unit circle.

Unfortunately, direct replacement of s in $H(s)$ as indicated in Equation 7.12a may lead to a digital filter with an undesirable response. This is readily shown by making the substitution $z = e^{j\omega T}$ and $s = j\omega'$ in Equation 7.12a. Simplifying, we find that the analogue frequency ω' and the digital frequency ω are related as

$$\omega' = k\tan\left(\frac{\omega T}{2}\right), \quad k = 1 \text{ or } \frac{2}{T} \tag{7.12b}$$

Equation 7.12b is sketched in Figure 7.10. It is seen that the relationship between the analogue frequency ω' and the digital frequency, ω, is almost linear for small values of ω, but becomes nonlinear for large values of ω, leading to a distortion (or warping) of the digital frequency response. Note, for example, that the passbands for the analogue filter on the left-hand side are of

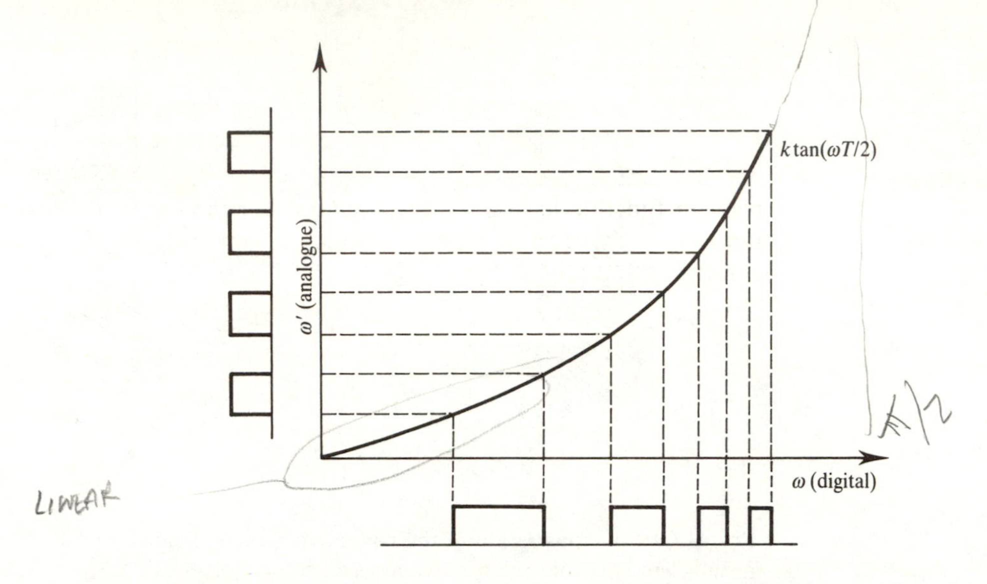

Figure 7.10 Relationship between analogue and digital frequencies showing the warping effect. Notice that the equally spaced analogue passbands are pushed together at the high frequency and, after transformation, in the digital domain.

constant width and centred at regular intervals, whereas the passbands for the digital equivalent are somewhat squashed up. This effect is normally compensated for by prewarping the analogue filter before applying the bilinear transformation.

7.4.7 Summary of the procedure for calculating digital filter coefficients by the BZT method

(1) Use the digital filter specifications to determine a suitable normalized transfer function, $H(s)$.

(2) Determine the cutoff frequency (or passband edge frequency) of the digital filter and call this ω_p.

(3) Obtain an equivalent analogue filter cutoff frequency (ω'_p) using the relation (prewarped)

$$\omega'_p = k \tan\left(\frac{\omega_p T}{2}\right), \quad k = 1 \text{ or } \frac{T}{2}$$

(4) Denormalize the analogue filter by frequency scaling $H(s)$. This is achieved by replacing s with s/ω'_p.

(5) Apply the bilinear transformation to obtain the desired digital filter transfer function $H(z)$ by replacing s by $(z-1)/(z+1)$.

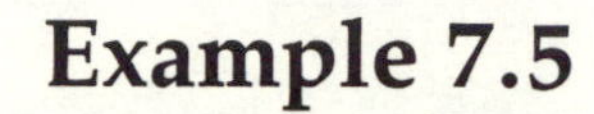

Example 7.5

An illustration of the BZT method Determine, using the BZT method, the transfer function and difference equation for the digital equivalent of the resistance–capacitance (RC) filter. Assume a sampling frequency of 150 Hz and a cutoff frequency of 30 Hz.

Solution

The normalized transfer function for the RC filter is

$$H(s) = \frac{1}{s+1}$$

The critical frequency for the digital filter is $\omega_p = 2\pi \times 30$ rad. The analogue frequency, after prewarping, is $\omega_p' = \tan(\omega_p T/2)$. With $T = 1/150$ Hz, $\omega_p' = \tan(\pi/5) = 0.7265$. The denormalized analogue filter transfer function is obtained from $H(s)$ as

$$H'(s) = H(s)|_{s=0.7265} = \frac{1}{s/0.7265 + 1} = \frac{0.7265}{s + 0.7265}$$

$$H(z) = H'(s)|_{s=(z-1)/(z+1)} = \frac{0.7265(1 + z)}{(1 + 0.7265)z + 0.7265 - 1}$$

$$= \frac{0.4208(1 + z^{-1})}{1 - 0.1584z^{-1}}$$

The difference equation is

$$y(n) = 0.1584y(n-1) + 0.4208[x(n) + x(n-1)]$$

The block diagram representation is shown in Figure 7.11.

Example 7.6

Further illustration of the BZT method It is required to design a digital filter to approximate the following analogue transfer function:

$$H(s) = \frac{1}{s^2 + \sqrt{2}s + 1}$$

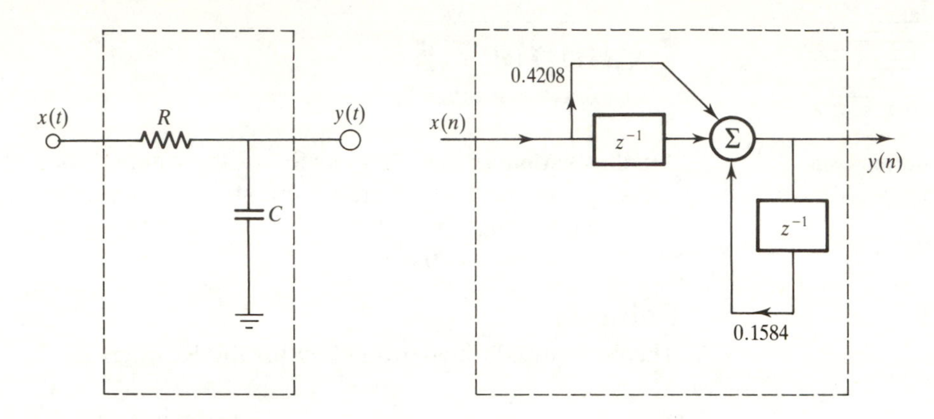

Figure 7.11 Simple RC filter and its digital equivalent.

Using the BZT method obtain the transfer function, $H(z)$, of the digital filter, assuming a 3 dB cutoff frequency of 150 Hz and a sampling frequency of 1.28 kHz.

Solution

The critical frequency is $\omega_{\mathrm{p}} = 2\pi \times 150$, giving the prewarped analogue frequency of

$$\omega'_{\mathrm{p}} = \tan(\omega_{\mathrm{p}} T/2) = 0.3857$$

The prewarped analogue filter is given by:

$$H'(s) = H(s)|_{s=s/\omega'_{\mathrm{p}}} = \frac{1}{(s/\omega'_{\mathrm{p}})^2 + \sqrt{2}s/\omega'_{\mathrm{p}} + 1}$$

$$= \frac{\omega'^{\,2}_{\mathrm{p}}}{s^2 + \sqrt{2}\omega'_{\mathrm{p}}s + \omega'^{\,2}_{\mathrm{p}}} = \frac{0.1488}{s^2 + 0.5455s + 0.1488}$$

Applying the BZT gives

$$H(z) = \frac{0.0878z^2 + 0.1756z + 0.0878}{z^2 - 1.0048z + 0.3561}$$

$$= \frac{0.0878(1 + 2z^{-1} + z^{-2})}{1 - 1.0048z^{-1} + 0.3561z^{-2}}$$

7.4.8 Comments on the bilinear transformation method

Essentially, the BZT method involves two separate transformations. First, the normalized analogue transfer function is frequency scaled by replacing s as follows:

$$s = \frac{s}{\omega_p'} \tag{7.13a}$$

where

$$\omega_p' = k \tan\left(\frac{\omega_p T}{2}\right), \quad k = 1 \text{ or } \frac{2}{T}$$

Second, the BZT is applied by replacing s in the new transfer function as

$$s = k\frac{z-1}{z+1} \tag{7.13b}$$

(1) It is common practice in many texts (for example, Rabiner and Gold, 1975) to use the factor $k = 2/T$ in the two operations above. It should be mentioned that both $k = 1$ and $k = 2/T$ lead to the same results because k is cancelled out anyway. To illustrate, consider the following simple filter:

$$H(s) = \frac{1}{s+1}$$

Assuming that the digital filter is to have a cutoff frequency of ω_p, then we must frequency scale $H(s)$ with the following frequency:

$$\omega_p' = k \tan\left(\frac{\omega_p T}{2}\right)$$

Thus the transfer function is

$$H'(s) = H(s)|_{s=s/\omega_p'} = \frac{1}{s/k \tan(\omega_p T/2) + 1}$$

Next, we replace s by $k(z-1)/(z+1)$:

$$H(z) = H'(s)|_{s=k(z-1)/(z+1)} = \frac{1}{[\not{k}(z-1)/(z+1)]/\not{k}\tan(\omega_p T/2) + 1}$$

From the above we see that the factor k is cancelled out, and it would not have mattered whether k was 1 or $2/T$.

(2) For computational efficiency, the two transformations can be combined into one:

$$s = \cot\left(\frac{\omega_p T}{2}\right)\frac{z-1}{z+1} \tag{7.14}$$

Example 7.7 illustrates this approach.

(3) For lowpass and highpass filters, the order of $H(z)$ is the same as the order of $H(s)$. For example, if $H(z)$ is derived from a second-order analogue filter $H(s)$ then $H(z)$ will also be a second-order system. For bandpass and bandstop filters, the order of $H(z)$ is twice the order of $H(s)$. This relationship is sometimes exploited to cut down on the algebraic manipulations in the BZT method (see for example Stanley *et al.*, 1984).

7.4.9 Use of classical analogue filters to design IIR digital filters

In real life, the analogue transfer function $H(s)$ from which $H(z)$ is obtained may not be available and will have to be determined from the filter specifications. For standard frequency selective filtering tasks, $H(s)$ can be derived from the classic Butterworth, Chebyshev or elliptic functions. We will provide only a brief summary of these classic filters here as detailed discussions are available in many other books. Only lowpass filters will be considered, since it is straightforward to obtain other filter types (bandpass, bandstop and so on) from normalized lowpass filters using standard transformations as will become evident.

7.4.9.1 Butterworth filter

The lowpass Butterworth filter is characterized by the following magnitude-squared frequency response:

$$|H(\omega')|^2 = \frac{1}{1 + (\omega'/\omega'_{\mathrm{p}})^{2N}} \qquad \textbf{(7.15)}$$

where N is the order of the filter and ω'_{p} is the 3 dB cutoff frequency. The magnitude–frequency response of a typical Butterworth lowpass filter is depicted in Figure 7.12(a), and is seen to be monotonic in both the passband and stopband. The response is said to be maximally flat because of its initial flatness (with a slope of zero at dc). The attenuation in decibels provided by a Butterworth filter and the filter order, N, required to provide the attenuation at a frequency ω'_{s}, are given by the following: the stopband attenuation is

$$-20\log_{10}(\delta_{\mathrm{s}}) \qquad \textbf{(7.16a)}$$

and the filter order is

$$N \geqslant \frac{\log_{10}[(1/\delta_{\mathrm{s}}) - 1]}{2\log_{10}(\omega'_{\mathrm{s}}/\omega'_{\mathrm{p}})} \qquad \textbf{(7.16b)}$$

The transfer function of the analogue Butterworth filter, $H(s)$, contains zeros at infinity and poles which are uniformly spaced on a circle of radius ω'_{p} in the

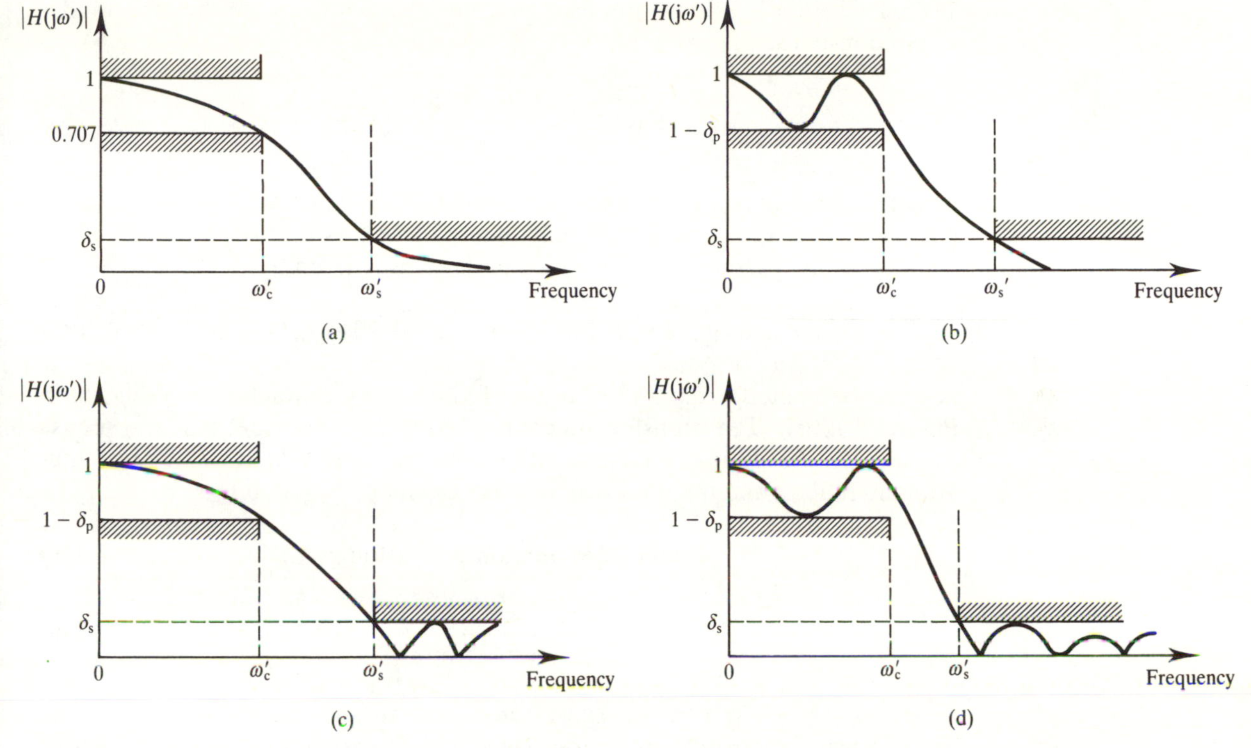

Figure 7.12 Sketches of frequency responses of some classical analogue filters: (a) Butterworth response; (b) Chebyshev type I; (c) Chebyshev type II; (d) elliptic.

s-plane at the following locations (Stearns and Hush, 1990; Jong, 1982):

$$s_k = \omega_p' e^{j\pi(2k+N-1)/2N} = \omega_p' \cos\left[\frac{(2k+N-1)\pi}{2N}\right] + j\omega_p' \sin\left[\frac{(2k+N-1)\pi}{2N}\right],$$

$$k = 1, 2, \ldots, N \quad \textbf{(7.17)}$$

The poles occur in complex conjugate pairs and lie on the left-hand side of the s-plane.

7.4.9.2 Chebyshev filter

The Chebyshev characteristic provides an alternative way of obtaining suitable analogue transfer function, $H(s)$. There are two types of Chebyshev filters, types I and II, with the following features (Figures 7.12(b) and 7.12(c)):

- Type I, with equal ripple in the passband, monotonic in the stopband;
- Type II, with equal ripple in the stopband, monotonic in the passband.

Type I Chebyshev filters, for example, are characterized by the magnitude-squared response

$$|H(\omega')|^2 = \frac{K}{1 + \varepsilon^2 C_N^2(\omega'/\omega'_p)} \quad \textbf{(7.18a)}$$

where $C_N(\omega'/\omega'_p)$ is a Chebyshev polynomial which exhibits equal ripple in the passband, N is the order of the polynomial as well as that of the filter, and ε determines the passband ripple, which in decibels is given by

$$\text{passband ripple} \leqslant 10\log_{10}(1 + \varepsilon^2) = -20\log_{10}(1 - \delta p) \quad \textbf{(7.18b)}$$

A typical amplitude response of a type I Chebyshev characteristic is shown in Figure 7.12(b). The transfer function, $H(s)$, for the Chebyshev response depends on the desired passband ripple and the filter order, N. The attenuation in decibels and the filter order N are given by

$$\text{stopband attenuation} \geqslant -20\log_{10}(\delta s) \quad \textbf{(7.19a)}$$

$$N \geqslant \frac{\cosh^{-1}(\delta/\varepsilon)}{\cosh^{-1}(\omega'_s/\omega'_p)} \quad \textbf{(7.19b)}$$

where $\delta_s = 1/(1 + \delta^2)^{1/2}$, ω'_s is the stopband frequency, ε and δ_s are the passband and stopband ripple parameters, and ω'_s specifies the frequency above which the stopband attenuation (Equation 7.19a) is satisfied.

The poles of the Chebyshev LPF lie on an ellipse in the s-plane and have coordinates given by (Stearns and Hush, 1990)

$$s_k = \omega'_p[\sinh(\alpha)\cos(\beta_k) + \mathrm{j}\cosh(\alpha)\sin(\beta_k)] \quad \textbf{(7.20)}$$

where

$$\alpha = \frac{1}{N}\sinh^{-1}\left(\frac{1}{\varepsilon}\right); \; \beta_k = \frac{(2k + N - 1)\pi}{2N}, \quad k = 1, 2, \ldots, N$$

7.4.9.4 Elliptic filter

The elliptic filter exhibits equiripple behaviour in both the passband and the stopband; see Figure 7.12(d). It is characterized by the following magnitude-squared response:

$$|H(\omega')|^2 = \frac{K}{1 + \varepsilon^2 G_N^2(\omega')} \quad \textbf{(7.21)}$$

where $G_N(\omega')$ is a Chebyshev rational function. Unlike the Butterworth and

Chebyshev filters, there is no simple expression for the poles of the elliptic filter. A procedure is available for computing locations of the poles (for example, in Antoniou, 1979; Jong, 1982, DeFatta, 1988). The zeros of the elliptic lowpass filter are entirely imaginary.

The elliptic characteristic provides the most efficient filters in terms of amplitude response. It yields the smallest filter order for a given set of specifications and should be the method of first choice in IIR filter design, except where the phase response is of concern when the Butterworth response may be preferred.

Tables of the polynomials of $H(s)$ for the Butterworth, Chebyshev and elliptic characteristics are available in most analogue design books in normalized form and can be used in the bilinear transformation. In practice, however, the computation of $H(z)$ from $H(s)$ is via a software package as we shall see later.

Example 7.7

Obtain the transfer function of a lowpass digital filter meeting the following specifications:

passband	0–60 Hz
stopband	$>$85 Hz
stopband attenuation	$>$15 dB

Assume a sampling frequency of 256 Hz and a Butterworth characteristic.

Solution

This example illustrates how to combine steps 4 and 5 in the BZT process into one, as suggested by Equation 7.14.

(1) The critical frequencies for the digital filter are

$$\omega_1 T = \frac{2\pi f_1}{F_s} = \frac{2\pi 60}{256} = 2\pi \times 0.2344$$

$$\omega_2 T = \frac{2\pi f_2}{F_s} = \frac{2\pi 85}{256} = 2\pi \times 0.3320$$

(2) The prewarped equivalent analogue frequencies are:

$$\omega_1' = \tan\left(\frac{\omega_1 T}{2}\right) = 0.906\,347; \ \omega_2' = \tan\left(\frac{\omega_1 T}{2}\right) = 1.715\,80$$

(3) Next we need to obtain $H(s)$ with Butterworth characteristics, a 3 dB cutoff frequency of 0.906 347, and a response at 85 Hz that is down by 15 dB. For an attenuation of 15 dB, $\delta_s = 0.1778$ and so from Equation 7.16b $N = 2.468$. We use $N = 3$, since it must be an integer. A normalized third-order filter is given by

$$H(s) = \frac{1}{(s+1)(s^2+s+1)} = \frac{1}{s+1}\,\frac{1}{s^2+s+1}$$

$$= H_1(s)\,H_2(s)$$

$$\cot\left(\frac{\omega_1 T}{2}\right) = \cot\left(\frac{2\pi \times 0.2344}{2}\right) = 1.103\,155$$

Performing the transform in two stages, one for each of the factors of $H(s)$ above, we obtain

$$H_2(z) = H_2(s)|_{s = \cot(\omega_1 T/2)[z-1)/(z+1)]}$$

$$= 0.3012\,\frac{1 + 2z^{-1} + z^{-2}}{1 - 0.1307z^{-1} + 0.3355z^{-2}}$$

which we have arrived at after considerable manipulation. Similarly, we obtain $H_1(z)$ as

$$H_1(z) = 0.4754\,\frac{1 + z^{-1}}{1 - 0.0490z^{-1}}$$

$H_1(z)$ and $H_2(z)$ may then be combined to give the desired transfer function, $H(z)$:

$$H(z) = H_1(z)H_2(z) = 0.1432\,\frac{1 + 3z^{-1} + 3z^{-2} + z^{-3}}{1 - 0.1801z^{-1} + 0.3419z^{-2} - 0.0165z^{-3}}$$

7.4.10 Designing highpass, bandpass, and bandstop filters

The previous discussions and examples have considered IIR filters with lowpass characteristics only. If highpass, bandpass or bandstop digital filters are required, three possible methods are as follows.

- *Method 1*. From the digital filter specifications, a suitable analogue prototype lowpass filter is derived. An appropriate analogue transformation is then used to convert the lowpass filter into an analogue filter with the desired characteristics (such as BPF or BSF). The resulting analogue filter is then digitized by applying the BZT (bilinear z transform).

- *Method 2*. From the digital filter specifications, a suitable analogue prototype lowpass filter is derived. The poles and zeros of the prototype lowpass filter are mapped, using appropriate transformations, into s-plane poles and zeros of an intermediate analogue filter. The s-plane poles and zeros of the intermediate analogue filter are in turn mapped into z-plane poles and zeros using BZT. The desired z-transfer function is then derived from the z-plane poles and zeros.
- *Method 3*. From the digital filter specifications, a suitable prototype analogue lowpass filter is derived. A prototype lowpass discrete filter transfer function $H(z)$ is obtained by applying the BZT. The digital LPF is then transformed into the desired filter type.

The first two methods are described below and illustrated with examples.

7.4.11 Calculating IIR filter coefficients: method 1

The steps involved in the first method are outlined below.

(1) Use the digital filter specifications to determine a suitable normalized lowpass filter, $H(s)$.

(2) Determine and prewarp the critical frequencies of the digital filter. For lowpass or highpass filters there is just one critical frequency – the band edge or cutoff frequency, ω'_p; for bandpass or bandstop filters, we have the lower and upper band edge frequencies, ω'_1 and ω'_2.

(3) Replace s in the transfer function, $H(s)$, using one of the following transformations, depending on the type of filter required:

$$s = \frac{s}{\omega'_p} \qquad \text{lowpass to lowpass} \qquad \textbf{(7.22a)}$$

$$s = \frac{\omega'_p}{s} \qquad \text{lowpass to highpass} \qquad \textbf{(7.22b)}$$

$$s = \frac{s^2 + \omega_0^2}{Ws} \qquad \text{lowpass to bandpass} \qquad \textbf{(7.22c)}$$

$$s = \frac{Ws}{s^2 + \omega_0^2} \qquad \text{lowpass to bandstop} \qquad \textbf{(7.22d)}$$

where $\omega_0^2 = \omega'_1\omega'_2$, $W = \omega'_2 - \omega'_1$.

(4) Apply the BZT to the new $H(s)$:

$$s = \frac{z - 1}{z + 1}$$

Alternatives to the BZT for the bandpass and bandstop filters are the following biquadratic transformations (Gold and Rader, 1969, Gray and Markel, 1976):

$$s = \cot\left[\frac{(\omega_2 - \omega_1)T}{2}\right]\left[\frac{z^2 - 2z\cos\gamma + 1}{z^2 - 1}\right] \quad \text{lowpass to bandpass} \quad \textbf{(7.23a)}$$

$$s = \tan\left[\frac{(\omega_2 - \omega_1)T}{2}\right]\left[\frac{z^2 - 1}{z^2\, 2z\cos\gamma + 1}\right] \quad \text{lowpass to bandstop} \quad \textbf{(7.23b)}$$

where

$$\cos\gamma = \cos\left[\frac{(\omega_2 + \omega_1)T}{2}\right]\Big/\cos\left[\frac{\omega_2 - \omega_1)T}{2}\right]$$

ω_1 and ω_2 are the lower and upper band edge frequencies (passband edge frequencies for BPFs, stopband edge frequencies for BSFs) and γ is the centre frequency.

Example 7.8

Highpass filter design Convert the simple lowpass filter in Example 7.5 into an equivalent highpass discrete filter. The s-plane transfer function is given by

$$H(s) = \frac{1}{s + 1}$$

Solution

$$\omega_{\text{p}}' = \tan(\omega_{\text{p}}T/2) = 0.7265$$

Using the LPF-to-HPF transformation of Equation 7.22b, the denormalized analogue transfer function is obtained as

$$H'(s) = H(s)|_{s = \omega_{\text{p}}'/s} = \frac{1}{\omega_{\text{p}}'/s + 1} = \frac{s}{s + 0.7265}$$

The z-plane transfer function is obtained by applying the BZT:

$$H(z) = H'(s)|_{s = (z-1)/(z+1)} = \frac{(z-1)/(z+1)}{(z-1)/(z+1) + 0.7265}$$

Simplifying, we have

$$H(z) = 0.5792 \frac{1 - z^{-1}}{1 + 0.1584z^{-1}}$$

The coefficients of the digital filter are

$$a_0 = \quad 0.5792 \qquad b_1 = 0.1584$$
$$a_1 = -0.5792$$

Example 7.9

Bandpass filter design A discrete bandpass filter with Butterworth characteristics meeting the following specifications is required. Obtain the coefficients of its transfer function, $H(z)$.

passband	200–300 Hz
sampling frequency	2000 Hz
filter order	2

Solution

The prewarped passband edge frequencies are given by

$$\omega_1' = \tan\left(\frac{\omega_1 T}{2}\right) = \tan(200\pi/2000) = 0.3249$$

$$\omega_2' = \tan\left(\frac{\omega_2 T}{2}\right) = \tan(300\pi/2000) = 0.5095$$

Thus $\omega_0^2 = 0.1655$ and $W = \omega_2' - \omega_1' = 0.1846$. A first-order normalized analogue lowpass filter is required (half the order of the bandpass filter). Thus we have

$$H(s) = \frac{1}{s + 1}$$

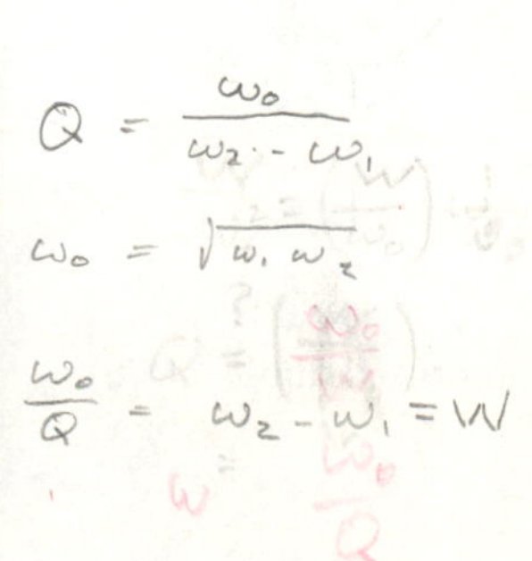

Using the lowpass-to-bandpass transformation (Equation 7.22c) we have

$$H'(s) = H(s)|_{s=(s^2+\omega_0^2)/Ws} = \frac{1}{(s^2 + \omega_0^2)/Ws + 1}$$

$$= \frac{Ws}{s^2 + Ws + \omega_0^2}$$

Applying the BZT to the analogue bandpass filter we have

$$H(z) = H'(s)|_{s=(z-1)/(z+1)} = \frac{W\,(z-1)/(z+1)}{[(z-1)/(z+1)]^2 + W(z-1)/(z+1) + \omega_0^2}$$

$$= \frac{W(z^2-1)/(1+W+\omega_0^2)}{z^2 + [2(\omega_0^2-1)/(1+W+\omega_0^2)]z + (1-W+\omega_0^2)/(1+W+\omega_0^2)}$$

Substituting the values of ω_0^2 and W and simplifying we have

$$H(z) = 0.1367\,\frac{1 - z^2}{1 - 1.2362z^{-1} + 0.7265z^{-2}}$$

The pole–zero diagrams of the normalized prototype LPF, the analogue bandpass filter and the discrete bandpass filter are depicted in Figure 7.13. Note that the lowpass-to-bandpass transformation has introduced a single zero at the origin of the s-plane and at infinity. The BZT has mapped the zeros to $z = \pm 1$. The zeros of the discrete bandpass filter are at $z = 1$ and $z = -1$. The poles are at $z = 0.6040 \pm 0.6015$j. The analogue bandpass zeros are at $s = 0$ and infinity (not shown) and the poles are at -0.0923 ± 0.3962j.

In practice, high-order IIR filters (for example, $N > 3$) are normally realized as cascades or parallel combinations of second- and/or first-order filter sections to reduce the effects of finite wordlength on filter performance (see later). Thus, after converting an analogue filter into discrete form the resulting z-transfer function $H(z)$, if it is of a high order, will need to be expressed in factored form (for cascade) or as the sum of second- and/or first-order terms (for parallel). To simplify this task we can express $H(s)$, at the outset, in factored form and then transform each factor separately. The resulting factors

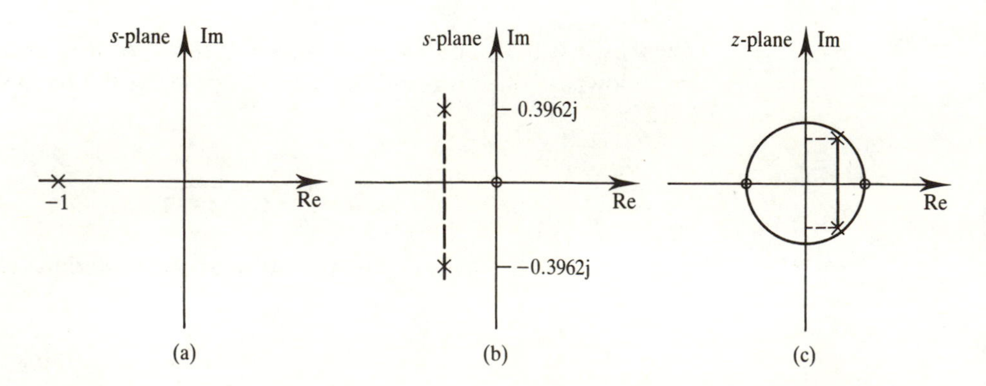

Figure 7.13 Pole–zero diagrams for (a) a prototype lowpass filter and those of (b) intermediate analogue bandpass and (c) discrete bandpass filters obtained by band transformation.

of $H(z)$ can then be combined or rearranged into a suitable format for the desired realization. This is basically the approach used in Example 7.7. We will illustrate the method with another example.

Example 7.10

An IIR digital bandpass filter meeting the specifications given below is required. Starting with a suitable normalized analogue LPF, (i) obtain the coefficients of $H(z)$, in factored form, using an appropriate band transformation and the BZT, and (ii) sketch the pole–zero diagrams for the analogue filters and the resulting digital bandpass filter.

passband	8–10 kHz
sampling frequency	32 kHz
bandpass filter order	4
filter type	Butterworth

Solution

The prewarped band edge frequencies are

$$\omega_1' = \tan(\omega_1 T/2) = \tan(8000\pi/32000) = 1$$

$$\omega_2' = \tan(\omega_2 T/2) = \tan(10000\pi/32000) = 1.4966$$

$$W = \omega_2' - \omega_1' = 0.4966$$

$$\omega_0^2 = \omega_1'\omega_2' = 1.4966$$

We require a normalized second-order LPF, that is $N = 4/2$:

$$H(s) = \frac{1}{s^2 + \sqrt{2}s + 1} \equiv \frac{1}{(s - s_{1,1})(s - s_{1,2})}$$
$$= H_1(s)H_2(s)$$

where

$$H_1(s) = \frac{1}{s - s_{1,1}}, \; H_2(s) = \frac{1}{s - s_{1,2}}$$

where $s_{1,1}$ and $s_{1,2}$ are the poles of $H(s)$, given by Equation 7.17 (with $\omega_p' = 1$):

$$s_{1,1} = \cos(3\pi/4) + \mathrm{j}\sin(3\pi/4) = \frac{-\sqrt{2}}{2} + \frac{\sqrt{2}\mathrm{j}}{2}$$

$$s_{1,2} = \cos(5\pi/4) + \mathrm{j}\sin(5\pi/4) = \frac{-\sqrt{2}}{2} - \frac{\sqrt{2}\mathrm{j}}{2}$$

We have expressed $H(s)$ in factored form to simplify the transformation process.

Next, each of the factors above is transformed using the lowpass-to-bandpass transformation. For the first factor we have

$$H_1'(s) = H_1(s)|_{s=(s^2+\omega_0^2/Ws}$$

$$= \frac{1}{(s^2 + \omega_0^2)/Ws - s_{1,1}}$$

$$= \frac{Ws}{(s^2 - s_{1,1}\, Ws + \omega_0^2)}$$

$$= \frac{Ws}{(s - s_{b,1})(s - s_{b,2})} \qquad \textbf{(7.24a)}$$

where

$$s_{b,1} = \tfrac{1}{2}\{s_{1,1}W + [(s_{1,1}W)^2 - 4\omega_0^2]^{1/2}\}$$
$$s_{b,2} = \tfrac{1}{2}\{s_{1,1}W - [(s_{1,1}W)^2 - 4\omega_0^2]^{1/2}\}$$

are two of the poles of the analogue bandpass filter. Using the values of $s_{1,1}$ and W, then

$$s_{1,1}W = -\ (0.3511 - 0.3511\mathrm{j})$$

and

$$(s_{1,1}W)^2 = -0.2466\mathrm{j}$$

Thus

$$s_{b,1} = \tfrac{1}{2}[-\ (0.3511 - 0.3511\mathrm{j}) + (-\ 5.9864 - 0.2466\mathrm{j})^{1/2}]$$
$$s_{b,2} = \tfrac{1}{2}[-\ (0.3511 - 0.3511\mathrm{j}) - (-\ 5.9864 - 0.2466\mathrm{j})^{1/2}]$$

In each case, the term whose square root is being taken is a complex number and can be evaluated using the following relationships (see the appendix for details):

$$x + \mathrm{j}y = u + \mathrm{j}v \qquad \textbf{(7.24b)}$$

where

$$u = \left[\frac{(x^2 + y^2)^{1/2} + x}{2}\right]^{1/2}$$

$$v = \frac{y}{2u}$$

Using these relationships, the transformed s-plane poles, $s_{b,1}$ and $s_{b,2}$, are given by

$$s_{b,1} = \tfrac{1}{2}[-(0.3511 - 0.3511j) + (0.0504 - 2.4472j)]$$
$$= -0.1504 - 1.0481j$$

$$s_{b,2} = \tfrac{1}{2}[-(0.3511 - 0.3511j) - (0.0504 - 2.4472j)]$$
$$= -0.2008 + 1.3992j$$

The transformation for $H_2(s)$ has a similar form to that of $H_1(s)$, since the two are a complex conjugate pair. Thus

$$H'_2(s) = H(s)|_{s=(s^2+\omega_0^2/Ws}$$
$$= \frac{Ws}{(s - s_{b,3})(s - s_{b,4})} \quad \textbf{(7.24c)}$$

where

$$s_{b,3} = s^*_{b,1},\ s_{b,4} = s^*_{b,2}$$

Next, we rearrange the poles of $H'_1(s)$ and $H'_2(s)$ by grouping complex conjugate poles together. This ensures that, when we apply the BZT, the resulting z-transfer functions will have real coefficients. Re-arranging the poles we have

$$H''_1(s) = \frac{Ws}{(s - s_{b,1})(s - s_{b,3})} = \frac{Ws}{(s - s_{b,1})(s - s^*_{b,1})} \quad \textbf{(7.24d)}$$

$$H''_2(s) = \frac{Ws}{(s - s_{b,2})(s - s_{b,4})} = \frac{Ws}{(s - s_{b,2})(s - s^*_{b,2})} \quad \textbf{(7.24e)}$$

Applying the BZT to $H''_1(s)$ we have

$$H_1(z) = H''_1(s)|_{s=(z-1)/(z+1)}$$
$$= \frac{W(z-1)/(z+1)}{[(z-1)/(z+1) - s_{b,1}][(z-1)/(z+1) - s^*_{b,1}]}$$
$$= \frac{W}{(1 - s_{b,1})(1 - s^*_{b,1})} \frac{(z-1)(z+1)}{(z - z_{b,1})(z - z^*_{b,1})} \quad \textbf{(7.25a)}$$

where $z_{b,1}$ is a z-plane pole given by

$$z_{b,1} = \frac{1 + s_{b,1}}{1 - s_{b,1}} \quad \textbf{(7.25b)}$$

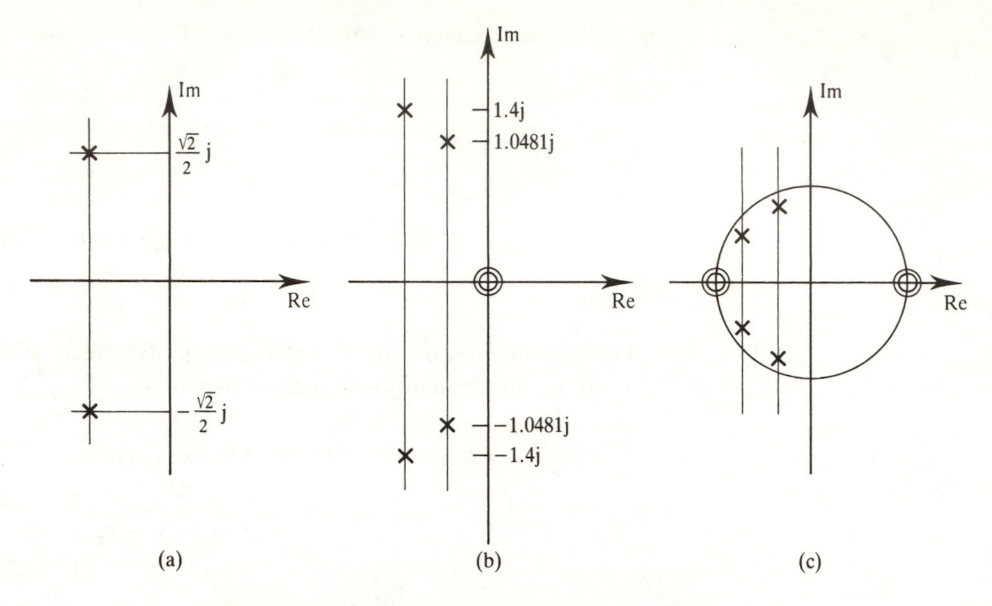

Figure 7.14 The pole–zero diagrams for Example 7.10.

and $z_{b,1}^*$ is its complex conjugate. Substituting the value of $s_{b,1}$ into Equation 7.25b, we find that $z_{b,1} = -0.05 + 0.8655j$. Substituting the values for $z_{b,1}$ and W into Equation 7.25a and simplifying, we obtain

$$H_1(z) = 0.20504 \frac{1 - z^{-2}}{1 + 0.1z^{-1} + 0.7516z^{-2}}$$

Similarly, the application of the BZT to $H_2''(s)$ leads to

$$\begin{aligned} H_2(z) &= H_2''(s)|_{s=(z-1)/(z+1)} \\ &= \frac{W}{(1 - s_{b,2})(1 - s_{b,2}^*)} \frac{(z - 1)(z + 1)}{(z - z_{b,2})(z - z_{b,2}^*)} \\ &= 0.14607 \frac{1 - z^{-2}}{1 + 0.5872z^{-1} + 0.7637z^{-2}} \end{aligned}$$

The overall transfer function, $H(z)$, is the product of $H_1(z)$ and $H_2(z)$:

$$H(z) = H_1(z)H_2(z)$$

The pole–zero diagram for the prototype LP filter, the analogue bandpass filter and the final digital (or discrete) filter are given in Figure 7.14. Two zeros at infinity in the LPF s-plane (not shown) are mapped to the origin and infinity in the bandpass filter s-plane (Figure 7.14(b)). After BZT, the two zeros at the origin in Figure 7.14(b) are mapped to $z = 1$, and those at infinity are mapped to $z = -1$ in the z-plane (Figure 7.14(c)).

7.4.12 Calculating IIR filter coefficients: method 2

An alternative, and perhaps more powerful and flexible, method of computing the coefficients of $H(z)$ for practical IIR filters is to map the individual poles and zeros of a suitable analogue filter from the s-plane into the z-plane and then to derive the digital filter coefficients from the z-plane poles and zeros. This approach is exploited by a number of commercial software implementations and is attractive when the filter is of a high order.

The procedure for calculating IIR coefficients by mapping s-plane poles and zeros into the z-plane is summarized below.

Step 1

As before, the designer starts with a normalized, Nth-order analogue lowpass filter of Butterworth, Chebyshev or elliptic type, depending on the design requirements. The poles of the normalized LPF are then obtained using Equation 7.17 for a Butterworth or Equation 7.20 for a Chebyshev filter. For an elliptic filter each pole is complex and, in general, has the form

$$s_{1,k} = \alpha_{p,k} + j\beta_{p,k} \tag{7.26}$$

For Butterworth and Chebyshev (type I) filters, the zeros for the prototype LPF are at infinity, but for elliptic filters they are entirely imaginary. In general, the locations of the zeros of the normalized LPF are easier to determine than those of the poles.

Step 2

Next, the poles and zeros of the normalized analogue LPF are converted into those of an LP, HP, BP or BSF using an appropriate transformation in Equations 7.22a–7.22d.

Lowpass and highpass filters

For a lowpass or highpass digital filter, the N poles of the normalized LP are transformed as follows (see Equations 7.22a and 7.22b):

$$s_{1,k} = \omega_p' s_{1,k} \quad k = 1, 2, \ldots, N \qquad \text{lowpass-to-lowpass} \tag{7.27a}$$

$$s_{h,k} = \omega_p' s_{1,k}^{-1} \quad k = 1, 2, \ldots, N \qquad \text{lowpass-to-lowpass} \tag{7.27b}$$

where ω_p' is the desired passband edge frequency, $s_{1,k}$ are the poles of the analogue lowpass filter and $s_{h,k}$ are the poles of the analogue highpass filter.

The similarity between Equations 7.27a and 7.27b is evident and is due to the duality between the lowpass and highpass characteristics. For N even, there will be $N/2$ complex pole pairs. For N odd there will be $(N-1)/2$ complex pole pairs and a single real pole.

For the classic filters – Butterworth, Chebyshev and elliptic – the transformations of Equations 7.27a and 7.27b map the zeros of the prototype LPF onto the imaginary axis in the s-plane. In the case of the Butterworth or Chebyshev filters, the zeros of the prototype filter are at infinity. In either case, the transformations map the zeros from infinity to infinity (for lowpass filters) or from infinity to the origin (for highpass filters), as shown in Figures 7.15(a)(ii) and 7.15(b)(ii).

Bandpass and bandstop filters

For bandpass digital filters, the analogue BPF poles are obtained from those of the normalized prototype LPF using the transformation

$$s_{1,k} = \frac{s_{b,k}^2 + \omega_0^2}{W s_{b,k}} \tag{7.28}$$

where $s_{1,k}$ are the poles of the prototype analogue LPF, $s_{b,k}$ are the pole pairs of the intermediate analogue BPF, $W = \omega_2' - \omega_1'$ is the width of the filter passband and $\omega_0^2 = \omega_1'\omega_2'$ the centre frequency of the passband. Equation 7.28 yields the following quadratic equation in $s_{b,k}$:

$$s_{b,k}^2 - W s_{1,k} s_{b,k} + \omega_0^2 = 0 \tag{7.29}$$

Solving for $s_{b,k}$ leads to the following expression for the bandpass analogue poles:

$$s_{b,k} = \frac{W}{2}\left[s_{1,k} \pm \left(s_{1,k}^2 - \frac{4\omega_0^2}{W^2}\right)^{1/2}\right] \tag{7.30}$$

It is seen from Equation 7.30 that each analogue LP pole, $s_{1,k}$ leads to a pair of analogue BPF poles as a result of the s^2 term in the transformation. In general, the analogue lowpass filter pole $s_{1,k}$ is complex and so the square-root term is complex. The evaluation of a complex square root using real arithmetic requires some care if the correct solution is to be obtained. Equation 7.24b may be used for this purpose.

For the bandstop digital filters, the following lowpass-to-bandstop transformation may be used:

$$s_{1,k} = \frac{W s_{r,k}}{s_{r,k}^2 + \omega_0^2} \tag{7.31}$$

where $s_{1,k}$ are the poles of the analogue prototype LPF, $s_{r,k}$ are the poles of the intermediate analogue bandstop filter, W is the width of the stopband and ω_0^2 is the centre frequency in the stopband. Equation 7.31 leads to the following

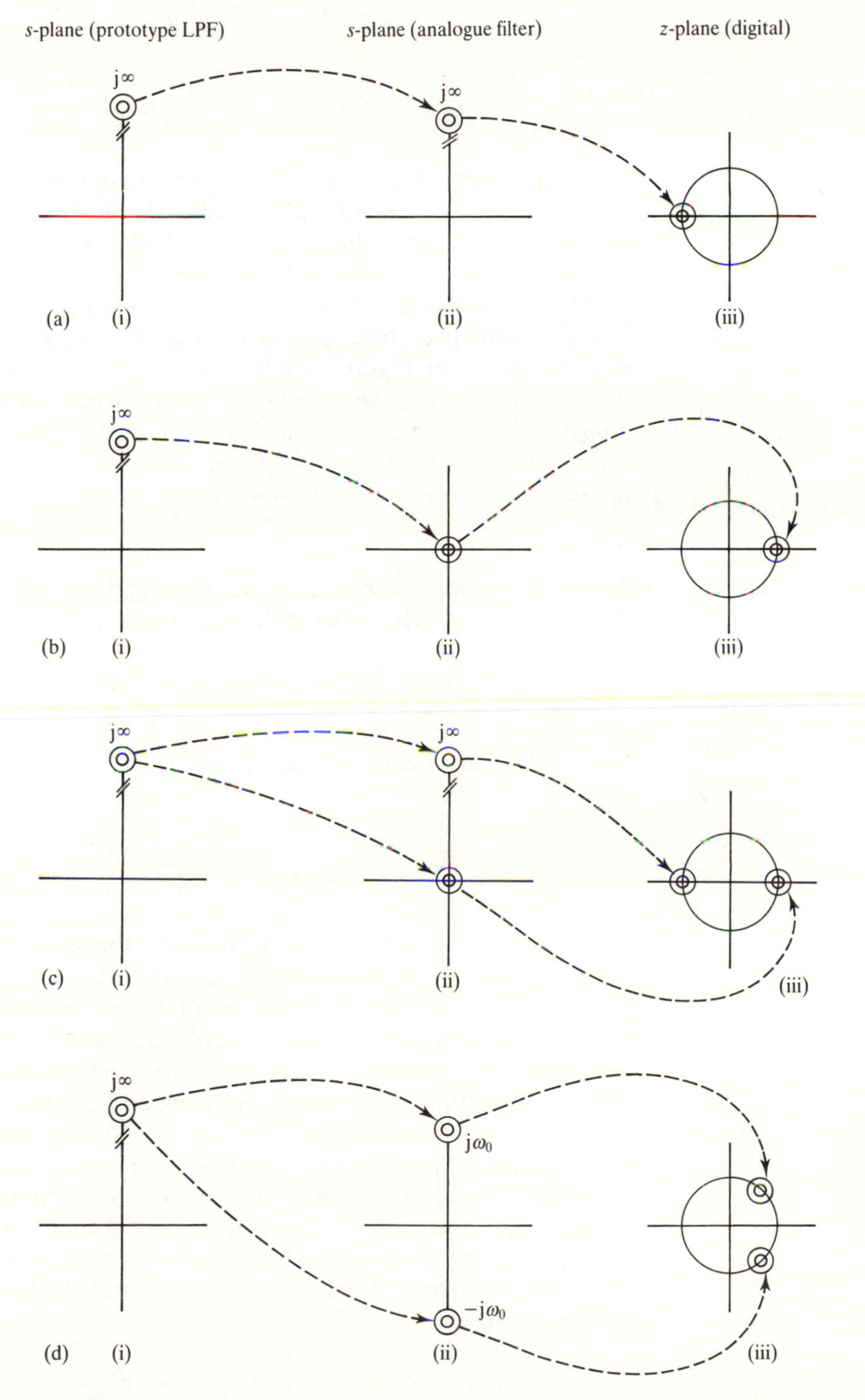

Figure 7.15 Mapping of zeros from a second-order prototype lowpass filter into (a) lowpass, (b) highpass, (c) bandpass and (d) bandstop.

expression for obtaining analogue bandstop poles from the prototype LPF:

$$s_{r,k} = \frac{W}{2}\left[s_{1,k}^{-1} \pm \left(s_{1,k}^{-2} - \frac{4\omega_0^2}{W^2}\right)^{1/2}\right] \quad \textbf{(7.32)}$$

Not surprisingly, Equation 7.32 has the same format as Equation 7.30 except for the inversion of the LPF poles due to the duality of the BPF and BSF.

As in the case of the lowpass and highpass filters, the transformations of Equations 7.30 and 7.32 map the zeros onto the imaginary axis. For Butterworth and Chebyshev filters, the lowpass-to-bandpass transformation maps the N zeros of the prototype lowpass filter from infinity to infinity and to the origin in the s-plane (see Figure 7.15(c)). The lowpass-to-bandstop transformation on the other hand maps the N zeros of the prototype lowpass filter from infinity to $\pm j\omega_0$ in the s-plane; see Figure 7.15(d). For elliptic bandpass and bandstop filters, the transformation maps the zeros of the prototype LPF which are entirely imaginary to other points on the $j\omega$ axis.

Step 3

The BZT is then used to map the poles and zeros from the s-plane into the digital z-plane. Each s-plane pole, $s_{p,k}$, is mapped as follows:

$$z_{p,k} = \frac{1 + s_{pk}}{1 - s_{pk}} \quad \textbf{(7.33)}$$

Similarly, each s-plane zero, $s_{z,k}$ of the transformed analogue filters is mapped onto the z-plane as follows:

$$z_{z,k} = \frac{1 + s_{zk}}{1 - s_{z,k}} \quad \textbf{(7.34)}$$

Figures 7.15(a)–7.15(d) illustrate the way the zeros are mapped from the s-plane to the z-plane via the BZT for the Butterworth and Chebyshev filters. Note, for example, that for lowpass filters, Figure 7.15(a)(ii), the zeros are at infinity in the s-plane and the BZT of Equation 7.34 maps these to the point $z = -1$ in the z-plane, Figure 7.15(a)(iii), whereas for the bandpass filter the s-plane zeros, Figure 7.15(c)(ii), are mapped from the origin to the point $z = 1$, and from infinity to $z = -1$.

For the elliptic lowpass, highpass, bandpass and bandstop filters, $s_{z,k}$ are imaginary and the BZT maps these onto the unit circle in the z-plane. In general, the z-plane zeros for all the classical filters (Butterworth, Chebyshev and elliptic) obtained via the BZT lie on the unit circle, regardless of the filter type. As a result, the numerator coefficients of $H(z)$ of classical filters are always integers (0, ±1, ±2).

Step 4

The final step is to determine the numerator and denominator coefficients of the second- and/or first-order filter sections. This is achieved by combining complex conjugate pole and zero pairs as follows:

$$H_i(z) = \frac{(z - z_{z,k})(z - z^*_{z,k})}{(z - z_{p,k})(z - z^*_{p,k})} \tag{7.35a}$$

$$= \frac{1 + a_{1i}z^{-1} + a_{2i}z^{-2}}{1 + b_{1i}z^{-1} + b_{2i}z^{-2}} \tag{7.35b}$$

In general, a complex pole or zero pair located at $\alpha \pm j\beta$ leads to a quadratic in z of the form

$$[z - (\alpha + j\beta)][z - (\alpha - j\beta)] = z^2 - 2\alpha z + \alpha^2 + \beta^2$$

$$= 1 - 2\alpha z^{-1} + (\alpha^2 + \beta^2)z^{-2} \tag{7.36a}$$

A single pole or zero on the real axis (that is, at $z = \pm\alpha$) leads to a first-order factor of the form

$$z \pm \alpha = 1 \pm \alpha z^{-1} \tag{7.36b}$$

Often we find that an Nth-order filter has N real zeros on the real axis in the z-plane. In this case, we may pair the z-plane zeros so that the numerator of each filter section is a quadratic of the form

$$1 \pm 2z^{-1} + z^{-2} \tag{7.36c}$$

The overall transfer function, $H(z)$, is given by

$$H(z) = KH_1(z)H_2(z) \ldots H_M(z)$$

where K is a gain factor which is used to adjust the magnitude response in the passband to a desired level. In most cases, K is set to a value to make the maximum response in the passband equal to unity.

7.4.13 Illustrative examples using method 2

Example 7.11

A second-order digital bandpass filter (BPF), with Butterworth characteristics and a passband between 200 and 300 Hz at a sampling frequency of 2 kHz, is required for a certain DSP application. Determine the transfer function of the digital filter by mapping the s-plane poles and/or zeros of a suitable prototype analogue lowpass filter into the z-plane.

Sketch and label the pole–zero diagrams of the prototype LPF, intermediate analogue BPF, and the digital BPF.

Solution

A first-order normalized LPF is required, since the order of the filter is doubled by the bandpass transformation. Thus

$$H(s) = \frac{1}{s + 1}$$

This transfer function has a single pole at $s_{1,1} = -1$. Now, $F_s = 2\text{ kHz} = 1/T$. Thus the prewarped band edge frequencies are

$$\omega_1' = \tan\left(\frac{\omega_1 T}{2}\right) = \tan\left(\frac{2\pi \times 200}{2 \times 2000}\right) = 0.3249$$

$$\omega_2' = \tan\left(\frac{\omega_2 T}{2}\right) = \tan\left(\frac{2\pi \times 300}{2 \times 2000}\right) = 0.5095$$

Thus ω_0 and W are given by

$$\omega_0^2 = \omega_1'\omega_2' = 0.1655, \; W = \omega_2' - \omega_1' = 0.1846$$

The single pole for the LPF is transformed into two poles for the BPF, using Equation 7.30, as follows:

$$s_{b,1} = \frac{0.1846}{2}\left\{-1 + \left[(-1)^2 - \frac{4 \times 0.1655}{(0.1846)^2}\right]^{1/2}\right\}$$

$$= -0.0923 + 0.3962\text{j}$$

$$s_{b,2} = \frac{0.1846}{2}\left\{-1 - \left[(-1)^2 - \frac{4 \times 0.1655}{(0.1846)^2}\right]^{1/2}\right\}$$

$$= -0.0923 - 0.3962\text{j} = s_{b,1}^*$$

From the BZT transformation, Equation 7.33 we have

$$z_{p,1} = \frac{1 - 0.0923 + 0.3962\text{j}}{1 + 0.0923 - 0.3962\text{j}} = 0.6181 + 0.569\text{j}$$

$$z_{p,2} = z_{p,1}^*$$

The prototype LPF has a single zero at infinity. This is mapped by the lowpass-to-bandpass transformation to the origin and to infinity in the bandpass s-plane. That is, $s_{z,1} = 0$, $s_{z,2} = \infty$. The BZT maps these zeros to the points $z = 1$ and $z = -1$ in the z-plane.

$$s_{z,1} \rightarrow z_{z,1} = 1; \quad s_{z,2} \rightarrow z_{z,2} = -1$$

We can now determine the discrete transfer function, $H(z)$, from the poles and zeros:

$$H(z) = \frac{(z-1)(z+1)}{(z - z_{p,1})(z - z_{p,2})}$$

$$= \frac{z^2 - 1}{z^2 - 1.236\,26z + 0.7265} = \frac{1 - z^{-2}}{1 - 1.236\,26z^{-1} + 0.7265z^{-2}}$$

The transfer function is identical to that of Example 7.9 to within a constant factor. The pole–zero diagrams are also identical to those given in Figure 7.13.

Example 7.12

Starting with a suitable analogue LPF, find the transfer function of a Chebyshev digital HPF in factored form to meet the following specifications:

passband edge frequency	15 kHz
attenuation at 18 kHz	>30 dB
passband ripple	1 dB
sampling frequency	48 kHz

Solution

The prewarped critical frequencies are

$$\omega'_p = \tan\left(\frac{15\,000\pi}{48\,000}\right) = 1.4966$$

$$\omega'_s = \tan\left(\frac{18\,000\pi}{48\,000}\right) = 2.4142$$

From the passband specifications, $\varepsilon = 0.3493$. The order of a suitable LP Chebyshev filter is 5 (the nearest integer).

With $\alpha = 1/N \sinh^{-1}(1/\varepsilon) = 0.3548$, $\sinh(\alpha) = 0.3623$, and $\cosh(\alpha) = 1.0636$, the left-hand poles of the normalized lowpass Chebyshev filter are located at ($\omega'_p = 1$)

$$s_{1,1} = 0.3623\cos\left[\frac{(2+5-1)\pi}{10}\right] + 1.0636\sin\left[\frac{(2+5-1)\pi}{10}\right]j$$
$$= -0.111\,96 + 1.0115j$$
$$s_{1,2} = 0.3623\cos\left[\frac{(4+5-1)\pi}{10}\right] + 1.0636\sin\left[\frac{(4+5-1)\pi}{10}\right]j$$
$$= -0.2931 + 0.6252j$$
$$s_{1,3} = 0.3623\cos\left[\frac{(6+5-1)\pi}{10}\right] + 1.0636\sin\left[\frac{(6+5-1)\pi}{10}\right]j = -0.3623$$
$$s_{1,4} = 0.3623\cos\left[\frac{(8+5-1)\pi}{10}\right] + 1.0636\sin\left[\frac{(8+5-1)\pi}{10}\right]j$$
$$= -0.2931 - 0.6252j$$
$$s_{1,5} = 0.3623\cos\left[\frac{(10+5-1)\pi}{10}\right] + 1.0636\sin\left[\frac{(10+5-1)\pi}{10}\right]j$$
$$= -0.111\,96 - 1.0115j$$

The reader should note the symmetry in the pole distribution and that $s_{1,1}$ and $s_{1,5}$ form a complex conjugate pair as do $s_{1,2}$ and $s_{1,4}$. Note also that each pole lies on the left-hand side of the s-plane, a necessary condition for stability.

Each prototype lowpass pole is transformed into the desired HP pole using Equation 7.27b:

$$s_{\mathrm{h},1} = -0.1618 + 1.4616j$$
$$s_{\mathrm{h},2} = -0.920\,13 + 1.9625j$$
$$s_{\mathrm{h},3} = -4.1306$$
$$s_{\mathrm{h},4} = s_{\mathrm{h},2} = -0.920\,13 - 1.9625j$$
$$s_{\mathrm{h},5} = s_{\mathrm{h},1} = -0.1618 - 1.4616j$$

Next, the poles are mapped from the s-plane into the z-plane using the BZT. From Equation 7.33 the z-plane poles after the BZT are given by (only the poles above the real axis are considered):

$$z_{\mathrm{h},1} = -0.3335 + 0.8386j$$
$$z_{\mathrm{h},2} = -0.4906 + 0.5207j$$
$$z_{\mathrm{h},3} = -0.6102j$$

All the zeros, $z_{z,k}$, are located at $z = 1$. The coefficients of the second- and first-order filter sections can then be obtained from the poles and zeros as (Equations 7.36a and 7.36b)

$$a_{11} = -2 \qquad a_{12} = -2 \qquad a_{13} = -1$$
$$a_{21} = 1 \qquad a_{22} = 1 \qquad a_{32} = 0$$

$$b_{11} = 0.6670 \qquad b_{12} = 0.9812 \qquad b_{13} = 0.6102$$
$$b_{21} = 0.8145 \qquad b_{22} = 0.5118 \qquad b_{32} = 0$$

Finally, the transfer function is given by

$$H(z) = KH_1(z)H_2(z)H_3(z)$$

where

$$H_1(z) = \frac{1 - 2z^{-1} + z^{-2}}{1 + 0.6670z^{-1} + 0.8145z^{-2}}$$

$$H_2(z) = \frac{1 - 2z^{-1} + z^{-2}}{1 + 0.9812z^{-1} + 0.5118z^{-2}}$$

$$H_3(z) = \frac{1 - z^{-1}}{1 + 0.6102z^{-1}}$$

7.4.14 Using an IIR filter design program

Whatever the method we adopt, it is evident that the bilinear transform method involves a considerable algebraic manipulation, with plenty of opportunity to make mistakes. Efficient computer programs now exist in the literature or commercially for calculating filter coefficients using the bilinear method, by merely specifying filter parameters of interest (IEEE, 1979; Gray and Markel, 1976; Parks and Burrus, 1987; Jong, 1982; DeFatta, 1988). Most of the computer programs in the literature are written in FORTRAN. There is a move away from such languages to more modern languages such as C or BASIC. A C language program for computing the filter coefficients using the BZT is on the PC disk for the book. The use of the program is illustrated in the following example.

Example 7.13

Illustrating the use of the IIR design program Obtain the coefficients of an audio digital filter with Chebyshev characteristics meeting the following specifications:

passband	0–2.5 kHz
stopband edge	2820 kHz
passband ripple	0.47 dB
sampling frequency	10 kHz
filter order	4

Solution

Using the program, the listing given below was obtained.

k	B_k	A_k
0	1.000000×10^{0}	1.934410×10^{-1}
1	-2.516884×10^{-1}	3.783311×10^{-1}
2	1.054118×10^{0}	5.241429×10^{-1}
3	-2.406030×10^{-1}	3.783311×10^{-1}
4	1.985861×10^{-1}	1.934410×10^{-1}

7.4.15 Choosing between the coefficient calculation methods

With the impulse invariant method, after digitizing the analogue filter, the impulse response of the original analogue filter is preserved, but not its magnitude–frequency response. Because of inherent aliasing, the method is inappropriate for highpass or bandstop filters. The bilinear method, on the other hand, yields very efficient filters and is well suited to the calculation of coefficients of frequency selective filters. It allows the design of digital filters with known classical characteristics such as Butterworth, Chebyshev and elliptic. Digital filters resulting from the bilinear transform method will, in general, preserve the magnitude response characteristics of the analogue filter but not necessarily the time domain properties. The impulse invariant method is good for simulating analogue systems with lowpass characteristics, but the bilinear method is best for frequency selective IIR filters.

For simple filtering applications, the pole–zero placement method provides a simple, but effective, way of obtaining the filter coefficients.

7.5 Stage 3: realization structures for IIR digital filters

Realization involves converting a given transfer function, $H(z)$, into a suitable filter structure. Flow or block diagrams are normally used to depict filter structures and they show the computational procedure for implementing the digital filter. The basic elements of realization structures are multipliers, adders, and delay elements; see Figure 7.16.

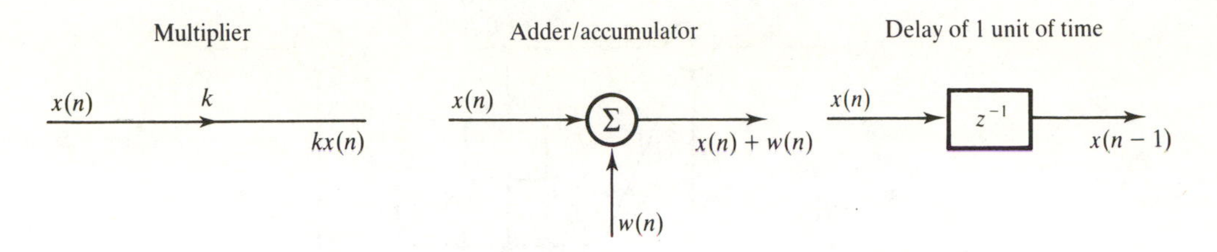

Figure 7.16 The basic elements of filter structures.

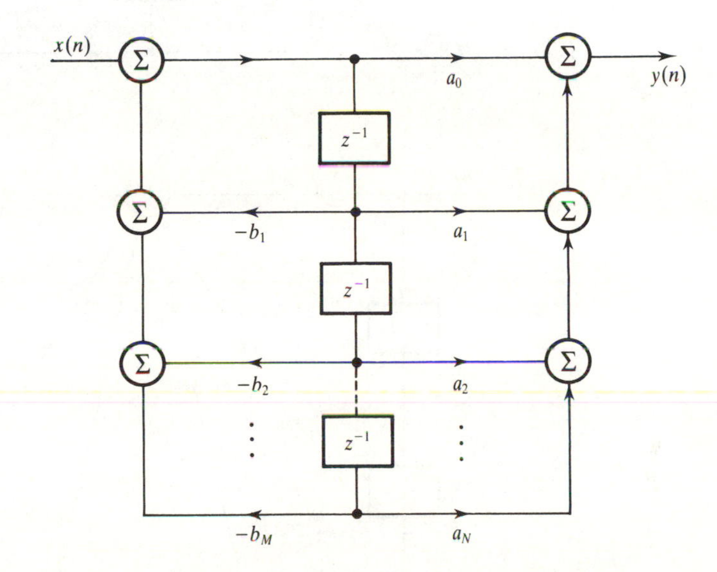

Figure 7.17 A direct form realization of an IIR filter.

Recall that the IIR filter is characterized by the following equations:

$$H(z) = \sum_{k=0}^{N} a_k z^{-k} \Bigg/ \left(1 + \sum_{k=1}^{M} b_k z^{-k}\right) \quad M \geqslant N \tag{7.37a}$$

$$y(n) = \sum_{k=0}^{N-1} a_k x(n-k) - \sum_{k=1}^{M} b_k y(n-k) \tag{7.37b}$$

A direct form realization of Equation 7.37 is shown in Figure 7.17, where $N = M$ for simplicity. Note that the coefficients used in the diagram are the same as in the transfer function, but with the signs reversed for the denominator coefficients. When the filter order is high, for example $M > 3$, direct realization of the filter as in Figure 7.17 is very sensitive to finite wordlength effects and should be avoided in these cases. In practice, $H(z)$ is normally broken down into smaller sections, typically second- and/or first-order blocks, which are then connected up in cascade or in parallel (see later).

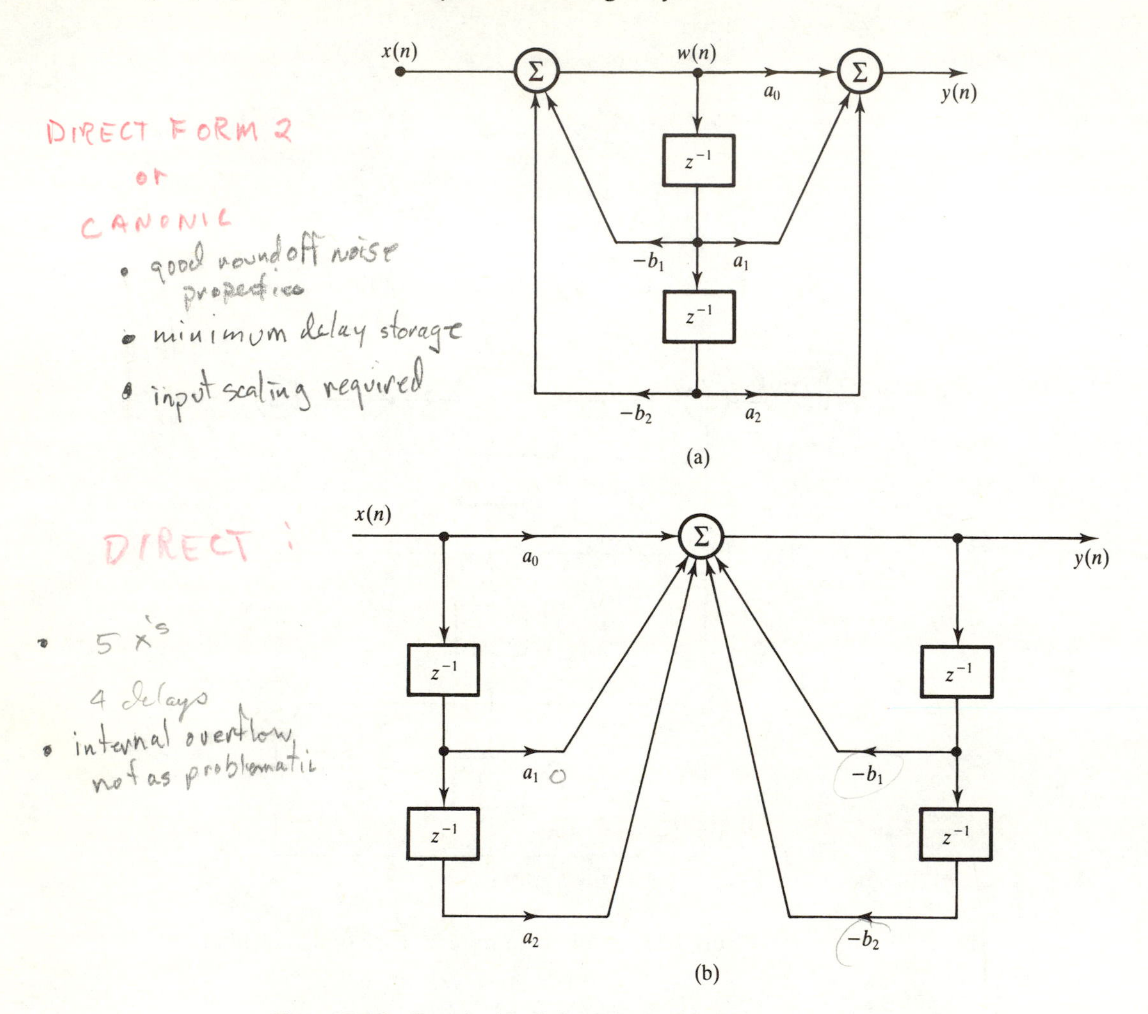

Figure 7.18 Practical building blocks for IIR realization structures: (a) canonic second-order section; (b) direct form second-order section.

7.5.1 Practical building blocks for IIR filters

Examples of practical second-order building blocks used in realizing higher-order IIR filters are depicted in Figure 7.18. The first (Figure 7.18(a)) is often called a canonic section (or direct form 2) because it has the minimum number of delay elements. This biquadratic section is characterized by the following equations:

$$w(n) = \sum_{k=0}^{2} a_k x(n) - \sum_{k=1}^{2} b_k w(n-k) \tag{7.38a}$$

$$y(n) = \sum_{k=0}^{2} a_k w(n-k) \tag{7.38b}$$

$$H(z) = \frac{a_0 + a_1 z^{-1} + a_2 z^{-2}}{1 + b_1 z^{-1} + b_2 z^{-2}} \tag{7.38c}$$

The second filter section (Figure 7.18(b)) is a direct realization of the second-order IIR equation. It is characterized by the following equations:

$$y(n) = \sum_{k=0}^{2} a_k x(n-k) - \sum_{k=1}^{2} b_k y(n-k) \tag{7.39a}$$

$$H(z) = \frac{a_0 + a_1 z^{-1} + a_2 z^{-2}}{1 + b_1 z^{-1} + b_2 z^{-2}} \tag{7.39b}$$

The canonic section (Figure 7.18(a)) is the most popular because it has a good roundoff noise property and requires a minimum number of storage elements, but it is susceptible to internal overflows. To avoid internal overflow it is necessary to scale the input to the filter section. Scaling is not mandatory for the direct form (Figure 7.18(b)) because it has only one adder, and may be preferred where scaling is not desired, for example high fidelity digital audio (Dattarro, 1988). Under certain conditions the direct form is superior to the canonic section, in terms of noise performance.

The coupled form has some desirable finite wordlength properties (Gold and Rader, 1969), but it requires more computational effort and cannot be readily used to implement transfer functions with second order numerator coefficient.

The filter blocks given in Figure 7.18 are general, second-order, sections. Several other filter blocks can be derived from them. For example, if the numerator coefficients a_1 and a_2 in Figure 7.18(a) are both zero we have a purely recursive structure. On the other hand, if the filter coefficients were obtained using elliptical functions the coefficient a_2 is unity. A first-order filter block is readily obtained by setting $a_2 = b_2 = 0$ in any of the structures above.

Figure 7.19 shows the transposes of the second-order canonic and the direct form filter sections. These are obtained from Figures 7.19(a) and 7.19(b) respectively, by interchanging all the adders and branch nodes, and reversing the directions of the arrows. Although the transfer functions of the sections in Figure 7.19 are identical to their transposes, their finite wordlength properties are quite different. Other structures that are less sensitive to finite wordlength effects are available, but these are usually more complicated. Examples include minimum-noise, state-variable, and lattice structure.

7.5.2 Cascade and parallel realization structures for higher IIR filters

In practice higher-order transfer functions are realized as cascades or parallel combinations of second- and/or first-order building blocks described above. Typically, in cascade realization the transfer function is factored into $N/2$

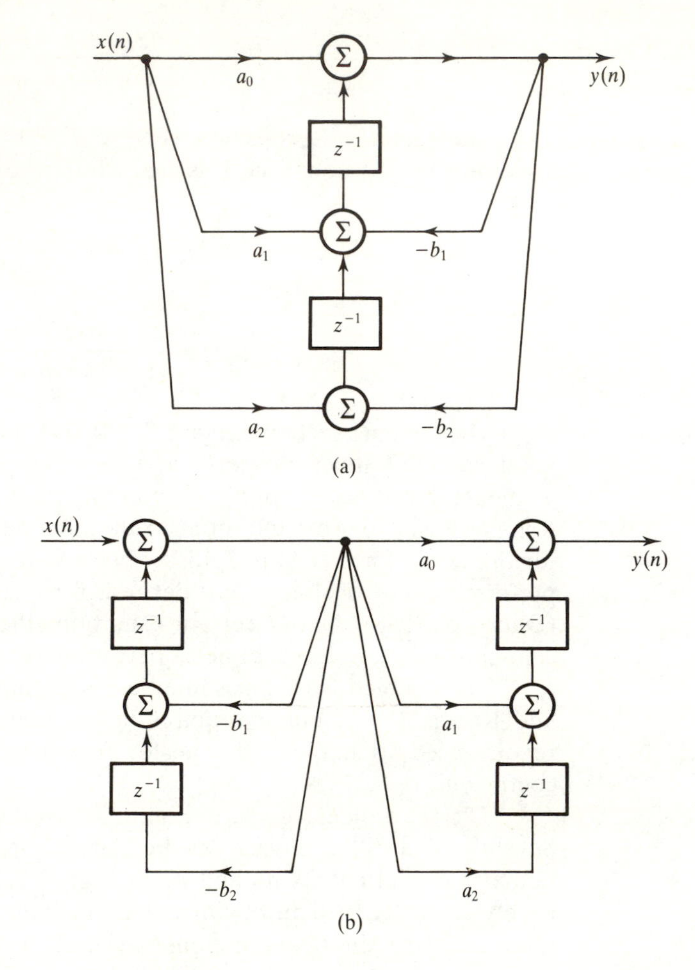

Figure 7.19 (a) Transpose canonic second-order section; (b) transpose direct form second-order section.

second-order factors:

$$H(z) = \prod_{k=1}^{N/2} \left[\frac{a_{0k} + a_{1k}z^{-1} + a_{2k}z^{-2}}{1 + b_{1k}z^{-1} + b_{2k}z^{-2}} \right]$$

$$= \prod_{k=1}^{N/2} \frac{N_k(z)}{D_k(z)} \tag{7.40a}$$

where

$$\begin{aligned} N_k(z) &= a_{0k} + a_{1k}z^{-1} + a_{2k}z^{-2} \\ D_k(z) &= 1 + b_{1k}z^{-1} + b_{2k}z^{-2} \end{aligned} \tag{7.40b}$$

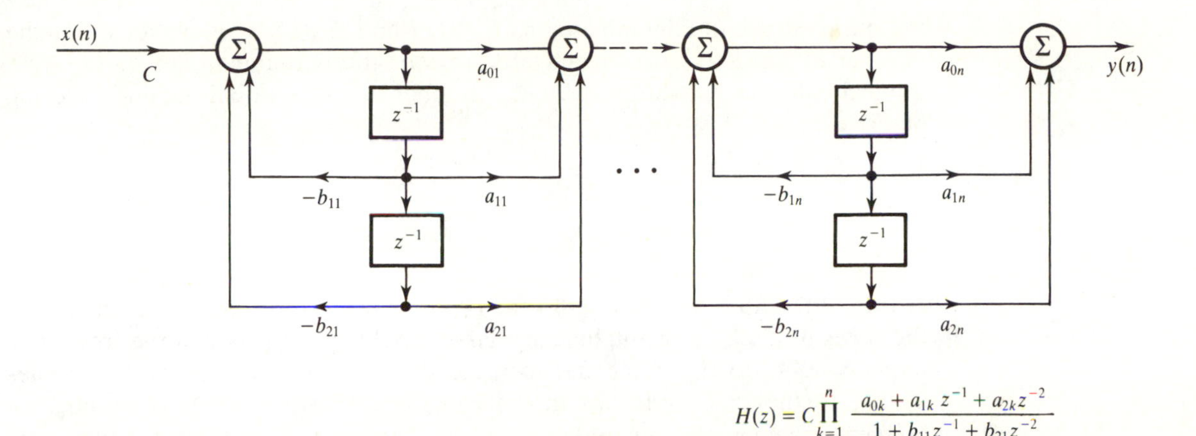

$$H(z) = C\prod_{k=1}^{n} \frac{a_{0k} + a_{1k}z^{-1} + a_{2k}z^{-2}}{1 + b_{11}z^{-1} + b_{21}z^{-2}}$$

Figure 7.20 Cascade realization.

and N, the filter order, is assumed to be even. If N is odd, then one of the $H_k(z)$ will be a first-order section.

Each second-order factor, $H_k(z)$, can be realized using one of the building blocks and the resulting blocks connected in cascade; see Figure 7.20. Three difficulties arise with the cascade realization: (1) how to pair the numerator factors with the denominator factors; (2) the order in which the individual sections should be connected; (3) the need to scale the signal levels at various points within the filter to avoid the levels becoming too large or too small.

The numerator and denominator factors can be ordered in a variety of ways. For example, a fourth-order filter can be factored into two second-order sections, and then paired and ordered in one of four different ways:

(1) $$H(z) = \frac{N_1(z)}{D_1(z)}\frac{N_2(z)}{D_2(z)}$$

(2) $$H(z) = \frac{N_2(z)}{D_2(z)}\frac{N_1(z)}{D_1(z)}$$

(3) $$H(z) = \frac{N_1(z)}{D_2(z)}\frac{N_2(z)}{D_1(z)}$$

(4) $$H(z) = \frac{N_2(z)}{D_1(z)}\frac{N_1(z)}{D_2(z)}$$

where each $N_k(z)$ and $D_k(z)$ is a second-order polynomial defined in Equation 7.40b. In the first case, the first filter section consists of the numerator and denominator pair $N_1(z)$ and $D_1(z)$, while the second filter section consists of the numerator and denominator pair $N_2(z)$ and $D_2(z)$. It is obvious that the number of possible pairings and orderings of the denominator is quite large. Typically, for an Nth-order filter, the number of different pairing and ordering possible is

$$\left(\frac{N}{2}!\right)^2 \tag{7.41}$$

A rule of thumb is to pair $N_i(z)$ with $D_k(z)$ if the zeros of $N_i(z)$ are closest to the poles of $D_k(z)$ to avoid having a large amplitude response at the frequency corresponding to the pole, and to place the second-order section with poles closest to the unit circle last in the cascade (Jackson, 1986). A number of efficient schemes have been developed for pairing and ordering the filter sections, based on which ordering gives the best signal-to-noise ratio, a topic intimately linked with pairing and ordering (see Section 7.6).

In parallel realization, an Nth-order transfer function $H(z)$ is expanded using partial fractions as

$$H(z) = C + \sum_{k=1}^{N/2} H_k(z) \tag{7.42}$$

where

$$C = \frac{a_N}{b_N}, \quad H_k(z) = \frac{a_{0k} + a_{1k}z^{-1}}{1 + b_{1k}z^{-1} + b_{2k}z^{-2}}$$

Again, each second-order section can be realized using the building blocks described previously as shown in Figure 7.21. It is worth noting that, in the parallel realization, the numerator coefficient for z^{-2} is zero. In the parallel structure, the order in which the sections is connected is not important. Furthermore, scaling is easier and can be carried out for each block independently (see later), and the SNRs are comparable with those of the best cascade realization (Jackson, 1986). However, the zeros of parallel structures are more sensitive to coefficient quantization errors. It should be mentioned that the sensitivity of the zeros of the parallel structure to coefficient quantization appears to be most serious when coefficient wordlength is down to as little as 5 bits or less. It appears that for coefficient wordlengths of 12 or more bits the differences between the parallel and cascade structures are small for most filters. However, an important advantage of the cascade method is that between 25% and 50% of the filter coefficients are simple integers (0, ±1 or ±2) when derived from classical analogue filters via the BZT. This is attractive in systems with only primitive arithmetic capability where the number of multiplications must be kept low. Further, most available software packages produce coefficients for cascade realization but not for the parallel structure. It is for these reasons that the cascade method has become very popular.

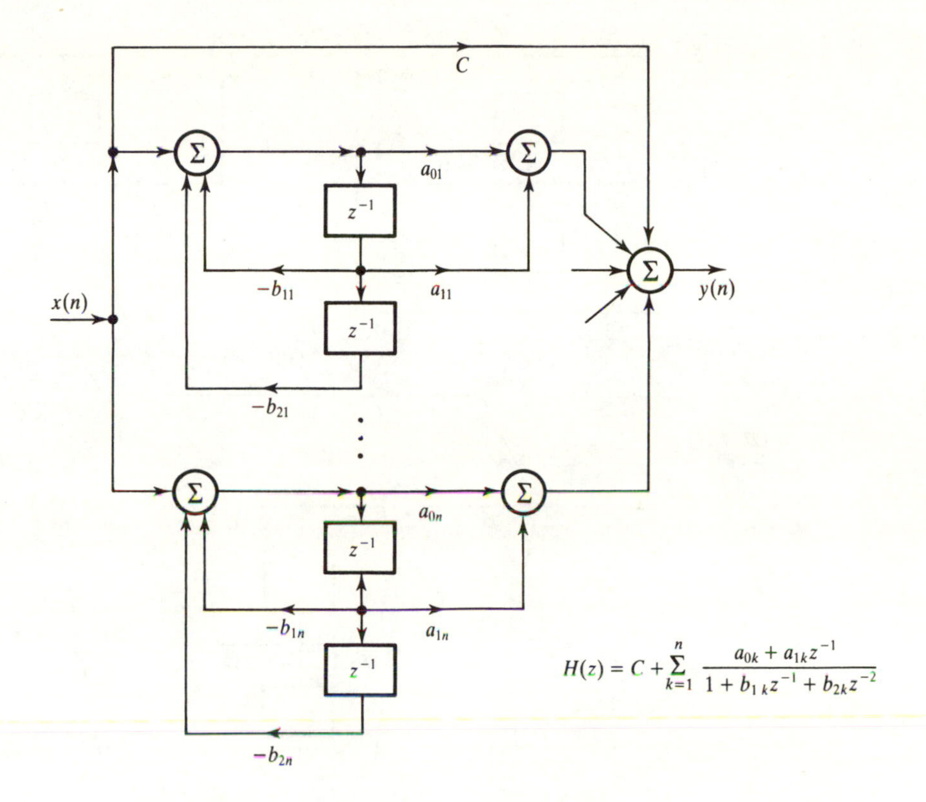

Figure 7.21 Parallel realization.

Example 7.14

Develop the transfer functions for (1) the cascade and (2) the parallel realization structures for the filter characterized by the following transfer function using second- and first-order sections:

$$H(z) = \frac{0.1432(1 + 3z^{-1} + 3z^{-2} + z^{-3})}{1 - 0.1801z^{-1} + 0.3419z^{-2} - 0.0165z^{-3}}$$

Solution

(1) For cascade realization, $H(z)$ is expressed in factored form:

$$H(z) = 0.1432 \frac{1 + 2z^{-1} + z^{-2}}{1 - 0.1307z^{-1} + 0.3355z^{-2}} \frac{1 + z^{-1}}{1 - 0.0490z^{-1}}$$

(2) For parallel realization $H(z)$ is expressed as the sum of second- and

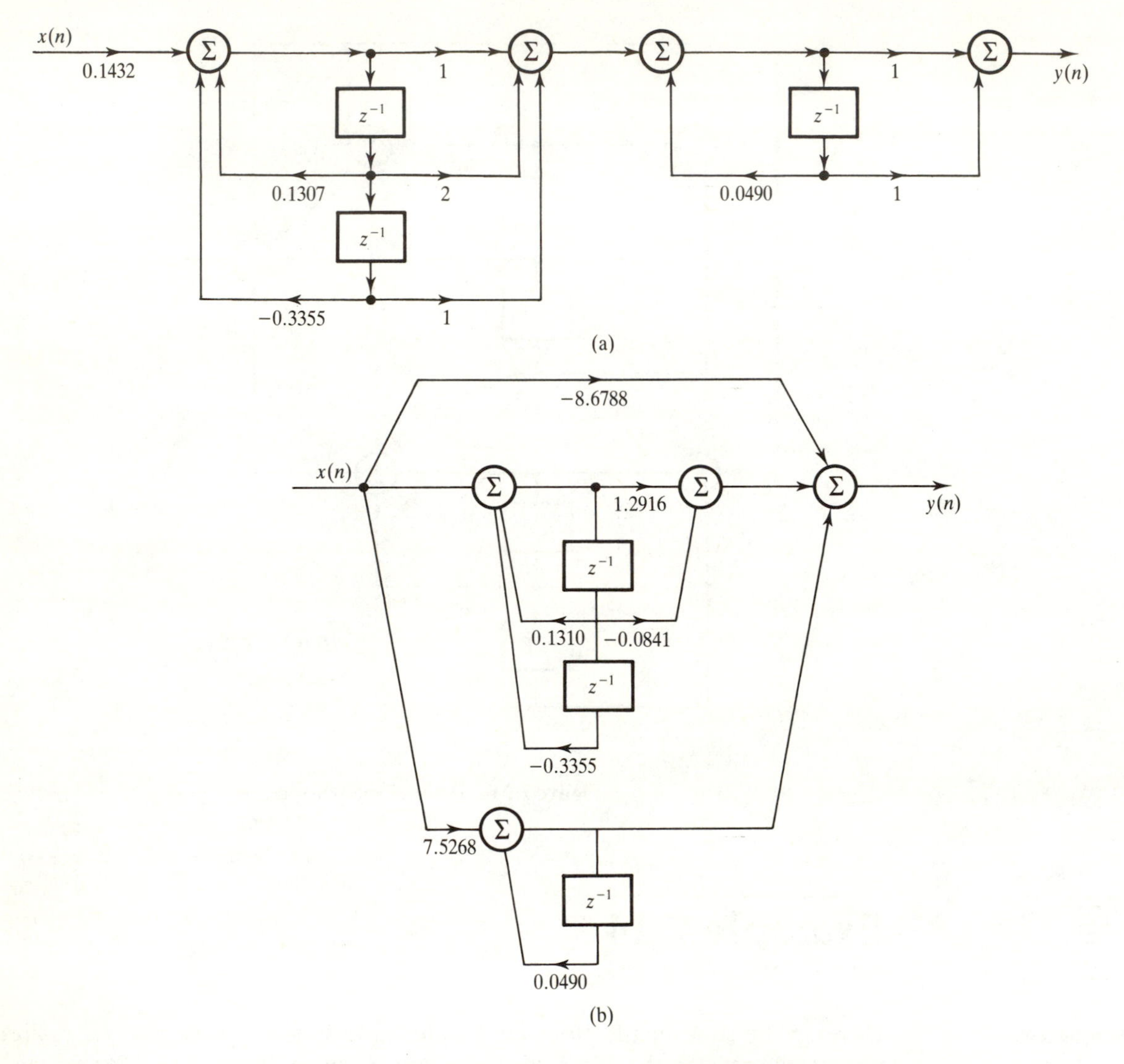

Figure 7.22 (a) Cascade and (b) parallel realizations for Example 7.14.

first-order sections using partial fractions:

$$H(z) = \frac{1.2916 - 0.08407z^{-1}}{1 - 0.131z^{-1} + 0.3355z^{-2}} + \frac{10.1764}{1 - 0.049z^{-1}} - 8.7107$$

The realization diagrams for the cascade and parallel realizations are given in Figures 7.22(a) and 7.22(b) respectively. The coefficients for the parallel realization were obtained using the C language program presented in Chapter 3.

7.6 Stage 4: analysis of finite wordlength effects

The coefficients, a_k and b_k, obtained from stage 2 are of infinite or very high precision, typically six decimal places. When an IIR digital filter is implemented in a small system, such as an 8-bit microcomputer, errors arise in representing the filter coefficients and in performing the arithmetic operations indicated by the difference equation. These errors degrade the performance of the filter and in extreme cases lead to instability.

Before implementing an IIR filter, it is important to ascertain the extent to which its performance will be degraded by finite wordlength effects and to find a remedy if the degradation is not acceptable. In general, the effects of these errors can be reduced to acceptable levels by using more bits but this may be at the expense of increased cost.

The main errors in digital IIR filters are as follows:

- ADC quantization noise, which results from representing the samples of the input data, $x(n)$, by only a small number of bits;
- coefficient quantization errors, caused by representing the IIR filter coefficients by a finite number of bits;
- overflow errors, which result from the additions or accumulation of partial results in a limited register length;
- product round-off errors, caused when the output, $y(n)$, and results of internal arithmetic operations are rounded (or truncated) to the permissible wordlength.

The extent of filter degradation depends on (i) the wordlength and type of arithmetic used to perform the filtering operation, (ii) the method used to quantize filter coefficients and variables, and (iii) the filter structure. From a knowledge of these factors, the designer can assess the effects of finite wordlength on the filter performance and take remedial action if necessary. Depending on how the filter is to be implemented some of the effects may be insignificant. For example, when implemented as a high level language program on most large computers, coefficient quantization and roundoff errors are not important. For real-time processing, finite wordlengths (typically 8 bits, 12 bits, and 16 bits) are used to represent the input and output signals, filter coefficients and the results of arithmetic operations. In these cases, it is nearly always necessary to analyse the effects of quantization on the filter performance.

The effects of finite wordlength on performance are more difficult to analyse in IIR filters than in FIR filters because of their feedback arrangements. However, the use of the PC-based program provided allows practical solutions to be obtained for specific filters. The effects of each of the four sources of errors listed above will be discussed, in turn, in the next few sections.

7.6.1 ADC quantization noise

The ADC quantizes the analogue input signal into a finite number of bits, typically 8, 12 or 16, which gives rise to quantization noise.

As discussed in Chapter 1, the quantization noise power (or variance, σ^2_{ADC}) is given by

$$\sigma^2_{\mathrm{A}} = \frac{q^2}{12} = \frac{2^{-2B}}{3} \qquad \textbf{(7.43)}$$

where q is the quantization step size and B is the number of ADC bits. Clearly, the level of the noise can be readily reduced by increasing the number of ADC bits. It is also possible to reduce it using multirate techniques (see Chapter 8). In general for values of B above 12 bits, the noise due to quantization error is insignificant, except for applications such as professional audio where at least 16 bits are required for acceptable performance.

The noise due to ADC quantization is fed into the IIR filter as an irreversible error. We will consider the effect of ADC noise on the performance of a simple IIR filter.

Example 7.15

An 8-bit ADC feeds a simple highpass IIR filter characterized by the following transfer function:

$$H(z) = \frac{1}{z + 0.5}$$

Estimate the steady state quantization noise power at the output of the filter.

Solution

The noise power, due to the ADC, at the system input is given by (Equation 7.43)

$$\sigma^2_{\mathrm{A}} = \frac{2^{-16}}{3}$$

The output noise, due to the ADC, is given by

$$\sigma^2_{\mathrm{oA}} = \sigma^2_{\mathrm{A}}\left[\frac{1}{2\pi \mathrm{j}}\oint_c H(z)H(z^{-1})\frac{\mathrm{d}z}{z}\right] = \sigma^2_{\mathrm{A}}\sum_{k=0}^{\infty} h^2(k) \qquad \textbf{(7.44)}$$

where $\oint_c$ indicates a contour integral and $h(k)$ is the impulse response of the system (see Chapter 3). The term in the square brackets can be viewed as the system power gain, serving to amplify (or alternate) the ADC noise.

For the simple transfer function in this example, the term in the square brackets can be readily obtained using the residue method discussed in Chapter 3. Using the results in Chapter 3 and the unit circle as the contour, the term inside the square brackets is simply equal to 4/3. Thus, the quantization noise power is

$$\sigma_{oA}^2 = \frac{2^{-16}}{3} \times \frac{4}{3}$$

7.6.2 Coefficient quantization errors

The primary effect of quantizing the filter coefficients into a finite number of bits is to alter the positions of the poles and zeros of $H(z)$ in the z-plane. For narrowband filters, for example, the poles will be close to the unit circle so that any significant deviation in their positions could make the filter unstable. The fewer the number of bits used to represent the coefficients the more will be the deviation in the pole and zero positions.

As well as potential instability, deviations in the locations of the poles and zeros also lead to deviations in the frequency response. The quantized filter should be analysed to ensure that its wordlength is sufficient for both stability and satisfactory frequency response. Thus, essentially, at this stage the designer should determine the number of bits that are required to represent the filter coefficients for stability and for the desired frequency response to be satisfied.

7.6.3 Coefficient wordlength to maintain stability

Our stability discussions will be restricted to second-order filter sections since these are the basic building blocks of any filter. Consider a second-order section characterized by the familiar equations

$$H(z) = \frac{a_0 + a_1 z^{-1} + a_2 z^{-2}}{1 + b_1 z^{-1} + b_2 z^{-2}}$$

$$y(n) = \sum_{k=0}^{2} a_k x(n-k) - \sum_{k=1}^{2} b_k y(n-k)$$

The poles (or the roots of the denominator) are located at

$$p_1 = \tfrac{1}{2}[-b_1 + (b_1^2 - 4b_2)^{1/2}] \quad \textbf{(7.45a)}$$

$$p_2 = \tfrac{1}{2}[-b_1 - (b_1^2 - 4b_2)^{1/2}] \quad \textbf{(7.45b)}$$

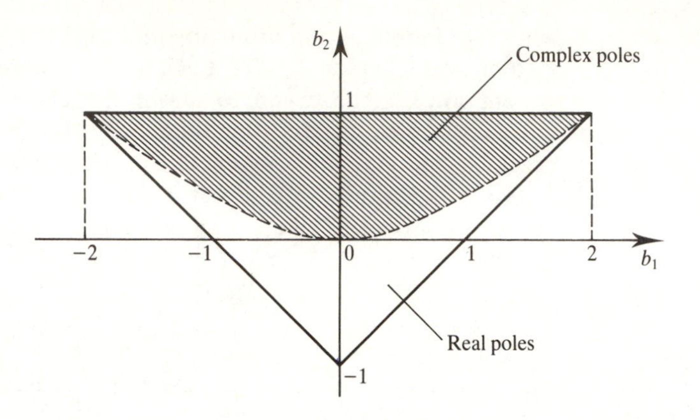

Figure 7.23 Stability triangle showing values of filter coefficients, b_1 and b_2, for which the filter is stable.

For each second-order section, three types of poles arise: complex conjugate poles, poles that are real and unequal, and real and equal (multiple-order) poles. Complex conjugate poles are the most common and occur if $b_1^2 < 4b_2$. For this case, the poles are each located at a radius, r, from the origin and at an angle θ given by:

$$p_1 = r\angle\theta,\ p_2 = r\angle-\theta \tag{7.46}$$

where

$$r = {b_2}^{1/2},\ \theta = \cos^{-1}\left(-\frac{b_1}{2r}\right)$$

Small changes in the coefficients b_1 and b_2, due to coefficient quantization, will lead to changes in both r and θ. For stability, the filter coefficients must lie inside the stability triangle (Figure 7.23) bounded by

$$0 \leqslant |b_2| < 1 \tag{7.47a}$$

$$|b_1| \leqslant 1 + b_2 \tag{7.47b}$$

The first bound specifies that the poles must lie inside the unit circle, since the pole radius is given by Equation 7.46. From Equations 7.46 and 7.47, a number of simple formulas can be derived for estimating the number of bits required to maintain stability, but these are applicable to a restricted set of cases. An alternative way of estimating suitable coefficient wordlength for stability is to analyse individual second-order blocks for various values of coefficient wordlengths (see example in Section 7.6.5).

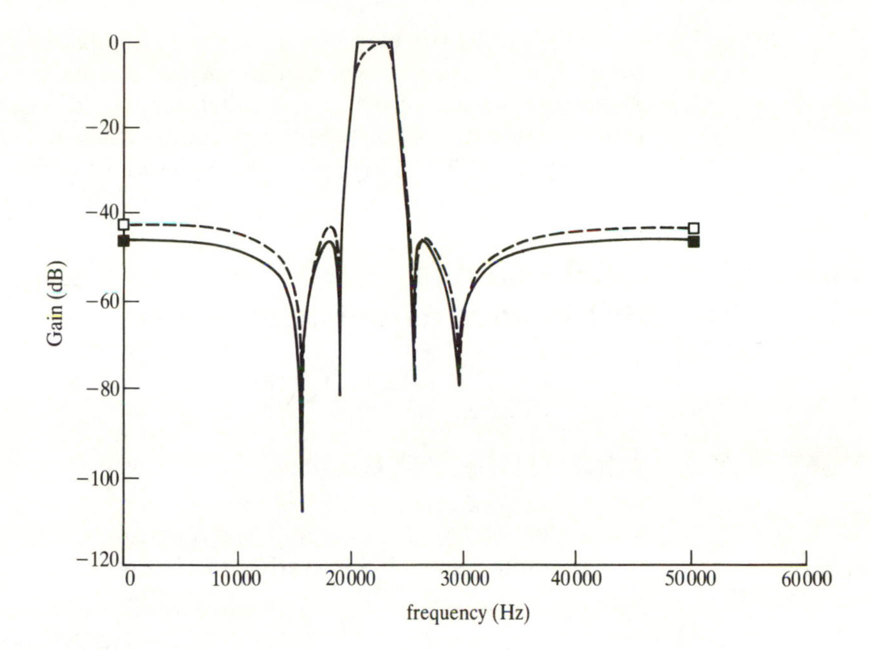

Figure 7.24 Practical effects of coefficient quantization on the frequency response: ■, unquantized; □, quantized, 5 bits.

7.6.4 Coefficient wordlength for desired frequency response

The number of bits required for stability may not guarantee a satisfactory response. The effect of representing the coefficients by too few bits is to alter the frequency response in the passband and stopband (see Figure 7.24). The changes in the passband are caused primarily by changes in the positions of the poles, and those in the stopband by changes in the locations of the zeros.

A procedure for determining a suitable coefficient wordlength for satisfactory frequency response is to find the minimum coefficient wordlength for which the passband and stopband requirements are satisfied. Although this approach may involve a lot of computation, the widespread availability of PCs and programs for finite wordlength analysis (FWA) makes it relatively straightforward to determine the wordlength for specific filters (an example program is on the PC disk for the book). An alternative approach employs a statistical method and is said to yield reasonably accurate estimates (Antoniou, 1979).

A smaller number of bits may be used to represent the filter coefficients if an optimization approach is used to quantize the filter coefficients, or by increasing the filter order, all other parameters remaining unchanged. This offers a trade-off between the coefficient wordlength and the filter order (Rabiner and Gold, 1975). For example, in a particular problem the designer may wish to use an existing processor or system with wordlengths already predetermined. If the designer finds that the required wordlength is greater than the processor wordlength, then it may be preferable to increase the order of the filter to bring the coefficient wordlength down to match that of the processor. However, the use of a higher-order filter would require more computational effort, which has some speed implications, and may be more susceptible to roundoff noise. The designer needs to consider the trade-off carefully.

7.6.5 A detailed example – coefficient wordlength requirements for stability and frequency response

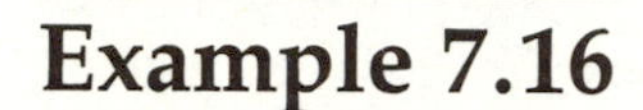

Example 7.16

A digital filter is required to satisfy the following frequency response specifications:

passband	20.5–23.5 kHz
stopband	0–19 kHz, 25–50 kHz
passband ripple	$\leqslant$ 0.25 dB
stopband attenuation	> 45 dB
sampling frequency	100 kHz

(1) Determine a suitable transfer function for the filter.

(2) Determine a suitable coefficient wordlength

(a) to maintain stability and

(b) to satisfy the frequency response specifications.

(3) Obtain and plot the frequency response of the unquantized filter and those of quantized filters corresponding to part (2).

Solution

(1) Using the design program (on the PC disk for the book) it was found that an elliptic filter characterized by the following transfer function is suitable:

$$H(z) = H_1(z)H_2(z)H_3(z)H_4(z)$$

where

$$H_1(z) = \frac{1 + 0.0339z^{-1} + z^{-2}}{1 - 0.1743z^{-1} + 0.9662z^{-2}}$$

$$H_2(z) = \frac{1 - 0.7563z^{-1} + z^{-2}}{1 - 0.5588z^{-1} + 0.9675z^{-2}}$$

$$H_3(z) = \frac{1 + 0.5331z^{-1} + z^{-2}}{1 - 0.2711z^{-1} + 0.9028z^{-2}}$$

$$H_4(z) = \frac{1 - 1.1489z^{-1} + z^{-2}}{1 - 0.4441z^{-1} + 0.9045z^{-2}}$$

(2) (a) The denominator coefficients of each of the second-order sections were each quantized by rounding to B bits ($B = 2, 3, \ldots, 29$) including the sign bits. For each value of B, the quantized coefficients and the pole location in polar form were computed. To illustrate, consider the first second-order filter section, $H_1(z)$. For $B = 8$ bits, the denominator coefficients are quantized by rounding as follows:

$$b_1 = -(0.1743 \times 2^7 + 0.5) = -22.8104 = -23$$

$$b_2 = 0.9662 \times 2^7 + 0.5 = 124.1736 = 124$$

In fractional notation, the coefficients are

$$b_1 = -23/128 = -0.179\,687\,5$$

$$b_2 = 124/128 = 0.968\,75$$

From Equation 7.46, the pole radius and angle for the section for $B = 8$ bits are given by

$$r = \sqrt{0.968\,75} = 0.9843,$$

$$\theta = \cos^{-1}\left(-\frac{b_1}{2r}\right) = \cos^{-1}(0.091\,28) = 84.76°$$

All the quantized coefficients and polar coordinates were computed using an analysis program. If, for any coefficient wordlength, the pole radial distance of a filter section is equal to or greater than unity then there is potential instability. It was found that for all the filter sections as few as $B = 5$ bits are required to maintain stability. In general, if the poles of an unquantized second-order section is at a radius $r < 0.9$, instability is unlikely if a coefficient wordlength of 8 bits or more is used.

(b) The coefficients of the second-order sections were each quantized to various wordlengths as described above. For each wordlength, the quantized coefficients were then combined to yield an overall quantized transfer function in direct form. Examples for wordlengths of 5 and 16 bits are given in Table 7.1. The passband ripples and stopband attenuation of the quantized filter for the various coefficient wordlengths were obtained. It was found that, to satisfy the frequency response specifications in both passband and stopband, 16 or more bits are required. We note that this is more than the wordlength required for stability.

Table 7.1 Coefficients for Example 7.16, showing wordlength effects.

	$A(k)$			$B(k)$		
k	*Ideal*	*5 bits*	*16 bits*	*Ideal*	*5 bits*	*16 bits*
0	1.000 000	1.000 000	1.000 000	1.000 000	1.000 000	1.000 000
1	−1.338 200	−1.250 000	−1.338 165	−1.448 300	−1.437 500	−1.448273
2	3.806 737	3.707 031	3.806 700	4.483 108	4.355 4690	4.483 071
3	−3.556 357	−3.288 574	−3.556 255	−4.220 527	−4.060 791	−4.220 431
4	5.629 177	5.443 726	5.629 105	6.647 162	6.261 536	6.647 087
5	−3.556 357	−3.288 574	−3.556 255	−3.945 450	−3.677 216	−3.945 354
6	3.806 737	3.707 031	3.806 700	3.918 3981	3.573 486	3.918 352
7	−1.338 200	−1.250 000	−1.338 165	−1.182 6020	−1.067 047	−1.182 575
8	1.000 000	1.000 000	1.000 000	0.763 3402	0.672 912	0.763 338

(3) The frequency responses, scaled to have a maximum of 0 dB, for the unquantized and quantized ($B = 5$ bits) filters are depicted in Figure 7.24. Visually, the response for the 16-bit quantized filter was the same as that of the unquantized filter and is therefore not shown.

7.6.6 Addition overflow errors

In 2's complement arithmetic, the addition of two large numbers of a similar sign may produce an overflow, that is a result that exceeds the permissible wordlength, which would cause a change in the sign of the output sample. Thus a very large positive number becomes a very large negative number and vice versa (Figure 7.25). Consider the canonic section in Figure 7.26. Because of the

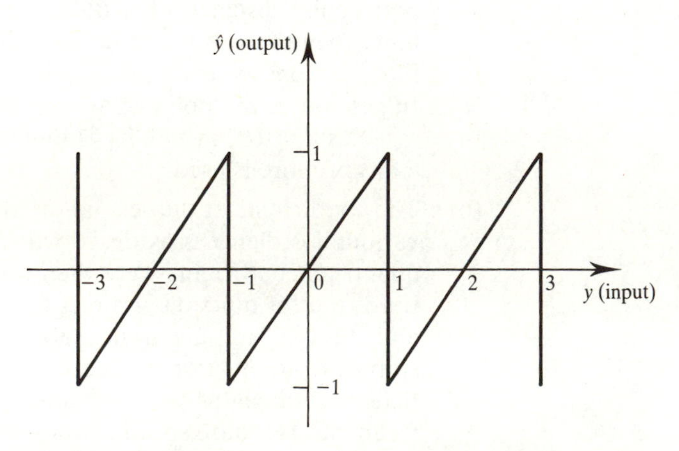

Figure 7.25 Overflow characteristics in 2's complement arithmetic. At instants when the input exceeds the permissible range $(-1, 1)$ overflow occurs.

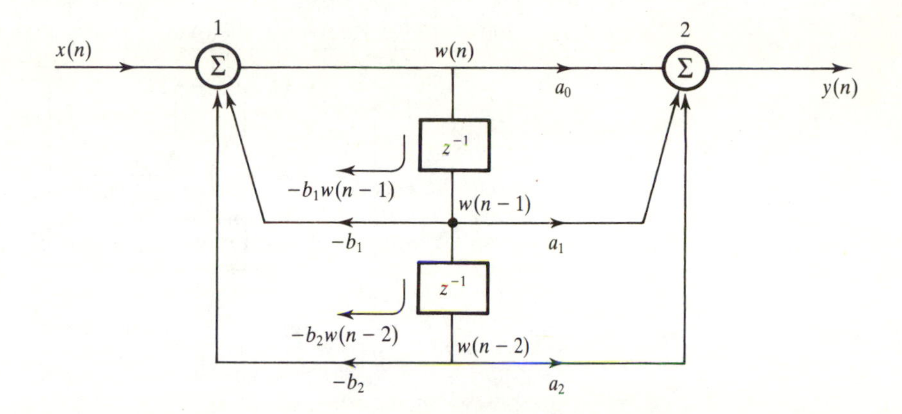

Figure 7.26 An illustration of the effects of addition overflow. Large inputs of the same sign at adder 1 will cause $w(n)$ to become too large. As $w(n)$ is fed back, the effect is self-sustaining.

recursive nature of the IIR filter, an overflow at $w(n)$ is fed back and used to compute the next output where it can cause further overflow, creating undesirable self-sustaining oscillations. Large-scale overflow limit cycles, as they are called, are difficult to stop once they start and may only be stopped by re-initializing the filter.

Large-scale overflow occurs at the outputs of the adders and may be prevented by scaling the inputs to the adders in such a way that the outputs are kept low, but this is at the expense of reduced signal-to-noise ratio (SNR). Thus it is important to select scale factors to prevent overflow while at the same time maintaining the largest possible SNR.

7.6.7 Principles of scaling

7.6.7.1 Canonic section

Consider the second-order canonic section in Figure 7.27(a). The scaling factor, s_1, at the filter input is chosen to avoid or reduce the possibility of overflow at the output of the left adder. To keep the overall filter gain the same, the numerator coefficients are multiplied by s_1.

There are three common methods of determining suitable scale factors for a filter. In method 1, often called the L_1 norm, the scale factor is chosen as follows:

$$s_1 = \sum_{k=0}^{\infty} |f(k)| \tag{7.48}$$

where $f(k)$ is the impulse response from the input to the output of the first

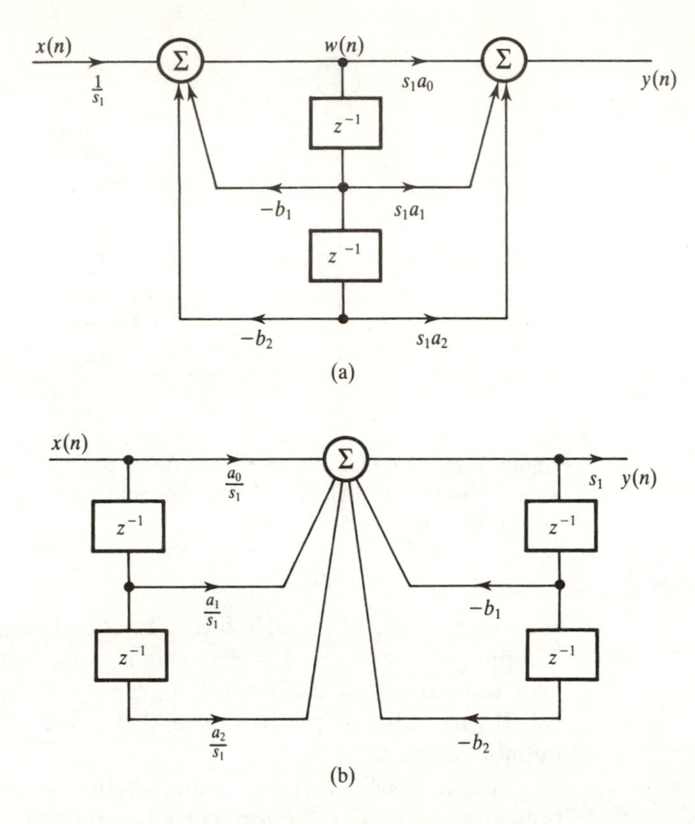

Figure 7.27 Principles of scaling in second-order filter sections: (a) canonic; (b) direct form.

adder, that is $w(n)$. This scale factor s_1 ensures that the overall gain of the filter from the input to $w(n)$ is unity and so there cannot be an overflow at $w(n)$. The impulse response, $f(k)$, may be obtained by first determining the corresponding transfer function $F(z)$ and then inverse z-transforming.

In the second method, often called the L_2 norm, the scale factor, s_1, is obtained as

$$s_1 = \left[\sum_{k=0}^{\infty} f^2(k)\right]^{1/2} \tag{7.49}$$

Alternatively, the L_2 norm scale factor may be obtained using contour integration via the relationship

$$\sum_{k=0}^{\infty} f^2(k) = \frac{1}{2\pi j}\oint F(z)F(z^{-1})\frac{dz}{z} \tag{7.50}$$

where $F(z)$ is the z-transform of $f(k)$ and $\oint$ indicates a contour integral around the unit circle $|z| = 1$. Evaluating this, we have (see Appendix 7C)

$$s_1^2 = \sum_{k=0}^{\infty} f^2(k) = \frac{1}{2\pi j}\oint \frac{1}{1 + b_1z^{-1} + b_2z^{-2}} \frac{1}{1 + b_1z + b_2z^2} \frac{dz}{z}$$

$$= \frac{1}{1 - b_2^2 - b_1^2(1 - b_2)/(1 + b_2)} \tag{7.51}$$

The use of Equation 7.51 avoids the evaluation of the infinite summation of Equation 7.49. In practice, however, $f(k)$ has only a finite number of significant terms and is readily evaluated with a suitable finite wordlength analysis program.

In method 3, known as the L_∞ norm, the scale factor is obtained as

$$s_1 = \max|F(w)| \tag{7.52}$$

where $F(w)$ is the peak amplitude of the frequency response between the input and $w(n)$.

The underlying assumption in method 1 is that the input is bounded, that is $|x(n)| < 1$. The scaling scheme is such that regardless of the type of input there will be no overflow. This is a somewhat drastic scaling scheme, as it caters for situations which are unlikely to happen in normal, real-world situations. The L_2 norm corresponds to placing an energy constraint on both the input and the transfer function. Its main attraction is that finite wordlength effect analysis requires the evaluation of L_2 norms (compare for example Equation 7.44 and Equation 7.50). It is also possible to derive closed form expressions for a variety of filter structures. Method 3 ensures that the filter does not overflow when a sine wave is applied and offers the best compromise. It is the scaling scheme often preferred, especially as it allows the effects of scaling to be verified experimentally using sine waves.

A compact way of expressing the ith scale factor is:

$$s_i = \|F\|_p$$

where the symbol $\|\cdot\|$ indicates the norm, and $p = 1, 2, \infty$ denotes the type of norm. Scale factors obtained by the three methods satisfy the following relationship:

$$L_2 < L_\infty < L_1$$

7.6.7.2 Direct structure

Consider the direct structure in Figure 7.27(b). Since the filter has one accumulator, internal overflows are not a problem, and so input scaling is not strictly necessary. This is one of the attractions of the direct structure. Intermediate overflows may occur in the output of the adder in the course of computing $y(n)$. Provided that the final output does not overflow, they do not matter. The scaling arrangement in Figure 7.27 may be used if scaling is required.

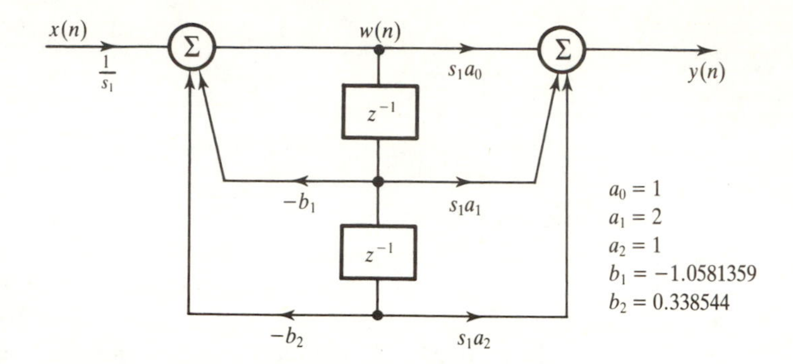

Figure 7.28 Block diagram representation for Example 7.16.

Example 7.17

Determine a suitable scale factor to prevent or reduce the possibility of overflow in an IIR lowpass filter characterized by the following transfer function:

$$H(z) = \frac{1 + 2z^{-1} + z^{-2}}{1 - 1.058\,135\,9z^{-1} + 0.338\,544z^{-2}}$$

Solution

The block diagram representation of the filter, using a second-order canonic section, is shown in Figure 7.28. Using the FWA program (on the PC disk) to evaluate Equations 7.48, 7.49 and 7.52, the scale factors for the three methods were computed. These are summarized below:

	L_1	L_2	L_∞
s_1	3.7112	1.7352	3.5663

Just as an illustration, we will also compute the L_2 norm using Equation 7.51:

$$s_1^2 = \frac{1}{1 - (0.3385)^2 - (1.058)^2[(1 - 0.3385)/(1 + 0.3385)]}$$

$$= 1/0.3322 = 3.01$$

$$s_1 = 1.7350$$

7.6.8 Scaling in cascade realization

In practice, filters are realized as cascades or parallel combinations of second–first-order sections. A scaling scheme for a sixth-order cascade realization is

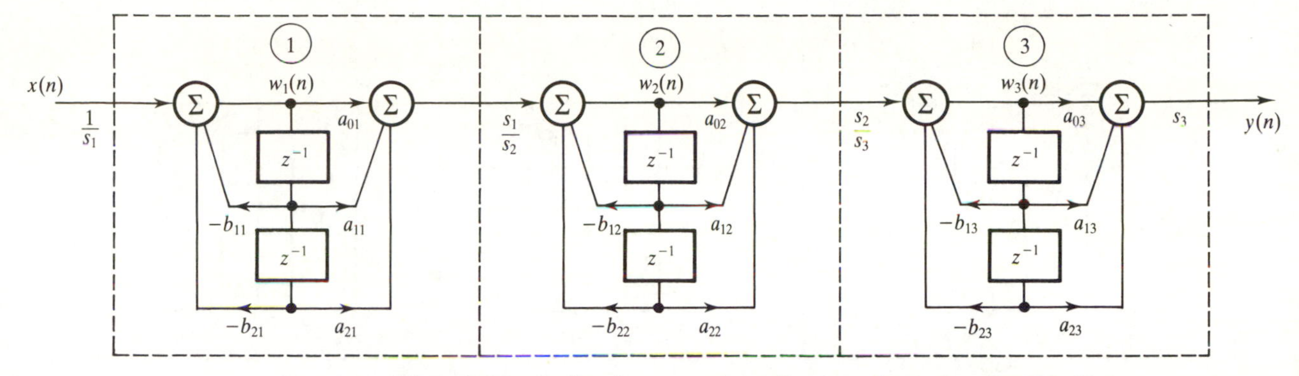

Figure 7.29 Scaling in a cascade realization of a sixth-order IIR filter.

shown in Figure 7.29. As before, the scale factors s_i, $i = 1, 2, 3$, are chosen to avoid or minimize overflows in the filter sections at the nodes labelled $w_i(n)$. The scaling scheme for each second-order section is essentially the same as for the single section considered before. The scale factors are obtained as

$$s_i = \|F_i(z)\|_p \tag{7.53}$$

where p denotes the type of norm: $p = 1, 2, \infty$. $F_i(z)$ is the transfer function from the input to the node $w_i(n)$ and given by

$$F_i(z) = \frac{\prod_{k=1}^{i-1} H_k(z)}{1 + b_{1i}z^{-1} + b_{2i}z^{-2}}, \quad i = 1, 2, 3$$

For the cascade realization, it is common practice to absorb the input scaling factor s_2 into the numerator of the first stage, s_3 into that of the second, and so on. Thus the scaling factors in Figure 7.29 can be rearranged as shown in Figure 7.30. It should be noted that the transfer function of the filter after scaling as discussed above is the same as that of the unscaled filter (theoretically at least).

Example 7.18

Compare the scale factors using the three methods above for the filter with the following transfer function, assuming cascade realization with second-order sections:

$$H(z) = H_1(z)H_2(z)H_3(z)$$

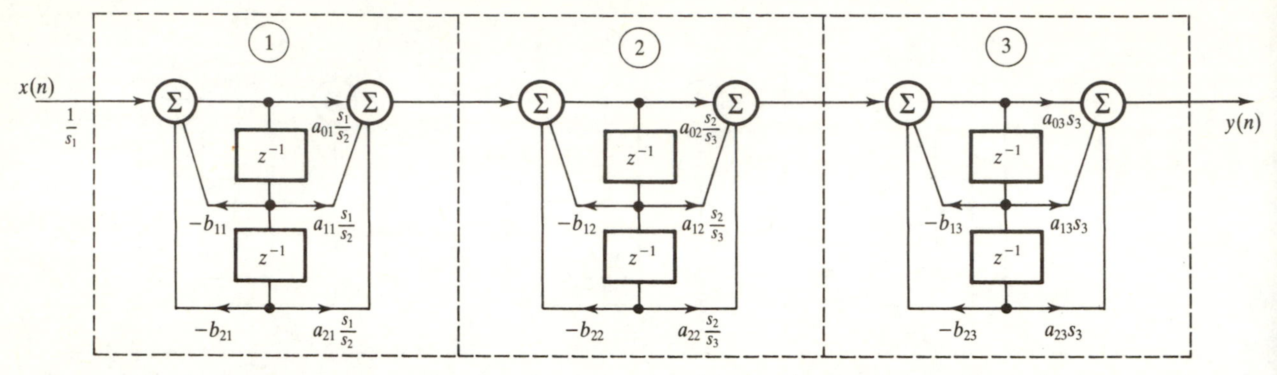

Figure 7.30 Scaling in a cascade realization of a sixth-order IIR filter (scaling factors absorbed into numerator, that is feedforward coefficients).

where

$$H_1(z) = \frac{1 + 0.2189z^{-1} + z^{-2}}{1 - 0.0127z^{-1} + 0.9443z^{-2}}$$

$$H_2(z) = \frac{1 - 0.5291z^{-1} + z^{-2}}{1 - 0.1731z^{-1} + 0.7252z^{-2}}$$

$$H_3(z) = \frac{1 + 1.5947z^{-1} + z^{-2}}{1 - 0.6152z^{-1} + 0.2581z^{-2}}$$

Using the FWA program, the scale factors s_1 to s_3 for the three methods were obtained. These are summarized below.

	L_1	L_2	L_∞
s_1	20.9608	3.0388	13.4098
s_2	19.0361	2.5358	10.1366
s_3	14.4467	2.9146	6.4087

As pointed out before, the L_1 norm is always the largest and the L_2 norm the smallest.

7.6.9 Scaling in parallel realization

Figure 7.31 depicts the scaling scheme for a parallel realization of a sixth-order IIR filter. It is seen that, in this case, the second-order sections are individually scaled as discussed previously. The scaling factor, s_i, at the input of each filter section ensures there is no overflow at the corresponding node $w_i(n)$. To keep the gain of the section the same as before scaling, the feedforward coefficients,

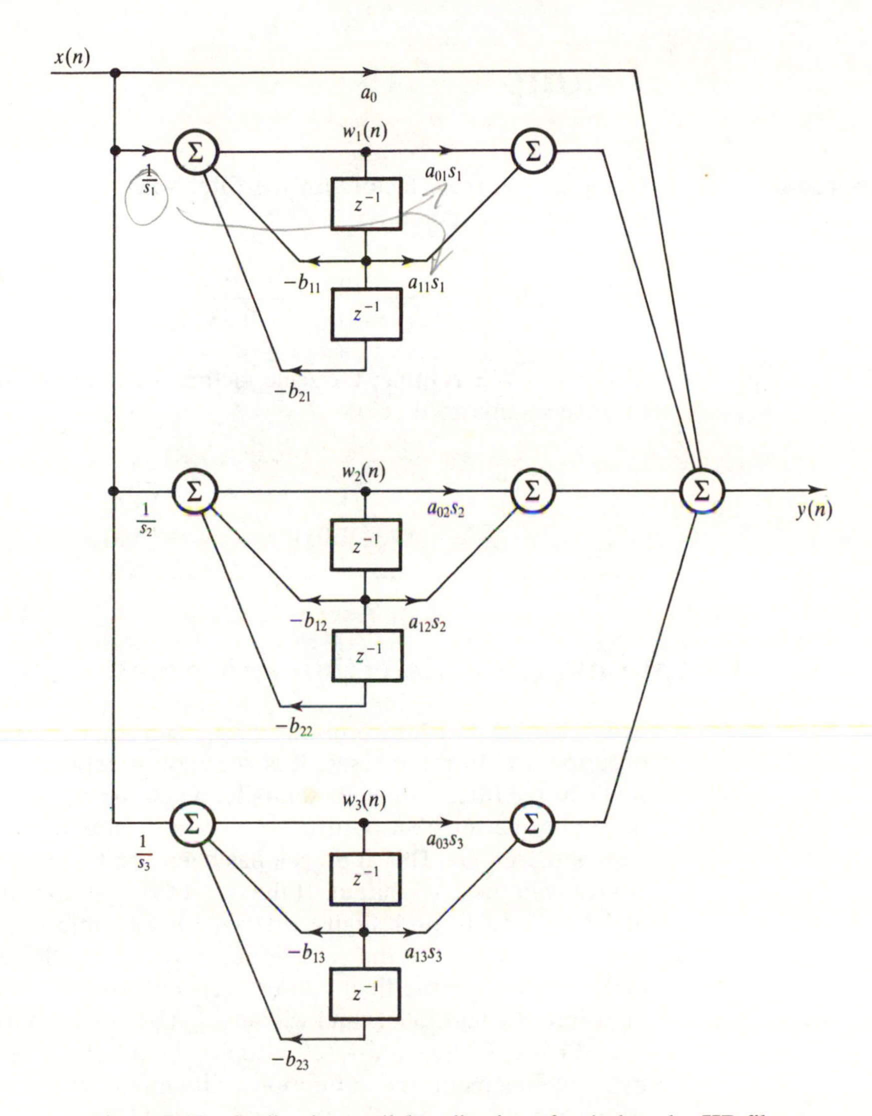

Figure 7.31 Scaling in parallel realization of a sixth-order IIR filter.

a_{ki}, are multiplied by s_i as shown in the diagram. The scaling factors are given by

$$s_i = \|F_i(z)\|_p$$

where $F_i(z)$, the transfer function from the input $x(n)$ to the node $w_i(n)$, is given by

$$F_i(z) = \frac{1}{1 + b_{1i}z^{-1} + a_{2i}z^{-2}}, \quad i = 1, 2, 3$$

Example 7.19

Compare the scale factors for the filter with the transfer function given below using the three methods of scaling:

$$H(z) = \frac{1.2916 - 0.08407z^{-1}}{1 - 0.131z^{-1} + 0.3355z^{-2}} + \frac{7.5268}{1 - 0.049z^{-1}} - 8.6788$$

Using an FWA routine, the scale factors for the three methods were computed and are summarized below.

	L_1	L_2	L_∞
s_1	1.7345	1.0667	1.5126
s_2	1.0515	1.0012	1.0515

7.6.10 Output overflow detection and prevention

When the L_2 and L_∞ norms are used, then output overflow is possible, albeit occasionally. In these cases, it is common practice to employ saturation arithmetic at the filter output. Essentially, when the output overflows, it is set to the maximum permissible positive or negative value depending on the sign of the true data sample. This approach has been shown to be effective in dealing with overflow in the final output. If the output is not saturated, then it will be wrong and may lead to undesirable effects, for example unpleasant sound in digital audio. In the case of the canonic section, that is all there is. In the case of the direct method, if the final output overflows without correction, this is also fed back into the multipliers and will affect subsequent output samples.

Figure 7.32(a) and 7.32(b) depict the overflow detection characteristics of two's complement and saturation arithmetic, respectively. In the figure, y is the correct output and $\hat{y}$ is the overflow output.

7.6.11 Product roundoff errors

Product roundoff error analysis is an extensive topic. Our presentation here will be brief and aims to make you aware of the nature of the errors, their effects and how to reduce them if necessary.

The basic operations in IIR filtering are defined by the familiar second-order difference equation:

$$y(n) = \sum_{k=0}^{2} a_k x(n-k) - \sum_{k=1}^{2} b_k y(n-k) \tag{7.54}$$

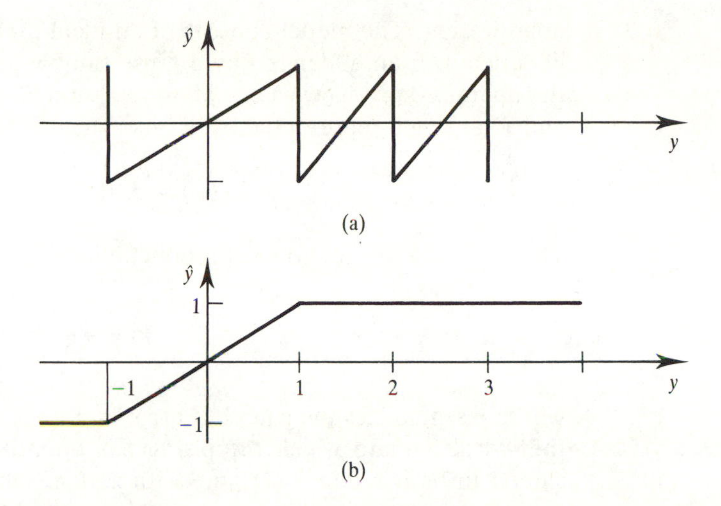

Figure 7.32 (a) Overflow characteristics; (b) saturation arithmetic.

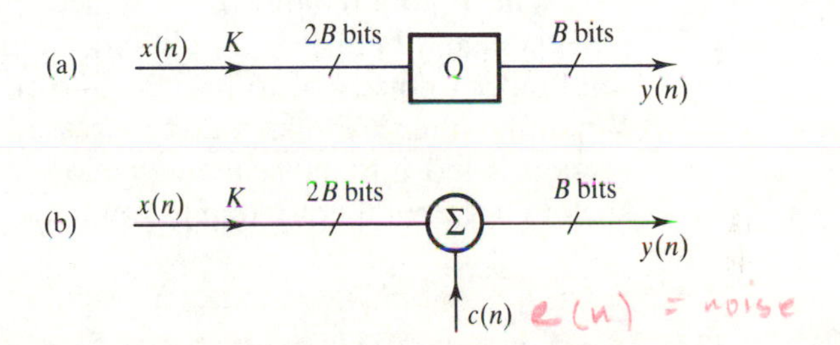

Figure 7.33 Representation of the product quantization error: (a) a block diagram representation of the quantization process; (b) a linear model of the quantization process.

where $x(n-k)$ and $y(n-k)$ are the input and output data samples, and a_k and b_k are the filter coefficients. In practice these variables are often represented as fixed point numbers. Typically, each of the products $a_k x(n-k)$ or $b_k y(n-k)$ would require more bits to represent than any of the operands. For example, the product of a B-bit data and a B-bit coefficient is $2B$ bits long. For recursive systems, if this result is not reduced subsequent computations will cause the number of bits to grow without limit.

Truncation or rounding is used to quantize the products back to the permissible wordlength. Quantizing the products leads to errors, popularly known as roundoff errors, in the output data and hence a reduction in the SNR. These errors can also lead to small-scale oscillations in the output of the digital filter, even when there is no input to the filter.

Figure 7.33(a) shows a block diagram representation of the product quantization process, and Figure 7.33(b) a linear model of the effect of product quantization. The model consists of an ideal multiplier, with infinite precision, in series with an adder fed by a noise sample, $e(n)$, representing the error in the quantized product, where we have assumed, for simplicity, that $x(n)$, $y(n)$, and K are each represented by B bits. Thus

$$y(n) = Kx(n) + e(n) \tag{7.55}$$

The noise power, due to each product quantization, is given by

$$\sigma_{\mathrm{r}}^2 = \frac{q^2}{12}$$

where r symbolizes the roundoff error and q is the quantization step defined by the wordlength to which the product is quantized. The roundoff noise is assumed to be a random variable with zero mean and a constant variance. Although this assumption may not always be valid (for example in the presence of narrowband, low level signals), it is useful in assessing the performance of the filter.

The roundoff noise may be fed into subsequent sections or stages of a DSP system where it is amplified, attenuated or modified in some way. The total output noise due to roundoff errors depends on the realization structure. When the filter is realized using a cascade structure, the noise produced by one section is fed into subsequent sections. Thus the sections should be arranged such that the total noise due to roundoff error is minimized.

7.6.12 Effects of roundoff errors on the signal-to-noise ratio

The effects of roundoff noise on filter performance depend on the type of filter structure used and the point at which the results are quantized. Figure 7.34(a) shows the quantization noise model for the direct form building block described earlier. It is assumed in the figure that the input data, $x(n)$, the output, $y(n)$, and the filter coefficients are represented as B-bit numbers (including the sign bit). The products are quantized back to B bits after multiplication by rounding (or truncation).

Since all five noise sources, e_1 to e_5 in Figure 7.34(a), feed to the same point (that is into the middle adder), the total output noise power is the sum of the individual noise powers (Figure 7.34(b)):

$$\sigma_{\mathrm{or}}^2 = \frac{5q^2}{12}\left[\frac{1}{2\pi j}\oint_{\mathrm{c}} F(z)F(z^{-1})\frac{\mathrm{d}z}{z}\right]s_1^2$$

$$= \frac{5q^2}{12}\left[\sum_{k=0}^{\infty} f^2(k)\right]s_1^2 \tag{7.56}$$

$$= \frac{5q^2}{12}\|F(z)\|_2^2 s_1^2 \tag{7.57}$$

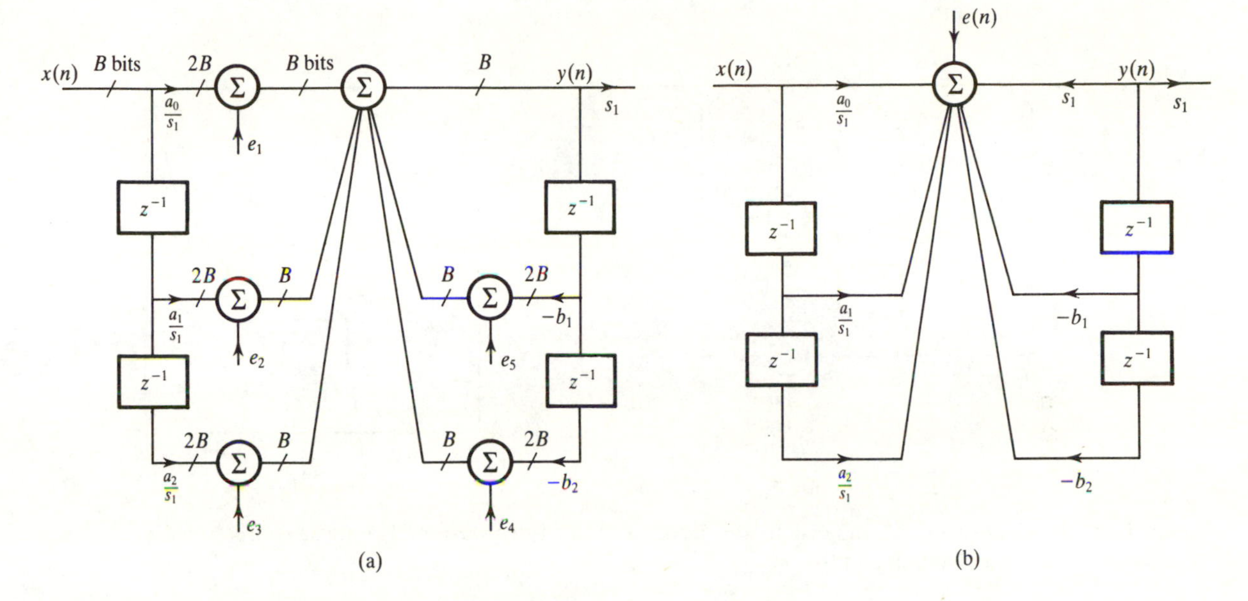

Figure 7.34 Product quantization noise model for the direct form filter section. All the noise sources in (a) have been combined in (b) as they feed to the same point.

where

$$F(z) = \frac{1}{1 + b_1 z^{-1} + b_2 z^{-2}}$$

$$f(k) = Z^{-1}[F(z)]$$

is the inverse z-transform of $F(z)$, which is also the impulse response from each noise source to the filter output, $\|\cdot\|_2^2$ is the L_2 norm squared and $q^2/12$ is the intrinsic product roundoff noise power. The total noise power at the filter output is the sum of the product roundoff noise and the ADC quantization noise (Equations 7.44 and 7.56):

$$\begin{aligned}\sigma_o^2 &= \sigma_{oA}^2 + \sigma_{or}^2\\ &= \frac{q^2}{12}\left[\sum_{k=0}^{\infty} h^2(k) + 5s_1^2 \sum_{k=0}^{\infty} f^2(k)\right]\\ &= \frac{q^2}{12}[\|H(z)\|_2^2 + 5s_1^2\|F(z)\|_2^2] \end{aligned} \qquad (7.58)$$

For the canonic section, Figure 7.35(a), the noise model again includes a scale factor as this generates a roundoff noise of its own. The noise sources $e_1(n)$ to $e_3(n)$ all feed to the left adder, whilst the noise sources $e_4(n)$ to $e_6(n)$ feed directly into the filter output. Combination of the noise sources feeding to the

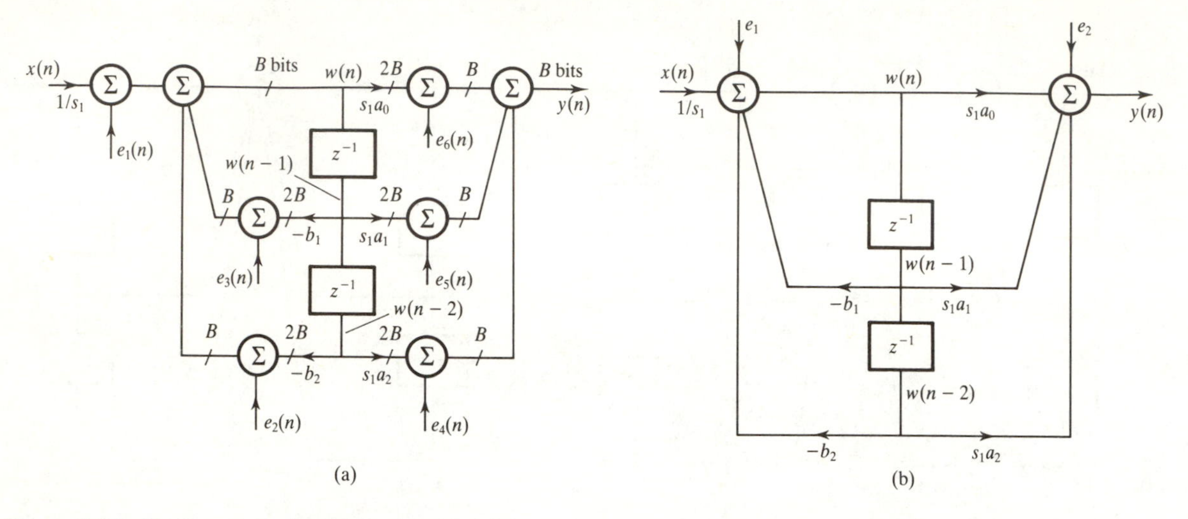

Figure 7.35 Product quantization noise model for the canonic filter section. The noise sources feeding the same point in (a) have been combined in (b).

same point leads to the noise model of Figure 7.35(b). Assuming uncorrelated noise sources, the total noise contribution is simply the sum of the individual noise contributions:

$$\sigma_{\text{or}}^2 = \frac{3q^2}{12}\sum_{k=0}^{\infty} f^2(k) + \frac{3q^2}{12}$$
$$= \frac{3q^2}{12}[\|F(z)\|_2^2 + 1] \qquad \textbf{(7.59)}$$

where $f(k)$ is the impulse response from the noise source e_1 to the filter output, and $F(z)$ the corresponding transfer function given by

$$F(z) = s_1\frac{a_0 + a_1z^{-1} + a_2z^{-2}}{1 + b_1z^{-1} + b_2z^{-2}} = s_1H(z) \qquad \textbf{(7.60)}$$

The total noise (ADC + roundoff noises) at the filter output is given by

$$\sigma_{\text{o}}^2 = \sigma_{\text{oA}}^2 + \sigma_{\text{or}}^2$$
$$= \frac{q^2}{12}\left\{3\left[1 + s_1^2\sum_{k=0}^{\infty} h^2(k)\right] + \sum_{k=0}^{\infty} h^2(k)\right\}$$
$$= \frac{q^2}{12}\{3[1 + s_1^2\|H(z)\|_2^2] + \|H(z)\|_2^2\} \qquad \textbf{(7.61)}$$

Equation 7.61 should be compared with Equation 7.58. Note also that the inclusion of the scale factor increases the output noise.

7.6.13 Illustrative example: assessment of the effect of roundoff errors on signal to noise ratio

Example 7.20

A filter is characterized by the following transfer function:

$$H(z) = \frac{0.1436 + 0.2872z^{-1} + 0.1436z^{-2}}{1 - 1.8353z^{-1} + 0.9747z^{-2}}$$

The filter is to be implemented in an 8-bit system with the input data, $x(n)$, the output data, $y(n)$, and filter coefficients represented as 8-bit fractional fixed point, 2's complement numbers.

Assuming that a second-order canonic section is used to realize the filter, and that each product (represented by 16 bits) is quantized to 8 bits immediately after multiplication,

(1) (a) sketch the realization diagram showing the sources of roundoff errors within the filter and determine a suitable scale factor for the system,

(b) estimate the total steady state output noise power and the degradation in the output SNR, in decibels, caused by roundoff errors, and

(2) repeat part (1) if direct form structure is used to realize the filter.

Solution

(1) (a) The realization diagram of the system with the noise sources is given in Figure 7.35(b). The noise source e_1 represents the sum of the errors resulting from rounding (or truncating) each of the products $b_1w(n-1)$, $b_2w(n-2)$ and $x(n)/s_1$ from 16 bits to 8 bits. The noise source e_2 is the sum of the three 16-bit inputs to the right adder, quantized to 8 bits.

Using the FWA program the scale factors for the three norms are $s_1 = 133.899\,66$ (L_1), $s_1 = 12.1395$ (L_2) and $s_1 = 102.088$ (L_∞).

(b) The output noise power due to roundoff error is obtained with the aid of the finite wordlength analysis program, based on Equation 7.59:

$$\sigma_{or}^2 = 1668.03q^2$$

where we have assumed L_2 scaling.

An estimate of the output noise due to the ADC, using Equation 7.44, is $3.7724q^2$. Thus total output noise power is

$$\sigma_o^2 = (1668.03 + 3.7724)q^2 = 1671.8024q^2$$

The output signal power, assuming a random signal input, is given by:

$$\sigma_x^2 = \tfrac{1}{3}\|H(z)\|_2^2 = 15.0896$$

The SNR (without roundoff error) is

$$\frac{15.0896}{3.7724q^2} = \frac{4}{q^2}$$

The SNR (with roundoff error) is

$$\frac{15.0896}{1671.8024q^2}$$

The degradation in SNR due to roundoff error is

$$10\log\left(\frac{4/q^2}{9.0260\times 10^{-3}/q^2}\right) = 26.47\ \text{dB}$$

(2) For the direct realization, the total output noise power due to roundoff error is $9048.82q^2$. The degradation in SNR due to roundoff error is 33.8 dB, with L_2 scaling. Without scaling, the degradation in SNR is about 1.11 dB. This low degradation in SNR when there is no scaling is why, in some applications, the direct realization is preferred as it offers the possibility of avoiding scaling (Dattoro, 1988).

7.6.14 Roundoff noise in cascade and parallel realizations

7.6.14.1 Cascade

Figure 7.36 shows the cascade realizations for a sixth-order IIR system using the second-order canonic section, where noise sources feeding the same point have been combined as suggested above and renumbered for simplicity. Thus e_1 is the sum of three noise sources, derived from the three multipliers feeding the leftmost adder. The composite noise source e_1 goes through the three filter sections $H_1(z)$, $H_2(z)$ and $H_3(z)$. The composite noise source e_2 goes through the transfer functions $H_2(z)$ and $H_3(z)$, and so on.

The total output noise due to roundoff error is the sum of all six noise sources:

$$\sigma_{\text{or}}^2 = \frac{3q^2}{12}\sum_{k=0}^{\infty} f_1^2(k) + \frac{3q^2}{12}\sum_{k=0}^{\infty} f_2^2(k) + \frac{2q^2}{12}\sum_{k=0}^{\infty} f_3^2(k) + \frac{3q^2}{12}\sum_{k=0}^{\infty} f_4^2(k)$$

$$+ \frac{2q^2}{12}\sum_{k=0}^{\infty} f_5^2(k) + \frac{3q^2}{12}$$

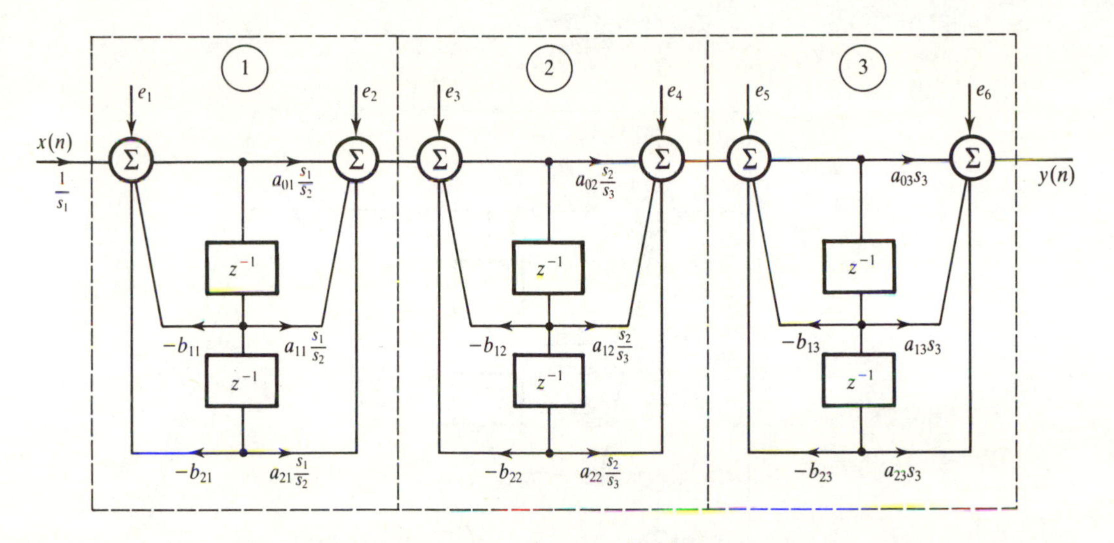

Figure 7.36 Noise model of a cascade realization of a sixth-order IIR filter.

$$= \frac{q^2}{12}\left[3\sum_{k=0}^{\infty} f_1^2(k) + 5\sum_{k=0}^{\infty} f_3^2(k) + 5\sum_{k=0}^{\infty} f_5^2(k) + 3\right]$$

$$= \frac{q^2}{12}[3\|F_1(z)\|_2^2 + 5\|F_3(z)\|_2^2 + 5\|F_5(z)\|_2^2 + 3] \quad \textbf{(7.62)}$$

where $f_i(k)$ is the impulse response between the noise source e_i and the output. The noise components due to e_2 and e_3 (Figure 7.36) each go through the same filter sections, that is through $H_2(z)$ and $H_3(z)$, and so their contributions at the output have been combined. The same is true of the contributions of the noise components e_4 and e_5.

7.6.14.2 Parallel

The roundoff noise model for parallel realization of a sixth-order filter is given in Figure 7.37. As before the noise sources due to individual product quantization have been combined. The noise sources e_1 to e_3 each go through a filter section to reach the output whereas the remaining noise sources feed directly to the output. The contribution of each of the noise sources, e_1 to e_3, to the output noise is

$$\sigma_{r,i}^2 = \frac{3q^2}{12}\sum_{k=0}^{\infty} f_i^2(k) = \frac{3q^2}{12}\|F_i(z)\|_2^2,\ i = 1, 2, 3$$

$$= \frac{3q^2}{12}s_i^2\sum_{k=0}^{\infty} h_i^2(k)$$

$$= \frac{3q^2}{12}s_i^2\|H_i(z)\|_2^2 \quad \textbf{(7.63))}$$

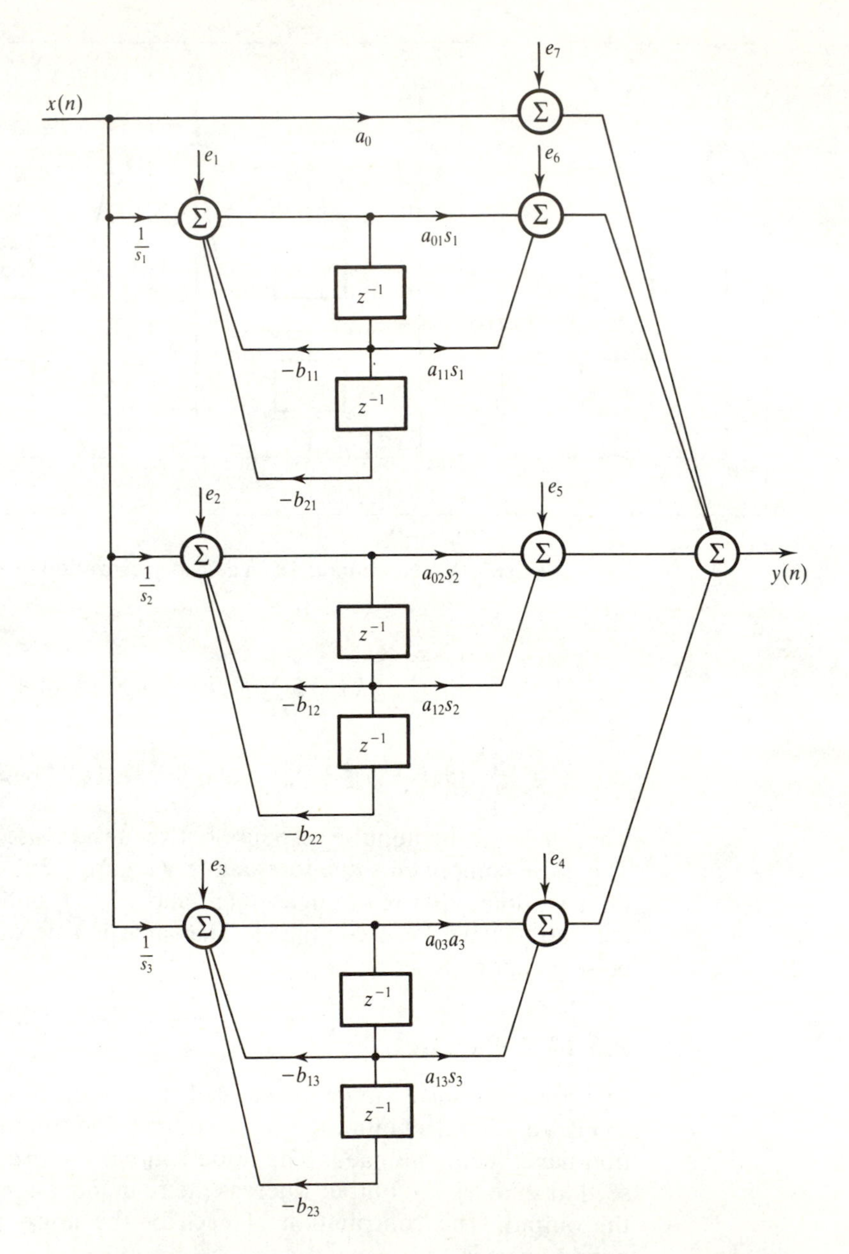

Figure 7.37 Noise model of a parallel realization of a sixth-order IIR filter.

where $H_i(z)$ and $h_i(k)$ are, respectively, the transfer function of filter section i and its corresponding impulse response. $F_i(z)$ and $f_i(k)$ are, respectively, the transfer function seen by the noise source i and the corresponding impulse response.

The noise sources e_i, $i = 4, 5, 6$, all feed directly into the output as does

e_7. Thus the total output noise power is given by

$$\sigma_{or}^2 = \frac{q^2}{12}\left\{7 + 3\sum_{i=1}^{3}\left[s_i^2\sum_{k=0}^{\infty} h_i^2(k)\right]\right\}$$

$$= \frac{q^2}{12}\left[7 + 3\sum_{i=1}^{3} s_i^2\|H_i(z)\|_2^2\right] \quad \textbf{(7.64)}$$

In general, for a parallel realization with L sections, the output power due to roundoff error is given by

$$\sigma_{or}^2 = \frac{q^2}{12}\left[2L + 1 + 3\sum_{i=1}^{L} s_i^2\|H_i(z)\|_2^2\right] \quad \textbf{(7.65)}$$

Estimates of all the equations above for roundoff noise power can be readily obtained using a suitable computer program (for example, the finite word-length analysis program described earlier).

Example 7.21

Given the following transfer function representing a fourth-order IIR filter (Mitra *et al.*, 1974),

$$H(z) = \frac{1 - 2z^{-1} + z^{-2}}{1 + 0.777z^{-1} + 0.3434z^{-2}} \frac{1 - 0.707z^{-1} + z^{-2}}{1 + 0.01877z^{-1} + 0.801z^{-2}} 0.093\,226$$

estimate the reduction in SNR, in decibels, if the filter is realized as

(1) a cascade of two second-order sections paired and ordered as given in $H(z)$, and
(2) parallel combinations of two second-order sections.

Assume in each case that the products are quantized to 8 bits before they are summed.

Solution

The noise models for the cascade and parallel realization structure are identical to Figures 7.36 and 7.37 respectively, if we ignore the third filter stage in each figure.

(1) For the cascade realization, the L_1 scale factors (using the PC-based programs) are

$$s_1 = 2.395\,746$$

$$s_2 = 15.703\,627$$

The roundoff noise power, the ADC noise power, and the signal power at the filter output are

$$\sigma_{\text{or}}^2 = 345.0391q^2$$
$$\sigma_{\text{oA}}^2 = 0.039\,76q^2$$
$$\sigma_y^2 = 0.1591$$

The SNR (without roundoff error) is

$$\frac{0.1591}{0.0397q^2}$$

and the SNR (with roundoff error) is

$$\frac{0.1591}{345.0789q^2}$$

Degradation in SNR due to roundoff error is

$$10\log(345.0789/0.039\,76) \approx 39.4 \text{ dB}$$

(2) For the parallel realization, using a partial fraction expansion (Mitra *et al.*, 1974), the transfer function becomes

$$H(z) = 0.093\,326\left(1 + \frac{-5.162 + 0.7867z^{-1}}{1 + 0.777z^{-1} + 0.3434z^{-2}} + \frac{1.657\,36 + 0.2759z^{-1}}{1 + 0.018\,77z^{-1} + 0.801z^{-2}}\right)$$

The roundoff noise power at the filter output is given by

$$\sigma_{\text{or}}^2 = \frac{q^2}{12}\left[5 + 3\sum_{i=1}^{2} s_i^{\,2}\|H_i(z)\|_2^2\right]$$

Using the FWA program we obtain the L_2 scale factors as $s_1 = 2.395\,746$, $s_2 = 5.450\,612$, $\|H_1(z)\|_2^2 = (7.378\,492)^2$, $\|H_2(z)\|_2^2 = (2.801\,937)^2$. Thus we have

$$\sigma_{\text{or}}^2 = \frac{q^2}{12}\{5 + 3[(2.395\,746)^2(7.378\,492)^2 + (5.450\,612)^2(2.801\,937)^2]\}$$
$$= 136.85q^2$$

Thus we have at the filter output, assuming the same output signal and ADC noise, an SNR (with roundoff error) of

$$0.1591/(136.85 + 0.039\,76)q^2 = 1.162\,28 \times 10^{-3}/q^2$$

The degradation in SNR due to roundoff error then becomes

$$10\log[(136.85 + 0.039\,76)/0.039\,76)] = 35.37 \text{ dB}$$

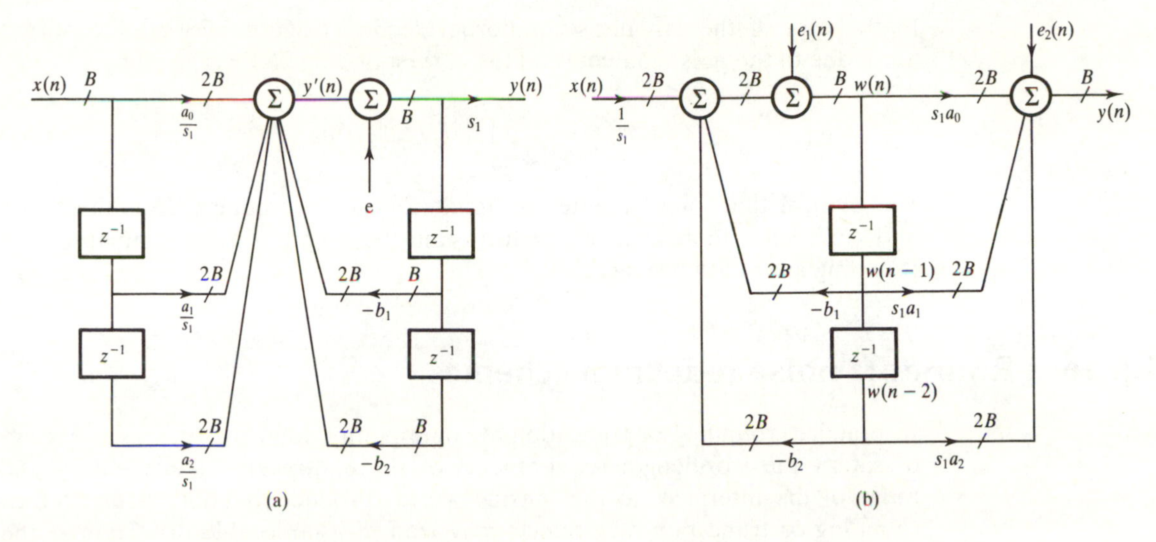

Figure 7.38 Noise models for IIR filter sections in a modern DSP system. The wordlengths at various points within the filter are shown. It is assumed that the input data and filter coefficients are each B bits long.

7.6.15 Effect of product roundoff noise in modern DSP systems

Early work on the effects of product roundoff errors on filter performance were based on a single, fixed internal wordlength, where it was mandatory to quantize each $2B$-bit product (strictly speaking, $2B-1$ bits) back to B bits before summation. With modern DSP processors such constraints do not exist because they support double-wordlength accumulation. All modern DSP processors feature, at least, a built-in 16×16-bit multiplier, 32-bit product register, and allow for the products to be accumulated as 32 bit numbers – the so-called 16/32 bit architecture.

Figure 7.38(a) shows the noise model for the direct second-order section when quantization is carried out after the products have been added. In the figure, the $2B$-bit sum of the products, $y'(n)$, is quantized to B bits. To distinguish this from the case where each product is separately quantized, we will refer to this as the post-accumulation quantization. It is clear that, in this case, there is only one noise source following the quantization. The output noise power in this case is given by

$$\sigma_{\text{or}}^2 = \frac{q^2}{12}\|F(z)\|_2^2 \tag{7.66}$$

where

$$F(z) = \frac{1}{1 + b_1 z^{-1} + b_2 z^{-2}}$$

In the case of the canonic second-order section (Figure 7.38(b)) the output noise due to the noise sources, and the corresponding SNR, is given by

$$\sigma_{\text{or}}^2 = \frac{q^2}{12}[s_1^2 \|H(z)\|_2^2 + 1] \tag{7.67}$$

It is evident that rounding after accumulation of the products (Equations 7.66 and 7.67) leads to a significant reduction in the roundoff noise compared with rounding after each product.

7.6.16 Roundoff noise reduction schemes

In practice, rounding or truncation at some point within the filter is necessary to satisfy the wordlength requirements of the multipliers, data memory and those of the interfaces to the outside world. Product roundoff errors due to rounding or truncation of products may lead to a noticeable distortion at the filter output with low level input signals and for high fidelity systems should be minimized. A number of schemes have been devised for reducing or eliminating the effects of roundoff errors in IIR filters. Effectively, these schemes shape the noise spectrum in such a way as to reduce or cancel their effects over certain bands of the filter. The schemes have been collectively called error spectral shaping (ESS).

Generalized noise reduction strategies for the direct and canonic filter sections are depicted in Figures 7.39(a) and 7.39(b), respectively. In both figures the feedback and feedforward coefficients, b_i' and a_i', are used to modify the transfer function seen by the roundoff errors such that the output roundoff noise is minimized without affecting the desired signal. In Figure 7.39(a), quantization of the sum of products at the output of the left adder generates an error, $e_1(n)$, which is the lower half of the double precision variable, $y(n)$. The products $b_1'e_1(n-1)$ and $b_2'e_2(n-2)$ do not have the same weight as the other inputs to the adder although they are all $2B+1$ bits long and so must be re-aligned or quantized before summing with the other inputs. The quantization error in this case is denoted by $e_2(n)$. Similarly, the terms $a_0'e(n)/s_1$, $a_1'e(n-1)/s_1$, and $a_2'e(n-2)/s_1$ need to be quantized before being summed with the other inputs to the right adder, giving rise to the error $e_3(n)$. Finally, the output of the right adder will be quantized from $2B+1$ bits to $B+1$ bits giving rise to the error $e_4(n)$. Similar considerations apply to the canonic section in Figure 7.39(b).

For the direct form realization with ESS (Figure 7.39(a)) the output noise is given by

$$\sigma_{\text{or}}^2 = \frac{q^2}{12}\left[\sum_{k=0}^{\infty} f_1^2(k) + \sum_{k=0}^{\infty} f_2^2(k) + 2s_1^2\right] \tag{7.68}$$

where $f_1(k)$ and $f_2(k)$ are, respectively, the impulse responses from the noise sources 1 and 2 to the output of the filter.

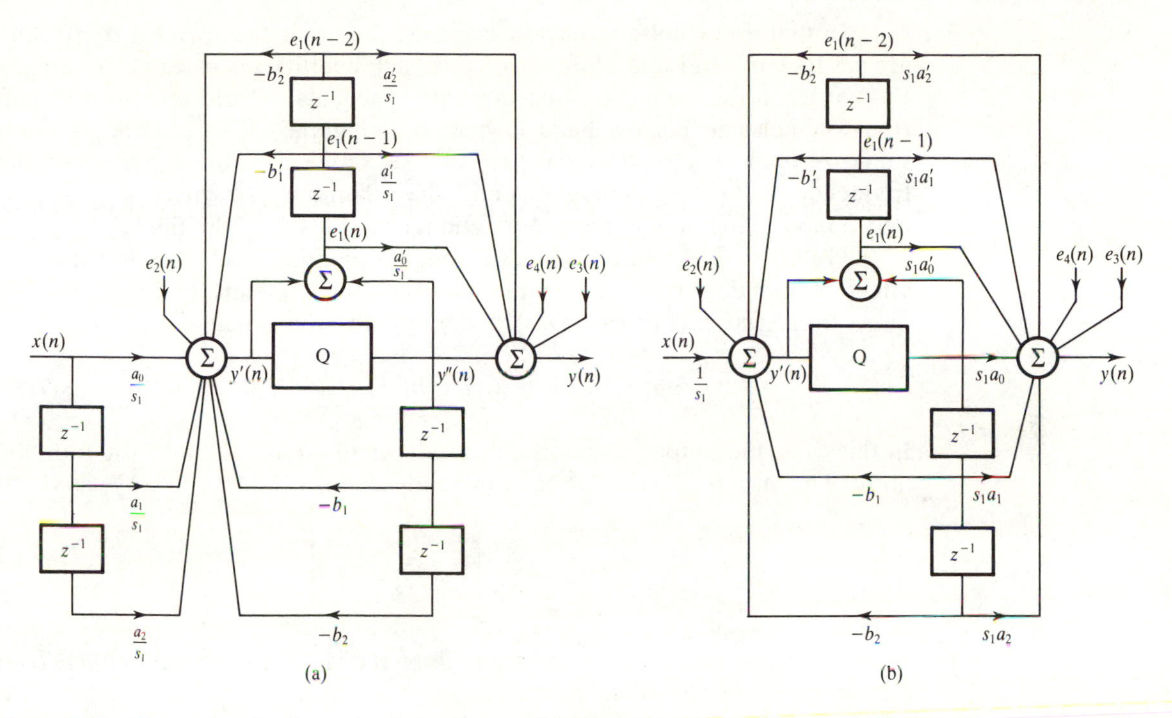

Figure 7.39 Generalized noise reduction schemes: (a) direct form filter; (b) canonic filter.

For the canonic section (Figure 7.39(b)) the output noise power due to roundoff is given by

$$\sigma_{\text{or}}^2 = \frac{q^2}{12}\left[\sum_{k=0}^{\infty} f_1^2(k) + \sum_{k=0}^{\infty} f_2^2(k) + 2\right] \tag{7.69}$$

$$= \frac{q^2}{12}[\|F_1(z)\|_2^2 + \|F_2(z)\|_2^2 + 2]$$

where

$$F_1(z) = (a_0' + a_1'z^{-1} + a_2'z^{-2})s_1 + \frac{(1 + b_1'z^{-1} + b_2'z^{-2})(a_0 + a_1z^{-1} + a_2z^{-2})s_1}{1 + b_1z^{-1} + b_2z^{-2}}$$

$$F_2(z) = \frac{(a_0 + a_1z^{-1} + a_2z^{-2})s_1}{1 + b_1z^{-1} + b_2z^{-2}}$$

The choice of the error spectral shaping (ESS) coefficients, a_i' and b_i', determines the effectiveness of the scheme in reducing noise. In practice, the ESS coefficients are often constrained to be integers to avoid further quantization processes. In practice, both first- and second-order ESS are normally used.

In first-order noise reduction schemes, the error feedforward coefficients are set to zero, and one of the error feedback coefficients is set to an integer. First-order noise reduction schemes are particularly effective in reducing roundoff noise in narrow band lowpass and highpass filters as they allow a single zero to be placed in the path of the roundoff error at either the low frequency or the high frequency end. The scheme is attractive as it involves only a modest increase in the computational complexity of the filter.

Optimum ESS can be achieved using second-order schemes in which the effect of roundoff noise at the filter output is completely nullified. For the direct form section (Figure 7.39(a)) the optimal ESS is obtained by setting

$$a_i' = 0,\ i = 0, 1, 2; \quad b_i' = b_i,\ i = 1, 2 \tag{7.70}$$

In this case, the output roundoff noise reduces to largely the intrinsic roundoff noise. The output roundoff noise power reduces to

$$\begin{aligned}\sigma_{or}^2 &= \frac{q^2}{12}\left[1 + \sum_{k=0}^{\infty} f^2(k)\right]\\ &= \frac{q^2}{12}[1 + \|F(z)\|_2^2]\end{aligned} \tag{7.71}$$

where

$$F(z) = \frac{s_1}{1 + b_1 z^{-1} + b_2 z^{-2}}$$

For the canonic section an optimal solution is obtained by setting

$$a_i' = -a_i,\ i = 0, 1, 2; \quad b_i' = b_i,\ i = 1, 2 \tag{7.72}$$

The output noise power is given by

$$\sigma_{or}^2 = \frac{q^4}{12}\sum_{k=0}^{\infty} f^2(k) = \frac{q^4}{12}\|F(z)\|_2^2$$

where

$$\frac{(a_0 + a_1 z^{-1} + a_2 z^{-2})s_1}{1 + b_1 z^{-1} + b_2 z^{-2}}$$

The optimal solution is computationally more expensive and, as pointed out by Mullis and Roberts (1982), effectively involves a double precision representation of the internal filter variables. A number of other suboptimal solutions, besides the integer ones considered previously, are available (for example in Higgins and Munson, 1982).

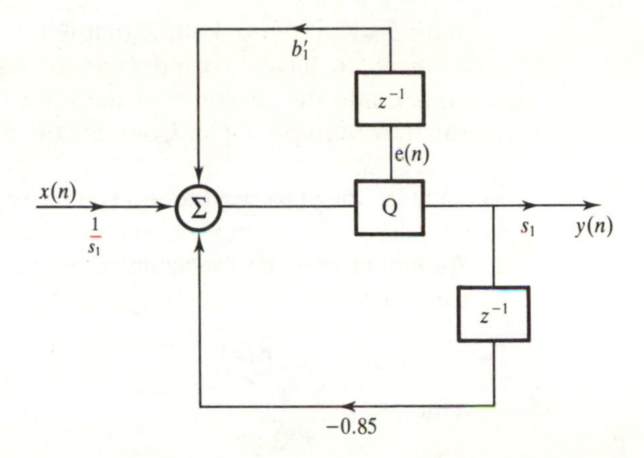

Figure 7.40 First-order IIR filter with error spectral shaping.

Example 7.22

Figure 7.40 shows a simple first-order IIR filter with an error spectral shaping scheme to minimize product roundoff noise at the filter output. Determine, analytically,

(1) a suitable L_2 scale factor to reduce the possibility of overflow, and

(2) the output noise power due to roundoff error for each of the following cases:

  (a) no error feedback, that is $b' = 0$;

  (b) the error feedback coefficient $b' = 1$.

Solution

(1) The L_2 scale factor is given by

$$\begin{aligned} s_1^2 &= \frac{1}{2\pi j}\oint F(z)F(z^{-1})\frac{dz}{z} \\ &= \frac{1}{2\pi j}\oint \frac{z^{-1}}{(1+bz^{-1})(1+bz)}dz \\ &= \frac{1}{2\pi j}\oint \frac{1}{(z+b)(1+bz)}dz \end{aligned}$$

$F(z)$ has only one pole inside the unit circle at $z = -b$. From the residue theory (see the appendix) we have

$$\begin{aligned} s_1^2 &= \lim_{z\to -b} \cancel{(z+b)}\frac{1}{\cancel{(z+b)}(1+bz)} \\ &= \frac{1}{1-b^2} \end{aligned}$$

With $b = 0.85$, $s_1^2 = 3.6036$, that is $s_1 = 1.8983$. The filter with ESS can be assumed to have two independent inputs, $x(n)$ and $e(n)$ (the error or noise), and the response of the filter to each input determined using the superposition principle. Consider the noise input alone:

$$y(n) = \hat{y}(n) + e(n) = -by(n-1) - b'e(n-1)$$

Assuming that the spectrum of $e(n)$ exists (since it can be replicated) we can write

$$\hat{Y}(z) + E(z) = -bz^{-1}Y(z) - b'z^{-1}E(z)$$

Thus

$$\hat{Y}(z) = -E(z)\frac{1 + b'z^{-1}}{1 + bz^{-1}}$$

It is seen that the transfer function from the noise source to the filter output is given by

$$F(z) = \frac{1 + b'z^{-1}}{1 + bz^{-1}}$$

The output noise is given by

$$\sigma_o^2 = \frac{q^2}{12}\frac{1}{2\pi j}\oint F(z)F(z^{-1})\frac{dz}{z}$$

(2) (a) With no error feedback, that is $b' = 0$,

$$F(z) = \frac{1}{1 + bz^{-1}} = \frac{1}{1 + 0.85z^{-1}}$$

and the output noise due to roundoff power is

$$\sigma_{o_r}^2 = \frac{q^2}{12}\frac{1}{1 - b^2} = \frac{1}{1 - (0.85)^2}\frac{q^2}{12} = 3.6036q^2/12$$

(b) With error feedback and $b' = 1$,

$$F(z) = \frac{1 + z^{-1}}{1 + 0.85z^{-1}}$$

and (Jury, 1964)

$$\sigma_{o_r}^2 = \frac{q^2}{12}\frac{1 + b'^2 - 2b'b}{1 - b^2} = \frac{q^2}{12}(2 - 2 \times 1 \times 0.85)/[1 - (0.85)^2]$$
$$= 1.0811q^2/12$$

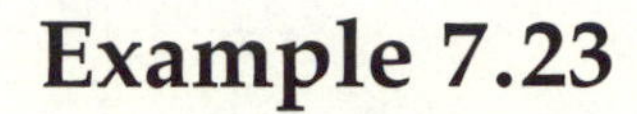

Example 7.23

Compare the roundoff noise performance of a second-order IIR filter which is characterized by the following transfer function:

$$H(z) = \frac{0.1436(1 + 2z^{-1} + z^{-2})}{1 - 1.8353z^{-1} + 0.9748z^{-2}}$$

if the filter is realized using (a) a canonic section or (b) a direct filter section for the following cases:

(1) $b'_i = 0$, $i = 1, 2$ (no error feedback)
(2) $b'_1 = -1$, $b'_2 = 0$
(3) $b'_1 = -2$, $b'_2 = 0$
(4) $b'_1 = -1$, $b'_2 = 1$
(5) $b'_1 = -2$, $b'_2 = 1$

Assume that all the feedforward error coefficients are zero in each case.

Solution

The realization structures, with ESS, are shown in Figure 7.41. The roundoff noise output power for the canonic and direct realization structures are, respectively,

$$\sigma_{or}^2 = \frac{q^2}{12}[\|F_1(z)\|_2^2 + 1]$$

and

$$\sigma_{or}^2 = \frac{q^2}{12}[\|F_2(z)\|_2^2 + 1]$$

where

$$F_1(z) = (1 + b'_1z^{-1} + b'_2z^{-2})\frac{(a_0 + a_1z^{-1} + a_2z^{-2})s_1}{1 + b_1z^{-1} + b_2z^{-2}}, s_1 = 12.1395 \text{ (L}_2\text{ scaling)}$$

$$F_2(z) = \frac{(1 + b_1z^{-1} + b_2z^{-2})s_1}{1 + b_1z^{-1} + b_2z^{-2}}, \qquad s_1 = 6.7282 \text{ (L}_2\text{ scaling)}$$

Using the FWA program, the output noise power for each case was obtained and summarized in Table 7.2. Note that for case 3 ($b'_1 = -2$, $b'_2 = 0$) the noise output has actually gone up instead of improving, emphasizing the importance of the choice of the feedback coefficients. The first order scheme, case 2, is quite effective in reducing the output noise.

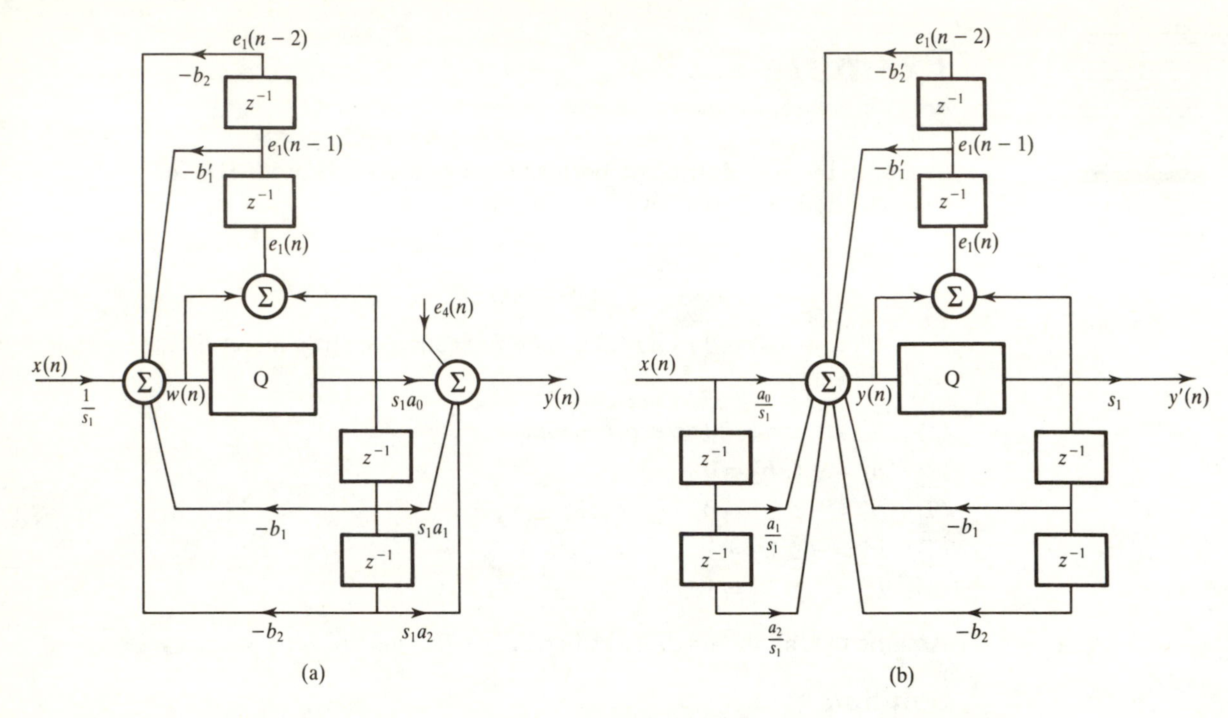

Figure 7.41 Two different realization structures of a second-order IIR filter.

Table 7.2 Output of the FWA program for Example 7.23.

	Noise power	
Case	*Canonic*	*Direct*
1	$556.0108q^2$	$556.0108q^2$
2	$77.1261q^2$	$78.6247q^2$
3	$710.0842q^2$	$713.0933q^2$
4	$414.1014q^2$	$413.845q^2$
5	$12.2659q^2$	$14.9999q^2$

7.6.17 Limit cycles due to product roundoff errors

In addition to the degradation in SNR, the error due to roundoff can cause an oscillation at the filter output or the output to remain stuck at a fixed nonzero value, even when there is no input. This effect is known as low level limit cycle. We will illustrate with an example.

Example 7.24

A first-order IIR filter is characterized by the difference equation

$$y(n) = x(n) + \alpha y(n-1) \quad n > 0$$

Given the initial condition $y(0) = 6$ and a zero input, that is $x(n) = 0$, $n = 0, 1, \ldots,$

(1) obtain and plot, assuming infinite precision, the first 10 output values for (i) $\alpha = -0.75$ and (ii) $\alpha = 0.75$,

(2) repeat part (1), but assume that the data and register lengths are each four bits long (that is 3 data bits and a sign bit) and that the products are rounded, and

(3) repeat parts (1) and (2), but assume truncation of products immediately after multiplication.

Solution

The values of the output samples for the three cases above are depicted in Figure 7.42 and listed in Table 7.3.

It is seen that if the input $x(n)$ is zero indefinitely the output $y(n)$, using infinite precision, decays exponentially to zero, regardless of the sign of α. However, if finite precision arithmetic is used, with the output rounded to the nearest integer, then for positive α the output remains fixed at a given level. The range of output levels over which the output is confined is known as the deadband. In this example, the deadband is the interval $[-2, 2]$. For the first-order filter the deadband interval is given by (Jackson, 1986)

$$k = \text{int}\left[\frac{0.5}{1 - \|\alpha\|}\right]$$

where $\text{int}[\cdot]$ is the integer part of the quantity in the square bracket. If α is negative, the output oscillates at a frequency of $F_s/2$ between two fixed levels of alternating sign. This results from the fact that when the input is removed the filter output decays to less than the quantization level, but then it is rounded up to the next level and the process repeats, creating a low level oscillation. These low level oscillations are undesirable in some applications. For example, they produce unpleasant noise in a telephone system during idle channel conditions when the speakers are silent. A way to reduce limit cycles is to increase the processor wordlength or to add a dither signal to the output before it is rounded. ESS, already discussed, has also been shown to reduce the amplitude of limit cycles and in some cases to eliminate them completely.

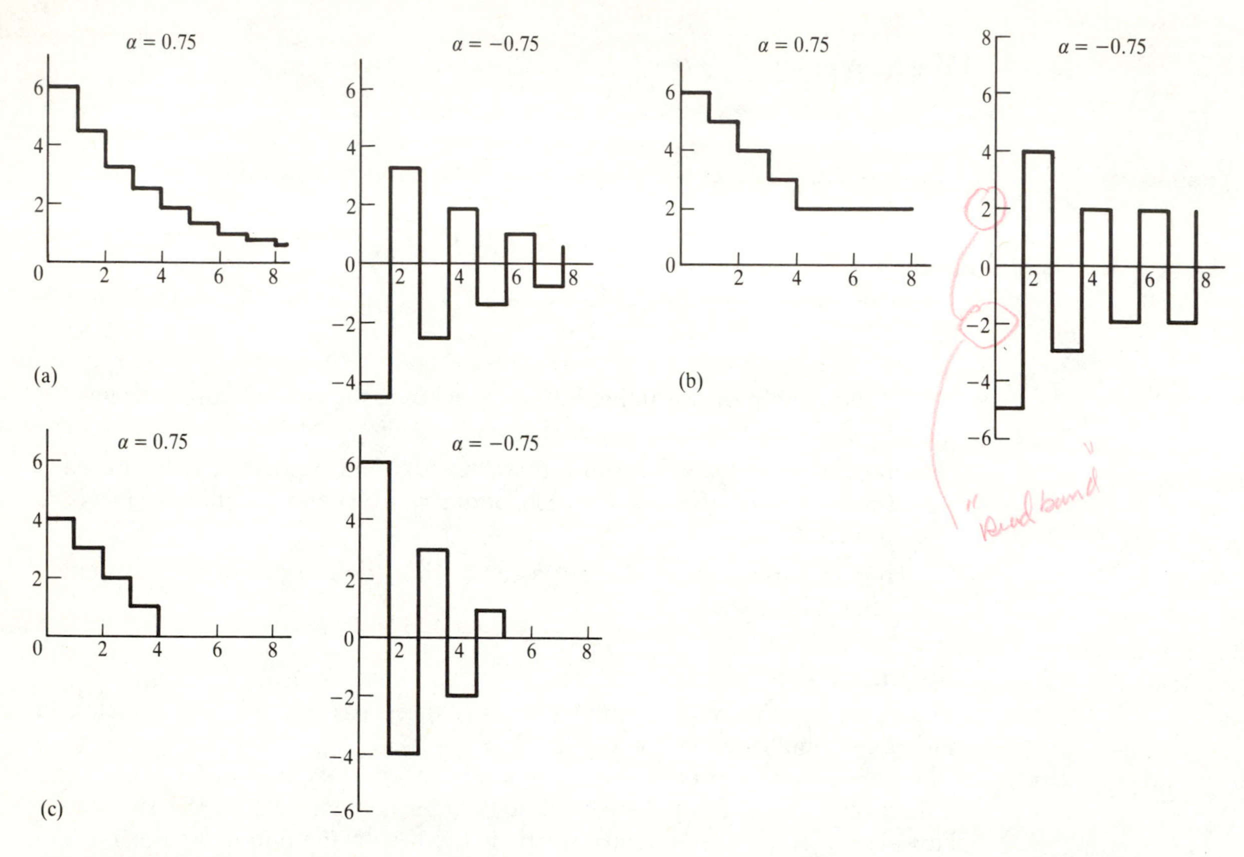

Figure 7.42 An illustration of low level limit cycle due to product quantization in a first-order IIR filter: (a) infinite precision; (b) quantization by rounding; (c) quantization by truncation.

Table 7.3 Results for Example 7.24.

	$y(n)$, $(\alpha = 0.75)$			$y(n)$, $(\alpha = -0.75)$		
n	*Infinite*	*Rounding*	*Truncation*	*Infinite*	*Rounding*	*Truncation*
0	6	6	6	6	6	6
1	4.5	5	4	−4.5	−5	−4
2	3.38	4	3	3.38	4	3
3	2.53	3	2	−2.53	−3	−2
4	1.90	2	1	1.90	2	1
5	1.42	2	0	−1.42	−2	0
6	1.07	2	0	1.07	2	0
7	0.80	2	0	−0.80	−2	0
8	0.60	2	0	0.60	2	0
9	0.45	2	0	−0.45	−2	0

In general roundoff limit cycles will not exist in a second-order filter if the coefficients lie inside the hatched area of the stability triangle.

7.6.18 Other nonlinear phenomena

As well as overflow and product limit cycles, other nonlinear effects that may influence the behaviour of an IIR digital filter include the following.

(1) *Jump phenomenon*. When the filter is fed by a sine wave, two possible output levels may exist for the same input signal. A small change in the amplitude or frequency of the input signal causes a jump from one output level to another. Several regions inside the stability triangle where such a phenomenon may exist have been identified. In these regions, the filter coefficients satisfy the condition $|b_1|b_2 < -1$. ESS has been shown to reduce the consequences of such nonlinear effects.

(2) *Subharmonic response*. For a sine wave input the output may contain subharmonics of the input (Claasen, 1974). Thus, for the same input signal but different initial conditions we can have outputs which are quite different. These effects are more serious for filters with poles close to the unit circle.

7.7 Stage 5: implementation of the filter

In the IIR filter, the output, $y(n)$, is computed for each input sample, $x(n)$. Assuming cascade realization using the second-order direct form, the key filtering equation is

$$y(n) = \sum_{k=0}^{2} a_k x(n-k) - \sum_{k=1}^{2} b_k y(n-k)$$

This equation clearly shows that to implement the filter we need the following components:

- memory (for example ROM) for storing filter coefficients;
- memory (for example RAM) for storing the present and past inputs and outputs $\{x(n), x(n-1), \ldots\}$ and $\{y(n), y(n-1), \ldots\}$;
- hardware or software multiplier(s);
- adder or arithmetic logic unit.

In modern real-time DSP, the filtering operations are efficiently performed with a DSP processor such as the TMS320C25. These processors have all the basic blocks on-board, including in-built hardware multiplier(s). In some applications, standard 8-bit or 16-bit microprocessors such as the Motorola 6800 or 68000 families offer attractive alternative implementations. In addition to the signal processing hardware, the designer must also provide suitable input–output (such as analogue–digital–analogue conversion) interfaces to the digital hardware, depending on the type of data source and sink. This approach may be described as hardware implementation.

In batch or off-line processing a suitable high level language is used to implement the filter. In this case, the filter is often implemented in a high level language (for example C or FORTRAN) and runs in a general purpose computer, such as a personal computer or a mainframe computer, where all the basic blocks are already configured. Thus, batch processing may be described as a purely software implementation.

Computational requirements

The designer must analyse the impact of the computational requirements of a digital filter on the processor that will be used. The primary requirements for digital filters are multiplication, additions, accumulation and delays or shifts. For example, a filter consisting of a second-order section would require typically four multiplications, four additions, and some shifts and storage. If the filtering is performed in real time, for example at 44.1 kHz (for digital audio), the arithmetic operations must be performed once every 1/(44.1 kHz). Allowance must also be made for other overheads such as fetching the input data or saving or outputting the filtered data samples as well as other housekeeping operations.

7.8 A detailed design example of an IIR digital filter

This example will be used to illustrate some of the many concepts presented in this chapter. In particular, we shall see how the five-stage design procedure is applied.

Stage 1: filter specifications

Design and implement a lowpass IIR digital filter using a software package and the TMS320C25-based target board to meet the following specifications:

sampling frequency	15 kHz
passband	0–3 kHz
transition width	450 Hz
passband ripple	0.5 dB
stopband attenuation	45 dB

Note: The target board has a 12-bit ADC and 12-bit DAC.

Stage 2: coefficient calculation

Using the software design program (on PC disk) for IIR filters, it was found that a fourth-order elliptic filter, via the bilinear transform method, would be required to satisfy the specifications. The output listing of the design program is summarized below.

	Denominator B_k	*Numerator* A_k
1	1.000000E+00	5.846399E−02
2	−1.325263E+00	1.359507E−01
3	1.480202E+00	1.820297E−01
4	−7.841098E−01	1.359506E−01
5	2.339270E−01	5.846398E−02

Poles		*Coefficients*	
Real	*Imaginary*	Z^{-1}	Z^{-2}
0.247967	0.836885	−0.495935	0.761864
0.414664	0.367559	−0.829328	0.307046

Zeros		*Coefficients*	
Real	*Imaginary*	Z^{-1}	Z^{-2}
−0.337859	0.941197	0.675718	1.000000
−0.824828	0.565383	1.649656	1.000000

From the listing, the transfer function of the filter, in direct form, is given by

$$H(z) = \frac{0.058\,463\,99 + 0.135\,950\,7z^{-1} + 0.182\,097\,9z^{-2} + 0.135\,950\,6z^{-3} + 0.058\,463\,98z^{-4}}{1 - 1.325\,263z^{-1} + 1.480\,202z^{-2} - 0.784\,109\,8z^{-3} + 0.233\,927z^{-4}}$$

Stage 3: realization

As explained earlier, direct form realization of $H(z)$ is very sensitive to many adverse effects of finite wordlength such as coefficient quantization errors, so it is important to break $H(z)$ down into smaller sections and then connect these up, for example in cascade or parallel structure. Assuming cascade structure, $H(z)$ is broken down into two second-order sections $H_1(z)$ and $H_2(z)$:

$$H(z) = H_1(z)H_2(z)$$

where

$$H_1(z) = \frac{a_{01} + a_{11}z^{-1} + a_{12}z^{-2}}{1 + b_{11}z^{-1} + b_{21}z^{-2}}$$

$$H_2(z) = \frac{a_{02} + a_{12}z^{-1} + a_{22}z^{-2}}{1 + b_{12}z^{-1} + b_{22}z^{-2}}$$

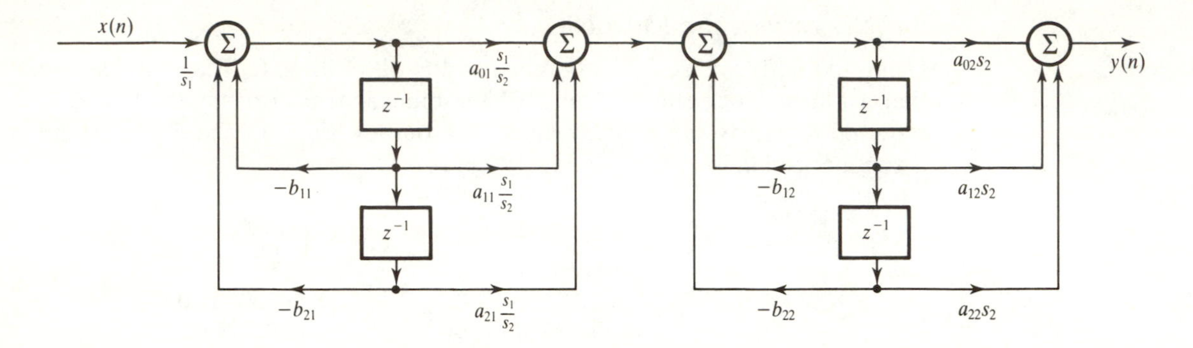

Figure 7.43 Realization diagram for the detailed design example.

The realization diagram is depicted in Figure 7.43, with each filter section realized using a standard biquad structure. The corresponding sets of difference equations, which define how the filtering operation will be carried out, are as follows.

Filter section 1

$$w_1(n) = x(n) - b_{11}w_1(n-1) - b_{21}w_1(n-2)$$

$$y_1(n) = a_{01}w_1(n)s_1/s_2 + a_{11}w_1(n-1)s_1/s_2 + a_{21}w_1(n-2)s_1/s_2$$

Filter section 2

$$w_2(n) = x(n) - b_{12}w_2(n-1) - b_{22}w_2(n-2)$$

$$y_2(n) = a_{02}w_2(n)s_2 + a_{12}w_2(n-1)s_2 + a_{22}w_2(n-2)s_2$$

The exact values of the coefficients, a_{ij} and b_{ij}, are dependent on how we pair the numerator and denominator polynomials of $H(z)$ and how the second-order filter sections used to realize the polynomials are ordered. The best pairing and ordering can only be determined from finite wordlength analysis.

Stage 4: analysis of finite wordlength effects

For the problem, based on the specifications given, we will assume that fixed point two's complement arithmetic will be used and that each coefficient will be quantized, by rounding, to a 16-bit wordlength.

The main objective here is to assess the effects of the various quantization errors on the filter performance and to determine the best filter configuration to implement, in terms of signal-to-noise ratios. The sources of errors of concern are

- overflow errors,
- roundoff errors, and
- coefficient quantization errors.

To avoid overflow at the output of the adders in Figure 7.43, suitable scale factors are introduced before the adders as shown in the figure.

Since $H(z)$ is fourth order to be realized as two second-order sections, its numerator and denominator factors can be paired and ordered in four possible ways:

$$H_A(z) = \frac{N_1(z)}{D_1(z)}\frac{N_2(z)}{D_2(z)}$$

$$H_B(z) = \frac{N_2(z)}{D_2(z)}\frac{N_1(z)}{D_1(z)}$$

$$H_C(z) = \frac{N_1(z)}{D_2(z)}\frac{N_2(z)}{D_1(z)}$$

$$H_D(z) = \frac{N_2(z)}{D_1(z)}\frac{N_1(z)}{D_2(z)}$$

where

$$N_1(z) = 1 + 0.675\,718z^{-1} + z^{-2}$$
$$N_2(z) = 1 + 1.649\,656z^{-1} + z^{-2}$$
$$D_1(z) = 1 - 0.495\,935z^{-1} + 0.761\,864z^{-2}$$
$$D_2(z) = 1 - 0.829\,328z^{-1} + 0.307\,046z^{-2}$$

Each of the four possible filter configurations will have different scale factors as well as different signal to roundoff noise performance. An objective at this stage is to determine the best pairing and ordering, in terms of signal-to-noise performance, for the filter. Overflow and roundoff errors are intimately linked, and so scaling and roundoff analysis should be performed simultaneously.

Using the finite wordlength analysis program, scale factors based on the L_1 norm, L_2 norm and L_∞ norm, for the four possible filters above were obtained and are summarized in Table 7.4. In this example, we have used the L_1 norm. For a fourth-order filter realized as a cascade of two second-order canonic sections, the roundoff noise at the output, after scaling, is given by

$$\sigma_o^2 = \frac{q^2}{12}[s_1^2\|H_1(z)H_2(z)\|_2^2 + 2s_2^2\|H(z)\|_2^2 + 1]$$

where q is the quantization stepsize or rounding, $\|\cdot\|_2^2$ symbolizes the L_2 norm squared, $H_1(z)$ is the transfer function for the first filter stage, $H_2(z)$ is the transfer function for the second filter stage, s_1 is the scale factor for the first filter stage and s_2 is the scale factor for the second filter stage.

The noise performances for each of the four possible filter configurations, with each coefficient quantized to 16 bits (after scaling), are shown in

Table 7.4 Scale factors for the four filter configurations.

Filter	*Scale factor*	L_1	L_2	L_∞
A	s_1	5.524 84	1.608 89	4.379 54
	s_2	11.821 57	3.677 381	7.262 393
B	s_1	2.479 158	1.359 467	2.175 539
	s_2	18.908 47	10.880 49	12.548 114
C	s_1	2.479 158	1.359 467	2.175 539
	s_2	11.821 571	10.880 490	7.262 393
D	s_1	5.524 844	1.608 890	4.379 544
	s_2	18.908 47	5.727 459	12.548 114

Table 7.5. It is evident that filter B has the best roundoff noise performance. The scaled filter transfer function is given by

$$H(z) = H'_{\mathrm{B}}(z) = \frac{s_1}{s_2}\frac{N_2(z)}{D_2(z)}\frac{N_1(z)}{N_2(z)}s_2$$

$$= 0.131\,113\,6\frac{1 + 1.649\,656z^{-1} + z^{-2}}{1 - 0.829\,328z^{-1} + 0.307\,046z^{-2}}$$

$$\times\frac{1 + 0.675\,718z^{-1} + z^{-2}}{1 - 0.495\,935z^{-1} + 0.761\,864z^{-2}}10.880\,490$$

Next, we analyse the effects of coefficient quantization errors. Specifically, we check that the specified coefficient wordlength is adequate for stability and to meet the frequency response specifications. As the poles are not very close to the unit circle a 16-bit coefficient wordlength is adequate for stability. For example, for the first filter section the FWA program shows that as few as 3 bits are adequate for stability, and quantizing the coefficients to 16 bits altered the poles radius by a mere 0.000 48%. The FWA program also showed that only 12 bits are required to maintain the frequency response within the tolerance limits. With 16-bit coefficient wordlength, the response of the filter is essentially the same as the unquantized filter response. The frequency response and pole–zero diagram for the unquantized filter are depicted in Figure 7.44.

Double length accumulation of sums of products and post-accumulation quantization are used to keep the effects of roundoff noise to a minimum.

Table 7.5 Comparison of roundoff noise performance of the four filter configurations.

Filter	*Noise power*
A	$703q^2$
B	$326.378q^2$
C	$382.32q^2$
D	$570.453q^2$

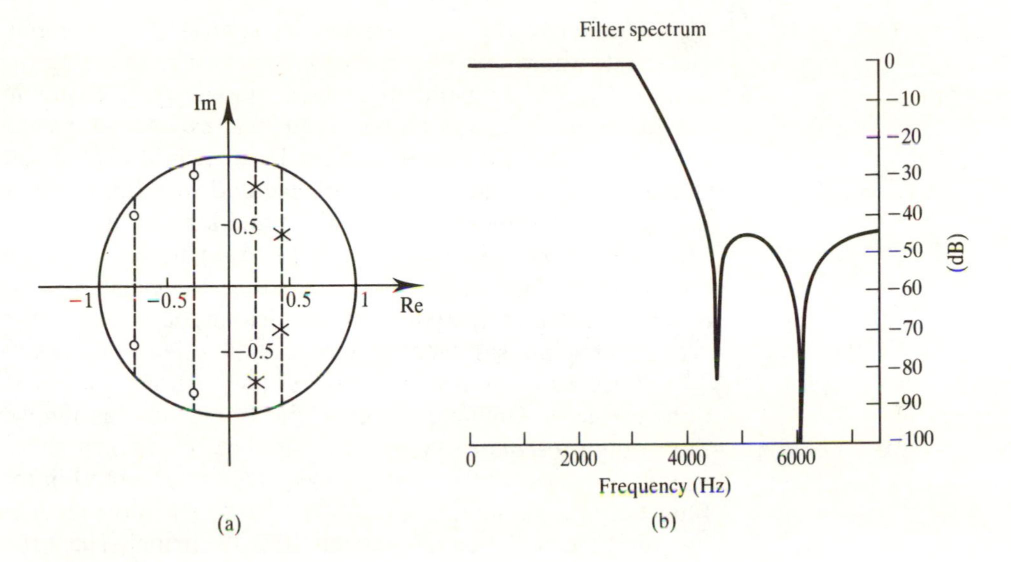

Figure 7.44 (a) The pole–zero diagram, and (b) the frequency response for the detailed design example.

Stage 5

The quantized coefficients (obtained by multiplying the scaled coefficients by 2^{15}), are entered into the TMS320C25 IIR filter program listed in the appendix. Chapter 11 gives a more detailed discussion of the IIR filtering routines and their development.

7.9 Summary

The design of IIR filters can be divided into five interrelated stages (see Figure 7.1). Filter specifications are often application dependent, but should include details such as band edges, tolerance limits for the amplitude response, sampling rate and I/O requirements. For filters with standard characteristics, the coefficients required to meet the amplitude response specifications can be efficiently obtained via the BZT. This approach as well as other useful coefficient calculation methods are described in the text and illustrated with lots of examples. Higher-order IIR filters are realized as cascades or parallel combinations of second- and first-order sections to keep changes in the positions of the poles and zeros due to the effects of finite wordlength small. Scaling the input to each section to prevent overflow in the internal nodes is necessary.

The performance of an IIR digital filter is limited by the number of bits used in its implementation. The four common sources of errors are those due to (1) input quantization, (2) coefficient quantization, (3) product roundoff and

(4) addition overflow. Techniques for analysing their effects on filter performance and, where possible, for eliminating or minimizing them have been presented. Coefficient wordlength must be adequate to minimize the effects of coefficient quantization on the frequency response and to prevent the possibility of instability. The stability of an IIR filter is always of concern. An IIR filter that is otherwise stable when implemented with infinite precision may become unstable if implemented with finite precision. In high fidelity audio work, for example, 24-bit coefficients are said to be necessary for processing low frequency audio signals. In most other cases, representing the coefficients with 16 or more bits and carrying out the arithmetic operations with double-length accumulators are sufficient to minimize the effects of finite wordlength.

Truncation or roundoff errors due to finite precision arithmetic operations create a nonlinear effect in the filter, such as limit cycles whereby the filter output oscillates even in the absence of or constant input. The effects of roundoff error on filter performance can be quantified in terms of SNRs at the filter output. Reduction in the SNR due to roundoff error can be offset by the use of the error spectral shaping (ESS) scheme. The primary effect of such schemes is to nullify the 'amplifying' effect of the poles of the filter on the roundoff errors. The price paid for this is an increase in the number of multiplications and additions, although first-order ESS with integer coefficients is computationally efficient.

Design programs are provided on the PC disks for the book to enable designers to calculate the filter coefficients and to analyse some of the effects of finite wordlength on filter performance.

7.10 Application examples

This section provides an overview of several applications where the IIR filter has either been used or is appropriate.

7.10.1 Digital audio

Digital filters have found use in many areas of digital audio, especially in systems with high quality digital sources such as the CD player and DAT, where it makes sense to carry out digitally as many of the signal processing operations as possible. DSP also makes it possible to generate acoustic properties of locations such as concert halls, jazz clubs and discos. Applications in digital audio where an IIR filter has been used include graphic equalization, tone control, channel equalization, noise shaping in ADC/DAC, and band splitting.

In digital graphic equalizers, for example, IIR filters are used to split the

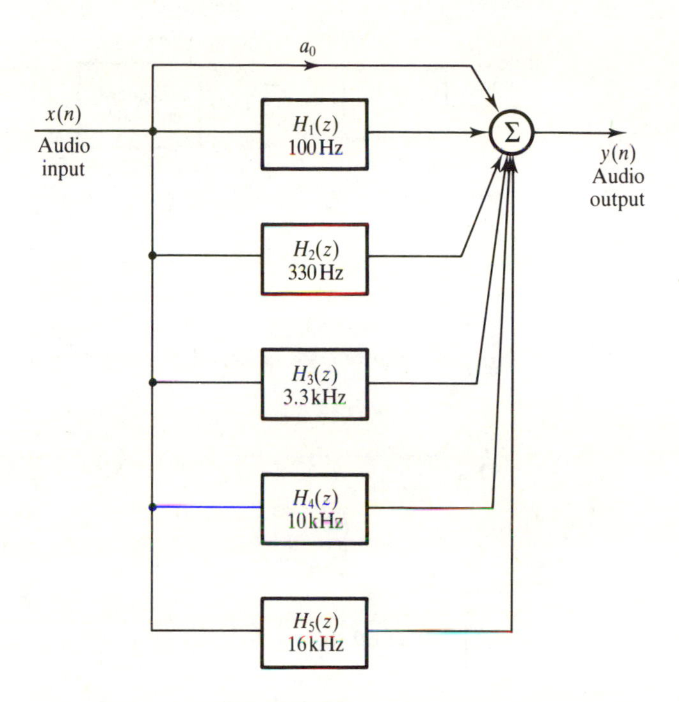

Figure 7.45 Simplified diagram of an all-digital graphic equalizer. The main component is a bank of parallel IIR filters with different centre frequencies. The gain of each filter is individually adjustable, for example with a sliding potentiometer, in the range ±10 dB, say.

entire audio frequency range into bands enabling a comprehensive tone adjustment of the reproduced sound to personal taste instead of the bass and treble controls. A typical five-band graphic equalizer would split the audio frequency range into five bands with centre frequencies of 100 Hz, 330 Hz, 3.3 kHz, 10 kHz and 16 kHz and allow adjustable signal level in each band in the range ±10 dB.

A simple filtering arrangement for graphic equalization is shown in Figure 7.45.

7.10.2 Digital control

With the increased awareness of the benefits of DSP and the availability of low cost processors, controllers are now being implemented digitally to achieve better accuracy and flexibility. Figure 7.46 shows the principles of digital control of an analogue plant, $H(s)$, which could be a car or motor, for example. In general, the digital controller has an IIR characteristic.

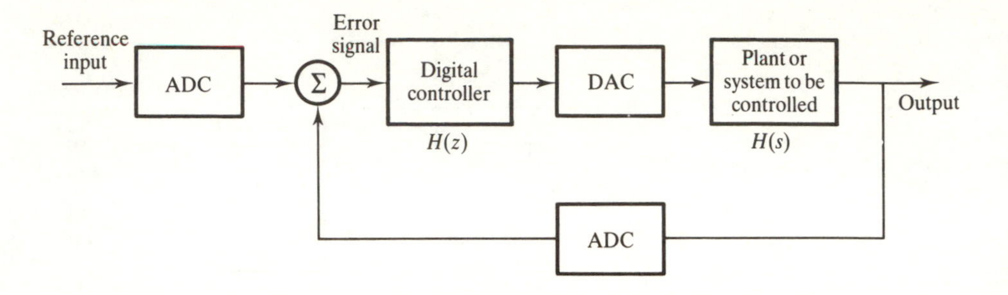

Figure 7.46 Principles of digital control of an analogue plant.

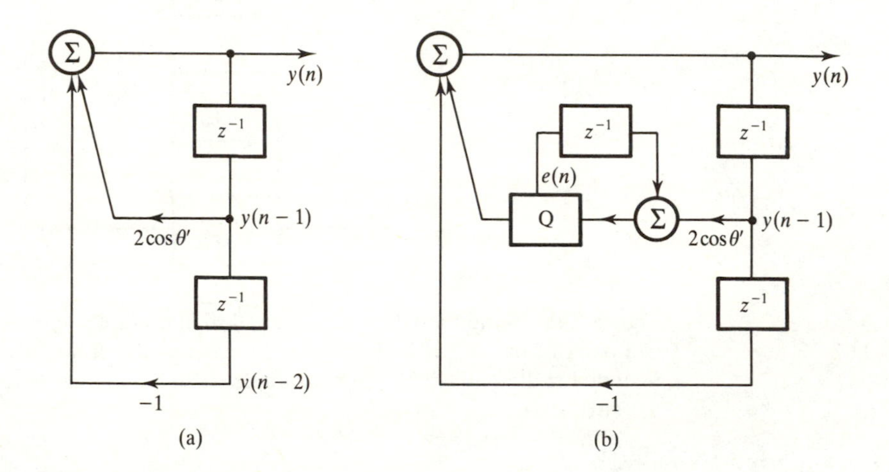

Figure 7.47 (a) A simple digital oscillator and (b) a simple digital oscillator with a first-order error spectral shaping.

7.10.3 Digital frequency oscillators

IIR filters have been used to generate accurate waveforms instead of the traditional look-up table approach. The approach exploits the fact that an IIR filter with poles on the unit circle is essentially unstable. A simple sine wave oscillator is depicted in Figure 7.47(a). The poles of the IIR filter are located at $e^{\pm j\theta}$ and the frequency of oscillation is given by

$$\theta' = w_0 T_{\mathrm{B}}$$

where T is the sampling period. The filter coefficient, $2\cos\theta'$, is constrained to be an integer by taking the integer part of $2^{\mathrm{B}} \times 2\cos\theta'$ (B is the number of bits).

A major issue in digital waveform generation using IIR filters concerns finite wordlength effects. For example, coefficient quantization would lead to unequally spaced frequencies, whereas product quantization leads to a buildup

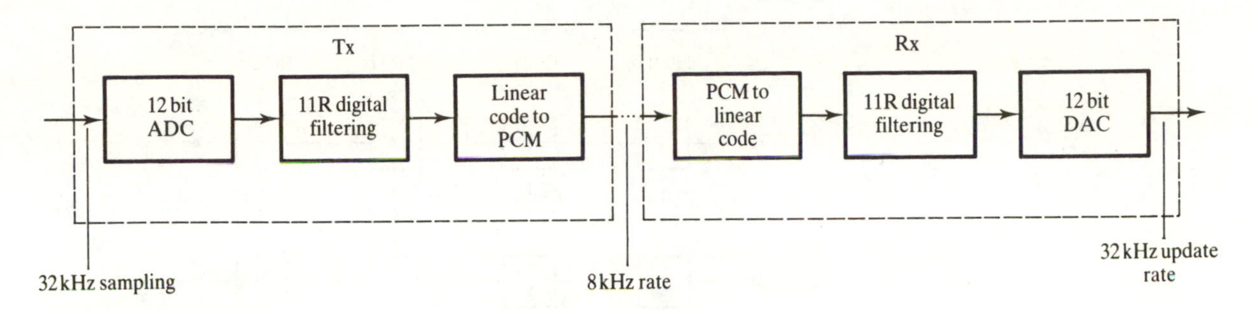

Figure 7.48 A PCM channel showing possible use of IIR filters for the main anti-aliasing filter (TX end) and anti-imaging filtering (RX end).

of roundoff errors which soon renders the waveform generator useless. However, using ESS techniques these errors can be reduced to a minimum. Figure 7.47(b) shows an oscillator utilizing the ESS technique (Abu-el-Haija and Al-Ibrahim, 1986), which offers a significant reduction in the roundoff noise effects.

7.10.4 Telecommunication

In digital telephony (Freeny *et al.*, 1971), PCM permits the simultaneous transmission of many voice channels. Each channel is sampled at 8 kHz after suitable bandlimiting and encoded using either A-law or μ-law. At the receiving end, the PCM data is converted back to analogue and anti-image filtered. Digital IIR filters can be used to provide the necessary filtering at the transmit and receive ends; Figure 7.48. In this case, the filtering is performed at a higher rate, for example 32 kHz, and then converted from linear to standard PCM codes.

7.10.5 Digital touch-tone generation and receiving

An excellent application of IIR filter is the all digital dual-tone multifrequency touch-tone receiver (Jackson *et al.*, 1968; Mock, 1985).

In modern telephone systems, the information required to establish communication, and for maintenance and charging, is normally provided by a multifrequency code. Typically, the telephone set generates two tones, one low frequency tone and another high frequency tone (see Figure 7.49).

The tone generator can be implemented using a pair of programmable second-order IIR oscillators; Figure 7.50. When a button is pressed the code for the dialled digit is used to select the appropriate filter coefficients and initializing conditions from ROMs to produce a pair of tones (one high frequency tone and one low frequency tone). The tones are added to produce the

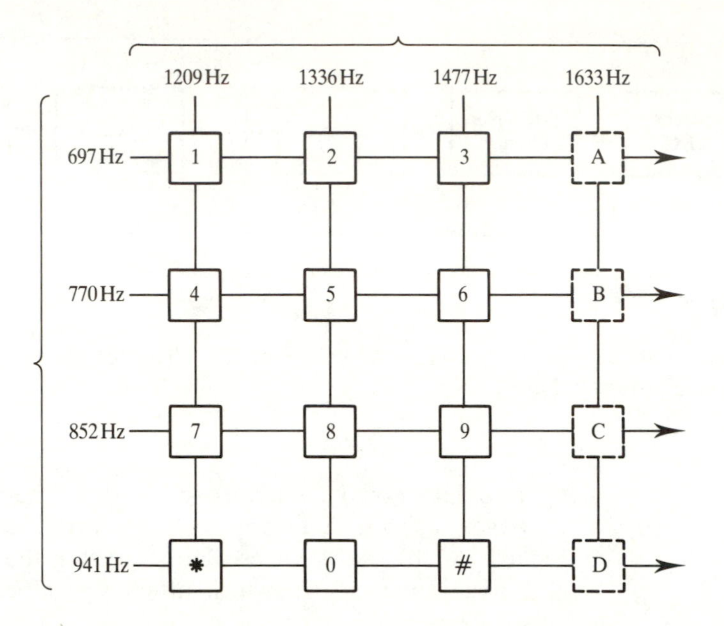

Figure 7.49 A simplified diagram of a 4×4 keypad for a touch-tone telephone. The buttons in broken lines are not available. Pressing a button generates a pair of tones, one from the low frequency group and the other from the high frequency group. For example, pressing 9 generates 852 Hz and 1477 Hz tones (after Mock, 1985).

touch-tone signal. As with the digital sine wave generator, the performance of the touch-tone generator can be improved by using an error feedback scheme.

At the receiving end, the information is digitized at a rate of 8 kHz and then separated into a low and a high frequency band by the front-end bandpass IIR filters. To detect the presence of a tone, level detection is carried out. This is performed by combined bandpass filtering and full wave rectification followed by lowpass filtering. To determine which of the low frequency tones is present the low frequency band is split into four bands by two sets of four BPFs. The same is true of the high frequency band. The resulting eight levels are passed to decision logic to determine the received code.

7.10.6 Clock recovery for data communication

A fundamental problem in most digital data communication over long distances is that of generating a clock at the receiving end at the correct frequency and phase so that the data may be correctly decoded. The clock is normally derived from the received data.

Traditionally, analogue circuits, for example using phase lock loops, are used to recover the clock, but these are susceptible to drift with age and temperature. Further, such circuits are unsuitable in applications that involve

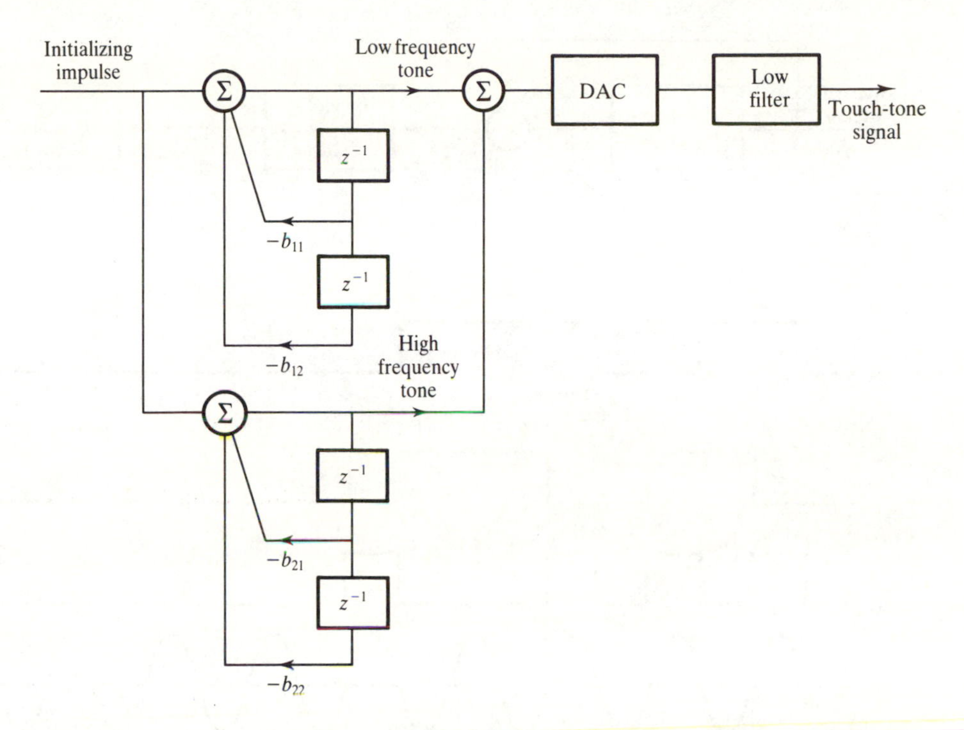

Figure 7.50 Touch-tone generator (after Mock, 1985). The code for the digit dialled is used to select filter coefficients and initial conditions and hence the oscillator frequency.

burst transmissions because of their slow response or where more than one data rate is involved (Smithson, 1992).

The input data stream is normally scrambled at the transmitting end (to provide clock information during idle periods) and then encoded, with each code representing a symbol. The codes are then transmitted at the so-called symbol rate. The problem at the receiving end is to recover the symbol clock.

The principles of symbol clock recovery using DSP are shown in Figure 7.51. The data stream is added modulo-2, that is exclusive ORed, with a version of itself delayed by one-half of a clock period to produce an output which contains level changes at the symbol rate (point C). The data is then applied to a marginally stable bandpass IIR filter. The impulse response of such a filter decays very slowly with time, producing a 'damped oscillation' at a frequency, w_0, the centre frequency of the filter. The use of a marginally stable filter ensures that there is an output even when there are no transitions in the input data stream for reasonably long periods. The sampling frequency of the filter is chosen to be a multiple of the symbol rate. The desired symbol clock is derived from the output of the filter by zero-crossing detection (point E in Figure 7.51). For a 2's complement representation, this is trivially implemented by examining the signs of the data samples at the output of the digital filter.

A simple all-pole, IIR filter of the form shown in Figure 7.52 may be used

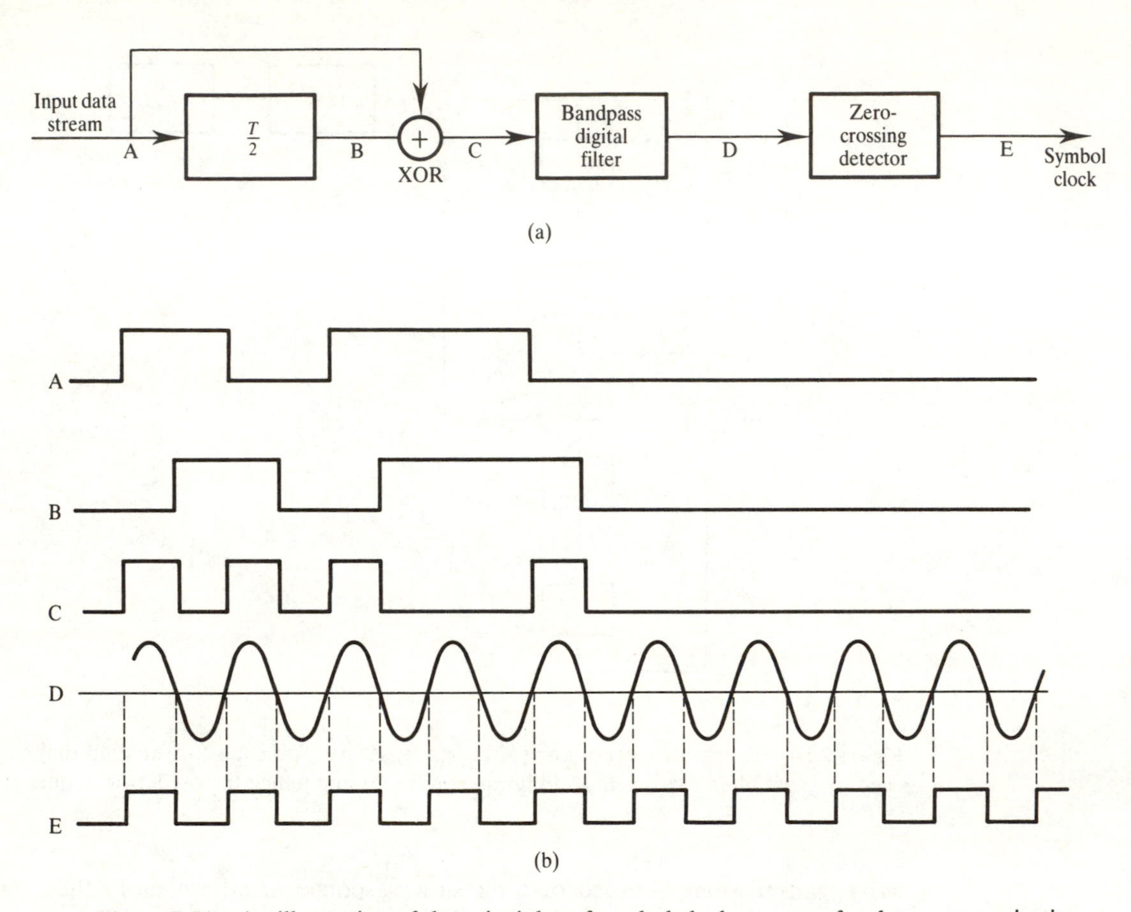

Figure 7.51 An illustration of the principles of symbol clock recovery for data communication.

for the symbol clock recovery. The filter is characterized by the following transfer function:

$$H(z) = \frac{1}{[z - r\exp(-jw_0T)][z - r\exp(jw_0T)]}$$

$$= \frac{1}{z^2 - 2r\cos(w_0T)z - r^2}$$

where w_0 is the centre frequency of the bandpass filter, r is the radius of the pole and T is the reciprocal of the sampling frequency. w_0 is normally chosen to be identical or very close to the symbol clock frequency to be recovered, and the sampling frequency is a multiple of the centre frequency. The bandwidth of the filter is determined by the radius of the pole (see Equation 7.4). To ensure that the impulse response decays slowly with time the pole is normally located very close to the unit circle, typically in the range $0.99 < r < 1$. As discussed in

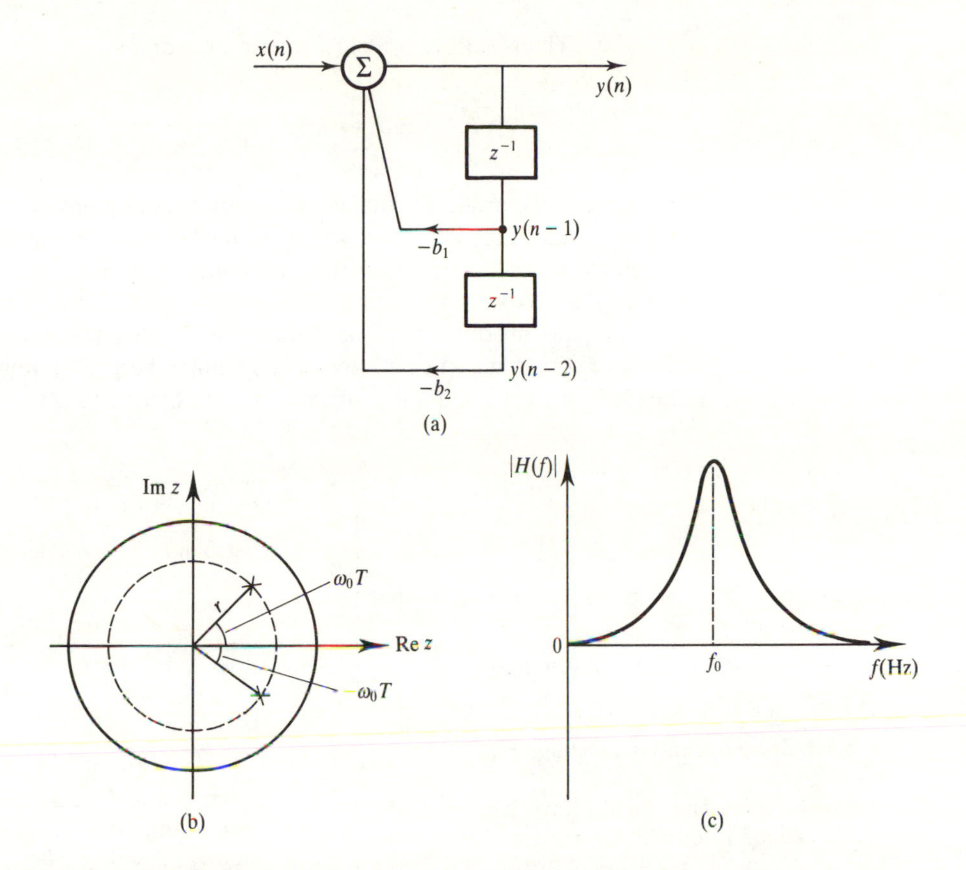

Figure 7.52 (a) Structure of the symbol clock recovery IIR filter, (b) pole diagram and (c) spectrum of filter.

Section 7.4.1 (Equation 7.5), the pole radius, r, and the bandwidth of the filter, bw, are related as

$$r \approx 1 - (\text{bw}/F_s)\pi$$

where $F_s = 1/T$ is the sampling frequency of the filter.

For example, to recover the symbol clock for a hypothetical 4800 baud modem, suitable filter parameters are

data rate	4.8 kbaud
filter centre frequency, f_0	4.8 kHz
sampling frequency	153.6 kHz
bandwidth, bw	100 Hz

In this case, the pole radius (from the above equation), $r = 0.997\,954\,69$, and the pole angle $w_0T = 2\pi f_0 T = (2\pi \times 4.8 \times 10^3/153.6 \times 10^3) = 0.195\,34$ rad

$\approx 11.25°$. The resulting transfer function becomes

$$H(z) = \frac{1}{z^2 - 1.957\,558z + 0.995\,913}$$

As discussed earlier in the chapter, finite wordlength effects must be considered if the filter is to operate as desired. In particular, the input to the filter needs to be scaled to avoid the possibility of self-sustaining oscillations at its output due to overflow, and the use of a simple roundoff noise shaping scheme may help to produce a 'clean' clock. In a practical clock recovery system, a second filter stage will be necessary to improve the system performance when the input data is a sequence of 1s or 0s (Smithson, 1992).

Problems

7.1 A lowpass filter has poles and a zero at the following locations:

zero, -0.5; poles, 0.370, $0.6 \pm 0.5j$

(1) Plot the pole–zero diagram.

(2) Obtain the transfer function, $H(z)$.

7.2 Digitize, using the impulse invariant method, the analogue filter with the transfer function

$$H(s) = \frac{\alpha}{s(s + \alpha)}, \quad \alpha = 0.5$$

Assume a sampling frequency of 1 (normalized).

7.3 A requirement exists to simulate in a digital computer an analogue system with the following normalized characteristic:

$$H(s) = \frac{1}{s^2 + \sqrt{2}s + 1}$$

Obtain a suitable transfer function using

(1) the impulse invariant method, and

(2) the bilinear transform method.

Assume a sampling frequency of 5 kHz and a 3 dB cutoff frequency of 1 kHz.

7.4 Obtain suitable coefficients for an IIR digital filter using the bilinear transform method and an elliptic characteristic to meet the following specifications:

passband	4 kHz–12 kHz
stopbands	0–3.4 kHz, 12.6 kHz–16 kHz
passband ripples	< 0.1 dB
stopband attenuation	> 30 dB
sampling frequency	32 kHz

Determine a suitable coefficient wordlength to ensure that the filter is stable and the frequency response lies within the specified limits.

7.5 Design and implement in software a digital lowpass filter to meet the following specifications:

passband edge	2.5 kHz
stopband edge	3 kHz
passband deviation	< 0.1 dB
stopband attenuation	> 60 dB
sampling frequency	15 kHz

7.6 Obtain, via the bilinear transform, the coefficients of a digital filter that is maximally flat in the passband, 0 to 4 kHz, and has an attenuation of at least 25 dB at frequencies over 10 kHz. Assume a sampling frequency of 32 kHz.

7.7 A requirement exists for a bandpass digital filter with a Chebyshev characteristic to meet the following specifications:

passband	1200–1800 Hz
stopband attenuation	> 30 dB
passband ripple	< 0.5 dB
transition width	400 Hz
sampling frequency	7.5 kHz

Using a suitable software program, obtain the filter coefficients.

7.8 The requirements for a certain lowpass filter are given below:

passband	0–30 Hz
stopband edge	50 Hz
stopband attenuation	> 40 dB, at $f > 50$ Hz
sampling frequency	256 Hz

Assuming that the filter has a Butterworth response, determine via the bilinear transformation the transfer function, $H(z)$, of the filter. Obtain the realization diagram for the filter in cascade form using second- and/or first-order sections.

7.9 A bandpass digital filter with a Butterworth response is required for a certain real-time digital signal processing system. The filter is to satisfy the following requirements:

passband	0.3–3.4 kHz
stopband	0–0.2 kHz and 4–8 kHz
stopband attenuation	25 dB
sampling frequency	32 kHz

Obtain a suitable transfer function for the filter using the bilinear transform method.

7.10 A digital filter is required to remove baseline wander and artefacts due to body movement in a certain biomedical application. The filter is required to meet the following requirements:

passband	1–30 Hz
stopband	0–0.5 Hz and 40–128 Hz
passband ripple	< 0.1 dB
stopband attenuation	> 30 dB
sampling frequency	256 Hz

Determine the order of a suitable IIR filter and its transfer function $H(z)$.

7.11 A narrowband reject digital filter is required to remove an interfering signal. The filter should satisfy the following specifications:

passband edges	45 Hz and 55 Hz
passband ripple	< 0.1 dB
stopband attenuation	> 50 dB
sampling frequency	500 Hz

Obtain the coefficients of the filter.

7.12 Determine, using the impulse invariant method, the transfer function and difference equation for the digital equivalent of a single-pole RC lowpass filter. Assume a sampling frequency of 150 Hz and a 3 dB cutoff frequency of 30 Hz.

7.13 A standard second-order analogue filter section, with simple poles, can be expressed as

$$\frac{A_0 + A_1 s}{B_0 + B_1 s + B_2 s^2} = \frac{C_1}{s + p_1} + \frac{C_2}{s + p_2}$$

where C_1 and C_2 are partial fraction coefficients and p_1 and p_2 are s-plane poles. Assume the impulse invariant transformation for the second-order section is given by

$$\frac{A_0 + A_1 s}{B_0 + B_1 s + B_2 s^2} \rightarrow \frac{c_1 + c_2 - (c_1 e^{-p_2 T} + c_2 p^{-p_1 T}) z^{-1}}{1 - (e^{-p_1 T} + e^{-p_2 T}) z^{-1} + e^{-(p_1 + p_2) T} z^{-2}} = \frac{a_0 - a_1 z^{-1}}{1 + b_1 z^{-1} + b_2 z^{-2}}$$

where T is the sampling interval.

(1) Find expressions for p_1, p_2, C_1 and C_2 in terms of A_0, A_1, B_0, B_1 and B_2.

(2) Obtain expressions for the coefficients a_0, a_1, b_1 and b_2 for the case when the poles are complex conjugates.

(3) Repeat part (2) when the poles are real and unequal.

(4) Given the following normalized analogue transfer function, use your results to obtain the coefficients of an equivalent discrete-time filter. Assume a sampling

frequency of 10 kHz and a cutoff frequency of 2 kHz.

$$H(s) = \frac{1}{s^2 + \sqrt{2}s + 1}$$

7.14 Starting with a suitable analogue Chebyshev LPF, obtain the transfer function of a digital bandstop filter to meet the following specifications:

stopband	10–15 kHz
sampling frequency	50 kHz
passband ripple	0.5 dB
filter order	6

7.15 Starting with a suitable analogue Chebyshev LPF and the biquadratic transformation given in Section 7.4.10 determine, using the s-plane to z-plane pole–zero mapping approach, the transfer function of a digital bandpass filter to meet the following specifications:

passband	10–15 kHz
sampling frequency	50 kHz
passband ripple	0.5 dB
bandpass filter order	6

7.16 An analogue lowpass filter is characterized by a pair of s-plane poles at

$$p_{1,2} = -1.4 \pm 1.2936j$$

It is desired to convert the filter into a digital bandpass filter with passband edges of 3 kHz and 5 kHz at a sampling frequency of 15 kHz. Given the digital lowpass to bandpass transformation and the BZT in Equation 7.31, determine

(1) the poles and zeros of the digital bandpass filter, and

(2) its transfer function in factored form.

7.17 An analogue filter is to be converted into an equivalent digital filter that will operate at a sampling frequency of 256 Hz. Assume that the analogue filter has the following transfer function:

$$H(s) = \frac{1}{s^3 + 2s^2 + 2s + 1}$$

(1) Obtain suitable coefficients for the digital filter.

(2) Assuming that the digital filter is to be realized using the cascade structure, draw a suitable realization block diagram and develop the difference equations.

(3) Repeat part (2) for the parallel structure.

7.18 Figure 7.53 shows a standard second-order filter section.

(1) Explain why overflow is permissible at nodes 1 and 3 but not at node 2.

(2) Find suitable scale factors to reduce the possibility of overflow at node 2.

(3) Assuming that the filter is implemented with pre-accumulator product quantization in an 8-bit system, determine the number of additional bits that will be required to achieve a reduction of roundoff noise by at least 20 dB.

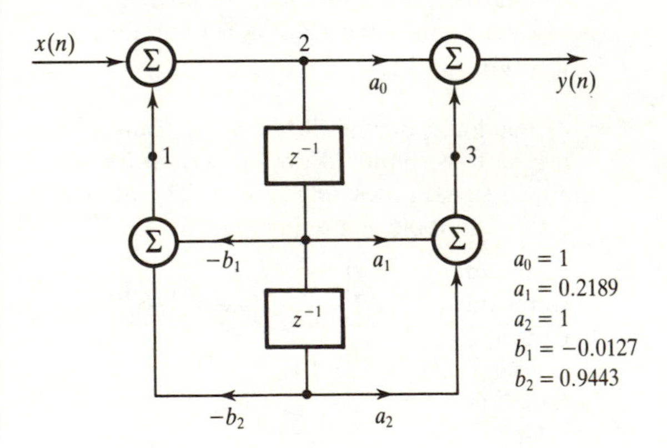

Figure 7.53 Standard second-order filter section for Problem 7.18.

7.19 An IIR filter has the following transfer function:

$$H(z) = \frac{0.1436 + 0.2872z^{-1} + 0.1436z^{-2}}{1 - 1.8353z^{-1} + 0.9748z^{-2}}$$

(1) Determine the positions of the poles and zeros and sketch the pole–zero diagram.

(2) Determine the radial distance of the pole from the origin.

(3) Estimate the number of bits required to represent each coefficient

(a) to maintain stability, and

(b) so that the amplitude response in the passband does not change by more than 1%.

7.20 The following transfer function is given:

$$H(z) = \frac{1 - 0.9631z^{-1} + z^{-2}}{1 - 1.5763z^{-1} + 0.9413z^{-2}}$$

(1) Find a suitable scale factor to avoid overflow if a canonic second-order section is used to realize it.

(2) Determine the minimum wordlengths required to achieve an output signal-to-noise ratio of 60 dB. State any assumptions made.

7.21 The following poles and zeros of an eighth-order IIR filter are given:

poles	zeros
0.2870 ± 0.9075j	0.0553 ± 0.9985j
0.7882 ± 0.5658j	0.8828 ± 0.4698j
0.4089 ± 0.7447j	−0.4816 ± 0.8764j
0.6479 ± 0.5975j	0.9617 + 0.2740j

(1) Sketch the pole–zero diagram and pair the poles and zeros, justifying your pairing scheme.

(2) Write down the transfer function of the filter from the diagram. Assuming the filter is to be realized in cascade form, decide on a suitable ordering scheme.

(3) Determine suitable scaling factors for the filter sections using the finite wordlength analysis program on the PC disks.

(4) Assume that the input data is digitized to 8 bits and that the degradation in SNR due to roundoff error is to be no more than 0.5 dB. Determine suitable wordlengths for the internal data, coefficients and data variables.

7.22 The following transfer function is given:

$$H(z) = \frac{1 - 1.4890z^{-1} + z^{-2}}{1 - 0.3724z^{-1} + 0.5119z^{-2}} \times \frac{1 - 1.9020z^{-1} + z^{-2}}{1 - 0.3779z^{-1} + 0.0851z^{-2}}$$

(1) Determine and plot the locations of the poles and zeros.

(2) Plot the magnitude and phase responses of the filter using a sampling frequency of 48 kHz.

(3) Write down the 2's complement fixed point representation of the filter coefficients using 8 bits (including the sign bit).

(4) Repeat parts (1) and (2) for the quantized filter and compare the two sets of results.

7.23 The following transfer function is given:

$$H(z) = 0.1436 \frac{1 + 2z^{-1} + z^{-2}}{1 - 0.67993z^{-1} + 0.49133z^{-2}}$$

(1) Determine a suitable scale factor to avoid overflow at the output of adder 1 and an output scale factor for the overall gain to be unity.

(2) Encode the filter coefficients and scale factors using 8-bit fixed point arithmetic.

(3) Determine the total roundoff noise.

7.24 A lowpass IIR filter is required in digital telephony to bandlimit the voice signal. The filter is required to satisfy the following specifications:

passband	0–3300 Hz
stopband	46 000–16 kHz
passband ripple	< 0.1 dB
stopband attenuation	> 30 dB
ADC	12 bits
coefficient wordlength	16 bits

Determine

(1) a suitable transfer for the filter, assuming the filter is realized in cascade form with second- and/or first-order sections,

(2) scale factors for each filter section,

(3) the change in passband and stopband ripples due to coefficient quantization, and

(4) the degradation in SNR due to roundoff error, assuming that post-accumulation quantization is employed.

7.25 Scaling is employed in IIR filters to prevent the adders overflowing. One scheme is to attenuate the input to each IIR filter section such that the scale factor is given by

$$s_1^2 = \frac{1}{2\pi j} \oint \frac{z^{-1}\,dz}{D(z)D(z^{-1})}$$

where $\oint$ indicates integration around the unit circle, that is $|z| = 1$, and

$$D(z) = 1 + b_1 z^{-1} + b_2 z^{-2}$$

(1) Find a general expression for s_1^2.

(2) Obtain the scale factor, s_1, for the filter with the following transfer function:

$$H(z) = \frac{1 + 1.2173z^{-1} + z^{-2}}{1 + 0.9140z^{-1} + 0.8793z^{-2}}$$

7.26 Design a Chebyshev lowpass IIR digital filter meeting the following specifications:

passband edge	12 kHz
stopband edge	16 kHz
passband ripples	0.5 dB
stopband attenuation	60 dB
sampling frequency	48 kHz

Assume that the filter will be implemented on a TMS320C25-based system with 12-bit ADC and DAC.

7.27 Design a Chebyshev highpass IIR digital filter meeting the following specifications:

passband edge	12 kHz
stopband edge	8 kHz
passband ripples	0.5 dB
stopband attenuation	60 dB
sampling frequency	48 kHz

Assume that the filter will be implemented on a TMS320C25-based system with 12-bit ADC and DAC.

References

Abu-el-Haija A. and Al-Ibrahim M.M. (1986). Improving performance of digital sinusoidal oscillators by means of error feedback circuits. *IEEE Trans. Circuits and Systems*, **33**(4), 373–80

Antoniou A. (1979). *Digital Filters Analysis and Design*. New York NY: McGraw-Hill

Claasen T. (1974). Improvement of overflow behaviour of 2nd-order digital filters by means of error feedback. *Electronics Lett.*, **10**(12), 240–1

Dattoro J. (1988). The implementation of recursive digital filters for high-fidelity audio. *J. Audio Engineering Society*, **36**(11), 851–78

DeFatta D.J., Lucas J.G. and Hodgkiss W.S. (1988). *Digital Signal Processing*. New York NY: Wiley

Feeney S.L., Kieburtz R.B., Mina K.V. and Tewksbury S.K. (1971). Design of digital filters for an all digital frequency division multiplex–time division multiplex translator. *IEEE Trans. Circuit Theory*, **18**, 702–11

Gold B. and Rader C.M. (1969). *Digital Processing of Signals*. New York NY: McGraw-Hill

Gray A.H. and Markel J.D. (1976). A computer program for designing elliptic filters. IEEE *Trans. Acoustics, Speech and Signal Processing*, **24**(6), 529–38

Higgins W.E. and Munson D. (1982). Noise reduction strategies for digital filters: error spectrum shaping versus the optimal linear state-space formulation. *IEEE Trans. Acoustics, Speech and Signal Processing*, **30**(6), 963–73

IEEE (1979). *Programs for Digital Signal Processing*. New York NY: IEEE Press

Jackson L.B. (1986). *Digital Filters and Signal Processing*. Boston MA: Kluwer

Jackson L.B., Kaiser J.F. and McDonald H.S. (1968). An approach to the implementation of digital filters. *IEEE Trans. Audio and Electroacoustics*, **16**(3), 413–21

Jong M.T. (1982). *Methods of Discrete Signal System Analysis*. New York NY: McGraw-Hill

Jury E.I. (1964). *Theory and Application of the Z-transform Method*. Huntington NY: Wiley

Lynn P.A. and Fuerst W. (1989). *Introductory Digital Signal Processing with Computer Applications*. Chichester: Wiley

Mitra S.K., Hirano K. and Sakaguchi H. (1974). A simple method of computing the input quantization and multiplication roundoff errors in a digital filter. *IEEE Trans. Acoustics, Speech and Signal Processing*, **22**(5), 326–9

Mock P. (1985) Add DTMF generation and decoding to DSP-μp designs. *EDN*, **30**

Mullis C.T. and Roberts R.A. (1982). An interpretation of error spectrum shaping in digital filters. *IEEE Trans. Acoustics, Speech and Signal Processing*, **30**(6), 1013–15

Parks T.W. and Burrus C.S. (1987). *Digital Filter Design*. New York NY: Wiley

Rabiner L.R. and Gold B. (1975). *Theory and Applications of Digital Signal Processing*. Englewood Cliffs NJ: Prentice-Hall

Rader C.M. and Gold B. (1967). Effects of parameter quantization on the poles of a digital filter. *Proc. IEEE*, **55**, 688–9

Smithson P. (1992). Clock recovery for a satellite data modem, University of Plymouth (personal communication)

Stanley W.D., Dougherty G.R. and Dougherty R. (1984). *Digital Signal Processing* 2nd edn. Reston VA: Reston Publishing Inc.

Stearns S.D. and Hush D.R. (1990). *Digital Signal Analysis* 2nd edn. Englewood Cliffs NJ: Prentice-Hall

Bibliography

Abu-el-Haija A.I. and Peterson A.M. (1979). An approach to eliminate roundoff errors in digital filters. *IEEE Trans. Acoustics, Speech and Signal Processing*, **27**, 195–8

Ahmed N. and Natarajan T. (1983). *Discrete-time Signals and Systems*. Reston VA: Reston Publishing Inc.

Allen J. (1975). Computer architecture for signal processing. *Proc. IEEE*, **63**(4), 624–48

Arjmand M. and Roberts R.A. (1981). On comparing hardware implementations of fixed point digital filters. *IEEE Circuits Systems Mag.*, **3**(2), 2–8

Avenhaus E. (1972). Filters with coefficients of limited word length. *IEEE Trans. Audio Electroacoustics*, **20**, 206–12

Barnes C.W., Tran B.N. and Leung S.H. (1985). On the statistics of fixed-point roundoff error. *IEEE Trans. Acoustics, Speech and Signal Processing*, **33**, 595–606

Bellanger M. (1984). *Digital Processing of Signals. Theory and Practice*. New York NY: Wiley

Chang T.L. (1978). A low roundoff noise digital filter structure. In *Proc. IEEE Int. Symp. on Circuits and Systems*, May 1978, pp. 1004–8

Chang T.L. (1979). Error-feedback digital filters. *Electronics Lett.* 348–9

Chang T.L. (1980). Comments on "An approach to eliminate roundoff errors in digital filters.". *IEEE Trans. Acoustics, Speech and Signal Processing*, **28**(2), 244–5

Chang T.L. (1981). Suppression of limit cycles in digital filters designed with one magnitude-truncation quantizer. *IEEE Trans. Circuits and Systems*, **28**(2), 107–11

Chang T.L. (1981). On low-roundoff noise and low-sensitivity digital filter structures. *IEEE Trans. Acoustics, Speech and Signal Processing*, **29**(5), 1077–80

Chang T.L. and White S.A. (1981). An error cancellation digital-filter structure and its distributed-arithmetic implementation. *IEEE Trans. Circuits and Systems*, **28**(4), 339–42

Charalambous C. and Best M.J. (1974). Optimization of recursive digital filters with finite word lengths. *IEEE Transactions Acoustics, Speech and Signal Processing*, **22**(6), 424–31

Chassaing R. and Horning D.W. (1990). *Digital Signal Processing with the TMS320C25*. New York NY: Wiley

Claasen T.A.C.M. and Kristiansson L.O.G. (1975). Necessary and sufficient conditions for the absence of overflow phenomena in a second order recursive digital filter. *IEEE Trans. Acoustics, Speech and Signal Processing*, **23**(6), 509–15

Claasen T.A.C.M., Mecklenbrauker W.F.G. and Peek J.B.H. (1973) Second-order digital filter with only one magnitude-truncation quantiser and having practically no limit cycles. *Electronics Lett.*, **9**, 531–2

Claasen T.A.C.M., Mecklenbrauker W.F.G. and Peek J.B.H. (1973). Some remarks on the classification of limit cycles in digital filters. *Philips Research Rep.*, **28**, 297–305

Claasen T., Mecklenbrauker W.F.G. and Peek J.B.H. (1975). Frequency domain criteria for the absence of zero-input limit cycles in nonlinear discrete-time systems, with applications to digital filters. *IEEE Trans. Circuits and Systems*, **22**, 232–9

Claasen T.A.C.M., Mecklenbrauker W.F.G. and Peek J.B.H. (1976). Effects of quantization and overflow in recursive digital filters. *IEEE Trans. Acoustics, Speech and Signal Processing*, **24**(6), 517–28

Crochiere R.E. (1975). A new statistical approach to the coefficient word length problem for digital filters. *IEEE Trans. Circuits and Systems*, **22**, 190–6

Crochiere R.E. and Oppenheim A.V. (1975). Analysis of linear digital networks. *Proc. IEEE*, **63**(4), 581–94

IEEE (1978). Digital Signal Processing II. IEEE

Diniz P.S.R. and Antoniou A. (1985). Low-sensitivity digital filter structures which are amenable to error-spectrum shaping. *IEEE Trans. Circuits and Systems*, **32**(10), 1000–7

Elliot D.F. ed. (1987). *Handbook of Digital Signal Processing*. London: Academic Press

Fethweis A. Roundoff noise and attenuation sensitivity in digital filters with fixed point filters. *IEEE Trans. Circuit Theory*, 174–5.

Jackson L.B. (1970). On the interaction of roundoff noise and dynamic range in digital filters. *BSTJ*, **49**(2), 159–84

Jackson L.B. (1976). Roundoff noise bounds derived from coefficient sensitivities for digital filters. *IEEE Trans. Circuits and Systems*, **23**(8), 481–5

Knowles J.B. and Olcayto E.M. (1968). Coefficient accuracy and digital filter response. *IEEE Trans. Circuit Theory*, **15**, 31–41

Liu B. (1971). Effect of finite word length on the accuracy of digital filters – a review. *IEEE Trans. Circuit Theory*, **18**, 670–7

Liu B. and Kaneko T. (1969). Error analysis of digital filters realized with floating-point arithmetic. *Proc. IEEE*, **57**(10), 1735–47

Markel J.D. and Gray A.H. (1975). Fixed-point implementation algorithms for a class of orthogonal polynomial filter structures. *IEEE Trans. Acoustics, Speech and Signal Processing*, **23**(5), 486–94

Markel J.D. and Gray A.H. (1975). Roundoff noise characteristics of a class of orthogonal polynomial structures. *IEEE Trans. Acoustics, Speech and Signal Processing*, **23**(5), 473–86

Motorola (1988). *Digital Stereo 10-band Graphic Equalizer Using the DSP56001*. Motorola Application Note

Mullis C.T. and Roberts R.A. (1976). Round-off noise in digital filters: frequency transformations and invariants. *IEEE Trans. Acoustics, Speech and Signal Processing*, **24**(6), 538–50

Munson D.C. and Liu B. (1980). Low-noise realization for narrow-band recursive digital filters. *IEEE Trans. Acoustics, Speech and Signal Processing*, **28**, 41–54

Nagle H.T. and Nelson V.P. (1981). Digital filter implementation on 16 bit microcomputers. *IEEE Micro*, **1**, 23–41

Oppenheim A.V. and Schafer R.W. (1975). *Digital Signal Processing*. Englewood Cliffs NJ: Prentice-Hall

Oppenheim A.V. and Weinstein, C.J. (1972). Effects of finite register length in digital filtering and the fast Fourier transform. *Proc. IEEE*, **60**, 957–76

Peled A., Liu B. and Steiglitz K. (1974). A new hardware realization of digital filters. *IEEE Trans. Acoustics, Speech and Signal Processing*, **22**, 456–62

Peled A., Liu B. and Steiglitz K. (1975). A note on implementation of digital filters. *IEEE Trans. Acoustics, Speech and Signal Processing*, **23**, 387–9

Rabiner L.R., Cooley J.W., Helms H.D., Jackson L.B., Kaiser J.F., Rader C.M., Schafer R.W., Steiglitz K. and Weinstein C.J. (1972). Terminology in digital signal processing. *IEEE Trans. Audio and Electroacoustics*, **20**, 322–37

Sandberg I.W. and Kaiser J.F. (1972). A bound on limit cycles in fixed-point implementations of digital filters. *IEEE Trans. Audio and Electroacoustics*, **20**, 110–12

Schmalzel J.L., Heine D.N. and Ahmed N. (1980). Some pedagogical considerations of digital filter hardware implementation. *IEEE Circuits and Systems Mag.*, **2**(1), 4–13

Sim P.K. and Pang K.K. (1985). Effects of input-scaling on the asymptotic overflow-stability properties of second recursive digital filters. *IEEE Trans. Circuits and Systems*, **32**(10), 1008–15

Steiglitz K. (1971). Designing short-word recursive digital. *Proc. 9th Ann. Allerton Conf. on Circuit and System Theory*, October 6–8, 1971, pp. 778–88

Steiglitz K., Bede L. and Liu B. (1976). An improved algorithm for ordering poles and zeros of fixed-point recursive digital filters. *IEEE Trans. Acoustics, Speech and Signal Processing*, **24**, 341–3

Taylor F.J. (1983). *Digital Filter Design Handbook*. New York NY: Dekker

Thong T. (1976). Finite wordlength effects in the ROM digital filter. *IEEE Trans. Acoustics, Speech and Signal Processing*, **24**, 436–7

Thong T. and Liu B. (1977). Error spectrum shaping in narrowband recursive digital filters. *IEEE Trans. Acoustics, Speech and Signal Processing*, **25**, 200–3

Williamson D. and Sridharan S. (1985). An approach to coefficient wordlength reduction in digital filters. *IEEE Trans. Circuits and Systems*, **32**(9), 893–903

Appendices

7A C programs for IIR digital filter design

Several C language programs for designing IIR digital filters have been developed for the book. Because of lack of space, only the program for the impulse invariant method of coefficient calculation is listed here. The following programs are not listed in the book, but are available on the PC disk for the book and can be obtained from the first author (see the preface for details of how to obtain a copy of the disk):

- bilinear transformation;
- coefficient calculation for classical IIR filters (Butterworth, Chebyshev and elliptic) via the bilinear transform method;
- finite wordlength analysis for IIR filters.

7A.1 C program for the impulse invariant method

The program for the impulse invariant method is listed in Program 7A.1. First, we will summarize the concepts on which the program is based and then illustrate its use by an example. The discussion will be based on the second-order filter section, since this is the basic building block for digital IIR filters.

Consider a general second-order s-plane transfer function which is to be converted into a discrete transfer function:

$$H(s) = \frac{A_0 + A_1 s}{B_0 + B_1 s + B_2 s^2} \tag{7A.1}$$

The equivalent z-transfer function is also a second-order section of the form

$$H(z) = \frac{a_0 + a_1 z^{-1}}{1 + b_1 z^{-1} + b_2 z^{-2}} \tag{7A.2}$$

Given the values of the coefficients of the analogue transfer function, $H(s)$, the C program listed in Program 7A.1 computes the coefficients of the equivalent z-transfer function, $H(z)$. To see how the program works, we will establish the relationship between the coefficients of $H(s)$ and those of $H(z)$.

Using a partial fraction expansion, the s-plane transfer function of Equation 7A.1 can be expressed as

$$\frac{A_0/B_2 + (A_1/B_2)s}{B_0/B_2 + (B_1/B_2)s + s^2} = \frac{c_1}{s - p_1} + \frac{c_2}{s - p_2} \tag{7A.3}$$

Program 7A.1 Impulse invariant method

```
/*..........................................................................*
*      Impulse invariant method                                             *
*                                                                           *
*      The analogue transfer function must be frequency-scaled              *
*      (normalized frequency) before using program                          *
*      30.10.92                                                             *
*...........................................................................*
*/

#include     <stdio.h>
#include     <math.h>
#include     <dos.h>

void         dfilter();
double       T;
double       a0, a1, a2, b0, b1, b2;
double       p1, p2, pr, pi;
double       c1, c2, cr, ci;
float        A0, A1, B0, B1, B2, temp;

main()
{
                                                       /* initialize coeffs*/
             A0=0; A1=0; B0=1; B1=0; B2=0;
             a0=0; a1=0; b0=1; b1=0; b2=0;
             c1=0; c2=0; p1=0; p2=0; a2=0;
```

```
                                                          /* read s-plane coefficients */
            printf("impulse invariant discrete filters \n");
            printf(" \n");
            printf("enter s-plane coefficients \n");
            printf("enter denominator coeffs: B0, B1, B2 \n");
            scanf("%f %f %f", &B0, &B1, &B2);
            printf("enter numerator coeffs: A0, A1 \n");
            scanf("%f, %f", &A0, &A1);
            T=1;
            dfilter();
            printf("\n");
            printf("press enter to continue\n");
            getch();
            exit(0);

}
/* .................................................................................................... */
void        dfilter()
{
/* Find the s-plane pole positions */
            temp = B1*B1 - 4*B0*B2;

            if(B2==0){                                    /* a single pole */
                  p1=-B0/B1;
                  a0=A0/B1;
                  b1=-exp(p1*T);
            }
            if(temp>0){                                   /* real and unequal poles */
                        pr=-B1/(2*B2);
                        pi=(pr*pr)-B0/B2;
                        pi=sqrt(pi);
                        p1=pr+pi;
                        p2=pr-pi;
                        c1=(A0+A1*p1)/((p1-p2)*B2);
                        c2=A1/B2-c1;
                        a0=c1+c2;
                        a1=-(c1*exp(p2*T) + c2*exp(p1*T));
                        b1=-exp(p1*T)-exp(p2*T);
                        b2=exp((p1+p2)*T);
            }
            if(temp<0){                                   /* complex conjugate poles */
                        pr=-B1/(2*B2);
                        pi=(pr*pr)-B0/B2;
                        pi=sqrt(-pi);
                        cr=A1/(B2*2);
                        ci=-(A0+A1*pr)/(2*pi*B2);
                        a0=2*cr;
                        a1=-(cr*cos(pi*T)+ci*sin(pi*T))*2*exp(pr*T);
                        b1=-2*exp(pr*T)*cos(pi*T);
                        b2=exp(2*pr*T);
            }
            printf("discrete filter coeffs: \n");
            printf("a0 a1 a2: \t%f %f %f \n", a0, a1, a2);
            printf("b0 b1 b2: \t%f %f %f \n", b0, b1, b2);
}
```

where p_1 and p_2 are the s-plane poles of $H(s)$ given by

$$p_{1,2} = \frac{-B_1}{2B_2} \pm \left[\left(\frac{B_1}{2B_2}\right)^2 - \frac{B_0}{B_2}\right]^{1/2} \tag{7A.4}$$

Multiplying both sides of Equation 7A.3 by $(s - p_1)(s - p_2)$ and equating coefficients of s and the constant terms we have

$$\frac{A_0}{B_2} = -(c_1 p_2 + c_2 p_1) \tag{7A.5a}$$

$$\frac{A_1}{B_2} = c_1 + c_2 \tag{7A.5b}$$

Solving for c_1 and c_2 we have

$$c_1 = \frac{A_0 + A_1 p_1}{(p_1 - p_2)B_2} \tag{7A.6a}$$

$$c_2 = \frac{A_1}{B_2} - c_1 \tag{7A.6b}$$

Applying the impulse invariant transformation to Equation 7A.3 gives the discrete transfer function, $H(z)$:

$$\begin{aligned} H(z) &= \frac{c_1 + c_2 - (c_1 e^{p_2 T} + c_2 e^{p_1 T})z^{-1}}{1 - (e^{p_1 T} + e^{p_2 T})z^{-1} + e^{(p_1 + p_2)T} z^{-2}} \\ &= \frac{a_0 + a_1 z^{-1}}{1 + b_1 z^{-1} + b_2 z^{-2}} \end{aligned} \tag{7A.7}$$

where

$$a_0 = (c_1 + c_2), \qquad a_1 = -(c_1 e^{p_2 T} + c_2 e^{p_1 T})$$
$$b_1 = -(e^{p_1 T} + e^{p_2 T}), \qquad b_2 = e^{(p_1 + p_2)T}$$

p_1 and p_2 are defined in Equation 7A.4, and c_1 and c_2 are defined in Equation 7A.6.

Thus, given the s-plane coefficients for a second-order filter (that is A_0, A_1, B_0, B_1 and B_2), the coefficients of the equivalent discrete filter can be obtained directly using the relationships above. Evaluation of the coefficients of $H(z)$ in Equation 7A.7 depends on the type of the s-plane poles, p_1 and p_2. In practice, three cases arise: these are when the two poles are (i) real and unequal, (ii) complex conjugate pair, or (iii) real and equal (that is, concident poles). Only the first two cases are considered as the third does not occur often and it is more involved.

In the first case, Equation 7A.7 can be used directly to obtain the coefficients of $H(z)$. In the second case, a simpler form of Equation A7.7 is used to avoid complex arithmetic. Exploiting the properties of the poles, Equation 7A.7 (for the second case) becomes

$$H(z) = \frac{(c_1 + c_1^*) - (c_1 e^{p_1^* T} + c_1^* e^{p_1 T})z^{-1}}{1 - (e^{p_1 T} + e^{p_1^* T})z^{-1} + e^{(p_1 + p_1^* T)} z^{-2}}$$

$$H(z) = \frac{2c_r - [c_r \cos(p_i T) + c_i \sin(p_i T)] 2e^{p_r T} z^{-1}}{1 - 2e^{p_r T} \cos(p_i T) z^{-1} + e^{2p_r T} z^{-2}} \tag{7A.8}$$

where p_r is the real part of p_1, p_i the imaginary part of p_1, c_r the real part of c_1 and c_i the imaginary part of c_1. From Equation 7A.4, the real and imaginary parts of p_1 are given by

$$p_r = -\frac{B_1}{2B_2}, \quad p_i = \left\{-\left[\left(\frac{B_1}{2B_2}\right)^2 - \frac{B_0}{B_2}\right]\right\}^{1/2}$$

and, from Equation 7A.6, the partial fraction coefficient, c_1, is given by

$$c_1 = \frac{A_1}{2B_2} - \frac{A_0 + A_1 p_r}{2p_i B_2} \mathrm{j} = c_r + c_i \mathrm{j}$$

Thus, in standard form, the coefficients of the second-order z-transfer function for the case where the poles of $H(s)$ are complex conjugates are

$$a_0 = 2c_r, \quad a_1 = -[c_r \cos(p_i T) + c_i \sin(p_i T)]2\mathrm{e}^{p_r T}$$
$$b_1 = -2\mathrm{e}^{p_r T}\cos(p_i T), \quad b_2 = \mathrm{e}^{2p_r T}$$

Example 7A.1

We will use the C program in Program 7A.1 to compute the coefficients for the discrete filter of Example 7.4.

The program expects the frequencies to be normalized. With a sampling frequency of 1280 Hz and a cutoff frequency of 150 Hz, the normalized cutoff frequency is 150/1280. The transfer function is first frequency scaled by replacing s by s/α where $\alpha = 2\pi \times 150/1280 = 0.736\,31$:

$$H'(s) = \frac{\alpha^2}{\alpha^2 + \sqrt{2}\alpha s + s^2} = \frac{1}{1 + (\sqrt{2}/\alpha)s + (1/\alpha^2)s^2}$$
$$= \frac{1}{1 + 1.920\,675s + 1.844\,96s^2}$$

The prompts and outputs of the program are given below. The discrete coefficients are identical to the values obtained in Example 7.4.

```
impulse invariant discrete filters

enter s-plane coefficients
enter denominator coeffs: B0, B1, B2
1  1.920675  1.84496
enter numerator coeffs: A0, A1
1  0
discrete filter coeffs:
a0  a1  a2:    0.000000 0.307718 0.000000
b0  b1  b2:    1.000000 -1.030953 0.353088
```

From the above listings, the z-transfer function can be written down directly:

$$H(z) = \frac{0.307\,718z^{-1}}{1 - 1.030\,953z^{-1} + 0.353\,088z^{-2}}$$

The coefficients are identical to the values obtained in Example 7.4.

7B Evaluation of complex square roots using real arithmetic

Given a complex number with rectangular coordinates $x + \mathrm{j}y$, let its square root be $u + \mathrm{j}v$:

$$u + \mathrm{j}v = (x + \mathrm{j}y)^{1/2} \tag{7B.1}$$

Squaring Equation 7.B1 we have

$$u^2 - v^2 + j2uv = x + jy$$

Equating real and imaginary parts gives

$$u^2 - v^2 = x \tag{7B.2a}$$

$$2uv = y \tag{7B.2b}$$

From Equation 7B.2b

$$v = y/2u \tag{7B.2c}$$

Substituting Equation 7B.2c in 7B.2a we have

$$u^2 - y^2/4u^2 = x \tag{7B.3}$$

and simplifying 7B.3 gives

$$4u^4 - 4xu^2 - y^2 = 0$$

This is a quadratic in u^2, which has the solution

$$u^2 = \frac{x}{2} + \frac{1}{8}(16x^2 + 16y^2)^{1/2}$$

$$= \frac{x + (x^2 + y^2)^{1/2}}{2}$$

(Note that the negative solution is not permitted since $u^2 \geqslant 0$.) Solving for u and using Equation 7B.2c, the solution for u is

$$u = \left[\frac{x + (x^2 + y^2)^{1/2}}{2}\right]^{1/2} \tag{7B.4a}$$

$$v = y/2u \tag{7B.4b}$$

or

$$u = -\left[\frac{x + (x^2 + y^2)^{1/2}}{2}\right]^{1/2} \tag{7B.5a}$$

$$v = -y/2u \tag{7B.5b}$$

The algorithm for computing the square root of $x + jy$ is then

- Step 1: let $u + jv = (x + jy)^{1/2}$
- Step 2: $u = \left[\frac{x + (x^2 + y^2)^{1/2}}{2}\right]^{1/2} = \left(\frac{x + |x + jy|}{2}\right)^{1/2}$
- Step 3: $v = y/2u$

For example

$$(-1 + j)^{1/2} = 0.455\,009 + j1.098\,68 \text{ and } -0.455\,09 - 1.098\,68j$$

$$(-1 - j)^{1/2} = 0.455\,009 - 1.098\,68j \text{ and } -0.455\,009 + 1.098\,68j$$

7C L_2 scaling factor equations

The standard canonic filter section is depicted in Figure 7.C1. The transfer function of the filter section is given by

$$H(z) = \frac{a_0 + a_1 z^{-1} + a_2 z^{-2}}{1 + b_1 z^{-1} + b_2 z^{-2}} \tag{7C.1}$$

The L_2 scale factor to reduce the possibility of overflow at the node labelled $w(n)$ is given by

$$s_1^2 = \frac{1}{2\pi j} \oint \frac{z^{-1}\,dz}{D(z)D(z^{-1})} = \frac{1}{2\pi j} \oint F(z)\,dz \tag{7C.2}$$

where

$$D(z) = 1 + b_1 z^{-1} + b_2 z^{-2}$$
$$F(z) = z^{-1}/D(z)D(z^{-1})$$

and $\oint$ is the contour integral around the circle $|z| = 1$.

Using the value of $D(z)$ in Equation 7C.2 we have

$$\begin{aligned} s_1^2 &= \frac{1}{2\pi j} \oint \frac{z^{-1}\,dz}{(1 + b_1 z^{-1} + b_2 z^{-2})(1 + b_1 z + b_2 z^2)} \\ &= \frac{1}{2\pi j} \oint \frac{z\,dz}{(z^2 + b_1 z + b_2)(1 + b_1 z + b_2 z^2)} \end{aligned}$$

The poles z_1 and z_2 of the integrand inside the unit circle are given by

$$z^2 + b_1 z + b_2 = (z - z_1)(z - z_2) = 0 \tag{7C.3}$$

From the calculus of residues, s_1^2 is the sum of the residues of $F(z)$:

$$\begin{aligned} s_1^2 &= \lim_{z \to z_1} \frac{(z - z_1)z}{(z^2 + b_1 z + b_2)(1 + b_1 z + b_2 z^2)} \\ &\quad + \lim_{z \to z_2} \frac{(z - z_2)z}{(z^2 + b_1 z + b_2)(1 + b_1 z + b_2 z^2)} \\ &= \frac{z_1}{(z_1 - z_2)(1 + b_1 z_1 + b_2 z_1^2)} - \frac{z_2}{(z_1 - z_2)(1 + b_1 z_2 + b_2 z_2^2)} \\ &= \frac{z_1(1 + b_1 z_2 + b_2 z_2^2 - z_2(1 + b_1 z_1 + b_2 z_1^2)}{(z_1 - z_2)(1 + b_1 z_1 + b_2 z_1^2)(1 + b_1 z_2 + b_2 z_2^2)} \\ &= \frac{1 - b_2 z_1 z_2}{(1 + b_1 z_1 + b_2 z_1^2)(1 + b_1 z_2 + b_2 z_2^2)} \\ &= \frac{1 - b_2 z_1 z_2}{1 + b_1(z_1 + z_2) + b_2(z_1^2 + z_2^2) + b_1^2 z_1 z_2 + b_1 b_2 z_1 z_2 (z_1 + z_2) + b_2^2 z_1^2 z_2^2} \\ &= \frac{1 - b_2 z_1 z_2}{1 + b_1(z_1 + z_2) + b_2[(z_1 + z_2)^2 - 2 z_1 z_2] + b_1^2 z_1 z_2 + b_1 b_2 z_1 z_2 (z_1 + z_2) + b_2^2 (z_1 z_2)^2} \end{aligned} \tag{7C.4}$$

Now, from Equation 7C.3,

$$z^2 + b_1 z + b_2 = (z - z_1)(z - z_2) = z^2 - (z_1 + z_2)z + z_1 z_2$$

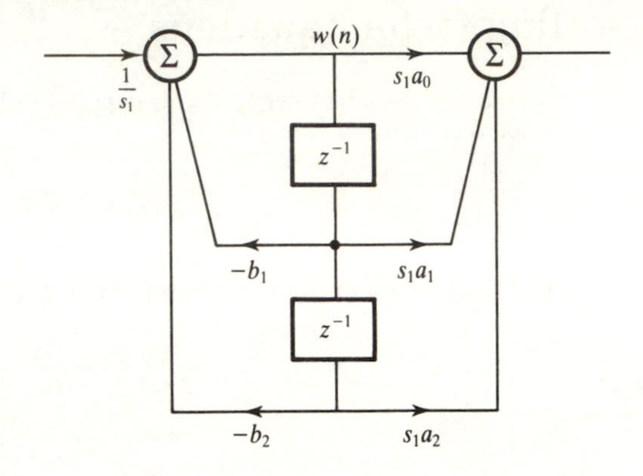

Figure 7C.1 Standard canonic filter section.

Thus

$$b_1 = -(z_1 + z_2)$$
$$b_2 = z_1 z_2$$

Using the values of b_1 and b_2 in Equation 7C.4 we have

$$\begin{aligned} s_1^2 &= \frac{1 - b_2^2}{1 - b_1^2 + b_2(b_1^2 - 2b_2) + b_1^2 b_2 - b_1^2 b_2^2 + b_2^4} \\ &= \frac{1 - b_2^2}{1 - b_1^2 - 2b_2^2 + 2b_1^2 b_2 - b_1^2 b_2^2 + b_2^4} \\ &= \frac{1 - b_2^2}{(1 - b_2^2)^2 - b_1^2(1 - 2b_2 + b_2^2)} \\ &= \frac{1 - b_2^2}{(1 - b_2^2)^2 - b_1^2(1 - b_2)^2} \\ &= \frac{1}{(1 - b_2^2) - b_1^2(1 - b_2)/(1 + b_2)} \end{aligned}$$

Thus

$$s_1^2 = \frac{1}{(1 - b_2^2) - b_1^2(1 - b_2)/(1 + b_2)} \tag{7C.5}$$

8

Multirate digital signal processing

8.1	Introduction	492
8.2	Concepts of multirate signal processing	493
8.3	Design of practical sampling rate converters	502
8.4	Software implementation of sampling rate converters–decimators	508
8.5	Software implementation of interpolators	514
8.6	Application examples	520
8.7	Summary	533
	References	534
	Bibliography	535
	Appendix	536

This chapter discusses, from a practical standpoint, the important topic of multirate processing. The basic concepts are explained and illustrated with fully worked examples. The design of actual multirate systems is presented to enable the reader to design his/her own system. A set of programs, in the C language, are provided for the design and software implementation of multirate processing on a personal computer.

8.1 Introduction

The increasing need in modern digital systems to process data at more than one sampling rate has led to the development of a new sub-area in DSP known as multirate processing (Crochiere and Rabiner, 1975, 1976, 1979, 1981, 1983, 1988). The two primary operations in multirate processing are decimation and interpolation and they enable the data rate to be altered in an efficient manner. Decimation reduces the sampling rate (that is, the sampling frequency), effectively compressing the data, and retaining only the desired information. Interpolation on the other hand increases the sampling rate. Often, the purpose of converting data to a new rate is to make it easier (for example, computationally more efficient) to process or to achieve compatibility with another system. To take an obvious example, if we reduce the sampling rate of a signal from 100 kHz to only 10 kHz, without loss of desired information, then at a stroke we reduce the computational burden in subsequent signal processing operations by a factor of 10. As another example, if we wish to play CD (compact disc) music which has a rate of 44.1 kHz in a studio which handles data at a 48 kHz rate, then the CD data must first be increased to 48 kHz using a multirate approach.

This chapter draws from many published materials, especially those of Rabiner and Crochiere who have made considerable contributions to multirate processing. We are indebted to them.

8.1.1 Some current uses of multirate processing in industry

Advantages of multirate processing are many and have been exploited in many and in an increasing number of modern systems. High quality data acquisition and storage systems are increasingly taking advantage of multirate techniques to avoid the use of expensive anti-aliasing analogue filters and to handle efficiently signals of different bandwidths which require different sampling frequencies. The basis for these applications is that, if an analogue signal is sampled at a rate much higher than specified by the sampling theorem, then a much simpler analogue anti-aliasing filter can be used to bandlimit it before it is digitized. Once in a digital form, the signal can be readily reduced to the desired rate using the multirate approach. A good example of such systems is the EDR8000 (Earth Data, UK) tape recorder.

In speech processing, multirate techniques are employed to reduce the storage space or the transmission rate of speech data. Estimates of speech parameters are computed at a very low sampling rate for storage or transmission. When required, the original speech is reconstructed from the low bit-rate representation at much higher rates using the multirate approach.

The need for inexpensive, high resolution analogue-to-digital converters (ADCs) in digital audio has led to the use of oversampling techniques in the design of such converters rather than the traditional successive-approximation technique (Adams, 1986; Agrawal and Shenoi, 1983; Claasen *et al*., 1980; Matsuya *et al*., 1987; Welland *et al*., 1989). For example, by oversampling the quantization noise inherent in ADCs is spread over a wider frequency range, so that the in-band noise is made small, effectively increasing the number of ADC bits. Further, these new ranges of high performance ADCs employ delta sigma modulation because of its simplicity, for example it requires no sample and hold amplifiers, and low cost. Most, if not all the inexpensive, high resolution ADCs (18, 20, 24 bits) in use today employ multirate processing. Examples include devices such as CS532X (from Crystal Semiconductor) and DSP56ADCx (from Motorola).

Multirate processing has found important application in the efficient implementation of DSP functions. For example, the implementation of narrow-band digital FIR filters using conventional DSP poses a serious problem because such filters require a very large number of coefficients to meet their tight frequency response specifications. The use of multirate techniques leads to very efficient implementation by allowing filtering to be performed at a much lower rate, which greatly reduces the filter order. Multirate techniques have also been used in many other applications including the well-known compact disc player. Details of the application of multirate processing in the CD player as well as many other applications are covered later in this chapter.

8.2 Concepts of multirate signal processing

A simply but naive approach to changing the sampling rate of a digital signal is to convert it back into analogue and then to re-digitize it at the new rate. Errors inherent in digital–analogue–digital conversion processes, such as quantization and aliasing errors, would degrade the signal. As the signal is already in a digital form, it is best to process it digitally throughout until conversion to analogue is mandatory, for example when the destination is the loudspeaker. Multirate processing is basically an efficient technique for changing the sampling frequency of a signal digitally. Its main attraction is that it allows the strengths of conventional DSP to be exploited. For example, much of the anti-aliasing and anti-imaging filtering in real-time DSP systems can be performed in the digital domain, enabling both sharp magnitude frequency as well as linear phase responses to be achieved.

The processes of decimation and interpolation are the fundamental operations in multirate signal processing, and they allow the sampling frequency to be decreased or increased without significant, undesirable effects of errors such as quantization and aliasing. We present details of these basic operations next.

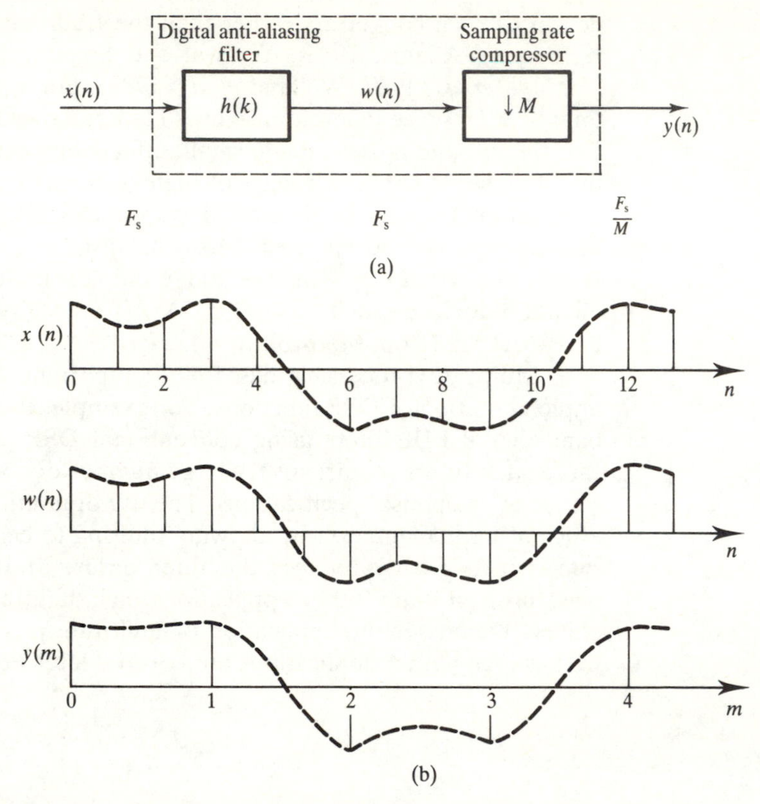

Figure 8.1 (a) Block diagram of decimation by a factor of M. (b) An illustration of decimation by a factor of $M = 3$. Note that only one out of every three samples of $w(n)$ appears at the output.

8.2.1 Sampling rate reduction: decimation by integer factors

A block diagram showing the process of decimating a signal $x(n)$ by an integer factor M is shown in Figure 8.1(a). It consists of a digital anti-aliasing filter, $h(k)$, and a sample rate compressor, symbolized by a down-arrow and the decimation factor, M. The rate compressor reduces the sampling rate from F_s to F_s/M. To prevent aliasing at the lower rate the digital filter is used to bandlimit the input signal to less than $F_s/2M$ beforehand. Thus the signal $x(n)$ is first bandlimited (the requirements of the decimating filter are discussed in detail in Section 8.3.1). Sampling rate reduction is achieved by discarding $M - 1$ samples for every M samples of the filtered signal, $w(n)$. The input–output relationship for the decimation process is

$$y(m) = w(mM) = \sum_{k=-\infty}^{\infty} h(k)x(mM - k) \tag{8.1a}$$

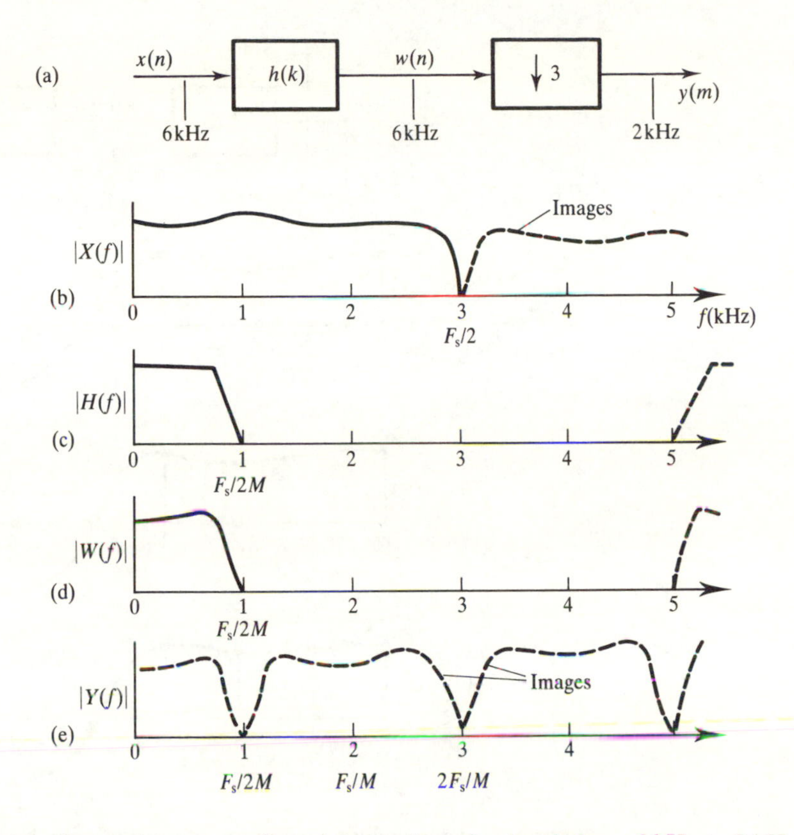

Figure 8.2 Spectral interpretation of decimation of a signal; from 6 kHz to 2 kHz. The image components would have aliased into the signal but for the digital filter.

where

$$w(n) = \sum_{k=-\infty}^{\infty} h(k)x(n-k) \tag{8.1b}$$

Figure 8.1(b) illustrates the process for the simple case where $M = 3$. In this case, two samples out of every three samples of $x(n)$ are discarded. In effect, decimation is a data compression operation.

The spectral description of the decimation process is shown in Figure 8.2, where we have assumed that the input $x(n)$ is a wideband signal. The broken lines in Figure 8.2(b) indicate the image components that would have caused aliasing had the input signal $x(n)$ not been bandlimited prior to decimation.

8.2.2 Sampling rate increase: interpolation by integer factors

In many ways, interpolation is the digital equivalent of the digital-to-analogue conversion process where the analogue signal is recovered by interpolating the

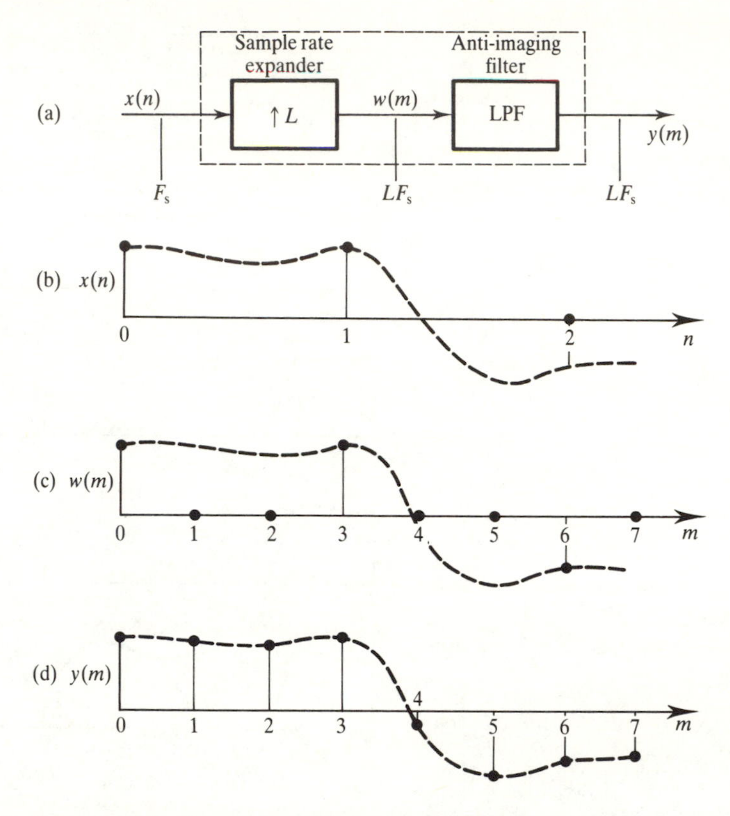

Figure 8.3 Time domain illustration of interpolation by a factor of $L = 3$. Note that for each sample of $x(n)$, three output samples $y(m)$ are obtained.

digital samples applied to the digital-to-analogue converter. In the case of digital interpolation, however, the process yields specific values.

Given a signal, $x(n)$, at a sampling frequency F_s, the interpolation process increases the sampling rate by L to LF_s. Figure 8.3(a) shows the interpolator. It consists of a sample rate expander, symbolized by an up-arrow and the interpolation factor, L, which indicates the amount by which the rate is increased. For each sample of $x(n)$ the expander inserts $L - 1$ zero-valued samples to form the new signal $w(m)$ at a rate of LF_s. This signal is then lowpass filtered to remove image frequencies created by the rate increase to yield $y(m)$. The insertion of $L - 1$ zeros spreads the energy of each signal sample over L output samples, effectively attenuating each sample by a factor of L. Thus it is necessary to compensate for this, for example by multiplying each sample of $y(n)$ by L.

The interpolation process is characterized by the following input–output relationship:

$$y(m) = \sum_{k=-\infty}^{\infty} h(k)w(m - k) \tag{8.2a}$$

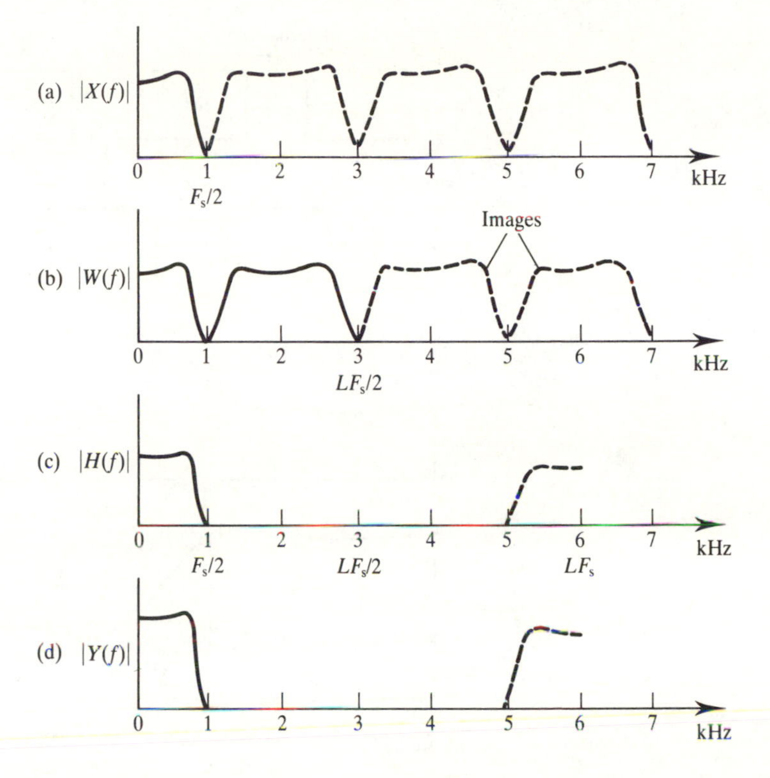

Figure 8.4 Spectral interpretation of interpolation of a signal from 2 kHz to 6 kHz.

where

$$w(m) = \begin{cases} x(m/L), & m = 0, \pm L, \pm 2L, \ldots \\ 0 \end{cases} \tag{8.2b}$$

Figures 8.3(b)–8.3(d) illustrate the interpolation process in the time domain, for the simple case of $L = 3$. Notice that, for each sample of $x(n)$, three output samples are produced. This is due to the two zero-valued samples inserted by the expander.

The frequency domain interpretation of the process is depicted in Figure 8.4. $X(f)$, $W(f)$, and $Y(f)$ are the frequency responses of the signals $x(n)$, $w(m)$ and $y(m)$, respectively. $H(f)$ is the amplitude response of the anti-imaging filter. The filter is necessary to remove the image components indicated by the broken lines in $W(f)$. It is worth pointing out at this stage, although the alert reader may have suspected this already, that decimation and interpolation processes are duals of each other, that is one is the inverse of the other. This duality property means that an interpolator can be readily derived from an equivalent decimator and vice versa.

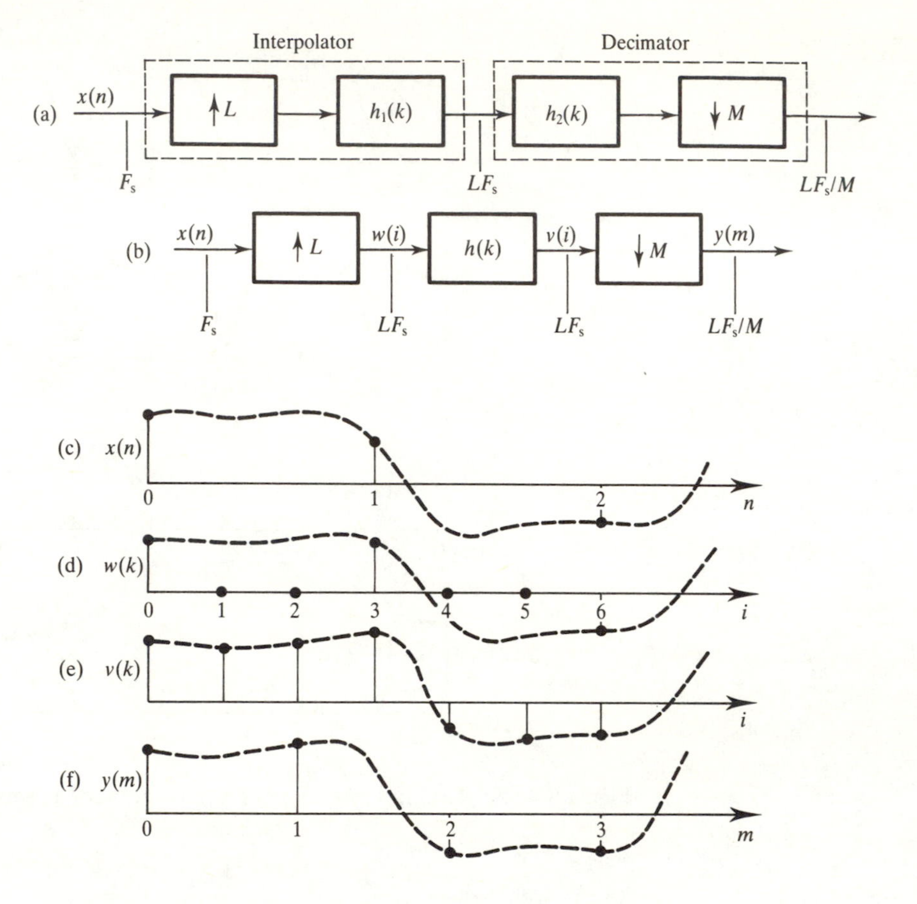

Figure 8.5 An illustration of interpolation by a rational factor ($L = 3$, $M = 2$).

8.2.3 Sampling rate conversion by non-integer factors

In some applications, the need often arises to change the sampling rate by a non-integer factor. An example is in digital audio applications where it may be necessary to transfer data from one storage system to another, where both systems employ different rates, possibly to discourage illegal copying of material. An example is transferring data from the compact disc system at a rate of 44.1 kHz to a digital audio tape (DAT) at 48 kHz. This can be achieved by increasing the data rate of the CD by a factor of 48/44.1, a non-integer.

In practice, such a non-integer factor is represented by a rational number, that is a ratio of two integers, say L and M, where L and M are integers such that L/M is as close to the desired factor as possible. The sampling frequency change is then achieved by first interpolating the data by L and then decimating by M (Figure 8.5(a)). It is necessary that the interpolation process precedes decimation otherwise the decimation process would remove some of the

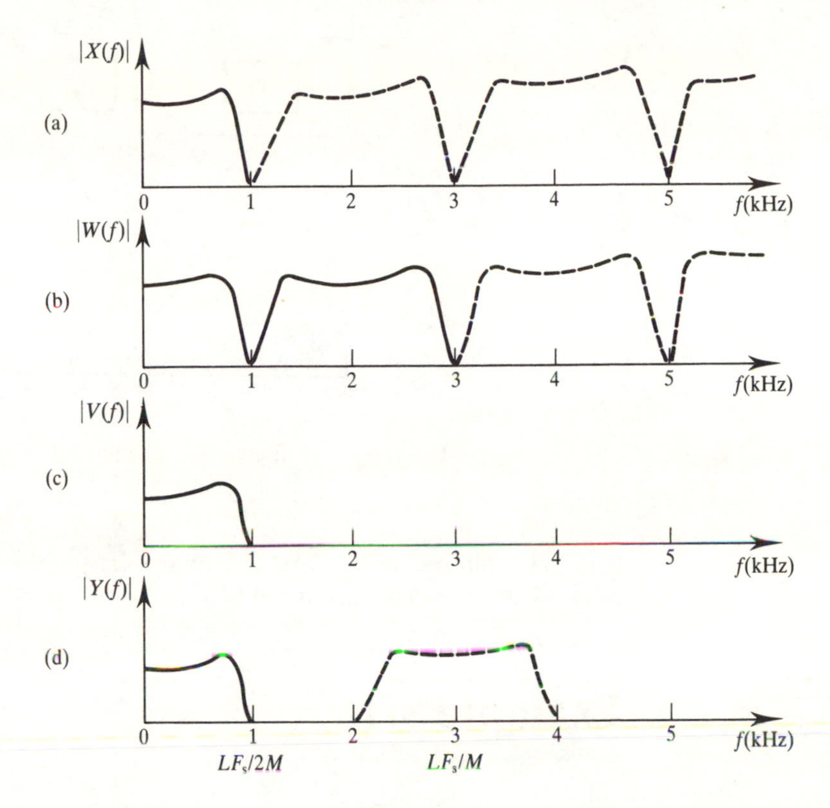

Figure 8.6 Spectral interpretation of sample rate increase of a signal at 2 kHz rate by a factor of 3/2. Signal rate is first increased by 3 to 6 kHz (a); after bandlimiting to avoid aliasing (c), the signal is reduced in rate by 2 to 3 kHz.

desired frequency components. In the CD–DAT example above, the rate conversion by 48/44.1 can be achieved by first interpolating by a factor $L = 160$ and then decimating by $M = 147$, that is we first increase the CD data rate by $L = 160$ to 7056 kHz and then reduce it by $M = 147$ to 48 kHz.

The two LPFs, $h_1(k)$ and $h_2(k)$, in Figure 8.5(a), can be combined into a single filter since they are in cascade and have a common sampling frequency to give the generalized sample rate converter in Figure 8.5(b). If $M > L$, the resulting operation is a decimation process by a non-integer, and when $M < L$ it is interpolation. If $M = 1$, the generalized system reduces to the simple integer interpolation described earlier, and if $L = 1$ it reduces to integer decimation.

Figure 8.5(c) illustrates interpolation by a factor of 3/2. The sample rate is first increased by 3, by inserting two zero-valued samples for each sample of $x(n)$ and then lowpass filtered to yield $v(i)$. The filtered data is then reduced by a factor of 2 by retaining only one sample for every two samples of $v(i)$. Figure 8.6 illustrates, in the frequency domain, the process of interpolation by 3/2. The input signal, $x(n)$, at a rate of 2 kHz, is first increased by a factor of 3

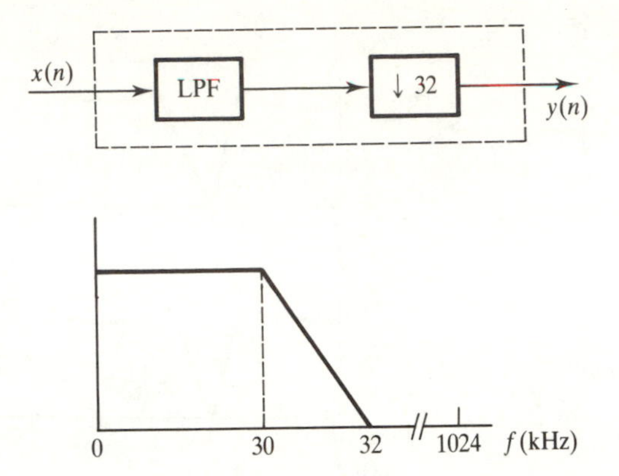

Figure 8.7 A single-stage decimator for Example 8.1.

to 6 kHz, filtered to remove the image frequencies which would otherwise cause aliasing, and then reduced by a factor of 2 to 3 kHz.

Example 8.1

A signal, $x(n)$, at a sampling frequency of 2.048 kHz is to be decimated by a factor of 32 to yield a signal at a sampling frequency of 64 Hz. The signal band of interest extends from 0 to 30 Hz. The anti-aliasing digital filter should satisfy the following specifications:

passband deviation	0.01 dB
stopband deviation	80 dB
passband	0–30 Hz
stopband	32–64 Hz

The signal components in the range from 30 to 32 Hz should be protected from aliasing. Design a suitable one-stage decimator.

Solution

The block diagram of the single stage decimator, and the specifications for the lowpass anti-aliasing filter are shown in Figure 8.7. From the specifications and the figure we can determine the following:

$$\Delta f = (32\text{–}30)/2048 = 9.766 \times 10^{-4}$$

$$\delta_p = 0.001\,15, \quad \text{from } 20\log(1 + \delta_p) = 0.01 \text{ dB}$$

$$\delta_s = 0.0001, \quad \text{from } -20\log(\delta_s) = 80 \text{ dB}$$

An estimate of the number of filter coefficients for the single-stage decimator is given by (see Chapter 6)

$$N \approx \frac{D_\infty(\delta_p, \delta_s)}{\Delta f} - f(\delta_p, \delta_s)\,\Delta f + 1 \tag{8.3}$$

where Δf is the width of the transition normalized to the sampling frequency,

$$D_\infty(\delta_p, \delta_s) = (\log_{10}\delta_s)[a_1(\log_{10}\delta_p)^2 + a_2(\log_{10}\delta_p) + a_3] + a_4(\log_{10}\delta_p)^2 + a_5(\log_{10}\delta_p) + a_6$$

$$f(\delta_p, \delta_s) = 11.012\,17 + 0.512\,44(\log_{10}\delta_p - \log_{10}\delta_s)$$

$$a_1 = 5.309 \times 10^{-3}; \qquad a_2 = 7.114 \times 10^{-2};$$

$$a_3 = -4.761 \times 10^{-1}; \qquad a_4 = -2.66 \times 10^{-3};$$

$$a_3 = -5.941 \times 10^{-1}; \qquad a_6 = -4.278 \times 10^{-1}.$$

δ_p is the passband ripple or deviation and δ_s is the stopband ripple or deviation.

Using the values of δ_p, δ_s, and Δf in Equation 8.3, we find that $N = 3947$. It is quite obvious that N is too large. In fact, none of the available filter design methods can be used to obtain the coefficients for such a filter because approximation errors would be too large. For all practical purposes, the design of the lowpass filter for the single-stage decimator is not possible. This example makes evident the need for an alternative, more efficient method of sampling rate conversion, especially when the rate change is large. Such an approach will be discussed in the next section.

8.2.4 Multistage approach to sampling rate conversion

In the previous sections, the changes in sampling rates were achieved in one fell swoop using one decimation or interpolation factor. When large changes in the sampling rate are required it is more efficient to change the rate in two or more stages than in one single stage. In fact, most practical multirate systems employ the multistage approach because it allows a gradual reduction or increase in the sampling rate, leading to a significant relaxation in the requirements of the anti-aliasing or anti-imaging filter at each stage.

Figure 8.8 shows an I-stage decimation process. The overall decimation factor, M, is expressed as the product of smaller factors:

$$M = M_1 M_2 \ldots M_I = N \tag{8.4}$$

where M_i, an integer, is the decimation factor for stage i. Each stage is an independent decimator as shown in dashed boxes. If $M \gg 1$, the multistage

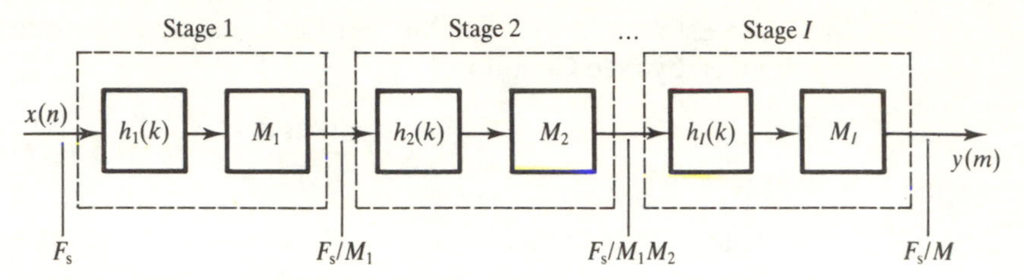

Figure 8.8 Multistage decimation process.

approach leads to much reduced computational and storage requirements, a relaxation in the characteristics of the filters used in the decimators, and consequently to filters that are less sensitive to finite wordlength effects.

These advantages are achieved at the expense of increased difficulty in the design and implementation of the system. Many examples are given later to illustrate the multirate approach after we have covered the design methodology.

8.3 Design of practical sampling rate converters

The design of a practical multistage sample rate converter can be broken down into four steps:

- specify the overall anti-aliasing or anti-imaging filter requirements and those for individual stages;
- determine the optimum number of stages of decimation or interpolation that will yield the most efficient implementation;
- determine the decimation or interpolation factors for each stage;
- design an appropriate filter for each stage.

8.3.1 Filter specification

From previous discussions, the need for a digital filter for anti-aliasing or anti-imaging filtering in sampling rate converters should be very clear. In fact, the performance of a multirate system depends critically on the type and quality of the filter used. Either FIR or IIR filters can be used for decimation or interpolation, but the FIR is the more popular.

In multirate processing, unlike conventional DSP, the computation efficiency of FIR filters is comparable with and, in some cases, exceeds that of IIR filters (Crochiere and Rabiner, 1975, 1976, 1983). Further, FIR filters have many desirable attributes (see Chapters 5 and 6 for details), such as linear phase response and low sensitivity to finite wordlength effects, as well as being simple to implement. For these reasons, only FIR filters will be considered in this chapter. All the coefficient calculation methods described in Chapter 6 for FIR filters can be used to design filters for multirate systems. In particular, the optimal and half-band filters are widely used.

The overall filter requirements for decimation, to avoid aliasing after rate reduction are,

passband	$0 \leqslant f \leqslant f_p$	**(8.5a)**
stopband	$F_s/2M \leqslant f \leqslant F_s/2$	**(8.5b)**
passband deviation	δ_p	**(8.5c)**
stopband deviation	δ_s	**(8.5d)**

where $f_p < F_s/2M$, and F_s is the original sampling frequency. Typically, f_p is the highest frequency of interest in the original signal.

In the case of interpolation, the anti-imaging filter must remove all but the useful information by bandlimiting the modified data to $F_s/2$ or less. Although the highest valid frequency after raising the rate to LF_s is $LF_s/2$, according to the sampling theorem, it is necessary to bandlimit to $F_s/2$ as this is the highest valid frequency in $x(n)$. The overall filter requirements for interpolation are

passband	$0 \leqslant f \leqslant f_p$	**(8.6a)**
stopband	$F_s/2 \leqslant f \leqslant LF_s/2$	**(8.6b)**
passband deviation	δ_p	**(8.6c)**
stopband deviation	δ_s	**(8.6d)**

where $f_p < F_s/2$. A gain of L is necessary in the passband to compensate for the amplitude reduction by the interpolation process.

8.3.2 Filter requirements for individual stages

The equiripple (optimal) filter is often used for sampling rate conversion, although filters obtained by the window method can also be used. The tolerance scheme for an equiripple lowpass filter is depicted in Figure 8.9(a).

For a multistage decimator (Figure 8.9(b)) the filter requirements for each stage to ensure that the overall filter requirements are met are (see also Figure 8.9(c))

passband	$0 \leqslant f \leqslant f_p$	**(8.7a)**
stopband	$(F_i - F_s/2M) < f < F_{i-1}/2, \quad i = 1, 2, \ldots, I$	**(8.7b)**
passband ripple	δ_p/I	**(8.7c)**
stopband ripple	δ_s	**(8.7d)**
filter length	$N \approx \dfrac{D_\infty(\delta_p, \delta_s)}{\Delta f_i} - f(\delta_p, \delta_s)\,\Delta f_i + 1$	**(8.7e)**

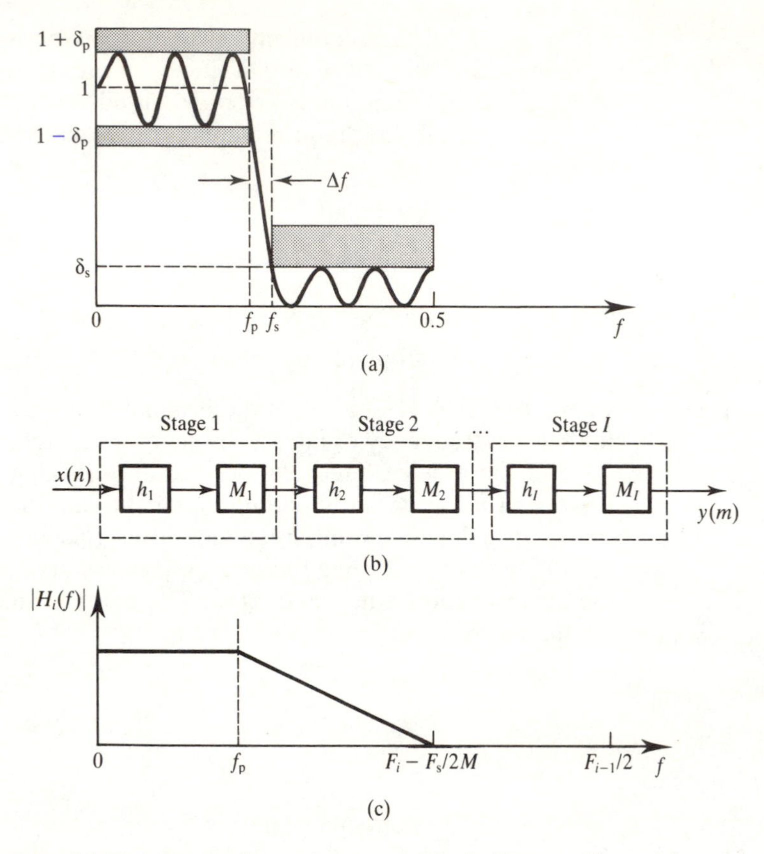

Figure 8.9 (a) Tolerance scheme for an equiripple lowpass filter; (b) multistage structure; (c) filter specifications for stage i, $i = 1, 2, \ldots, I$.

where F_i, N_i and Δf_i are, respectively, the output sampling frequency, the filter length and the normalized transition width for the ith-stage decimator. The parameters $D_\infty(\delta_p, \delta_s)$ and $f(\delta_p, \delta_s)$ have the same meaning as in Equation 8.3. The output sampling frequency for stage i is given by

$$F_i = F_{i-1}/M_i, \quad i = 1, 2, \ldots, I \tag{8.8}$$

where M_i is the decimation factor for the stage. The initial and final sampling rates are F_0 and F_I, respectively. To relate to previous discussions, $F_0 = F_s$ and $F_I = F_s/M$.

For multistage decimation, a lower passband deviation is necessary for each stage to ensure that the overall passband deviation is δ_p. The stopband deviation for each stage is the same as the overall stopband deviation because as the signal goes from stage to stage the stopband components are attenuated further. If anything, the overall stopband deviation will be better than the overall stopband requirements. For a one-stage decimator, the filter requirements are the same as specified in Equation 8.5.

8.3.3 Determining the number of stages and decimation factors

The multistage design offers significant savings in computation and storage requirements over a single-stage design. The extent of this saving depends on the number of stages used and the choice of decimation factors for the individual stages. A major problem is determing the optimum number of stages, I, and the decimation factors for each stage. An optimum number of stages is one which leads to the least computational effort, for example as measured by the number of multiplications per second (MPS) or the total storage requirements (TSR) for the coefficients:

$$\text{MPS} = \sum_{i=1}^{I} N_i F_i \qquad \textbf{(8.9a)}$$

$$\text{TSR} = \sum_{i=1}^{I} N_i \qquad \textbf{(8.9b)}$$

where N_i is the number of filter coefficients for stage i, and we have ignored any symmetry in the filter coefficients.

The choice of the number of stages, I, and the decimation factors is not a trivial problem. However, in practice the number of stages, I, is rarely more than 3 or 4. Further, for a given value of M, there are only a limited set of possible integer factors. Thus a viable approach is to determine all the possible factors of M, that is all the set of M_i values, and their corresponding MPS or total storage requirements. The most efficient or preferred solution can then be chosen by inspection. The algorithm for this approach is summarized in Table 8.1. A C language implementation is provided in the disk for this book, and a spreadsheet implementation is described by DeFatta *et al*. (1988).

In general, for optimum MPS or TSR the decimation factors satisfy the following relationship (Crochiere and Rabiner, 1975, 1976):

$$M_1 > M_2 > \ldots M_I \qquad \textbf{(8.10)}$$

where M_i are continuous. However, when the factors are integers it is not always possible to satisfy Equation 8.10 for some values of I, for example if $I = 3$ and $M = 32$ (see the comments in Example 8.2).

For $I = 2$, that is two stages of decimation, the optimal values of the decimation factors for which TSR is minimized are

$$M_{1_{\text{opt}}} = \frac{2M}{2 - \Delta f + (2M\,\Delta f)^{1/2}} \qquad \textbf{(8.11a)}$$

$$M_{2_{\text{opt}}} = \frac{M}{M_{1_{\text{opt}}}} \qquad \textbf{(8.11b)}$$

For $I > 2$, no simple closed-form expression exists and it becomes necessary to use a computer-aided optimization routine or the algorithm given in Table 8.1 to find the optimum decimation factors, M_i.

Table 8.1 Algorithm for finding optimum values of I and M_i values.

- Specify overall filter parameters (F_s, M, f_p, f_s, δ_s, δ_p)
- For each value of I, ($I = 1, 2, \ldots, I_{max}$), obtain all the possible set of integer decimation factors of M.
- For each set of decimation factors determine the filter requirements, the MPS and storage from Equation 8.9.
- For each value of I, select the decimation factors giving the most efficient design in terms of storage requirements.
- Select the most efficient or desired solution.

8.3.4 Illustrative design example

Example 8.2

The sampling rate of a signal $x(n)$ is to be reduced, by decimation, from 96 kHz to 1 kHz. The highest frequency of interest after decimation is 450 Hz. Assume that an optimal FIR filter is to be used, with an overall passband ripple, $\delta_p = 0.01$, and passband deviation, $\delta_s = 0.001$. Design an efficient decimator.

Solution

We will start by finding the most efficient design for each value of I, $I = 1$, 2, 3, 4. We will then compare these designs and select the best.

(1) First let us consider a one-stage design ($I = 1$). The block diagram and filter specifications for the stage are given in Figure 8.10(a).

(2) Next, we consider a two-stage design. Using the design program referred to in the text, the optimum integer decimation factors for $I = 2$ are $M_1 = 32$, $M_2 = 3$. The two-stage system, including its specifications, is shown in Figure 8.10(b). At the first stage, the sampling rate is reduced by 32 to 3 kHz, and, at the second stage, this is further reduced by 3 to 1 kHz.

(3) For the three-stage case ($I = 3$), the optimum integer decimation factors, in terms of storage, are $M_1 = 8$, $M_2 = 6$, $M_3 = 2$. The system for this is depicted in Figure 8.10(c).

(4) For the four-stage design, the optimum integer decimation factors are $M_1 = 4$, $M_2 = 4$, $M_3 = 3$, $M_4 = 2$. The system and filter specifications for this are depicted in Figure 8.10(d).

The results are summarized below:

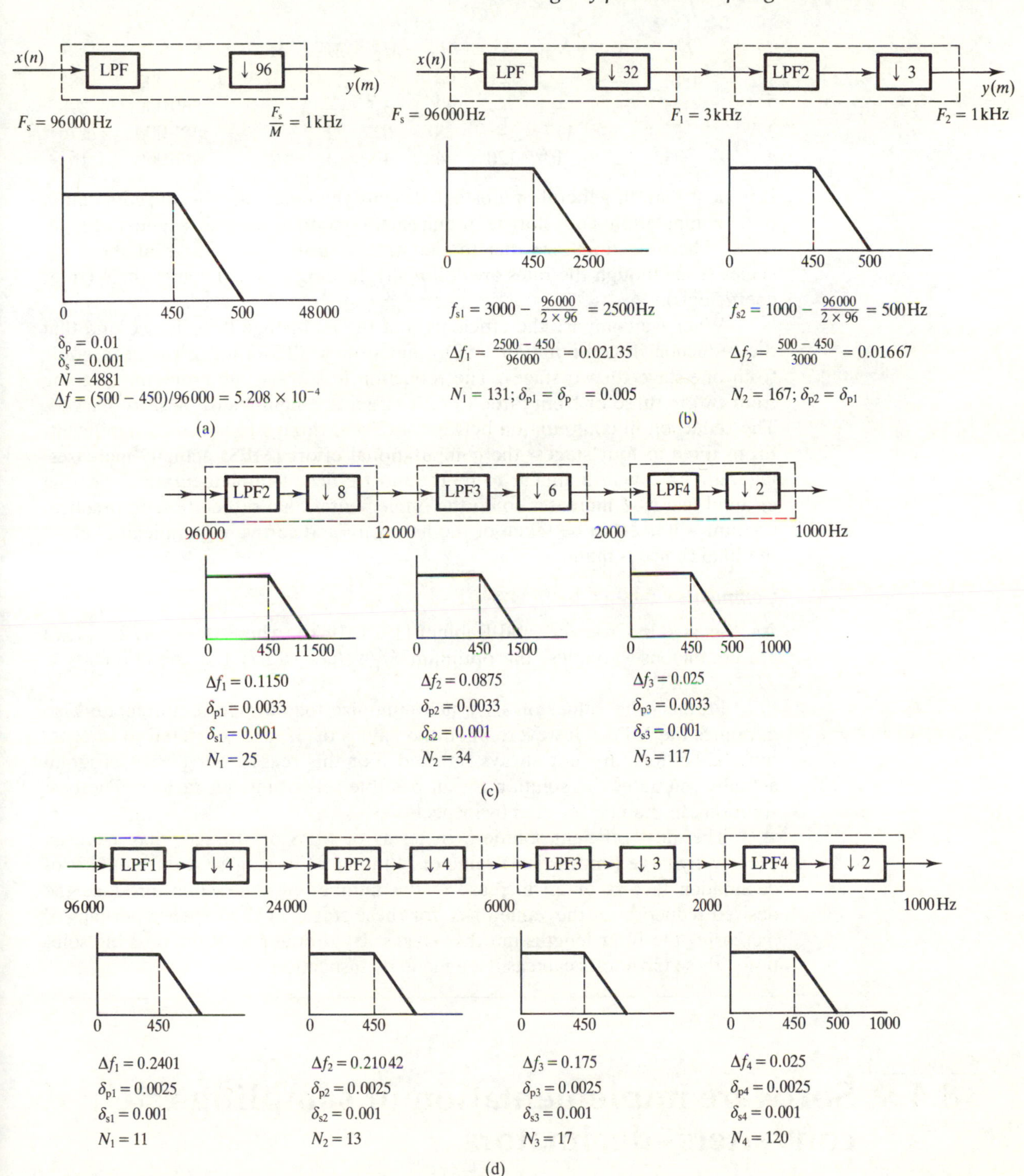

Figure 8.10 (a) Single-stage decimator. (b) Two-stage decimator. (c) Three-stage decimator. (d) Four-stage decimator (Example 8.2).

I	N_1	N_2	N_3	N_4	M_1	M_2	M_3	M_4	MPS	TSR
1	4881	–	–	–	96	–	–	–	48 881 000	4881
2	131	167	–	–	32	3	–	–	560 000	298
3	25	34	117	–	8	6	2	–	485 000	176
4	11	13	17	120	4	4	3	2	496 000	161

It is clear that, in general, multistage designs yield very significant reduction in both computation and storage requirements compared with single-stage designs. The reductions are due to the wide transitions of filters at the early stages (even though the rates are still high), leading to small values of N (filter coefficients).

When we compare the efficiencies of the multistage designs, we find that the reduction in computation (MPS) and storage (TSR) are greatest in going from one stage to two stages. The reduction in storage requirements in going from two to three or from three to four stages are significant, but less marked. The reduction in computation between two and three stages is also significant. From three to four stages, the computational effort (MPS) actually increases. Overall, $I = 3$ would appear to be the most efficient implementation, bearing in mind that as I increases so do the implementational difficulties. In practice, account will need to be taken of the hardware and software complexity before the final choice is made.

Comments

As discussed in Crochiere and Rabiner (1975, 1976), when M_i ($i = 1, 2, \ldots, I$) are continuous variables, the optimum M_i values satisfy the condition $M_1 > M_2 > \ldots M_I$.

Further, the values of M_i which minimize the storage requirements also minimize the MPS. However, when the values of M_i are restricted to integers these conditions are not always satisfied. For this reason our design program actually computes the solution for all possible sets of integer factors. The best solution can then be selected by inspection.

The most efficient solution, in terms of MPS or storage, may have an excessively large (impractical) value of N in one of its stages. A different set of decimation factors or an increase in the number of stages may produce the desired reduction in the value of N for these stages at the expense perhaps of increasing the filter lengths in other stages. By producing all the possible solutions, these trade-offs can easily be made by inspection.

8.4 Software implementation of sampling rate converters–decimators

A simple block diagram representation of the decimator is given in Figure 8.11(a), where $h(k)$ is an anti-aliasing digital filter. Assuming a direct form

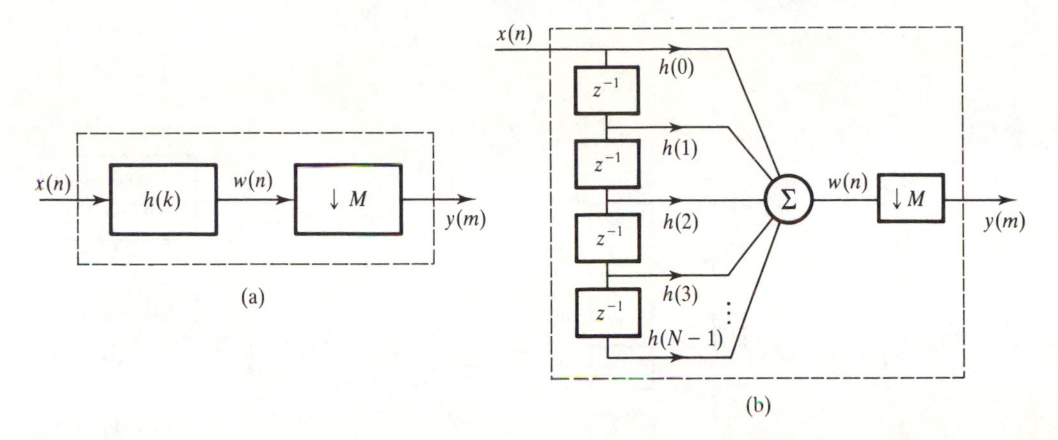

Figure 8.11 (a) A simple block diagram of the decimator; (b) signal flowgraph for the decimator.

implementation (that is, the use of tapped-delay lines), the output of the filter, $w(n)$, and the input, $x(n)$, are related as

$$w(n) = \sum_{k=0}^{N-1} h(k)x(n-k) \tag{8.12a}$$

where N is the number of FIR filter coefficients. The output of the decimator $y(m) = w(mM)$, which on combining with Equation 8.12a leads to the decimator equation

$$y(m) = \sum_{k=0}^{N-1} h(k)x(Mm-k) \tag{8.12b}$$

The signal flow diagram for the decimator is given in Figure 8.11(b). The input $x(n)$ is fed into the delay line one sample at a time. For every M samples of $x(n)$ applied to the delay line, one output sample $y(m)$ is computed. This involves keeping the first sample of $w(n)$, discarding the next $M-1$ samples, keeping the next sample, and discarding the next $M-1$ samples, and so on. Since for each sample that is kept, the next $M-1$ samples of $w(n)$ are discarded, it is unnecessary to perform Equation 8.12(a) for those samples of $w(n)$ that are discarded. Thus, the down-sampling (discarding of samples) operation can be performed before the multiplication of the input samples by the coefficients (Figure 8.12(a)). The multiplications and additions involving the filter coefficients are now performed at the lower sampling frequency F_s/M, leading to a reduction in the computational effort by a factor of M.

The operation of a single-stage decimator is summarized in the flowchart of Figure 8.12(b). The flowchart for a three-stage decimation is given in Figure 8.13, which is a straightforward extension of the single-stage operation.

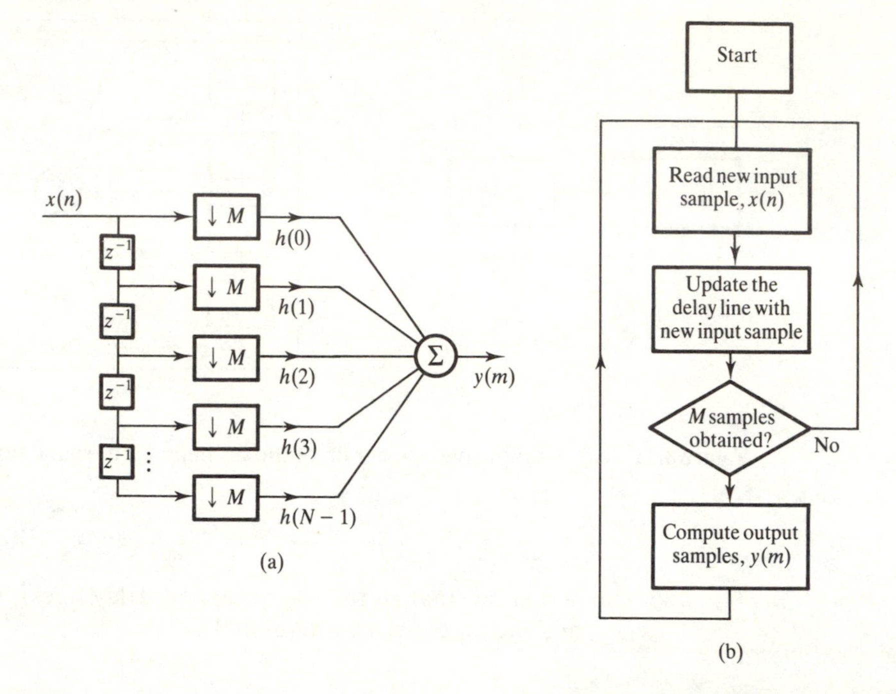

Figure 8.12 (a) A more efficient flowgraph for decimation; (b) flowchart of the decimation process.

8.4.1 Program for multistage decimation

A self-contained, interactive C language program based on the above methods that runs on a PC (personal computer) is provided in the appendix. The program decimates input data using up to three decimation stages; see the flow chart of Figure 8.13. Each stage of decimation requires an integer decimation factor and a set of N-point filter coefficients representing a linear phase FIR digital filter.

The input data is read from a data file in the PC and the decimated data is written to a user-specified output data file one sample at a time. Assuming a three-stage decimation (see Figure 8.13), for every M_1 input samples fed into stage 1, one output sample is computed. For every M_2 output samples computed for stage 1, one output sample is computed at stage 2. Finally, for every M_3 output samples from stage 2 one output sample is computed from stage 3. Thus at the end of a decimation cycle for each M input samples of $x(n)$, where $M = M_1 M_2 M_3$, one output sample is computed and stored in the output data file. The process is repeated until all the input samples have been processed.

The program is self-contained. To use the program, the user must specify the number of stages, the overall decimation factor, the decimation factor for each stage, and a set of direct form FIR linear phase filter coefficients for each stage. The user specifies the names of files containing the data to be decimated,

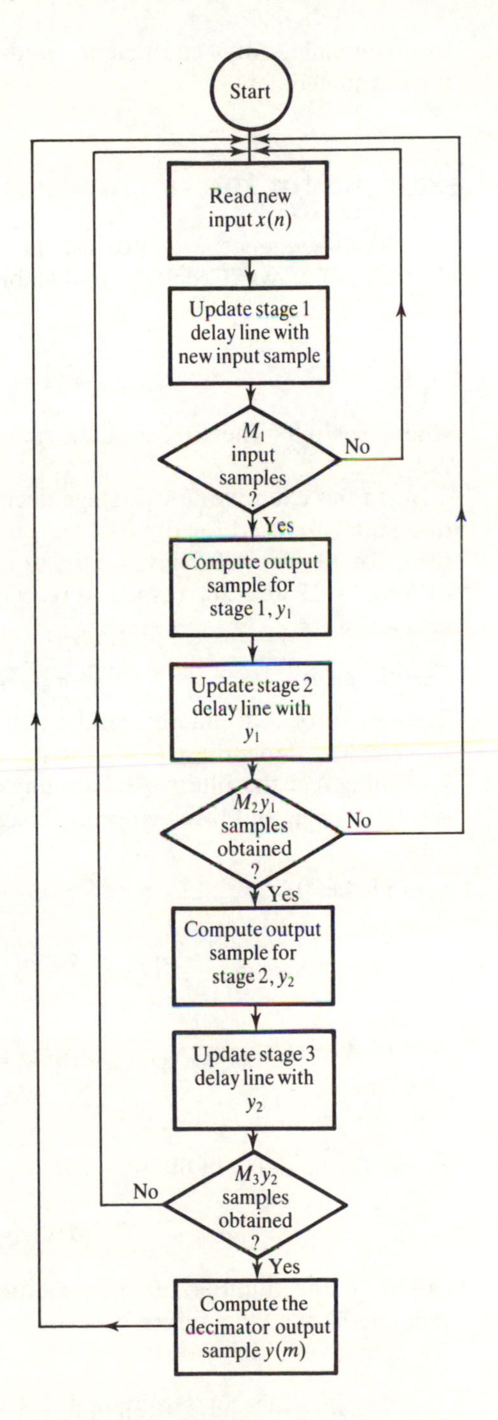

Figure 8.13 Flowchart for a three-stage decimator.

files containing filter coefficients, and the name of the output file which holds the decimated data.

8.4.2 Test example for the decimation program

The input sequence used to test the decimation process is the allpass signal derived as follows (Crochiere and Rabiner, 1979):

$$x(n) = -\alpha, \quad n = 0$$
$$= (1 - \alpha^2)\alpha^{n-1}, \quad n = 1, 2, \ldots \tag{8.13}$$

where $\alpha = 0.9$. The first 29 data samples of the sequence are listed in Table 8.2.

In this example, a two-stage decimation is performed. Decimation factors of 5 and 2 are used for the first and second stages, giving an overall decimation factor of 10. Table 8.2 gives a list of the FIR filter coefficients used which are of lengths 25 and 28, respectively. The result of the decimation operation is also given.

Output delay

The output of a decimator will be delayed from the input by a certain number of samples, depending on the type of filter used at the decimator stages. Assuming that the filters used are linear phase FIR filters, the group delays for single, double, and three-stage decimation are, respectively,

$$T(1\text{ stage}) = \frac{1}{M}[T_1 - (M - 1)] \text{ samples} \tag{8.14a}$$

$$T(2\text{ stages}) = \frac{1}{M_1 M_2}[T_1 + M_1 T_2 - (M_1 M_2 - 1)] \text{ samples} \tag{8.14b}$$

$$T(3\text{ stages}) = \frac{1}{M_1 M_2 M_3}[T_1 + M_1 T_2 + M_1 M_2 T_3 - (M_1 M_2 M_3 - 1)] \text{ samples} \tag{8.14c}$$

where T_i, the delay in the ith-stage filter, is given by

$$T_i = (N_i + 1)/2 \text{ samples}$$

and N_i is the number of filter coefficients for stage i. In the two-stage test example above, the filters have delays of 14.5 and 13 samples, respectively, given an overall delay of

$$T(2\text{ stages}) = (1/5 \times 2)[14.5 + 5 \times 13 - (5 \times 2 - 1)] = 7.05 \text{ samples}$$

If an integer delay is desired then N_i should be chosen such that the value of T in the appropriate equation above is an integer. Such is the case when the input

Table 8.2 Data for test example on decimation.

n	$x(n)$	$y(m)$	$h_1(k)$	$h_2(k)$
0	−0.9000		−0.000 174	−0.000 303
1	0.1900		−0.002 682	0.001 807
2	0.1710		−0.006 346	0.003 120
3	0.1539		−0.011 033	−0.001 169
4	0.1385		−0.014 156	−0.009 267
5	0.1247		−0.012 024	−0.007 792
6	0.1122		−0.000 775	0.011 124
7	0.1010		0.021 904	0.027 651
8	0.0909		0.055 181	0.007 674
9	0.0818	0.000 040	0.094 397	−0.045 444
10	0.0736		0.131 836	−0.064 816
11	0.0662		0.158 866	0.022 946
12	0.0596		0.168 728	0.202 371
13	0.0537		0.158 866	0.352 610
14	0.0483		0.131 836	0.352 610
15	0.0435		0.094 397	0.202 371
16	0.0391		0.055 181	0.022 946
17	0.0352		0.021 904	−0.064 816
18	0.0317		−0.000 775	−0.045 444
19	0.0285	−0.000 286	−0.012 024	0.007 674
20	0.0257		−0.014 156	0.027 651
21	0.0231		−0.011 033	0.011 124
22	0.0208		−0.006 346	−0.007 792
23	0.0187		−0.002 682	−0.009 267
24	0.0168		−0.000 174	−0.001 169
25	0.0152			0.003 120
26	0.0136			0.001 807
27	0.0123			−0.000 303
28	0.0110			
29	0.0099	0.001 116		
30	0.0089			
31	0.0081			
32	0.0072			
33	0.0065			
34	0.0059			
35	0.0053			
36	0.0048			
37	0.0043			
38	0.0039			
39	0.0035	−0.001 659		
40	0.0031			
41	0.0028			
42	0.0025			
43	0.0023			
44	0.0020			
45	0.0018			
46	0.0017			
47	0.0015			
48	0.0013			
49	0.0012	−0.000 402		
50	0.0011			

$x(n)$ and $y(m)$ are the input and decimated data. $h_1(k)$ and $h_2(k)$ are decimating filter coefficients.

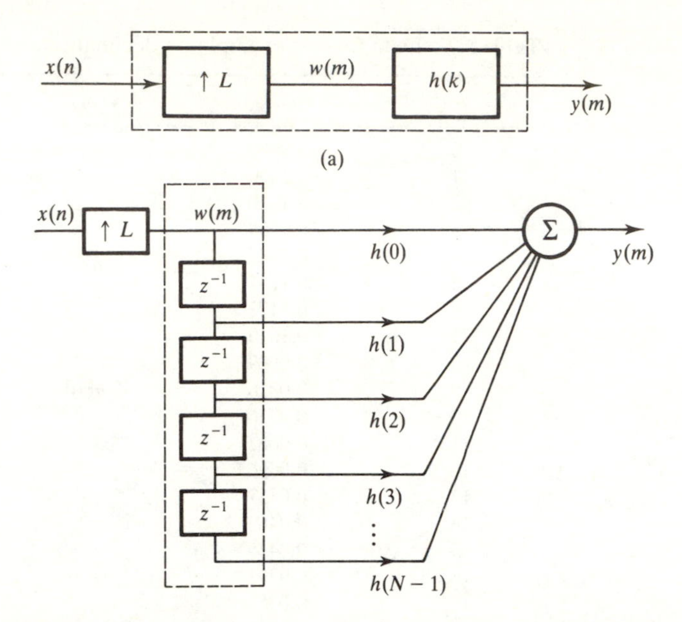

Figure 8.14 (a) Block diagram of the interpolator; (b) signal flow diagram of the interpolator.

and output samples are to be compared, for example in multirate highpass filtering the output sample needs to be corrected for the delay through the decimator and interpolator.

8.5 Software implementation of interpolators

A block diagram representation of the interpolator is shown in Figure 8.14(a) and its signal flow diagram is depicted in Figure 8.14(b). For every input sample, $x(n)$, fed into the interpolator, the rate expander (the box with an up-arrow) inserts $L-1$ zero-valued samples after the input sample. These are then filtered to yield $y(m)$. Thus for each input sample of $x(n)$ we have L samples of $y(m)$. Effectively, the input sampling frequency is increased from F_s to LF_s by the interpolator. One implication of inserting $L-1$ zeros after each sample is that the energy of each input sample is spread across L output samples. Thus the interpolator has a gain of $1/L$. After interpolation, each output sample should be multiplied by L to restore it to its proper level.

The interpolation equations are

$$y(m) = \sum_{k=0}^{N-1} h(k)w(m-k) \tag{8.15a}$$

$$w(m-k) = \begin{cases} x[(m-k)/L], & m-k = 0, L, 2L, \ldots \\ 0 \end{cases} \tag{8.15b}$$

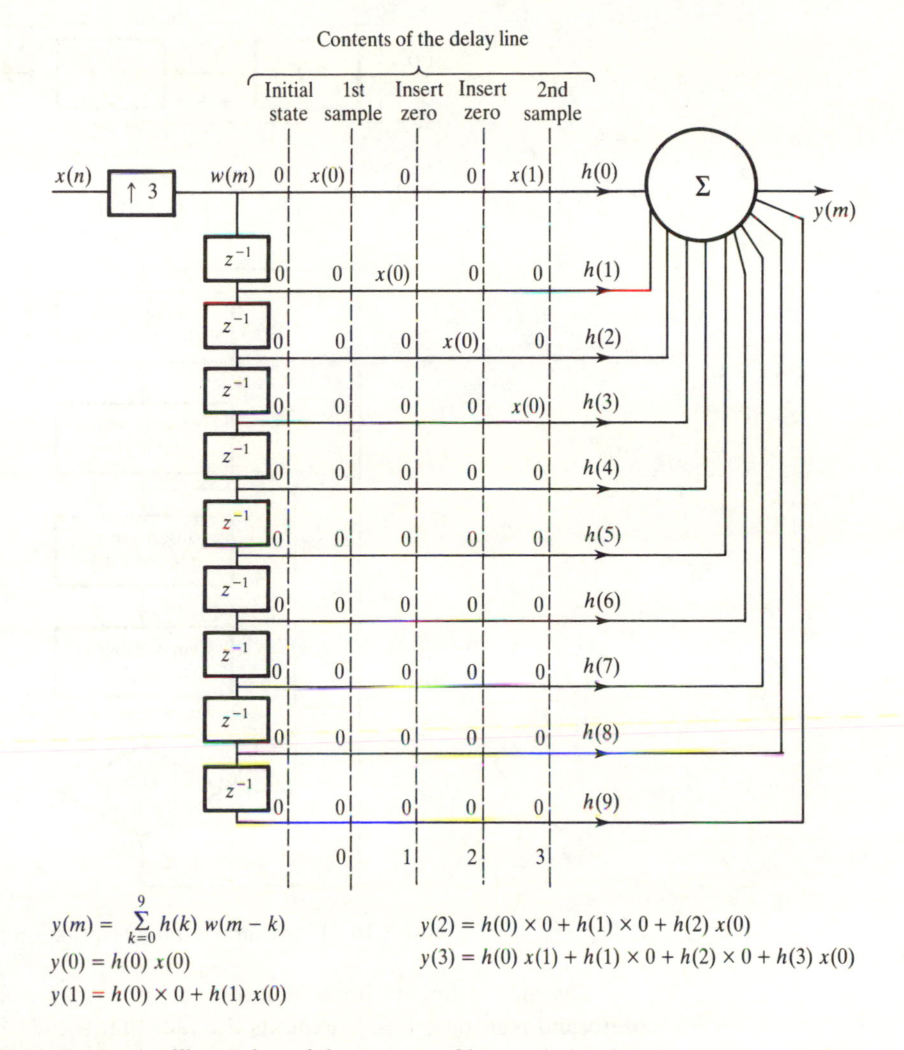

Figure 8.15 An illustration of the process of interpolation for the simple case of $L = 3$.

Figure 8.15 illustrates the process for the simple case where $L = 3$ and the filter length is 10. The delay line is fed with an input sample followed by two zeros, then the next input sample followed by two zeros and so on. An output sample is computed for each sample (data or zero) fed into the delay line. The contents of the delay line after two samples of $x(n)$ have been fed into the interpolator are shown. We see that for each input sample fed in, three samples are computed. The nonzero samples (that is the actual samples of $x(n)$) in the delay line are separated by $L - 1$ zeros (two in this example). Clearly, multiplication operations by the zero-valued samples are unnecessary.

Figure 8.16 gives the flowchart of a single-stage interpolator. In this implementation, only the nonzero-valued samples are fetched and used in the computation of the output samples. The flowchart for up to three stages of interpolation is given in Figure 8.17.

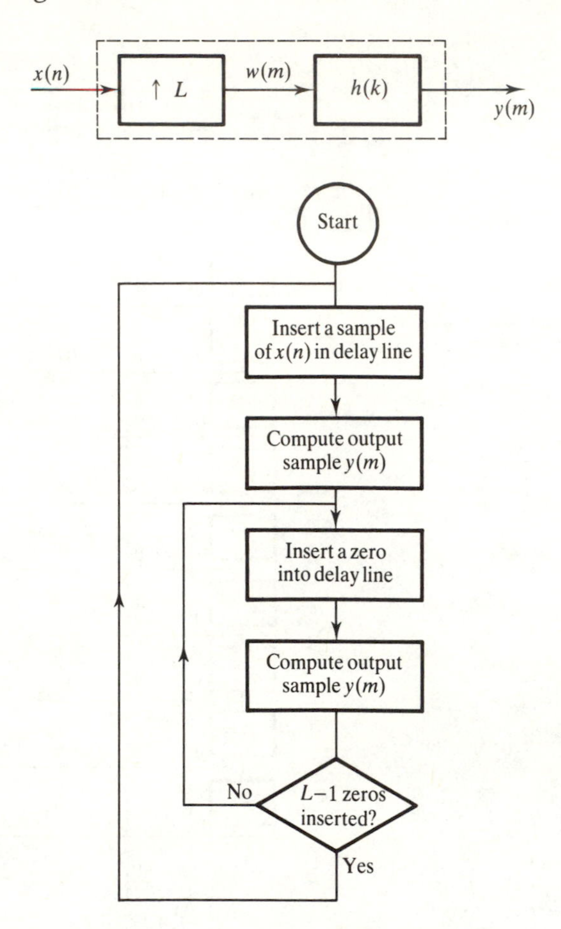

Figure 8.16 Flowchart of the interpolation process.

Another efficient implementation, known as polyphase filtering (Crochiere and Rabiner, 1983), exploits the fact that some of the delay line samples are zero. In this case the rate expander is removed altogether to eliminate the need to store zero-valued samples. The delay line is then shortened to N/L locations. In this approach, for each input sample fed into the delay line, the N/L delay line samples are used to compute L output samples, with each output sample computed with a different set of filter coefficients (that is, those filter coefficients that correspond to zero-valued samples are skipped). The limitation of the polyphase filter approach is that the ratio of N to L must be an integer. In the next section, we describe a C language implementation of a three-stage interpolator.

8.5.1 Program for multistage interpolation

The program is based on the method described in the last section. The program interpolates input data using up to three interpolation stages; see the flowchart

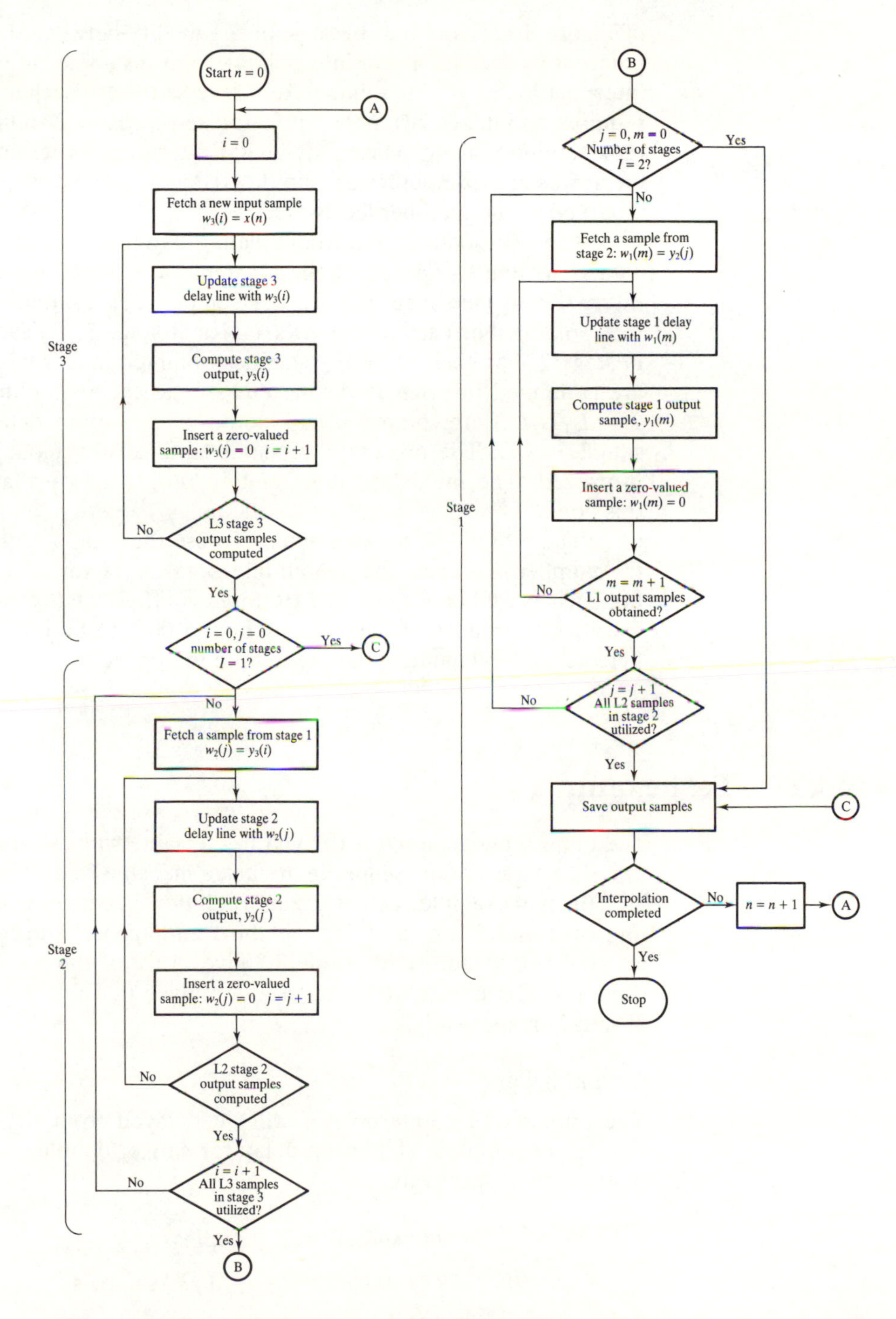

Figure 8.17 Interpolation using up to three stages. If, for example, $I = 1$ then only the sections labelled stage 3 are executed.

of Figure 8.17. Note that because of the duality between decimation and interpolation (a decimator and interpolator form a sample rate converter pair) the interpolation stages are numbered in reverse order. Each stage of interpolation requires an integer interpolation factor and a set of N-point filter coefficients representing a linear phase FIR digital filter. The program is intended to run on a personal computer such as the IBM or compatible, but can be readily modified to run on other hardware.

The input data is read from a data file in a computer and the interpolated data is written to an output file a sample at a time. Assuming a three-stage interpolation (see Figure 8.17), L_3 samples are computed at stage 3 of the interpolator. For each of the L_3 samples at stage 3, L_2 samples are computed in stage 2. For each of the L_2 samples obtained in stage 2, L_1 output samples are computed in stage 1. At the end of the cycle, L output samples (where $L = L_1L_2L_3$) are computed and stored in the output data file for each input sample, $x(n)$. The process is repeated until all the input samples have been processed. The multistage interpolation program is available on the disk for this book.

The program is self-contained. To use the program, the user must specify the number of stages, the overall interpolation factor, the interpolation factor for each stage, and a set of FIR filter coefficients for each stage. The user specifies the names of files containing the data to be interpolated, filter coefficients, and the name of the file to hold the results.

8.5.2 Test example

The input sequence used is derived in the same way as for the decimator. The first six samples of the sequence are given in Table 8.3.

In this example, two-stage interpolation is performed. Interpolation factors of 2 and 5 are used for the third and second stages, giving an overall interpolation factor of 10. Table 8.3 gives a list of the results of the interpolation. The FIR filter coefficients used are given in Table 8.3, and are of lengths 25 and 28 respectively.

Output delay

The output of the interpolator will be delayed from the input by a certain number of samples. The group delay for single, double, and three-stage decimation are, respectively,

$$T(1\text{ stage}) = T_1 \text{ samples} \quad \textbf{(8.16a)}$$

$$T(2\text{ stages}) = T_1 + M_1T_2 \text{ samples} \quad \textbf{(8.16b)}$$

$$T(3\text{ stages}) = T_1 + M_1T_2 + M_1M_2T_3 \text{ samples} \quad \textbf{(8.16c)}$$

Table 8.3 Data for test example on interpolation.

n	$x(n)$	$y(m)$	$h_2(k)$	$h_1(k)$
0	−0.9000	-4.74498×10^{-7}	−0.000303	−0.000174
		-7.313815×10^{-6}	0.001807	−0.002682
		-1.730554×10^{-5}	0.003120	−0.006346
		-3.0087×10^{-5}	−0.001169	−0.011033
		-3.860341×10^{-5}	−0.009267	−0.014156
		-2.995969×10^{-5}	−0.007792	−0.012024
		4.150394×10^{-5}	0.011124	−0.000775
		1.629372×10^{-4}	0.027651	0.021904
		3.299083×10^{-4}	0.007674	0.055181
		4.876397×10^{-4}	−0.045444	0.094397
1	0.1900	5.600492×10^{-4}	−0.064816	0.131836
		5.226861×10^{-4}	0.022946	0.158866
		2.857456×10^{-4}	0.202371	0.168728
		-1.480226×10^{-4}	0.352610	0.158866
		-7.700115×10^{-4}	0.352610	0.131836
		−0.0015445	0.202371	0.094397
		-2.448377×10^{-3}	0.022946	0.055181
		-3.40052×10^{-3}	−0.064816	0.021904
		-4.320959×10^{-3}	−0.045444	−0.000775
		-5.079387×10^{-3}	0.007674	−0.012024
2	0.1710	-5.534875×10^{-3}	0.027651	−0.014156
		-5.728923×10^{-3}	0.011124	−0.011033
		-5.466501×10^{-3}	−0.007792	−0.006346
		-4.756987×10^{-3}	−0.009267	−0.002682
		-3.522772×10^{-3}	−0.001169	−0.000174
		-1.715353×10^{-3}	0.003120	
		5.557998×10^{-4}	0.001807	
		3.324822×10^{-3}	−0.000303	
		6.400164×10^{-3}		
		9.565701×10^{-3}		
3	0.1539	0.0125976		
		1.544317×10^{-2}		
		1.774401×10^{-2}		
		1.933819×10^{-2}		
		1.984539×10^{-2}		
		1.894878×10^{-2}		
		1.681832×10^{-2}		
		1.303225×10^{-2}		
		7.845689×10^{-3}		
		1.357867×10^{-3}		
4	0.1385	-6.262392×10^{-3}		
		-1.462789×10^{-2}		
		-2.343143×10^{-2}		
		-3.207272×10^{-2}		
		-3.972186×10^{-2}		
		-4.567938×10^{-2}		
		-4.993166×10^{-2}		
		-5.142782×10^{-2}		
		-5.009625×10^{-2}		
		-4.527419×10^{-2}		

$x(n)$ and $y(m)$ are the input and interpolated data. $h_2(k)$ and $h_1(k)$ are interpolating filters.

where T_i is the delay in the ith-stage filter: $T_i = (N_i + 1)/2$ samples. N_i is the number of filter coefficients for the stage. In the test example above, the filters have delays of 13 and 14.5 samples, giving an overall delay of $13 + 2 \times 14.5 = 42$ samples.

If an integer delay is desired then N_i should be chosen such that the appropriate equation of Equations 8.16 is an integer. Thus if the input and output of the interpolator are to be compared (for example highpass narrowband filtering where the filtering operation is performed as the inverse of a lowpass), the output signal should be corrected or adjusted for this delay.

8.6 Application examples

Digital audio engineering is an area that has benefited significantly from multirate techniques. For example, they are used in the compact disc player to simplify the D/A conversion processes, while at the same time maintaining the quality of the reproduced sound. At the front-end of digital audio systems, efforts have been directed at using delta modulation techniques, combined with multirate processing, to obtain high quality digital data from analogue audio signals.

Other areas where multirate techniques have been used are in the acquisition of high quality data, high resolution spectral analysis, and the design and implementation of narrowband digital filtering.

In the next few sections we will describe a number of these applications.

8.6.1 High quality analogue-to-digital conversion for digital audio

The constant demand in digital audio for better quality, higher resolution and higher speed ADC has led to the development of single-bit ADCs using delta sigma modulation techniques. This offers the possibility of eliminating altogether from the conversion process most of the analogue circuitry at the front-end of a digital audio system, including the analogue anti-aliasing filters and sample-and-hold circuits.

A simplified block diagram of a fast single bit ADC process is shown in Figure 8.18 (Adams, 1986; Matsuya *et al.*, 1987; Welland *et al.*, 1989). The analogue audio signal is first converted into a single bit stream, using delta sigma modulation at a 3.072 MHz rate. The single bit stream is then downsampled to 48 kHz, using a multistage decimator, to yield 16-bit PCM words. Many ADCs utilizing multirate techniques are now available, off-the-shelf. Examples are 16- and 18-bit stereo ADCs by Crystal Semiconductor (CS5326, CS5327, CS5328, CS5329), and by Motorola Semiconductor (DSP56ADC16).

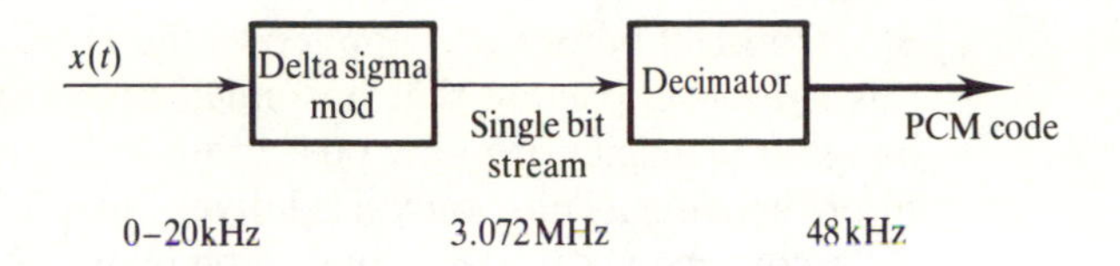

Figure 8.18 Simplified block diagram of single-bit ADC scheme.

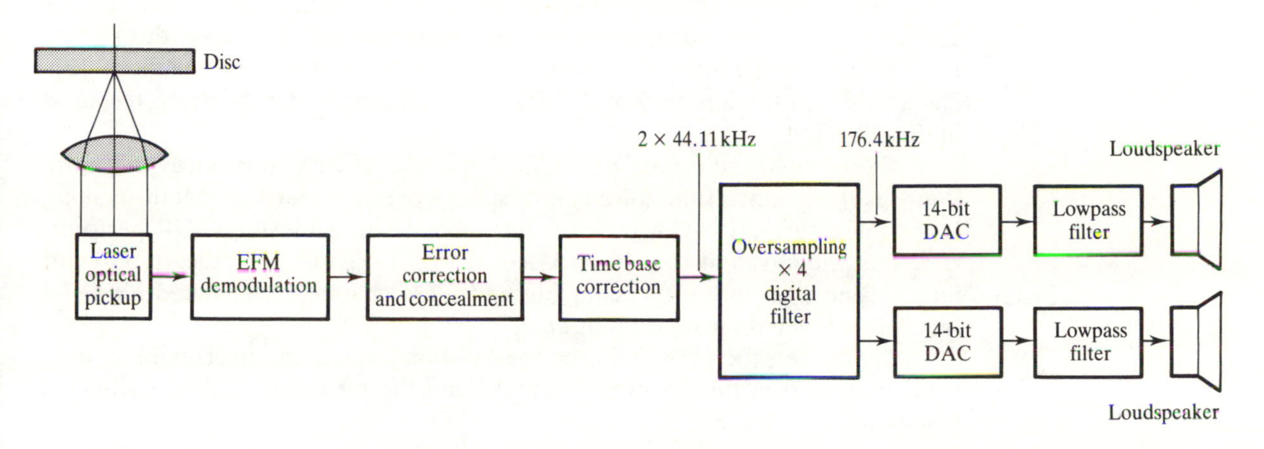

Figure 8.19 Audio signal reproduction in the compact disc system.

8.6.2 Efficient digital-to-analogue conversion in compact hi-fi systems

One of the first serious applications of multirate techniques was in the reproduction of sound and music in the compact disc (CD) player.

Figure 8.19 depicts the process of reconstituting the analogue audio signal from the digital signal read from the CD. After decoding and error correction the digital signals are in 16-bit words, representing the acoustic information at a 44.1 kHz sampling rate.

If these digital codes were converted directly into analogue image frequency bands centred at multiples of the sampling frequency of 44.1 kHz would be produced. Although the image frequencies would be inaudible as they are above the baseband of 0–20 kHz, they could cause overloading if passed on to the player's amplifier and loudspeaker, or they could set up intermodulation distortion. Thus the frequency components above the baseband need to be attenuated by at least 50 dB. Analogue filters that can provide this level of attenuation will have to meet very tight specifications, and would require trimming to ensure that the filters for the two stereo channels are matched.

To avoid the problems with analogue filters, multirate filtering is employed in the compact disc player. This is achieved by increasing, by interpolation, the sampling frequency of the data by 4 to 176 kHz (4 × 44.1 kHz

= 176.4 kHz) before it is applied to the DAC. In the time domain, the effect of this is to give a signal that has much finer steps. In the frequency domain, the image frequencies are now pushed up to higher frequencies, making it easier to filter them out. Thus, only a relatively simple lowpass filter is required after the D/A conversion. In the actual implementation, the digital filter incorporates a $\sin x/x$ correction (see Chapter 1) to compensate for the effects of the holding circuit following the DAC. The $\sin x/x$ correction has a beneficial effect in that it attenuates the signals on either side of 174 kHz by more than 18 kB, further simplifying the analogue image filtering requirements. A simple third-order Bessel filter is used after interpolation to provide additional attenuation. This has a 3 dB cutoff frequency of 30 kHz and a reasonably linear phase response in the passband.

Oversampling the data has other beneficial effects. It reduces the noise floor, as the quantization noise is now spread over a wider bandwidth, making it possible to use a DAC with fewer bits, and still to achieve an SNR performance that is equivalent to a 16-bit DAC. Thus, in Figure 8.19, the 16-bit-word interpolated data, after oversampling and noise shaping, is rounded off to 14 bits before it is fed into the 14-bit DAC.

There are other DACs in the market that exploit the oversampling concepts presented in this chapter. Examples are the bit stream DACs by Philips Components (SAA7322, SAA7323, SAA7350).

Example 8.3

A digital audio system exploits oversampling techniques to relax the requirements of the analogue anti-imaging filter. The overall filter specifications for the system is given below:

baseband	0 to 20 kHz
input sampling frequency F_s	44.1 kHz
output sampling frequency	176.4 kHz
stopband attenuation	50 dB
passband ripple	0.5 dB
transition width	2 kHz
stopband edge frequency	22.05 kHz

Design a suitable interpolator.

Solution

Using the multirate design program on the PC disk for the book the interpolation factors and filter characteristics of the possible interpolators (with integer factors) are summarized below.

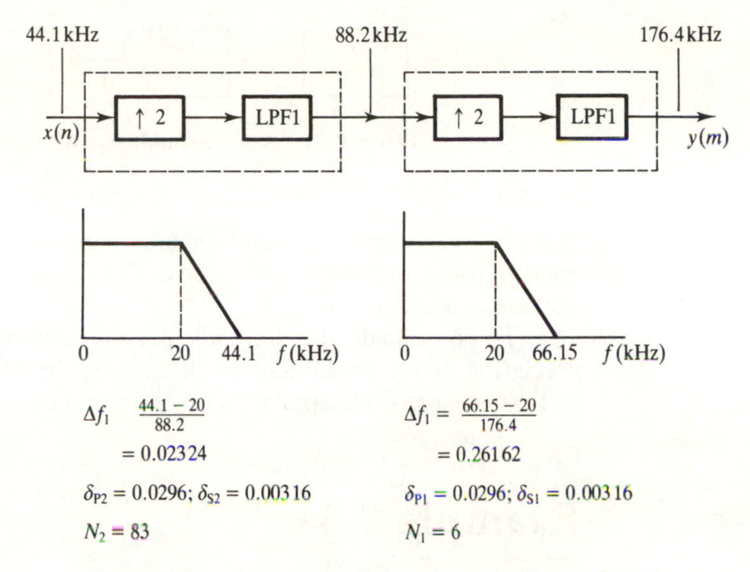

Figure 8.20 A two-stage interpolator for Example 8.3.

Number of stages	*Interpolation factor, L_i*	*Filter length N_i*	*Normalized transition width Δf_i*	*Passband ripple, δ_p*	*Stopband ripple, δ_s*
1	4	146	0.045 35	0.059 25	0.003 16
2	2	6	0.261 62	0.0296	0.003 16
	2	83	0.023 24	0.0296	0.003 16

For the two-stage interpolator, the system has the form shown in Figure 8.20.

8.6.3 Application in the acquisition of high quality data

In the acquisition of almost any real-life data, the need to keep aliasing low often dictates the use of a relatively complex analogue anti-aliasing filter. In a multichannel system, each analogue channel must be fitted with a separate anti-aliasing filter as such filters cannot be readily multiplexed. Where there is a large number of analogue channels (for example, in biomedicine as many as 32 channels may be required), the use of analogue anti-aliasing filters can become expensive. By using anti-aliasing digital filters, the complex analogue filters on each channel can be replaced by a much simpler filter, leading to a substantial cost reduction. Further, the phase matching problems of tightly specified analogue filters can be avoided. The difficulty of supporting multiple sampling frequencies (each sampling frequency would require a different cutoff frequency) when analogue anti-aliasing filters are used is also overcome.

Figure 8.21 shows a block diagram of a multirate data acquisition system (Quarmby, 1984). The desired aliasing level is achieved by the front-end RC

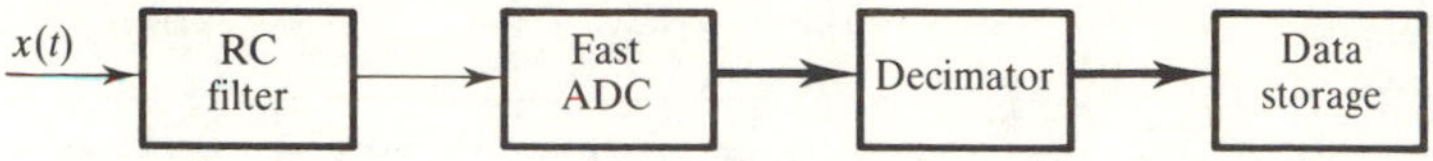

Figure 8.21 A simple multirate data acquisition system.

filters oversampling the input signal, then downsampling to the desired frequency using multirate techniques. The main price is that the ADC must operate at a faster rate.

To consolidate the materials presented here and to provide a better appreciation of the benefits of implementing the anti-aliasing filter digitally, we will discuss a real-life application by way of an example.

Example 8.4

A requirement exists for a general purpose, multichannel (up to 32 channels) data acquisition system for collecting physiological data. Each analogue channel is individually configurable, by the user, to have a cutoff frequency of between 0.5 and 200 Hz, and a selectable sampling frequency in the range from 1 to 1000 Hz. The overall filter requirements for each channel are

passband ripple	$\leqslant 0.5$ dB
signal-to-aliasing ratio	$\geqslant 45$ dB (in the passband)
passband edge frequency	$0.5 \text{ Hz} \leqslant f_p \leqslant 200 \text{ Hz}$
stopband edge frequency	$\leqslant 3f_p$

Both amplitude and phase distortion should be kept as low as possible. To reduce component count and cost and the size of the PCB for the system only simple analogue filters should be used at the front-end.

Solution

To provide the anti-aliasing filtering using analogue filters alone would require a very high-order filter. An alternative approach is to fit each channel with a simple, identical filter, to oversample at a common, fixed rate, and then to decimate to the desired rate(s). At each stage we must ensure that the specifications are met.

A simple, one-pole RC filter could be used for each channel, but this would require a very high sampling frequency to satisfy the specification. We will use a second-order Butterworth filter, as we have found it satisfactory in our biomedical engineering work.

The amplitude response of a second-order Butterworth filter is given by

$$A(f) = \frac{1}{[1 + (f/f_c)^4]^{1/2}}$$

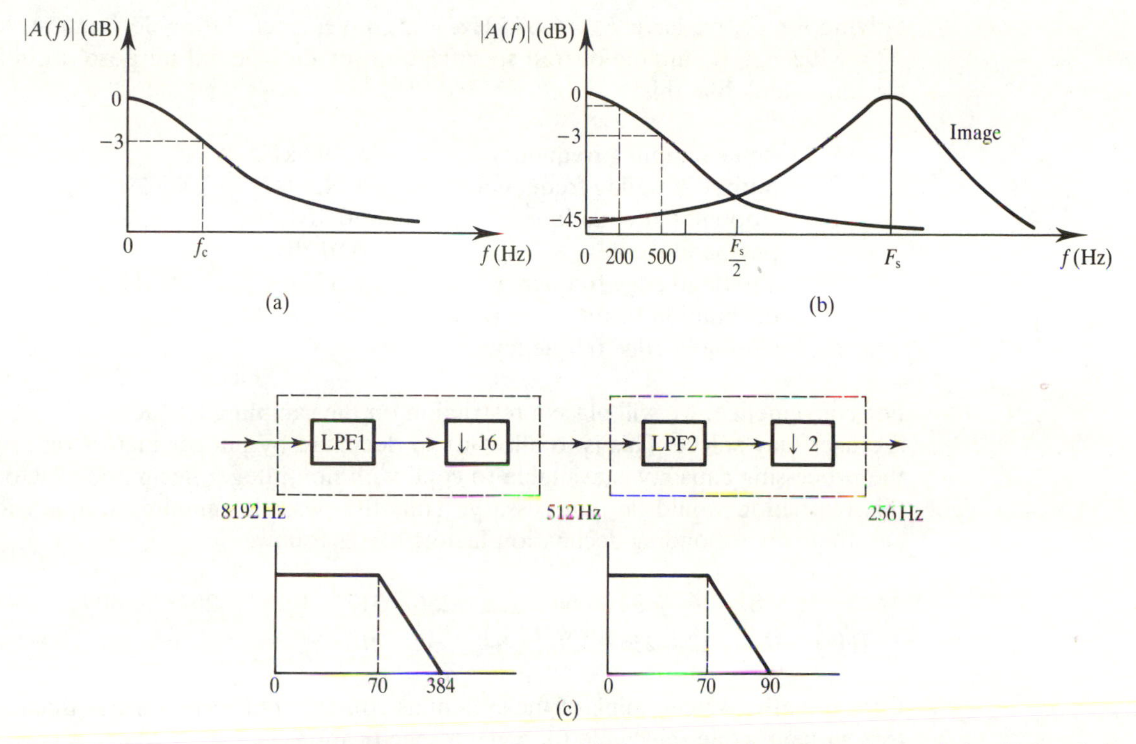

Figure 8.22 (a) A second-order Butterworth response. (b) Spectrum of bandlimited, sampled, wideband signal. (c) A two-stage decimator for Example 8.3.

This is depicted in Figure 8.22(a). It is evident from the figure that a significant amplitude error exists in the band from 0 to f_c. To keep the error within the specification, the highest frequency of interest (200 Hz in our case) should be well below f_c. A suitable value for f_c can be obtained from

$$20\log[1 + (200/f_c)^4]^{1/2} \leqslant 0.5 \text{ dB}$$

Solving for f_c gives $f_c = 338.39$ Hz. For convenience and to allow for additional errors that would be introduced by subsequent stages $f_c = 500$ Hz would be used. An f_c of 500 Hz gives a response that is down, at 200 Hz, by about 0.11 dB.

Next, we establish a common sampling frequency for all the channels. After bandlimiting each channel with a second-order Butterworth filter and sampling, the spectrum of the signal is depicted in Figure 8.22(b) (assuming a wideband signal). Referring to the figure, it should be clear that we need a sampling frequency, F_s, such that the aliasing level is at least 45 dB down on the signal level at f_p (where $f_p = 200$ Hz):

$$20\log\{1 + [(F_s - 200)/500]^4\}^{1/2} \geqslant 45 \text{ dB}$$

Solving for F_s, we have $F_s \approx 6.67$ kHz. For convenience during decimation, let $F_s = 8192$ Hz. A suitable overall specification for the general purpose decimator might look like this:

input sampling frequency	8.192 kHz
output sampling frequency	1 Hz $< F_s <$ 1000 Hz
stopband attenuation	50 dB
passband ripple	0.01 dB
passband edge frequency	0.5 Hz $< f_p <$ 200 Hz
decimation factor	$8.192 < M < 8192$
stopband edge frequency	$< 2f_p$

For convenience, we will place a restriction on the sampling frequency, F_s, that the user may select. This is to allow us to decimate by integer factors only. If the processing capacity is available to cope with non-integer decimation factors this restriction would be unnecessary. Thus the possible sampling frequencies and their corresponding decimation factors are as follows:

M	8	16	32	64	128	256	512	1024	2048	4096	8192
F_s (Hz)	1024	512	256	128	64	32	16	8	4	2	1

Conceptually, we can think of the system as consisting of 11 multistage decimators with only one selectable for a given specification.

As an illustration, let us consider a system for collecting EEG (electroencephalography) signals. The user specification for each channel could be

sampling frequency	256 Hz
stopband attenuation	45 dB
passband ripple	0.5 dB
passband	0–70 Hz

This specification is translated into that for a rate converter, consistent with that for the general purpose decimator above:

input sampling frequency	8192 kHz
output sampling frequency	256 Hz
decimation factor	32
stopband attenuation	50 dB
passband ripple	0.01 dB
passband	0–70 Hz
stopband	90–128 Hz

Using the design program (provided on the disk), an efficient decimator (in terms of computation and system complexity) is the two-stage system depicted in Figure 8.22(c).

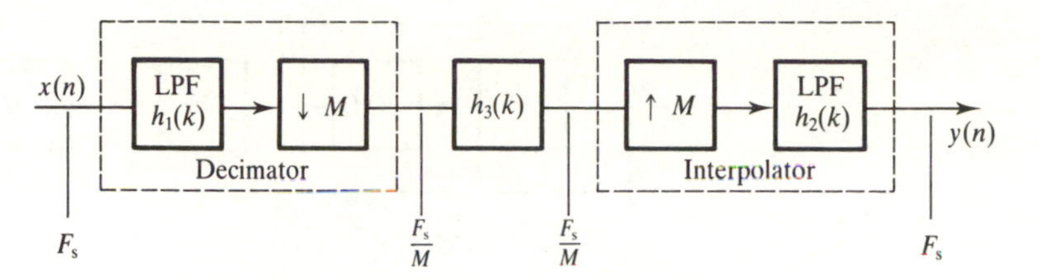

Figure 8.23 Multirate narrow band filtering.

8.6.4 Efficient implementation of narrowband digital filters

Narrowband digital filters are characterized by sharp transitions between the passband and stopbands, and by passbands which are very small compared with the sampling frequency. As a result narrowband FIR filters require a very large number of coefficients. This poses a problem in the design and implementation of such filters because they are highly susceptible to finite wordlength effects (for example roundoff noise and coefficient quantization errors). Further, large coefficient storage requirements and computational effort are required. The multirate approach overcomes these problems, and leads to FIR filters with a computational performance that is comparable with that of elliptic IIR filters.

Figure 8.23 shows a simple arrangement for multirate filtering. The sampling frequency of the input sequence is first reduced as far as possible by decimation, the desired filtering operation is then performed at the low sampling frequency, and finally the sampling frequency of the filtered data is restored back to its original rate by interpolation. The use of the same sampling rate conversion factor at the decimator and the interpolator ensures that the input signal, $x(n)$, and the output signal, $y(n)$, are at the same sampling rate.

Narrowband lowpass and bandpass filtering

For narrowband lowpass filtering, the filters $h_1(k)$ and $h_2(k)$ in Figure 8.23 would be lowpass filters and there may be no need for $h_3(k)$. A design aim would be for the overall input–output characteristics of the structure in Figure 8.23 to be equivalent to those of the desired conventional lowpass filter. The characteristics that result would, however, not be identical to those of a conventional lowpass filter because of the effects of aliasing and imaging. In practice, to guarantee that the overall filter meets the desired specifications, the filters $h_1(k)$ and $h_2(k)$ are identical, each with a passband ripple of $\delta_p/2$ and a stopband ripple of δ_s, where δ_p and δ_s are, respectively, the passband and stopband deviations of the equivalent lowpass filter.

The design of multirate bandpass filters is somewhat more involved, except when one wishes to design the so-called integer-band bandpass filters where the band edges are exact multiples of the lowest sampling frequency in the system, that is $F_s/2M$. In these cases, the decimation/interpolation factor, M, and the filter band edges satisfy the following conditions (Crochiere and Rabiner, 1983):

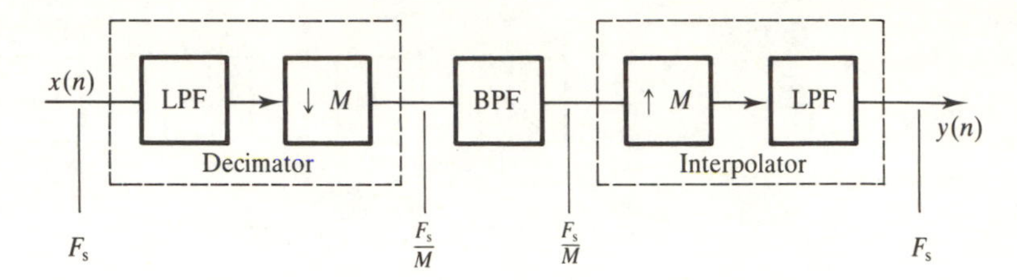

Figure 8.24 An approach to narrowband multirate bandpass filtering.

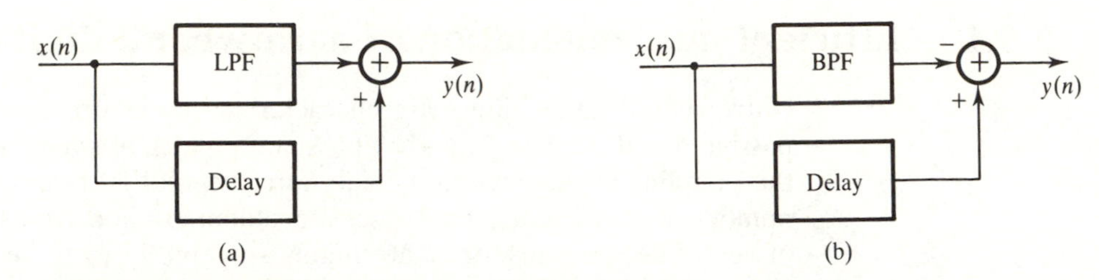

Figure 8.25 Multirate realization of highpass and bandstop filtering as duals of lowpass and bandpass filters: (a) multirate highpass filtering; (b) multirate bandstop filtering.

$$M = F_s/2(f_{su} - f_{sl}) \tag{8.17a}$$

$$f_{sl} = kF_s/M,\ k \text{ an integer},\quad 0 < k < M - 1 \tag{8.17b}$$

$$f_{su} = (k + 1)F_s/M \tag{8.17c}$$

where f_{sl} and f_{su} are the lower and upper stopband edge frequencies respectively. Equation 8.17(a) gives the maximum decimation factor possible, and Equations 8.17(b) and 8.17(c) specify the lower stopband edge and upper stopband edge and the band number k.

A simple alternative, but less efficient, multirate bandpass filtering scheme is to decimate the data as far as possible as described previously using suitable lowpass filters, to bandpass filter the low rate signal, and then to interpolate back to the desired rate. This is illustrated in Figure 8.24. Quite clearly, care must be taken to ensure that the desired passband is protected from the effects of aliasing and imaging during decimation and interpolation.

Narrowband highpass and bandstop filters

Narrowband highpass and bandstop filters can be realized as the duals of the lowpass and bandpass filters, respectively:

$$H_{hp}(w) = 1 - H_{1p}(w) \tag{8.18a}$$

$$H_{bs}(w) = 1 - H_{bp}(w). \tag{8.18b}$$

The structures for highpass and bandstop filters are depicted in Figure 8.25. In the case of a highpass filter, for example, the signal, $x(n)$, is first lowpass filtered. The filtered signal is then subtracted from the unfiltered signal to yield the desired signal. The signal $x(n)$ must be delayed by an amount equal to the delay of the lowpass filter before subtraction. Clearly, it is necessary for the delay through the lowpass filter to be an integer number of samples for this to be possible. Correct passband and stopband specifications must be used for the lowpass filter so that the desired highpass filter is obtained.

Example 8.5

In connection with a research project in foetal monitoring, a need arose to assess the effects of the measurement system on the electrical activity of the foetal heart (the ECG or electrocardiogram) (Westgate *et al.*, 1990). To achieve this required that certain features of the ECG be quantified including the mains frequency content of the signal. Because of the existence of signal energy in the neighbourhood of the mains frequency (50 Hz), it was necessary to use a very narrowband filter. The specifications of the filter are

passband	49 to 51 Hz
stopband edge frequencies	47 and 53 Hz
stopband attentuation	30 dB ($\delta_p = 0.031\,62$)
passband ripple	0.1 dB ($\delta_s = 0.011\,579$)
sampling frequency	500 Hz

Solution

If a direct design of the above filters were attempted, Equation 8.3 suggests that we would require 4018 coefficients, which is rather too long.

Using the multirate approach, a number of options exist. One option is to decimate the data to as low a rate as possible (consistent with the specifications above) (see also Problem 8.4). In this case, the lowest rate would be 125 Hz as this would still allow us to have available the band from 0 to 62.5 Hz. The overall specifications for the decimator are

passband ripple	0.05 dB ($\delta_p = 0.005\,789\,5$)
stopband attentuation	30 dB ($\delta_s = 0.031\,62$)
passband	0–53 Hz
input sampling frequency	500 Hz
output sampling frequency	125 Hz

The decimation factor is 4. Using the multirate design program on the disk for the book a two-stage decimator was designed (see Figure 8.26). The optimal method (see Chapter 6) was used to obtain the coefficients of the two filters in Figure 8.26. The ECG data were then decimated using the filters and the

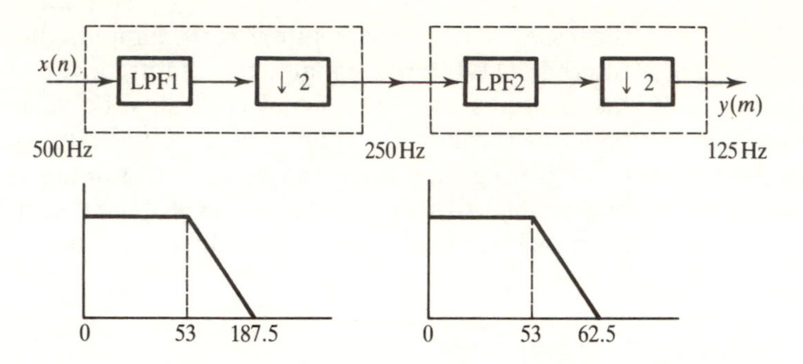

Figure 8.26 Decimator for reducing the rate of the ECG data.

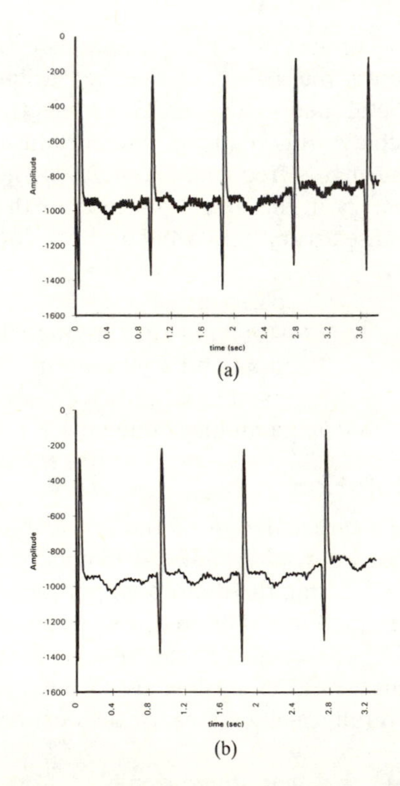

Figure 8.27 (a) Raw ECG data. (b) ECG data after decimation (adjusted for delay).

multistage decimation program described in Section 8.4. Examples of the data before and after decimation are shown in Figure 8.27.

A mains filter, satisfying the specifications given in the problem (but at the new, reduced rate) was designed. In this case, the number of filter coefficients was 113.4 for the mains filter ($\delta_p = 0.005\,789\,5$, $\delta_s = 0.031\,62$). After filtering, the data can be restored to its original rate by interpolation.

Example 8.6

Design a suitable multirate lowpass filter for extracting the baseline shifts in the foetal ECG. The filter should satisfy the following specifications:

passband	0 to 0.4 Hz
stopband	0.5 to 250 Hz
passband ripple	0.01
stopband ripple	0.001
sampling frequency	500 Hz

Solution

The approach we will use is to decimate down to a 1 Hz rate and then interpolate back up to 500 Hz. In this case, the overall filter specifications for the decimator are

passband ripple	0.01
stopband ripple	0.001
passband	0–0.4 Hz
stopband edge	0.5 Hz
sampling frequency	500 Hz
decimation factor	500

Using the multirate design program on the disk for the book various practical decimators with up to four stages of decimation were obtained. The characteristics of the most interesting ones are summarized in Table 8.4. Taking into account implementational complexity, the three-stage decimator depicted in Figure 8.28 was selected as the best solution. Again using the optimal design program (Chapter 6) the filter coefficients of the filters of the decimator were computed. Using the multistage decimation program, the ECG data was decimated down to 1 Hz and then interpolated back up to 500 Hz.

8.6.5 High resolution narrowband spectral analysis

As discussed in Chapter 10, an important application of the FFT is in the estimation of the spectrum of signals. The FFT provides spectral components of a signal at uniform intervals between 0 and one-half the sampling frequency. In many applications such as sonar, seismology, radar, biomedicine and vibration analysis, the desired signal may occupy only a narrow band in the spectrum of the acquired data. In such cases, direct use of the FFT would require a significant and unnecessarily high computational effort. The multirate technique can be used to isolate and translate the frequency band of interest to a

Table 8.4 Summary of the efficient decimators.

Number of stages	*MPS*	*TSR*	M_i	*Filter length* N_i	*Passband edge* f_p (Hz)	*Stopband edge* (Hz)	*Normalized transition width* Δf_i	*Passband ripple*	*Stopband ripple*
1			500	12 707	0.4	0.5	0.002 2	0.01	0.001
2	1807	430	50	153	0.4	9.5	0.018 20	0.005	0.001
			10	277	0.4	0.5	0.01	0.005	0.001
3	1705	189	25	77	0.4	19.5	0.038 20	0.0033	0.001
			10	53	0.4	1.5	0.055	0.0033	0.001
			2	59	0.4	0.5	0.05	0.0033	0.001
4	1444	172	2	2	0.4	249.5	0.498 2	0.0025	0.001
			25	83	0.4	9.5	0.036 40	0.0025	0.001
			5	27	0.4	1.5	0.110 00	0.0025	0.001
			2	60	0.4	0.5	0.050 0	0.0025	0.001
4	1724	169	25	79	0.4	19.5	0.038 20	0.0025	0.001
			2	3	0.4	9.5	0.455 00	0.0025	0.001
			5	27	0.4	1.5	0.110 0	0.0025	0.001
			2	60	0.4	0.5	0.050 0	0.0025	0.001

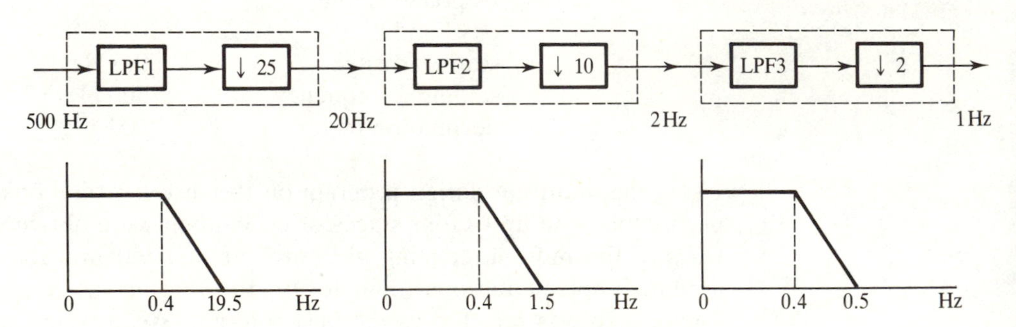

Figure 8.28 The three-stage decimator for Example 8.6.

lower frequency before the FFT is applied, leading to a significant reduction in computation, as well as permitting a trade-off between resolution and computational effort.

The FFT performed on the downsampled data allows an equivalent resolution for a much reduced computation or a greater resolution for about the same amount of computation as direct FFT on the original sequence. Effectively, downsampling allows us to see the spectrum of the narrowband on an expanded scale.

Narrowband spectral analysis using multirate techniques is essentially an extension of narrowband bandpass filtering discussed earlier, and has the same restrictions. The signal is first bandpass filtered to isolate the band of interest. The sampling frequency of the filtered signal is then reduced by decimation to F_s/M, where F_s is the sampling frequency of $x(n)$. The spectrum of the much reduced sequence $y(n)$ is then computed using the FFT. A correction factor is used to compensate for the errors in the spectrum due to the ripples in the passband of $h(n)$. When the band of interest does not meet the conditions, this can be circumvented by using a slightly wider frequency band encompassing the desired band. Another approach is to use the method described in Liu and Mintzer (1978), which involves a computer search to find permissible decimation factors.

8.7 Summary

Digital sytems that handle more than one sampling rate are known as multirate systems. The two key elements of a multirate system are the decimator and interpolator. The decimator allows us to reduce, efficiently, the rate of a signal by an integer factor M or a rational factor L/M $(L < M)$. The interpolator allows us to increase the sampling rate by an integer factor L or a rational factor L/M $(L > M)$.

In practice, sample rate changes are implemented in two or more stages for maximum computational and/or storage efficiency. The individual digital filters used in multistage designs have relaxed specifications, leading to fewer coefficients and hence to lower sensitivity to finite wordlength effects, both being directly related to the number of filter coefficients. A practical method of designing sample rate converters was described in detail.

The main strength of multirate systems lies in their ability to exploit the advantages of DSP, in particular the ability to use DSP to bandlimit a signal almost to the Nyquist frequency, with substantial attenuation, and without violating the sampling theorem requirements. The advantages have been exploited in many applications including the compact disc, digital filtering, data acquisition and high resolution data acquisition systems. Many of these systems have been described in detail and their multirate elements designed.

A set of C language programs are provided on the PC disk for the book which allow the design and software implementation of multirate systems.

Problems

8.1 A one-stage decimator is characterized by the following:

decimation factor 3
anti-aliasing filter coefficients

$$h(0) = -0.06 = h(4)$$
$$h(1) = 0.30 = h(3)$$
$$h(2) = 0.62$$

Given the data, $x(n)$, with successive values $\{6, -2, -3, 8, 6, 4, -2\}$, calculate and list the filtered output, $w(n)$, and the output of the decimator, $y(m)$.

8.2 Design a decimator for a high quality data acquisition system with the following overall specification for the decimation filter:

audio band	0 to 20 kHz
input sampling frequency	3.072 MHz

output sampling frequency	48 kHz
passband ripple	< 0.001 dB
stopband attenuation	> 86 dB

8.3 A requirement exists to compute the spectrum of a narrowband signal embedded in a wideband signal. The band of interest is 49–51 Hz, but the composite signal occupies the band from 0 to 100 Hz. An N-point sequence $x(n)$ is obtained by sampling the composite signal at a rate of 1 kHz.

(1) Illustrate how the desired signal spectrum would be obtained using the multirate approach.

(2) Estimate the computational advantage of the multirate approach over the conventional FFT approach. Compare the resolution of the spectrum for both methods.

8.4 A high quality, efficient narrowband filter is required to extract and assess the mains component in a signal. The filter should satisfy the following specifications:

passband	49 to 51 Hz
stopband edge frequencies	48 and 52 Hz
stopband attenuation	60 dB
passband ripple	0.01 dB
sampling frequency	1000 Hz

Using the multirate approach, design a suitable filter.

8.5 There is a need to interpret the activity in a certain physiological signal, captured at a rate of 256 Hz. To achieve this requires the extraction and analysis of both time and frequency domain features in each band. As a first step towards this, design a suitable multirate system for splitting the signal into the following bands:

0.5–4 Hz
4–8 Hz
8–13 Hz
13–16 Hz

The multirate system should introduce no more than 0.01 dB ripple in the bands and the out-of-band signals should be attenuated by at least 50 dB.

References

Adams R.W. (1986). Design and implementation of an audio 18 bit analog-to-digital converter using oversampling techniques. *J. Audio Engineering Society*, **34**(3), 153–66

Agrawal B.P. and Shenoi K. (1983). Digital methodology for ΣΔM. *IEEE Trans. Communications*, **31**(3), 360–70

Claasen T.A.C.M., Mecklenbrauker W.F.G., Peek J.B.H. and Hurck Van N. (1980). Signal processing method for improving the dynamic range of A/D and D/A converters. *IEEE Trans. Acoustics, Speech and Signal Processing*, **28**(5), 529–38

Crochiere R.E. and Rabiner L.R. (1975). Optimum FIR digital filter implementations for decimation, interpolation, and narrow-band filtering. *IEEE Trans. Acoustics, Speech and Signal Processing*, **23**(5), 444–56

Crochiere R.E. and Rabiner L.R. (1976). Further considerations in the design of decimators and interpolators. *IEEE Trans. Acoustics, Speech and Signal Processing*, **24**, 296–311

Crochiere R.E. and Rabiner L.R. (1983). *Multirate Digital Signal Processing*. Englewood Cliffs NJ: Prentice-Hall

Crochiere R.E. and Rabiner L.R. (1979). A program for multistage decimation, interpolation, and narrow band filtering. In *IEEE Programs for DSP*. IEEE

Crochiere R.E. and Rabiner L.R. (1981). Interpolation and decimation of digital signals – a tutorial review. *Proc. IEEE*, **69**(3), 300–31

Crochiere R.E. and Rabiner L.R. (1988). Multirate processing of digital signals. In *Advanced Topics in Signal Processing* (Lim J.S. and Oppenheim A.V., eds.). Englewood Cliffs NJ: Prentice-Hall

DeFatta D.J., Lucas J.G. and Hodgkiss W.S. (1988). *Digital Signal Processing: A System Design Approach*. New York NY: Wiley

Liu B. and Mintzer F. (1978). Calculation of narrow-band spectra by direct decimation. *IEEE Trans. Acoustics, Speech and Signal Processing*, **26**(6), 529–34

Matsuya Y., Uchimura K., Iwata A., Kobayashi T., Ishikawa M. and Yoshitome T. (1987). A 16-bit oversampling A-to-D conversion technology using triple integration noise shaping. *IEEE J. Solid State Circuits*, **22**(6), 921–8

Quarmby D., ed. (1984). *Signal Processor Chips*, Chapter 5. London: Granada

Welland D.R., Del Signore B.P., Swanson E.J., Tanaka T., Hamashita, K., Hara S. and Takasuka K. (1989). A stereo 16-bit delta-sigma A/D converter for digital audio. *J. Audio Engineering Society*, **37**(6), 476–86

Westgate J.A., Keith R.D.F., Gurnow J.S.K., Ifeachor E.C. and Greene K.R.G. (1990). Suitability of fetal scalp electrodes for fetal electrocardiogram during labour. *J. Clin. Physics & Physiological Measurement*, **11**(4), 297–306

Bibliography

Analog Devices (1988). *ADSP-2100 Family Applications Handbook*, Volume 2, Chapter 3. Analog Devices Inc.

Bellanger M.G. (1977). Computation rate and storage estimation in multirate digital filtering with halfband filters. *IEEE Trans. Acoustics, Speech and Signal Processing*, **25**, 344–6

Bellanger M.G., Daquet J.L. and Lepagnol G.P. (1974). Interpolation, extrapolation and reduction of computation speed in digital filters. *IEEE Trans. Acoustics, Speech and Signal Processing*, **22**, 231–5

Brown Jr. J.L. (1981). Multichannel sampling of lowpass signals. *IEEE Trans. Circuits Systems*, **28**, 101–6

Cox R.V., Bock D.E., Bauer K.B., Johnston J.D. and Snyder J.H. (1987). The analogue voice privacy system. *AT&T Technical J.*, **66**, 119–31

Elliot D.F., ed. (1987). *Handbook of Digital Signal Processing*. New York NY: Academic Press

Goodman D.J. and Carey M.J. (1977). Nine digital filters for decimation and interpolation. *IEEE Trans. Acoustics, Speech and Signal Processing*, **25**(2), 121–6

Goodman D.J. and Flanagan J.L. (1971). Direct digital conversion between linear and adaptive delta modulation formats. In *Proc. IEEE Int. Communications Conf.*, Montreal, PQ, Canada, June 1971

Goedhart D., van der Plassche R.J. and Stikvoort E.F. (1982). Digital-to-analog conversion in playing a compact disc. *Philips Technical Rev.*, **40**(6), 174–9

Huber A., De Man E., Schiller E. and Ulbrich W. (1986). FIR lowpass filter for signal decimation with 15 MHz clock frequency. In *IEEE Int. Conf. Acoustics, Speech and Signal Processing*, Tokyo, April 7–11, pp. 1533–6

Jerri A.J. (1977). The Shannon sampling theorem – its various extensions and applications: a tutorial review. *Proc. IEEE*, **65**(11), 1565–96

Linden D.A. (1959). A discussion of sampling theorems. *Proc. IRE.*, **47**, 1219–26

Mintzer F. (1982). On half-band, third-band and *N*th-band FIR filters and their design. *IEEE Trans. Acoustics, Speech and Signal Processing*, **30**, 734–8

Mintzer F. and Liu B. (1978). The design of optimal multirate bandpass and bandstop filters. *IEEE Trans. Acoustics, Speech and Signal Processing*, **26**(6), 534–43

Mintzer F. and Liu B. (1978). Aliasing error in the design of multirate filters. *IEEE Trans. Acoustics, Speech and Signal Processing*, **26**, 76–88

Montijo B.A. (1988). Digital filtering in a high-speed digitizing oscilloscope. *Hewlett Packard J.*, (June), 70–6

Mou Z.J. and Duhamel P. (1987). Fast FIR filtering: algorithms and implementations. *Signal Processing*, **13**, 377–84

Plassche Van De R.J. and Dijkmans E.C. (1983). A monolithic 16-bit D/A conversion system for digital audio. In *Digital Audio* (Blesser B., ed.), pp. 54–60. Audio Engineering Inc

Princen J.P. and Bradley A.B. (1986). Analysis/synthesis filter banks design based on time domain aliasing cancellation. *IEEE Trans. Acoustics, Speech and Signal Processing*, **23**, 1153–61

Rabiner L.R. and Crochiere R.E. (1975). A novel implementation for narrow-band FIR digital filters. *IEEE Trans. Acoustics, Speech and Signal Processing*, **23**(5), 457–64

Regalia P.A., Fujii N., Mitra S.K. and Neuvo Y. (1987). Active RC crossover networks with adjustable characteristics. *J. Audio Engineering Society*, (January–February), **35**(1/2), 24–30

Rorabacher D.W. (1975). Efficient FIR filter design for sample rate reduction and interpolation. In *Proc. IEEE International Symposium on Circuits and Systems*, April 21–23, 1975, pp. 396–9

Schafer R.W. and Rabiner R.L. (1973). A digital signal processing approach to interpolation. *Proc. IEEE*, **61**, 692–702

Scheuermann H. and Gockler H. (1981). A comprehensive survey of digital transmultiplexing methods. *Proc. IEEE*, **69** 1419–50

Shannon C.E. (1949). Communications in the presence of noise. *Proc. IRE*, **37**, 10–21

Thong T. (1989). Practical consideration for a continuous time digital spectrum analyzer. In *Proc. IEEE International Symposium on Circuits and Systems*, Portland OR, May 1989, 1047–50

Tuffs D.W., Rorabacher D.W. and Mosier W.E. (1970). Designing simple, effective digital filters. *IEEE Trans. Audio Electro-Acoustics*, **18**, 142–58

Vaidyanathan P.P. (1990). Multirate digital filters, filter banks, polyphase networks, and applications: a tutorial. *Proc. IEEE*, **78**(1), 56–93

Vaidyanathan P.P. and Nguyen T.Q. (1987). A trick for the design of FIR half-band filters. *IEEE Trans. Circuits and Systems*, **34**, 297–300

Zobel R.N. and Tang P.S. (1985). A high performance multichannel decimating FIR digital filter system for microprocessor based data acquisition. *Proc. ISCAS*, 1149–52

Appendix

A8 C programs for multirate processing and systems design

The following C language programs are available on the PC disk for the book (see the preface for details of how to obtain the disk):

(1) decimate.c, which decimates the data using up to three stages of decimation;

(2) interpol.c, which interpolates the data using up to three stages of interpolation;

(3) moptimum.c, which determines the characteristics of an I-stage decimator (or interpolator) for $I = 1, 2, 3$, or 4 (the characteristics include decimation factor and filter characteristics for each stage, and the measures of efficiency (such as number of multiplications per second) for various configurations).

To limit the size of the book, only the first program, decimate.c, is listed here (see Program 8A.1). The prompts, responses, and output from the program when used to decimate the data described in Section 8.4.2 are given below. The reader should refer to Section 8.4.2 and Table 8.2 for further details on this example.

Program 8A.1

```
/* ......................................................................... *
*                                                                            *
*       Sample rate conversion - general, 3 stage decimation                 *
*       program performs 1, 2 or 3-stage decimation on data in a             *
*       user-specified data file and saves results in dout.dat               *
*                                                                            *
*       program name: decimate.c
*                                                                            *
*       Manny Ifeachor, 15.4.90                                              *
*                                                                            *
*                                                                            *
* ........................................................................... *
*/
#include     <stdio.h>
#include     <math.h>
#include     <dos.h>

#define npt      150

void    init_arrays();
void    decimator_spec();
void    read_coeffs();
void    decimator();

float   y1[npt], h1[npt], h2[npt], h3[npt];
float   x1[npt], x2[npt], x3[npt], xn;
long    n1, n2, n3, m1, m2, m3, m, mm;
long    count1, count2, count3, ndata2=0;
int     nstage;
FILE    *in, *out, *fopen();
char    din[30], fname1[30], fname2[30], fname3[30];
main()
{
        init_arrays();                          /* initialize coeffs and data arrays */
        decimator_spec();                       /* ask user for decimator specifications */
        read_coeffs();                          /* read decimator filter coefficients */

        while(fscanf(in,"%f",&xn)!=EOF){        /* read and decimate data */
              decimator();
              ++ndata2;
        }
        printf("number of data points in input file \t%ld \n",ndata2);
        printf("number of data points stored in output file \t%ld \n",mm);
        fclose(in);
        fclose(out);
        exit(0);
}
/* ......................................................................... */
void    init_arrays()
{
        long    i;
        for(i=0; i<npt; ++i){
```

```
                x1[i]=0;   h1[i]=0;
                x2[i]=0;   h2[i]=0;
                x3[i]=0;   h3[i]=0;
                y1[npt]=0;
        }
}
/* ........................................................................................................ */
void    decimator_spec()
{
        printf("enter name of file holding data to be decimated \n");
        scanf("%s",din);
        printf("the decimated data will be stored in dout.dat \n");
        printf("\n");
        printf("enter decimation factor and number of decimation stages\n");
        scanf("%ld %ld",&m,&nstage);
        if(nstage==1){
                printf("enter coefficient file name \n").
                scanf("%s",fname1);
                m1=m;
        }
        if(nstage==2){
                printf("enter coefficient filename and decimation factor for stage 1\n");
                scanf("%s %ld", fname1,&m1);
                printf("enter coefficient filename and decimation factor for stage 2\n");
                scanf("%s %ld", fname2,&m2);
        }
        if(nstage==3){
                printf("enter coefficient filename and decimation factor for stage 1\n");
                scanf("%s %ld", fname1,&m1);
                printf("enter coefficient filename and decimation factor for stage 2\n");
                scanf("%s %ld", fname2,&m2);
                printf("enter coefficient filename and decimation factor for stage 3\n");
                scanf("%s %ld", fname3,&m3);
        }
        count1=m1; count2=m2; count3=m3;
        mm=0;

}
/* ........................................................................................................ */
void    read_coeffs()
{

        n1=0; n2=0; n3=0;

        if((in=fopen(fname1,"r"))==NULL){
                printf("first coeff file not found\n");
                exit(1);
        }
        while(fscanf(in,"%f",&h1[n1])!=EOF){
                ++n1;
        }
        --n1;
        fclose(in);
        if(nstage>=2){
                if((in=fopen(fname2,"r"))==NULL){
```

```
                printf("second coeff file not found\n");
                exit(1);
            }
            while(fscanf(in,"%f",&h2[n2])!=EOF){
                ++n2;
            }
            --n2;
            fclose(in);
        }
        if(nstage==3){
            if((in=fopen(fname3,"r"))==NULL){
                printf("third coeff file not found\n");
                exit(1);
            }
            while(fscanf(in,"%f",&h3[n3])!=EOF){
                ++n3;
            }
            --n3;
            fclose(in);
        }
        if((in=fopen(din,"r"))==NULL){              /* open file holding input data */
            printf("error reading input data file\n");
            exit(1);
        }
        if((out=fopen("dout.dat","w"))==NULL){  /* open file to hold result */
            printf("cannot open output data file\n");
            exit(1);
        }
}
/* ........................................................................................ */
void    decimator()
{
        float       xnew, y;
        double      y1;
        long        j, k;
                                                    /* stage of decimation starts here */
        xnew=xn;
        for(j=1; j<=n1; ++j){                       /* update stage1 delay line */
            k=n1-j+1;
            x1[k]=x1[k-1];
        }
        x1[0]=xnew;
        count1--;
        if(count1>0)
            return;
        y1=0;
        for(j=0; j<=n1; ++j){                       /* 1st stage FIR filtering */
            y1=y1+h1[j]*x1[j];
        }
        count1=m1;
        if(nstage==1){
            y=y1;
            fprintf(out,"%f \n",y);
            ++mm;
            return;
```

```
        }
                                        /* second stage of decimation starts here */

        for(j=1; j<=n2; ++j){           /* stage2 delay line */
                k=n2-j+1;
                x2[k]=x2[k-1];
        }
        x2[0]=y1;
        count2--;
        if(count2>0)
                return;
        y1=0;
        for(j=0; j<=n2; ++j){           /* 2nd stage FIR filtering */
                y1=y1+h2[j]*x[j];
        }
        count2=m2;
        if(nstage==2){
                y=y1;
                fprint(out,"%f \n",y);
                ++mm;
                return;
        }
                                        /* third stage of decimation starts here */
        for(j=1; j<=n3; ++j){           /* updata stage3 delay line */
                k=n3-j+1;
                x3[k]=x3[k-1];
        }
        x3[0]=y1;
        count3--;
        if(count3>0)
                return;
        y1=0;
        for(j=0; j<=n3; ++j){           /* 3rd stage FIR filtering */
                y1=y1+h3[j]*x3[j];
        }
        count3=m3;
        y=y1;
        fprint(out,"%f \n",y);
        ++mm;
}
```

```
enter name of file holding data to be decimated
dtest.dat
the decimated data will be stored in dout.dat

enter decimation factor and number of decimation stages
10   2
enter coefficient filename and decimation factor for stage 1
tcoef1.dat 5
enter coefficient filename and decimation factor for stage 2
tcoef2.dat 2
number of data points in input file       60
number of data points stored in output file      6
```

9

Adaptive digital filters

9.1	When to use adaptive filters and where they have been used	542
9.2	Concepts of adaptive filtering	543
9.3	Basic Wiener filter theory	547
9.4	The basic LMS adaptive algorithm	550
9.5	Recursive least squares algorithm	557
9.6	Application example 1 – adaptive filtering of ocular artefacts from the human EEG	561
9.7	Application example 2 – adaptive telephone echo cancellation	563
9.8	Other applications	565
	References	569
	Bibliography	570
	Appendix	571

An adaptive filter is essentially a digital filter with self-adjusting characteristics. It adapts, automatically, to changes in its input signals. Adaptive filters are the central topic in the sub-area of DSP known as adaptive signal processing. This chapter describes key aspects of this important topic based on the LMS (least mean square) and RLS (recursive least squares) algorithms which are two of the most widely used algorithms in adaptive signal processing. The treatment is practical with only the essential theory included in the main text. C language implementations of a variety of LMS and RLS based adaptive filters are included in the appendix and on the PC disk for the book. A number of real-world applications are presented.

9.1 When to use adaptive filters and where they have been used

The contamination of a signal of interest by other unwanted, often larger, signals or noise is a problem often encountered in many applications. Where the signal and noise occupy fixed and separate frequency bands, conventional linear filters with fixed coefficients are normally used to extract the signal. However, there are many instances when it is necessary for the filter characteristics to be variable, adapted to changing signal characteristics, or to be altered intelligently. In such cases, the coefficients of the filter must vary and cannot be specified in advance. Such is the case where there is a spectral overlap between the signal and noise (see Figure 9.1) or if the band occupied by the noise is unknown or varies with time. Typical applications where fixed coefficient filters are inappropriate are the following.

(1) Electroencephalography (EEG), where artefacts or signal contamination produced by eye movements or blinks is much larger than the genuine electrical activity of the brain and shares the same frequency band with signals of clinical interest. It is not possible to use conventional linear filters to remove the artefacts while preserving the signals of clinical interest.
(2) Digital communication using a spread spectrum, where a large jamming signal, possibly intended to disrupt communication, could interfere with the desired signal. The interference often occupies a narrow but unknown band within the wideband spectrum, and can only be effectively dealt with adaptively.
(3) In digital data communication over the telephone channel at a high rate. Signal distortions caused by the poor amplitude and phase response characteristics of the channel lead to pulses representing different digital codes to interfere with each other (intersymbol interference), making it difficult to detect the codes reliably at the receiving end. To compensate for the channel distortions which may be varying with time or of unknown characteristics at the receiving end, adaptive equalization is used.

An adaptive filter has the property that its frequency response is adjustable or modifiable automatically to improve its performance in accordance with some criterion, allowing the filter to adapt to changes in the input signal characteristics. Because of their self-adjusting performance and in-built flexibility, adaptive filters have found use in many diverse applications such as telephone echo cancelling, radar signal processing, navigational systems, equalization of communication channels, and biomedical signal enhancement.

In summary we use adaptive filters

- when it is necessary for the filter characteristics to be variable, adapted to changing conditions,

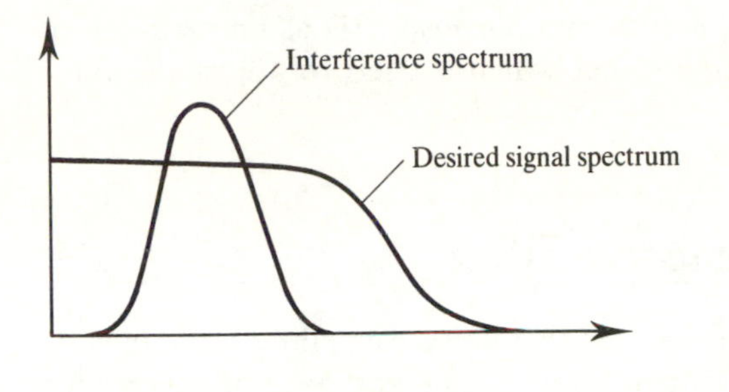

Figure 9.1 An illustration of spectral overlap between a signal and a strong interference.

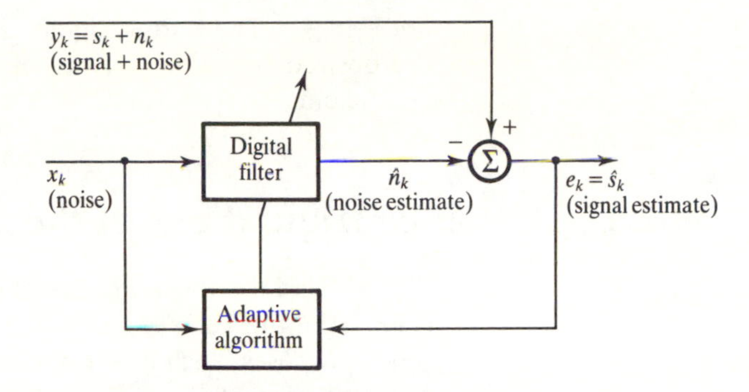

Figure 9.2 Block diagram of an adaptive filter as a noise canceller.

- when there is spectral overlap between the signal and noise (see Figure 9.1), or
- if the band occupied by the noise is unknown or varies with time.

The use of conventional filters in the above cases would lead to unacceptable distortion of the desired signal. There are many other situations, apart from noise reduction, when the use of adaptive filters is appropriate (see later).

9.2 Concepts of adaptive filtering

9.2.1 Adaptive filters as a noise canceller

An adaptive filter consists of two distinct parts: a digital filter with adjustable coefficients, and an adaptive algorithm which is used to adjust or modify the coefficients of the filter (Figure 9.2). Two input signals, y_k and x_k, are applied simultaneously to the adaptive filter. The signal y_k is the contaminated signal containing both the desired signal, s_k, and the noise, n_k, assumed uncorrelated with each other. The signal, x_k, is a measure of the contaminating signal which is correlated in some way with n_k. x_k is processed by the digital filter to produce an estimate, $\hat{n}_k$, of n_k. An estimate of the desired signal is then obtained by subtracting the digital filter output, $\hat{n}_k$, from the contaminated signal, y_k:

$$\hat{s}_k = y_k - \hat{n}_k = s_k + n_k - \hat{n}_k \tag{9.1}$$

The main objective in noise cancelling is to produce an optimum estimate of the noise in the contaminated signals and hence an optimum estimate of the desired signal. This is achieved by using $\hat{s}_k$ in a feedback arrangement to adjust the digital filter coefficients, via a suitable adaptive algorithm, to minimize the noise in $\hat{s}_k$. The output signal, $\hat{s}_k$, serves two purposes: (i) as an estimate of the desired signal and (ii) as an error signal which is used to adjust the filter coefficients.

9.2.2 Other configurations of the adaptive filter

The discussions above are based on the adaptive noise cancelling principles. It is important to keep in mind that adaptive filters can be and have been used for other purposes, such as for linear prediction, adaptive signal enhancement and adaptive control. In general, the meaning of the signal x_k, y_k, and e_k or the way they are derived are application dependent, a fact which should be borne in mind. Figure 9.3 shows different configurations of the adaptive filter.

9.2.3 Main components of the adaptive filter

In most adaptive systems, the digital filter in Figure 9.2 is realized using a transversal or finite impulse response (FIR) structure (Figure 9.4). Other forms are sometimes used, for example the infinite impulse response (IIR) or the lattice structures, but the FIR structure is the most widely used because of its simplicity and guaranteed stability. For the N-point filter depicted in Figure 9.4, the output is given by

$$\hat{n}_k = \sum_{i=0}^{N-1} w_k(i) x_{k-i} \tag{9.2}$$

where $w_k(i)$, $i = 0, 1, \ldots$, are the adjustable filter coefficients (or weights), and $x_k(i)$ and $\hat{n}_k$ are the input and output of the filter. Figure 9.4 illustrates the single-input, single-output system. In a multiple-input single-output system, the x_k may be simultaneous inputs from N different signal sources.

9.2.4 Adaptive algorithms

Adaptive algorithms are used to adjust the coefficients of the digital filter (in Figure 9.2) such that the error signal, e_k, is minimized according to some criterion, for example in the least squares sense. Common algorithms that have found widespread application are the least mean square (LMS), the recursive least squares (RLS), and the Kalman filter algorithms. In terms of computation and storage requirements, the LMS algorithm is the most efficient. Further, it does not suffer from the numerical instability problem inherent in the other two algorithms. For these reasons, the LMS algorithm has become the algorithm of first choice in many applications. However, the RLS algorithm has superior convergence properties.

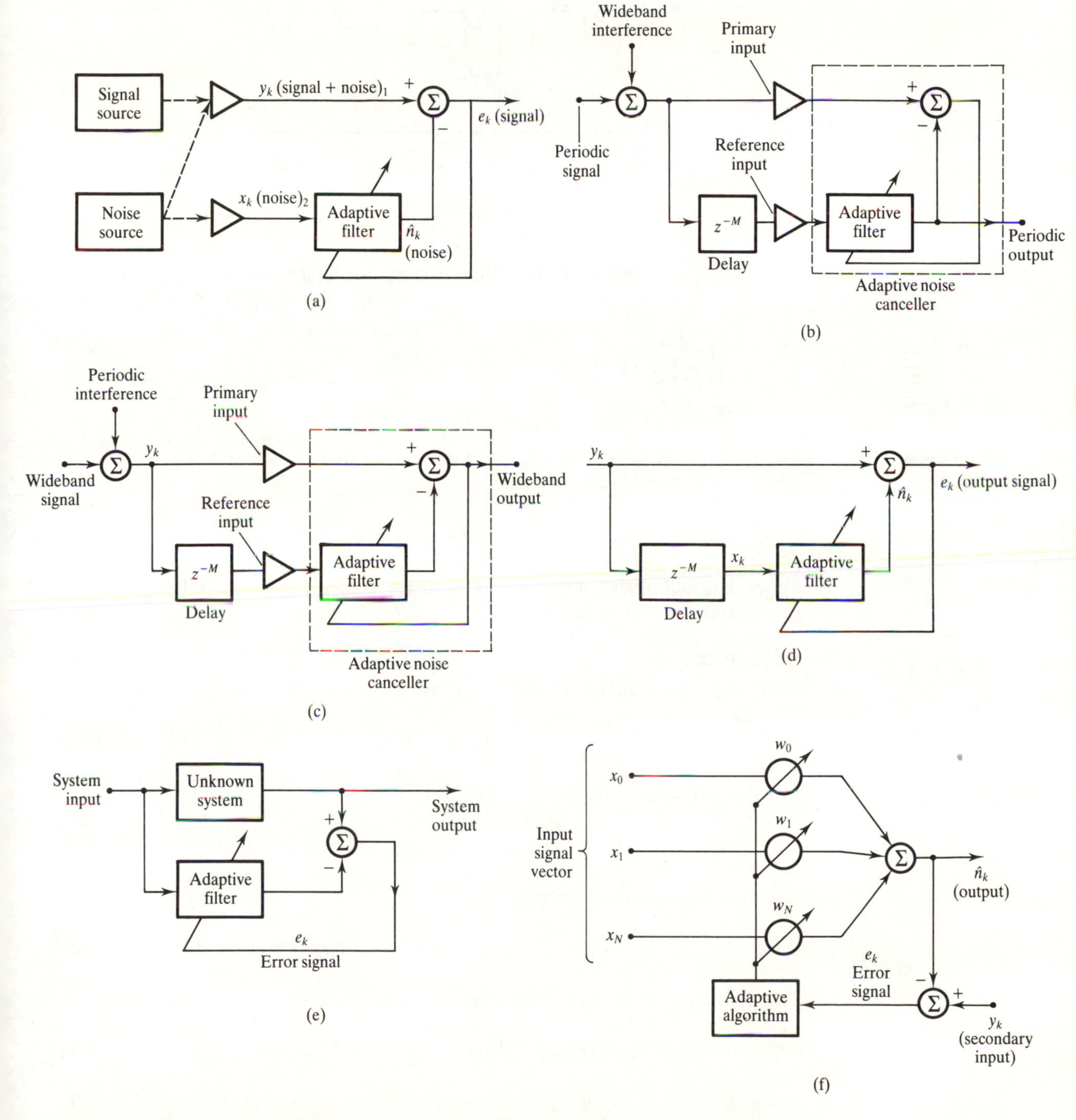

Figure 9.3 Some configurations of the adaptive filter (after Widrow and Winter, 1988): (a) adaptive noise canceller; (b) adaptive self-tuning filter; (c) cancelling periodic interference without an external reference source; (d) adaptive line enhancer; (e) system modelling; (f) linear combiner.

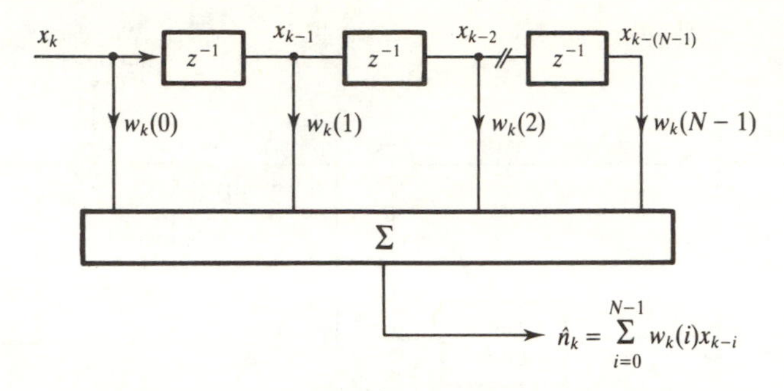

Figure 9.4 Finite impulse response filter structure.

Example 9.1

The estimate of the desired signal at the output of an adaptive noise canceller is given by (Widrow *et al.*, 1975a)

$$\hat{s}_k = y_k - \hat{n}_k = s_k + n_k - \hat{n}_k$$

Show that minimizing the total power at the output of the canceller maximizes the output signal-to-noise ratio.

Solution

Now, the contaminated signal is given by

$$y_k = s_k + n_k \tag{9.3}$$

and the estimate of the desired signal is given by

$$\hat{s}_k = y_k - \hat{n}_k = s_k + n_k - \hat{n}_k \tag{9.4}$$

Squaring Equation 9.4 we have:

$$\hat{s}_k^2 = s_k^2 + (n_k - \hat{n}_k)^2 + 2s_k(n_k - \hat{n}_k) \tag{9.5}$$

Taking the expectations of both sides of Equation 9.5,

$$E[\hat{s}_k^2] = E[s_k^2] + E[(n_k - \hat{n}_k)^2] + 2E[s_k(n_k - \hat{n}_k)] \tag{9.6}$$

Since the desired signal, s_k, is uncorrelated with n_k or with $\hat{n}_k$ the last term in Equation 9.6 is zero and we have

$$E[\hat{s}_k^2] = E[s_k^2] + E[(n_k - \hat{n}_k)^2] \tag{9.7}$$

where $E[s_k^2]$ represents the total signal power, $E[\hat{s}_k^2]$ represents the estimate of the signal power (it also represents the total output power) and $E[(n_k - \hat{n}_k)^2]$ represents the remnant noise power which may still be in s_k. It is evident in Equation 9.7 that if the estimate $\hat{n}_k$ is the exact replica of n_k, the output power will contain only the signal power. By adjusting the adaptive filter towards the optimum position, the remnant noise power and hence the total output power are minimized. The desired signal power is unaffected by this adjustment since s_k is uncorrelated with n_k. Thus

$$\min E[\hat{s}_k^2] = E[s_k^2] + \min E[(n_k - \hat{n}_k)^2] \tag{9.8}$$

It is clear in Equation 9.8 that the net effect of minimizing the total output power is to maximize the output signal-to-noise ratio. When the filter setting is such that $\hat{n}_k = n_k$, then $\hat{s}_k = s_k$. In this case, the output of the adaptive noise canceller is noise free. When the signal y_k contains no noise, that is when $n_k = 0$, the adaptive filter turns itself off (in theory at least) by setting all the weights to zero.

9.3 Basic Wiener filter theory

Many adaptive algorithms can be viewed as approximations of the discrete Wiener filter (Figure 9.5). Two signals, x_k and y_k, are applied simultaneously to the filter. Typically, y_k consists of a component that is correlated with x_k and another that is not. The Wiener filter produces an optimal estimate of the part of y_k that is correlated with x_k which is then subtracted from y_k to yield e_k.

Assuming an FIR filter structure with N coefficients (or weights – the popular phrase in the literature), the error, e_k, between the Wiener filter output and the primary signal, y_k, is given by

$$e_k = y_k - \hat{n}_k = y_k - \mathbf{W}^{\mathrm{T}}\mathbf{X}_k = y_k - \sum_{i=0}^{N-1} w(i)x_{k-i} \tag{9.9}$$

where $\mathbf{X}_k$ and $\mathbf{W}$, the input signal vector and weight vector, respectively, are given by

$$\mathbf{X}_k = \begin{bmatrix} x_k \\ x_{k-1} \\ \vdots \\ x_{k-(N-1)} \end{bmatrix} \quad \mathbf{W} = \begin{bmatrix} w(0) \\ w(1) \\ \vdots \\ w(N-1) \end{bmatrix} \tag{9.10}$$

The square of the error is given as

$$e_k^2 = y_k^2 - 2y_k\mathbf{X}_k^{\mathrm{T}}\mathbf{W} + \mathbf{W}^{\mathrm{T}}\mathbf{X}_k\mathbf{X}_k^{\mathrm{T}}\mathbf{W} \tag{9.11}$$

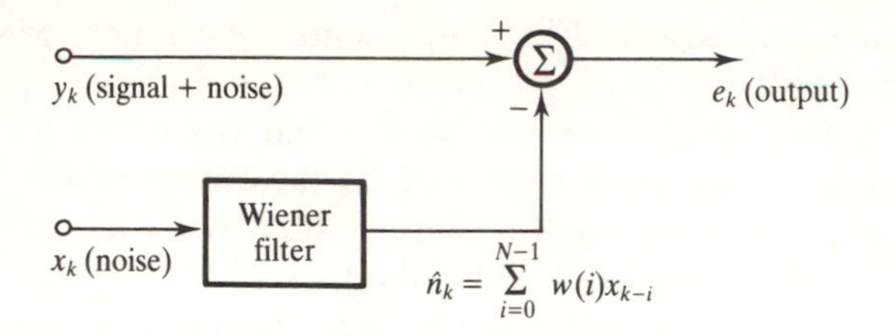

Figure 9.5 The basic Wiener filter.

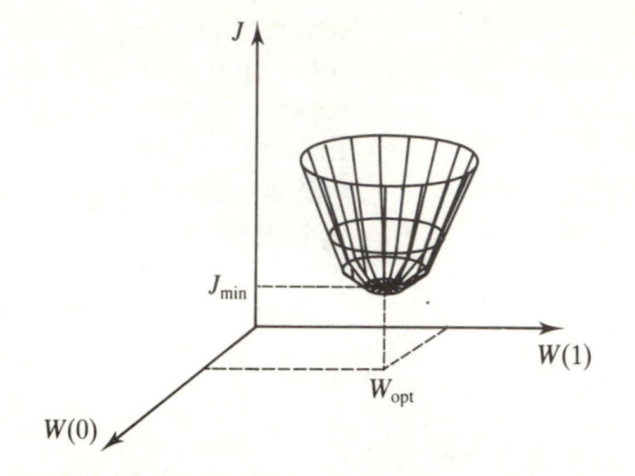

Figure 9.6 Error–performance surface.

The mean square error (MSE), J, is obtained by taking the expectations of both sides of Equation 9.11, assuming that the input vector $\mathbf{X}_k$ and the signal y_k are jointly stationary:

$$J = E[e_k^2] = E[y_k^2] - 2E[y_k\mathbf{X}_k^T\mathbf{W}] + E[\mathbf{W}^T\mathbf{X}_k\mathbf{X}_k^T\mathbf{W}]$$
$$= \sigma^2 + 2\mathbf{P}^T\mathbf{W} + \mathbf{W}^T\mathbf{R}\mathbf{W} \quad \textbf{(9.12)}$$

where $E[\cdot]$ symbolizes expectation, $\sigma^2 = E[y_k^2]$ is the variance of y_k, $\mathbf{P} = E[y_k\mathbf{X}_k]$ is the N length cross-correlation vector and $\mathbf{R} = E[\mathbf{X}_k\mathbf{X}_k^T]$ is the $N \times N$ autocorrelation matrix. A plot of the MSE against the filter coefficients, $\mathbf{W}$, is bowl shaped with a unique bottom (see Figure 9.6). This figure is known as the performance surface and is non-negative. The gradient of the performance surface is given by

$$\nabla = \frac{dJ}{d\mathbf{W}} = -2\mathbf{P} + 2\mathbf{R}\mathbf{W} \quad \textbf{(9.13)}$$

Each set of coefficients, $w(i)$ $(i = 0, 1, \ldots, N-1)$, corresponds to a point on the surface. At the minimum point of the surface, the gradient is zero and the filter weight vector has its optimum value, $\mathbf{W}_{opt}$ (see Example 9.2):

$$\mathbf{W}_{opt} = \mathbf{R}^{-1}\mathbf{P} \quad \textbf{(9.14)}$$

Equation 9.14 is known as the Wiener–Hopf equation or solution. The task in adaptive filtering is to adjust the filter weights, $w(0)$, $w(1)$, . . ., using a suitable algorithm, to find the optimum point on the performance surface.

The Wiener filter has a limited practical usefulness because

- it requires the autocorrelation matrix, $\mathbf{R}$, and the cross-correlation vector, $\mathbf{P}$, both of which are not known *a priori*,
- it involves matrix inversion, which is time consuming, and

- if the signals are nonstationary, then both **R** and **P** will change with time and so $\mathbf{W}_{\text{opt}}$ will have to be computed repeatedly.

For real-time application, a way of obtaining $\mathbf{W}_{\text{opt}}$ on a sample-by-sample basis is required. Adaptive algorithms are used to achieve this without having to compute **R** and **P** explicitly or performing a matrix inversion.

Example 9.2

Starting with the equation for the mean square error (Equation 9.12), derive the Wiener–Hopf equation.

Solution

The MSE is given by

$$\text{MSE} = J = \sigma^2 + 2\mathbf{P}^{\mathrm{T}}\mathbf{W} + \mathbf{W}^{\mathrm{T}}\mathbf{R}\mathbf{W} \tag{9.15}$$

The gradient, ∇, of the MSE is obtained by differentiating the MSE with respect to the weight vector **W**, and setting the result to zero (Haykin, 1986):

$$\nabla = \frac{\mathrm{d}J}{\mathrm{d}\mathbf{W}} = \frac{\mathrm{d}\sigma^2}{\mathrm{d}\mathbf{W}} + \frac{\mathrm{d}(\mathbf{P}^{\mathrm{T}}\mathbf{W})}{\mathrm{d}\mathbf{W}} + \frac{\mathrm{d}(\mathbf{W}^{\mathrm{T}}\mathbf{R}\mathbf{W})}{\mathrm{d}\mathbf{W}} \tag{9.16}$$

Now,

$$\frac{\mathrm{d}\sigma^2}{\mathrm{d}\mathbf{W}} = \mathbf{0}$$

$$\frac{\mathrm{d}(2\mathbf{P}^{\mathrm{T}}\mathbf{W})}{\mathrm{d}\mathbf{W}} = -2\mathbf{P}$$

$$\frac{\mathrm{d}(\mathbf{W}^{\mathrm{T}}\mathbf{R}\mathbf{W})}{\mathrm{d}\mathbf{W}} = 2\mathbf{R}\mathbf{W}$$

Using these results, and setting $\nabla = \mathbf{0}$, Equation 9.16 becomes

$$\nabla = \frac{\mathrm{d}J}{\mathrm{d}\mathbf{W}} = -2\mathbf{P} + 2\mathbf{R}\mathbf{W} = \mathbf{0} \tag{9.17}$$

The optimum coefficient vector is then given by

$$\mathbf{W}_{\text{opt}} = \mathbf{R}^{-1}\mathbf{P} \tag{9.18}$$

9.4 The basic LMS adaptive algorithm

One of the most successful adaptive algorithms is the LMS algorithm developed by Widrow and his coworkers (Widrow *et al.*, 1975a). Instead of computing $\mathbf{W}_{opt}$ in one go as suggested by Equation 9.18, in the LMS the coefficients are adjusted from sample to sample in such a way as to minimize the MSE. This amounts to descending along the surface of Figure 9.6 towards its bottom.

The LMS is based on the steepest descent algorithm where the weight vector is updated from sample to sample as follows:

$$\mathbf{W}_{k+1} = \mathbf{W}_k - \mu \boldsymbol{\nabla}_k \tag{9.19}$$

where $\mathbf{W}_k$ and $\boldsymbol{\nabla}_k$ are the weight and the true gradient vectors, respectively, at the kth sampling instant. μ controls the stability and rate of convergence.

The steepest descent algorithm in Equation 9.19 still requires knowledge of $\mathbf{R}$ and $\mathbf{P}$, since $\boldsymbol{\nabla}_k$ is obtained by evaluating Equation 9.17. The LMS algorithm is a practical method of obtaining estimates of the filter weights $\mathbf{W}_k$ in real time without the matrix inversion in Equation 9.18 or the direct computation of the autocorrelation and cross-correlation. The Widrow–Hopf LMS algorithm for updating the weights from sample to sample is given by

$$\mathbf{W}_{k+1} = \mathbf{W}_k + 2\mu e_k \mathbf{X}_k \tag{9.20a}$$

where:

$$e_k = y_k - \mathbf{W}_k^{\mathrm{T}} \mathbf{X}_k \tag{9.20b}$$

Clearly, the LMS algorithm above does not require prior knowledge of the signal statistics (that is the correlations $\mathbf{R}$ and $\mathbf{P}$), but instead uses their instantaneous estimates (see Example 9.3). The weights obtained by the LMS algorithm are only estimates, but these estimates improve gradually with time as the weights are adjusted and the filter learns the characteristics of the signals. Eventually, the weights converge. The condition for convergence is:

$$0 < \mu > 1/\lambda_{max} \tag{9.21}$$

where λ_{max} is the maximum eigenvalue of the input data covariance matrix. In practice, $\mathbf{W}_k$ never reaches the theoretical optimum (the Wiener solution), but fluctuates about it (see Figure 9.7).

9.4.1 Implementation of the basic LMS algorithm

The computational procedure for the LMS algorithm is summarized below.

(1) Initially, set each weight $w_k(i)$, $i = 0, 1, \ldots, N-1$, to an arbitrary fixed value, such as 0.

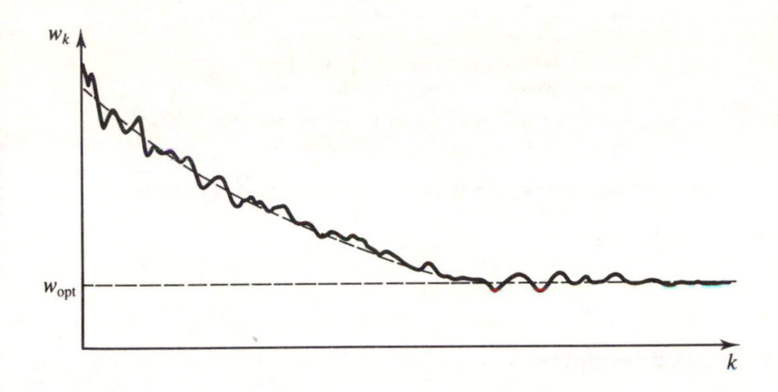

Figure 9.7 An illustration of the variations in the filter weights.

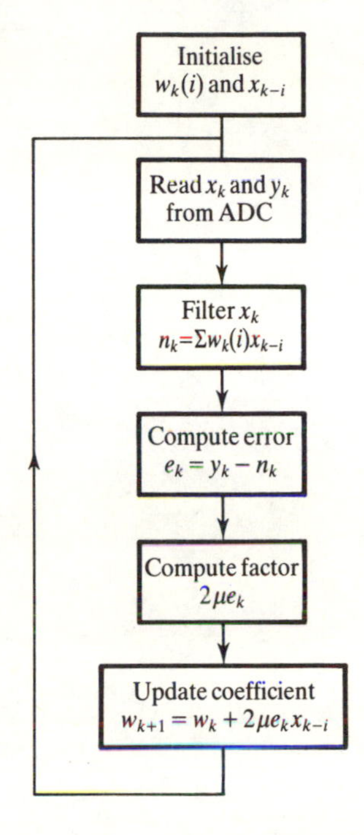

Figure 9.8 Flowchart for the LMS adaptive filter.

For each subsequent sampling instants, $k = 1, 2, \ldots$, carry out steps (2) to (4) below:

(2) compute filter output

$$\hat{n}_k = \sum_{i=0}^{N-1} w_k(i)x_{k-i}$$

(3) compute the error estimate

$$e_k = y_k - \hat{n}_k$$

(4) update the next filter weights

$$w_{k+1}(i) = w_k(i) + 2\mu e_k x_{k-i}$$

The simplicity of the LMS algorithm and ease of implementation, evident from above, make it the algorithm of first choice in many real-time systems. The LMS algorithm requires approximately $2N + 1$ multiplications and $2N + 1$ additions for each new set of input and output samples. Most signal processors are suited to the mainly multiply–accumulate arithmetic operations involved, making a direct implementation of the LMS algorithm attractive.

The flowchart for the LMS algorithm is given in Figure 9.8. Figures 9.9

```
Inputs:     xk(i)   vector of the latest input samples
            yk      current contaminated signal sample
            wk(i)   vector of filter coefficients
Outputs:    ek      current desired output (or error) sample
            wk(i)   vector of updated filter coefficients

      /* compute the current error estimate       */

            ek = yk
            for i=1 to N do
               ek = ek − xk(i) * wk(i)
            end

      /* update filter coefficients                */

            gk = 2u * ek
            for i = 1 to N do
               wk(i) = wk(i) + xk(i) * gk
            end

            return
```

Figure 9.9 Coding of the LMS adaptive filter.

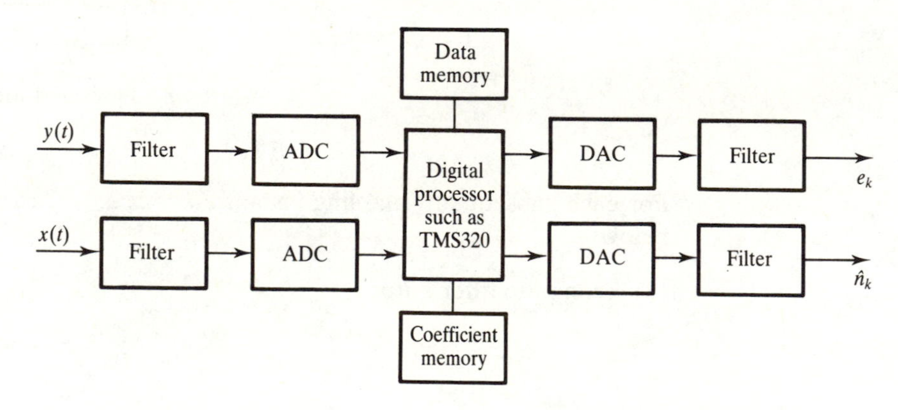

Figure 9.10 Hardware implementation for real-time LMS adaptive filtering.

and 9.10, respectively, show a pseudocode for the software and hardware implementations. A C language implementation of the LMS algorithm is given in the appendix.

Example 9.3

Starting with the steepest descent algorithm

$$\mathbf{W}_{k+1} = \mathbf{W}_k - \mu \nabla_k$$

where $\mathbf{W}_k$ is the filter weight vector at the kth sampling instant, μ controls stability and rate of convergence and $\boldsymbol{\nabla}_k$ is the true gradient of the error–performance surface, derive the Widrow–Hopf LMS algorithm for adaptive noise cancelling, stating any reasonable assumptions made.

Solution

The steepest descent algorithm is given by

$$\mathbf{W}_{k+1} = \mathbf{W}_k - \mu\boldsymbol{\nabla}_k \tag{9.22}$$

The gradient vector, $\boldsymbol{\nabla}$, the cross-correlation between the primary and secondary inputs, $\mathbf{P}$, and the autocorrelation of the primary input, $\mathbf{R}$, are related as

$$\boldsymbol{\nabla} = -2\mathbf{P} + 2\mathbf{RW} \tag{9.23}$$

In the LMS algorithm, instantaneous estimates are used for $\boldsymbol{\nabla}$. Thus

$$\begin{aligned}\boldsymbol{\nabla}_k &= -2\mathbf{P}_k + 2\mathbf{R}_k\mathbf{W}_k = -2\mathbf{X}_k y_k + 2\mathbf{X}_k\mathbf{X}_k^{\mathrm{T}}\mathbf{W}_k \\ &= -2\mathbf{X}_k(y_k - \mathbf{X}_k^{\mathrm{T}}\mathbf{W}_k) = -2e_k\mathbf{X}_k\end{aligned} \tag{9.24}$$

where

$$e_k = y_k - \mathbf{X}_k^{\mathrm{T}}\mathbf{W}_k$$

Substituting Equation 9.24 in the equation for the steepest descent algorithm we have the basic Widrow–Hopf LMS algorithm:

$$\mathbf{W}_{k+1} = \mathbf{W}_k + 2\mu e_k\mathbf{X}_k \tag{9.25a}$$

where

$$e_k = y_k - \mathbf{W}_k^{\mathrm{T}}\mathbf{X}_k \tag{9.25b}$$

9.4.2 Practical limitations of the basic LMS algorithm

In practice, several practical problems are encountered when using the basic LMS algorithm, leading to a lowering of performance. Some of the more important problems are discussed here.

Effects of nonstationarity

In a stationary environment, the error performance surface of the filter has a constant shape and orientation, and the adaptive filter merely converges to and operates at or near the optimum point. If the signal statistics change after the weights have converged, the filter responds to the change by re-adjusting its

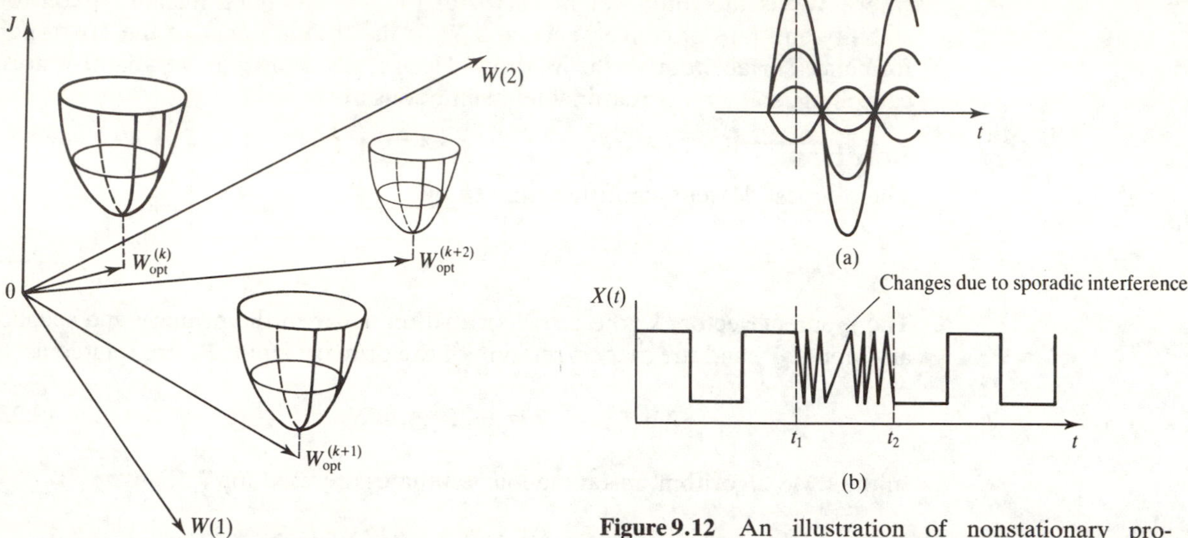

Figure 9.11 Time-varying error–performance surface.

Figure 9.12 An illustration of nonstationary processes: (a) modulated waveform; (b) sporadic interference.

weights to a new set of optimal values, provided that the change in signal statistics is sufficiently slow for the filter to converge between changes. In a nonstationary environment, however, the bottom or minimum point continually moves, and its orientation and curvature may also be changing (see Figure 9.11). Thus the algorithm in this case has the task not only of seeking the minimum point of the surface but also of tracking the changing position, leading to a significant lowering of performance. (Note that a variable is said to be stationary if its statistics (such as mean, variance, autocorrelation) change with time. Such changes can result from, for example, sudden changes due to sporadic interference of short duration (Figure 9.12) or bad data, and often upset the filter weights.)

A number of schemes have been developed to overcome this problem but these in general tend to increase the complexity of the basic LMS algorithm. One such scheme is the time-sequenced adaptive filter (Ferrara and Widrow, 1981).

Effects of signal component on the interference input channel

The performance of the algorithm relies on the measured interference signal, $x_k(i)$, being highly correlated with the actual interference, but weakly correlated (theoretically zero) with the desired signal. In most cases, this condition is not met. In some applications, the contaminating input may contain both the undesired interference as well as low level signal components. This leads to a cancellation of some of the desired signal components. Such a situation is illustrated in Figure 9.13. It is shown in Widrow *et al.* (1975a) that the adaptive noise cancelling process still leads to a significant improvement in the desired signal-to-noise ratio in these cases but only at the expense of a small signal

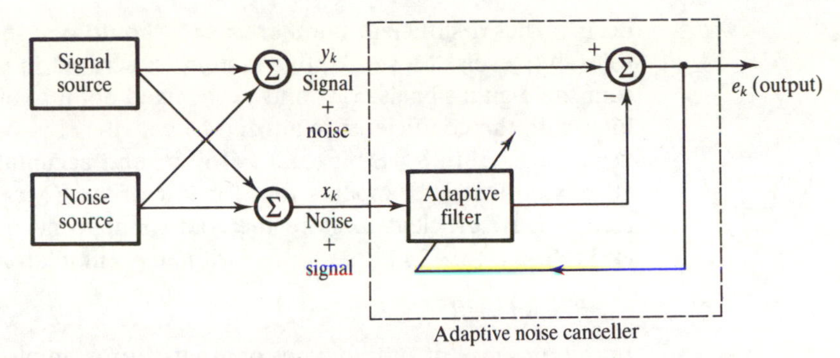

Figure 9.13 Adaptive noise cancelling with some signal components in both the desired signal and interference input channels.

distortion. However, if x_k contains only signals and no noise component whatsoever, the desired signal in y_k may be completely obliterated. Our work in biomedical signal processing confirms their results (Ifeachor *et al*. 1986).

Computer wordlength requirements

The LMS-based FIR adaptive filter is characterized by the following equations: for the digital filter,

$$\hat{n}_k = \sum_{i=0}^{N-1} w_k(i)x_{k-i} \tag{9.26a}$$

for the adaptive algorithm,

$$\mathbf{W}_{k+1} = \mathbf{W}_k + 2\mu e_k \mathbf{X}_k \tag{9.26b}$$

where

$$e_k = y_k - \mathbf{W}_k^{\mathrm{T}}\mathbf{X}_k$$

When adaptive filters are implemented in the real world, the filter weights, w_k, and the input variables, x_k and y_k, are of necessity represented by a finite number of bits. Similarly, the numerical operations involved are carried out using a finite precision arithmetic. The recursive nature of the LMS algorithm means that the wordlength will grow without limit and so some of the bits must be discarded before each updated weight is stored. Thus the y_k, e_k and $w_k(i)$ may differ significantly from their true values. The use of filter weights and results of arithmetic operations with limited accuracy may introduce errors into the adaptive filter whose effects may include (i) possible nonconvergence of the adaptive filter to the optimal solution, leading to an inferior performance. For example, if the filter is used as an interference canceller some residual interference may remain, (ii) the filter outputs may contain noise which will cause it to fluctuate randomly, and (iii) a premature termination of the algorithm may occur. Thus a sufficient number of bits should be used to keep these errors at tolerable levels. Most adaptive systems described in the open literature represent the digital signals, x_{k-i} and y_k, as fixed point numbers of between 8 and 16 bits, with the coefficients quantized to between 16 and 24 bits. The multipliers used range from 8×8 bit to 24×16 bit, and accumulators of between 16 and 40 bits are used. It appears that for low order filters (up to about 100 coefficients) it is sufficient to store the coefficient to no more than 16-bit accuracy and to use a 16×16 bit multiplier with an accumulator of length 32 bits.

Coefficient drift

In the presence of certain types of inputs (for example narrowband signals), the filter coefficients may drift from the optimum values and grow slowly, eventually exceeding the permissible wordlength. This is an inherent problem in the LMS algorithm and leads to a long-term degradation in performance. In practice, coefficient drift is counteracted by introducing a leakage factor which gently nudges the coefficients towards zero. Two such schemes are given in Equations 9.27:

$$w_{k+1}(i) = \delta w_k(i) + 2\mu e_k x_{k-i} \qquad 0 < \delta < 1 \tag{9.27a}$$

$$w_{k+1}(i) = w_k(i) + 2\mu e_k x_{k-i} \pm \delta \qquad 0 < \delta < 1 \tag{9.27b}$$

Small δ, the leakage factor, ensures that drift is contained, but introduces bias in the error term, e_k.

The usefulness of the basic LMS algorithm has been extended by more sophisticated LMS-based algorithms as mentioned before. These include

(1) the complex LMS algorithm which allows the handling of complex data,
(2) the block LMS algorithm which offers substantial computational advantages and in some cases faster convergence, and
(3) time-sequenced LMS algorithms to deal with particular types of nonstationarity.

9.4.3 Other LMS-based algorithms

Complex LMS algorithm

The complex LMS algorithm for updating the filter weights is given by (Widrow *et al* ., 1975b)

$$\widetilde{\mathbf{W}}_{k+1} = \widetilde{\mathbf{W}}_k + 2\mu \tilde{e}_k \widetilde{\mathbf{X}}_{k-i} \tag{9.28}$$

where the symbol $\sim$ denotes a complex variable. The GEC–Plessey PDSP16000 processors (GEC–Plessey, 1990) are ideally suited to the complex LMS algorithm as they can perform arithmetic operations directly on complex data, which is a distinct advantage over conventional processors.

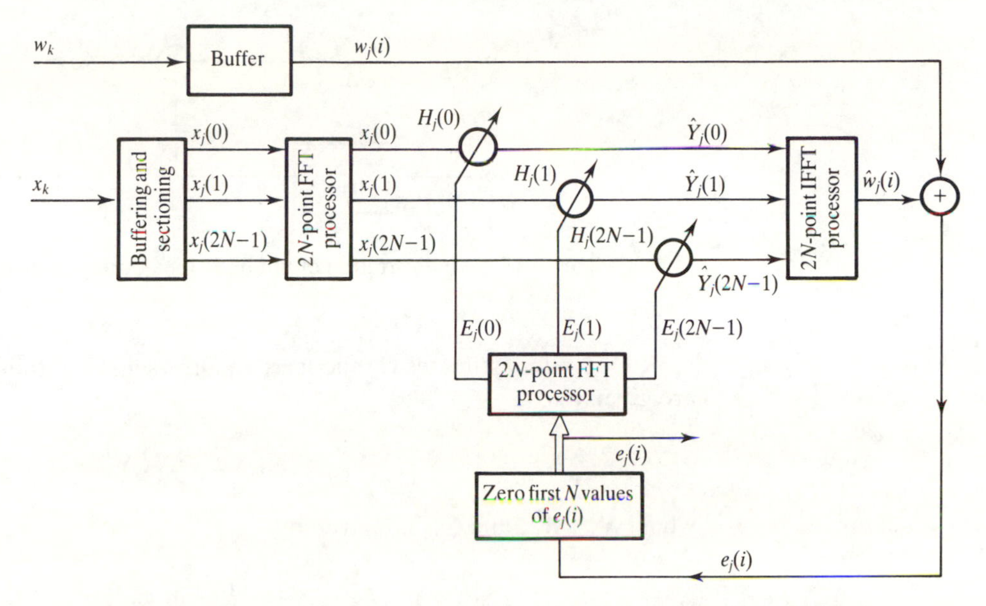

Figure 9.14 Simplified block diagram of a frequency domain LMS filter.

Fast LMS algorithms

A number of block LMS algorithms have been proposed which offer substantial computational savings especially when the number of filter coefficients is large. The computational savings result from processing the data in blocks instead of one sample at a time. Frequency domain implementations of the block LMS exploit the computational advantages of the fast Fourier transform (FFT) in performing convolutions (Mansour and Gray, 1982).

An efficient frequency domain filter is depicted in Figure 9.14.

9.5 Recursive least squares algorithm

The RLS algorithm is based on the well-known least squares method (Figure 9.15). An output signal, y_k, is measured at the discrete time, k, in response to a set of input signals, $x_k(i)$, $i = 1, 2, \ldots, n$. The input and output signals are related by the simple regression model

$$y_k = \sum_{i=0}^{n-1} w(i)x_k(i) + e_k \tag{9.29}$$

where e_k represents measurement errors or other effects that cannot be accounted for, and $w(i)$ represents the proportion of the ith input that is contained in the primary signal, y_k. The problem in the LS method is, given the $x_k(i)$ and y_k above, to obtain estimates of $w(0)$ to $w(n-1)$.

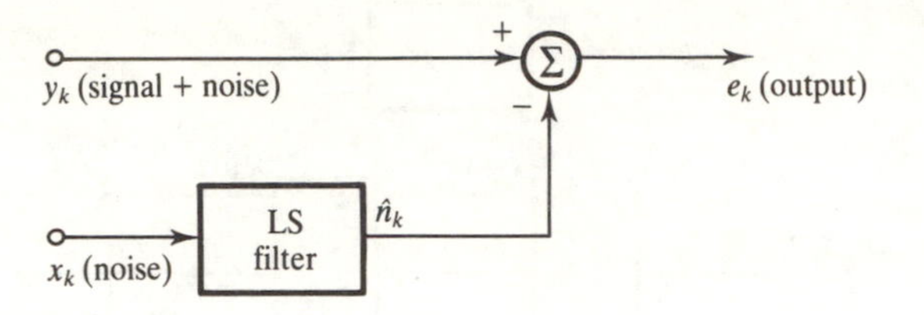

Figure 9.15 An illustration of the basic idea of the least-squares method.

Optimum estimates (in the least squares sense) of the filter weights, $w(i)$, are given by

$$\mathbf{W}_m = [\mathbf{X}_m^{\mathrm{T}}\mathbf{X}_m]^{-1}\mathbf{X}_m^{\mathrm{T}}\mathbf{Y}_m \tag{9.30}$$

where $\mathbf{Y}_m$, $\mathbf{W}_m$ and $\mathbf{X}_m$ are given by

$$\mathbf{Y}_m = \begin{bmatrix} y_0 \\ y_1 \\ y_2 \\ \vdots \\ y_{m-1} \end{bmatrix} \quad \mathbf{X}_m = \begin{bmatrix} \mathbf{x}^{\mathrm{T}}(0) \\ \mathbf{x}^{\mathrm{T}}(1) \\ \mathbf{x}^{\mathrm{T}}(2) \\ \vdots \\ \mathbf{x}^{\mathrm{T}}_{(m-1)} \end{bmatrix} \quad \mathbf{W}_m = \begin{bmatrix} w(0) \\ w(1) \\ w(2) \\ \vdots \\ w(n-1) \end{bmatrix}$$

$$\mathbf{x}^{\mathrm{T}}(k) = [x_k(0)\ x_k(1)\ \ldots\ x_k(n-1)], \quad k = 0, 1, \ldots, m-1$$

The suffix m indicates that each matrix above is obtained using all m data points and T indicates transposition. Equation 9.30 gives the OLS estimate of $\mathbf{W}_m$ which can be obtained using any suitable matrix inversion technique. The filter output is then obtained as

$$\hat{n}_k = \sum_{i=0}^{n-1} \hat{w}(i)x_{k-i}, \quad k = 1, 2, \ldots, m \tag{9.31}$$

9.5.1 Recursive least squares algorithm

The computation of $\mathbf{W}_m$ in Equation 9.30 requires the time-consuming computation of the inverse matrix. Clearly, the LS method above is not suitable for real-time or on-line filtering. In practice, when continuous data is being acquired and we wish to improve our estimate of $\mathbf{W}_m$ using the new data, recursive methods are preferred. With the recursive least squares algorithm the estimates of $\mathbf{W}_m$ can be updated for each new set of data acquired without repeatedly solving the time-consuming matrix inversion directly.

A suitable RLS algorithm can be obtained by exponentially weighting the data to remove gradually the effects of old data on $\mathbf{W}_m$ and to allow the tracking of slowly varying signal characteristics. Thus

$$\mathbf{W}_k = \mathbf{W}_{k-1} + \mathbf{G}_k e_k \tag{9.32a}$$

$$\mathbf{P}_k = \frac{1}{\gamma}[\mathbf{P}_{k-1} - \mathbf{G}_k \mathbf{x}^{\mathrm{T}}(k)\mathbf{P}_{k-1}] \tag{9.32b}$$

where

$$\mathbf{G}_k = \frac{\mathbf{P}_{k-1}\mathbf{x}(k)}{\alpha_k}$$

$$e_k = y_k - \mathbf{x}^{\mathrm{T}}(k)\mathbf{W}_{k-1}$$

$$\alpha_k = \gamma + \mathbf{x}^{\mathrm{T}}(k)\mathbf{P}_{k-1}\mathbf{x}(k)$$

$\mathbf{P}_k$ is essentially a recursive way of computing the inverse matrix $[\mathbf{X}_k^{\mathrm{T}}\mathbf{X}_k]^{-1}$.

The argument k emphasizes the fact that the quantities are obtained at each sample point. γ is referred to as the forgetting factor. This weighting scheme reduces to that of the LS when $\gamma = 1$. Typically, γ is between 0.98 and 1. Smaller values assign too much weight to the more recent data, which leads to wildly fluctuating estimates. The number of previous samples that significantly contribute to the value of $\mathbf{W}_k$ at each sample point is called the asymptotic sample length (ASL) given by

$$\sum_{k=0}^{\infty} \gamma^k = \frac{1}{1-\gamma} \tag{9.33}$$

This effectively defines the memory of the RLS filter. When $\gamma = 1$, that is when it corresponds to the LS, the filter has an infinite memory.

9.5.2 Limitations of the recursive least squares algorithm

The RLS method is very efficient and involves exactly the same number of arithmetic operations between samples as $\mathbf{W}_k$ and $\mathbf{P}_k$ in Equation 9.32 have fixed dimensions. This is an important requirement for efficient real-time filtering. There are, however, two main problems that may be encountered when the RLS algorithm is implemented directly. The first, referred to as 'blow-up', results if the signal $x_k(i)$ is zero for a long time, when the matrix $\mathbf{P}_k$ will grow exponentially as a result of division by γ (which is less than unity) at each sample point:

$$\lim_{k\to\infty} \mathbf{P}_k = \lim_{k\to\infty} \left(\frac{\mathbf{P}_{k-1}}{\gamma_{k-1}}\right) \tag{9.34}$$

The second problem with the RLS is its sensitivity to computer roundoff errors, which results in a negative definite $\mathbf{P}$ matrix and eventually to instability. For successful estimation of $\mathbf{W}$, it is necessary that the matrix $\mathbf{P}$ be positive semidefinite which is equivalent to requiring in the LS method that the matrix $\mathbf{X}^{\mathrm{T}}\mathbf{X}$

be invertible, but, because of differencing of terms in Equation 9.32b, positive definiteness of $\mathbf{P}$ cannot be guaranteed. This problem can be worse in multi-parameter models, especially if the variables are linearly dependent and when the algorithm is implemented on a small system with a finite wordlength. When the algorithm has iterated for a long time the two terms in the parentheses in Equation 9.32b are very nearly equal and subtraction of such terms in a finite wordlength system may lead to errors and a negative definite $\mathbf{P}_k$ matrix.

The problem of numerical instability may be solved by suitably factorizing the matrix $\mathbf{P}$ such that the differencing of terms in Equation 9.32b is avoided. Such factorization algorithms are numerically better conditioned and have accuracies that are comparable with the RLS algorithm that uses double precision. Two such algorithms are the square root and the UD factorization algorithms. In terms of storage and computation the UD algorithm is more efficient, and is thus preferred. In fact, the UD algorithm is a square-root-free formulation of the square root algorithm and thus shares the same properties as the latter.

9.5.3 Factorization algorithms

Square root algorithm

In the square root method, the matrix $\mathbf{P}_k$ is factored as (Peterka, 1975)

$$\mathbf{P}_k = \mathbf{S}_k\mathbf{S}_k^{\mathrm{T}} \tag{9.35}$$

where $\mathbf{S}_k$, an upper triangular matrix, and $\mathbf{S}_k^{\mathrm{T}}$, its transpose, are square roots of $\mathbf{P}_k$. Thus if $\mathbf{S}_k$ instead of $\mathbf{P}_k$ is updated the positive definiteness of $\mathbf{P}_k$ is guaranteed, since the product of two square roots is always positive. $\mathbf{S}_k$ is updated as

$$\mathbf{S}_k = \frac{1}{\gamma^{1/2}}\mathbf{S}_{k-1}\mathbf{H}_{k-1} \tag{9.36}$$

where $\mathbf{H}_k$ is an upper triangular matrix.

UD factorization algorithm

In the UD method $\mathbf{P}_k$ is factored as (Bierman, 1976)

$$\mathbf{P}_k = \mathbf{U}_k\mathbf{D}_k\mathbf{U}_k^{\mathrm{T}}$$

where $\mathbf{U}_k$ is a unit upper triangular matrix, $\mathbf{U}_k^{\mathrm{T}}$ is its transpose and $\mathbf{D}_k$ is a diagonal matrix. Thus instead of updating $\mathbf{P}_k$ as in the RLS, its factors $\mathbf{U}$ and $\mathbf{D}$ are updated. A C language code for the UD algorithm is given on the disk for this book.

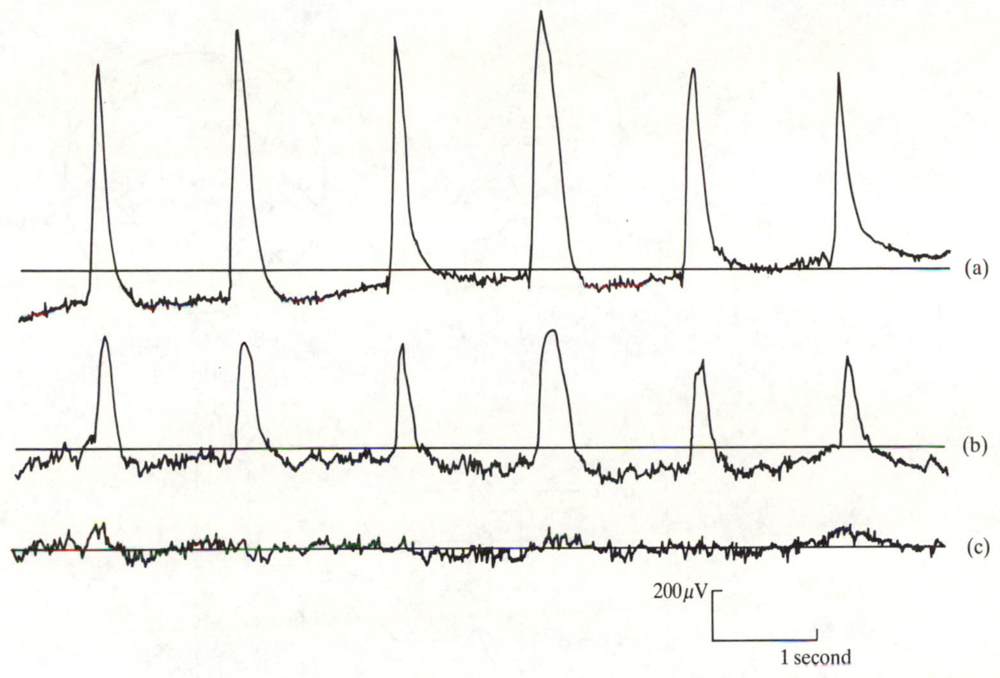

Figure 9.16 The problem of ocular artefacts in electroencephalography: (a) measured EOG, (b) corresponding contaminated EEG signal and (c) EEG signal corrected for artefact.

9.6 Application example 1 – adaptive filtering of ocular artefacts from the human EEG

9.6.1 The physiological problem

The human electroencephalogram (EEG) is the electrical activity of the brain and contains useful diagnostic information on a variety of neurological disorders. Normal EEG signals are measured from electrodes placed on the scalp, and are often very small in amplitude, of the order of 20 μV. The EEG, like all biomedical signals, is very susceptible to a variety of large signal contaminations or artefacts which reduce its clinical usefulness. For example, blinking or moving the eyes produces large electrical potentials around the eyes called the electrooculogram (EOG). The EOG spreads across the scalp to contaminate the EEG, when it is referred to as an ocular artefact (OA). Examples of measured EOG and the corresponding contaminated EEG are given in Figure 9.16.

Ocular artefacts are a major source of difficulty in distinguishing normal brain activities from abnormal ones. In some cases, for example brain-damaged babies and patients with frontal tumours, it is difficult to distinguish between the associated pathological slow waves in the EEG and OAs. The similarity between the OAs and signals of clinical interest also makes it difficult to

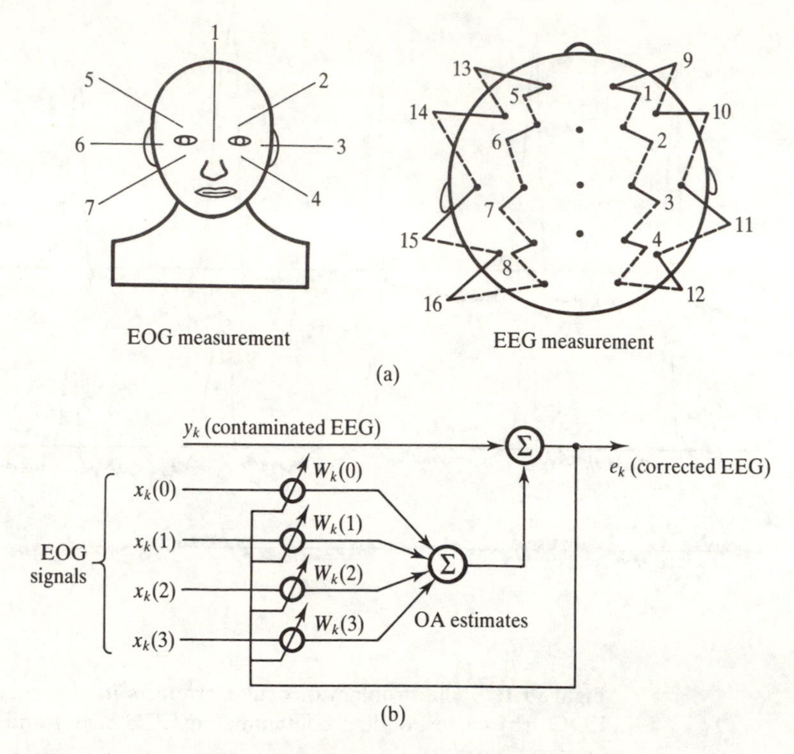

Figure 9.17 Adaptive ocular artefact removal method: (a) possible electrode positions for EOG (ocular movement) and EEG measurements; (b) adaptive ocular artefact filter.

automate the analysis of the EEG by computer. In general, neurological disorders often manifest themselves in the EEG as slow waves which unfortunately not only have appearance similar to OAs but share the same frequency bands as OAs. The problem then is to remove the OAs while preserving the signals of clinical interest.

9.6.2 Artefact processing algorithm

Several methods have been proposed for processing OAs. However, factors such as the requirements of the clinical laboratory, constraints of real-time applications, costs, the random nature of OAs and the spectral overlap between OAs and some signals of cerebral origin dictate that OA processing should be adaptive and in real time.

An adaptive ocular artefact filtering scheme is depicted in Figure 9.17. In this method estimates of OAs are obtained by suitably scaling the EOGs. The OA estimates are then subtracted from the contaminated EEGs to yield 'artefact-free' EEG signals. To illustrate this, consider the simple problem of correcting a single EEG channel for ocular artefact using four EOG signals (Figure 9.17(b)). The information contained in the contaminated EEG, y_k, and the EOGs, $x_k(0)$ to $x_k(3)$, is used to obtain an estimate of the ocular artefact, $\sum_{i=0}^{3} w_k(i)x_k(i)$. The OA estimate is then subtracted from the contaminated EEG to yield an 'artefact-free' EEG, e_k:

$$e_k = y_k - \sum_{i=0}^{n-1} w_k(i)x_k(i) \tag{9.37}$$

where $w_k(i)$, $i = 0, 1, \ldots, n-1$, are the coefficients of the adaptive filter and represent the fractions of the EOGs that reach the EEG as artefacts. e_k is also used to adjust the coefficients (weights) of the adaptive filter, using a numerically stable recursive least squares algorithm, so that optimal estimates of OAs are obtained. Continuous adjustment of $w_k(i)$ is necessary to account for changes in OAs due, for example, to changes in ocular movements.

The adaptive filtering algorithm used to remove the OAs is the UD algorithm described previously. This numerically stable formulation of the RLS algorithm was preferred to the LMS because of its superior convergence time, enabling it to cope better with different OAs each of which requires a different optimum set of coefficients for effective removal. An example of an EEG signal adaptively corrected for artefacts is shown in Figure 9.16(c).

9.6.3 Real-time implementation

An on-line microprocessor-based ocular artefact removal system that uses the UD algorithm described above has been developed (Ifeachor *et al.*, 1986). The system implements a variety of user-selectable models. The system has been tested on several normal and patient subjects. Good results were obtained for various categories of patient subjects.

However, it was found that when pathological waves, such as slow waves, epileptic spike and wave complexes, were picked up at both the EEG and EOG electrodes, the waves in the corrected EEG were reduced in amplitude. This is because the fraction of the EOG subtracted depends on the degree of correlation between the EOG and its component in the EEG, and the presence of slow waves of similar shape to the OA can lead to the subtraction of a fraction which depends on slow waves as well as the EOG. Thus, it is necessary to distinguish between the OA and slow waves, using a knowledge-based system, for example (Ifeachor *et al.*, 1990).

9.7 Application example 2 – adaptive telephone echo cancellation

Echoes arise primarily in communication systems when signals encounter a mismatch in impedance. Figure 9.18(a) shows a simplified long distance telephone circuit. The hybrid circuit at the exchange converts the two-wire circuit

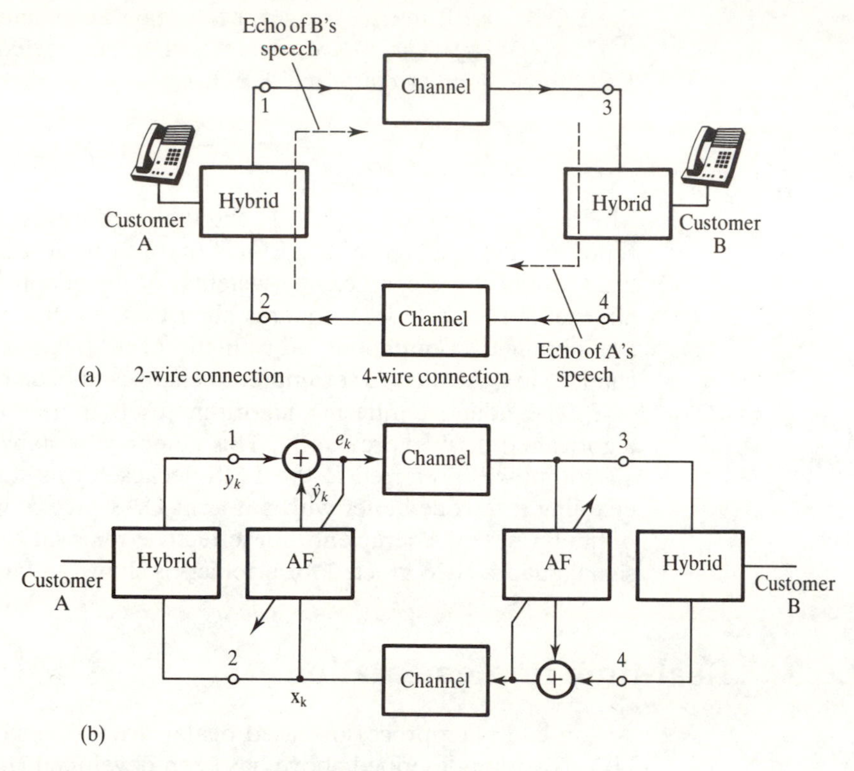

Figure 9.18 (a) Simplified long-distance telephone circuit; (b) echo cancellation in long-distance voice telephony.

from the customer's premises to a four-wire circuit, and provides separate paths for each direction of transmission. This is largely for economic reasons, for example to allow multiplexing, that is simultaneous transmission of many calls.

Ideally, the speech signal originating from customer A travels along the upper transmission path to the hybrid on the right and from there to customer B, while that from B travels along the lower transmission path to A. The hybrid network at each end should ensure that the speech signal from the distant customer is coupled into its two-wire port and none to its output port. However, because of impedance mismatches the hybrid network allows some of the incoming signals to leak into the output path and to return to the talker as an echo. When the telephone call is made over a long distance (for example using geostationary satellites) the echo may be delayed by as much as 540 ms and represents an impairment that can be annoying to the customers. The impairment increases with distance. To overcome this problem, echo cancellers are installed in the network in pairs, as illustrated in Figure 9.18(b).

At each end of the communication system (Figure 9.18(b)), the incoming signal, x_k, is applied to both the hybrid and the adaptive filter (Duttweiller, 1978). The cancellation is achieved by making an estimate of the echo and subtracting it from the return signal, y_k. The underlying assumption here is that the echo return path (through the hybrid) is linear and time invariant. Thus the return signal at time k may be expressed as

$$y_k = \sum_{i=0}^{N-1} w_k(i)x_{k-i} + s_k \tag{9.38}$$

where x_k are samples of the incoming signal (from the far-end speaker), s_k is the near-end speaker plus any additive noise and w_k is the impulse response of the echo path. The echo canceller makes an estimate of this impulse response and produces a corresponding estimate, $\hat{y}_k = \sum w_k(i)x_{k-j}$, of the echo which is then subtracted from the normal return signal, y_k. Economic considerations place a limit on the sampling rate and the wordlengths of filter coefficients and input data which in turn limits the canceller's performance. Fundamental limits come from misadjustment in the adaptive filter and from nonlinearities in the echo path.

9.8 Other applications

Loudspeaking telephones

- The hybrid network is used to separate the transmit and receive paths (that is, the loudspeaker from the microphone), but there is a significant acoustic coupling between the loudspeaker and the microphone because of their proximity as well as a leakage across the imperfectly matched hybrid network (South *et al.*, 1979).
- The difficulty then is how to provide adequate gain for the receive and transmit directions without causing instability.
- The conventional solution to the problems is to use a voice-activated switch to select the transmit and receive paths, but this is not satisfactory because it does not allow full duplex communication.
- A better solution is to use adaptive filtering techniques to estimate and control the acoustic and hybrid echoes (Figure 9.19(b)). The number of filter coefficients here can be quite large, for example 512, making the use of a fast algorithm attractive.
- In teleconferencing networks (or public address systems) acoustic feedback leads to problems similar to those described above. Adaptive filters used for these may require large numbers of coefficients (250 to 1000), especially in rooms with long reverberation times, and must converge rapidly.

Multipath compensation

- In a type of spread spectrum system each data bit is transmitted as one of two orthogonal M-length pseudorandom sequences of bits. The sequence

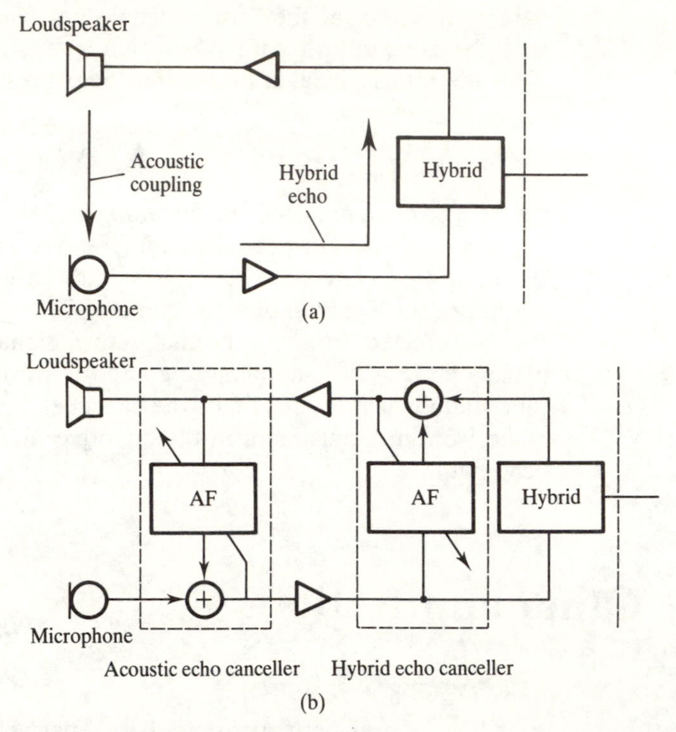

Figure 9.19 (a) Loudspeaking telephone; (b) acoustic and hybrid echo cancellation in loudspeaking telephone.

transmitted depends on whether the data bit is a logic 0 or 1. At the receiver two sequences identical to those used at the transmitter are cross-correlated with the received sequence to determine whether a 1 or 0 is received.

- In the presence of a multipath, the signal travels through separate paths to the receiver. Such effects could occur in mountainous or urban regions through reflection. The received signal is the sum of a number of components whose amplitudes and phases may differ (Figure 9.20). This reduces the performance of the receiver.
- The adaptive filter is used to estimate the overall multipath response and to compensate for its effects.

Adaptive jammer suppression

- In direct sequence spread spectrum a need often arises to suppress the effects of a jamming signal at the receiver to improve the performance of the receiver. Adaptive filtering may be used for this purpose (Figure 9.21). In such a system, use is made of the fact that the jammer is highly correlated whereas the pseudorandom code is weakly correlated. Thus the output of the filter, y_k, is an estimate of the jammer. This is sub-

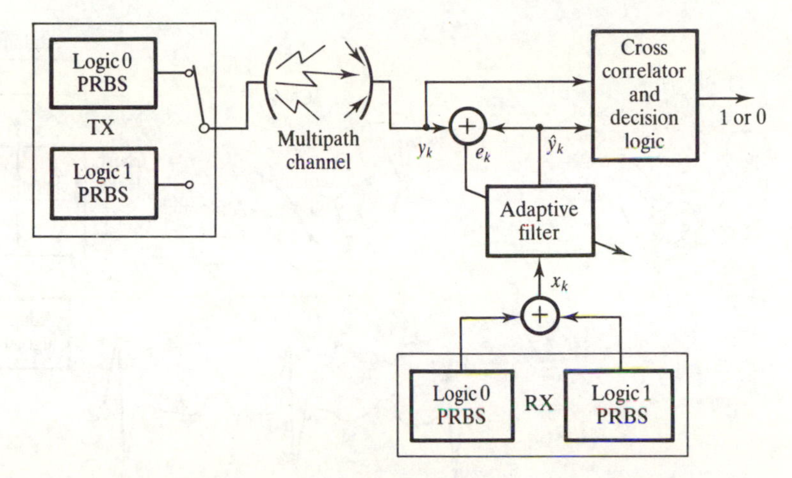

Figure 9.20 An adaptive spread spectrum communication system with multipath-effect compensation.

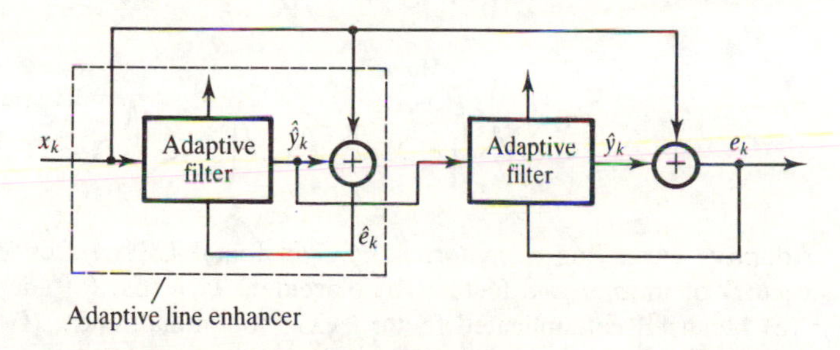

Figure 9.21 Jammer suppression in direct sequence spread spectrum receiver.

tracted from the received signal, x_k, to yield an estimate of the spread spectrum signal.

- To enhance the performance of the system a two-stage jammer suppressor is used. The adaptive line enhancer, which is essentially another adaptive filter, counteracts the effects of finite correlation which leads to partial cancellation of the desired signal. The number of coefficients required for either filter is moderate (about 16), but the sampling frequency may be well over 400 kHz.

Radar signal processing

Adaptive signal processing techniques are widely used to solve a number of problems associated with radar. For example, adaptive filters are used in monostatic radar systems to remove or cancel clutter components from the desired target signals. In HF ground wave radar, adaptive filters are used to reduce co-channel interference which is a major problem in the HF band.

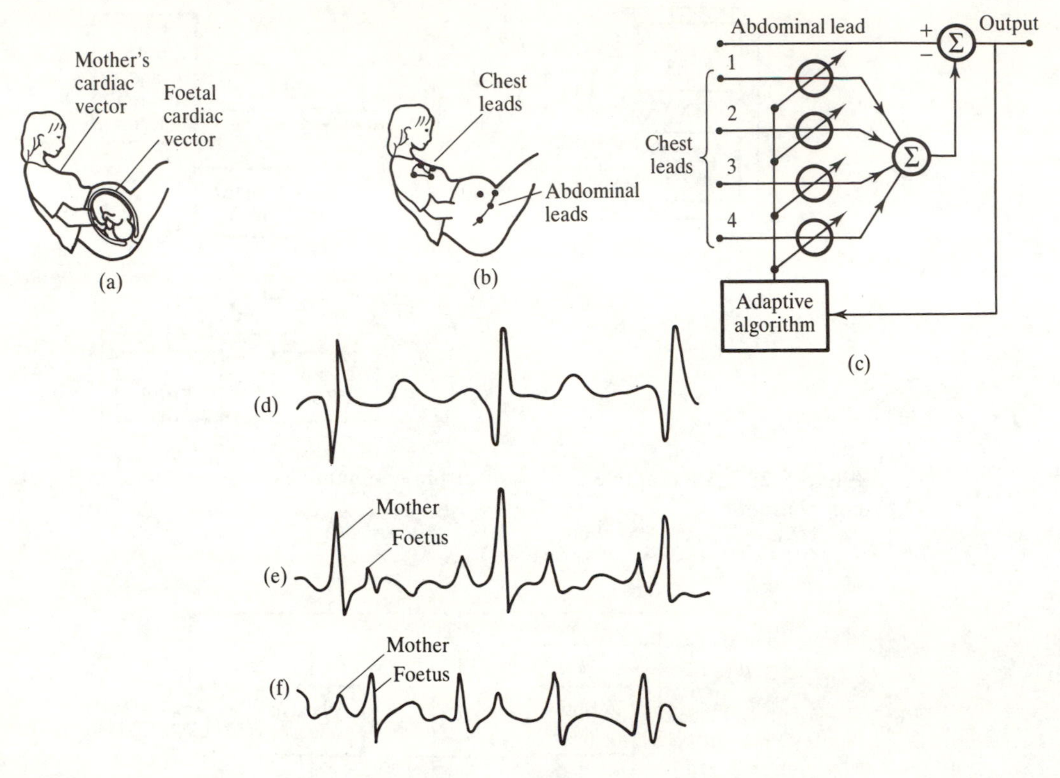

Figure 9.22 Adaptive cancelling of maternal ECG in foetal ECG (after Widrow *et al.*, 1975a): (a) cardiac electric field vectors of mother and foetus; (b) placement of leads; (c) adaptive; (d) idealized mother's ECG (chest leads); (e) idealized contaminated foetal ECG (abdominal lead); (f) output of noise canceller showing reduced mother's ECG.

Separation of speech signals from background noise

Acoustic background noise is a serious problem in speech processing. An adaptive filter may be used to enhance the performance of speech systems in noisy environments (for example in fighter aircrafts, tanks, cars) to improve both intelligibility and recognition of speech.

Foetal monitoring – cancelling of maternal ECG during labour

- Information derived from the foetal electrocardiogram (ECG), such as the foetal heart rate pattern, is valuable in assessing the condition of the baby before or during childbirth.
- The ECG derived from electrodes placed on the mother's abdomen is susceptible to contamination from much larger background noise (for example muscle activity and foetal motion) and the mother's own ECG.
- Adaptive filters have been used to derive a 'noise-free' foetal ECG. Figure 9.22 illustrates the concept.

- Four chest leads are used to detect the baby's ECG, and one or more leads to detect the combined mother and baby's ECG. A four-channel adaptive filter, with 32 coefficients per channel, is used to cancel the mother's heartbeat as shown.

Problems

9.1 Justify the use of adaptive filters instead of conventional filters in applications such as

(1) the removal of ocular artefacts from human EEGs,

(2) echo cancellation in long distance telephony, and

(3) supression of jammer signal in spread spectrum communication.

Starting with the steepest descent algorithm:

$$\mathbf{W}_{k+1} = \mathbf{W}_k - \mu\boldsymbol{\nabla}_k$$

where $\mathbf{W}_k$ is the filter weight vector at the discrete time k, μ controls stability and rate of convergence, and $\boldsymbol{\nabla}_k$ is the true gradient vector of the error–performance surface at the discrete time k, derive the Widrow–Hopf LMS algorithm for adaptive noise cancelling, stating any reasonable assumptions made. Comment on the practical significance of the LMS algorithm.

Comment on two major practical limitations of the LMS algorithm and how they lower the performance of the algorithm. Suggest how these limitations may be overcome.

9.2 The output signal from an adaptive noise canceller is given by

$$e_k = y_k - \mathbf{X}_k^{\mathrm{T}}\mathbf{W}_k$$

where $\mathbf{W}_k$ is the adaptive filter weight vector and the other variables have the usual meaning. Starting with this equation, derive

(1) the discrete Wiener–Hopf equation and

(2) the basic LMS adaptive algorithm.

State any assumptions made.

9.3 Show that the adaptive filter turns itself off when there is no correlation between the interference signal, x_k, and the contaminated signal y_k.

9.4 Explain briefly, with the aid of a block diagram, the basic concepts of adaptive noise cancelling. Discuss critically the benefits and limitations of adaptive noise cancelling in a real-time application of your choice and suggest ways of overcoming the limitations.

References

Bierman G.J. (1976). Measurement updating using the UD factorization. *Automatica*, **12**, 375–82

Duttweiller D.L. (1978). A twelve-channel digital echo canceler. *IEEE Trans. Communications*, **26**, 647–53

Ferrara E.R. and Widrow B. (1981). The time-sequenced adaptive filter. *IEEE Trans. Acoustics, Speech and Signal Processing*, **29**(3), 766–70

GEC–Plessey (1990). *Digital Signal Processing IC Handbook*. GEC–Plessey Semiconductor

Haykin S. (1986). *Adaptive Filter Theory*. Englewood Cliffs NJ: Prentice-Hall

Ifeachor E.C., Jervis B.W., Morris E.L., Allen E.M. and Hudson N.R. (1986). A new microcomputer-based on-line ocular artefact removal (OAR) system. *IEE Proc.*, **133**, 291–300

Ifeachor E.C., Hellyar M.T., Mapps D.J. and Allen E.M. (1990). Knowledge based enhancement of EEG signals. *Proc. IEE (Part F)*, **137**(5), 302–10

Mansour D. and Gray A.H. (1982). Unconstrained frequency domain adaptive filter. *IEEE Trans. Acoustics, Speech and Signal Processing*, **30**, 726–34

Peterka P. (1975). A square root filter for real-time multivariate regression. *Kybernetika*, **11**, 53–67

South C.R., Hoppitt C.E. and Lewis A.V. (1979). Adaptive filters to improve loudspeaker telephone. *Electronics Lett*, **15**, 673–4

Widrow B. and Winter R. (1988). Neural nets for adaptive filtering and adaptive pattern recognition. *IEEE Computer*, 25–30

Widrow B., Glover J.R., McCool J.M., Kaunitz J., Williams C.S., Hearn R.H., Zeidler J.R., Dong E. and Goodlin R.C. (1975a). Adaptive noise cancelling: principles and applications. *Proc. IEEE*, **63**, 1692–76

Widrow B., McCool J.M. and Ball M. (1975b). The complex LMS algorithm. *Proc. IEEE*, 719–20

Bibliography

Clark G.A., Mitra S.K. and Parker S.R. (1981). Block implementation of adaptive digital filters. *IEEE Trans. Acoustics, Speech and Signal Processing*, **29**, 744–52

Clark G.A., Parker S.R. and Mitra S.K. (1983). A unified approach to time- and frequency-domain realization of FIR adaptive digital filters. *IEEE Trans. Acoustics, Speech and Signal Processing*, **31**, 1073–83

Cowan C.F.N. and Grant P.M. (eds.) (1985). *Adaptive Filters*. Englewood Cliffs NJ: Prentice-Hall

Dentino M., McCool J.M. and Widrow B. (1978). Adaptive filtering in the frequency domain. *Proc. IEEE*, **66**, 1658–9

Dudek M.T. and Robinson J.M. (1981). A new adaptive circuit for spectrally efficient digital microwave-radio-relay systems. *Electronics and Power*, 397–401

Falconer D.D. (1982). Adaptive reference echo cancellation. *IEEE Trans. Communications*, **30**, 2083–94

Ferrara E.R. and Widrow B. (1981). Multichannel adaptive filtering for signal enhancement. *IEEE Trans. Acoustics, Speech and Signal Processing*, **29**, 766–70

Harrison W.A., Lim J.S. and Singer E. (1986). A new application of adaptive noise cancellation. *IEEE Trans. Acoustics, Speech and Signal Processing*, **34**, 21–7

Holte N. and Stueflotten S. (1981). A new digital echo canceler for two-wire subscriber lines, *IEEE Trans. Communications*, **29**, 1573–81

Lappage R., Clarke J., Palma G.W.R. and Huizing A.G. (1987). The Byson research radar. In *International Conf. Radar 87*, October 1987, London: IEE, 453–61

Madden (1987). The adaptive suppression of interference in HF ground wave radar. In *International Conf. Radar 87*, October 1987. London: IEE, 98–102

Messerschmitt D.G. (1984). Echo cancellation in speech and data transmission. *IEEE J. Selected Areas in Communications,* **2**, 283–97

Mikhael W.B. and Wu F.H. (1987). Fast algorithms for block FIR adaptive digital filtering. *IEEE Trans. Circuits and Systems,* **34**, 1152–60

Mueller K.H. (1976). A new digital echo canceler for two-wire full-duplex data transmission. *IEEE Trans. Communications*, **24**, 956–62
Ochia K., Araseki T. and Ogihara T. (1977). *IEEE Trans. Communications,* **25**, 589–94
Ogue J.C., Saito T. and Hoshiko Y. (1983). A fast convergence frequency domain adaptive filter. *IEEE Trans. Acoustics, Speech and Signal Processing,* **31**, 1312–14
Reed F.A., Feintuch P.L. and Bershad N.J. (1985). The application of the frequency domain LMS adaptive filter to split array bearing estimation with a sinusoidal signal. *IEEE Trans. Acoustics, Speech and Signal Processing*, **33**, 61–9
Saulnier G.J., Das P.K. and Milstein L. (1985). *IEEE J. Selected Areas in Communications*, **3**, 676–86
Sethares W.A., Lawrence D.A., Johnson C.R. and Bitmead R.R. (1986). Parameter drift in LMS adaptive filters. *IEEE Trans. Acoustics, Speech and Signal Processing*, **34**, 868–79
Sondhi M.M. and Berkley D.A. (1980). Silencing echoes on the telephone network. *Proc. IEEE,* **68**, 948–63
Tao Y.G., Kolwicz K.D., Gritton C.W.K. and Duttweiler D.L. (1986) A cascadable VLSI echo canceller. *IEEE Trans. Acoustics, Speech and Signal Processing*, **34**, 297–303
Thornton C.L. and Bierman G.J. (1978). Filtering and error analysis via the UDU covariance factorization. *IEEE Trans. Automatic Control*, **23**, 901–7
Widrow B. (1966). *Adaptive filters 1: Fundamentals*. Report SU-SEL-66-126, Stanford Electronics Laboratory, Stanford University CA
Widrow B. (1971). Adaptive filters. In *Aspects of Network and System Theory* (Kalman R. and DeClaris N., eds.), pp. 563–87. New York: Holt, Rinehart and Winston
Widrow B. (1976). Stationary and nonstationary learning characteristics of the LMS adaptive filter. *Proc. IEEE*, **64**, 1151–62
Widrow B. and Stearns S.D. (1985). *Adaptive Signal Processing*. Englewood Cliffs NJ: Prentice-Hall
Widrow B., Mantey P., Griffiths L. and Goode B. (1967). Adaptive antenna systems. *Proc. IEEE*, **55**, 2143–59

Appendix

9A C language programs for adaptive filtering

Several adaptive algorithms have been implemented in the C language for this chapter. These are

(1) lmsflt.c, the LMS algorithm,
(2) uduflt.c, the UD algorithm,
(3) sqrflt.c, the square root algorithm and
(4) rlsflt.c, the recursive least squares algorithm.

Only the first program is listed here to limit the size of the book (see Program 9A.1). However, all the programs are available on the computer disk for the book (see the preface for details of how to obtain a copy).

To illustrate how to implement adaptive filters, we will use the program listed in Program 9A.1 to detect a tone in broadband noise.

Program 9A.1 C language implementation of the LMS algorithm (lmsflt.c).

```
/* ........................................................................ */
/*        implementation of the LMS algorithm                               */
/*                                                                          */
/*        manny 6.11.92                                                     */
/*                                                                          */
/*        inputs:                                                           */
/*        x[]           input data vector                                   */
/*        dk            latest input data value                             */
/*        w[]           coefficient vector                                  */
/*                                                                          */
/*        outputs:                                                          */
/*        ek            error value                                         */
/*        yk            digital filter output                               */
/*        w[]           updated coefficient vector                          */
/* ........................................................................ */

double      lmsflt()
{
            int       i;
            double    uek,yk;

            yk=0;
            for(i=0; i<N; ++i){                         /* digital filtering */
                      yk=yk+w[i]*x[i];
            }
            ek=dk-yk;                                   /* compute output error*/

            uek=2*mu*ek;                                /* update the weights */
            for(i=0; i<N; i++){
                      w[i]=w[i]+uek*x[i];
            }
            return(yk);
}
```

Adaptive enhancement of narrowband signals buried in noise

Adaptive filters are often used to detect or enhance narrow-band signals buried in wideband noise. The structure that is commonly used for this purpose is depicted in Figure 9A.1. It consists of a delay element, symbolized by z^{-M}, and an adaptive predictor. The delay element removes any correlation that may exist between the samples of the noise component. The adaptive predictor is essentially an FIR filter with adjustable coefficients and its output, y_k, gives the enhanced narrowband signal. In some applications, the second output of the adaptive filter, e_k, and not y_k is the desired output. The prediction coefficients, $w_k(i)$, are optimized by a suitable adaptive algorithm, which in our case is an LMS algorithm (see Section 9.4 for details).

In the case of the LMS algorithm, the adaptive filter is characterized by the following equations:

$$y_k = \sum_{i=0}^{N-1} w_k(i)x_k(i) \qquad \textbf{(9A.1)}$$

$$e_k = d_k - y_k \qquad \textbf{(9A.2)}$$

$$w_{k+1}(i+1) = w_k(i) + 2\mu e_k x_k(i), \quad i = 0, 1, \ldots, N-1 \qquad \textbf{(9A.3)}$$

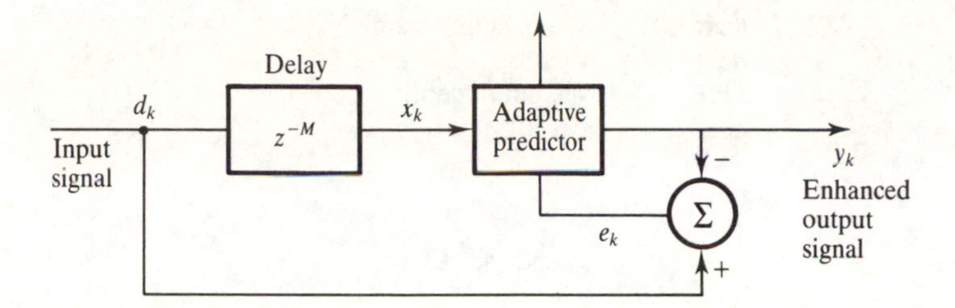

Figure 9A.1 Adaptive signal enhancement.

where d_k is the noisy narrowband signal sample, $x_k(i)$ is the input data vector, derived from delayed values of d_k, $w_k(i)$ is the prediction coefficient vector at the kth sampling instant, μ is the stability factor, and y_k is the enhanced, narrowband signal.

The function, lmsflt.c, in Program 9A.1 is a C language implementation of the above equations. The program listed in Program 9A.2 illustrates how to use the function lmsflt.c for signal enhancement. To simulate the problem a broadband noise was added to a 500 Hz sine wave signal, and the composite data stored in ASCII format in a file din.dat. The noisy sine wave was then applied to the adaptive filter. To simulate real-time adaptive filtering, the input data is read from the file and applied to the adaptive filter one sample at a time. For very long lengths of data the

Program 9A.2 Program for adaptive signal enhancement.

```
/* ......................................................................... */
/*                                                                           */
/*      program to illustrate adaptive filtering using                       */
/*      the LMS algorithms                                                   */
/*                                                                           */
/*      program name: adfilter.c                                             */
/*                                                                           */
/*      manny, 7.11.92                                                       */
/* ......................................................................... */

#include        <stdio.h>
#include        <math.h>
#include        <dos.h>

/* constant definitions */

#define N       30          /* filter length */
#define M       1           /* delay */
#define w0      0           /* initial value for adaptive filter coefficients */
#define npt     N+M
#define SF      2048        /* factor for reducing the data samples - 11 bit ADC
                               assumed */
#define mu      0.04

double      lmsflt();
void        initlms();
void        update_data_buffers();
void        initfiles();
```

```
float       x[npt], d[npt], dk, ek;
double      w[npt];
FILE        *in,*out,*fopen();
char        din[30];

main()
{
            double yk, yk1;

            initfiles();
                                                   /* lms-based adaptive filter */
            initlms();
            while(fscanf(in,"%f",&dk)!=EOF){
                  dk=dk/SF;
                  update_data_buffers();
                  yk=lmsflt();
                  yk1=SF*yk;
                  fprintf(out,"%lf  \n",yk1);
            }
            fcloseall();
}

/* ................................................................................ */
void        initfiles()
{
            clrscr();
            printf("enter name of file holding data to be filtered \n");
            scanf("%s",din);
            printf("\n");
            printf("the filtered data will be stored in dout.dat \n");

            if((in=fopen(din,"r"))==NULL){
                  printf("cannot open input data file \n");
                  exit(1);
            }
            if((out=fopen("dout.dat","w"))==NULL){
                  printf("cannot open output data file \n");
                  exit(1);
            }
            return;
}
/* ................................................................................ */

void        update_data_buffers()
{
            long      j, k;

            for(j=1; j<N; ++j){                      /* update x-data buffer */
                  k=N-j;
                  x[k]=x[k-1];
            }
            x[0]=dk;
            if(M>0)
                  x[0]=d[M-1];
```

```
        for(j=1; j<M; ++j){                    /* update d-data buffer */
                k=M-j;
                d[k]=d[k-1];
        }
        d[0]=dk;
}

/* ................................................................ */
void    initlms()
{
        long    i;
        for(i=0; i<npt; ++i){
                x[i]=0;
                d[i]=0;
                w[i]=w0;
        }
}
/* ................................................................ */

#include        "lmsflt.c";
```

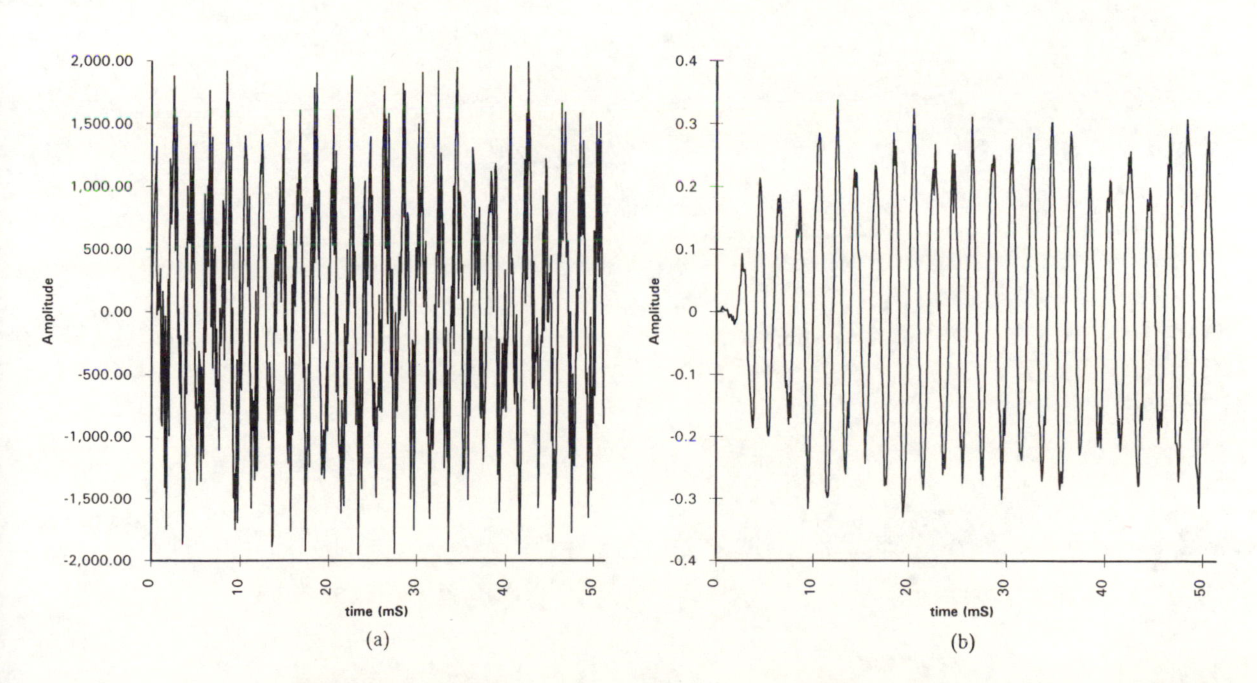

Figure 9A.2 Adaptive enhancement of a narrowband signal: (a) noisy signal; (b) enhanced signal.

user may need to read the data in blocks for efficiency. Figure 9A.2 shows the results for the LMS based filters.

As may be evident from Figure 9A.2, to use the adaptive algorithms, the user needs to specify the parameters of the adaptive filters, for example the length of the FIR filter, N, the delay factor, M, and the stability factor, μ. Attention must also be paid to the format of the input data. For example, in some applications the input data may come from a multichannel source, with each element of $x_k(i)$ representing the data value from a channel. In this case, the input data array to the adaptive algorithm will need to be suitably modified.

10

Spectrum estimation and analysis

10.1	Introduction	578
10.2	Principles of spectrum estimation	580
10.3	Traditional methods	583
10.4	Modern parametric estimation methods	603
10.5	Comparison of estimation methods	604
10.6	Application examples	604
10.7	Summary	608
10.8	Worked example	609
	References	612

The following topics are covered in this chapter: the basic concepts of spectrum analysis, the pitfalls to be avoided especially in the nonparametric methods, the properties of data windows, pre-processing of the data and choice of the data window function. Nonparametric methods described and compared include the modified periodogram, the Blackman–Tukey method based on fast Fourier transformation of the autocorrelation function of the data and the fast correlation method. An introductory review of parametric methods of estimation, based on time domain parametric models of the data, is included together with recommendations and references to the literature. Applications described are the processing of electrical waveforms of the brain to differentiate between patients with brain diseases and normal subjects, and the analysis of human electroencephalograms based on autoregressive modelling.

10.1 Introduction

The transformation of data from the time domain to the frequency domain was introduced in Section 2.1. The principles and practice of the estimation and analysis of spectra in the frequency domain are the subject of this chapter. Plots of harmonic amplitude or phase versus frequency often result in a more comprehensible presentation of the data or waveform, particularly when the latter is of a random nature. By selecting certain harmonics on some suitable criterion and rejecting others significant data reduction may be possible. Spectrum analysis has found various applications such as the study of communication engineering signals, of event-related or stimulated responses of the human electroencephalogram (EEG) in the diagnosis of brain diseases (Jervis *et al.*, 1993), of other biological signals, of meteorological data, and in industrial process control and the measurement of noise spectra for the design of optimal linear filters.

The spectrum estimation techniques available may be categorized as nonparametric and parametric. The nonparametric methods include the periodogram, the Bartlett and Welch modified periodogram, and the Blackman–Tukey methods. All these methods have the advantage of possible implementation using the fast Fourier transform, but with the disadvantage in the case of short data lengths of limited frequency resolution. Also, considerable care has to be exercised to obtain meaningful results. Parametric methods on the other hand can provide high resolution in addition to being computationally efficient. The disadvantage of these methods, however, is that considerable time and effort may be necessary to form a sufficiently accurate model of the process from which to estimate the spectrum. The most common parametric approach is to derive the spectrum from the parameters of an autoregressive model of the signal. Moving average and autoregressive moving average models have also been employed.

A number of pitfalls have to be avoided in performing nonparametric spectral analysis and the associated topics of aliasing, scalloping loss, finite data length, spectral leakage, and spectral smearing should be understood and are discussed in Section 10.3.

The deleterious effects of spectral leakage and smearing may be minimized by windowing the data by a suitable window function. The sampled data values are multiplied point by point by the sampled values of the selected window function. The topic of windowing is discussed in Section 10.3.2. The equivalent noise bandwidth, processing gain, worst-case processing loss, and minimum resolution bandwidth are all properties of window functions which are considered when choosing a suitable window (see Section 10.3.2.1). In the section on overlap correlation it is shown that averaging the spectra of a number of windowed sections of the data instead of computing the spectrum of the windowed data directly leads to a significantly improved estimate of the spectrum. The part of Section 10.3.2.1 on the biasing effect of data windows shows how the loss of signal energy and the dc biasing effect of data windows may be overcome by pre-processing the data prior to windowing.

Judgements of the quality of spectral estimates are based on estimation theory, and so some basic concepts of that theory are now presented. Statistical estimation involves determining the expected values of statistical quantities derived from samples of the population. However, in time series analysis, discrete data obtained as a function of time is usually available rather than samples of the population taken simultaneously. This difficulty is commonly avoided by assuming that the process is ergodic, that is the properties of the time series data are the same as would be those of the hypothetical samples. Some statistical definitions are now in order.

The mean value of a time series with data values $x(n)$, $n = 0, 1, \ldots, N-1$, is the expected value of $x(n)$, $E[x(n)]$, given by

$$E[x(n)] = \frac{1}{N} \sum_{n=0}^{N-1} x(n) \tag{10.1}$$

where E denotes expectation. The variance of the same time series is

$$\operatorname{var}[x(n)] = E\{[x(n) - \bar{x}(n)]^2\} \tag{10.2}$$

The autocovariance of $x(n)$ is given by

$$c_{xx}(m) = E\{[x(n) - \bar{x}(n)]\,[x(n+m) - \bar{x}(n)]\} \tag{10.3}$$

where m denotes the lag in the data points and $\bar{x}(n)$ denotes $E[x(n)]$. The power spectral density estimated from the finite realization is

$$P_{\mathrm{E}}(\omega) = \sum_{m=-\infty}^{\infty} c_{xx}(m) \exp(-\mathrm{j}\omega m) \tag{10.4}$$

Note that if finite duration waveforms as opposed to infinitely long stochastic processes are being analysed it would more appropriate to use the energy spectral density. The power spectral density has units of $V^2\mathrm{Hz}^{-1}$. If the statistical attribute α is being estimated then the bias of the estimate is defined to be the difference between the true (population) value and the estimate:

$$\text{bias} = \alpha - E[\alpha] \tag{10.5}$$

If the bias is zero then the estimate gives the true value; if the bias is not zero it represents the error in α and the estimated value of α is said to be biased. Clearly good estimators will not be biased. The variance of α is a measure of the width of the peak of the probability density distribution function of α. Small variances correspond to narrow peaks, and as the variance tends to zero the estimated value approaches the population (true) value if the estimate is unbiased. If the variance tends to zero as the number of data, N, increases, the estimate is said to be consistent. If the estimate is inconsistent then the estimates will fluctuate more and more wildly from realization to realization as the number of data is increased. It is therefore desirable that statistical estimates be both unbiased and consistent.

The properties of the periodogram methods of spectrum estimation are discussed in Sections 10.3.3 and 10.3.4. It is shown that the spectrum estimates derived as periodograms are inconsistent, that is successive realizations yield fluctuating estimates and are only unbiased for a large number of data. Stable and more accurate estimates are obtained by windowing sections of the data and averaging their spectra. The associated methods are known as the Bartlett and Welch modified periodogram methods. The final nonparametric method described is known as the Blackman–Tukey method. The windowed autocorrelation function of the data is first computed and the energy spectrum is then obtained from its FFT. The Blackman–Tukey spectrum estimate is characterized by a larger quality factor than the other periodogram methods.

It is also desirable to identify a quality factor for the estimates of power spectral density to allow a comparison of the different estimations. The ratio of the square of the mean of the power spectral density to its variance has been proposed as a suitable quality factor (Proakis and Manolakis, 1989):

$$Q = \frac{\{E[P_{\mathrm{E}}(f)]\}^2}{\mathrm{var}\,[P_{\mathrm{E}}(f)]} \tag{10.6}$$

10.2 Principles of spectrum estimation

In this section a voltage waveform, plotted against time, will be considered initially. The shape of the waveform may provide useful information. For example, it may be a sine wave which can obviously be characterized by its amplitude, frequency, and phase angle. To be more specific, the waveform would be describable as consisting of a single component with a certain amplitude and phase at the known frequency. As an alternative to representing the waveform as a plot of voltage versus time, it could be represented by two plots: one of amplitude versus frequency and the other of phase versus frequency. Because the sine wave only has one amplitude, one phase, and one frequency the amplitude and phase plots would each consist of a single point only. It can be shown by Fourier analysis (see Chapter 2) that all waveforms can be represented mathematically as the summation of a number of sinusoidal waveforms, each with a specific amplitude and phase at its specific frequency. Thus any waveform can be alternatively represented by a plot of amplitude versus frequency together with a plot of phase versus frequency. These plots are known as the amplitude and phase spectra. These spectra are important because they provide a complementary way of representing the waveform which more clearly reveals information about the frequency content of the waveform. The observed shapes of the spectra and changes in them are often helpful in the understanding and interpretation of the waveforms. Amplitude and phase spectra very often provide more useful information than the waveforms. The topic of transformation from the time to the frequency domain, and the converse, was described in Chapter 2. The transformation of periodic waveforms to the frequency domain by the Fourier series and the complex Fourier series was reviewed. It was shown that the frequencies of the sinusoidal frequency components of the periodic waveform, known as the Fourier components, are harmonically related to each other, i.e. each is an integral multiple of the first harmonic frequency, f, where

$$f = 1/T_p$$

where T_p is the repetition period of the waveform. It follows that adjacent harmonic components are all equally spaced by the frequency interval $f = 1/T_p$, which in this sense of separation is known as the frequency resolution. The amplitude spectrum has an amplitude measured in volts. As an example Figure 2.1(a) shows a periodic voltage pulse waveform while Figures 2.1(b) and 2.1(c) display the amplitude spectrum and the phase spectrum of this waveform respectively. Various uses of the amplitude and phase spectra are mentioned in the introduction to Chapter 2.

Non-periodic but continuous signals can be transformed from the time to the frequency domain using the Fourier transform described in Section 2.1.2. The 'amplitude' of this transform was shown to have the dimension of $\mathrm{V\,Hz^{-1}}$ and when plotted against frequency therefore represents the amplitude spectral density. Thus the area under the curve between two frequencies yields the 'average' voltage of the waveform for those frequency components lying between the two frequencies. Squaring the calculated Fourier transform 'amplitudes' gives the energy spectral density of the voltage waveform in $\mathrm{J\,Hz^{-1}}$. The term 'spectrum' is often used to refer to plots of the energy spectral density versus frequency. Figures 2.2(a) and 2.2(b) show the amplitude and energy spectral densities of a rectangular pulse respectively. It was also shown in Chapter 2 how the spectra of sampled and nonperiodic voltage waveforms may be calculated using the discrete Fourier transform (DFT). The DFT components were shown to be harmonically related with the first harmonic angular frequency being $\Omega = 2\pi/NT$ so that the first harmonic frequency is f where

$$f = 1/NT \tag{10.7a}$$

where N is the number of data, and T is the sampling interval. Since $NT \approx T_p$, the duration of the sampled waveform, for $N \approx N - 1$ the first harmonic frequency is expressible as

$$f = 1/T \tag{10.7b}$$

Again, owing to the harmonic relationship the frequency resolution of the spectra is also given by $1/T$. Thus the longer the waveform, or the more data samples, the better will be the frequency resolution of the spectra. This will be seen to be a reason for the addition of additional zeros to the data from short data sequences.

An example calculation of the DFT of a data sequence $\{1, 0, 0, 1\}$ is given in Section 2.2. The DFT of the data sequence was shown to be $\{2, 1+j, 0, 1-j\}$. Thus, the second harmonic component, $1+j$, is of magnitude $\sqrt{(1^2+1^2)} = \sqrt{2}$. If the data sequence represented sampled voltages the amplitude of this second harmonic would be $\sqrt{2}$ V, and its energy would be $(\sqrt{2}\ \mathrm{V})^2$, i.e. 2 Joules. The corresponding phase angle is given by $\tan^{-1}$ (imaginary component/real component) $= \tan^{-1} 1 = 45°$. The data sequence is plotted in Figure 2.3(a) and its amplitude and phase spectra in Figures 2.3(b) and 2.3(c) respectively. The amplitude spectrum has the dimension of volts. The DFT and the Fourier transform were shown in Section 2.2 to be related by $F(j\omega) = TX(k)$. The decimation-in-time fast Fourier transform (FFT) algorithm for accelerating the computation of the DFT was introduced in Section 2.5 and an example of its use given in Section 2.5.1 for calculating the DFT of the above sequence $\{1, 0, 0, 1\}$.

If the waveform under investigation is long compared with the time interval over which it may be regarded as having constant statistical moments, then the spectrum estimate is likely to be inaccurate. This will also be the case when there is a large noise component in the waveform. It is then desirable to smooth the estimated spectrum to obtain an improved estimate. The purpose of spectrum smoothing is essential to remove randomness. The signal-to-noise ratio in random waveforms can be improved by averaging the waveforms, the signal-to-noise ratio being improved by a factor of $\sqrt{K}$ if K waveforms are averaged. Thus one method of improving the accuracy of the estimated spectrum consists of dividing the data into K equal length sections, determining the spectrum of each section, and then averaging the spectra. In this way the average amplitude and average phase of each harmonic frequency component of the K spectra are obtained and plotted to yield the average amplitude and phase spectra. The accuracy of the spectra may be obtained in terms of their variance. For example the smaller the variance of the power spectral density the more accurate its estimate. It is therefore important to know what is the effect of a spectrum estimation method on the variance of the spectrum. Spectrum estimation by averaging is discussed in Section 10.3.2.1 in which it is explained that spectra estimated as the averages of the spectra of K data sections have a lower variance than those computed directly, with the variance decreasing in proportion to the number of sections taken. Even if the noise content of the waveform is low, that is high signal-to-noise ratio, the averaged result from K sections still results in a significant improvement in accuracy by this modified periodogram method. However, sectioning the data results in fewer data per FFT and consequently a reduction in frequency resolution which may be overcome by the addition of augmenting zeros (Section 10.3.1.2). It is therefore always necessary to bear in mind the opposing claims of the accuracy of the estimate and the required frequency resolution, and to design for the best compromise. Another approach to smoothing the periodogram is to calculate it from the DFT of the windowed autocorrelation function of the data. This is the Blackman–Tukey method as described in Section 10.3.5. Since the autocorrelation function of the data consists of the average of the sums of products of the data by themselves at different lags (Section 4.2), the signal-to-noise ratio is improved (Section 4.2.2). The Blackman–Tukey method results in a spectrum with a larger quality factor than the modified periodogram methods.

Data windows also exert a smoothing effect on spectra. In particular, windows with small side lobes in the frequency domain filter out noise which falls outside the main lobe and will offer improved smoothing. In fact this type of spectral smoothing is sometimes performed by convolving the spectrum of the data with that of the chosen window function.

Parametric and more recent methods of spectrum estimation are discussed briefly in Section 10.4. While these methods are less of an art than the nonparametric methods and may be automated, they are mathematically too complex to be fully described in this text and the interested reader is referred to the references. The concepts of forming models of the data in terms of model parameters and of obtaining the spectra from the frequency response functions of linear systems in terms of these models are mentioned as are more recent methods.

10.3 Traditional methods

In Section 10.3.1 various pitfalls found in spectral analysis are detailed and explanations of how they may be overcome are provided. The technique of windowing and the properties of data windows are described in Section 10.3.2. Some properties of spectra and spectral smoothing are described in Section 10.3.3.

10.3.1 Pitfalls

10.3.1.1 Sampling rate and aliasing

Before the spectral analysis can be carried out the first procedure must be to pass the analogue signals through an anti-aliasing filter, the function of which is to prevent aliasing of the sampled signal after the following stage of analogue-to-digital conversion. Aliasing refers to distortion of the signal spectrum by the introduction of spurious low frequency components owing to a combination of an inadequate anti-aliasing filter and of too low a sampling rate. This topic is fully discussed and explained in Section 1.4.2.

10.3.1.2 Scalloping loss or picket-fence effect

As explained in Section 2.2, the discrete Fourier transform (DFT) consists of harmonic amplitude and phase components regularly spaced in frequency. The spacing of the spectral lines depends on the number of data samples, decreasing with the number of data. The latter is usually restricted by both sampling rate and realization length limitations. If, therefore, there is a signal component which falls between two adjacent harmonic frequency components in the spectrum then it cannot be properly represented. Its energy will be shared between neighbouring harmonics and the nearby spectral 'amplitudes' will be distorted. The amplitude density spectrum for a uniform spectral density signal is shown in Figure 10.1. Note the finite width of the main lobes occurring at the harmonic frequencies and that a signal component at a nonharmonic frequency such as f_{nh} cannot be properly represented. A solution to this difficulty lies in arranging for the harmonic components to be more closely spaced and coincident with the signal frequencies. This may be achieved by adding additional data in the form of zeros to the true data. The added zeros are termed augmenting zeros and serve to increase the fidelity of the estimated spectrum to the true spectrum without adding additional information. Sufficient augmenting zeros, N', are added to the N data to satisfy the simultaneous requirements that

$$N + N' = 2^m \tag{10.8}$$

for a radix-2 fast Fourier transform (FFT) algorithm, where m is integer, and that $1/(N + N' - 1)T$ in which T represents the sampling interval gives sufficient resolution.

The scalloping loss, SL, is defined to represent the maximum reduction in processing gain which occurs mid-way between the harmonically related frequencies (Harris, 1978):

$$\text{SL} = \frac{|W(\omega_s/2N)|}{W(0)} = \frac{\left|\sum_{n=0}^{N-1} w(nT)\exp(-\text{j}\pi n/N)\right|}{\sum_{n=0}^{N-1} w(nT)} \tag{10.9}$$

where W represents the DFT of the window function (Section 10.3.2), $\omega_s = 2\pi f_s$ is the angular sampling frequency, f_s is the sampling frequency, N is the number of data, n is the datum number, and $w(nT)$ is the time domain sampled window function.

As already stated, the effect of finite data length is to limit the achievable frequency resolution to $1/(N - 1)T$ (Hz). This results in a coarse spectrum which may be rendered smooth and continuous by supplementing the data with augmenting zeros. The process is simply one of interpolation of the spectral curve between adjacent harmonics. A real improvement in resolution can only be achieved if a longer realization is available. The frequency interval between the lines of the spectrum becomes $1/(N + N' - 1)T$ (Hz) on addition of N' augmenting zeros.

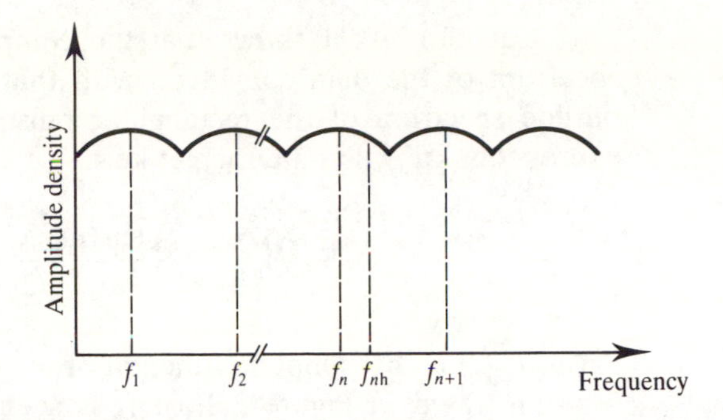

Figure 10.1 The amplitude density spectrum for a signal of uniform spectral density.

10.3.1.3 Trend removal

Any trends in the data must be removed prior to computation of the spectrum because error terms owing to addition of the trend to the data will be integrated and may produce large errors in the estimated spectrum.

10.3.1.4 Spectral leakage and spectral smearing

The FFT of a set of sampled data is not the true FFT of the process from which the data was obtained. This is because the process is continuous whereas the data is the sampled values of a realization which is truncated at its beginning and end. The data which represents a length T_s (s) of the signal is effectively obtained by multiplying all the sampled values in the interval T_s by unity, while all values outside this interval are multiplied by zero. This is equivalent to multiplying, or windowing, the signal by a rectangular pulse, or window, of width T_s and height 1. In this case the sampled data values $v(n)$ are given by the product of the data values, $s(n)$, and the values of the window function, $w(n)$:

$$v(n) = w(n)s(n) \qquad \textbf{(10.10)}$$

This time-domain product is equivalent to a convolution in the frequency domain; see Sections 2.3 and 4.3. Thus the value of the FFT for the nth harmonic is

$$V(\omega_n) = \sum_{k=-N}^{N} W(\omega_n - \omega_k)S(\omega_k) \qquad \textbf{(10.11)}$$

where ω_n is the angular frequency of the nth harmonic, $V(\omega_n)$ is the complex DFT component at frequency ω_n, $W(\omega_n)$ is the DFT of the window at frequency ω_n, and $S(\omega_k)$ is the true DFT component of the signal at frequency ω_k.

Equation 10.11 shows that the computed spectrum consists of the true spectrum of the data convolved with that of the window function. The amplitude spectrum of the rectangular pulse $S_{\mathrm{R}}(\omega_n)$ is given by the following expression, known as a Dirichlet kernel:

$$S_{\mathrm{R}}(\omega_n) = \frac{T_{\mathrm{s}} \sin(\omega_n T_{\mathrm{s}}/2)}{\omega_n T_{\mathrm{s}}/2} = \mathrm{Sa}\left(\frac{\omega_n T_{\mathrm{s}}}{2}\right) \tag{10.12}$$

$\mathrm{Sa}(\omega_n T_{\mathrm{s}}/2)$ is the sampling function of $\omega_n/2$ (see Sections 2.1.1 and 2.1.2) and was illustrated in Figure 2.2(a). It is seen to consist of a main lobe and an infinite number of side lobes peaking at 0 Hz and $(n+0.5)/T_{\mathrm{s}}$ Hz respectively. Now the amplitude spectrum of a single sine wave component of the signal at frequency f_n comprises two impulses at frequencies $\pm f_n$. Convolution by the sampling function produces the spectrum of Figure 10.2. The two impulses have been transformed into two overlapping sampling functions. The effect of the rectangular window has been to introduce spurious peaks into the computed spectrum owing to the effect of the side lobes. This will be true of each frequency component of the signal, and so the amplitude spectrum of the signal will be distorted by the addition and subtraction of the large number of window side lobes and main lobes. The effect may be to introduce spurious peaks, or to conceal true peaks in the spectrum, the phenomenon being known as spectral leakage. To avoid this it is necessary to modify the data by multiplying them by a window shaped to reduce the side lobe effect. Suitable windows have a value of 1 at the mid-data point and are tapered to zero at points $n = 0$ and $n = N - 1$. At least 23 such windows have been developed and their relative suitabilities have been investigated by Harris (1978).

In order to minimize spectral leakage the window shape is chosen to minimize its side lobe levels. Unfortunately, this has the effect of increasing the main lobe width, causing it to spread into the adjacent side lobes. This therefore aliases the side lobes. This is repeated at each harmonic frequency, the overall result being an aliasing of the signal spectrum, known as smearing. Windows and their parameters therefore have to be carefully chosen in order to strike the optimum balance between frequency resolution and the statistical accuracy of the spectral estimate.

10.3.2 Windowing

Various properties of windows are described in this section, basically in the time domain. However, it should again be noted that windowing may be performed either in the time domain (data windows) or in the frequency domain (frequency windows) because of the equivalence between multiplication in the time domain and convolution in the frequency domain (Section 10.3.1.4). Frequency domain windowing may thus be carried out by convolving the frequency domain window with the signal spectrum. This procedure may be achieved by applying Equation 4.91.

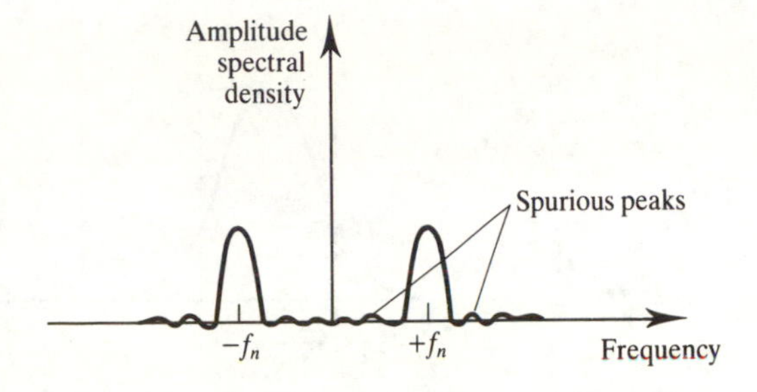

Figure 10.2 Amplitude spectral density of a sinusoidal signal convolved with the sampling function.

10.3.2.1 Properties of windows

Equivalent noise bandwidth

It was seen in Section 10.3.1.4 that owing to the phenomenon of spectral leakage the impulse functions which theoretically represent the amplitude spectral densities become sampling functions. That is, the infinitely narrow spectral components are replaced by the broader bandwidth sampling function. The side lobes of these functions which bias the signal components may be regarded as contributing unwanted noise to the signal, and the frequency window may be regarded as a broadband filter. It is therefore desirable from this point of view to design the windows to have a low noise bandwidth by reducing the side lobe amplitudes. This noise bandwidth is measured for comparison purposes between different windows by the equivalent noise bandwidth. This is defined to be the bandwidth of an ideal rectangular filter which passes the same amount of applied white noise as the spectral filter in question (see Figure 10.3). With this definition it is possible to compare the side lobe properties of different windows by comparing their equivalent noise bandwidths. Thus equivalent noise bandwidth is an important window parameter. The smaller it is then the better the window from this point of view.

The equivalent noise bandwidth is given by

$$\text{ENBW} = \frac{\sum_{n=0}^{n=N-1} w^2(nT)}{\left[\sum_{n=0}^{n=N-1} w(nT)\right]^2} \tag{10.13}$$

Overlap correlation

When data is windowed the ends of the data sequence are tapered to zero. This represents a loss of information. In particular, short duration events occurring in the tapered region may be missed. One approach to solving this problem is to partition the data sequence into overlapping sections and to window and

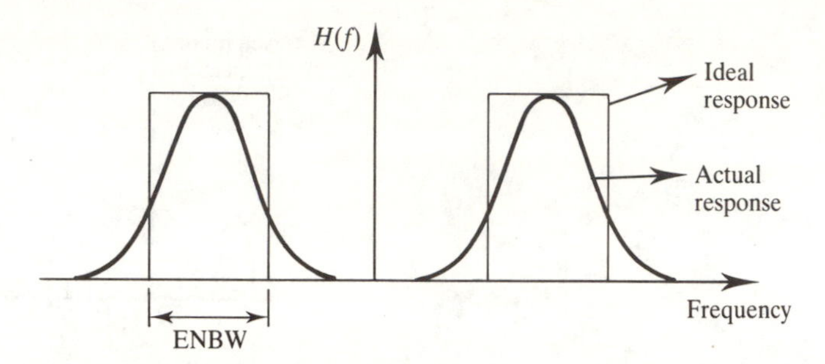

Area under ideal response curves = area under actual response curves

Figure 10.3 Equivalent noise bandwidth of a filter.

transform each section separately. If the overlap is about 50% to 75%, then most features of the data will be included in the sequences. The resulting spectra are then averaged to obtain an estimate of the true spectrum. The partitioning of the data is illustrated in Figure 10.4. This procedure is called redundancy or overlap processing. Generally 50–75% overlap processing provides 90% of the possible performance improvement for most weighting functions (DeFatta *et al.*, 1988). By averaging the spectra of the sections the variance of the spectrum is reduced. For K statistically identical but independent measurements the variance of the average is $1/K$ times the variance of the individual spectrum. However, this is not true when the spectra of overlapped sections are averaged because of the correlation between them. For the cases of 50% and 75% overlap the variance of the averaged spectrum to the individual spectrum is given in Harris (1978), from which it can be shown that, for example, the averaging of four spectra reduces the variance to 25% of that of the original spectrum. This represents a significant improvement in the estimate of the spectrum.

Processing gain

The processing gain, PG, is defined as the ratio of the output signal-to-noise ratio subsequent to windowing to the input signal-to-noise ratio prior to windowing:

$$\mathrm{PG} = \frac{(S/N)_{\mathrm{O/P}}}{(S/N)_{\mathrm{I/P}}} \tag{10.14}$$

The processing gain depends on the shape of the window since this determines its equivalent noise bandwidth (see the previous section on this subject). The taper of the window reduces the signal power causing a processing loss, while the side lobes enhance the noise bandwidth.

Worst-case processing loss

The worst-case processing loss (WCPL) is defined to be the sum (in decibels) of the maximum scalloping loss of a window and its processing loss (PL). It

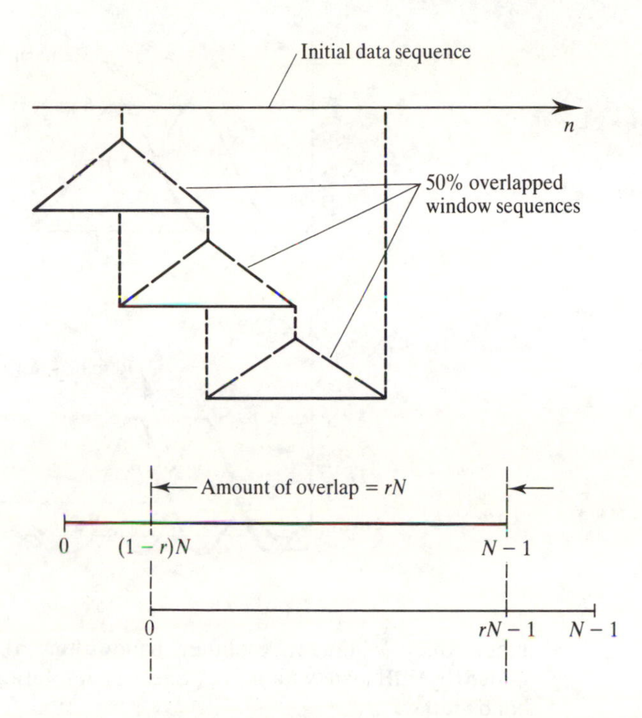

Figure 10.4 Partitioning of the data for overlap processing.

represents the reduction in output signal-to-noise ratio owing to windowing and worst-case frequency location. It always lies between 3.0 and 4.3 dB. Windows for which it exceeds 3.8 dB should be avoided. These include the rectangular, Poisson ($\alpha = 4$), Hanning–Poisson ($\alpha = 2.0$), Cauchy ($\alpha > 4$), and minimum four-sample Blackman–Harris windows.

Minimum resolution bandwidth

Normally when two identical spectral peaks overlap they may be resolved provided that they do not overlap beyond their 3 dB points, (Figure 10.5(a)). However, in the case of DFT components the adjacent spectral components are weighted by the window and then summed coherently, that is the side lobes are included in the summation. The gain of each component at cross-over must not exceed 0.5. This means that the spectral resolution is determined by the 6 dB bandwidth of the components rather than the 3 dB bandwidth (Figure 10.5(b)).

Biasing effect of data windows

Multiplication of the data by a tapered window reduces the sample magnitudes at the taper and hence the total signal power. It can be shown that each frequency component is affected identically by the window and that the multiplying factor is proportional to the square root of the coherent power gain, which represents the normalized power of the data window regarded as a voltage waveform. Thus the reduction in signal power may be restored without

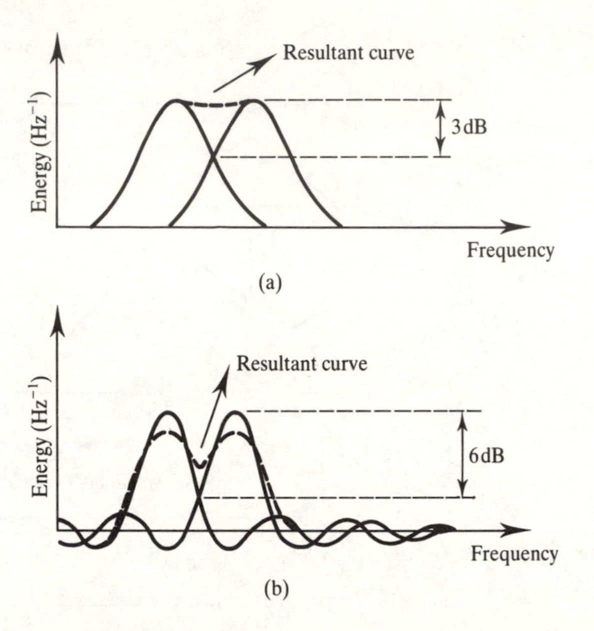

Figure 10.5 Minimum resolution bandwidth. (a) Resolution of spectral peaks determined by 3dB bandwidths. (b) Spectral resolution of DFT peaks determined by 6dB bandwidths.

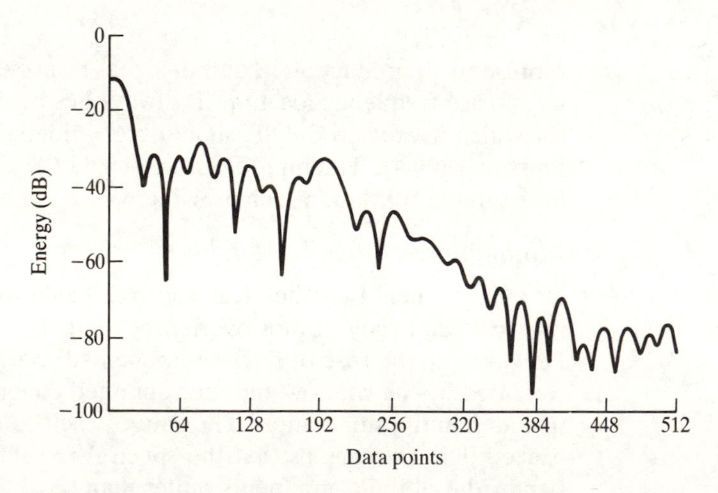

Figure 10.6 Energy-spectrum of 1 s interstimulus interval (ISI) CNV, Kaiser–Bessel window. 64 data points, 1024-point FFT.

distorting the power density spectrum. The window also adjusts the mean level of the data, thereby increasing the apparent energy of the lower frequency components in the spectrum. Some means of compensating for this is required but the straightforward subtraction of the mean of the windowed data results in pronounced high frequency side lobes.

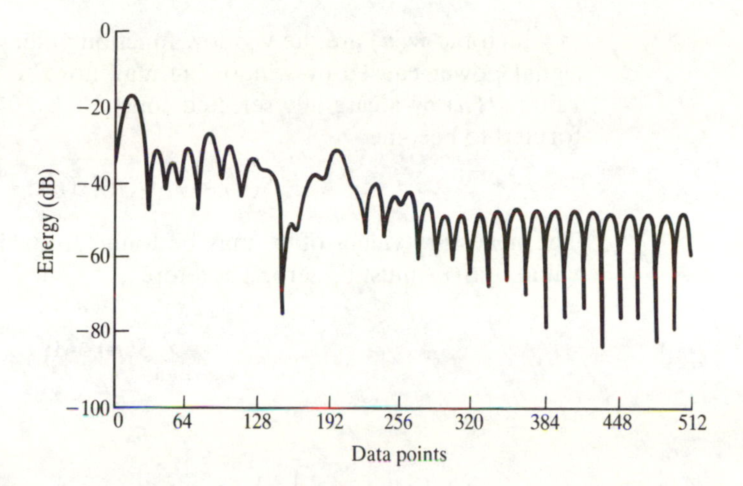

Figure 10.7 Energy spectrum of 1 s ISI CNV, Kaiser–Bessel window. 64 data points, 1024-point FFT. Mean level of windowed data removed by subtraction of mean from data.

Figure 10.6 shows the energy density spectrum of 64 data points from which the mean level of the data had first been subtracted after which the data had then been multiplied by the corresponding values of a Kaiser–Bessel window. It can be seen that the spectrum contains low frequency components introduced by the windowing process. Figure 10.7 shows the spectrum obtained when the mean level of the windowed data had first been subtracted. The low frequency components are seen to have been removed but at high frequencies a marked side lobe structure has become apparent. The following example shows how the deleterious effects of windowing may be overcome.

Example 10.1

Show that the twin effects of the window in reducing the signal energy and in introducing low frequency components into the spectrum may be overcome by windowing a linear function of the data rather than the data (Jervis *et al.*, 1989a).

Solution

Assume the initial data $s(n)$ has zero mean. Let the mean level introduced into the data by windowing be removed by subtracting a constant k_1 from $s(n)$. The new windowed data values are now given by $s^1(n)$ where

$$s^1(n) = w(n)[s(n) - \mathrm{k}_1] \qquad \textbf{(10.15)}$$

in which the $w(n)$ are the window function values or weights. The reduction in signal power caused by windowing may now be restored by multiplying each value $s^1(n)$ by a carefully selected constant k_2. Thus the data points are transformed to become

$$S(n) = k_2 w(n)[s(n) - k_1] \tag{10.16}$$

The necessary values of k_1 may be found from the condition that the average value of $S(n)$ must be zero. Therefore

$$\sum_{n=0}^{N-1} S(n) = 0$$

Hence

$$k_2\left[\sum_{n=0}^{N-1} w(n)s(n) - \sum_{n=0}^{N-1} w(n)k_1\right] = 0$$

Therefore

$$k_1 = \frac{\sum_{n=0}^{N-1} w(n)s(n)}{\sum_{n=0}^{N-1} w(n)} \tag{10.17}$$

The normalized ac power of the data prior to windowing is

$$E[(s(n) - k_1)^2] = \sigma_{sN}^2 \tag{10.18}$$

where E denotes the expected value and σ_{sN}^2 is the variance of $s(n)$ with mean k_1. The normalized ac power of the windowed data is

$$E\{k_2^2 w^2(n)[s(n) - k_1]^2\}$$

where $w(n)$ and $s(n)$ are mutually independent. Now

$$\begin{aligned} E\{k_2^2 w^2(n)[s(n) - k_1]^2\} &= E[k_2^2 w^2(n)]E\{[s(n) - k_1]^2\} \\ &= k_2^2 E[w^2(n)]\sigma_{sN}^2 \end{aligned} \tag{10.19}$$

The value of k_2 required to make the powers of the windowed and unwindowed data the same may be obtained by equating Equations 10.18 and 10.19:

$$\sigma_{sN}^2 = k_2 E[w^2(n)]\sigma_{sN}^2$$

so

$$k_2^2 = \frac{1}{E[w^2(n)]} = \frac{1}{(1/N)\sum_{n=0}^{N-1} w^2(n)} = \frac{N}{\sum_{n=0}^{N-1} w^2(n)}$$

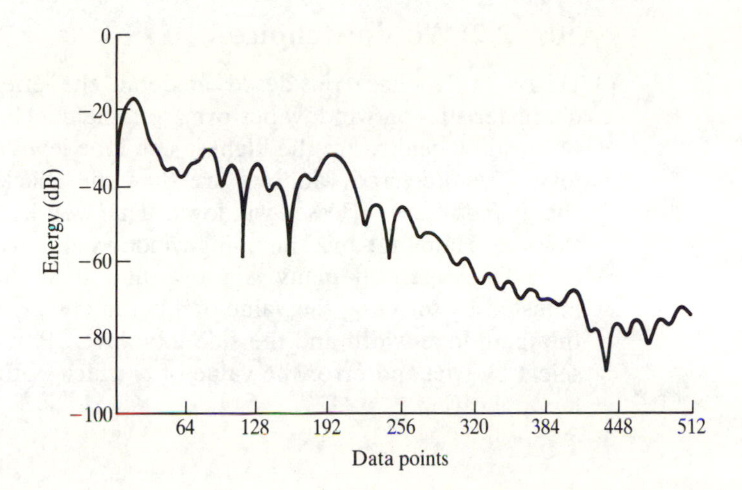

Figure 10.8 Energy spectrum of 1 s ISI CNV, Kaiser–Bessel window. 64 data points, 1024-point FFT. Data pre-processed to remove windowed mean level and to maintain mean power.

Therefore

$$k_2 = \left[\frac{N}{\sum_{n=0}^{N-1} w^2(n)} \right]^{1/2} \tag{10.20}$$

Substituting Equations 10.17 and 10.20 into Equation 10.16 finally gives

$$S(n) = w(n) \left[s(n) - \frac{\sum_{n=0}^{N-1} w(n)s(n)}{\sum_{n=0}^{N-1} w(n)} \right] \left[\frac{N}{\sum_{n=0}^{N-1} w^2(n)} \right]^{1/2} \tag{10.21}$$

Figure 10.8 shows the resulting energy spectrum when the mean of the windowed data is removed and the signal energy restored using Equation 10.21. It can be seen that the dc level owing to windowing and the side lobe effect have both been removed.

The recommended procedure is therefore to modify the data $s(n)$ according to Equation 10.21 prior to computation of its DFT. This is equivalent to first subtracting k_1 from the data and then multiplying the difference by k_2 prior to windowing.

10.3.2.2 Window choice

Harris (1978) has considered in detail the effects of the different window characteristics on window performance and concluded that the major influences on window quality are the highest side lobe level and the worst-case processing loss. The preferred windows are then the Blackman–Harris, Dolph–Chebyshev, and Kaiser–Bessel windows. The well-known Tukey (cosine-tapered), Poisson, Hanning, and Hamming windows all have inferior performance.

The taper on many windows and their shape may in many cases be adjusted by selecting the value of a parameter, α. The effect of this is to adjust the main lobe width and the side lobe level. Part of the art of windowing is to select by trial and error the value of α which optimizes the results in a particular application.

Example 10.2

The effect of different windows on an amplitude spectrum Figure 10.9(a) shows the DFT components of two sine waves, differing in amplitude by 40 dB, and of frequencies $100f$ and $120f$, where f is the first harmonic frequency corresponding to a record length of T_s(s), and which were obtained without windowing, that is with a rectangular window. When the signal is harmonically related to the window length in this way the signal appears periodic and infinite and is faithfully reproduced even by a rectangular window. This periodic relationship was broken to yield the results of Figure 10.9(b) by changing the frequency of the larger signal to $102.5f$ which is not a harmonic frequency. It can be seen that the result has been a big increase in side lobe level which almost masks out the smaller signal. This effect may be largely overcome by the choice of a suitable window. Figures 10.10(a) and 10.10(b) show tapered cosine (Tukey) windows with $\alpha = 0.1$ and $\alpha = 0.5$ respectively while Figures 10.11(a) and 10.11(b) show the respective spectra. Figures 10.12(a) and 10.12(b) show a Hamming window ($\alpha = 0.54$) and the resulting spectrum respectively. Figures 10.13(a)–10.13(c) show Kaiser–Bessel windows with $\alpha =$ 2.0, 3.0, and 4.0 respectively while Figure 10.14 illustrates the corresponding spectra. The result nearest the true spectrum is given by the Kaiser–Bessel window with $\alpha = 4.0$. However, it should be noted that as α is increased to decrease the side lobe level the main lobe width increases. A compromise value of α must be sought. Of course in a real situation all the harmonics will be affected so that window selection with respect to multitone performance is very important if spurious results are to be avoided.

The Dolph–Chebyshev window gives the best results with regard to low side lobes and worst-case processing loss but the coherent addition of its side

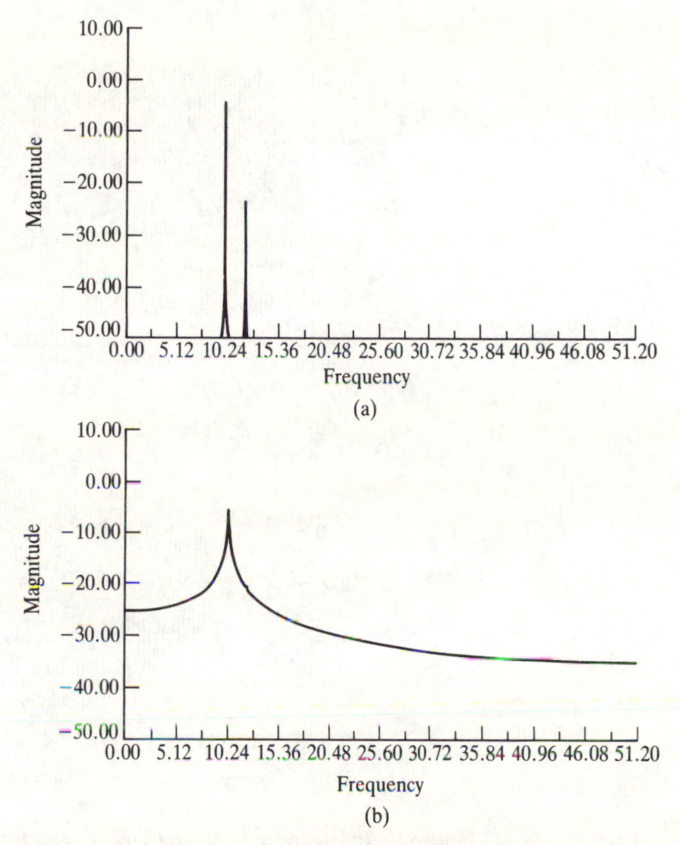

Figure 10.9 Amplitude spectra of two sine waves differing in amplitude by 40dB. (a) Window length is an integer multiple of both periods. (b) Window length is not an integer multiple of one of the periods (the larger period).

lobes detracts from its multitone detection performance. Also, its side lobe structure exhibits extreme sensitivity to coefficient errors. Therefore the Blackman–Harris or Kaiser–Bessel windows are to be preferred. The Kaiser–Bessel window has the advantages that its coefficients are easier to generate and the side lobe level versus main lobe width compromise is easier to adjust by varying α. The expression for the Kaiser–Bessel window is (Kuo and Kaiser, 1966)

$$w(n_{\mathrm{KB}}) = I_0\left\{\pi\alpha\left[1.0 - \left(\frac{n_{\mathrm{KB}}}{N/2}\right)^2\right]^{1/2}\right\}/I_0(\pi\alpha), \quad 0 \leqslant |n_{\mathrm{KB}}| \leqslant N/2 \qquad \textbf{(10.22)}$$

in which n_{KB} is the window function sample number, α is a numerical parameter which may be adjusted to select the best side lobe level versus main lobe width compromise, N is the number of sample points in the window, and

$$I_0(x) = \sum_{k=0}^{K}\left[\frac{(x/2)^k}{k!}\right]^2 \qquad \textbf{(10.23)}$$

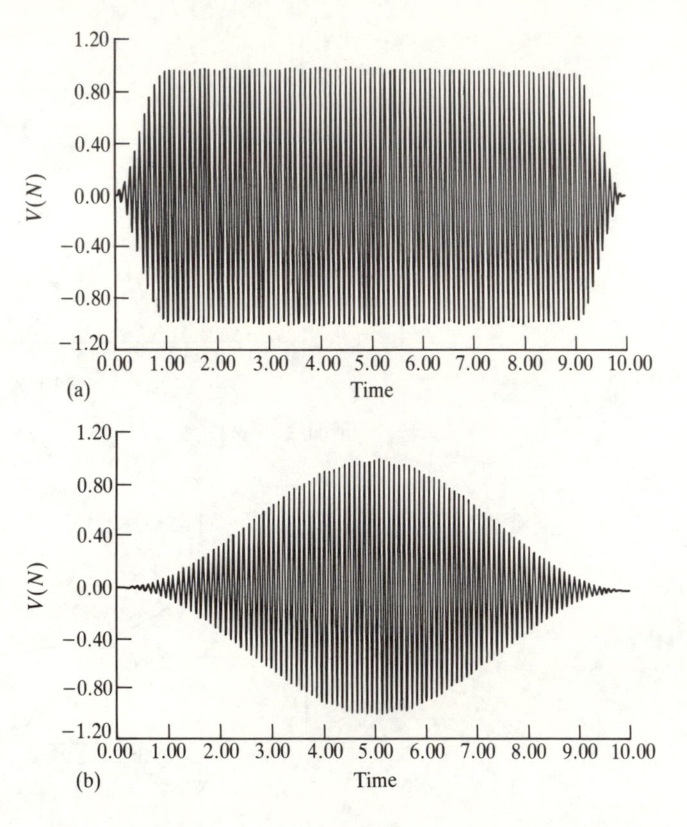

Figure 10.10 Tapered cosine (Tukey) windows. (a) $\alpha = 0.1$. (b) $\alpha = 0.5$.

is a zero-order modified Bessel function of the first kind, and K is theoretically infinite but, because the magnitude of the Bessel function decreases rapidly with k, it is usually adequate to set $K = 32$.

Equation 10.22 defines the window between the points $-N/2$ and $N/2 - 1$. The DFT is normally required to extend from $n_{\mathrm{DFT}} = 0$ to $n_{\mathrm{DFT}} = N - 1$ where n_{DFT} is the DFT datum number. Therefore for use with the DFT the Kaiser–Bessel window has to be shifted right $N/2$ places to satisfy

$$n_{\mathrm{DFT}} = n_{\mathrm{KB}} + N/2$$

or

$$n_{\mathrm{KB}} = n_{\mathrm{DFT}} - N/2 \qquad \textbf{(10.24)}$$

Hence Equation 10.22 must be modified to

$$w(n_{\mathrm{DFT}}) = I_0\left\{\pi\alpha\left[1.0 - \left(\frac{n_{\mathrm{DFT}} - N/2}{N/2}\right)^2\right]^{1/2}\right\}/I_0(\pi\alpha), \quad 0 \leqslant n_{\mathrm{DFT}} \leqslant N - 1 \qquad \textbf{(10.25)}$$

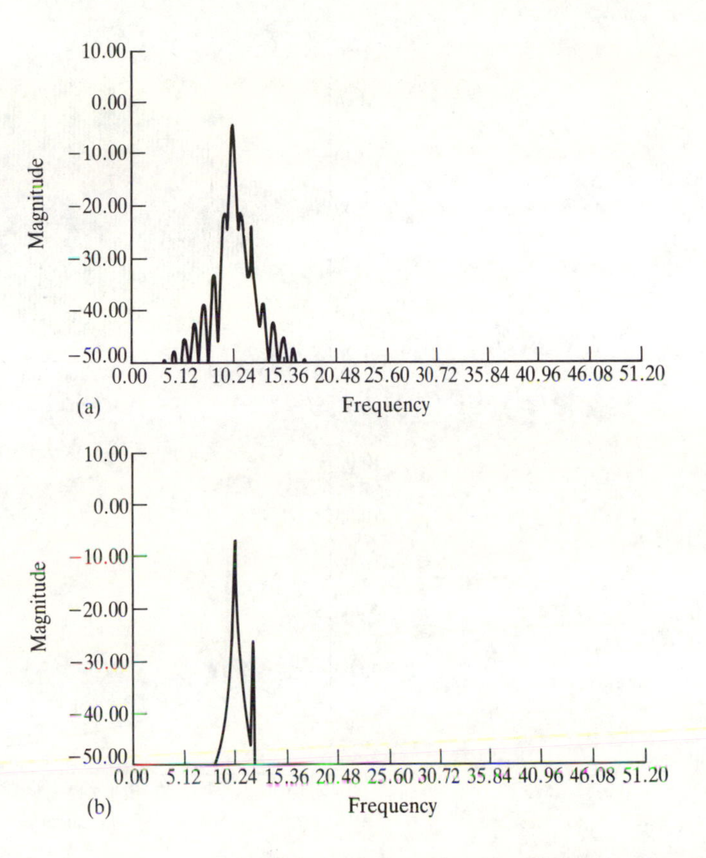

Figure 10.11 Amplitude spectra of the two sine waves multiplied by the tapered cosine windows.

10.3.3 The periodogram method and periodogram properties

The square of the modulus of the Fourier transform, $|F(j\omega)|^2$, is the estimated power spectral density, $E[P(f)]$, and is also known as the periodogram. $E[P(f)]$ can be shown (Proakis and Manolakis, 1989; DeFatta *et al.*, 1988) to be given by

$$E[P(f)] = \sum_{m=-\infty}^{m=\infty} w_{\mathrm{B}}(m) c_{xx}(m) \exp(-\mathrm{j}2\pi f m) \qquad \mathbf{(10.26)}$$

in which $c_{xx}(m)$ is the autocovariance of $x(n)$ evaluated at lag m, f is the frequency, and $w_{\mathrm{B}}(m)$ is the triangular (Bartlett) window defined by

$$w_{\mathrm{B}}(m) = 1 - \frac{|m|}{N} \qquad |m| \leq N - 1 \qquad \mathbf{(10.27)}$$

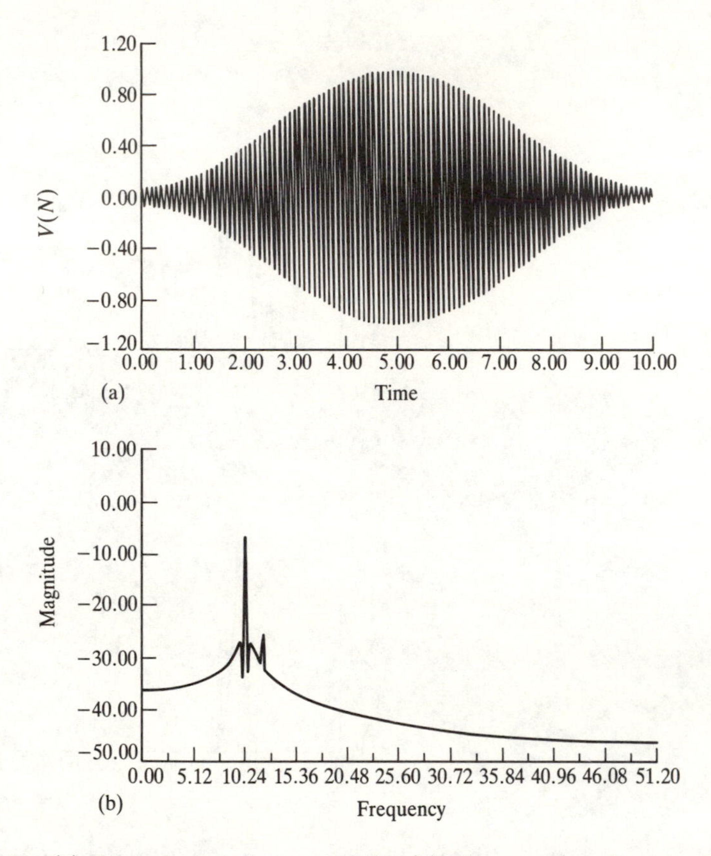

Figure 10.12 (a) Hamming window, $\alpha = 0.54$. (b) The corresponding amplitude spectrum of the two sine waves.

By comparison the true power spectral density, $P(f)$, of $x(n)$ is

$$P(f) = \sum_{m=-\infty}^{\infty} c_{xx}(m) \exp(-j2\pi fm) \tag{10.28}$$

The power spectral density given by the periodogram is therefore biased with the bias given by

$$\begin{aligned} P(f) - \mathrm{E}[P(f)] &= \sum_{m=-\infty}^{\infty} [1 - w_{\mathrm{B}}(m)] c_{xx}(m) \exp(-j2\pi fm) \\ &= \frac{|m|}{N} P(f) \end{aligned} \tag{10.29}$$

For $N \gg |m|$ the bias becomes small and $E[P(f)] \rightarrow P(f)$, that is the periodogram is asymptotically unbiased. Also, for large N the variance of the periodogram becomes

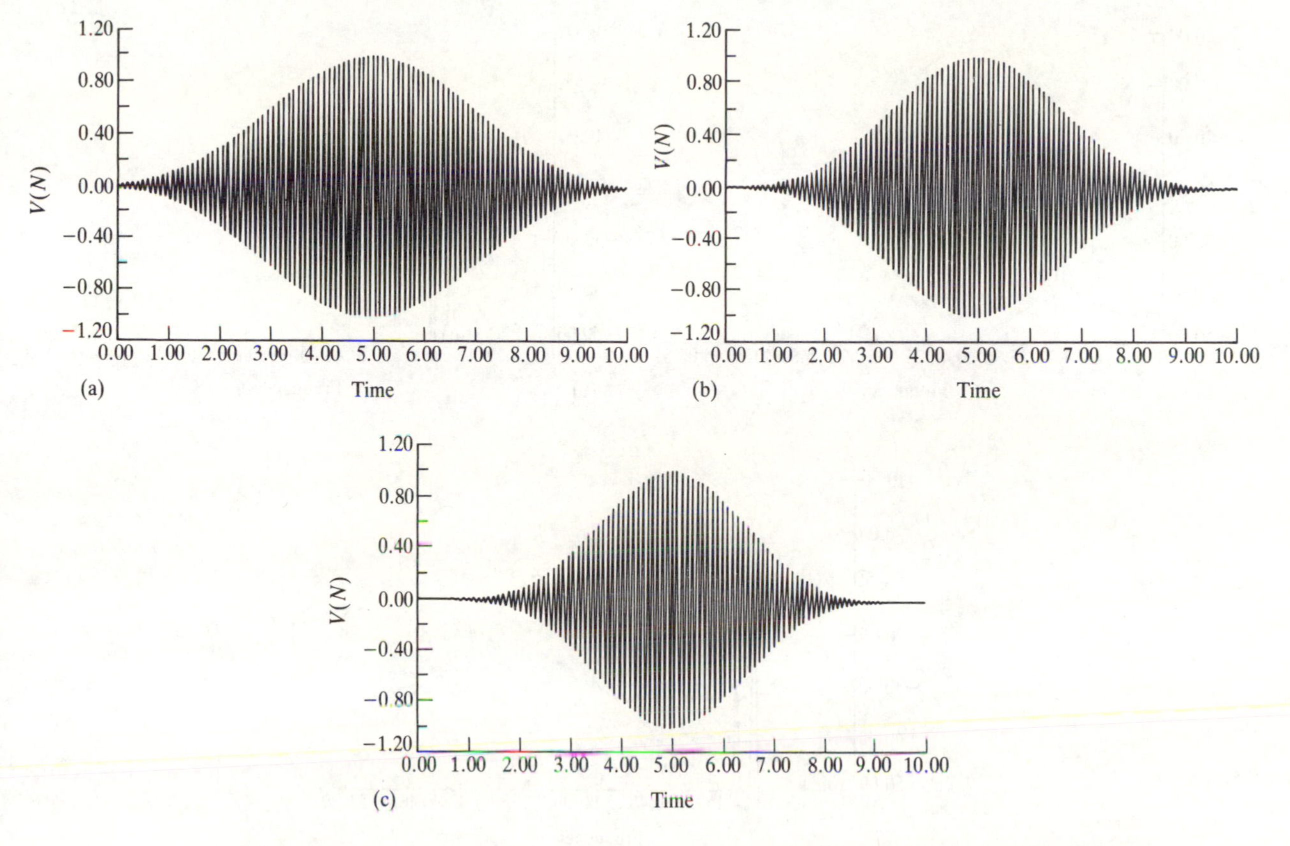

Figure 10.13 Kaiser–Bessel windows. (a) $\alpha = 2.0$. (b) $\alpha = 3.0$. (c) $\alpha = 4.0$.

$$\operatorname{var}[P(f)] \approx FP^2(f) \tag{10.30}$$

where F depends on the window function used, that is the variance depends on the square of the power spectral density and does not converge towards zero with increasing N. This means that power spectral density estimates obtained from the periodogram are inconsistent and yield fluctuating estimates of $P(f)$ from successive realizations.

Note that the autocovariance is usually determined by averaging over N terms rather than the alternative $N - |m|$ terms. Both estimates are consistent and asymptotically unbiased, but the former has a smaller variance which is why it is preferred.

Furthermore, when the DFT is applied to obtain the spectrum, the corresponding periodogram is defined as $(1/N)|X(k)|^2$ and has the dimension of normalized energy although the reader may note that some authors still refer to the function $(1/N)|X(k)|^2$ as the power spectral density.

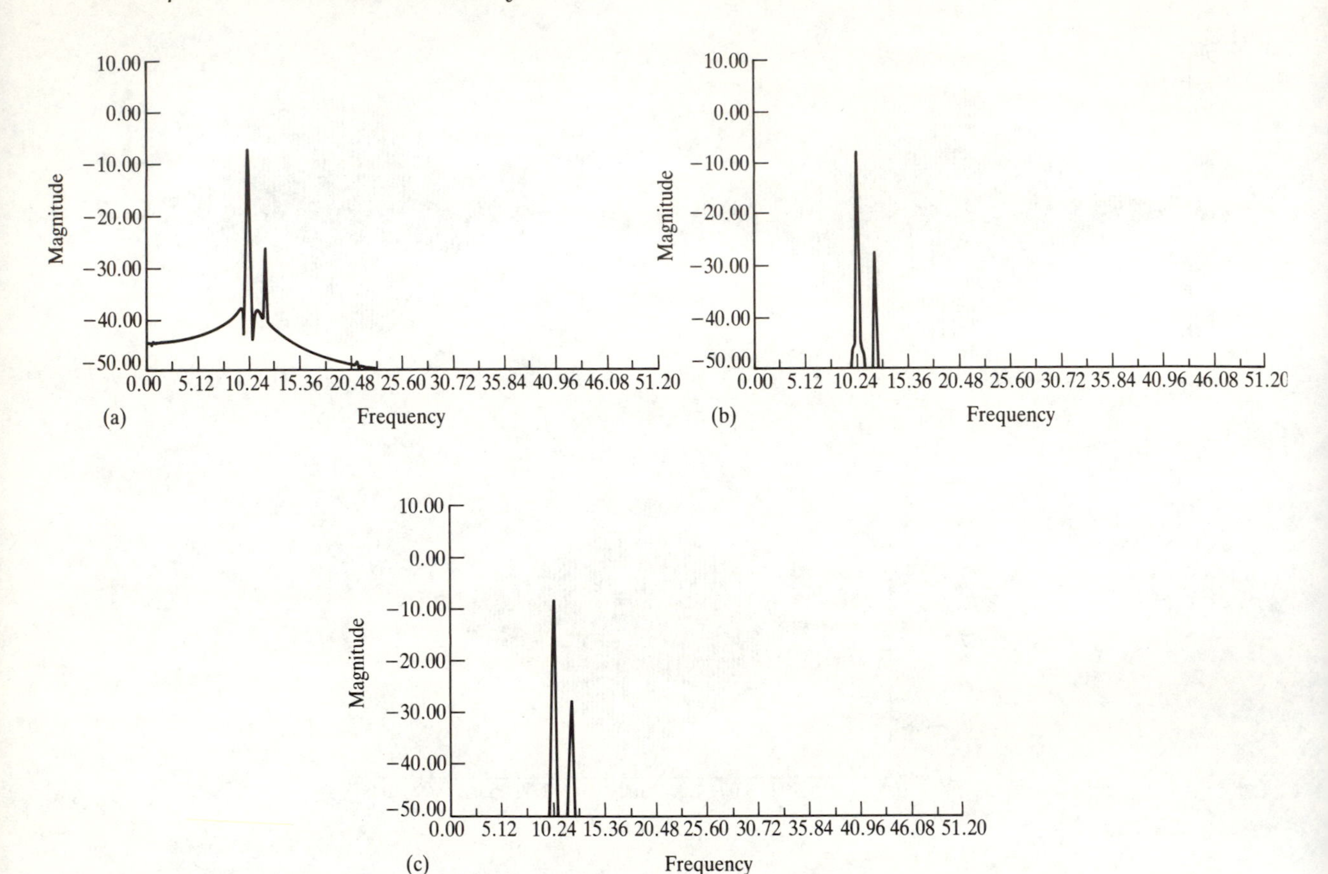

Figure 10.14 The amplitude spectra of the two sine waves calculated using the corresponding Kaiser–Bessel windows of Fig 10.13. (a) $\alpha = 2.0$. (b) $\alpha = 3.0$. (c) $\alpha = 4.0$.

10.3.4 Modified periodogram methods

Welch (1967) suggested that the inconsistency of the periodogram method could be overcome by averaging a number of modified periodograms. Each of the latter consists of a section of the data. These sections may be sequential (the Bartlett method) or overlapped (the Welch method). The methods also reduce the variance of the power spectral density estimates. This is demonstrated for the Bartlett method in Example 10.3.

10.3.4.1 The Welch method

An advantage of the Welch method over the Bartlett method is that the variance of the power spectral density is further reduced. However, this is at the expense of a further decrease in spectral resolution. In the Welch method L data sections of length M are overlapped and the periodograms are computed from the L windowed data sections. Also, the periodograms are normalized by the factor U to compensate for the loss of signal energy owing to the windowing procedure. In fact U equates to $1/k_2^{1/2}$ where k_2 is the factor derived in part of Section 10.3.2.1 on the biasing effect of data windows as necessary to compensate for this reduction in signal energy. Thus,

$$U = \frac{1}{M} \sum_{n=0}^{M-1} w^2(n) \tag{10.31}$$

The Welch power density spectral estimate, $P_{\mathrm{WE}}(f)$, is therefore

$$P_{\mathrm{WE}}(f) = \frac{1}{L} \sum_{j=0}^{L-1} P_{\mathrm{j}}(f) \tag{10.32}$$

The expected value of the Welch estimate is

$$E[P_{\mathrm{WE}}(f)] = \frac{1}{L} \sum_{j=0}^{L-1} E[P_{\mathrm{j}}(f)] = E[P_{\mathrm{j}}(f)] \tag{10.33}$$

that is the same as the expected value of the modified periodogram. It can be shown (Proakis and Manolakis, 1989) that as $N \to \infty$ and $M \to \infty$ the value converges to that of the true power spectral density, $P(f)$. Thus for large N and M the Welch power spectral density estimate is unbiased. Under the same conditions the variance of the Welch estimate converges to zero, that is the estimate is consistent. Welch showed that for the case of no overlap ($L = K$)

$$\mathrm{var}\,[P_{\mathrm{WE}}(f)] \approx (1/K)P^2(f)$$

which is equal to that of the Bartlett variance under the same conditions. For 50% overlap ($L = 2K$)

$$\mathrm{var}\,[P_{\mathrm{WE}}(f)] \approx (9/8L)P^2(f)$$

which is less than that for the Bartlett case by the factor $9/16 = 0.56$.

10.3.5 The Blackman–Tukey method

It was established in Chapter 2 that the power density spectrum is given by the DFT of the autocorrelation function of the data. One might query the usefulness of this approach knowing that the periodogram can be computed directly from the data as the square of the DFT. Well, first it is noteworthy that the Blackman–Tukey method was introduced in 1958 (Blackman and Tukey, 1958) while the FFT algorithm for fast computation of the DFT was not published by Cooley and Tukey until 1965 (Cooley and Tukey, 1965). Second, it is possible that the Blackman–Tukey approach might contain some advantage over the periodogram method. Indeed, it will be shown in the next section that the Blackman–Tukey method is characterized by a larger quality factor. In addition autocorrelation functions may now be computed using DFTs by the fast correlation method (Section 10.3.6). The Blackman–Tukey procedure then is

(1) to calculate the autocorrelation function of the data,
(2) to apply a suitable window function to the data, and
(3) to compute the FFT of the resulting data to obtain the power density spectrum.

By comparison with the periodogram method we see that the smoothing is achieved by the averaging effect of the autocorrelation process rather than by the averaging of several periodograms.

The autocorrelation function is windowed to taper it towards its extremes because at larger lags fewer data points enter the computation so these estimates are less accurate. Tapering has the effect of attaching less weight to these estimates.

The Blackman–Tukey estimate, $P_{\mathrm{BTE}}(f)$, is

$$P_{\mathrm{BTE}}(f) = \sum_{m=-(M-1)}^{M-1} r_{xx}(m)w(m)\exp(-\mathrm{j}2\pi fm) \qquad \textbf{(10.34)}$$

where $r_{xx}(m)$ is the autocorrelation function of the data and $w(n)$ is the window function of length $2M-1$ and is zero for $|m| \geqslant M$.

In order to obtain real estimates $w(n)$ must be symmetrical about $m=0$, and for the estimates to be positive its transform must be positive. Not all windows satisfy these criteria. The Hanning and Hamming windows are examples of two which do not.

It may be shown that the expected value of the Blackman–Tukey estimate is

$$E[P_{\mathrm{BTE}}(f)] = \sum_{m=-(M-1)}^{M-1} c_{xx}(m)w_{\mathrm{B}}(m)\exp(-\mathrm{j}2\pi fm) \qquad \textbf{(10.35)}$$

in which $w_{\mathrm{B}}(m)$ is the triangular Bartlett window.

The condition $M<N$ has to be satisfied to obtain additional smoothing of the spectrum. If $N \gg m$ the estimate will be asymptotically unbiased. Also, if $W(k)$, the DFT of the window, $w(n)$, is narrower than $P(f)$, the true power-density spectrum, then

$$\mathrm{var}[P_{\mathrm{BTE}}(f)] \approx P^2(f)\left[\frac{1}{N}\sum_{m=-(M-1)}^{M-1} w^2(m)\right] \qquad \textbf{(10.36)}$$

and as $N/M \to \infty$, $\mathrm{var}[P_{\mathrm{BTE}}(f)] \to 0$, showing that under these conditions the Blackman–Tukey estimate is consistent.

10.3.6 The fast correlation method

It was pointed out in Section 4.3.4 that if in excess of 128 data are to be correlated the computation is quicker if use is made of the correlation theorem

Table 10.1 Quality factors Q for power spectral density estimates.

Estimation method	*Conditions*	Q	*Comments*
Periodogram	$N \to \infty$	1	Inconsistent, independent of N
Bartlett	$N, M \to \infty$	$1.11Nf$	Quality improves with data length
Welch	$N, M \to \infty$, 50% overlap	$1.39Nf$	Quality improves with data length
Blackman–Tukey	$N, M \to \infty$, triangular window	$2.34Nf$	Quality improves with data length

(Equation 4.50) to implement the calculations using FFTs. For example, this yields a tenfold speed increase if $N = 1024$. In addition, if large amounts of input data are involved such as may exceed the memory capacity of the system then the overlap–add or overlap–save sectioning techniques may be applied (Sections 4.3.5–4.3.7). When the autocorrelation in the Blackman–Tukey method is computed using FFTs in these ways the method is known as the fast correlation method for spectral estimation.

10.3.7 Comparison of the power spectral density estimation methods

A quality factor for estimates of power spectral density was given in Equation 10.6. It can be shown (Proakis and Manolakis, 1989) that the quality factors for the four nonparametric spectral analysis methods are as given in Table 10.1 where f is the 3 dB main lobe width of the associated windows. It is seen that the Blackman–Tukey method is superior for quality, and that, with the exception of the periodogram method, the quality may be maintained as the frequency resolution is increased (decrease in f) by increasing N.

Great care and some trial computations are required to ensure satisfactory results. On balance the Blackman–Tukey method would seem to be marginally the best, but other considerations may lead to a preference for one of the other methods.

10.4 Modern parametric estimation methods

The nonparametric methods described in the previous sections of this chapter which utilize periodograms and FFTs are subject to the aforementioned limitations of low spectral resolution in the case of short records and the requirement for windowing to reduce the spectral leakage. These difficulties may be overcome by parametric methods (Burg, 1968; Nuttall, 1976; Ulrych and Clayton, 1976; Marple, 1980; Cadzow, 1979, 1982; Graupe *et al.*, 1975; Kay, 1980; Friedlander, 1982). The price to be paid is an extensive investigation of an appropriate model for each process, a determination of the necessary order of the chosen model for adequate representation of the data (Whittle, 1965; Jenkins and Watts, 1968; Box and Jenkins, 1976; Chatfield, 1979; Akaike, 1969, 1973, 1974, 1978, 1979; Shibata, 1976; Rissanen, 1983), and computation of the model parameters (Proakis and Manolakis, 1989; Makhoul, 1975; Levinson, 1947; Durbin, 1959; Priestley, 1981; Wold, 1954; Chatfield, 1984). The advantages gained are increased spectral resolution, applicability to short data lengths, and avoidance of spectral leakage, scalloping loss, spectral smearing, and window biasing effects. Nevertheless, although an improvement over the nonparametric methods described, these parametric methods do have some disadvantages which may be avoided by alternative modern approaches such as the sequential or adaptive (Friedlander, 1982; Kalouptsidis and Theodoridis, 1987) and the maximum likelihood methods (Capon, 1969; Lacoss, 1971).

To summarize, the parametric approach calls for parametric modelling of the data, a well-established branch of time series analysis (Jenkins and Watts, 1968; Box and Jenkins, 1976; Priestley, 1981), combined with an interpretation of the data as being the output of a linear system excited by white noise. This system is represented by a polynomial transfer function expressed in terms of the model parameters. The spectrum of the data is computed from this transfer function.

10.5 Comparison of estimation methods

Of the nonparametric methods the Blackman–Tukey method has the larger quality factor and is therefore to be preferred, although for convenience one of the other approaches may be employed.

The parametric methods give greater frequency resolution and avoid the use of window functions. The Capon, maximum likelihood, method yields a minimum variance unbiased estimate with a spectral resolution intermediate between the Burg or the unconstrained least squares methods, and the nonparametric methods. The adaptive filtering methods emphasize the more recent data and are suitable for nonstationary data.

10.6 Application examples

10.6.1 Use of spectral analysis by a DFT for differentiating between brain diseases

Amplitude and phase spectra derived using an FFT have been utilized in a procedure to differentiate between Huntington's disease, schizophrenia, Parkinson's disease, and normal subjects by analysing selected harmonics in the

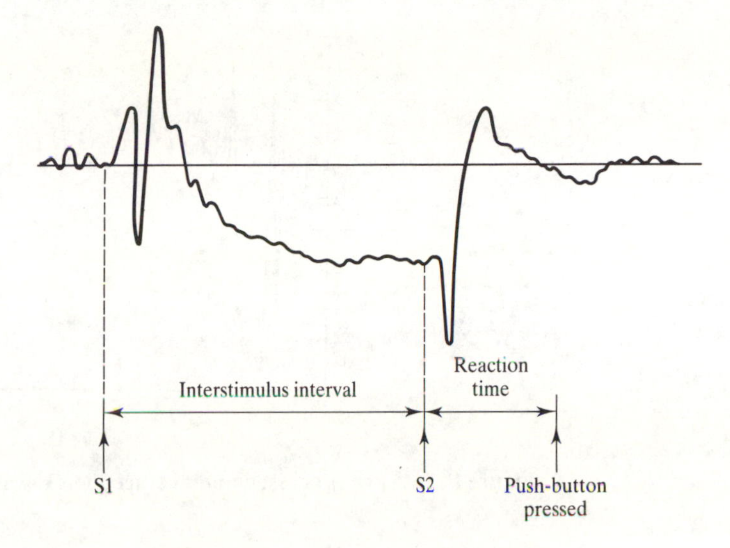

Figure 10.15 Schematic CNV waveform.

spectrum of the contingent negative variation (CNV) in the subjects' electro-encephalogram (EEG) (Jervis *et al.*, 1993).

The CNV is an event-related potential (ERP) manifesting as a negative electrical potential shift on the scalp on elicitation by a suitable auditory stimulus paradigm. A schematic drawing of a CNV waveform is shown in Figure 10.15. The CNV is the negative waveform found between the points of onset of the auditory stimuli at S_1 and S_2. There are reasons to believe that the CNV waveform is affected in people who suffer from one of the above-mentioned diseases. By determining the spectra of appropriate parts of the waveform, and treating them statistically, it is possible to differentiate between the subject categories.

Several CNVs were recorded from each subject using a purpose-designed signal processing instrumentation system (Jervis and Saatchi, 1990; Saatchi and Jervis, 1991). The data was then pre-processed to reduce the effect of the background EEG and of ocular artefacts on the CNV waveform. The mean level of the signals was removed so that a comparison over time could be made and to ensure that the ocular artefact removal algorithm functioned properly. The mean level removal caused a positive shift of the pre- and post-stimulus baseline. Therefore the baseline was corrected by subtracting the means of the different sections of the response from the corresponding sections. Digital lowpass filtering was then applied to filter out the unwanted high frequency components of the EEG. An FIR filter was used rather than an IIR filter because it would not distort the waveform. An ocular artefact removal algorithm was then applied to remove ocular artefacts by the method of proportional subtraction (Jervis *et al.*, 1989b). The averages of eight pre-processed CNV waveforms for one each of the subject categories are shown in Figures

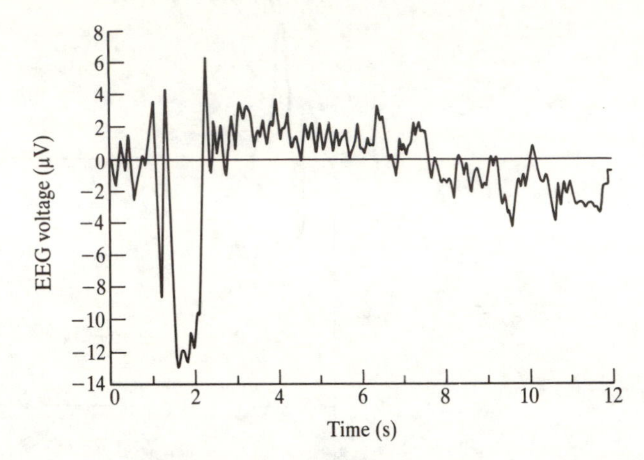

Figure 10.16 Pre-processed and averaged CNV waveform of a normal subject.

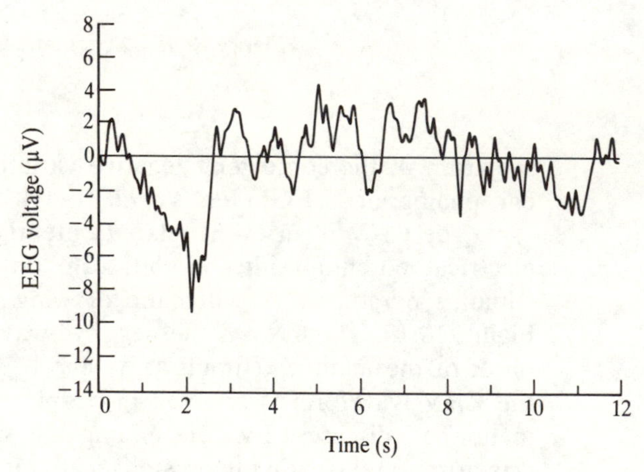

Figure 10.17 Pre-processed and averaged CNV waveform of a Huntington's Disease patient.

10.16–10.19. These are the averaged CNVs of a normal, a Huntington's disease, a schizophrenic, and a Parkinson's disease subject respectively.

Two 512 ms (64 sample) segments of each individual CNV waveform were then windowed using a Kaiser–Bessel window. Experiments indicated that a value of the window parameter, α, of 0.75 offered a satisfactory compromise between side lobe level and main lobe width. 960 augmenting zeros were added to the 64 data samples to reduce the scalloping loss. The DFTs of the sets of 1024 data were then calculated. Four statistical tests were applied to the first 96 harmonic components of the spectra thus produced. These tests are described in reference (Jervis *et al.*, 1983). The tests were entitled the nearest and furthest mean amplitude test, the pre- and post-stimulus mean amplitude difference test, the Rayleigh test of circular variance, and the modified Rayleigh test of circular variance. These tests resulted in a number of test statistics.

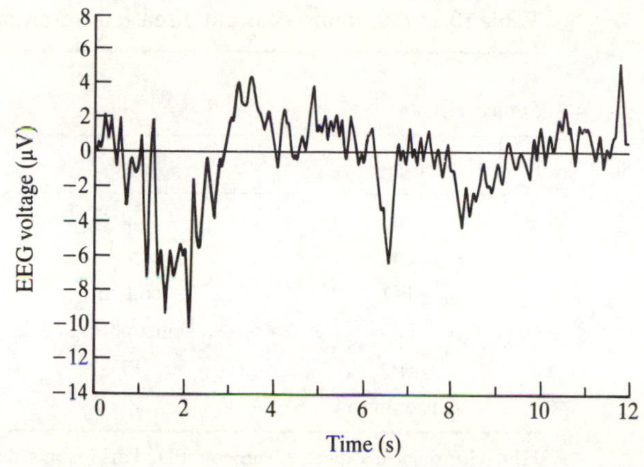

Figure 10.18 Pre-processed and averaged CNV waveform of a schizophrenic patient.

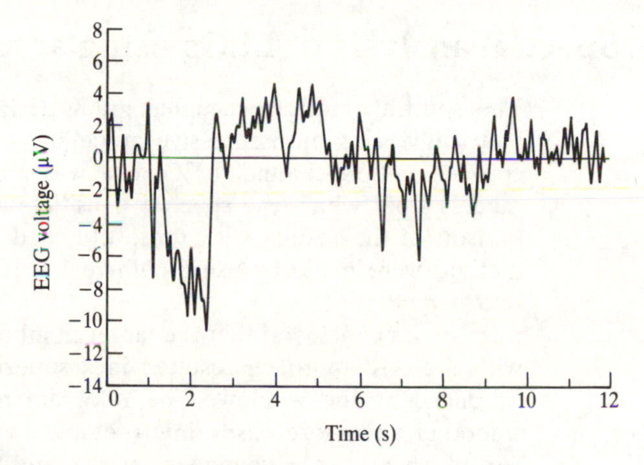

Figure 10.19 Pre-processed and averaged CNV waveform of a Parkinson's Disease patient.

In order to reduce the number of test statistics by selecting the more discriminatory ones they were subjected to the univariate test, the *t*-test, and stepwise discriminant analysis. This procedure was described in Jervis *et al.*, (1993) and was implemented by means of the statistical programs package SAS (SAS, 1982).

The classification of the individuals was now carried out using discriminant analysis (Morrison, 1976). Again, further details are given in Jervis *et al.* (1993) and the SAS package was used for implementation. The results are summarized in Table 10.2.

These results show the usefulness of spectral analysis applied to the CNV waveform when combined with statistical techniques in successfully differentiating to a high degree of accuracy between Huntington's disease, schizophrenic, Parkinson's disease, and normal subjects.

Table 10.2 Summary of brain disease differentiation results.

Subject types		*Differentiation success rates (%) in test domain*	
Type 1	*Type 2*	*Type 1*	*Type 2*
HD	control	100	100
schizophrenic	control	95	100
PD	control	93.8	87.5
HD	schizophrenic	100	90.9
HD	PD	90.9	81.8
schizophrenic	PD	81.3	93.8

HD, Huntington's disease subject; PD, Parkinson's disease subject; control, age- and sex-matched control subject.

10.6.2 Spectral analysis of EEGs using autoregressive modelling

The spectra of electroencephalograms (EEGs) have been determined parametrically using autoregressive modelling (Gersch, 1970). Simultaneously recorded multiple-channel EEG data was represented by autoregressive (AR) models from which the spectral densities were derived parametrically. Comparison of the results with those obtained using the windowed periodogram method were made by using both real EEG data and data simulated from a known model.

It was concluded that the larger number of degrees of freedom associated with the AR approach resulted in a superior statistical performance to that produced by the windowed periodogram method, and appeared to provide smoother and more easily interpretable results. The parametric method was less of an art since windowing was unnecessary and selection of the model order was automated. While more recent improvements in determining model order now exist, this particular application illustrates some advantages of parametric methods.

10.7 Summary

It has been suggested in this chapter that parametric methods of spectrum estimation may yield more reliable results. Since they may be automated they are probably preferable to the non-parametric periodogram-based methods. The latter are less reliable and are more of an art in that they require the application of significant expertise to secure meaningful results. However, their usefulness was illustrated particularly well by the clinical application to the differentiation of brain diseases.

10.8 Worked example

Example 10.3

Show that the Bartlett estimate of the power spectral density is asymptotically unbiased, that the variance of the estimate decreases with the number of data sections, and that the spectrum estimates are consistent. What effect does the modified periodogram method have on the frequency resolution?

Solution

In this method the periodograms of K non-overlapping sections of M data are computed. Given a total of N data this means $K = N/M$. The average of the K periodograms then gives the Bartlett power density spectral estimate, $P_{\mathrm{BE}}(f)$, where

$$P_{\mathrm{BE}}(f) = \frac{1}{K} \sum_{j=0}^{K-1} P_{\mathrm{j}}(f) \tag{10.37}$$

and j denotes the sequence number and $P_{\mathrm{j}}(f)$ is the corresponding jth estimate of the power spectral density.

Examining the expected values of $P_{\mathrm{BE}}(f)$ gives

$$\begin{aligned} E[P_{\mathrm{BE}}(f)] &= \frac{1}{K} \sum_{j=0}^{K-1} E[P_{\mathrm{j}}(f)] \\ &= E[P_j(f)] \\ &= \sum_{m=-(M-1)}^{M-1} \left(1 - \frac{|m|}{M}\right) c_{xx}(m) \exp(-\mathrm{j}2\pi fm) \end{aligned} \tag{10.38}$$

For $M \gg |m|$, the Bartlett window term $1 - |m|/M$ approaches unity and $E[P_{\mathrm{BE}}(f)]$ becomes the true power spectral density, $P(f)$. The Bartlett estimate of $P(f)$ is therefore asymptotically unbiased.

Turning now to the variance,

$$\mathrm{var}\,[P_{\mathrm{BE}}(f)] = \frac{1}{K^2} \sum_{j=0}^{K-1} \mathrm{var}\,[P_j(f)] = \frac{1}{K}\,\mathrm{var}\,[P_j(f)] \tag{10.39}$$

Thus the variance decreases in inverse proportion to the number of sections, K, into which the N data sequence has been divided. Because the variance decreases with K the Bartlett estimate of $P(f)$ is consistent. However, because the number of data from which the periodograms are calculated has been reduced by the factor K from N to $M = N/K$ the spectral resolution has also been reduced by the same factor. Thus the main lobe width obtained by the Bartlett method is K times that obtained using the full N data set.

Problems

10.1 (1) A waveform is sampled at the rate of 30 kHz. The first 524 288 samples are fast Fourier transformed. Calculate the frequency of the first harmonic and the frequency resolution of the spectrum.

(2) If the true spectrum of the waveform contained a sinusoidal component at 5.7505 Hz how would you modify the data to ensure this component was properly represented in the estimated spectrum?

10.2 A waveform is sampled at 8 kHz, the sampled voltages being 1.0, 1.0, 1.0, 1.0, 1.0, 1.0, 1.0, 1.0 V. The data is then windowed by a window function, the corresponding sampled values of which are 0, 0.5, 1.0, 1.0, 1.0, 1.0, 0.5, 0.0. Determine the DFT and hence the Fourier transform of the windowed data.

10.3 Calculate the scalloping loss and equivalent noise bandwidth associated with the data of Problem 10.2.

10.4 By referring to Figures 10.10, 10.12 and 10.13 estimate the equivalent noise bandwidth and processing gain of the following windows by using eight sampled values in each case:

(1) rectangular;

(2) Tukey (tapered cosine), $\alpha = 0.1$;

(3) Tukey, $\alpha = 0.5$;

(4) Hamming, $\alpha = 0.54$;

(5) Kaiser–Bessel, $\alpha = 2.0$;

(6) Kaiser–Bessel, $\alpha = 4.0$.

10.5 A telecommunications pulse waveform is sampled every 0.167 μs giving the values 0, 0, 1, 1, 1, 1, 1, 0 V. The nonzero values are windowed with a Kaiser–Bessel window with $\alpha = 4.0$. Compute the energy spectrum of the windowed pulse using an eight-point FFT (or DFT). (Obtain estimated window data from Figure 10.13c.)

10.6 Determine the scalloping loss, the processing loss, and the worst-case processing loss of the window in Problem 10.5.

10.7 For the data of Problem 10.5 determine the worst-case processing loss and the amplitude of the first side lobe for the Kaiser–Bessel window with $\alpha = 4.0$, the Hamming window with $\alpha = 0.54$, the Tukey window with $\alpha = 0.5$, and the rectangular window. Tabulate the results and select the most suitable window. (Use Figures 10.13c, 10.12a and 10.10b to obtain estimates of the window functions.)

10.8 The sampled voltages obtained from a waveform sampled at 8 kHz are 0, 4.0, 2.4, 1.0, −1.0, −3.8, −1.3, 0 V. Compute and plot the energy spectra:

(1) applying a Kaiser–Bessel window, $\alpha = 2.0$, to the data;

(2) after modification of the data according to the formula of Equation 10.21,

$$S(n) = w(n)\left[s(n) - \frac{\sum_{n=0}^{N-1} w(n)s(n)}{\sum_{n=0}^{N-1} w(n)}\right] \times \left[\frac{N}{\sum_{n=0}^{N-1} w^2(n)}\right]^{1/2}$$

where the $w(n)$ are the sampled values of the Kaiser–Bessel window with $\alpha = 2.0$;

(3) explain the reasons for any differences between the results obtained in parts (1) and (2).

10.9 Obtain the energy spectrum of the sampled data sequence {0, 1, 0, 1, 0, 1, 0, 1} and compare it with what you would expect for a square waveform.

10.10 Apply the Bartlett modified periodogram method to obtain the energy spectrum of the data of Problem 10.9 by sectioning the data into the two non-overlapping sequences {0, 1, 0, 1} and {0, 1, 0, 1}.

10.11 Apply the Welch modified periodogram method to obtain the energy spectrum of the data of Problem 10.9 by subdividing the data sequence into three equal length sections with 50% overlap.

10.12 The data sequence of Problem 10.9 is now assumed to contain a random noise component such that the new data sequence becomes {0.763, 1.656, 0.424, 1.939, 0.133, 1.881, 0.328, 1.348}. Calculate the energy spectrum of this noisy data and compare it with that of the original data. Estimate the signal-to-noise ratio from the sampled data.

10.13 Now repeat the Bartlett modified periodogram method of obtaining the energy spectrum by using the two non-overlapping sequences {0.763, 1.656, 0.424, 1.939} and {0.133, 1.881, 0.328, 1.348}.

10.14 Now repeat Problem 10.11 using the noisy data sequence of Problem 10.12.

10.15 Now assume that the original data sequence of Problem 10.9 is more highly contaminated with noise, so that the noisy data sequence is {6.03, 6.18, 3.35, 8.42, 1.05, 7.96, 2.59, 3.75}. Calculate the energy spectrum of this data and estimate the signal-to-noise ratio from the sampled data.

10.16 Obtain an improved estimate of the energy spectrum of the data of Problem 10.15 by the Bartlett method using two equal length sections.

10.17 Enhance the quality of the estimated energy spectrum of the data of Problem 10.15 by using the Welch modified periodogram method with three equal sections with 50% overlap.

10.18 Draw up a table to compare the results of Problems 10.9–10.17 for the different methods of spectrum estimation and for the different signal-to-noise ratios. Discuss the effects of the signal-to-noise ratio on the different methods and select your preferred method.

10.19 Obtain the power density spectrum of the data of Problem 10.9, that is {0, 1, 0, 1, 0, 1, 0, 1} using the Blackman–Tukey method and compare the result with that obtained for Problem 10.9. Use the window function {0, 0.5, 1, 1, 1, 1, 0.5, 0}.

10.20 Repeat Problem 10.19 using the noisy data of Problem 10.12.

10.21 Repeat Problem 10.19 using the noisy data of Problem 10.15.

10.22 Now compare the results obtained as answers to Problems 10.12–10.17 and ascertain whether a periodogram method or the Blackman–Tukey method gives the best result in the presence of noise.

10.23 Determine the energy spectrum of a square wave of period 1.0 μs which alternates between 0 V and 5 V in amplitude by using the Blackman–Tukey method. Compare your result with the theoretical one.

10.24 Write computer programs to generate waveforms of interest to yourself and design suitable spectrum analysis procedures using the techniques described in the chapter to evaluate the energy and phase spectra.

10.25 Devise some amplitude and phase spectra which show some interesting features which would challenge the spectrum estimation techniques, such as close peaks. Transform them to the time domain and then add noise to achieve a low, a unity, and a high signal-to-noise ratio. Now determine and criticize the amplitude and phase spectra.

References

Akaike H. (1969). Fitting autoregressive models for prediction. *Ann. Institute of Statistical Mathematics*, **21**, 243–7

Akaike H. (1973). Information theory and an extension of the maximum likelihood principle. In *2nd International Symposium on Information Theory* (Petrov B. N. and Csaki F. eds.), pp. 267–81. Budapest: Akademiai Kiade

Akaike H. (1974). A new look at the statistical model identification. *IEEE Trans. Automatic Control*, **19**, 716–22

Akaike H. (1978). A Bayesian analysis of the minimum AIC procedure. *Ann. Institute of Statistical Mathematics*, **30A**, 9–14

Akaike H. (1979). A Bayesian extension of the minimum AIC procedure of autoregressive model fitting. *Biometrika*, **66**, 237–42

Blackman R.B. and Tukey J.W. (1958). *The Measurement of Power Spectra*. New York NY: Dover

Box G.E.P. and Jenkins G.M. (1976). *Time-Series Analysis, Forecasting, and Control*. San Francisco CA: Holden-Day

Burg J.P. (1967). Maximum entropy spectral analysis. In *Proc. 37th Meeting Society Exploration Geophysicists*, Oklahoma City, October. Reprinted in Childers D. G., ed. (1968). *Modern Spectrum Analysis*. New York NY: IEEE Press

Burg J.P. (1968). A new analysis technique for time series data. In *NATO Advanced Study Institute on Signal Processing with Emphasis on Underwater Acoustics*, August 12–23, 1968. Reprinted in Childers D. G., ed. (1968). *Modern Spectrum Analysis*. New York NY: IEEE Press

Cadzow J.A. (1979). ARMA spectral estimation: an efficient closed-form procedure. In *Proc. RADC Spectrum Estimation Workshop*, Rome NY: October 1979, pp. 81–97

Cadzow J.A. (1982). Spectral estimation: an overdetermined rational model equation approach. *Proc. IEEE*, **70**, 907–38

Capon J. (1969). High-resolution frequency–wavenumber spectrum analysis. *Proc. IEEE*, **57**, 1408–18

Chatfield C. (1979). Inverse autocorrelation. *J. Royal Statistical Society A*, **142**, 363–77

Chatfield C. (1984). *The Analysis of Time Series* 3rd edn. London: Chapman and Hall

Cooley J.W. and Tukey J.W. (1965). An algorithm for the machine calculation of complex Fourier series. *Mathematics of Computation*, **19**, 297–301

DeFatta D.J., Lucas J.G. and Hodgkiss W.S. (1988). *Digital Signal Processing: A System Design Approach*, Section 6.6.5, p. 263. New York NY: Wiley

Durbin J. (1959). Efficient estimation of parameters in moving-average models. *Biometrika*, **46**, 306–16

Friedlander B. (1982). Lattice methods for spectral estimation. *Proc. IEEE*, **70**, 990–1017

Gersch W. (1970). Spectral analysis of EEGs by autoregressive decomposition of time series. *Mathematical Biosciences*, **7**, 205–22

Graupe D., Krause D.J. and Moore J.B. (1975). Identification of autoregressive-moving average parameters of time series. *IEEE Trans. Automatic Control*, **20**, 104–7

Harris F.J. (1978). On the use of windows for harmonic analysis with the discrete Fourier transform. *Proc. IEEE*, **66**(1), 51–84

Jenkins G.M. and Watts D.G. (1968). *Spectral Analysis and its Applications*. San Francisco CA: Holden-Day

Jervis B.W. and Saatchi M.R. (1990). An integrated system for process control and the acquisition, storage and processing of data. In *IEE Colloq. on PC-Based Instrumentation*,

January 31, 1990, IEE Digest No. 1990/025

Jervis B.W., Nichols M.J., Johnson T.E., Allen E.M. and Hudson N.R. (1983). A fundamental investigation of the composition of auditory evoked potentials. *IEEE Trans. Biomedical Engineering*, **30**(1), 43–50

Jervis B.W., Coelho M. and Morgan G.W. (1989a). Spectral analysis of EEG responses. *Medical and Biological Engineering and Computing*, **27**, 230–8

Jervis B.W., Coelho M. and Morgan G.W. (1989b). Effect on EEG responses of removing ocular artefacts by proportional EOG subtraction. *Medical and Biological Engineering and Computing*, **27**, 484–90

Jervis B.W., Saatchi M.R., Allen E.M., Hudson N.R., Oke S. and Grimsley M. (1993). A pilot study of computerised differentiation of Huntington's disease, schizophrenia, and Parkinson's disease patients using the contingent negative variation. *Medical and Biological Engineering and Computing*, **31** (January), 31–8

Kalouptsidis N. and Theodoridis S. (1987). Fast adaptive least-squares algorithms for power spectral estimation. *IEEE Trans. Acoustics, Speech and Signal Processing*, **35**, 661–70

Kay S.M. (1980). A new ARMA spectral estimator. *IEEE Trans. Acoustics, Speech and Signal Processing*, **28**, 585–8

Kuo F.F. and Kaiser J.F. (1966). *System Analysis by Digital Computer*, Chapter 7, pp. 232–8. New York NY: Wiley

Lacoss R.T. (1971). Data adaptive spectral analysis methods. *Geophysics*, **36**, 661–75

Levinson N. (1947). The Weiner RMS error criterion in filter design and prediction. *J. Mathematical Physics*, **25**, 261–78

Makhoul J. (1975). Linear prediction: a tutorial review. *Proc. IEEE*, **63**, 561–80

Marple S.L. (1980). A new autoregressive spectrum analysis algorithm. *IEEE Trans. Acoustics, Speech and Signal Processing*, **28**, 441–54

Morrison D.F. (1976). *Multivariate Statistical Methods* 2nd edn. New York NY: McGraw-Hill

Nuttall A.H. (1976). *Spectral Analysis of a Univariate Process with Bad Data Points via Maximum Entropy and Linear Predictive Techniques*. NUSC Technical Report TR-5303, New London CN

Priestley M.B. (1981). *Spectral Analysis and Time Series*, Volume 1, *Univariate Series*, Chapters 6 and 7. New York NY: Academic Press

Proakis J.G. and Manolakis D.G. (1989). *Introduction to Digital Signal Processing*, Sections 1.3.2, 11.2.4, 11.3 and 11.3.4 and Appendix 6A. Basingstoke: Macmillan

Rissanen J. (1983). A universal prior for the integers and estimation by minimum description length. *Ann. Statistics*, **11**, 417–31

Saatchi M.R. and Jervis B.W. (1991). PC-based integrated system developed to diagnose specific brain disorders. *Computing and Control Engineering J.*, **2**(2), 61–8

SAS (1982). *SAS User Guide*. SAS Institute

Shibata R. (1976). Selection of the order of an autoregressive model by Akaike's information criterion. *Biometrika*, **63**, 117–26

Ulrych T.J. and Clayton R.W. (1976). Time series modelling and maximum entropy. *Physics Earth and Planetary Interiors*, **12**, 188–200

Welch P.D. (1967). The use of fast Fourier transform for the estimation of power spectra. *IEEE Trans. Audio and Electroacoustics*, **15**, 70–3

Whittle P. (1965). *Prediction and Regulation*. London: English Universities Press

Wold H. (1954). *A Study of the Analysis of Stationary Time Series*. Stockholm: Almquist and Wichsells

11

General- and special-purpose hardware for DSP

11.1	Introduction	615
11.2	Computer architectures for signal processing	615
11.3	General-purpose digital signal processors	628
11.4	Implementation of DSP algorithms on general-purpose digital signal processors	636
11.5	Special-purpose DSP hardware	662
11.6	Summary	668
	References	671
	Bibliography	671
	Appendix	672

The main objectives of this chapter are to provide an understanding of the key issues underlying general- and special-purpose processors for DSP, the impact of DSP algorithms on the hardware and software architectures of these processors, and how key DSP algorithms are implemented for real-time execution on general-purpose digital signal processors or realized as a piece of special-purpose hardware.

Real-time often implies 'as soon as possible' but within specified time limits. Real-time processing may be divided into two broad categories (although further subdivision is possible): stream processing, for example digital filtering, where data is processed one sample at a time, and block processing, for example FFT and correlation, where fixed blocks of data points are processed at a time. The implementation of DSP algorithms in real time requires both hardware and software. The hardware may be an array of processors, standard microprocessors, DSP chips or microprogrammed special-purpose devices. The software is often low level assembly codes or microcodes native to the hardware, although the trend now is to write software codes in an efficient high level language, such as C.

11.1 Introduction

For convenience, DSP processors can be divided into two broad categories: general purpose and special purpose. General-purpose DSP processors include devices such as the Texas Instruments TMS320C25, and ADSP56000 from Motorola. There are two types of special-purpose hardware.

(1) Hardware designed for efficient execution of specific DSP algorithms such as digital filters, fast Fourier transform. This type of special-purpose hardware is sometimes called an algorithm-specific digital signal processor.

(2) Hardware designed for specific applications, for example for telecommunications, digital audio, or control applications. This type of hardware is sometimes called an application-specific digital signal processor.

In most cases application-specific digital signal processors execute specific algorithms, such as PCM encoding/decoding, but are also required to perform other application-specific operations. Examples of special-purpose processors are DSP56200, an FIR digital filter from Motorola, and A100, an FIR digital filter from INMOS. Both general-purpose and special-purpose processors can be designed with single chips or with individual blocks of multipliers, ALUs, memories, and so on.

First, we will discuss the architectural features of digital signal processors that have made real-time DSP in many areas possible.

11.2 Computer architectures for signal processing

Most processors available today are based on the von Neumann concepts, where operations are performed sequentially. Figure 11.1 shows a simplified architecture for a standard von Neumann processor. When an instruction is

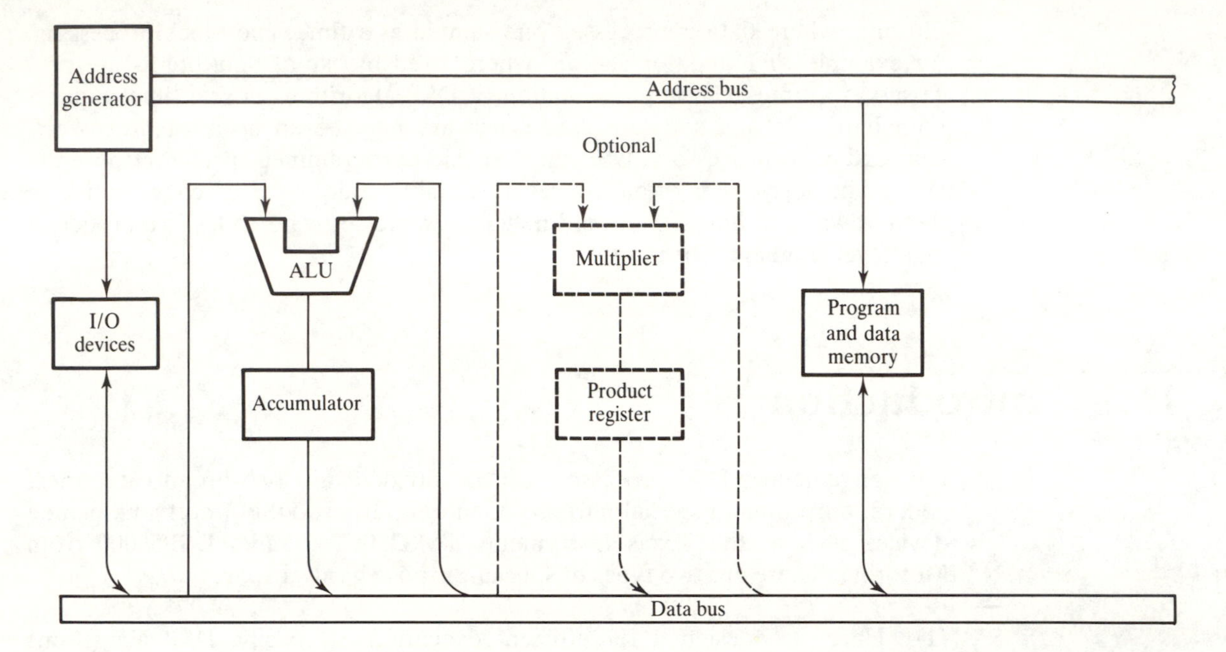

Figure 11.1 A simplified architecture for standard microprocessors.

processed in such a processor, units of the processor not involved at each instruction phase wait idly until control is passed on to them. Increase in processor speed is achieved by making the individual units operate faster, but there is a limit on how fast they can be made to operate.

If it is to operate in real time, a DSP processor must have its architecture optimized for executing DSP functions. Figure 11.2 shows a generic hardware architecture suitable for real-time DSP. It is characterized by the following:

- Multiple bus structure with separate memory space for data and program instructions. Typically the data memories hold input data, intermediate data values, output samples, as well as fixed coefficients for, for example, digital filters or FFTs. The program instructions are stored in the program memory.
- The I/O port provides a means of passing data to and from external devices such as the ADC and DAC or for passing digital data to other processors. Direct memory access (DMA), if available, allows for rapid transfer of blocks of data directly to or from data RAM, typically under external control.
- Arithmetic units for logical and arithmetic operations, which include an ALU, a hardware multiplier.

Why is such an architecture necessary? Most DSP algorithms (such as filtering, correlation and fast Fourier transform) involve repetitive arithmetic operations such as multiply, add, memory accesses, and heavy data flow through the CPU.

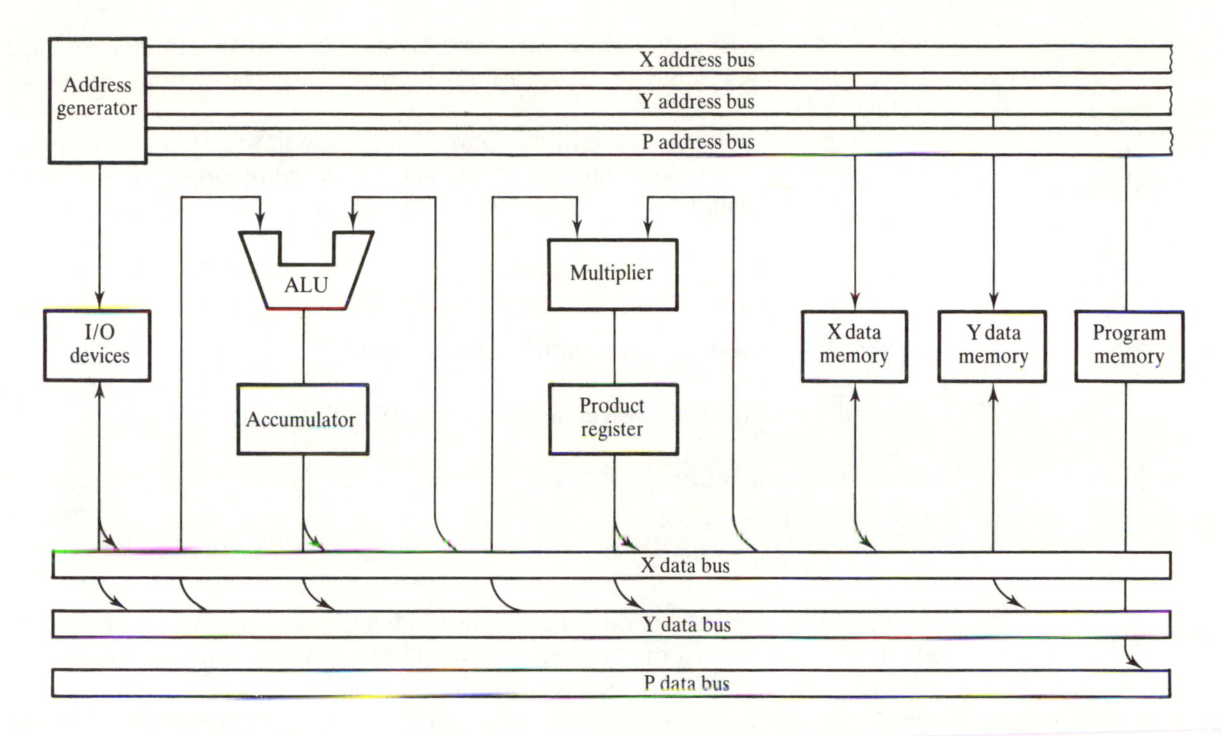

Figure 11.2 Hardware architecture for signal processing.

The architecture of standard microprocessors is not suited to this type of activity. An important goal in DSP hardware design is to optimize both the hardware architecture and the instruction set for DSP operations. In digital signal processors, this is achieved by making extensive use of the concepts of parallelism. In particular, the following techniques are used:

- Harvard architecture;
- pipelining;
- fast, dedicated hardware multiplier/accumulator;
- special instructions dedicated to DSP;
- replication;
- on-chip memory/cache.

For successful DSP design, it is important to understand these key architectural features.

11.2.1 Harvard architecture

The principal feature of the Harvard architecture is that the program and data memories lie in two separate spaces, permitting a full overlap of instruction fetch and execution. Standard microprocessors, such as the Intel 6502, are characterized by a single bus structure for both data and instructions, as shown in Figure 11.1.

Suppose that in a standard microprocessor we wish to read a value op1 at address ADR1 in memory into the accumulator and then store it at two other addresses, ADR2 and ADR3. The instructions could be

```
LDA    ADR1    load the operand op1 into the accumulator from ADR1
STA    ADR2    store op1 in address ADR2
STA    ADR3    store op1 in address ADR3
```

Typically, each of these instructions would involve three distinct steps:

- instruction fetch;
- instruction decode;
- instruction execute.

In our case, the instruction fetch involves fetching the next instruction from memory, and instruction execute involves either reading or writing data into memory. In a standard processor, without Harvard architecture, the program instructions (that is, the program code) and the data (operands) are held in one memory space; see Figure 11.3. Thus the fetching of the next instruction while the current one is executing is not allowed, because the fetch and execution phases each require memory access.

In a Harvard architecture (Figure 11.4), since the program instructions and data lie in separate memory spaces, the fetching of the next instruction can overlap the execution of the current instruction; see Figure 11.5. Normally, the program memory holds the program code, while the data memory stores variables such as the input data samples.

Strict Harvard architecture is used by some digital signal processors (for example, Motorola DSP56000), but most use a modified Harvard architecture (for example, the TMS320 family of processors). In the modified architecture used by the TMS320, for example, separate program and data memory spaces are still maintained, but communication between the two memory spaces is permissible, unlike in the strict Harvard architecture.

11.2.2 Pipelining

Pipelining is a technique which allows two or more operations to overlap during execution. In pipelining, a task is broken down into a number of distinct subtasks which are then overlapped during execution. It is used extensively in digital signal processors to increase speed. A pipeline is akin to a typical production line in a factory, such as a car or television assembly plant. As in the production line, the task is broken down into small, independent subtasks

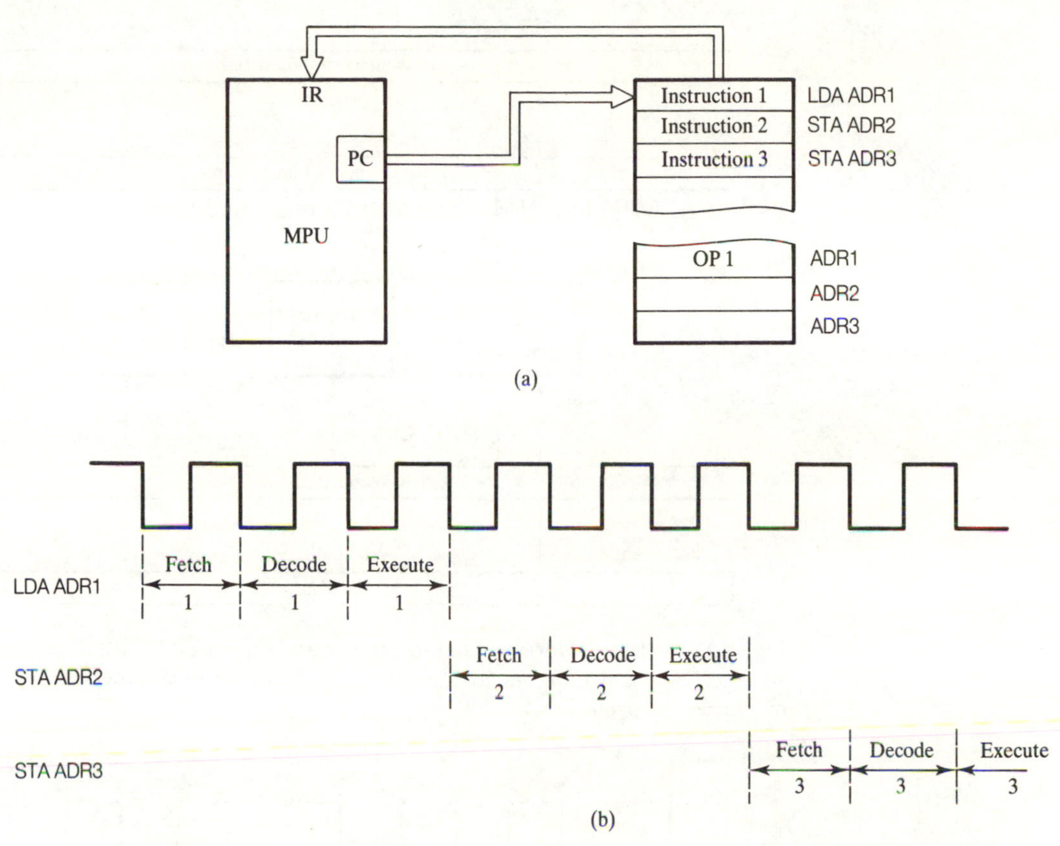

Figure 11.3 An illustration of instruction fetch, decode and execute in a non-Harvard architecture with single memory space: (a) instruction fetch from memory; (b) timing diagram.

called pipe stages. The pipe stages are connected in series to form a pipe and the stages executed sequentially.

As we have seen in the last section, an instruction can be broken down into three steps. Each step in the instruction can be regarded as a stage in a pipeline and so can be overlapped. By overlapping the instructions, a new instruction is started at the start of each clock cycle (Figure 11.6(a)).

Figure 11.6(b) gives the timing diagram for a three-stage pipeline, drawn to highlight the instruction steps. Typically, each step in the pipeline takes one machine cycle. Thus during a given cycle up to three different instructions may be active at the same time, although each will be at a different stage of completion. The key to an instruction pipeline is that the three parts of the instruction (that is, fetch, decode and execute) are independent and so the execution of multiple instructions can be overlapped. In Figure 11.6(b), it is seen that, at the ith cycle, the processor could be simultaneously fetching the ith instruction, decoding the $(i-1)$th instruction and at the same time executing the $(i-2)$th instruction.

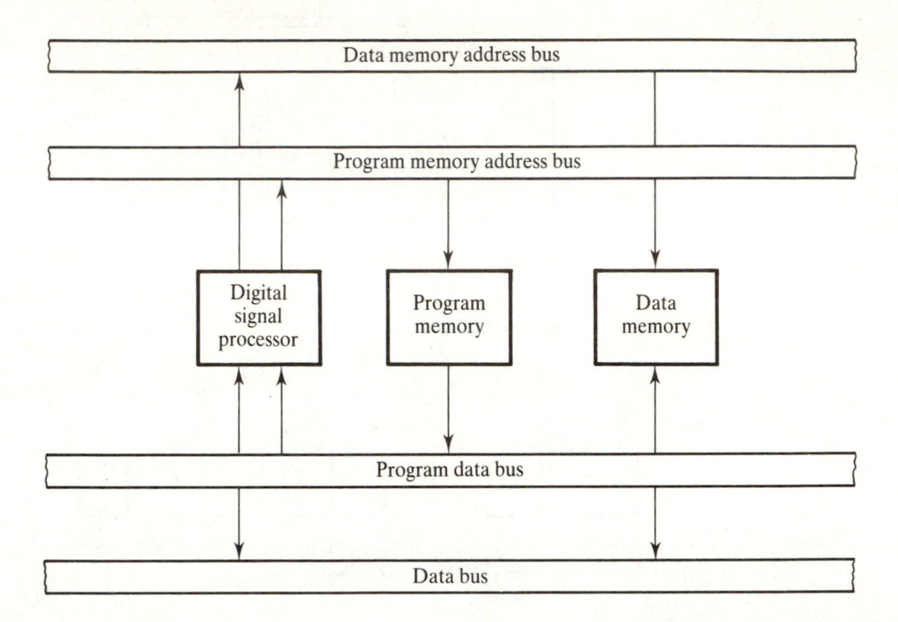

Figure 11.4 Basic Harvard architecture with separate data and program memory spaces. Data and program instruction fetches can be overlapped as two independent memories are used.

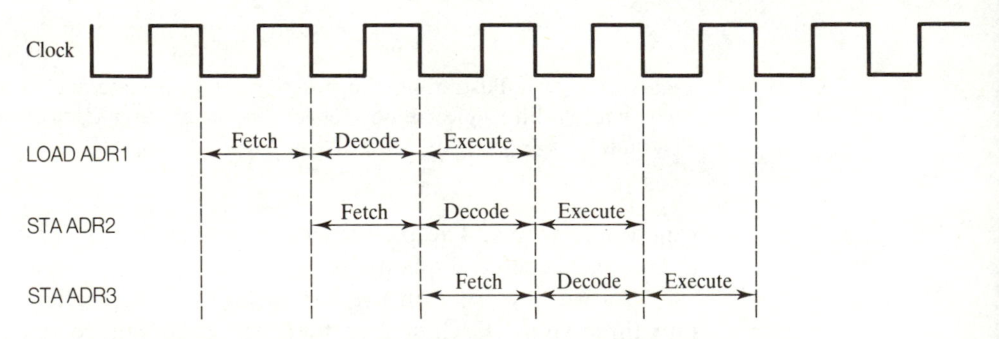

Figure 11.5 An illustration of instruction overlap made possible by the Harvard architecture.

The three-stage pipelining discussed above is based on the technique used in the Texas Instruments TMS320C25. As in other applications of pipelining, in the TMS320C25 a number of registers are used to achieve the pipeline: a prefetch counter holds the address of the next instruction to be fetched, an instruction register (IR) holds the instruction to be executed, a queue instruction register (QIR) stores the instructions to be executed if the instruction in the IR is still executing. The program counter contains the address of the next instruction to execute. On completion of the execution of the instruction in the IR, the instruction in the QIR is transferred to the IR for execution.

By exploiting the inherent parallelism in the instruction stream, pipelining

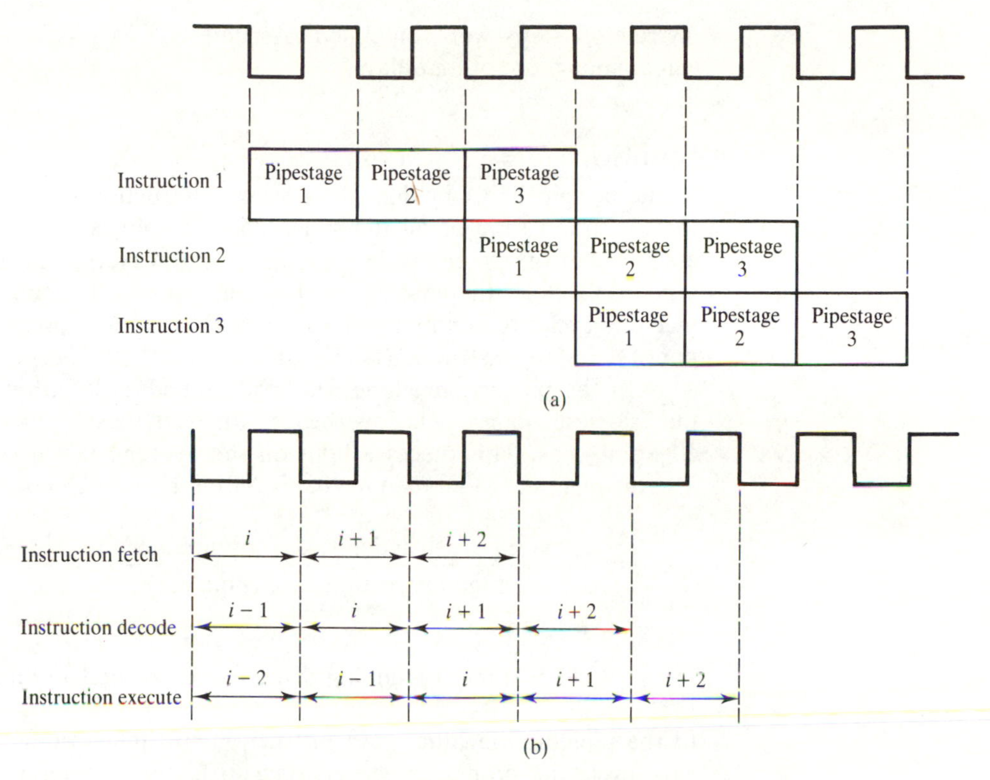

Figure 11.6 An illustration of the concept of pipelining.

leads to a significant reduction, on average, of the execution time per instruction. The throughput of a pipeline machine is determined by the number of instructions through the pipe per unit time. As in a production line, all the stages in the pipeline must be synchronized. The time for moving an instruction from one step to another within the pipe (see Figure 11.6(a)) is one cycle and depends on the slowest stage in the pipeline. In a perfect pipeline, the average time per instruction is given by (Hennessy and Patterson, 1990)

$$\frac{\text{time per instruction (nonpipeline)}}{\text{number of pipe stages}} \quad \textbf{(11.1)}$$

In the ideal case, the speed increase is equal to the number of pipe stages. In practice, the speed increase will be less because of the overheads in setting up the pipeline, and delays in the pipeline registers, and so on.

Example 11.1

In a nonpipeline machine, the instruction fetch, decode and execute take 35 ns, 25 ns, and 40 ns, respectively. Determine the increase in throughput if the instruction steps were pipelined. Assume a 5 ns pipeline overhead at each stage, and ignore other delays.

Solution

In the nonpipeline machine, the average instruction time is simply the sum of the execution time of all the steps: $35 + 25 + 40$ ns $= 100$ ns. However, if we assume that the processor has a fixed machine cycle with the instruction steps synchronized to the system clock, then each instruction would take three machine cycles to complete: 40 ns $\times 3 = 120$ ns. This corresponds to a throughput of 8.3×10^6 instructions s^{-1}.

In the pipeline machine, the clock speed is determined by the speed of the slowest stage plus overheads. In our case, the machine cycle is $40 + 5 = 45$ ns. This places a limit on the average instruction execution time. The throughput (when the pipeline is full) is 22.2×10^6 instructions s^{-1}. Then

$$\begin{aligned} \text{speedup} &= \frac{\text{average instruction time (nonpipeline)}}{\text{average instruction time (pipeline)}} \qquad (11.2) \\ &= 120/45 \\ &= 2.67 \text{ times (assuming nonpipeline executes in three cycles)} \end{aligned}$$

In the pipeline machine, each instruction still takes three clock cycles, but at each cycle the processor is executing up to three different instructions. Pipelining increases the system throughput, but not the execution time of each instruction on its own. Typically, there is a slight increase in the execution time of each instruction because of the pipeline overhead.

Pipelining has a major impact on the system memory. The number of memory accesses in a pipeline machine increases, essentially by the number of stages. In DSP the use of Harvard architecture, where data and instructions lie in separate memory spaces, promotes pipelining.

When a slow unit, such as a data memory, and an arithmetic element are connected in series, the arithmetic unit often waits idly for a good deal of the time for data. Pipelining may be used in such cases to allow a better utilization of the arithmetic unit. The next example illustrates the concept.

Example 11.2

Most DSP algorithms are characterized by multipy-and-accumulate operations typified by the following equation:

$$a_0x(n) + a_1x(n-1) + a_2x(n-2) + \ldots + a_{N-1}x(n-(N-1))$$

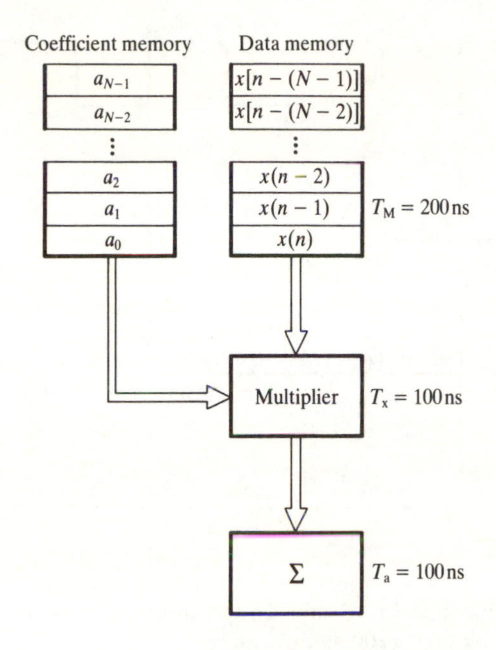

Figure 11.7 Non-pipelined MAC configuration. Products are clocked into the accumulator every 400 ns.

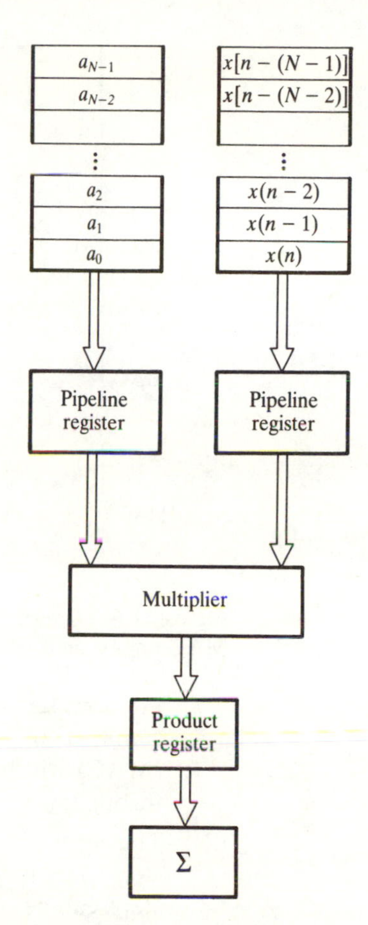

Figure 11.8 Pipelined MAC configuration. The pipeline registers serve as temporary store for coefficient and data sample pair. The product register also serves as a temporary store for the product.

Figure 11.7 shows a nonpipeline configuration for an arithmetic element for executing the above equation. Assume a transport delay of 200 ns, 100 ns, and 100 ns, respectively for the memory, multiplier and the accumulator.

(1) What is the system throughput?

(2) Reconfigure the system with pipelining to give a speed increase of 2:1. Illustrate the operation of the new configuration with a timing diagram.

Solution

(1) The coefficients, a_k, and the data arrays are stored in memory as shown in the Figure 11.7(a). In the nonpipelined mode, the coefficients and data

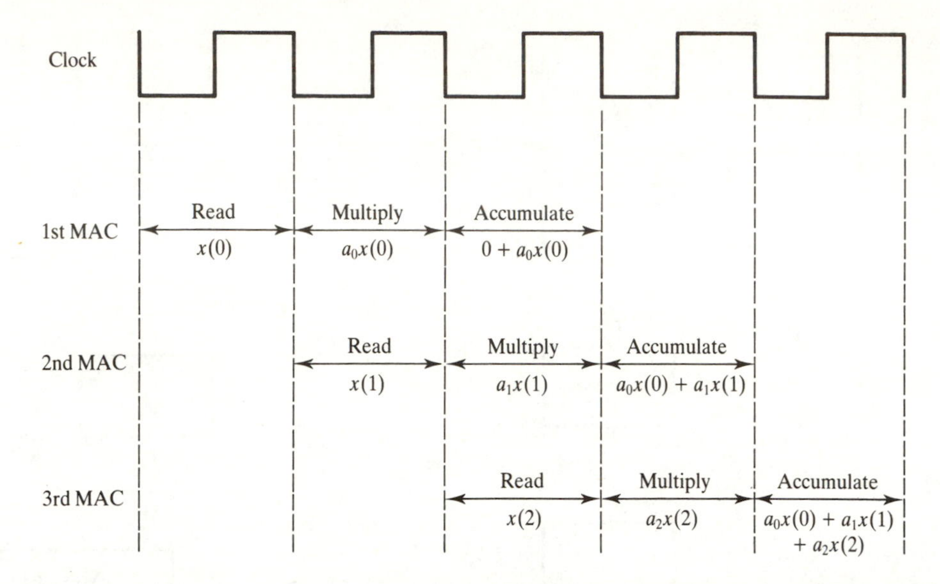

Figure 11.9 Timing diagram for a pipelined MAC unit. When the pipeline is full, a MAC operation is performed every clock cycle (200 ns).

are accessed sequentially and applied to the multiplier. The products are summed in the accumulator. Successive multiplication–accumulation (MAC) will be performed once every 400 ns (200 + 100 + 100), that is a throughput of 2.5×10^6 operations s^{-1}.

(2) The arithmetic operations involved can be broken up into three distinct steps: memory read, multiply, and accumulate. To improve speed these steps can be overlapped. A speed improvement of 2:1 can be achieved by inserting pipeline registers between the memory and multiplier and between the multiplier and accumulator as shown in Figure 11.8. The timing diagram for the pipeline configuration is shown in Figure 11.9. As is evident in the timing diagram, the MAC is performed once every 200 ns. The limiting factor is the basic transport delay through the slowest element, in this case the memory. Pipeline overheads have been ignored.

DSP algorithms are often repetitive but highly structured, making them well suited to multilevel pipelining. For example, FFT requires the continuous calculation of butterflies. Although each butterfly requires different data and coefficients the basic butterfly arithmetic operations are identical. Thus arithmetic units such as FFT processors can be tailored to take advantage of this. Pipelining ensures a steady flow of instructions to the CPU, and in general leads to a significant increase in system throughput. However, on occasions pipelining may cause problems. For example, in some digital signal processors, pipelining may cause an unwanted instruction to be executed, especially near branch instructions, and the designer should be aware of this possibility.

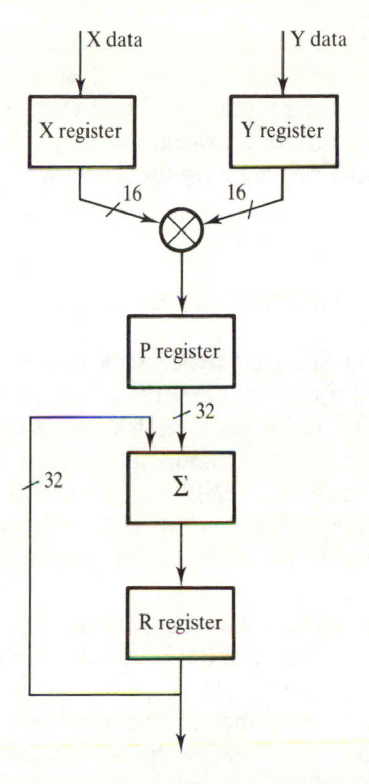

Figure 11.10 A typical MAC configuration in DSPs.

11.2.3 Hardware multiplier–accumulator

The basic numerical operations in DSP are multiplications and additions. Multiplication, in software, is notoriously time consuming. Additions are even more time consuming if floating point arithmetic is used. To make real-time DSP possible a fast, dedicated hardware multiplier–accumulator (MAC) using fixed or floating point arithmetic is mandatory. Fixed or floating hardware MAC is now standard in all digital signal processors. In a fixed point processor, the hardware multiplier typically accepts two 16-bit 2's complement fractional numbers and computes a 32-bit product in a single cycle (100 ns typically). The average MAC instruction time can be significantly reduced through the use of special repeat instructions.

A typical DSP hardware MAC configuration is depicted in Figure 11.10. In this configuration, the multiplier has a pair of input registers that hold the inputs to the multiplier, and a 32-bit product register which holds the result of a multiplication. The output of the P (product) register is connected to a double-precision accumulator, where the products are accumulated.

The principle is very much the same for hardware floating point multiplier–accumulators, except that the inputs and products are normalized floating point numbers. Floating point MACs allow fast computation of DSP results with minimal errors. As discussed in Chapters 6 and 7 DSP algorithms such as FIR and IIR filtering suffer from the effects of finite wordlength (coefficient quantization and arithmetic errors). Floating point offers a wide dynamic range and reduced arithmetic errors, although for many applications the dynamic range provided by the fixed-point representation is adequate.

11.2.4 Special instructions

Digital signal processors provide special instructions optimized for DSP. The benefits of these special instructions are twofold: they lead to a more compact code which takes up less space in memory (nearly as compact as a code written in a high level language such as C) and they lead to an increase in the speed of execution of DSP algorithms. Special instructions provided by DSP chips include (i) instructions that support general DSP operations, (ii) instructions that reduce the overhead in instruction loops and (iii) application-oriented instructions.

Many key algorithms in DSP, such as digital filtering and correlation, require data shifts or delays to make room for new data samples. Digital signal processors provide special instructions that allow a data sample to be copied to the next higher memory address as it is being fetched from memory or operated on, all in one cycle. For example, the instruction pair LTD and MPY in the TMS320C25 permits the simultaneous execution of the loading of data into the temporary register for the multiplier, data shifting (to implement the unit delay symbolized by z^{-1}), and accumulation of products.

Special instructions are provided to speed up DSP operations that are often repeated. For example, in the TMS320C25 an instruction is provided that allows the next instruction to be repeated $N+1$ times. As the repeat instruction requires only a single instruction fetch a piece of code that ordinarily requires a multicycle loop effectively becomes a single-cycle instruction. The repeat instruction is especially useful in block moves and FIR filtering.

For example, it may be necessary to move a block of code or coefficients from slow external program memory to an on-chip memory for faster execution. The following TMS320C25 program fragment illustrates moving a set of N coefficients from program memory to data memory:

```
LARP      AR7
LRLK      AR2, dest     ;destination
RPTK      NM1
BLKD      coeff,*-      ;move coeffs.
```

FIR filters can be efficiently implemented using the instruction pair

```
RPTK     NM1
MACD     HNM1, XNM1
```

The first instruction RPTK NM1 loads the filter length minus 1 ($N - 1$) into the repeat instruction counter, and causes the MACD instruction following it to be repeated N times. The MACD instruction performs a number of operations in one cycle:

(1) loads the program counter with the address of the coefficients in the on-chip program memory (RAM block B0, configured as program memory);
(2) multiplies the data sample, $x(n - k)$, in the on-chip data memory by the coefficient, $h(k)$;
(3) adds the previous product to the accumulator;
(4) implements the unit delay, symbolized by z^{-1}, by shifting the input data sample up to the next location in memory;
(5) increments the PC to point to the next filter coefficient in memory.

The TMS320C30, a floating point digital signal processor, takes the concept of instruction repeat further. It provides an instruction that allows a block of code, not just a single instruction as in the C25, to be repeated N times, eliminating the overhead in executing instruction loops. The format is

```
        RPTB    loop
        :
loop    (last instruction)
```

Such zero-overhead looping and single-cycle branching instructions are becoming quite common in DSP devices.

Application-oriented instructions include those that support the update of an LMS-based adaptive filter. In fast Fourier transformation there is always a need for scrambling the input data sequence before FFT or unscrambling the data after FFT to ensure the data points appear in the correct sequence. The more advanced digital signal processors provide special instructions for bit-reversal addressing which performs the required scrambling/unscrambling at the same time as a data sample is being moved or fetched.

11.2.5 Replication

In DSP, replication involves using two or more basic units, for example using more than one ALU, multiplier or memory unit. Often the units are arranged to work simultaneously. For example, in the INMOS A100, 32 16×16-bit multipliers are used. In DSP, the norm is to have one CPU, with one or more arithmetic elements replicated.

However, full-blown parallel processing concepts where for example a number of independent processors work on a given task, or several processors under one control unit work simultaneously on a single problem, are now being extended to DSP. At least one parallel DSP chip, the TMS320C40, is now available. The INMOS Transputer (a parallel processor), although not a DSP device, can be combined with special-purpose processors (for example, the A100) to provide an efficient parallel digital signal processing machine.

11.2.6 On-chip memory/cache

In most cases, DSP chips operate so fast that slow inexpensive memories are unable to keep up. The common practice is to slow the processor down by adding wait states. In some processors, wait states are software programmable, but in others a piece of external hardware is necessary to slow the processor down. Wait states mean of course that the processor cannot operate at full speed.

To alleviate this problem many DSP chips contain fast on-chip data RAMs and/or ROMs (see for example, Table 11.1). In such processors, slow external memories may be used to hold the program code. At initialization, the code may be transferred to the fast, internal memory for full-speed execution. Fast on-chip EPROMs are useful for real-time development and for final prototyping. Some chips provide an on-chip program cache which may be used to hold often repeated sections of a program. Execution of codes in the cache avoids further memory fetches and speeds up program execution.

Provision of an on-chip memory is now a norm.

11.3 General-purpose digital signal processors

General-purpose DSP chips are basically high speed microprocessors, with hardware architectures and instruction sets optimized for DSP operations. These chips make extensive use of parallelism, Harvard architecture, pipelining and dedicated hardware whenever possible to perform time-consuming operations, such as shifting/scaling, multiplication, and so on. DSP chips available today differ in their detailed architecture and the on-board peripherals provided.

Table 11.1 summarizes the key features of three popular general-purpose DSP processor families. We will describe, briefly, their architectural features. The TMS320 family will be used to illustrate the implementation of a number of real-time DSP algorithms on general-purpose DSP chips.

11.3.1 Texas Instruments TMS320 family

The TMS320 family consists of a number of single-chip, fixed and floating point DSP processors. The TMS320 architecture is based on the Harvard architecture, modified to allow more flexibility, for example transfers between program

Table 11.1 Comparison of features of general-purpose digital signal processors.

Manufacturer	*Device*	*Arithmetic and wordlength*	*Accumulator wordlength*	*On-chip memory, RAM (words)*	*On-chip memory, ROM (words)*	*Extended memory data (words)*	*Wait state type*	*Bit reverse capability*	*Zero-over-head loop*	*Cycle time (ns)*[a]	*Cache memory (words)*
Texas Instruments	TMS320C10	16-bit fixed point	32 bit	144	1.5 K	4 K	None	None	No	114	0
	TMS320C25	16-bit fixed point	32 bit	544	4 K	2 × 64 K	HW	Yes	No[b]	78	0
	TMS320C30	32-bit floating point	40 bit	2 K	4 K	16 M	HW/SW	Yes	Yes	50	64
Motorola	DSP56000/1	24-bit fixed point	56 bit	1.5 K	544	3 × 64 K	HW/SW	Yes	Yes	60	0
	DSP96002	32-bit floating point	96 bit	2 K	1088	3 × 4 G	HW/SW	Yes	Yes	50	0
Analog Devices	ADSP2100	16-bit fixed point	40 bit	0	0	32 K + 16 K	HW	Yes	Yes	60	16
	ADSP21000	32-bit floating point	80 bit	0	0	4 G + 16 M	HW/SW	Yes	Yes	40	32

[a]Fastest quoted cycle time.
[b]Next-instruction-repeat looping available.
HW, hardware wait state; SW, software wait state.

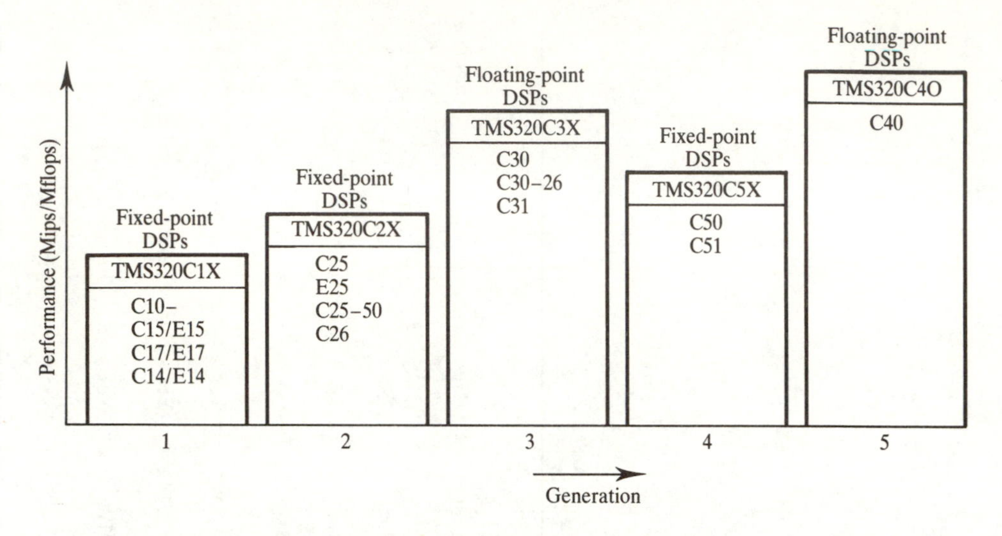

Figure 11.11 The TMS320 family of digital signal processors: generations.

and data spaces. At the present, the family consists of five generations of processors and several offspring (see Figure 11.11). The first and second generations have a lot in common, architecturally, but the second has more features. Later generations have more peripheral devices built in to reduce chip count. The TMS320C30 is a floating point device. The C40 is also a floating point device with similar architecture to the C30, but designed to support full parallel processing operations as in the well known INMOS Transputer. The TMS320C25 and TMS320C30 are among the most important members of the TMS320 family.

The TMS320C25, introduced in 1986, is a second-generation member of the TMS320 family of signal processors from Texas Instruments. It is a CMOS version of its predecessor, the TMS32020, but with more features and increased speed. It is said to have a performance of about 4 times that of the TMS320C10. The internal architecture is shown in Figure 11.12 in a simplified form to emphasize the dual internal memory spaces which are characteristic of the Harvard architecture.

The C25 has a number of special instructions dedicated to DSP operations. In particular, it has a repeat instruction which when combined with a multiply–accumulate with data move instruction can execute an FIR instruction with considerable time savings. Its bit-reversed addressing capability is very useful in FFTs. There is also a set of special instructions for adaptive filtering.

Unlike the first generation, C10, which has a very limited memory space and external I/O support, the C25 provides more on-chip and off-chip memory spaces and facilities. The on-chip data memory consists of three data blocks, B0, B1 and B2, each divided into 128-word pages as shown on the right-hand

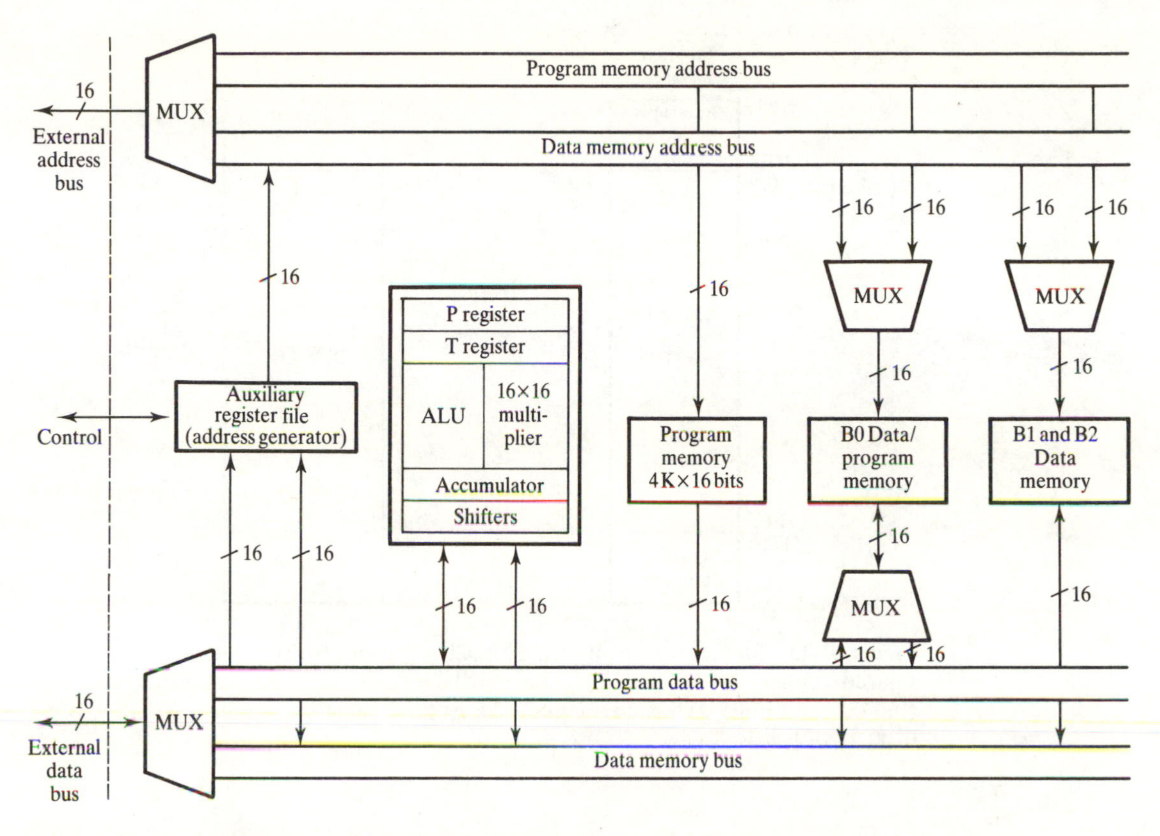

Figure 11.12 A simplified block diagram of the TMS320C25 showing internal memory buses. Note that internally the processor has four buses, but two buses externally.

side of Figure 11.13. The 256-word RAM block B0 can be configured as either data or program memory, but blocks B1 and B2 can only be used as data memory. The addressable RAM can be extended off-chip to 64K words. The microcomputer version of the TMS320C25 has 4096 16-bit word, factory maskable, on-chip program ROM. Program code in the off-chip program memory will execute at full speed if the external memories used are fast enough (about 45 ns) or at reduced speed with wait states. A useful feature of the C25 is the ability to download program code from slow off-chip program memory to the on-chip data memory block B0 and then to execute the code at full speed.

The TMS320C30 is a 32-bit single-chip digital signal processor. It supports both integer and floating point arithmetic operations. It has a large memory space (16M × 32-bit words) and is equipped with many on-chip peripheral facilities to simplify system design. These include a program cache to improve the execution of commonly used codes, and on-chip dual access memories. The large memory spaces cater for memory intensive applications, for example

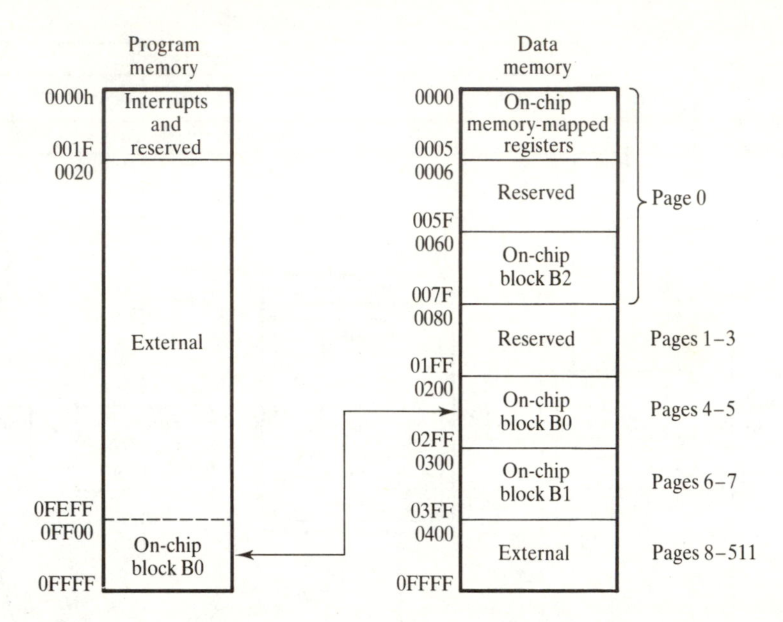

Figure 11.13 Memory map of the TMS320C25. The on-chip block B0 can be configured either as a data memory (0200h–02FFh) or as a program memory (0FF00h–0FFFFh). When configured as a program memory, the data memory locations 0200–02FFh cease to exist.

graphics and image processing, which are difficult to implement on the earlier generations.

The ability of DSP chips to perform DSP operations in floating point is a welcome development. This minimizes finite wordlength effects such as overflows, roundoff errors, and coefficient quantization errors inherent in DSP. It also facilitates algorithm development, as a designer can develop an algorithm on a large computer in a high level language and then port it to a DSP device.

In the TMS320C30, floating point multiplication requires 32-bit operands and produces a 40-bit normalized floating point product. Integer multiplication requires 24-bit inputs and yields 32-bit results. Three floating point formats are supported. The first is a 16-bit short floating point format, with 4-bit exponents, 1 sign bit and 11 bits for the mantissa. This format is for immediate floating point operations. The second is a single-precision format with an 8-bit exponent, 1 sign bit and 23-bit fractions (32 bits). The third is a 40-bit extended precision format which has an 8-bit exponent, 1 sign bit, and 31-bit fractions. The floating point representation differs from that of standard IEEE, but facilities are provided to allow conversion between the two formats.

The TMS320C30 combines the features of Harvard architecture (separate buses for program instructions, data and I/O) and von Neumann processor (unified address space). Special instructions provided by the C30 include block repeat, bit reversal addressing, and those that can execute in parallel.

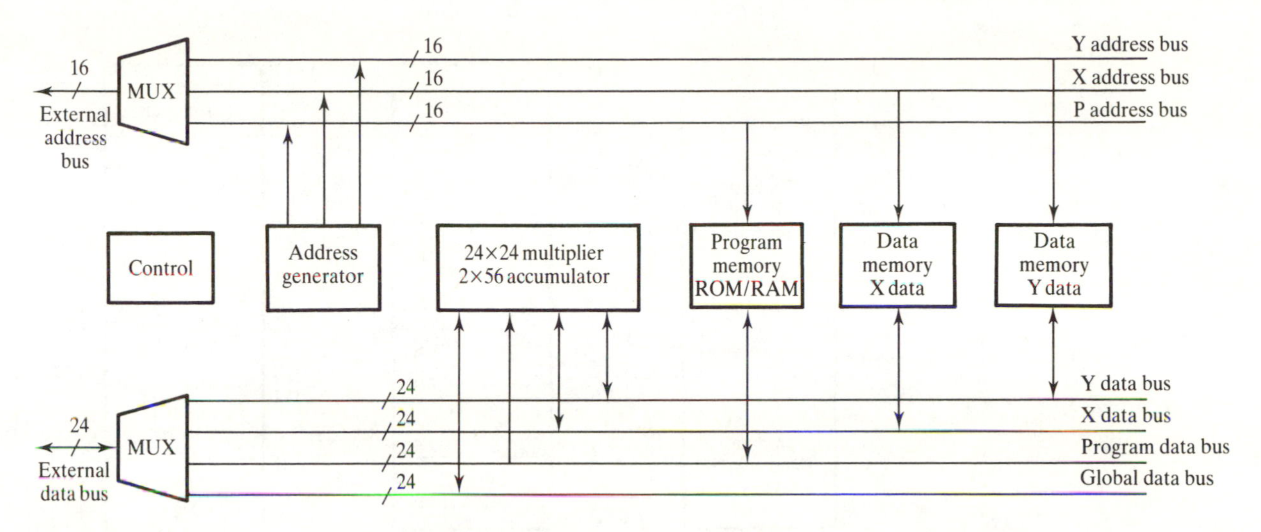

Figure 11.14 Simplified architecture of the DSP56000/1/2.

11.3.2 Motorola DSP56000 family

The Motorola DSP56000 is a high precision fixed point digital signal processor. Its architecture is depicted in Figure 11.14. Internally, it has two independent data memory spaces, the X-data and Y-data memory spaces, and one program memory space.

Having two separate data memory spaces allows a natural partitioning of data for DSP operations and facilitates the execution of the algorithm. For example, in graphics applications data can be stored as X and Y data, in FIR filtering as coefficients and data, and in FFT as real and imaginary. During program execution, pairs of data samples can be fetched or stored in internal memory simultaneously in one cycle. Externally, the two data spaces are multiplexed into a single data bus, reducing somewhat the benefits of the dual internal data memory. The data memories are expandable off chip to 128K words and the program memory is expandable to 64K words. The memory map for the DSP56000 is shown in Figure 11.15.

The arithmetic units consists of two 56-bit accumulators and a parallel fixed point hardware multiplier–accumulator (MAC). The MAC accepts 24-bit inputs and produces a 56-bit product. The 24-bit wordlength provides sufficient accuracy for representing most DSP variables while the 56-bit accumulator wordlength prevents arithmetic overflows. These wordlengths are adequate for most applications, including digital audio, which impose stringent requirements.

The DSP56000 has an excellent host interface port that simplifies interfacing to other systems. A useful feature in the 56000 is that the I/O port can be programmed to insert wait states when slow memory or other peripherals are used. Up to 15 wait states, from half a machine cycle to 7.5 cycles,

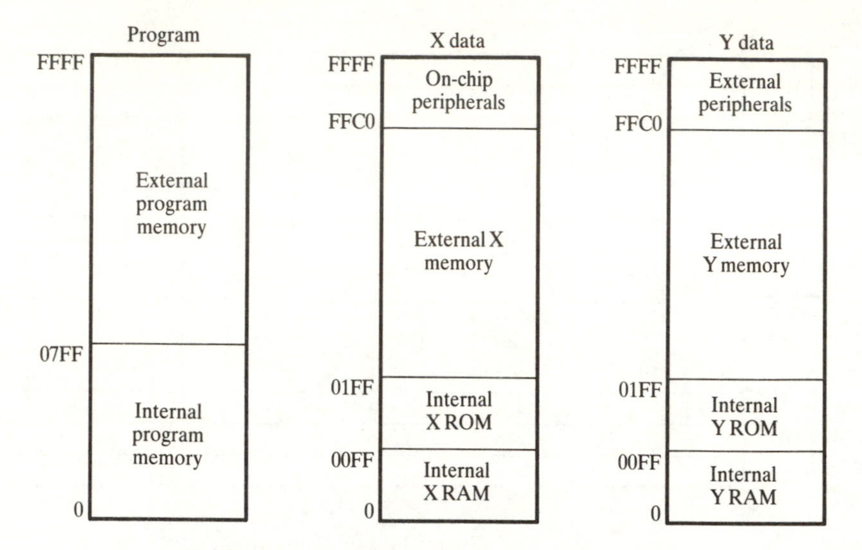

Figure 11.15 Memory map for DSP 56000.

can be programmed. The DSP56000 provides special instructions that allow zero-overhead looping and bit-reversed addressing capability for scrambling input data before FFT or unscrambling the fast Fourier transformed data.

There are other derivatives of the DSP56000 family. For example, the DSP56156 is a 16-bit version with built-in codec (coder–decoder) aimed at the telecommunication applications. The DSP96002 is a 32-bit floating/fixed point digital signal processor. It is architecturally similar to the fixed point DSP56000, but with considerable enhancement and greater precision. Like the DSP56000, it has two separate on-chip data memory spaces, X-data and Y-data memory, and one program memory. Each memory space is expandable to 4G words, divided into 0.5G word areas. The wordlength is 32 bits. The multiplier takes 32-bit floating point inputs and produces a 44-bit product, or 32-bit fixed point inputs and produces a 64-bit product. It is equipped with 96-bit/32-bit accumulators. The instruction set supports zero-overhead looping, circular and bit reversal addressing capabilities. Like the DSP56000, wait states can be configured in hardware or programmable in software.

11.3.3 Analog Devices ADSP2100 family

The Analog Devices ADSP2100 is one of the few general-purpose DSP chips with no on-chip memory, but it is also unique in having two separate external memory spaces; one holds data only, and the other holds program code as well as data.

The block diagram of the ADSP2100 internal architecture is depicted in Figure 11.16. The main components are the ALU, multiplier–accumulator, and shifters. The MAC accepts 16×16-bit inputs and produces a 32-bit pro-

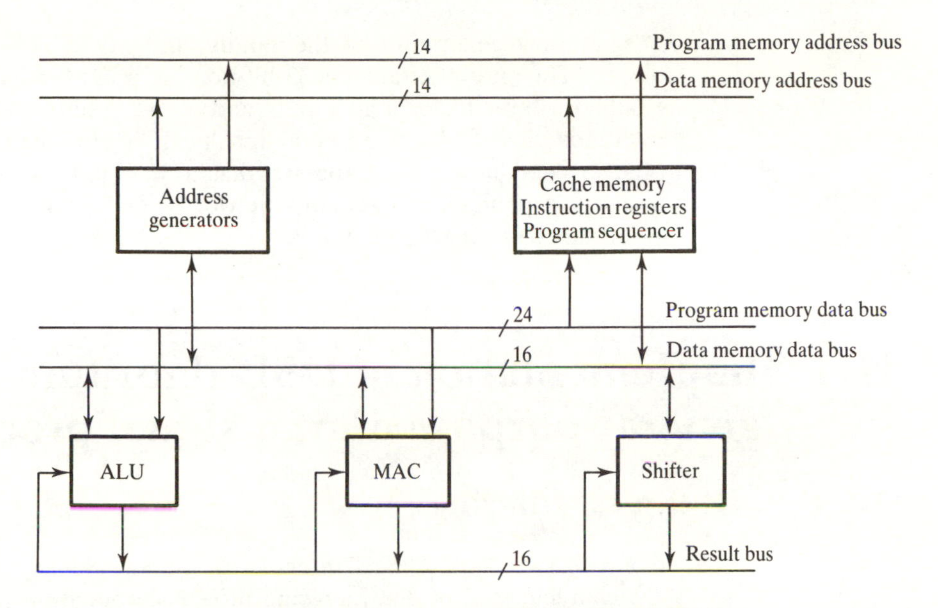

Figure 11.16 Simplified architecture of ADSP2100.

duct in 1 cycle. A useful feature of the ADSP2100 is that all the arithmetic and logic units (MAC, ALU and shifter) are connected to a common 16-bit result (R) bus. Thus the result of an arithmetic operation from one unit can be used immediately as an input for the next operation by any of the units.

The ADSP2100 departs from the strict Harvard architecture, as it allows the storage of both data and program instructions in the program memory. A signal line (data access signal) is used to indicate when data and not program instructions are being fetched from the program memory. Storage of data in the program memory inhibits a steady data flow through the CPU as data and instruction fetches cannot occur simultaneously. To avoid a bottleneck, the ADSP2100 has an on-chip program memory cache which holds the last 16 instructions executed. This eliminates the need, especially when executing program loops, for repeated instruction fetches from the program memory.

The ADSP2100 provides special instructions for zero-overhead looping and supports a bit-reversing addressing facility for FFT. It has only hardware wait states. The processor also provides facilities for context switching, that is on interrupt a fast exchange of working registers and shadow registers is performed. After interrupt servicing, the registers are exchanged again, restoring the CPU to its original state.

The lack of on-chip memory in the ADSP2100 is a severe restriction, especially in low budget projects. To run at full speed fast memories are necessary, which may be too expensive for low budget applications. Later derivatives of the ADSP2100 have on-chip memory, but combine program and data buses externally much like the DSP56000. They also have software programmable wait states.

A new generation of the family, the ADSP21000, is a floating point device. The multiplier accepts 32-bit floating point inputs and produces a 32-bit result or 40-bit fixed point inputs and 40-bit results. Also, 32-bit fixed point operands yield 64-bit fixed point results. It is equipped with two 80-bit fixed point accumulators. The processor has a 32×48-bit instruction cache, a data memory expandable to $4G \times 40$-bit words, and program memory expandable externally to $16M \times 48$-bit words.

11.4 Implementation of DSP algorithms on general-purpose digital signal processors

11.4.1 FIR digital filtering

Nonrecursive N-point FIR filters, with the structure given in Figure 11.17(a), are characterized by the following difference equation (see Chapter 6 for details):

$$y(n) = \sum_{k=0}^{N-1} h(k)x(n-k) \tag{11.3}$$

A fragment of a C language implementation of the general FIR filter is given in Program 11.1. For real-time FIR filtering, the data and coefficients are stored in memory, conceptually, as shown in Figure 11.17(b). To appreciate how the FIR filter works, consider the simple case of $N = 3$, with the following difference equation:

$$y(n) = h(0)x(n) + h(1)x(n-1) + h(2)x(n-2) \tag{11.4}$$

$x(n)$ represents the latest input sample, $x(n-1)$ the last sample, and $x(n-2)$ the sample before that.

Suppose the three-coefficient digital filter is fed from an ADC. The first thing to do is to allocate two sets of contiguous memory locations (in RAM), one for storing the input data ($x(n)$, $x(n-1)$, $x(n-2)$) and the other for the filter coefficients ($h(0)$, $h(1)$, $h(2)$) as depicted below:

Data RAM	Coefficient memory
0	$h(0)$
0	$h(1)$
0	$h(2)$

At initialization, the RAM locations where the data samples are to be stored are set to zero since we always start with no data. The following operations are then performed.

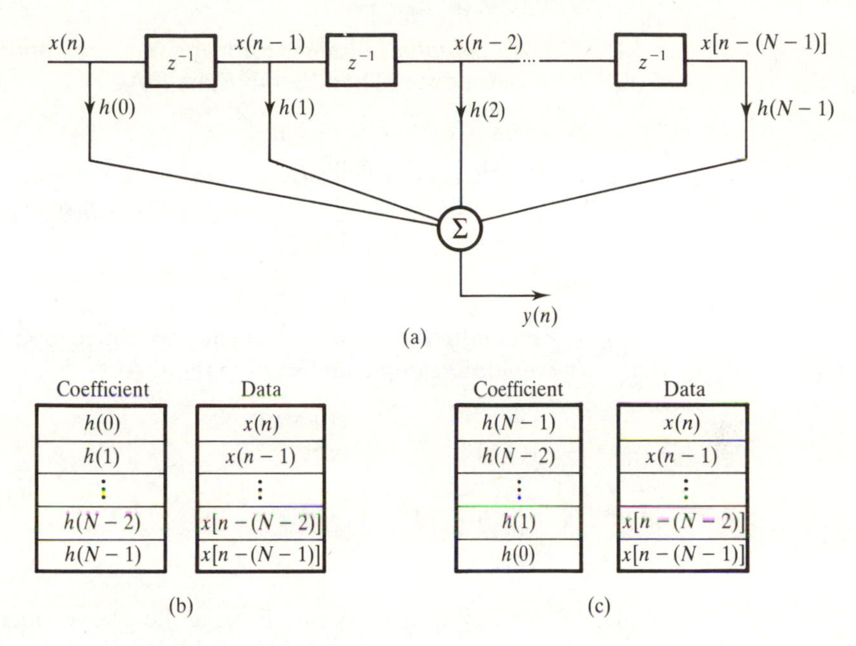

Figure 11.17 Implementation of FIR filter: (a) filter structure; (b) coefficient and data memory map; (c) alternative memory map.

Program 11.1 A C language pseudo-code for FIR filtering.

```
nm1=N-1;
yn=0;
for(k=0; k<nm1;++k){                  /*shift data to make room for new sample */
        x[nm1-k]=x[nm1-k-1];
        x[0]=xn;
}
for(k=0; k<N;++k){
        yn=yn+h[k]*x[k];              /* filter data and compute output sample */
}
return(yn);                           /* filter output sample */
```

(1) *First sampling instant*. Read data sample from the ADC, shift data RAM one place (to make room for the new data), save the new input sample, compute output sample from Equation 11.4 and then send the computed output sample to the DAC:

Data RAM	Coefficient memory	
$\rightarrow x(1)$	$h(0)$	$y(1) = h(0)x(1) + h(1)x(0) + h(2)x(-1)$
0	$h(1)$	
0	$h(2)$	

(2) *Second sampling instant*. Repeat the above operation and work out the new output sample and send to the DAC:

Data RAM	Coefficient memory	
$\rightarrow x(2)$	$h(0)$	$y(2) = h(0)x(2) + h(1)x(1) + h(2)x(0)$
$x(1)$	$h(1)$	
0	$h(2)$	

(3) *Third sampling instant*. Repeat the above operation and work out the new output sample and send to the DAC:

Data RAM	Coefficient memory	
$\rightarrow x(3)$	$h(0)$	$y(3) = h(0)x(3) + h(1)x(2) + h(2)x(1)$
$x(2)$	$h(1)$	
$x(1)$	$h(2)$	

(4) *Fourth sampling instant*. Repeat the above operation and work out the new output sample and send to the DAC:

Data RAM	Coefficient memory	
$\rightarrow x(4)$	$h(0)$	$y(4) = h(0)x(4) + h(1)x(3) + h(2)x(2)$
$x(3)$	$h(1)$	
$x(2)$	$h(2)$	

Note that the oldest data sample has now fallen off the end.

(5) *nth sampling instant*. Repeat the above operation and work out the new output sample and send to the DAC:

Data RAM	Coefficient memory	
$\rightarrow x(n)$	$h(0)$	$y(n) = h(0)x(n) + h(1)x(n-1) + h(2)x(n-2)$
$x(n-1)$	$h(1)$	
$x(n-2)$	$h(2)$	

A TMS320C10 implementation of the three-point FIR filter is given in Program 11.2. In this case, the computation of the products starts at the bottom of the data and coefficients to exploit the TMS320C10 data move instructions. The instruction pair LTD and MPY are central to the TMS320C10-based FIR filter implementation. For example, the multiplication operation, $h(1)x(n-1)$, and the shift implied in Equation 11.4 or represented by z^{-1} in Figure 11.17(a) is carried out by the instruction pair

```
LTD     XNM1
MPY     H1
```

Program 11.2 Straight-line code for three-point FIR filter.

```
NXTPT   IN      XN, ADC
        ZAC
        LT      XNM2
        MPY     H2          ;h(2)x(n−2)

        LTD     XNM1        ;0+h(2)x(n−2); x(n−2)=x(n−1)
        MPY     H1          ;h(1)x(n−1); x(n−1)=x(n−2)

        LTD     XN          ;h(2)x(n−2)+h(1)x(n−1); x(n−1)=x(n)
        MPY     H0          ;h(0)x(n)

        APAC                ;h(2)x(n−2)+h(1)x(n−1)+h(0)x(n)

        SACH    YN,1        ;save output sample
        OUT     YN,DAC      ;output sample to DAC

        B       NXTPT
```

The instruction LTD XNM1 loads the T (temporary) register with the data sample $x(n-1)$ (held in data RAM address XNM1), adds the previous product, $h(2)x(n-2)$, which is still in the P (product) register to the accumulator, and shifts $x(n-1)$ up to the next address, that is $x(n-2) = x(n-1)$. The second instruction MPY multiplies the contents of the T register with $h(1)$ and leaves the result in the product register. The shifting scheme ensures that the input data samples are in the right locations when the next sample is to be computed.

Straight-line coding of the FIR filter, such as Program 11.2, leads to a fast implementation, but is not general purpose, and for large N-point filters will not yield a compact program.

Example 11.3

A digital FIR notch filter satisfying the specifications given below is to be implemented on the TMS320C10 target board.

notch frequency	1.875 kHz
attenuation at notch frequency	>60 dB
passband edge frequencies	1.575 and 2.175 kHz
passband ripple	<0.01 dB
sampling frequency	7.5 kHz
number of coefficients	61

A 61-point, optimal FIR filter satisfies the above specifications. The design of this filter was discussed in detail in Section 6.6.5. Here, we will concentrate only on the implementation. The coefficients of the filter are quantized to 16 bits

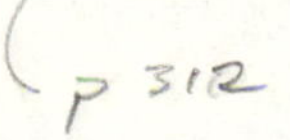

Table 11.2 Filter coefficients for Example 11.3.

	Quantized coefficients
FILTER LENGTH = 61	
***** IMPULSE RESPONSE *****	
H(1) = 0.12743640E−02 = H(61)	42
H(2) = 0.26730640E−05 = H(60)	0
H(3) = −0.23681110E−02 = H(59)	−78
H(4) = −0.17416350E−05 = H(58)	0
H(5) = 0.43428480E−02 = H(57)	142
(H 6) = 0.53579250E−05 = H(56)	0
H(7) = −0.71570240E−02 = H(55)	−235
H(8) = −0.49028620E−05 = H(54)	0
H(9) = 0.10897540E−01 = H(53)	357
H(10) = 0.89629280E−05 = H(52)	0
H(11) = −0.15605960E−01 = H(51)	−511
H(12) = −0.85508990E−05 = H(50)	0
H(13) = 0.21226410E−01 = H(49)	695
H(14) = 0.12250150E−04 = H(48)	0
H(15) = −0.27630130E−01 = H(47)	−905
H(16) = −0.11091200E−04 = H(46)	0
H(17) = 0.34579770E−01 = H(45)	1133
H(18) = 0.13800660E−04 = H(44)	0
H(19) = −0.41774130E−01 = H(43)	−1369
H(20) = −0.11560390E−04 = H(42)	0
H(21) = 0.48832790E−01 = H(41)	1600
H(22) = 0.12787590E−04 = H(40)	0
H(23) = −0.55359840E−01 = H(39)	−1814
H(24) = −0.90065860E−05 = H(38)	0
H(25) = 0.60944450E−01 = H(37)	1997
H(26) = 0.88997300E−05 = H(36)	0
H(27) = −0.65232190E−01 = H(35)	−2137
H(28) = −0.38167120E−05 = H(34)	0
H(29) = 0.67925720E−01 = H(33)	2226
H(30) = 0.27041150E−05 = H(32)	0
H(31) = 0.93115220E+00 = H(31)	30512

(Q15 format), by multiplying each coefficient by 2^{15}, and then rounding to the nearest integer. The quantized and unquantized coefficients are listed in Table 11.2.

A TMS320C10 program for the notch filter is given on the disk for the book. The flowchart for real-time implementation using the TMS320C10 is given in Figure 11.18. The TMS320C10 implementation makes use of the indirect addressing features of the TMS320, and contains a loop controlled by the BANZ instruction as illustrated in Figure 11.18. We have used indirect addressing to reduce program size and to make the program general purpose. Only the filter length, N, and the coefficients need to be changed to use the program

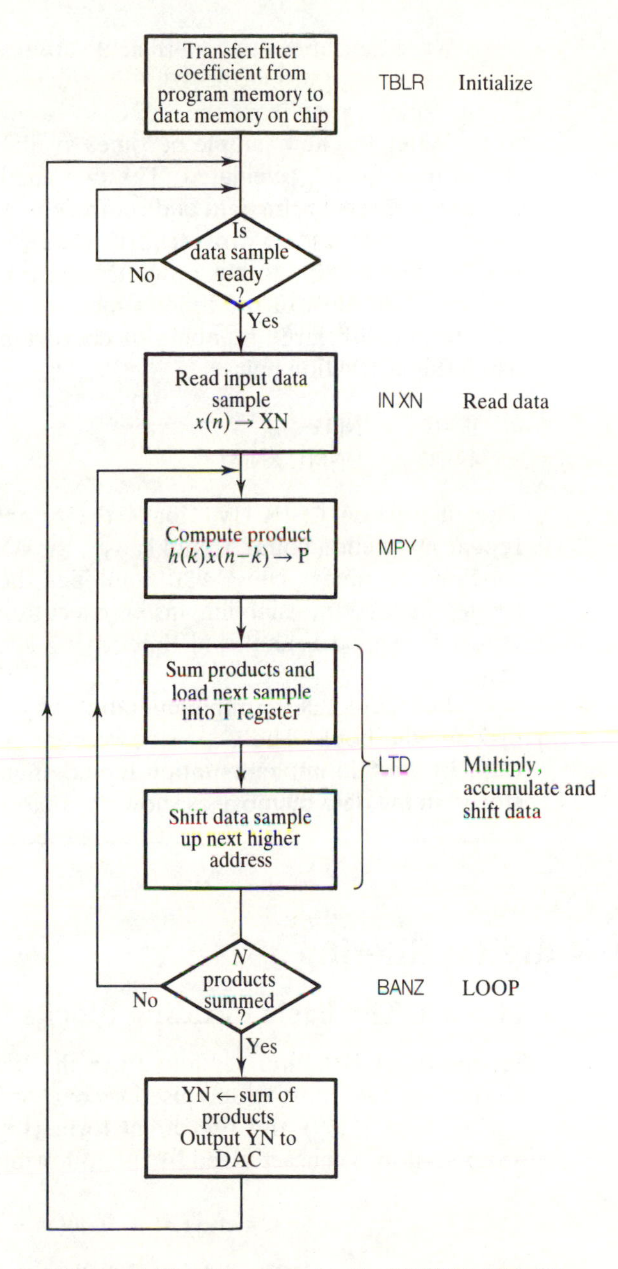

Figure 11.18 Flowchart for the TMS320 FIR filter.

for a different FIR filter. Program 11A.1 in the appendix which is a TMS320C10 implementation of the design example in Section 6.11 illustrates how the notch filter program (on the disk) can be used for other filters. However, because of the overhead associated with the loop control this approach is slower in execution than if it was coded in straight line.

At initialization, the coefficients are transferred from program memory to the data memory using the TBLR instruction. The processor waits for the new input sample $x(n)$ from the ADC to become available (the BIOZ line goes low). When the new sample becomes available, it is read into the memory and the output sample calculated. The two auxiliary registers, AR0 and AR1, are used as pointers to the data and coefficient. One is used as a loop counter.

The TMS320C25 provides a single-cycle multiply–accumulate instruction, MACD, which helps to cut down the time to execute an FIR filter. With the TMS320C10, most of the time is spent in the BANZ loop. In the TMS320C25, FIR filters with large numbers of coefficients can be efficiently implemented using the instruction pair

```
RPTK      NM1
MACD      HNM1, XNM1
```

The instruction RPTK NM1 loads the filter length minus 1 $(N-1)$ into the repeat instruction counter, and causes the MACD instruction following it to be repeated N times. The MACD combines the instruction pair LTD–MPY into a single instruction, enabling faster execution. The instruction pair RPTK and MACD is a good example of time-saving special instructions available in DSP chips.

The TMS320C25 implementation of the 61-point FIR filter is also on the disk for the book. The C25 code is more compact and faster than that of the C10. In the C25 implementation the coefficients and data sample sequence are stored in the data memory as shown in Figure 11.17(c).

11.4.2 IIR digital filtering

11.4.2.1 The basic building blocks for IIR filters

Second-order IIR filter sections form the basic building blocks for digital IIR filters. The two most widely used second-order structures are the canonic section (Figure 11.19) and the direct form (Figure 11.20). The canonic second-order section is characterized by the following equations:

$$w(n) = \mathrm{sf}_1 x(n) - b_1 w(n-1) - b_2 w(n-2) \quad \textbf{(11.5a)}$$

$$y(n) = a_0 w(n) + a_1 w(n-1) + a_2 w(n-2) \quad \textbf{(11.5b)}$$

where $x(n)$ represents the input data, $w(n)$ represents the internal node, $y(n)$ is the filter output sample and sf_1 is a scale factor, equal to $1/s_1$.

A TMS320C10 implementation of this IIR filter section, in straight line code, is given in Program 11.3. The memory map showing the storage of the filter coefficients and internal input data sequence, $w(n)$, is depicted in Figure 11.19(b). (For the TMS320, the feedback coefficients, b_1 and b_2, are in fact

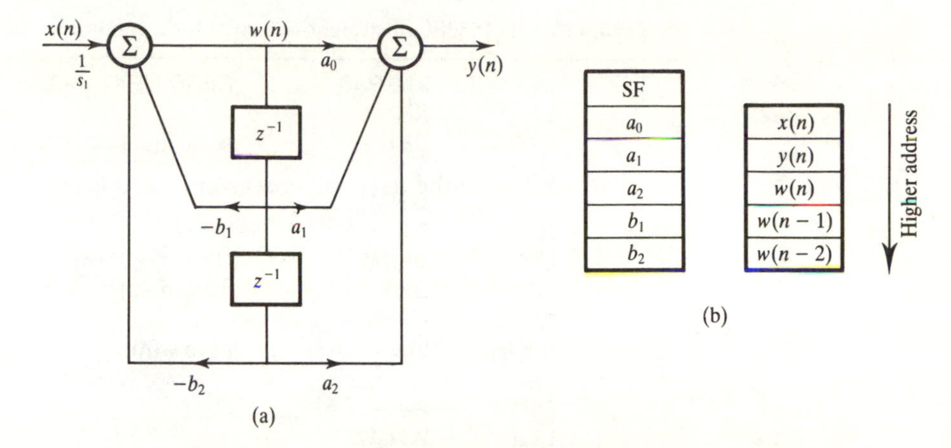

Figure 11.19 (a) Second-order canonic section; (b) coefficient–data storage.

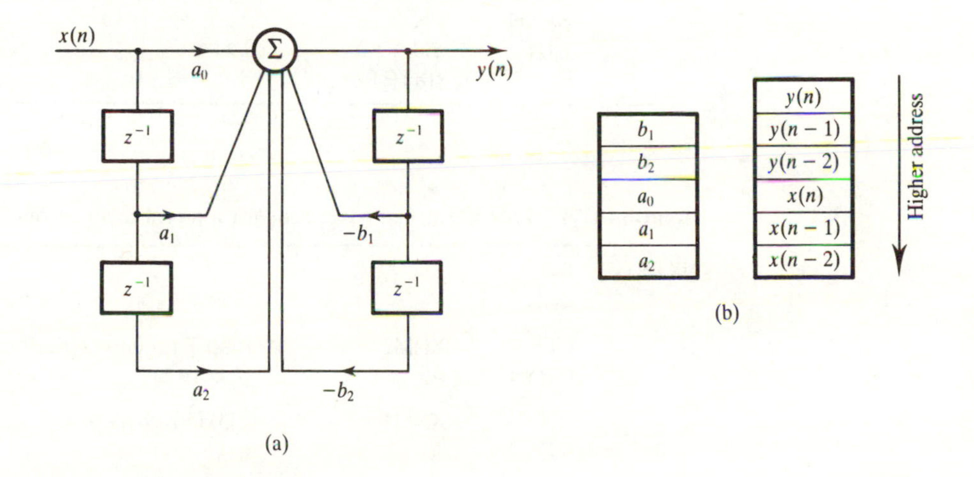

Figure 11.20 Implementation of the direct form second-order section: (a) realization diagram; (b) data and coefficient storage.

stored with their signs reversed.) As in the case of FIR filter implementation, the multiply-and-add operations implicit in Equations 11.5 are performed with the instruction pair LTD and MPY.

The difference equation for the direct form second-order IIR filter (Figure 11.20(b)) section is given by

$$y(n) = a_0x(n) + a_1x(n-1) + a_2x(n-2) - b_1y(n-1) - b_2y(n-2) \tag{11.6}$$

where $x(n-k)$ are the input data sequence and $y(n-k)$ are the output data

Program 11.3 TMS320C10 straight-line code for a canonic section.

```
NXTPT   IN      XN, PA0     ;READ NEXT DATA
        LT      XN
        MPY     SF1         ;scale input sample sf1*x(n)
        PAC
        LT      WNM1
        MPY     B1          ;b1*w(n-1)
        LTA     WNM2        ;sf1*x(n)+b1*w(n-1)
        MPY     B2          ;b2*w(n-2)
        APAC
        SACH    WN          ;save w(n)
        ZAC
        MPY     A2
        LTD     WNM1
        MPY     A1
        LTD     WN
        MPY     A0          ;a0*w(n)
        APAC                ;a1*w(n-1)+a0*w(n)
        SACH    YN          ;save output of
        OUT     YN, PA0
        B       NXTPT
```

Program 11.4 TMS320 straight-line code for a direct form second-order section.

```
NXTPT   IN      XN, ADC
        ZAC

        LT      XNM2        ;load T register with data sample x(n-2)
        MPY     A2          ;a2x(n-2)

        LTD     XNM1        ;SUM=a2x(n-2)
        MPY     A1          ;a1x(n-1)

        LTD     XN          ;SUM=a2x(n-2)+a1x(n-1)
        MPY     A0          ;a0x(n)

        LTA     YNM2        ;SUM=a2x(n-2)+a1x(n-1)+a0x(n)
        MPY     B2          ;b2y(n-2)

        LTD     YNM1        ;SUM=a2x(n-2)+a1x(n-1)+a0x(n)+
                            ;b2y(n-1)
        MPY     B1          ;b1y(n-1)

        LTA     YN          ;SUM=a2x(n-2)+a1x(n-1)+a0x(n)+
                            ;b2y(n-2)+b1y(n-1)
        SACH    YN,1        ;save the upper 16 bits in data memory
                            location YN
        OUT     YN,DAC
        B       NXTPT
```

sequence. The data and coefficient storage for the direct structure is depicted in Figure 11.20(b) and the TMS32010 code given in Program 11.4.

The direct form filter is simpler to program and can lead to a somewhat faster implementation than the canonic section because of the simpler indexing involved: compare, for example, Equations 11.5 and 11.6.

11.4.2.2 Higher-order IIR filters

Higher-order IIR filters are realized as either a cascade or a parallel combination of the second-order filter sections (see Chapter 7 for more details).

Cascade realization

The transfer function, $H(z)$, of an Nth-order IIR filter, using second-order sections in cascade, is given by

$$H(z) = \prod_{k=1}^{N/2} \frac{a_{0k} + a_{1k}z^{-1} + a_{2k}z^{2}}{1 - b_{1k}z^{-1} - b_{2k}z^{-2}} \quad \textbf{(11.7)}$$

The cascade realization of a fourth-order ($N = 4$) IIR filter using second-order canonic sections is shown in Figure 11.21(a). The storage of the filter variables (data and coefficients) is shown in Figure 11.21(b). The set of difference equations for the fourth-order IIR filter, using canonic sections, is given by

$$w_1(n) = \mathrm{sf}_1 x(n) - b_{11}w_1(n-1) - b_{21}w_1(n-2) \quad \textbf{(11.8a)}$$

$$y_1(n) = a_{01}w_1(n) + a_{11}w_1(n-1) + a_{21}w_1(n-2) \quad \textbf{(11.8b)}$$

$$w_2(n) = y_1(n) - b_{12}w_1(n-1) - b_{22}w_2(n-2) \quad \textbf{(11.8c)}$$

$$y_2(n) = a_{02}w_2(n) + a_{12}w_2(n-1) + a_{22}w_2(n-2) \quad \textbf{(11.8d)}$$

A C language pseudo-code for an IIR filter realized as a cascade of second-order canonic sections is given in Program 11.5.

Example 11.4

Design and implement a lowpass IIR digital filter using the TMS320C10-based target board (see Chapter 12) to meet the following specifications:

sampling frequency	15 kHz
passband	0–3 kHz
transition width	450 Hz
passband ripple	0.5 dB
stopband attenuation	45 dB

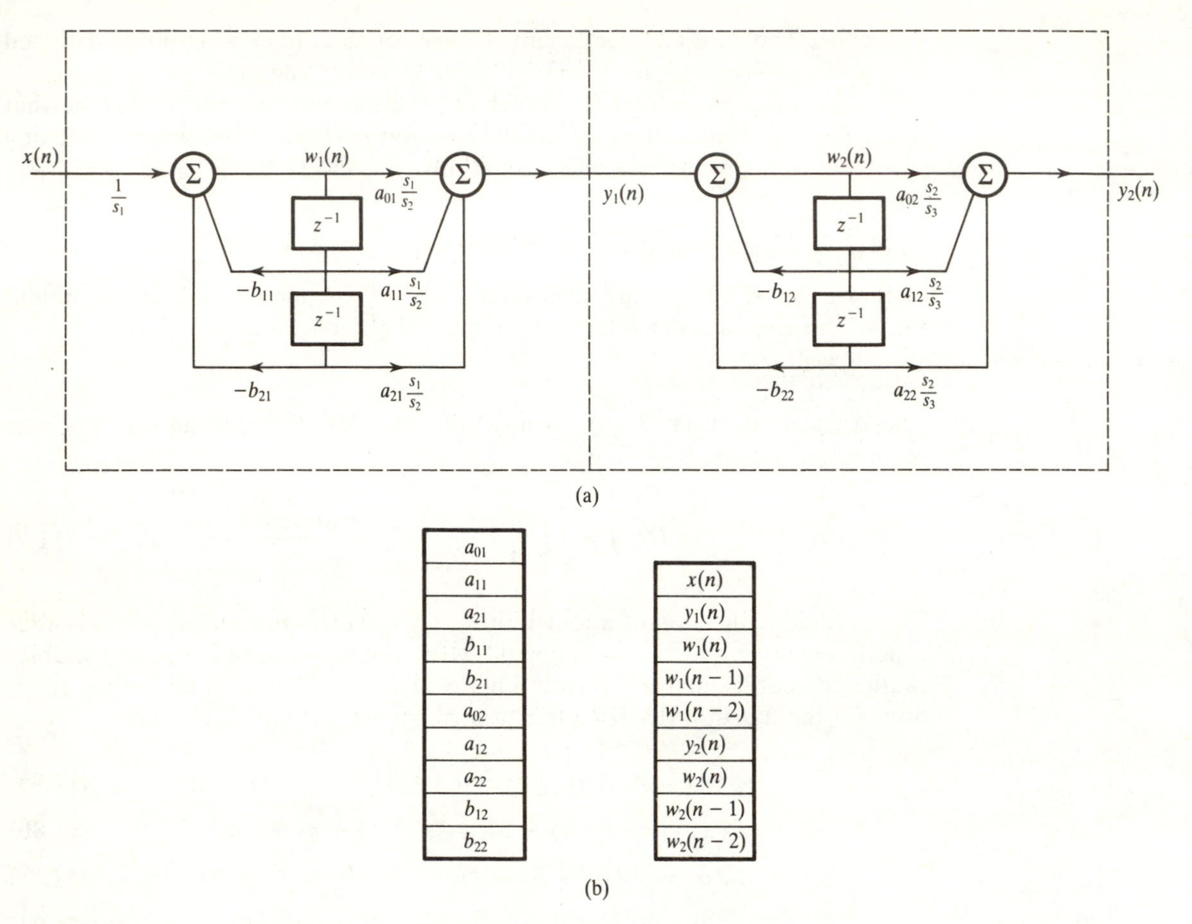

Figure 11.21 Cascade realization of an IIR filter: (a) realization diagram; (b) data and coefficient storage.

Program 11.5 C language pseudo-code for a cascade IIR filter.

```
for(n=0; n<(Nsamples-1); ++n){          /* Nsamples no of data samples */

    xn=x[n];
    for(k=1; k<N; ++k){
        wk=sk[k]*xn-b1[k]*w1[k]-b2[k]*w2[k];
        yk=(a0[k]*wk+a1[k]*w1[k]+a2[k]*w2[k]);
                                        /*output of 1st section*/
        w2[k]=w1[k];
                                        /*shift and save delay node data */
        w1[k]=wk;
        xn=yk;                          /*kth section feeds next section */
    }
    y[n]=yk;                            /*nth output sample */
}
```

Table 11.3 Filter coefficients before and after quantization to 16 bits.

	Coefficient	*Scaled*	*Quantized*
a_{02}	1	0.999 969 5	32 767
a_{12}	0.675 718	0.675 718	22 142
a_{22}	1	0.999 969 5	32 767
b_{12}	−0.495 935	−0.495 935	−16 251
b_{22}	0.761 864	0.761 864	24 965
a_{01}	1	0.131 113 6	4 296
a_{11}	1.649 656	0.216 292 4	7 087
a_{21}	1	0.131 113 6	4 296
b_{11}	−0.829 328	−0.829 328	−27 175
b_{21}	0.307 046	0.307 046	10 061

$s_1 = 2.479\,158\ (L_1)$; sf $= 0.403\,362\,7$; $s_2 = 18.908\,47\ (L_1)$.

A detailed design of this filter was given in Chapter 7 (Section 7.8). It was shown there that a fourth-order elliptic filter with the following transfer function would meet the specifications:

$$H(z) = \frac{1 + 0.675\,718z^{-1} + z^{-2}}{1 - 0.495\,935z^{-1} + 0.761\,864z^{-2}} \times \frac{1 + 1.649\,656z^{-1} + z^{-2}}{1 - 0.829\,328z^{-1} + 0.307\,046z^{-2}}$$

The difference equations for the cascade realization using canonic sections are the same as Equation 11.8.

The coefficients, scaled to avoid overflow and quantized to 16 bits, are listed in Table 11.3. The TMS320C10 codes for the fourth-order filter using the canonic form filters are available on the PC disk for this book. The filter was also implemented on the TMS320C25 in the SWDS (software development system) environment. The complete C25 codes are given in Program 11A.2.

Parallel realization

The transfer function of an Nth-order IIR filter for parallel realization is given by

$$H(z) = \sum_{k=1}^{N/2} \frac{a_{0k} + a_{1k}z^{-1}}{1 + b_{1k}z^{-1} + b_{2k}z^{-2}} + C \tag{11.9}$$

The realization diagram, using second-order canonic sections, for $N = 4$, is given in Figure 11.22. For the canonic section, the difference equation is given by

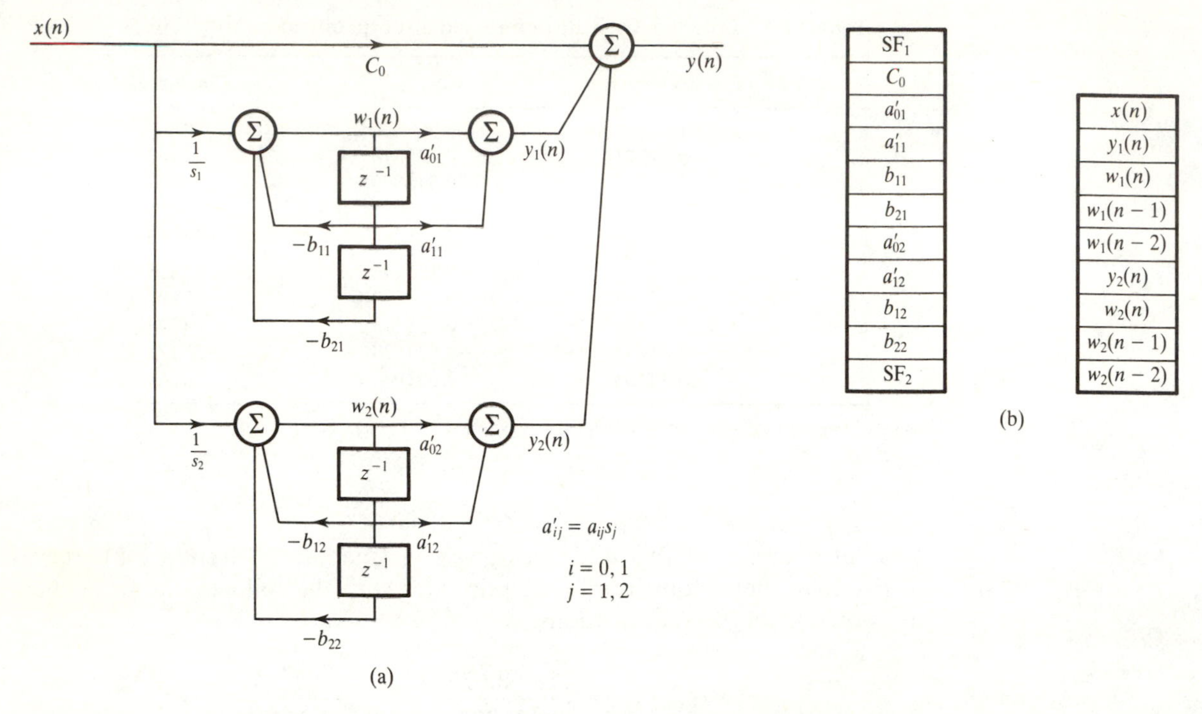

Figure 11.22 Implementation of a fourth-order IIR filter: (a) realization diagram; (b) coefficient and data storage.

$$w_1(n) = \text{sf}_1 x(n) - b_{11} w_1(n-1) - b_{21} w_1(n-2)$$

$$y_1(n) = a_{01} w_1(n) + a_{12} w_1(n-1)$$

$$w_2(n) = \text{sf}_2 x(n) - b_{12} w_2(n-1) - b_{22} w_1(n-2)$$

$$y_2(n) = a_{02} w_2(n) + a_{12} w_2(n-1)$$

$$y(n) = a_0 x(n) + y_1(n) + y_2(n)$$

A simple C language code for an IIR filter realized as a parallel combination of second-order canonic sections is given in Program 11.6.

Example 11.5

Represent the transfer function of Example 11.4 in a parallel form using second-order canonic sections as building blocks. Implement the filter using the same hardware as the last example.

Using the partial fraction expansion program discussed in Chapter 3, the coefficients for parallel realization were obtained from those of the cascade

Program 11.6 C language pseudo-code for parallel realization.

```
for(n=0; n<(Nsamples-1); ++n){                    /*Namples no of data samples */

        y[n]=c*x[n];                              /*output through constant path */
        for(k=1; k<N; ++k){
                wk=sk[k]*x[n]-b1[k]*w1[k]-b2[k]*w2[k];
                yk=(a0[k]*wk+a1[k]*w1[k])/sk[k];  /*output of 1st section*/
                w2[k]=w1[k];                      /*shift and save delay node data */
                w1[k]=wk;
                y[n]=yk+y[n];
        }
}
```

Table 11.4 Implementation of fourth-order IIR filter of Example 11.5: filter coefficients before and after quantization to 16 bits.

	Unquantized coefficients	*Quantized coefficients*
sf1	0.181 00	5 931
c_0	0.249 923 79	8 190
a_{01}	−0.132 922 5	−24 063
a_{11}	−0.180 523 2	−32 670
b_{11}	0.028 994	16 251
b_{21}	−0.044 541 6	−24 965
a_{02}	−0.058 534	−4 756
a_{12}	0.508 420 5	20 653
b_{12}	0.048 489 9	27 178
b_{22}	−0.017 951	−10 061
sf_2	0.403 32	13 216

realization. The transfer function becomes

$$H(z) = \frac{-0.132\,922\,5 - 0.180\,523\,2z^{-1}}{1 - 0.028\,994z^{-1} + 0.044\,541\,6z^{-2}} + \frac{-0.058\,534 + 0.508\,420z^{-1}}{1 - 0.048\,489\,9z^{-1} + 0.017\,951\,1z^{-2}} + 0.249\,923\,79$$

$$s_1 = 5.524\,484\,4, \quad s_2 = 2.4794$$

Table 11.4 gives the coefficient values before and after quantization to 16 bits. The TMS320C10 and TMS320C25 codes for the filter are available on the PC disk for this book.

Extension of the implementation techniques discussed above for both the cascade and parallel structures to higher-order IIR filters is relatively easy.

However, a more compact code may be obtained by implementing the second-order building block as a subroutine.

11.4.3 FFT processing

The discrete Fourier transform (DFT) of a finite data sequence, $x(n)$, is defined as

$$X(k) = \sum_{n=0}^{N-1} x(n) W_N^{nk}$$

where W_N, often called the twiddle factor, is a set of complex coefficients.

Direct computation of the DFT coefficients, $X(k)$, is time consuming when N is large. FFT algorithms provide efficient ways of computing $X(k)$ with significant reduction in computation time. As discussed in Chapter 2, the butterfly and twiddle factor are central to FFT algorithms.

11.4.3.1 Implementation of the butterfly

Figures 11.23(a) and 11.23(b) depict the two types of butterflies used in the radix-2 FFT. FFTs based on these butterflies lead to the same result. For the decimation in time (Figure 11.23(a)) the butterfly takes a pair of input data, A and B, and produces a pair of outputs:

$$A' = A + BW_N^k \tag{11.10a}$$

$$B' = A - BW_N^k \tag{11.10b}$$

In general the input and output data samples as well as the twiddle factors are all complex and can be expressed as

$$A = A_r + jA_i \tag{11.11a}$$

$$B = B_r + jB_i \tag{11.11b}$$

$$W_N^k = e^{-j2\pi k/N} = \cos(2\pi k/N) - j\sin(2\pi k/N) \tag{11.11c}$$

where the suffix r indicates the real part and i the imaginary part of the data. The butterfly operation in Equations 11.10 involves complex arithmetic, but in practice it is often carried out using real arithmetic. To express the operation in a form suitable for real arithmetic, we note that the product of B and W in Equations 11.10 has the form

$$BW_N^k = B_r\cos(X) + B_i\sin(X) + j[B_i\cos(X) - B_r\sin(X)] \tag{11.12}$$

where $X = 2\pi k/N$. Using Equations 11.11 and 11.12 in Equations 11.10a and 11.10b we have

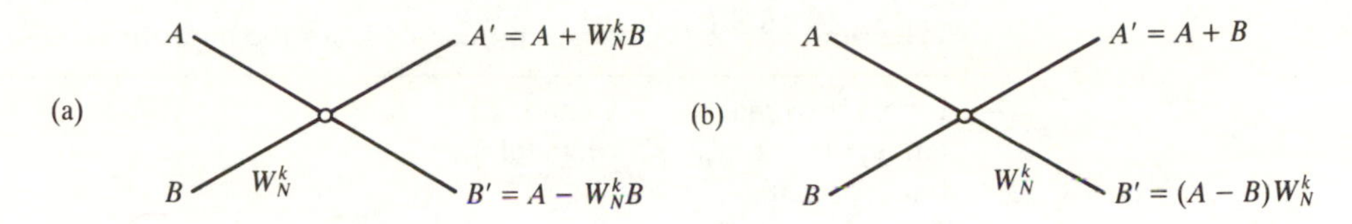

Figure 11.23 The two types of butterflies used in radix-2 FFT algorithm: (a) butterfly for the decimation in time radix-2 FFT; (b) butterfly for the decimation in frequency radix-2 FFT.

Program 11.7 A C language pseudo-code for pre-calculating the twiddle factor values.

```
pi=6.28315307179586/N;
for(k=0; k<N/2; ++k){
        X=k*pi;
        w.real[k]=cos[X];
        w.imag[k]=sin[X];
}
```

$$A' = A_r + B_r \cos(X) + B_i \sin(X) + j[A_i + B_i \cos(X) - B_r \sin(X)] \tag{11.13a}$$

$$B' = A_r - [B_r \cos(X) + B_i \sin(X)] + j\{A_i - [B_i \cos(X) - B_r \sin(X)]\} \tag{11.13b}$$

The outputs of the butterfly, A' and B', are now in the desired form. Thus, given a pair of complex data points, A and B, in rectangular form, Equations 11.13a and 11.13b are used to compute the output of the butterfly using real arithmetic.

The computation of the sine and cosine terms in Equations 11.13 is time consuming. In real-time FFT, a more efficient approach is to pre-calculate the real and imaginary parts of the twiddle factor (Equation 11.11c), and to store these values in a look-up table. The C language pseudo-code in Program 11.7 illustrates how the twiddle factor values are pre-calculated.

A C language pseudo-code for the radix-2 butterfly, with twiddle factor values pre-calculated and stored in a look-up table, is given in Program 11.8.

A TMS320C25 pseudo-code is shown in Program 11.9. The pre-calculated twiddle factor values are stored in Q15 format. The input data is assumed complex, with the real and imaginary parts stored in consecutive locations in data RAM. For complex inputs, the values of A' or B' can attain a maximum value of 2.414 42, and 2 for real data input. In fixed point arithmetic this will cause overflow. To avoid overflow the input data to a butterfly should be scaled. In the C25 implementation, the scaling is dynamic, advantage being taken of the fact that the product of two fixed point numbers produces an extra sign bit. The extra sign bit is normally removed by a left shift, but by leaving it the result is effectively scaled by 2.

Program 11.8 A C language pseudo-code for the butterfly.

```
t.real=br*w.real[k]+bi*w.imag;
t.imag=bi*w.real[k]-br*w.imag[k];
b.real[j]=a.real-t.real;
b.imag[j]=a.imag-t.imag;
a.real[j]=a.real+t.real;
a.imag[j]=a.imag+t.imag;
```

Program 11.9 TMS320C25 code for the butterfly.

```
*
*       compute terms common to the two butterfly outputs, A' and B'
*
*
        LT      BR        ;compute 1/2*[b.real*cos(X)+b.imag*sin(X)]
        MPY     WREAL     ;1/2*b.real*cos(X)
        LTP     BI
        MPY     WIMAG     ;1/2*b.imag*sin(X)
        APAC              ;1/2[b.real*cos(X)+b.imag*sin(X)]
        MPY     WREAL     ;1/2*b.imag*cos(X)
        LT      BR
        SACH    BR        ;1/2[b.real*cos(X)+b.imag*sin(X)]

        PAC               ;compute [q.imag*cos(X)-q.real*sin(X)]
        MPY     WIMAG
        SPAC
        SACH    BI
*
*       compute and save the butterfly outputs
*
        LAC     AR, 14    ;compute and save the real parts of the output
        ADD     BR, 15
        SACH    AR, 1     ;save a.real
        SUBH    BR
        SACH    BR, 1     ;save b.real

        LAC     AI,14     ;compute and save the imaginary parts of the
                          ;output
        ADD     BI, 15
        SACH    AI, 1     ;save a.imag
        SUBH    BI
        SACH    BI,1      ;save b.imag
```

11.4.3.2 In-place computation and constant geometry

The signal flowgraph of an eight-point FFT is shown in Figure 11.24. From the figure, it is evident that to obtain the DFT coefficients, $X(k)$, shown on the

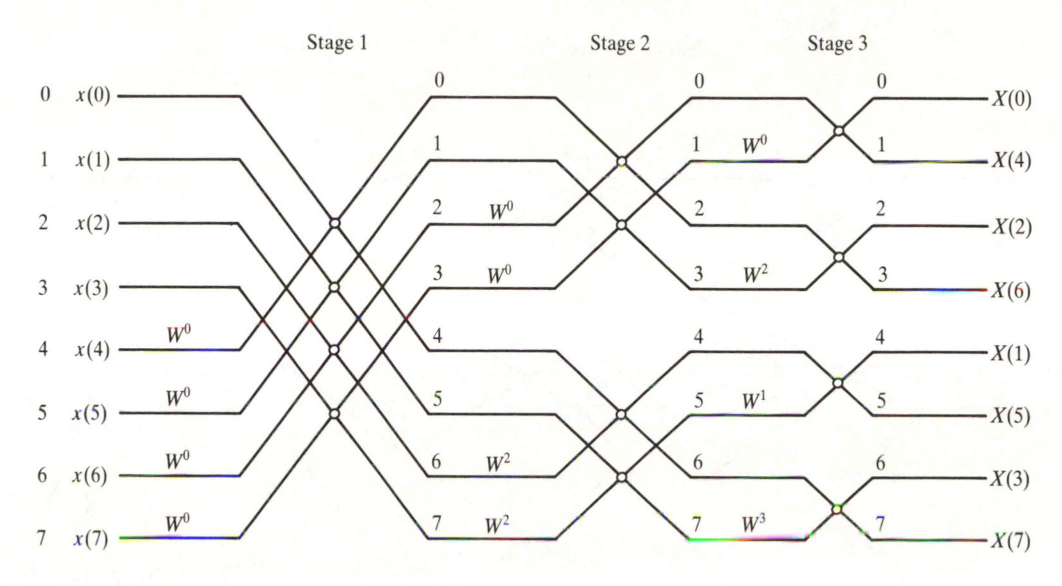

Figure 11.24 An in-place DIT FFT flowgraph, with input in natural order but output in bit-reversed order.

right-hand side, given the input, a series of butterfly computations will have to be performed. The radix-2 FFT algorithm is a method of carrying out the series of butterfly computations in an orderly manner. In the flowgraph, the data flows from left to the right. Thus, once the outputs of a butterfly, A' and B', are computed the inputs, A and B, are no longer required and so can be overwritten by the outputs. This is the basis of the concept of in-place computation.

The in-place algorithm makes efficient use of available memory as the transformed data overwrites the input data. In the past, when memory was very expensive, this was an important consideration. However, with in-place computation the indexing required to determine where in memory to fetch the input data to each butterfly is quite complex. For example, in Figure 11.24 the top butterfly in stage 1 takes its inputs from addresses 0 and 4 and writes the outputs back to the same addresses. On the other hand, the top butterfly in stage 2 takes its inputs from addresses 0 and 2. In general, in the in-place FFT the input/output addresses vary from stage to stage. Further, for high speed FFTs, the use of the same memory for input and output slows down computations because of long memory access (except, for example, if dual-port RAMs are used). As both memory and multipliers are cheap, the trend now is to optimize the whole FFT processor to push speed up.

An alternative FFT implementation, known as the non-in-place or constant geometry, reads the input data to a butterfly from a pair of addresses and stores the output in another pair of addresses as shown in Figure 11.25. Unlike the in-place FFT, where the input/output addresses for each butterfly vary from stage to stage, in this case the addressing for each butterfly is fixed and much simpler. For an N-point FFT, the inputs of the nth butterfly at each

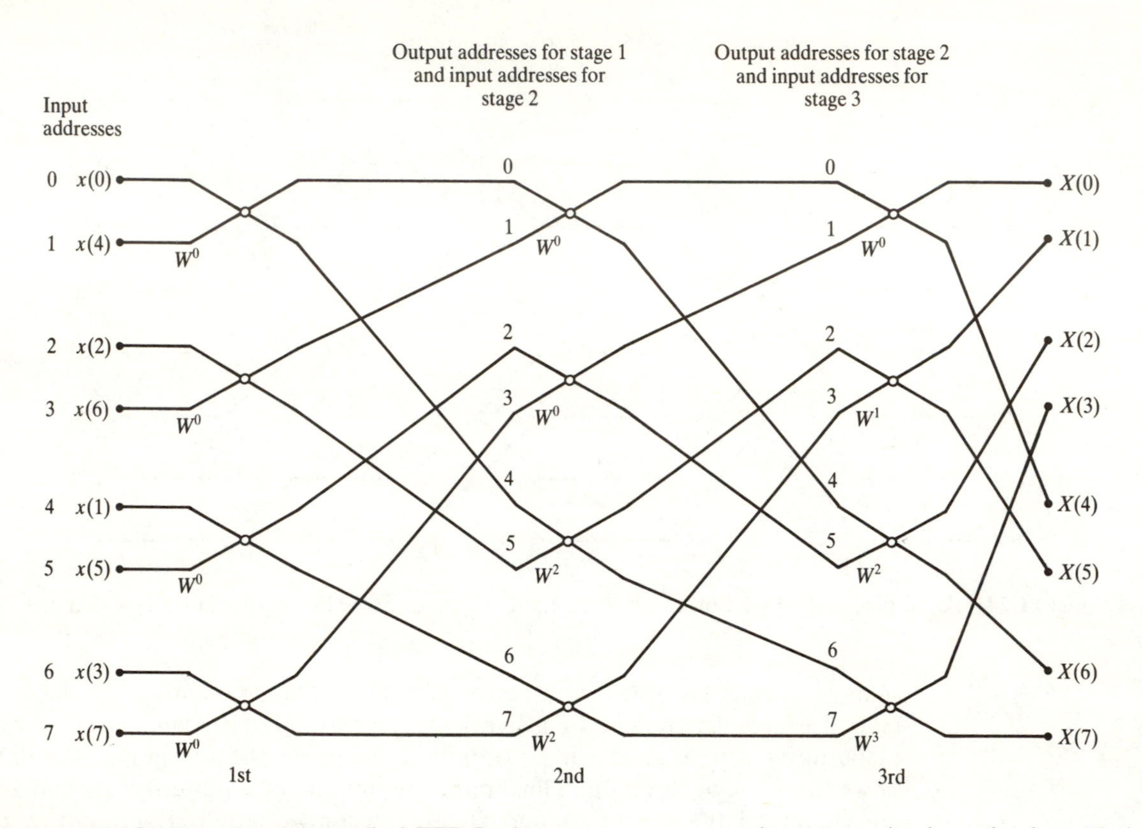

Figure 11.25 Constant geometry radix-2 FFT. In the constant geometry, the computation is not in place. Notice that for each butterfly the input and output span is constant.

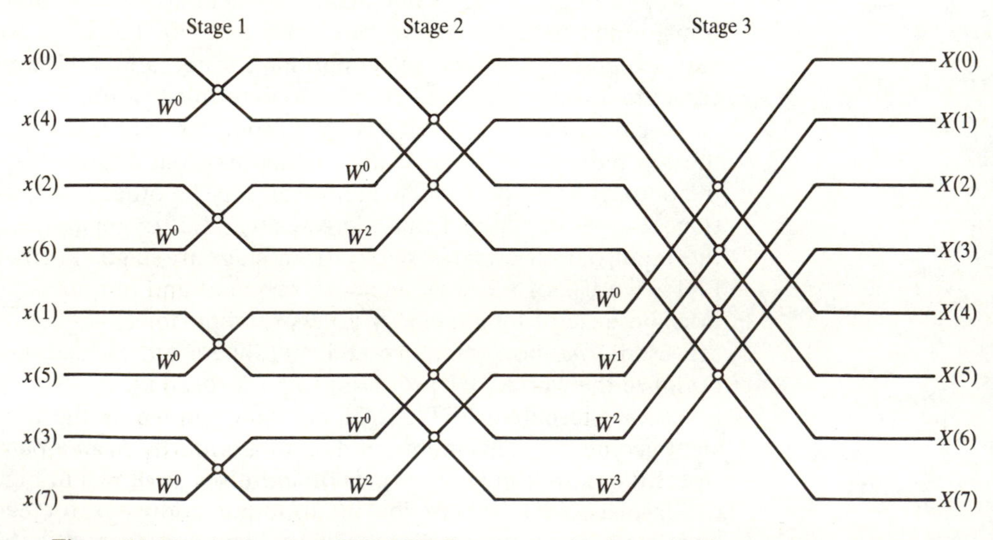

Figure 11.26 An in-place DIT FFT flowgraph with input in bit-reversed order and the output in natural order.

Table 11.5 Data for eight-point FFT showing the concepts of bit reversing.

Input sequence, natural order	*Binary code of sequence*	*Input sequence, bit reversed*	*Binary code of sequence (bit reversed)*
$x(0)$	000	$x(0)$	000
$x(1)$	001	$x(4)$	100
$x(2)$	010	$x(2)$	010
$x(3)$	011	$x(6)$	110
$x(4)$	100	$x(1)$	001
$x(5)$	101	$x(5)$	101
$x(6)$	110	$x(3)$	011
$x(7)$	111	$x(7)$	111

stage are $2n$ and $2n+1$, $n = 0, 1, \ldots, N/2 - 1$. The outputs of the nth butterfly are stored at addresses n and $N/2 + n$. For example, at the second stage of Figure 11.25 the top butterfly takes its inputs from addreses 0 and 1 and stores its outputs at addresses 0 and 4. Clearly, for the non-in-place FFT to work two separate memories or arrays are required; one holds the inputs and the other the outputs. After each stage the roles of the memories are reversed.

11.4.3.3 Data scrambling and bit reversal

In the DIT (decimation-in-time) FFT, if the input data sequence is applied to the FFT processor in natural order, the output of the FFT appears scrambled (see Figure 11.24). To ensure the output appears in the correct order (that is, as $X(0)$, $X(1), \ldots, X(N-1)$), we either scramble the input data sequence before taking the FFT (see Figures 11.25 and 11.26) or unscramble the output after taking the FFT (see Figure 11.24).

For radix-2 FFT, input data scrambling is achieved by storing the input sequence in bit-reversed order. Assuming the input data has already been stored in natural order (that is, as $x(0)$, $x(1)$, $x(2)$, $\ldots$, $x(N-1)$), the bit-reversed order is achieved by representing the indices of the input data in binary as shown in the second column of Table 11.5 for an eight-point FFT, and then swapping the bits about the centre (fourth column in Table 11.5). Notice for example, that in the table the index of data sample $x(3)$ has a binary representation of 011. Swapping the first and third bits about the middle bit gives 110 (that is, bit 1 which is 0 becomes 1 and bit 3 which is 1 becomes 0 while the middle bit about which we perform the operation remains unchanged). The bit-reversed code, 110, is decimal number 6. To effect scrambling, we swap the locations of the data samples $x(3)$ and $x(6)$. Applying the same principle to the remainder of the inputs, we obtain the bit-reversed sequence given in Table 11.5, third column. Notice that, after scrambling, the locations of the first and last data samples remain unchanged. This is because bit reversal of the indices 000 and 111 does not have any effect. In general, the first and last data points are not affected by scrambling in radix-2 FFT. The

Program 11.10 In-place bit reversal for radix-2 FFT.

```
/* perform in-place bit reversal */

j=1;
for(i=1; i<N; ++i){
	if(i<j){
		tr=x.real[j];	/* swap x[j] and x[i] */
		ti=x.imaj[j];
		x.real[j]=x.real[i];
		x.imag[j]=x.imag[i];
		x.real[i]=tr;
		x.imag[i]=ti;
		k=N/2;
		while(k<j){
			j=j-k;
			k=k/2;
		}
	}
	else {
		k=N/2;
		while(k<j){
			j=j-k;
			k=k/2;
		}
	}
j=j+k;
}
```

alert reader will have spotted other samples of the input sequence immune to bit reversal.

When the input data is held in a memory or an array, scrambling the input data involves identifying pairs of input data locations and interchanging or swapping the data in those locations. A bit reversal algorithm due to Rader (Rabiner and Gold, 1975) is the most widely used to determine the indices of the memory locations to swap. A C language pseudo-code for the bit reversing algorithm is given in Program 11.10.

Advanced DSP chips now provide instructions to perform bit reversal on the input data samples as they are fetched from memory in readiness for FFT or on the transformed data as they are being stored in memory after FFT. In the TMS320C25, for example, the instruction

```
IN	*BRO+, PA2
```

reads a data sample from an external memory, via port PA2, and stores it in the correct scrambled location using the so-called reverse carry propagation. Two auxiliary registers are involved in performing the reverse carry propaga-

Program 11.11 Fragment of TMS320C25 codes for bit reversal.

```
LARK    AR0, N        ;load FFT length in AR0
LRLK    AR1           ;load base address in AR1
LARP    1             ;select auxiliary register 1
RPTK    N-1           ;read data, scramble and save in memory
IN      *BR0+, PA2
```

tion. For an N-point FFT, auxiliary register AR0 holds the value of the transform length, N, and another register, for example, AR1, holds the base or start address of the data. The instruction above is then combined with the repeat instruction to fetch and save data in memory in bit-reversed order. Program 11.11 gives a fragment of the C25 assembly code for bit reversing data stored in an external memory in natural order. It is assumed that the input data is complex with each input data sample stored in memory in pairs with the real part in the even address location and the imaginary part in the odd address location.

11.4.4 Multirate processing

As discussed in Chapter 10, multirate processing involves performing DSP operations at more than one sampling rate. The two fundamental operations in multirate processing are decimation (sample rate reduction) and interpolation (sampling rate increase). We will illustrate the implementation of a real-time decimator by an example.

Example 11.6

The sampling rate of a signal is to be reduced by a three-stage decimation process from 30 kHz to 1 kHz. Assume the highest frequency of interest after decimation is 400 Hz, an in-band ripple of 0.08 dB and stopband rejection of 50 dB. The decimator is to be implemented on the TMS320C25 in the SWDS environment (Texas Instruments, 1988).

Solution

Using the multirate design program (on the PC disk for this book), parameters of the three-stage decimator were obtained; see Figure 11.27. The coefficients of the three filters, using the optimal FIR design program, are given in Table 11.6. The filter lengths are slightly longer than the estimates predicted by the decimator design program (12, 13, 48 instead of 13, 12, 46) to ensure the specifications are met.

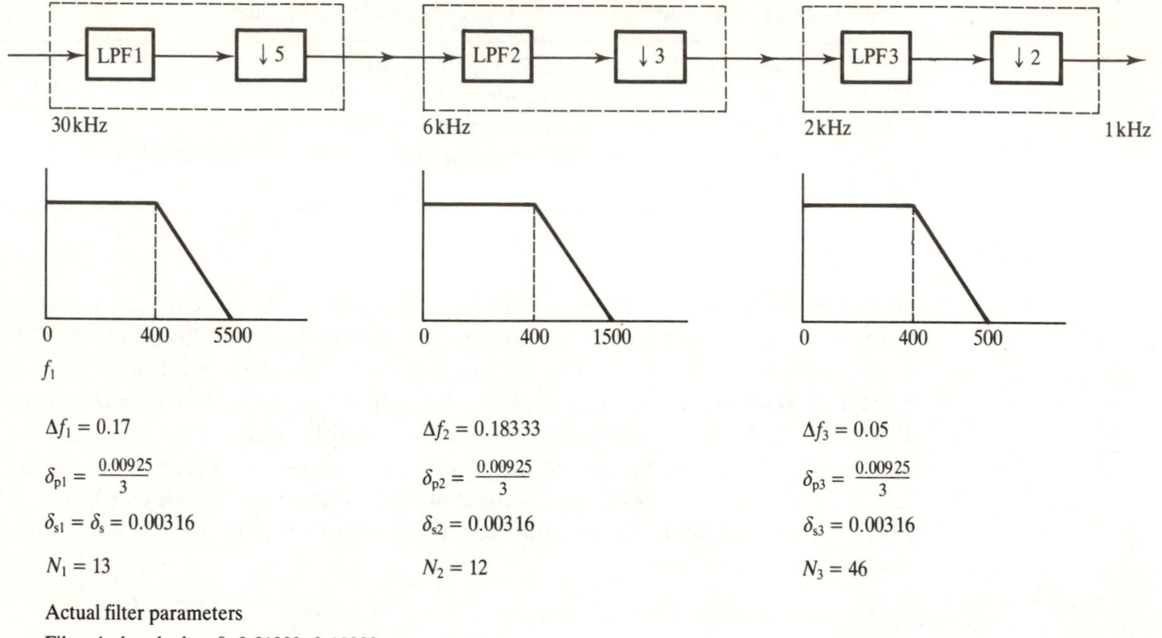

Actual filter parameters

Filter 1: band edge, 0, 0.01333, 0.18333
filter length, 12; weight, 1, 3

Filter 2: band edge, 0, 0.06666, 0.25, 0.5
filter length, 13; weight, 1, 3

Filter 3: band edge, 0, 0.2, 0.25, 0.5
filter length, 48; weight, 1, 3

Figure 11.27 Parameters of the three-stage decimator.

The flowchart for a general three-stage decimator is given in Figure 8.13. The coefficient and data storage map for the TMS320C25-based decimator is shown in Figure 11.28. RAM block B0, configured as a program memory, is divided into three and holds the coefficients of the three filters. The coefficients for filter 1, $h_1(k)$, are allocated 32 memory locations starting from address FF00h; coefficients for the stage 2 filter start at location FF20h and can have a length of up to 96; the third-stage filter starts at FF80h and can have a length of 128. This allocation reflects the fact that in general the filter lengths of practical decimators increase from stage to stage. RAM block B1 holds the data for the delay lines associated with the three FIR filters; see Figure 11.28. $x_1(n)$ is associated with $h_1(k)$, $x_2(n)$ with $h_2(k)$, and $x_3(n)$ with $h_3(k)$.

The filter coefficients are quantized to 16 bits by multiplying each coefficient by 2^{15} and then rounding up to the nearest integer. The decimation program, in TMS320C25 assembly language, is general purpose, and can be modified to implement one-stage, two-stage or three-stage decimation by replacing the coefficients, specifying the filter lengths, number of decimation stages and decimation factors.

Table 11.6 Filter coefficients for the three-stage decimator.

FILTER LENGTH = 12

***** IMPULSE RESPONSE *****

H(1) =	0.73075550E−02 = H(12)	239
H(2) =	0.27123260E−01 = H(11)	889
H(3) =	0.59286430E−01 = H(10)	1943
H(4) =	0.10198970E+00 = H(9)	3342
H(5) =	0.14187870E+00 = H(8)	4649
H(6) =	0.16675770E+00 = H(7)	5464

***** IMPULSE RESPONSE *****

H(1) =	−0.86768190E−02 = H(13)	−284
H(2) =	−0.25476870E−01 = H(12)	−835
H(3) =	−0.25468170E−01 = H(11)	−834
H(4) =	0.24184320E−01 = H(10)	792
H(5) =	0.13238570E+00 = H(9)	4338
H(6) =	0.24907950E+00 = H(8)	8162
H(7) =	0.30075170E+00 = H(7)	9855

FILTER LENGTH = 48

***** IMPULSE RESPONSE *****

H(1) =	0.17780220E−02 = H(48)	585
H(2) =	−0.17396640E−02 = H(47)	−57
H(3) =	−0.49461790E−02 = H(46)	−162
H(4) =	−0.25451430E−02 = H(45)	−83
H(5) =	0.40843330E−02 = H(44)	134
H(6) =	0.42773070E−02 = H(43)	140
H(7) =	−0.45042640E−02 = H(42)	−148
H(8) =	−0.80385180E−02 = H(41)	−263
H(9) =	0.29002500E−02 = H(40)	95
H(10) =	0.12193670E−01 = H(39)	400
H(11) =	0.92281120E−03 = H(38)	30
H(12) =	−0.16199860E−01 = H(37)	−531
H(13) =	−0.76966970E−02 = H(36)	−252
H(14) =	0.18898710E−01 = H(35)	619
H(15) =	0.17966280E−01 = H(34)	589
H(16) =	−0.18756490E−01 = H(33)	−615
H(17) =	−0.32451860E−01 = H(32)	−1063
H(18) =	0.13458800E−01 = H(31)	441
H(19) =	0.52945520E−01 = H(30)	1735
H(20) =	0.17620600E−02 = H(29)	58
H(21) =	−0.86433440E−01 = H(28)	−2832
H(22) =	−0.44585360E−01 = H(27)	−1461
H(23) =	0.18176500E+00 = H(26)	5956
H(24) =	0.41039480E+00 = H(25)	13448

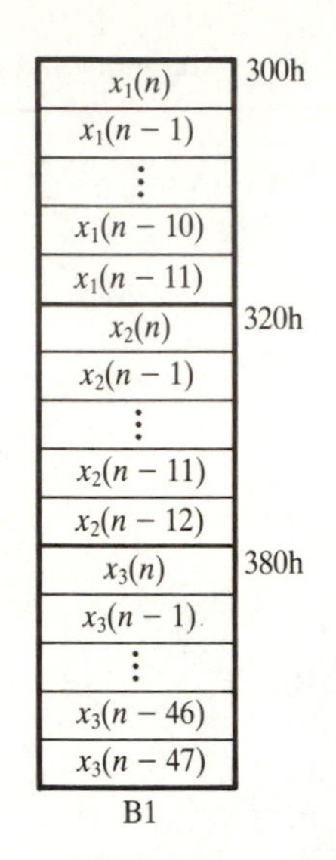

Figure 11.28 Coefficient and data storage map for a three-stage decimator.

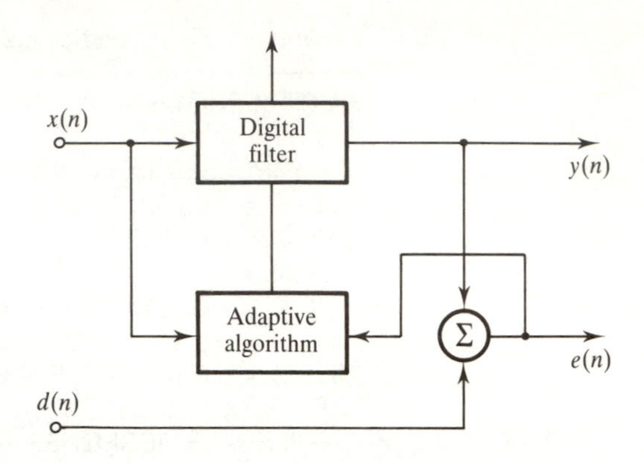

Figure 11.29 General structure of an adaptive filter: a pair of inputs and a pair of outputs.

11.4.5 Adaptive filtering

The general structure for an adaptive filter is depicted in Figure 11.29. As discussed in Chapter 9, adaptive filtering involves two processes.

(1) *Digital filtering*. The coefficients of the digital filter in Figure 11.29 are used to extract appropriate information from the input signal, $x(n)$, to yield $y(n)$. Assuming a transversal FIR structure, the filter is given by

$$y(n) = \sum_{k=0}^{N-1} w_n(k)x(n-k) \tag{11.14}$$

where $w_n(k)$, $k = 0, 1, \ldots, N-1$, are the digital filter coefficients (often called weights) and $x(n-k)$, $k = 0, 1, \ldots, N-1$, is a sequence of the input data.

The implementation of the digital filter given in Equation 11.14 is very similar to that of a standard FIR filter discussed earlier. Thus a C language implementation of the filter would have the familiar form

```
y[n]=0;
for(k=0; k<N; k++){
      y[n]=y[n]+wn[k]*xn[k];
}
```

Program 11.12 C pseudo-code for the LMS adaptive filter coefficient update.

```
uen=2*u*e[n]
for(k=0; k<N; k++){
      wn[k]=wn[k]+uen*xn[k];
}
```

Program 11.13 TMS320C25 pseudo-code for the LMS adaptive filter coefficient update.

```
          LT      ERR
          MPY     U                ;COMPUTE U*e(n)
          PAC
          ADD     ONE, 15
          SACH    ERRF
          LRLK    AR1, N-1         ;LOAD FILTER LENGTH
          LRLK    AR2, WN          ;POINT TO LAST COEFF
          LRLK    AR3, XN+1        ;POINT TO LAST DATA X(n-(n-1))
          LT      ERRF
          MPY     *-,AR2           ;U*e(n)*x(n-k)
ADAPT     ZALR    *,AR3
          MPYA    *-,AR2           ;wk(n+1)=wk(n)+P
          SACH    *+,0,AR1
          BANZ    ADAPT,*-,AR2
```

In TMS320C25 assembly language the filtering part has the form

```
LRLK   AR1, XNM1      ;point to data x[n-(N-1)]
RPTK   N-1            ;multiply and accumulate
MACD   COEFF,*-
```

The MACD–RPTK instruction pair allows the multiply and accumulation necessary to compute the filter output, $y(n)$, in full precision. Only the final output is quantized, minimizing the roundoff error.

(2) *Adaptive process*. This process involves updating, that is adjusting the filter coefficients towards their optimal values. When the basic LMS algorithm is used, the coefficients are updated as follows:

$$w_{n+1}(k) = w_n(k) + 2\mu e(n)x(n-k), \quad k = 0, 1, 2, \ldots, N-1 \qquad \textbf{(11.15)}$$

where $w_n(k)$ is the kth coefficient of the digital filter at the nth sampling instants, μ is the stability factor, and $x(n-k)$ is the kth input data sample at the kth delay line. A C language implementation of the basic LMS update equation is given by Program 11.12. The term 2ue[n] is a scalar and is the same for all coefficients and so it is computed once and placed outside the loop. A TMS320C25 implementation of the adaptation process is given in Program 11.13.

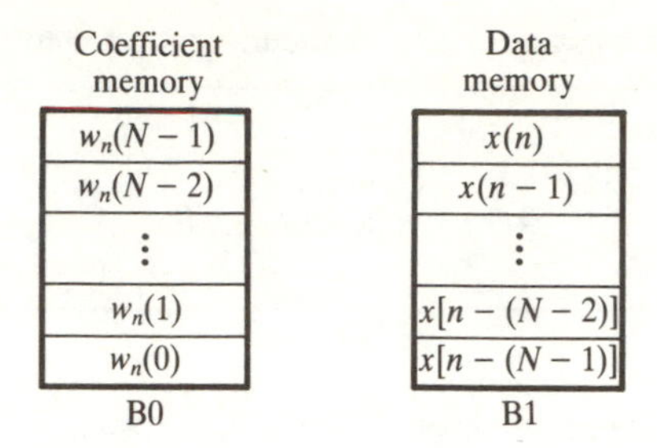

Figure 11.30 Data and coefficient storage for adaptive filtering.

In the TMS320C25 implementation, the filter coefficients are stored in RAM block B0 and the input data, $x(n-k)$, in RAM block B1 as shown in Figure 11.30. During the filtering process, RAM block B0 is configured as a program memory to allow the use of the multiply–accumulate and repeat instructions but, during coefficient update, RAM block B0 is configured as a data memory. The complete C25 code for an LMS adaptive filter is given on the PC disk for this book.

11.5 Special-purpose DSP hardware

Why special purpose?

Digital signal processing operations are computationally intensive. In wide bandwidth applications where the input/output data rates are high, most general-purpose digital signal processors (DSPs) cannot perform the required computations fast enough. This is of course the reason that general-purpose DSPs are often found in audio frequency applications. Further, for a given application most general-purpose DSPs contain many on-chip resources that are either redundant or underutilized, for example addressing modes, instruction set, I/O peripherals. In special-purpose DSPs, the hardware is optimized to execute a specific algorithm or to perform certain functions in a specific application. This leads to greater utilization of on-chip resources and increased speed of operation.

Special-purpose hardware can be implemented as a single-chip product or realized as blocks of individual components. The building block approach using individual components is more flexible and leads to increased speed, but the hardware is difficult to develop and may be more expensive. Single-chip DSPs, if they exist for the task, have lower chip counts, do not require knowledge of an obscure assembly language, or have problems of software debugging.

Basic requirements of special-purpose DSPs

The most common arithmetic operation in DSP algorithms, such as digital filtering, correlation and transformations, is the sum of products:

$$y = \sum_{k=0}^{N-1} a_k x_k \tag{11.16}$$

where a_k is a set of coefficients or variables and x_k a data sequence.

The characteristic Equation 11.16 can be written in a recursive form to allow for a more efficient computation of the sums of products:

$$y_k = a_k x_k + y_{k-1}, \quad k = 0, 1, \ldots, N-1 \tag{11.17}$$

where

$$y_{-1} = 0$$
$$y = y_{N-1}$$

In special-purpose DSPs Equation 11.17 is computed with a multiplier–accumulator (MAC) at a very fast rate, for example 40 ns per MAC.

Like the general-purpose DSPs, the architecture of a special-purpose DSP include data memory, RAM and/or ROM, for storing data and variables (such as filter or FFT coefficients), fast hardware multiplier–accumulator, and temporary registers to store data or intermediate results. Extensive use is made of parallelism, multiplexing and pipelining to achieve maximum speed.

In the next few sections, we will discuss some basic issues involved in the design of special-purpose hardware for DSP.

11.5.1 Hardware digital filters

11.5.1.1 FIR digital filters

The direct form FIR filter is characterized by the following equation:

$$y(n) = \sum_{k=0}^{N-1} h(k)x(n-k)$$

Figure 11.31 shows a basic architecture for an FIR digital filter using blocks of individual components. The main components are coefficient and data memories, analogue input/output units (ADC and DAC), multiplier–accumulator (MAC), and a controller (not shown). The components of the FIR filter can be implemented with fast, off-the-shelf products.

At each sampling instant, a new data sample, $x(n)$, is read from the ADC and saved in the data memory. Each input data sample and the corresponding coefficient are fetched from the memory simultaneously and applied to the multiplier. The products are then accumulated to yield the output sample. The computation of each output sample, $y(n)$, would require N data–coefficient fetches from memory and N MACs.

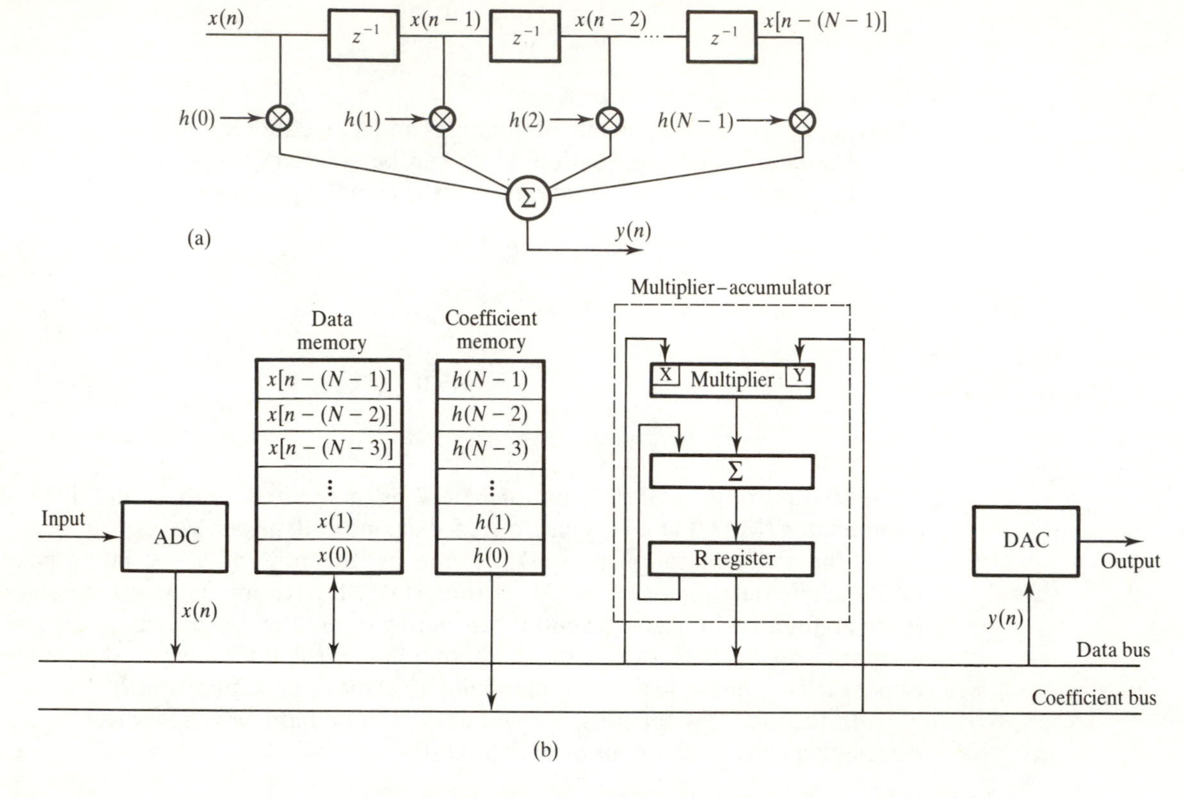

Figure 11.31 Architecture of a hardware FIR digital filter.

The FIR filtering operation is regular and well structured, and can be readily implemented in a single IC. Special-purpose single-chip FIR filters are now available. Examples are the Motorola DSP56200 and the INMOS A100. They are both algorithm-specific digital signal processors. The DSP56200 is special-purpose hardware designed to perform sum of product operations, typified by the FIR.

A simplified block diagram of the internal architecture of the DSP56200 is shown in Figure 11.32. The main elements, as would be expected, are a coefficient RAM (256 × 16 bit) and a data RAM (256 × 16 bit), and an arithmetic unit. The device operates as either an FIR filter (single or dual FIR) or an LMS-based adaptive filter. The A100 contains effectively 32 fast, 16 × 16-bit MACs in a single chip. It is essentially an efficient hardware implementation of the transversal filter (Figure 11.33(a)). In the A100, the basic transversal structure is modified so that the multipliers are fed in parallel (Figure 11.33(b)).

11.5.1.2 IIR digital filter

The architecture for a second-order canonic IIR filter is shown in Figure 11.34. In this case the data memory holds the internal node data $w(n)$. The standard

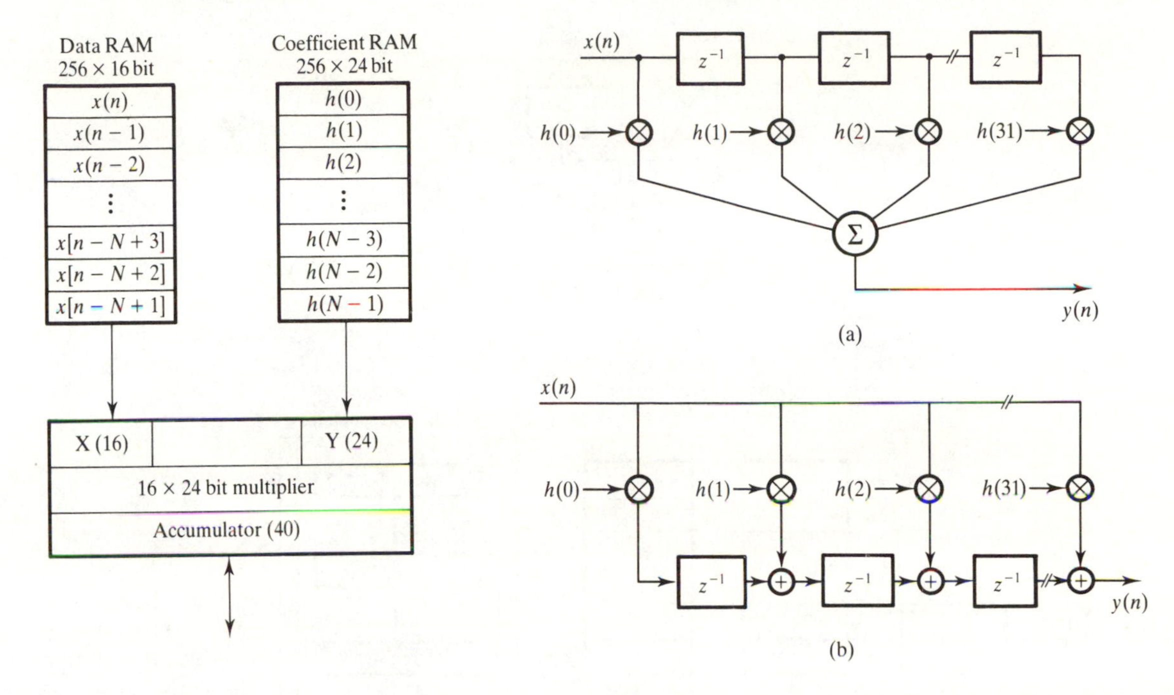

Figure 11.32 A simplified block diagram of the internal architecture of the DSP56200.

Figure 11.33 (a) Transversal FIR architecture. (b) Modified transversal FIR architecture used by the INMOS A100.

second-order canonic section in Figure 11.34(a) is characterized by the following equations:

$$w(n) = s_1x(n) - b_1w(n-1) - b_2w(n-2)$$

$$y(n) = a_0w(n) + a_1w(n-1) + a_2w(n-2)$$

where $x(n)$ represents the input data, $w(n)$ represents the internal node, $y(n)$ is the filter output sample and s_1 is a scale factor.

11.5.2 Hardware FFT processors

The DFT takes a set of N time domain samples and transforms these into a set of N frequency domain samples, $X(k)$. The FFT is an efficient way of computing the DFT coefficients, $X(k)$. Butterfly arithmetic is the basic operation in the FFT. A butterfly is characterized by the following equations:

$$A' = A + W_N^kB$$

$$B' = A - W_N^kB$$

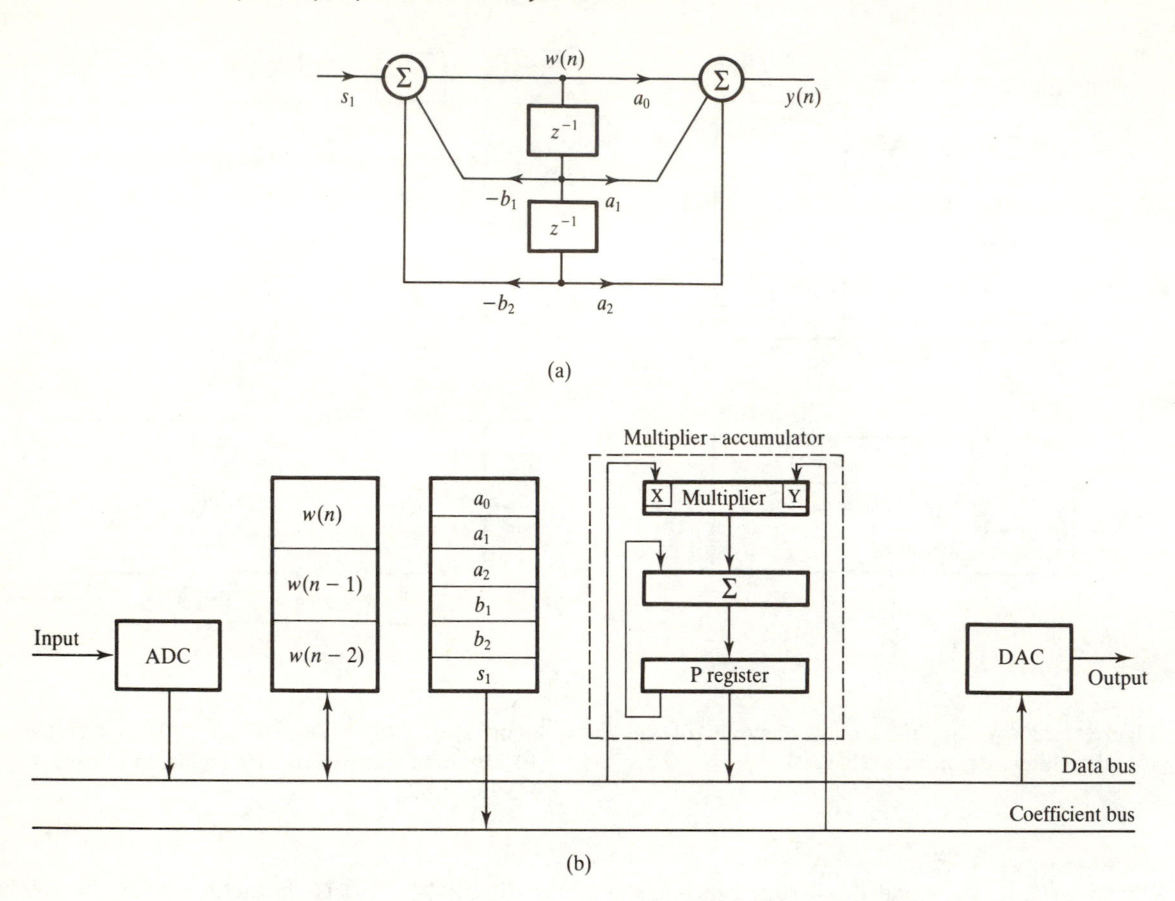

Figure 11.34 (a) IIR filter structure. (b) Hardware architecture for an IIR filter biquad section.

where A and B are a pair of complex-valued input data samples to the butterfly, A' and B' are the outputs of the butterfly, and W_N is the twiddle factor, also complex valued.

Each butterfly operation requires a complex multiplication, that is $W_N^k B$, a complex addition and a complex subtraction. Figure 11.35 shows a direct hardware implementation of a butterfly processor using individual blocks of complex arithmetic units: a complex multiplier and a pair of complex accumulators. The complex multiplier calculates the common terms, $W_N^k B$. The two complex accumulators calculate the two outputs of the butterfly, A' and B'.

A 50 ns, single-cycle butterfly processor can be implemented using the Plessey Semiconductor complex multiplier (PDSP16112A) and a pair of complex accumulators (PDSP16318A). With standard real arithmetic units, an equivalent butterfly processor would consist of four multipliers and six adders. The use of complex arithmetic units clearly leads to a lower chip count and possibly to enhanced system performance. A hardware FFT processor built around the butterfly processor is shown in Figure 11.36.

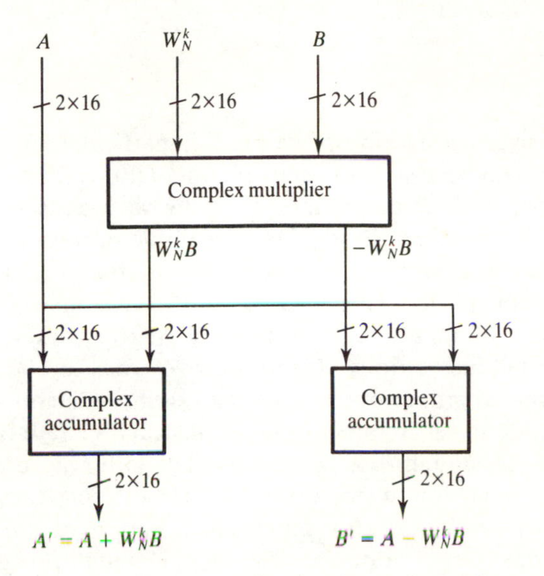

Figure 11.35 Concepts of a hardware butterfly processor using individual blocks of complex arithmetic units.

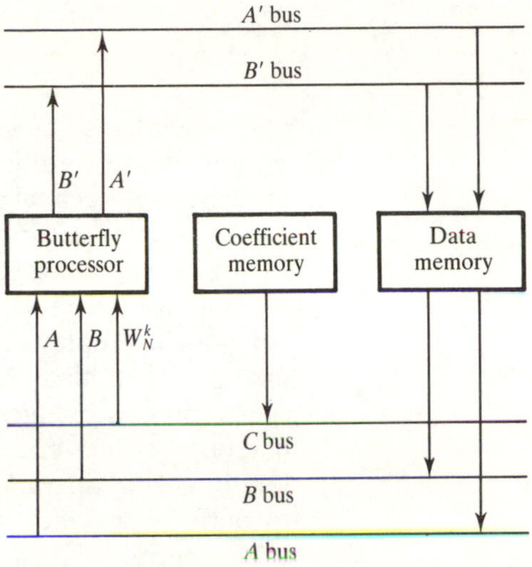

Figure 11.36 A simplified architecture of a hardware FFT processor. Controller and address generator are not shown.

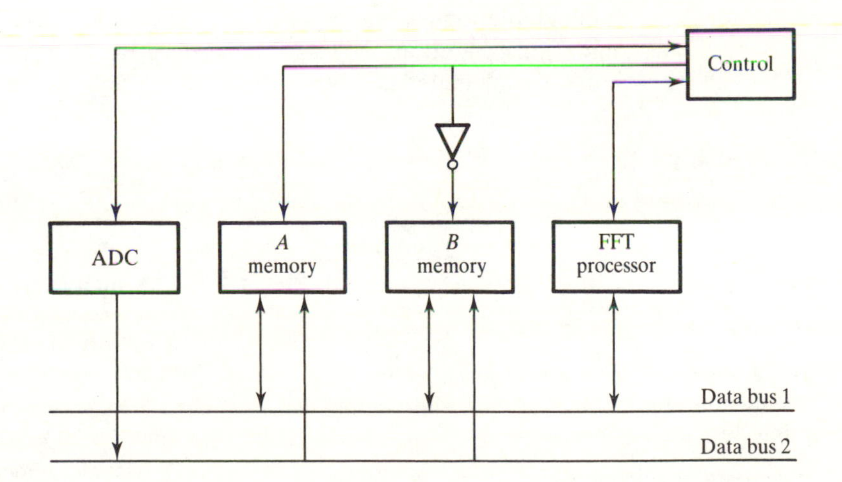

Figure 11.37 Double buffering in real-time FFT.

A real-time, double-buffered, FFT configuration is depicted in Figure 11.37. N-point FFT is performed alternately from each of the two buffers. FFT is performed on the N-point data in buffer A, while buffer B is being filled with new data. The double buffering allows real-time continuous FFT without loss of data. The maximum time to complete the N-point FFT is the interval $T_f = NT$ (s).

11.6 Summary

DSP algorithms involve extensive arithmetic operations, in particular multiplications and additions with heavy data flow through the CPU. Efficient execution of such algorithms in real time requires a hardware architecture and instruction set radically different from those of standard microprocessors. In digital signal processors, this is achieved by making use of the concepts of Harvard architecture, pipelining, dedicated hardware: for example fast hardware multiplier–accumulator and shifters, and by providing fast internal memories and many DSP-oriented instructions. There are two types of digital signal processors: general-purpose processors (which are akin to standard microprocessors, except that they have architectures and instruction sets tailored for DSP operations), and special-purpose processors. The latter are used to perform specific DSP algorithms, for example digital FIR filtering (algorithm-specific digital signal processors), or for efficient execution of some application-dependent operations (application-specific digital signal processors). Compared with general-purpose digital signal processors, special-purpose DSPs offer speed, but are less flexible.

The basic ideas underlying DSP hardware and the impact of DSP algorithms on the architecture of DSP processors were discussed in detail. The implementation of some key DSP algorithms using general-purpose digital signal processors as well as special-purpose DSP processors was discussed to illustrate the issues involved.

Problems

11.1 Write short critical notes on each of the following concepts, using diagrams where appropriate to illustrate your answer:

- Harvard architecture;
- pipelining;
- multiplier–accumulator;
- special instructions;
- data and program memory.

Explain how Harvard architecture as used by the TMS320 family differs from the strict Harvard architecture. Compare this with the architecture of a standard von Neumann processor.

11.2 (1) A multiplier–accumulator, with three pipe stages, is required for a digital signal processor. Sketch a block diagram of a suitable configuration for the MAC. Explain, briefly, and with the aid of a timing diagram, how your MAC works.

(2) Assume a memory access time of 150 ns, multiplication time of 100 ns, addition time of 100 ns, and overhead of 5 ns at each pipe stage. Determine the throughput of the MAC. Comment on your answer.

(3) The DSP system is required to execute the following algorithm in real time:

$$\begin{aligned} y(n) = {} & a_0x(n) + a_1x(n-1) \\ & + a_2x(n-2) + \dots \\ & + a_{N-1}x[n-(N-1)] \end{aligned}$$

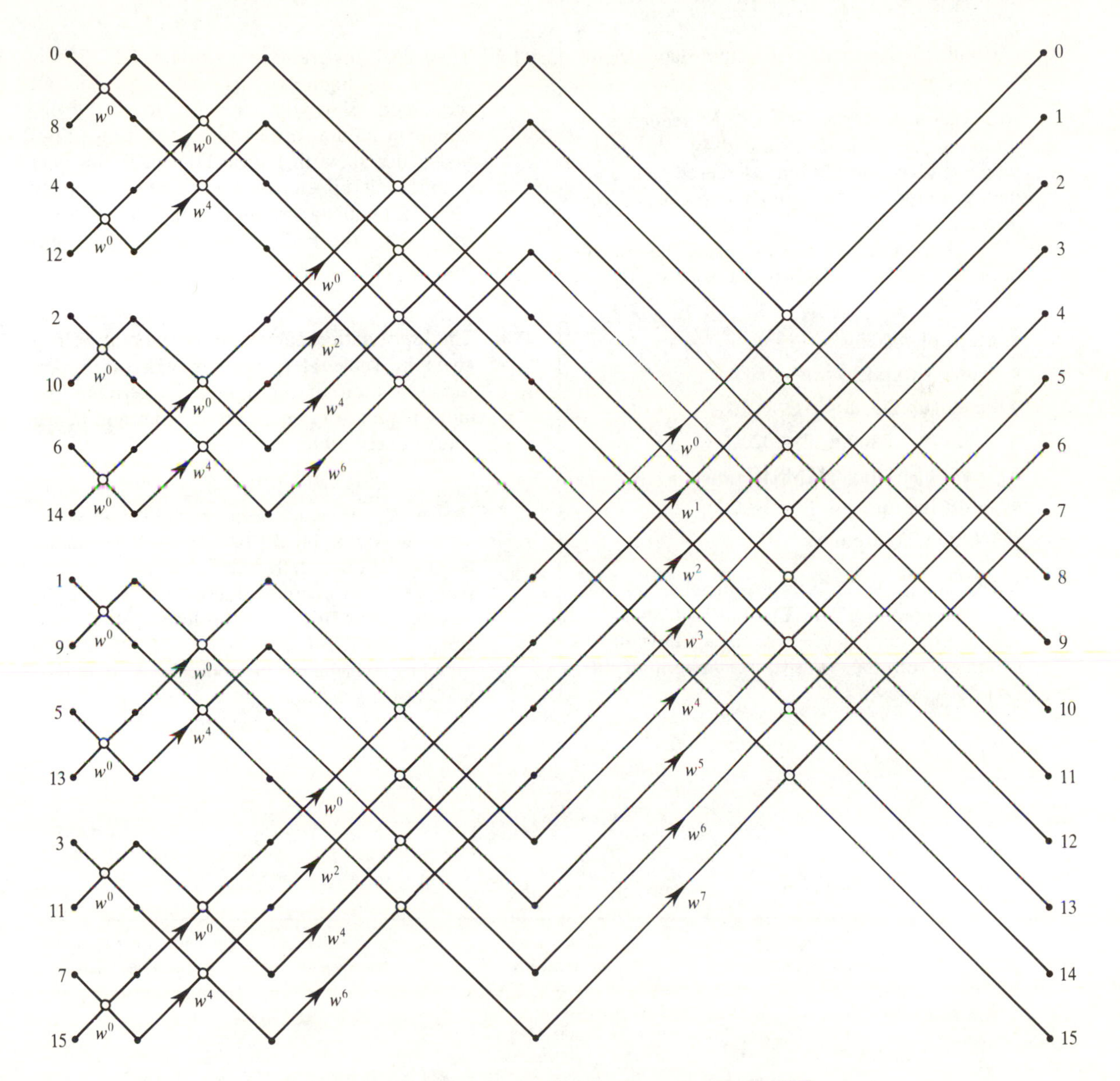

Figure 11.38 Flowgraph of a 16-point radix-2 DIT FFT.

How long will it take the MAC to compute each output sample?

11.3 M. J. Flynn in his paper (Flynn, 1966) divides high speed computers into the following four categories:

(1) single instruction stream, single data stream (SISD);

(2) single instruction stream, multiple data stream (SIMD);

(3) multiple instruction stream, single data stream (MISD);

(4) multiple instruction, multiple data stream (MIMD).

where an instruction stream is the sequence of program instructions executed by the computer, and a data stream is the sequence of data required by the computer to execute the instructions.

Determine, with justification, the appropriate category for each of the following processors:

- Motorola 68000;
- Motorola DSP56000;
- Analogue Devices ADSP2100;
- Texas Instruments TMS320C25;
- Texas Instruments TMS320C30;
- Texas Instruments TMS320C40;
- INMOS Transputer.

11.4 Given in Figure 11.38 is the signal flowgraph for a 16-point DIT FFT. Construct an equivalent constant geometry signal flowgraph. Comment on the relative advantages of the two flowgraphs.

11.5 The signal flowgraph of an 8-point DIT FFT is shown in Figure 11.24 with the output scrambled. Show that the output can be obtained in natural order by bit reversing $X(k)$ and hence show that, in a DIT FFT, the final output will appear in the correct order by either scrambling the input data sequence before taking the FFT or unscrambling the output after FFT.

11.6 The input data sequence for a 16-point FFT is given in Table 11.7 together with the binary representation of its indices. Determine the input sequence in bit-reversed order and hence complete the table.

11.7 Design an efficient special-purpose hardware, using individual blocks of arithmetic elements, for a real-time Nth-order IIR digital filter realized as a cascade of second-order sections. Assume an ADC/DAC resolution of 12 bits, and a coefficient wordlength of 16 bits. A sampling frequency of 100 kHz is required. State any assumptions made.

Table 11.7 Input data sequence for Problem 11.6.

Input sequence, natural order	*Binary code of sequence*	*Input sequence, bit reversed*	*Binary code of sequence (bit reversed)*
$x(0)$	0000		
$x(1)$	0001		
$x(3)$	0011		
$x(5)$	0101		
$x(6)$	0110		
$x(7)$	0111		
$x(8)$	1000		
$x(9)$	1001		
$x(10)$	1010		
$x(11)$	1011		
$x(12)$	1100		
$x(13)$	1101		
$x(14)$	1110		
$x(15)$	1111		

References

Flynn M.J. (1966). Very high-speed computing systems. *Proc. IEEE*, **54**(12), 1901–9

Hennessy J.L. and Patterson D.A. (1990). *Computer Architecture: A Quantitative Approach*, San Mateo CA: Morgan Kaufmann

Rabiner L.R. and Gold B. (1975). *Theory and Applications of Digital Signal Processing*. Englewood Cliffs NJ: Prentice-Hall

Texas Instruments (1988). *TMS320C2x Software Development System User's Guide*. Dallas TX: Texas Instruments

Bibliography

Analogue Devices (1989). *DSP Products Handbook*. Analogue Devices

Casey P.E. and Simmers L. (1986). Digital signal processing IC helps to shed new light on image processing applications. *Electronic Design*, (March 20, 1986), 135

Chassaing R. and Horning D.W. (1990). *Digital Signal Processing with the TMS320C25*. New York NY: Wiley

Cragon H. (1980). The elements of single-chip microcomputer architecture. *Computing Mag.*, **13**(10), 27–41

Croisier A., Estaban D.J., Levilion M.E. and Rizo W. (1973). *Digital Filter for PCM Encoded Signal*. US Patent 3777130, December 3, 1973

Er M.H., Wong D.J., Sethu A.A.-L. and Ngeow K.S. (1991). Design and implementation of RSA cryptosystem using multiple DSP chips. *Microprocessors and Microsystems*, **15**(7), 369–78

Gallant J. (1990). Plug-in DSP boards. *EDN*, (April 26, 1990), **35**, 142–60

Ganesan S. (1991). A dual-DSP microprocessor system for real-time digital correlation. *Microprocessor and Microsystem*, **15**(7), 379–84

Gore A.E. (1986). Cascadable digital signal processor. *New Electronics*, (October 14, 1986), 39–40

Hillman G.D. (1987). DSP56200: an algorithm-specific digital signal processor peripheral. *Proc. IEEE*, **75**(9), 1185–91

Jouppi P. and Wall D.W. (1989). Available instruction-level parallelism for superscalar and superpipelined machines. In *Proc. Third Conf. on Architectural Support for Programming Languages and Operating Systems*. IEEE/ACM (April), Boston MA, pp. 272–82

Kogge P.M. (1981). *The Architecture of Pipelined Computers*. New York NY: McGraw-Hill

Leary K. and Morgan D. (1986). Fast and accurate analysis with LPC gives a DSP chip speech-processing power. *Electronic Design*, (April 17, 1986), 153

Lin K.S. ed. (1988). *Digital Signal Processing Applications with the TMS320 Family* Vol. 1. Englewood Cliffs NJ: Prentice-Hall

Lin K.S., Frantz G.A. and Simar R. (1987). The TMS320 family of digital signal processors. *Proc. IEEE*, **75**(9), 1143–59

McKee D. (1990). TMS32010 routine finds phase. *EDN*, (May 10, 1990), **35**, 148

Mennen P. (1991). DSP chips can produce random numbers using proven algorithm. *EDN*, (January 21, 1991), **36**, 141–6

Messer D.D. (1991). Convolutional encoding and viterbi decoding using the DSP56001. *Microprocessors and Microsystems*, **15**(1), 54–62

Mitchell P. (1988). Floating point and CMOS up data rates. *New Electronics*, (May 1988), 23–9
Motorola (1986). *DSP56000 Digital Signal Processor User's Manual*. Motorola
Papamichalis P. (1990). *Digital Signal Processing Applications with the TMS320 Family. Theory, Algorithms, and Implementations* Vol. 3. Dallas TX: Texas Instruments
Papamichalis P. and Simar R. (1988). The TMS320C30 floating-point digital signal processor. *IEEE Micro Mag.*, **8**(6), 10–28
Plessey Semiconductors (1988). *Digital Signal Processing IC Handbook*. Plessey Semiconductors
Roesgen J. (1986). Fast modem designs benefit from DSP chip's versatility. *Electronic Design*, (June 12, 1986)
Roesgen J. and Tung S. (1986). Moving memory off chip, DSP μP squeezes in more computational power. *Electronic Design*, (February 20, 1986), 131
Rosen S. (1969). Electronic computers: a historical survey. *Computer Survey*, **1**(1), 7–36
Schmalzel D., Hein and Ahmed N. (1980). Some pedagogical considerations of digital filter hardware implementation. *IEEE Circuits and Systems Mag.*, **2**(1), 4–13
Shear D. (1991). EDN's DSP-chip directory. *EDN*, (October 1, 1991), **36**, 104–33
So J. (1983). TMS 320 – step forward in digital signal processing. *Microprocessors and Microsystems*, **7**(10), 451–60
Stokes J. and Sophie G.R.L. (1991). Implementation of PID controllers on the Motorola DSP56000/DSP56001. Part 1. *Microprocessors and Microsystems*, **15**(6), 321–31
Stokes J. and Sophie G.R.L. (1991). Implementation of PID controllers on the Motorola DSP56000/DSP56001. Part 2. *Microprocessors and Microsystems*, **15**, 385–92
Texas Instruments (1989). *Second-Generation TMS320 User's Guide*. Dallas, TX: Texas Instruments
Titus J. (1986). DSP ICs. *EDN*, (October 16, 1986), 163–76

Appendix

11A TMS320 assembly language programs for real-time signal processing and a C language program for constant geometry radix-2 FFT

The following TMS320C10/C25 programs are available on the PC disk (see the Preface for details of how to obtain the disk):

(1) a TMS320C10-based FIR digital notch filter;

(2) a TMS320C10 implementation of an FIR digital bandpass filter;

(3) a TMS320C25-based FIR digital notch filter;

(4) a TMS320C10 fourth-order digital IIR filter realized with second-order sections in cascade;

(5) a TMS320C25 fourth-order digital IIR filter realized with second-order sections in cascade;

(6) a TMS320C25 fourth-order digital IIR filter realized with second-order sections in parallel;

Program 11A.1 A TMS320C10-based implementation of an FIR digital bandpass filter.

```
METAi Assembler 4.00 ©1988 Crash Barrier    Thu Nov 19 00:37:40 1992
Page 1 Assembler

                                      targbpf.asm

00000000                          1   c:\metai\32010.tab/
                                  2
00000000                          3            .ctrl       27, 15
00000000                          4            SEGMENT     word at 0000 'ram'
00000000                          5   ;;;;;;;;;;;;;;;;;;;;;;;;;;;;;;;;;;;;;;;;;;;;;;;;;;;;;;;;;;;;;;;;;;;;;;;;;;;;;;;;
00000000                          6            ;                                                              ;
00000000                          7            ;    FIR BANDPASS FILTER                                       ;
00000000                          8            ;                                                              ;
00000000                          9            ;    Filter specification:                                     ;
00000000                         10            ;                                                              ;
00000000                         11            ;    filter type                          bandpass filter      ;
00000000                         12            ;    sampling frequency                   15 kHz               ;
00000000                         13            ;    passband                             900–1100 Hz          ;
00000000                         14            ;    transition width                     450 Hz               ;
00000000                         15            ;    passband ripple                      < 0.87 dB            ;
00000000                         16            ;    stopband attenuation                 > 30 dB              ;
00000000                         17            ;    filter length                        41                   ;
00000000                         18            ;                                                              ;
00000000                         19            ;    Hardware: TMS320C10 Target board with 8-bit ADC/DAC       ;
00000000                         20            ;                                                              ;
00000000                         21   ;;;;;;;;;;;;;;;;;;;;;;;;;;;;;;;;;;;;;;;;;;;;;;;;;;;;;;;;;;;;;;;;;;;;;;;;;;;;;;;;
00000000                         22            ;
                                 23
00000000  F900002B               24            B           START
                                 25
00000028                         26   NM1      EQU         40          ;N-1
00000000                         27   XN       EQU         0           ;CURRENT I/P SAMPLE
00000028                         28   XNM1     EQU         NM1
00000029                         29   H0       EQU         NM1+1
00000051                         30   HNM1     EQU         H0+NM1
0000007B                         31   YN       EQU         123
0000007C                         32   ONE      EQU         124
00000000                         33   PA0      EQU         0           ;address of I/O for D/A IN TARGET BOARD
00000001                         34   PA1      EQU         1
00000002                         35   PA2      EQU         2
00000003                         36   PA3      EQU         3
00000002                         37   COEFF    EQU         2           ;START ADDRESS OF COEFFS.
00000000                         38   R0       EQU         0
00000001                         39   R1       EQU         1
00000002                         40   ;
00000002                         41   ;TABLE OF COEFFS. THESE ARE INITIALLY
00000002                         42   ;STORED IN PROGRAM MEMORY.
                                 43
                                 44
00000002  FE09FFFEFF5B019F       45
00000002  02B3038E03D9           45            DC.W        -503,-2,-165,415,691,910,985
00000009  035001D9FF97FCE8       46
00000009  FA57F881               46            DC.W        848,473,-105,-792,-1449,-1919
0000000F  F7EAF8DDFB52FEE8       47
```

```
0000000F  02F406A8          47            DC.W      -2070,-1827,-1198,-280,756,1704
00000015  093F0A2E093F06A8  48
00000015  02F4FEE8          48            DC.W      2367,2606,2367,1704,756,-280
0000001B  FB52F8DDF7EAF881  49
0000001B  FA57              49            DC.W      -1198,-1827,-2070,-1919,-1449
00000020  FCE8FF9701D90350  50
00000020  03D9038E02B3      50            DC.W      -792,-105,473,848,985,910,691
00000027  019F00A5FFFEFE09  51            DC.W      415,165,-2,-503

METAi Assembler 4.00 ©1988 Crash Barrier    Thu Nov 19 00:37:40 1992
Page 2 Assembler
                                targbpf.asm
                            52
                            53
0000002B                    54  ; ============= START OF MAIN PROGRAM==============
0000002B                    55  ;
0000002B                    56  ; INITIALIZATION
0000002B                    57  ;
0000002B  7E01              58  START   LACK    1
0000002C  507C              59          SACL    ONE
0000002D  6E00              60          LDPK    0          ;POINT TO PAGE ZERO OF DATA
                                                           ;MEMORY
0000002E                    61  ;
0000002E                    62  ;TRANSFER COEFFICIENTS TO DATA MEMORY FROM PROGRAM MEMORY
                                ;IN EPROM SPACE
0000002E                    63  ;
0000002E  7E02              64          LACK    COEFF      ;LOAD COEFF ADDRESS INTO
                                                           ;ACCUMULATOR
0000002F  7028              65          LARK    AR0,NM1    ;NO OF COEFF INTO AUXILIARY
                                                           ;REGISTER 0
00000030  7129              66          LARK    AR1,H0     ;AND DATA MEMORY ADDRESS OF
                                                           ;COEFFICIENTS INTO AR1
00000031  6881              67  LOAD    LARP    R1         ;SELECT AR1 AND BEGIN TO TRANSFER
                                                           ;COEFF.
00000032  67A0              68          TBLR    *+,R0      ;INTO DATA MEMORY, THEN INCREMENT
                                                           ;THE CONTENTS OF AR1

00000033  007C              69          ADD     ONE        ;INCREMENT THE ACCUMULATOR
00000034  F4000031          70          BANZ    LOAD       ;DEC AR0, AND BRANCH IF NOT ZERO
00000036                    71  ;
00000036                    72  ;WAIT FOR NEW INPUT SAMPLE
00000036                    73  ;
00000036  6880              74          LARP    R0
00000037  F600003B          75  WAIT    BIOZ    NXTPT      ;SEE IF SAMPLE IS RDY
00000039  F9000037          76          B       WAIT       ;IF NOT GO WAIT
                            77
0000003B  4000              78  NXTPT   IN      XN,PA0     ;IF READY THEN READ SAMPLE . . . was
                                                           ;PA2 for EVM
0000003C                    79  ;
0000003C                    80  ;CALCULATE FILTER OUTPUT IN YN AND OUTPUT TO DAC
0000003C                    81  ;
0000003C  7028              82  skip    LARK    AR0,XNM1
0000003D  7151              83          LARK    AR1,HNM1
0000003E  7F89              84          ZAC
                            85
0000003F  6A91              86          LT      *-,R1      ;LOAD XN(N-1) SAMPLE
```

```
00000040  6D90        87          MPY     *-,R0       ;COMPUTE H(N-1)*XN(N-1)
00000041  6B81        88   LOOP   LTD     *,R1        ;COMPUTE SIG[H(K)*X(N-K)]
00000042  6D90        89          MPY     *-,R0
00000043  F4000041    90          BANZ    LOOP
00000045  7F8F        91          APAC                ;ADD H(N-1)*X(N-1)
00000046  597B        92          SACH    YN,1        ;OUTPUT SAMPLE
00000047  487B        93          OUT     YN,PA0      ;was PA3 for EVM
                      94
00000048  F6000048    95   onhi   BIOZ    onhi        ;wait here until BIO line goes high before
                                                      ;going for next
                      96
0000004A  F9000037    97          B       WAIT
                      98
0000004C              99          end
No errors on assembly of 'targbpf.asm'
```

Program 11A.2 A TMS320C25 fourth-order digital IIR filter realized with second-order sections in cascade.

```
;;;;;;;;;;;;;;;;;;;;;;;;;;;;;;;;;;;;;;;;;;;;;;;;;;;;;;;;;;;;;;;;;;;;;;;;
; Fourth order Elliptic filter, connected                              ;
; as a cascade of 2 biquad canonic sections                            ;
'Manny Ifeachor, Jan., 1992                                            ;
;                                                                      ;
;        FILTER SPECIFICATIONS:                                        ;
;                                                                      ;
;        Filter Type              lowpass                              ;
;        Sampling frequency       15 kHz                               ;
;        Passband                 0-3 kHz                              ;
;        transition width         450 Hz                               ;
;        Passband ripple          0.5 dB                               ;
;        Stopband attenuation     45 dB                                ;
;                                                                      ;
;        Hardware: TMS320C25 SWDS with AIB                             ;
;        1-bit ADC/DAC (filter B)                                      ;
; -------------------------------------------------------------------- ;
;

XN       .set    0
YN1      .set    1
W1N      .set    2
W1NM1    .set    3
W1NM2    .set    4
YN2      .set    5
W2N      .set    6
W2NM1    .set    7
W2NM2    .set    8
SF1      .set    9
A01      .set    10
A11      .set    11
A21      .set    12
B11      .set    13
B21      .set    14
A02      .set    15
A12      .set    16
```

```
A22      .set     17
B12      .set     18
B22      .set     19
SF2      .set     20
ONE      .set     21
RATED    .set     22
MODED    .set     23
WONE     .set     24
TEMP     .set     25
PBM1     .set     0300h
;
         .sect    "IRUPTS"
START    B        INIT
;
         .text
COEFFS   .word      13217          ; SF1=0.4033627 IN Q15 FORMAT
                                   ;(SCALE FACTOR)
;
         .word       4296          ;a01=0.1311136
         .word       7087          ;a11=0.2162924
         .word       4296          ;a21=0.1311136
         .word      27175          ;-b11=0.829328
         .word     -10061          ;-b21=-0.307046
;
         .word      32767          ;a02=0.9999695 (largest +ve number)
         .word      22142          ;a12=0.675718
         .word      32767          ;a22=0.9999695
         .word      16251          ;-b12=0.495935
         .word     -24965          ;-b22=-0.761864
;
         .word      29769          ;SF2=0.90847
;
MODEP    .word    0Ah
RATEP    .word    0299h
;
*
**       INITIALIZE THE AIB  **
*
INIT     LDPK     6
         SSXM
         LACK     MODEP
         TBLR     MODED
         OUT      MODED,PA0
         LACK     RATEP            ;SET UP AIB SAMPLING FREQ
         TBLR     RATED
         OUT      RATED,PA1
         OUT      RATED,PA3        ;LET GO AIB
*
**       TRANSFER COEFFS FROM PROG MEMORY TO DATA MEMORY
*
         LARP     AR0
         LRLK     AR0,PMB1+SF1
         RPTK     11
         BLKP     COEFFS,*+
```

```
*
**        INITIALIZE DMA FOR INTERNAL NODE DATA
*
INITWN    ZAC
          SACL      W1N
          SACL      W1NM1
          SACL      W1NM2
          SACL      W2N
          SACL      W2NM1
          SACL      W2NM2
*
**        WAIT FOR NEW DATA SAMPLE TO BE READY
*
RDATA     BIOZ      NXTPT           ;FETCH THE NEW SAMPLE
          B         RDATA
NXTPT     IN        XN,PA2
          LT        XN
*
**        START OF FILTER BLOCK 1
*
BLOCK1    MPY       SF1             ;SCALE INPUT DATA SAMPLE: SF*X(N)
          PAC
          LT        W1NM1           ;LOAD T-REGISTER WITH W(N-1)
          MPY       B11             ;B11W1(N-1)
          LTA       W1NM2           ;SFX(N)+B11W1(N-1)
          MPY       B21             ;B21W1(N-2)
          APAC                      ;SFX(N)+B11W1(N-1)+B21W1(N-2)
          SACH      W1N
          ZAC
          MPY       A21             ;A21W1(N-2)
          LTD       W1NM1           ;SUM=A21W(N-2); W1(N-2)=W1(N-1)
          MPY       A11             ;A11W1(N-1)
          LTD       W1N             ;SUM=A21W1(N-2)+A11W1(N-1);W1(N-1)=W1(N)
          MPY       A01             ;A01W1(N)
*
**        START OF FILTER BLOCK 2
*
BLOCK2    LTA       W2NM1           ;Y1(N)=A21W(N-2)+A11W1(N-1)+A01W1(N)
          MPY       B12             ;B12W2(N-1)
          LTA       W2NM2           ;Y1(N)+B12W2(N-1)
          MPY       B22             ;B22*W2(N-2)
          APAC                      ;Y1(N)+B12*W2(N-1)+B22*W2(N-2)
;                                   ;SUM=Y1(N)+B12*W2(N-1)+2*B22*W2(N-2)
          SACH      W2N             ;STORE HIGH 16 BIT WORD OF SUM IN W2(N)
          MPY       A22             ;A22*W2(N-2)
          ZAC                       ;SUM=SUM+A22*W2(N-2)
          LTD       W2NM1           ;W2(N-2)=W2(N-1); SUM=A22*W2(N-2)
          MPY       A12             ;A12*W2(N-1)
          APAC                      ;SUM=A22*W2(N-2)+A12*W2(N-1)
          LTD       W2N             ;W2(N-1)=W2(N)
          MPY       A02             ;A02*W2(N)
          APAC
          SACH      YN2             ;
*
**        SCALE OUTPUT SAMPLE AND SEND IT TO DAC
*
```

```
        LT      YN2         ;SCALE OUTPUT BACKUP
        MPY     SF2         ;SF2*YN2
        PAC                 ;
        APAC
        APAC
        APAC
        APAC
        APAC
        APAC
        APAC
        APAC
        APAC
        APAC
        APAC
        APAC
        APAC
        APAC
        APAC
        APAC
        APAC
        APAC
        APAC
        APAC
        SACH    YN2         ;21*SF2*YN2
;
        OUT     YN2,PA2
        B       RDATA
;
        END
;@\\\000300AB1800001100AB18;

/* This is the link command file */
MEMORY
{                                                        /*Program Memory*/
    PAGE 0: VECTORS:origin=0h, length=01Fh
            CODE:origin=20h, length=0F90h
                                                         /* Data Memory*/
    PAGE 1: RAMB2:origin=60h, length=020h
            RAMB0:origin=200h, length=0FFh
            RAMB1:origin=300h, length=0FFh
}
SECTIONS
{   IRUPTS   :{}    > VECTORS    PAGE 0
    .text    :{}    > CODE       PAGE 0
    .data    :{}    > RAMB2      PAGE 1
    .bss     :{}    > RAMB0      PAGE 1
}
```

(7) a C language program for constant geometry radix-2 FFT;

(8) a TMS320C25-based radix-2 FFT algorithm;

(9) a TMS320C25 adaptive filter.

Only programs (2) and (5) are listed in this appendix because of lack of space.

12

Applications and case studies

12.1	TMS320C10 target board for real-time DSP	680
12.2	TMS320C25 target board for real-time DSP	685
12.3	TMS320C25 Software Development System (SWDS)	687
12.4	FFT spectrum analyser	689
12.5	Detection of foetal heartbeats during labour	697
12.6	Real-time adaptive removal of ocular artefacts from human EEGs	706
12.7	Fixed- and floating point implementation of DSP systems	723
12.8	Equalization of digital audio signals	733
12.9	Adaptive ocular artefact filter	736
12.10	Summary	744
	References	745
	Bibliography	747
	Appendices	748

The objectives of this chapter are twofold. The first is to describe the hardware and development environment used to implement some of the DSP algorithms described in previous chapters. Two low cost target boards are described: a

TMS320C10 target board and a TMS320C25 target board which we have developed for demonstrating the principles of DSP to large groups of students. Commercial DSP hardware is still an expensive platform on which to base the teaching of DSP, especially when required to demonstrate DSP to large groups. Also, they are 'invisible', often plugged into the back of the PC, out of sight. It is helpful for the hardware used for teaching DSP to be simple and visible to emphasize that, in principle, both the DSP hardware and software are very simple and flexible. We have found the two target boards valuable in demonstrating the principles of DSP in real time and hope that readers will benefit from our experience.

Our second objective is to describe a number of real-world applications of DSP in the form of case studies. Applications described here include FFT spectrum analysis, removal of signal contamination from the human electroencephalogram (electrical activity of the brain), and the detection of foetal heartbeats from the foetal electrocardiogram (electrical activity of the heart) which is necessary for assessing the condition of the baby during childbirth. The presentation draws on many concepts discussed in earlier chapters.

12.1 TMS320C10 target board for real-time DSP

12.1.1 Background and system specifications

As in other areas of engineering, a practical experience of designing and implementing DSP algorithms is necessary for gaining a proper appreciation of the issues involved in DSP. Engineering students with a background in only analogue signal processing have genuine difficulty getting to grips with the new techniques involved in DSP, especially if they do not have the necessary mathematical background to understand the concepts from a theoretical point of view. Often they are puzzled by, for example, how the numerical operations used in FIR or IIR filtering can lead to filtering. They are reasonably comfortable with analogue filters and the concepts of how the frequency-dependent characteristics of capacitors and inductors combined with resistance achieve filtering. In DSP, there are no obvious frequency-dependent parameters. How does a digital filter actually work, some will ask.

We became convinced that what we needed was a simple stand-alone piece of hardware, not connected to a PC (to avoid the impression that you are playing some tricks), which students could play with and use to design and implement simple DSP functions. We also wanted to demonstrate some of the practical issues involved in real-time DSP, such as the concepts of aliasing, imaging, $\sin x/x$, overflows and so on.

First, we will set out target specifications for a suitable system. Next, a system we have developed to achieve these is described at block diagram level.

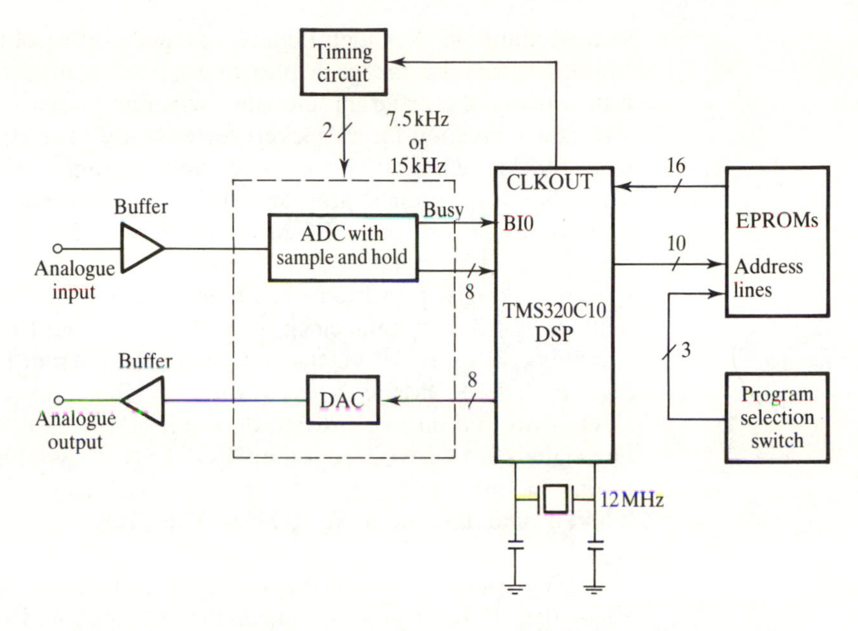

Figure 12.1 A simple block diagram of the TMS320C10 target board.

The hardware needed to meet the specifications and the detailed design of the system are then given.

Specifications for the system are as follows:

- stand-alone board capable of executing simple DSP algorithms in real time;
- single analogue input/output digitized to 8 bits;
- system should be flexible and inexpensive;
- allow codes for given DSP algorithms to be modified;
- be capable of operating at two different sampling frequencies;
- allow the study of aliasing and imaging.

12.1.2 System description

The system block diagram, shown in Figure 12.1, consists of four main units: a TMS320C10 digital signal processor which is the heart of the system, an ADC/DAC unit, the timing circuit, and the memory units. The TMS320C10 is ideally suited to the task: it is inexpensive, simple to design with (provided that one is not intending to do a lot of clever things), and we already have experience of using it.

To minimize chip count and costs, the ADC/DAC unit is an 8-bit chip, Analog Devices AD7569. The ADC has a conversion of 2 μs and an on-chip sample and hold. For high fidelity systems 8-bit resolution is not adequate, but for demonstrating the principles of DSP we found it sufficient. The memory unit consists of a program selection switch and a pair of EPROMs, mounted on ZIF (zero insertion force) sockets for easy use. The EPROM is partitioned into eight blocks each of 1K words, selectable through the program selection switch. This allows up to eight programs to be held in the EPROMs. For stand-alone operation, the use of EPROMs is necessary. The processor derives its clock from a 12 MHz crystal, instead of the 20 MHz clock required to operate at full speed, making it possible to use inexpensive EPROMs with slow access time. With a 12 MHz system clock, any 8K × 8-bit EPROMs with access time of <165 ns can be used with the system. In the current implementation, a pair of 8K × 8-bit EPROMs with access time of 150 ns are used. The 12 MHz system clock is divided internally to produce a 3 MHz CLOCKOUT signal. The sampling signals are derived from the CLOCKOUT signal by dividing it down in the timing circuit which generates two user-selectable sampling signals, one at 7.5 kHz and the other at 15 kHz. The ADC is controlled by the sampling signals.

The complete circuit diagram of the system is shown in Figure 12.2. To allow the study of aliasing and imaging, input and output filters are not included. If required, the input signal can be passed through an appropriate anti-aliasing filter before being applied to the system and/or the output of the system can be fed into an anti-imaging filter. The system accepts an input signal between ±2.5 V, samples and digitizes it to 8 bits at a rate of 7.5 kHz or 15 kHz. At the end of each conversion, the busy line of the ADC pulls the BIO pin of the processor low to signal that a new piece of data is ready. Alternatively, the busy line can be used to interrupt the processor when the data is ready. After processing, the data is fed to the DAC.

Input/output data transfers to or from the ADC/DAC unit by the processor are performed with the IN and OUT instructions. The IN instruction reads data from the ADC and places it in the internal data memory of the processor. The data enable ($\overline{\text{DEN}}$) signal of the processor combined with the address line A0 allows the 8-bit data from the ADC to be strobed into the upper half of the 16-bit data bus (D8–D15) and stored in the data memory location XN. The OUT instruction transfers data from a data memory location to the DAC. The write enable signal ($\overline{\text{WE}}$) combined with the address line A0 is used to place the data onto the data bus D0–D15 and to transfer it to the DAC. The program memory is controlled by the memory enable ($\overline{\text{MEN}}$) signal from the processor and allows the program instructions to be fetched from the EPROMs. The most significant three address lines of the EPROMs are connected to a selection switch, which effectively partitions the program memory into eight blocks. With the 8K × 8-bit EPROMs we have used, each block is 1K words long, as shown in Table 12.1. Typically, each block contains a separate program that executes a specific DSP function.

The TMS320C10 target board is constructed in a double-sided PCB. The PCB layout was carried out in house, but the board was professionally made at less than $75 per board in quantities of 5.

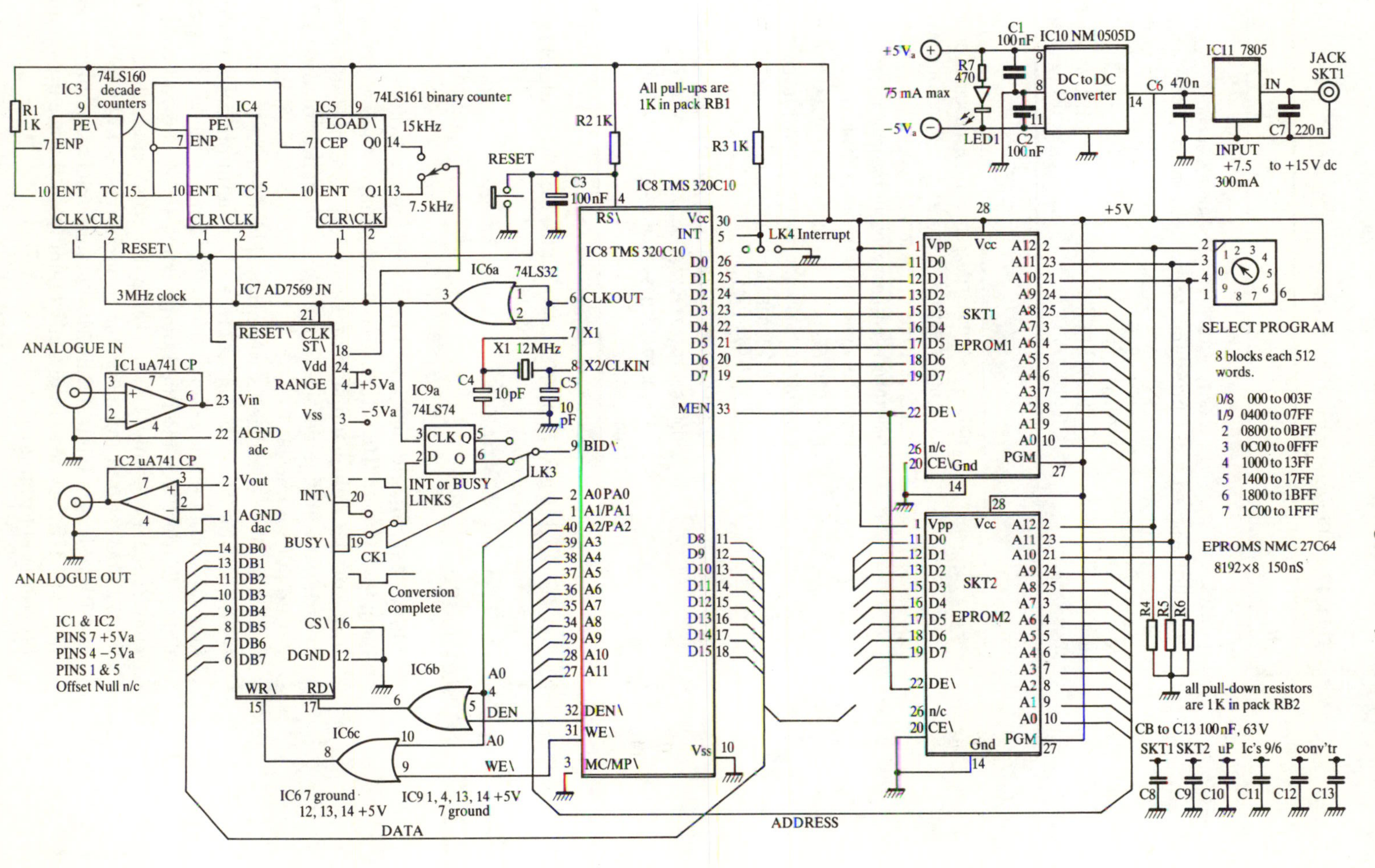

Figure 12.2 TMS 32010 target board.

Table 12.1 Program memory selection.

Block	*Address*	*DSP programs*
0	0000–03FF	Input–output loop
1	0400–07FF	Noise generator
2	0800–0BFF	Square wave generator
3	0C00–0FFF	41-point bandpass FIR filter
4	1000–13FF	61-point FIR notch filter
5	1400–17FF	fourth-order IIR lowpass filter in cascade
6	1800–1BFF	fourth-order IIR lowpass filter in parallel
7	1C00–1FFF	fourth-order IIR bandpass filter in cascade

```
:10000000F900003F002A0000FFB20000008E00004F
:10001000FF15000001650000FE01000002B70000AE
:10002000FC780000046A0000FAAA0000063F000005
:10003000F8EC000007CC0000F7A8000008B10000B1
:1000400077300000008B10000F7A8000007CC0000DE
:10005000F8EC0000063F0000FAAA0000046A000065
:10006000FC78000002B70000FE01000001650000FE
:10007000FF150000008E0000FFB20000002A7E0184
:10008000507C6E007E02703C713D688167A0007CF0
:10009000F40000456880F600004FF900004B400076
:1000A000703C71797F896A916D906B816D90F400DD
:1000B00000557F8F597B487BF600005CF900004BB0
:00000001FF
```

Figure 12.3 Hex codes for the FIR filter in Intel format.

12.1.3 System use

Many DSP algorithms have been implemented on the target board. These include FIR and IIR filters, noise and square wave generators. Examples are listed on the right-hand side of Table 12.1. To illustrate the use of the target board, we will give details of the implementation of the FIR notch filter discussed in Example 11.3.

The filter coefficients were first quantized to 16 bits, by multiplying each coefficient by 2^{15} (see Table 11.2). A general-purpose TMS320C10 assembly language program developed using the commercial cross-assembler, Metai, by Crash Barrier, was modified by inserting the quantized coefficients, and filter length; see Program 11A.1. Using the cross-assembler, a HEX file in Intel format was created to enable the program codes to be programmed into a pair of EPROMs; see Figure 12.3. Most commercial assemblers would create such a file. The HEX codes are directly taken from the second column of the assembly listing in Program 11A.1.

Programming the EPROMs with the HEX codes depends on the type of EPROM programmer and associated software. Our simple set-up uses the Jabal programmer. With the 'odd' EPROM inserted into the socket of the programmer, the odd bytes in the HEX file are programmed by entering a command such as this:

```
podtalk targbsf.hex 2764a ia1000o
```

The 'even' EPROM is programmed similarly by replacing the last character, 'o' (odd), in the above command, by an 'e' (even). The lower byte of each instruction is held in the odd EPROM and the upper byte in the even EPROM. The 'i' in the command line indicates to the EPROM programmer to ignore the fact that the EPROM may not be empty, allowing up to eight different programs to be held in one EPROM. The '1000' is the offset address indicating the start of program codes (in this case address block 4 in Table 12.1).

With the even and odd EPROM secured in the ZIF socket, the selection switch in one of eight positions, and the reset button pressed the system automatically executes the selected program. For the digital filter the selection switch is set on position 4.

12.2 TMS320C25 target board for real-time DSP

12.2.1 Background and system specifications

The TMS320C10 target board described in the last section has proved a useful and an inexpensive platform for demonstrating simple DSP algorithms in real time and given students the opportunity to implement their own DSP algorithms. However, having used it for about three years, we found that additional facilities are sometimes necessary, such as communication with a PC, for example to download or upload code or data to the PC, dual analogue input /output channels to allow operations such as adaptive filtering, and correlation.

We decided to design an entirely new target board based on the more powerful processor, the TMS320C25, to take advantage of the improved performance it offers. The features of the new target board are as follows:

- two analogue input/output channels;
- 12-bit ADC and DAC for each channel, supporting sampling rates of up to 100 kHz;
- sampling frequency to be software changeable;
- one serial port and one parallel port for communication with a PC;
- jumper selectable anti-aliasing and anti-imaging filters;
- 8K words of data RAM and 8K words of program RAM, 8K words of program EPROM;
- no wait state RAMS to allow full speed execution.

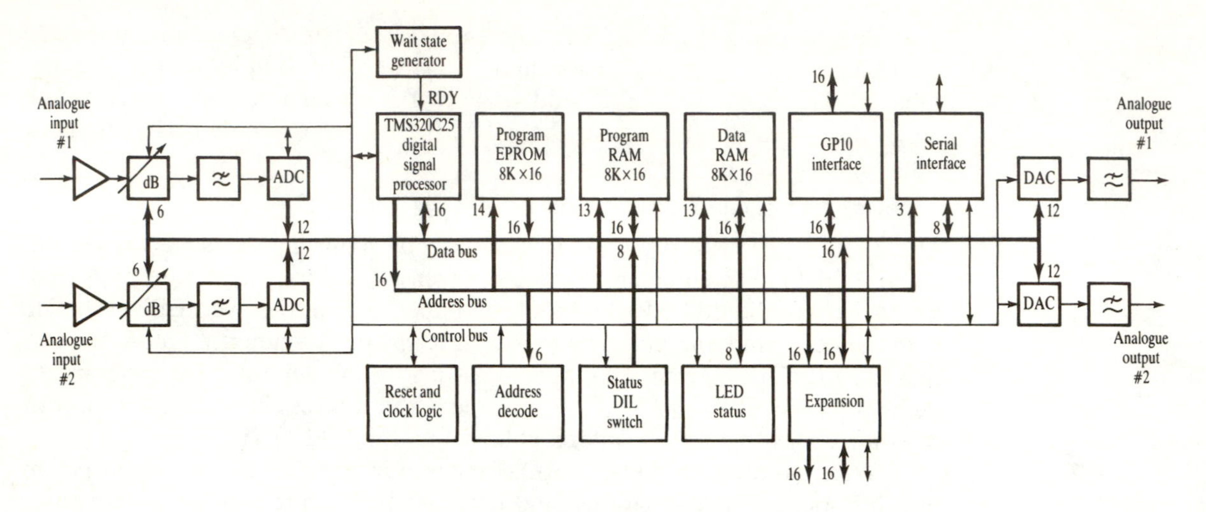

Figure 12.4 TMS320C25 target board.

12.2.2 System description

A simplified block diagram of the target board is shown in Figure 12.4. The system communicates with a PC via either a parallel port or an RS232 serial port. The parallel port goes via a general-purpose input/output (GPIO) port.

Programs can be downloaded from the PC into the external program RAM and executed at full speed. The program EPROM contains a small boot program which allows limited communication with the PC. Data can be uploaded to the PC for storage on a disk file.

Two analogue I/O channels each fitted with a 12-bit ADC and 12-bit DAC are provided. The ADCs can each support sampling at up to a 100 kHz rate. The sampling signals for the two ADCs are simultaneous and are derived from the XF line of the TMS320C25 or from an external source. When the internal signal XF is used, the on-chip timer on the TMS320C25 generates an interrupt once every T seconds, where:

$$T = F_{ck}/N$$

and N is the content of the period register (PRD) internal to the TMS320C25, F_{ck} is the clockout signal of the C25, which is 40 MHz at full speed, and T is the sampling period.

The anti-aliasing filter and the anti-imaging filter have Butterworth characteristic. Each consists of four second-order sections mounted on headers to enable the user to configure the filters for different applications. The filters can also be bypassed using jumpers, to allow the study of aliasing or imaging effects.

The TMS320C25 has three external address spaces: program memory, data memory, and I/O spaces. These three spaces are selectable by the pro-

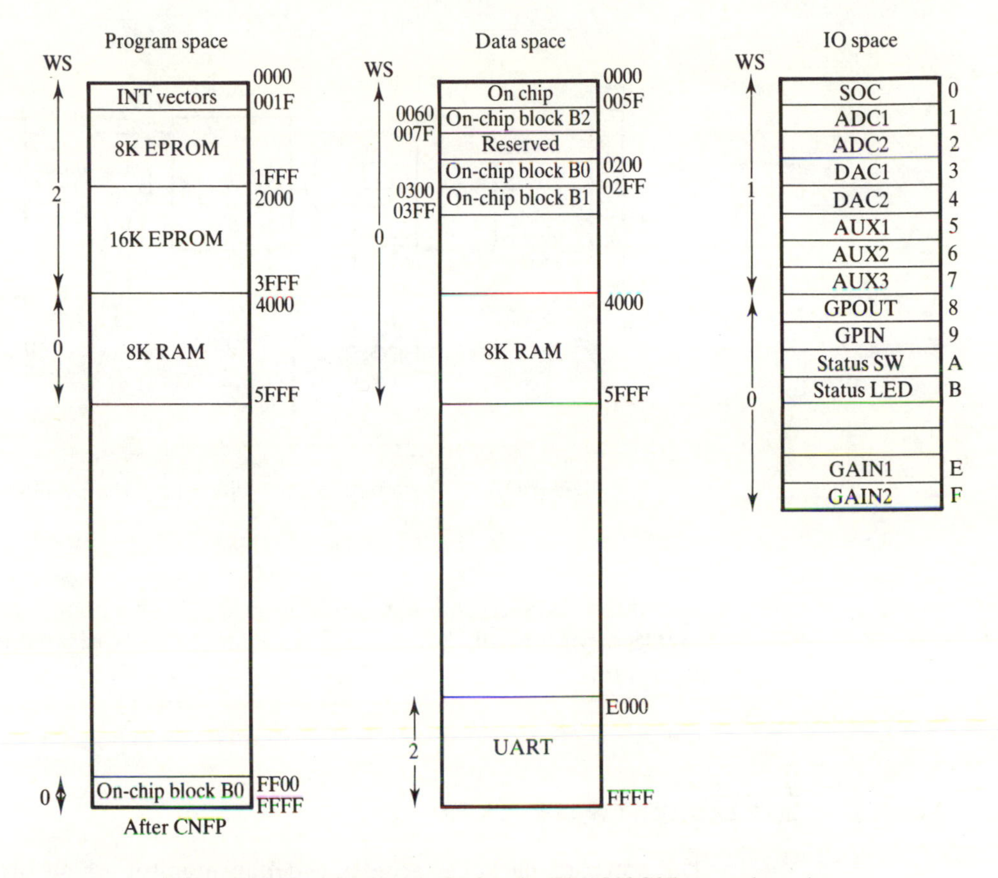

Figure 12.5 Memory maps for the TMS320C25 target board.

gram space (PS), data space (DS) and I/O space (IS) lines respectively. The program memory space contains a pair of 8K × 8-bit EPROMs and a pair of 8-bit × 8-bit RAMs. The RAMs are 25 ns memories, requiring no wait states. The EPROMs, mounted on ZIF sockets, require one wait state to execute (access time of 140 ns or less). The memory maps for the TMS320C25 target board are shown in Figure 12.5. The detailed circuit diagram for the target board is available on request from the first author. In Section 12.4 we will see an example application of the TMS320C25 target board.

12.3 TMS320C25 Software Development System (SWDS)

The SWDS is a PC-resident development tool for the Texas Instruments TMS32020/C25 processor-based systems that allows real-time software simula-

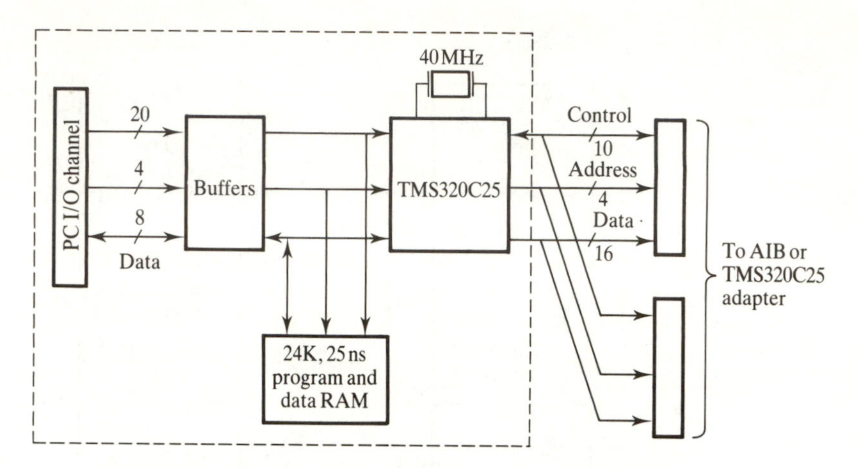

Figure 12.6 A simplified block diagram of the SWDS hardware.

tion of the TMS320 codes. Several programs in this book were developed in the SWDS environment. The SWDS package consists of a software package and hardware.

12.3.1 SWDS software

The software package includes a debug monitor which provides an environment that allows assembly language source files to be assembled, linked, debugged (for example using single stepping, breakpoints, memory and register display/modify) and execute. Data logging facilities are also available and permit some I/O operations to disk.

12.3.2 SWDS hardware

The main hardware is a single card that plugs into a full-size slot in the IBM PC or compatible. The card contains a TMS320C25 processor with 24K words of user-configurable, external, no wait states, program and data RAM. A simple block diagram of the SWDS board is given in Figure 12.6.

Two 40-way ribbon cables connect the SWDS card, via one of two adapters, to a user target hardware system for I/O operations: the TMS32020 adapter allows connection between the SWDS card and a TMS32020/25-based target system, via the ribbon cables, allowing in-circuit emulation of the TMS320C25/20 in the target system. The AIB adapter allows connection between the SWDS and the analogue interface board (AIB).

12.3.3 Using the SWDS with the analogue interface board

The AIB consists of a 12-bit ADC (with a sample and hold), 12-bit DAC and analogue input and output filters to allow analogue I/O signals to the TMS320. The AIB adapter allows the SWDS to use the AIB as a target system. The adapter converts the C25 I/O structure to allow access to the ADC, DAC and control registers on the AIB.

The PC, SWDS and the AIB together form a complete DSP software and hardware development system for the TMS320C25.

The SWDS communicates with the AIB via the following ports PA0 to PA2:

PA0	AIB mode control register
PA1	sample rate register
PA2	ADC output register (IN)
PA2	DAC input register (OUT)

Port PA0 is used to set up the modes of the ADC and DAC, and the sample rate register is accessed by an OUT instruction to port 1. The sampling frequency is determined as follows:

$$N = F_{clk}/F_s - 1$$

where F_{clk} is the system clock, which is 40 MHz for the TMS320C25, F_s is the sampling frequency, and N is a user-defined value in the sample rate register.

The reader is referred to Chapter 11 for examples of programs written to run under the SWDS–AIB environment.

12.4 FFT spectrum analyser†

The availability of low cost and powerful computers (such as the IBM PC or compatible) has led to the development of economical laboratory instruments based around the PCs. Our aim here is to combine the graphics capability of the IBM PC with the high speed numerical processing power of the Texas Instruments TMS320C25 digital signal processor, to produce an economical laboratory spectrum analyser. The analyser is required to take an analogue input, to digitize and calculate its spectrum using an FFT algorithm, and to display the spectrum on the IBM PC monitor. In future, it is intended to develop the system further and use it as a general-purpose real-time signal analyser.

A conceptual block diagram of the spectrum analyser is depicted in Figure 12.7. The analyser performs N-point radix-2 FFTs on one of two input data, calculates the power spectrum density and transmits this to the IBM PC

†This section is based, in part, on the final year project of one of our previous students, Mr Ali Mohammed.

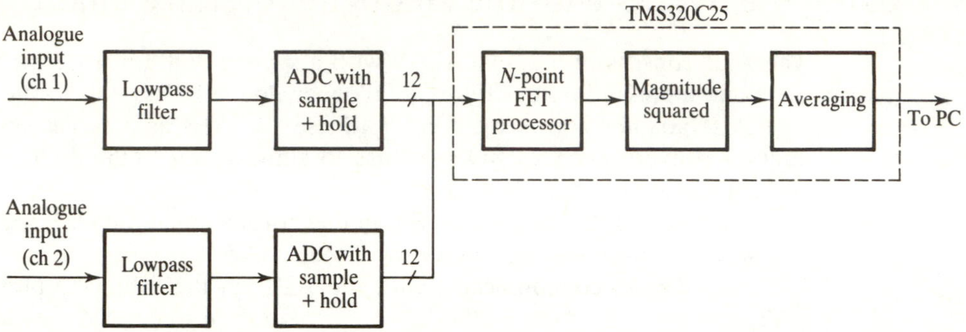

Figure 12.7 Conceptual block diagram of the FFT spectrum analyser.

through an RS232C link. A C language program on the PC receives and displays the spectrum. All the numerical computations associated with spectrum estimation are performed by the TMS320C25 processor.

In the next few sections, we will describe the features of the analyser, the spectrum estimation techniques used, and the analyser hardware and software.

12.4.1 Features of the analyser

The FFT spectrum analyser consists of a TMS320C25-based board connected to an IBM PC or compatible, and a PC-resident software package for displaying the signal spectrum. The main features of the analyser are as follows:

- dual analogue input channels, each with a 12-bit ADC;
- sampling rate per channel <40 kHz;
- estimates power spectrum density of signals in the frequency range 0–10 kHz using the FFT algorithm;
- serial communication with a host personal computer via an RS232 port;
- display of signal spectrum on the PC using the host's graphics facilities.

Although not implemented at the present, the analyser can readily support multivariable analysis (for example, cross-spectrum, cross-correlation), and the display of input signals in time.

12.4.2 Spectrum estimation in the analyser

The spectrum analyser uses the radix-2 FFT algorithm to convert a time domain data into frequency. The FFT algorithm takes N data samples, $x(n)$, $n = 0, 1, \ldots, N-1$, and produces N-point complex frequency samples, $X(k)$, $k = 0, 1, \ldots, N-1$. The raw power spectrum density (PSD) is then obtained as the scaled magnitude squared of complex frequency samples:

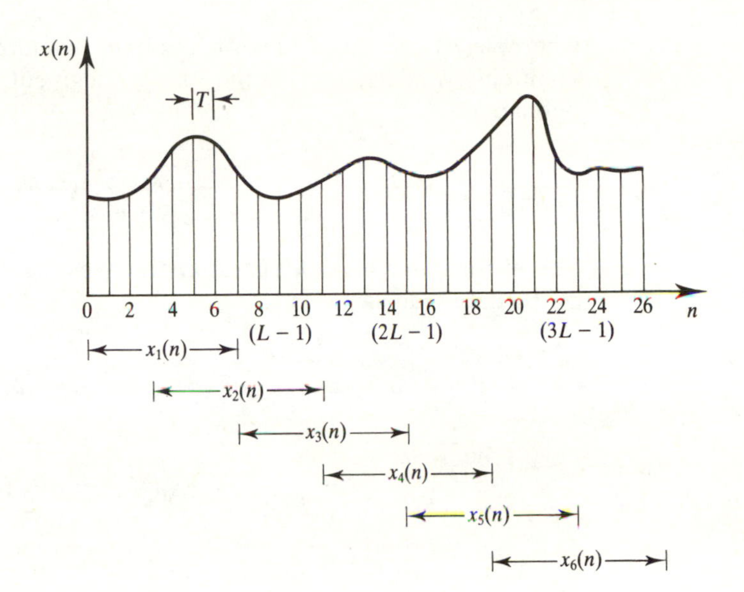

Figure 12.8 An illustration of subsectioning of data with 50% overlap and section length, $L = 8$.

$$S(k) = (1/N)|X(k)|^2 = (1/N)\{\mathrm{Re}[X(k)^2] + \mathrm{Im}[X(k)^2]\} \qquad k = 0, 1, \ldots, N/2 \quad \textbf{(12.1)}$$

where N is the number of data points. The PSD gives a measure of the distribution of the average power of a signal over frequency. For random or noisy data, direct computation of the power spectrum as in Equation 12.1 may not yield good spectrum estimates because of random errors. To obtain consistent estimates of the PSD some form of averaging is normally used at some stage in the estimation process, for example correlation or frequency averaging. In the analyser, we employed a method due to Welch (1967).

In the Welch method, the input data sequence $(x(n),\ n = 0, 1, \ldots, N-1)$ whose spectrum is to be estimated is first divided into K overlapping segments each of length L. For example, for a 50% overlap the data segments are (see Figure 12.8)

$$x_i(n) = x[n + iL/2], \quad n = 0, 1, \ldots, L-1; i = 0, 1, \ldots, K-1 \quad \textbf{(12.2)}$$

The transform, $X_i(k)$, and the raw power spectrum, $S_i(k)$, of each data segment are then obtained as

$$X_i(k) = \sum_{n=0}^{L-1} w(n)x_i(n)\exp(-j2\pi nk), \quad k = 0, 1, \ldots, L-1 \quad \textbf{(12.3)}$$

$$S_i(k) = \frac{1}{LQ}|X_i(k)|^2, \quad k = 0, 1, \ldots, L/2 \quad \textbf{(12.4)}$$

where $w(n)$ are the values of a suitable window function, for example the Hamming window, $x_i(n)$ is the ith data segment, and Q is the window energy defined as

$$Q = \frac{1}{L} \sum_{n=0}^{L-1} w^2(n) \tag{12.5}$$

The average of the raw PSD estimates gives the desired 'smooth' PSD estimate of the data sequence, $x(n)$:

$$\begin{aligned} S(k) &= \frac{1}{K} \sum_{i=0}^{K-1} S_i(k), \quad k = 0, 1, 2, \ldots, L/2 \\ &= \frac{1}{KLQ} \sum_{i=0}^{K-1} |X_i(k)|^2, \quad k = 0, 1, 2, \ldots, L/2 \end{aligned} \tag{12.6}$$

The values of $X_i(k)$ in Equation 12.3 are of course obtained using the FFT for efficiency. To improve the spectral clarity of $S(k)$, the length, L, of each data segment may be extended, after windowing, to say $L + R$ by adding R zeros, and the FFT of the zero-padded sequences then obtained.

In the analyser eight data segments were used. The segment length, L, was fixed at 128 data points (that is no zeros were added) and the overlap between segments was kept constant at half the segment length.

The input data sequence $x(n)$ whose PSD is to be estimated is derived by sampling an analogue signal at a frequency, F_s, consistent with the sampling theorem:

$$F_s > 2f_{max}; \quad F_s = 1/T$$

where T is the sampling interval and f_{max} the highest frequency in the analogue signal. When we transform $x(n)$ into frequency as described above, the frequency spacing, Δf, between the frequency components is given by:

$$\Delta f = 1/LT \tag{12.7}$$

This is the effective resolution of the FFT. For the analyser, $L = 128$, and the maximum sampling frequency, $F_s = 40$ kHz, giving a resolution <312.5 Hz.

12.4.3 Analyser hardware

The block diagram of the system hardware is shown in Figure 12.9. Anti-aliasing filtering at the input is achieved using a second-order Butterworth filter for simplicity. Using this low-order anti-aliasing filter means that the sampling frequency should be sufficiently high to keep the effects of aliasing low (see Chapter 1 for details).

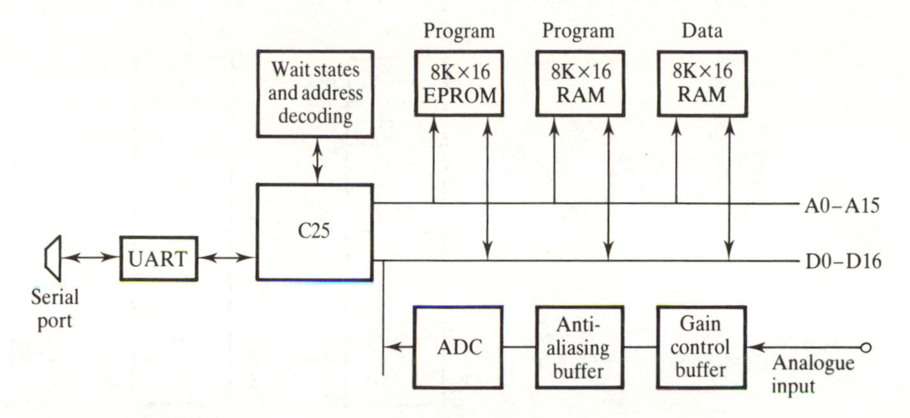

Figure 12.9 PC-based FFT spectrum analyser hardware.

The program memory stores the FFT twiddle factors, precomputed sine and cosine values, as well as the window coefficients, $w(n)$, in Equation 12.3. The input data samples, $x(n)$, are stored first in the external data memory, and each block of data to be transformed is then transferred to the internal memory and stored in a complex format (that is, as real and imaginary); see the map for data storage in Figure 12.10. The FFT is performed in place. The spectrum analyser board passes the spectrum estimates to the IBM PC for display through the serial port.

The TMS320C25 target board described in Section 12.2 meets all the analyser hardware specifications and so was used to implement the analyser.

12.4.4 Analyser software

The analyser software consists of a set of TMS320C25 assembly language programs for computing the PSD values and transferring these to the PC, and a C language program resident on the PC for displaying the power spectrum.

12.4.4.1 TMS320C25 programs

The TMS320C25 programs carry out the following operations.

(1) Collect N data points, $x(n)$, from the ADC (at the present, $N = 576$) and save in the external data RAM.

(2) Divide the N data points into eight overlapping segments, 128 data points each (Equation 12.2).

(3) For each data segment perform the following operations:

- window the data using a Hamming window

$$x_i(n) = x_i(n)w(n), \quad n = 0, 1, \ldots, 127$$

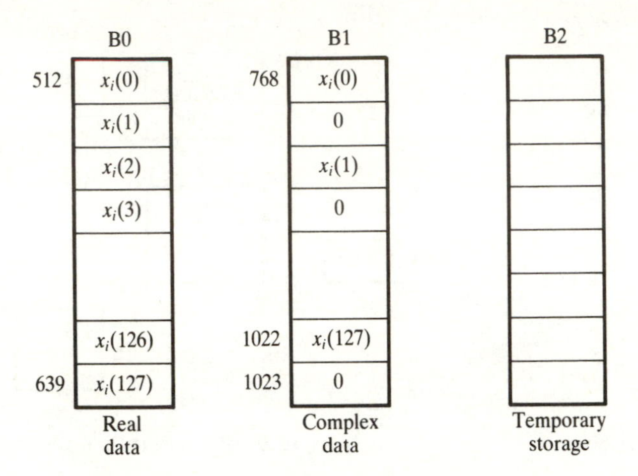

Figure 12.10 On-chip data storage for the analyser.

- transfer the windowed data into the internal RAM block B1, and store as real and imaginary;
- perform a 128-point, in-place FFT on the data segment;
- compute the raw PSD (Equation 12.4);
- save raw PSD in external RAM.

(4) Average the raw PSD for all segments (Equation 12.6).

(5) Transfer averaged PSD to the PC.

The TMS320C25 code, written for the Texas Instruments assembler, consists of a main program, four subroutines, and an interrupt service routine:

- main routine;
- PSD initialization;
- interrupt service routine;
- data collection;
- PSD estimation;
- data transfer.

Some of the programs are described briefly below.

PSD initialization

The PSD initialization routine supplements the system initialization in the boot program referred to in Section 12.2. Its major role is to prepare the hardware for PSD computations. It performs the following operations:

- configure on-chip memory block B0 as RAM;
- initialize internal data memories blocks B0, B1, and B2 with zero;

- program the timer for the correct sampling frequency;
- unmask timer interrupt;
- enable the timer interrupt.

Programming the timer involves loading the 16-bit period register (PRD), data memory location 3, with an appropriate value to set up the sampling frequency:

$$F_s = F_{clk}/N$$

where F_s is the sampling frequency, F_{clk} is the TMS320C25 CLKOUT frequency, which is 10 MHz, and N is a 16-bit unsigned integer loaded into the PRD register. To generate a 40 kHz sampling frequency, the PRD is loaded with 250.

Data collection

After PSD initialization, the processor branches to the data collection subroutine where it goes into the idle state and waits until the timer interrupt occurs. On interrupt, the processor jumps to the interrupt service routine and issues a start of conversion (SOC) command to the ADCs by toggling the external flag, XF. The processor then returns to the data collection subroutine and waits for the end of the conversion (EOC) signal from the ADCs by testing for the BIO line to go low.

When the EOC is received the data is read from the 12-bit ADC and stored in the external data RAM in Q15 format by left shifting the data four places before it is stored in external data memory (only the channel 1 ADC is used for the PSD estimation). The processor returns to the main routine when 576 consecutive data samples have been collected.

Power spectrum density estimation

The FFT is the heart of the PSD estimation. It is a radix-2, 128-point decimation in frequency (DIF) algorithm based on the implementation in Texas Instruments (1986). For each data segment, the PSD routine initializes memory blocks B0 and B1 with zero and then moves the data from the external memory to the internal memory block B0. The data is windowed by multiplying each data sample by an appropriate window coefficient and then transferred to the on-chip memory, block B1, where it is stored in a complex form. The real data samples are stored in even addresses, starting at address 768. The odd address locations, which contain zeros, represent the imaginary parts; see Figure 12.10. The complex data is then FFTed, in place, and its raw PSD computed. The raw PSDs for the segments are then averaged to obtain a smooth PSD.

Data transfer

The averaged PSD is transferred to the PC, via the serial interface, with each averaged PSD value represented as 2 bytes.

12.4.4.2 IBM PC display software

The simple C language program on the PC reads the averaged PSD values from the analyser hardware and displays it on the screen. When the program is first executed a menu is displayed requesting the user to select an option:

(1) display power spectrum density;
(2) save and display PSD;
(3) read PSD from disk and display;
(4) zoom;
(5) exit.

If option 1 or 2 is selected, control is passed to the main program in the analyser, which computes the PSD and transfers it to the PC as described above. For option 1, the PSD is then displayed. For option 2, the PSD values are first saved on the disk in a user-specified file, before the PSD is displayed. For option 3, the C program first reads the PSD from disk and then displays it. For option 4, the user is asked to specify the frequency range of interest, and the selected region of the current display is then expanded accordingly. Selecting option 5 returns the user to DOS.

12.4.5 Using the analyser

Figures 12.11(a) and 12.11(b) show examples of the PSD of, respectively, 1 kHz square wave and a 2 kHz sine wave.

At present the analyser is used only for power spectrum estimation. It can be extended to perform other signal analysis or used in other applications, for example

- frequency response estimation,
- auto- and cross-spectral analysis of noisy or random signals,
- convolution in the frequency domain,
- correlation in the frequency domain,
- cross-spectral density,
- magnitude coherence, and
- time domain display.

The data from the analyser hardware may also be sent to the PC in various forms for further processing and/or display, for example as

- raw data,
- real and imaginary, and
- magnitude and phase.

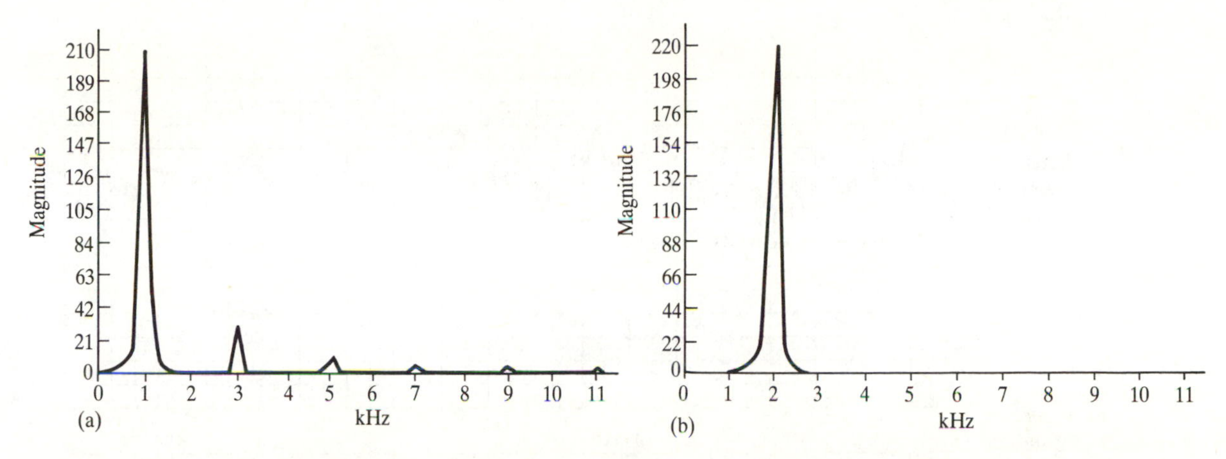

Figure 12.11 (a) Frequency spectrum for 1 kHz square wave signal. (b) Frequency spectrum for 2 kHz sine wave signal.

12.5 Detection of foetal heartbeats during labour†

Worldwide, the standard method of monitoring the foetus during labour is the display of continuous foetal heart rate (FHR) and the uterine activity which together constitute the cardiotocogram (CTG) (Figure 12.12). By analysis and appropriate interpretation of changes in the CTG obstetricians hope to prevent the delivery of dead or impaired babies who had suffered as a result of a lack of oxygen during labour and delivery.

12.5.1 The foetal electrocardiogram

The foetal heart rate is routinely obtained during labour from the electrocardiogram (ECG), the electrical activity of the heart (see Figure 12.13) or ultrasound. Like the adult ECG, normal foetal ECG is characterized by five peaks and valleys labelled with successive letters of the alphabet P, Q, R, S, and T (Greene, 1987). Thus, the ECG is said to consist of the P wave, QRS complex and T wave (Greene, 1987).

As shown in Figure 12.13, the reciprocal of the heart period, that is the

†This section is based, in part, on the final year project of one of our previous students, Mr Iead Rezek.

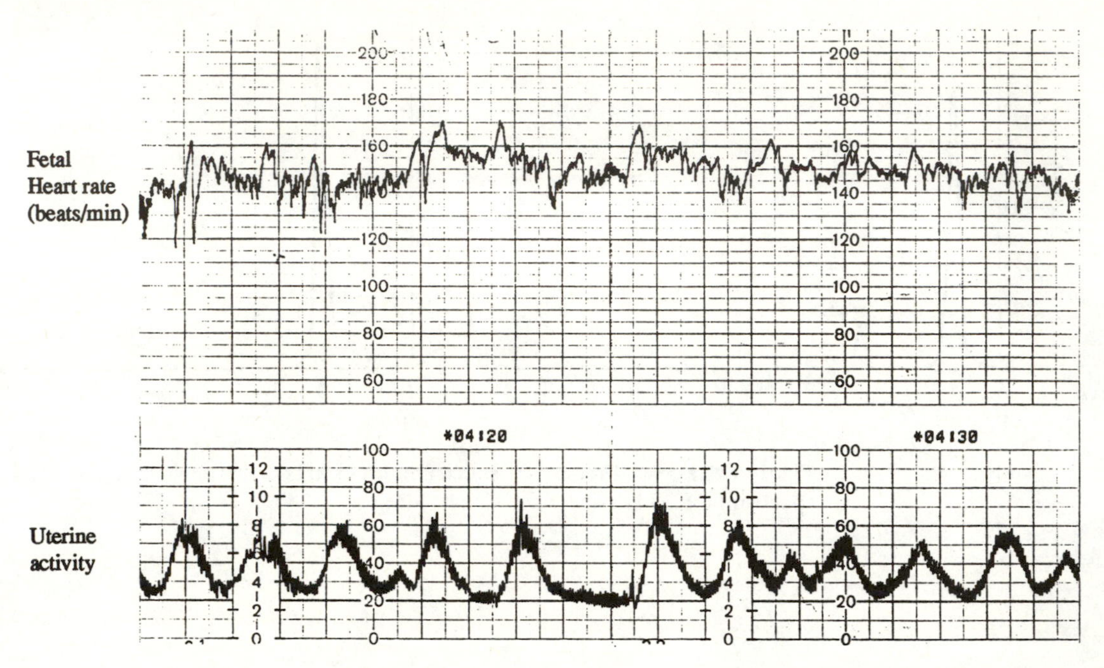

Figure 12.12 An example of a cardiotocogram (CTG). The CTG consists of the foetal heart rate pattern and the uterine activity.

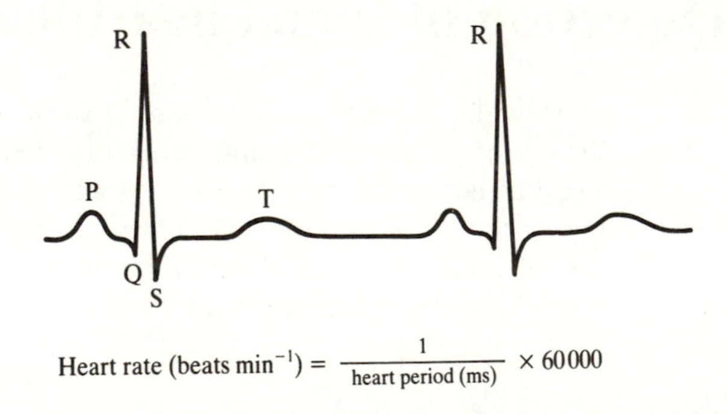

$$\text{Heart rate (beats min}^{-1}) = \frac{1}{\text{heart period (ms)}} \times 60\,000$$

Figure 12.13 The electrocardiogram.

time interval between the R-to-R peaks (in milliseconds), multiplied by 60 000 gives the instantaneous heart rate. The FHR pattern in the upper half of Figure 12.12 is a plot of successive instantaneous heart rates.

In practice, to measure the foetal heart rate a suitable DSP algorithm is employed to detect, in hardware or software, successive QRS complexes and from these to calculate the R-to-R intervals and the corresponding FHR. Most QRS detection methods assume that the shape of the foetal QRS complex is known *a priori*, but that its time of occurrence is unknown. This assumption is

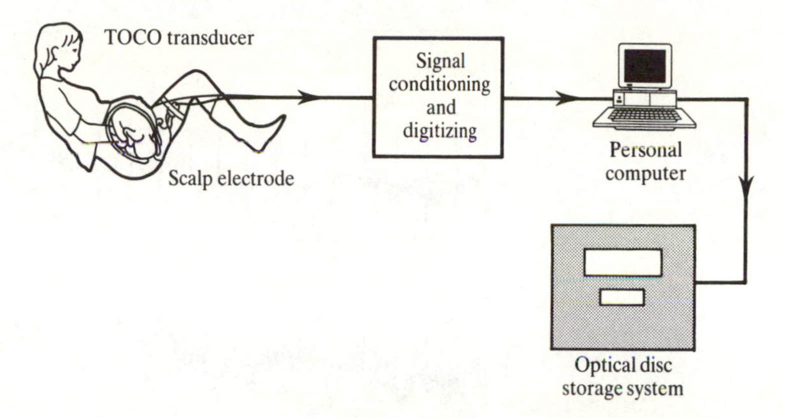

Figure 12.14 Measurement of foetal electrocardiogram.

reasonable, although not always valid as the shape of the QRS complex may change from patient to patient and indeed within the same patient. Thus by comparing the ECG signal against a known, representative QRS template the locations of the QRS complexes in the ECG can be determined based on some measure of similarity, for example a high value of cross-correlation.

A fundamental problem is the reliable detection of the QRS complexes. Signal degradation due, for example, to baseline wander, mains interference, uterine contractions, ADC saturation, and movements of the baby or mother will lead to false detection or missed QRS complexes. The aim in this case study is to investigate and compare two QRS detection methods that may be of practical value in real-time foetal monitoring. This work is a small part of an ongoing research initiative with local hospitals to develop intelligent systems to assist the busy clinicians in managing labour (Ifeachor *et al*., 1991).

The foetal ECG data used in the case study was taken from our foetal research database. The ECG signal was obtained by measuring differentially between an electrode on the foetal scalp and a standard skin electrode placed on the maternal thigh, and a second maternal electrode was used as earth (Figure 12.14). This lead system has a sensitivity vector in the longitudinal plane of the foetus compared with the sagittal plane of the standard foetal scalp electrode connection and should reduce ECG vector change resulting from rotation of the foetus (Lindecrantz *et al*., 1988). The FECG was fed through a patient, isolation box amplified, analogue bandpass filtered (passband 0.07–100 Hz), and digitized at 500 samples s^{-1} with a resolution of 8 bits.

Examples of measured foetal ECGs are shown in Figures 12.15(a) to 12.15(c). From visual examination, it is seen that the data in Figure 12.15(a) has a relatively high SNR, compared with Figures 12.15(b) and 12.15(c), with large amplitude R waves (seen as spikes). The data in Figure 12.15(b) on the other hand has a relatively high noise content and significant base line shifts, although the R waves are still discernible. The data in Figure 12.15(c) contains ADC errors, seen as large amplitude swings between the maximum and minimum ADC values at the start of the record, due perhaps to ADC saturation,

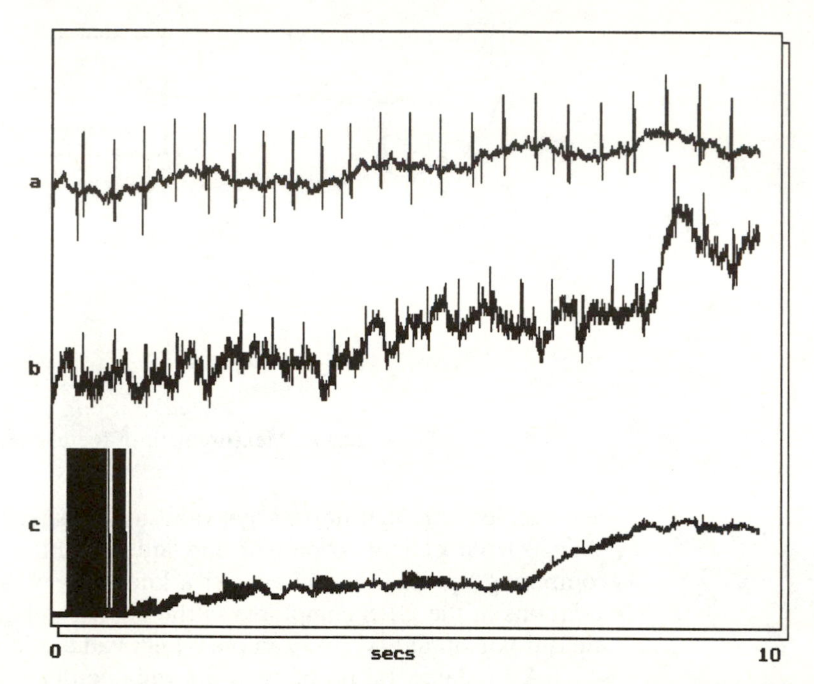

Figure 12.15 Examples of grades of ECG data: (a) grade 1 (good); (b) grade 2 (average); (c) grade 3 (poor).

as well as severe baseline shifts and high frequency noise including mains contamination. The three data sets in Figures 12.15(a), 12.15(b) and 12.15(c) have been subjectively classified as grade 1 (good), grade 2 (average) and grade 3 (poor) ECGs, respectively.

12.5.2 Foetal ECG signal pre-processing

For grade 2 and 3 data, noise levels and baseline shifts make detection of QRS complexes from the raw ECG more difficult. For a reliable QRS detection, it is necessary to pre-process the raw ECG to minimize the influence of these sources of signal degradation before attempting to detect the QRS complexes. It is known that significant frequency components of the QRS complex lie between about 4 and 45 Hz. Baseline shifts in the ECG are normally of a low frequency, typically less than about 3 Hz, although, for grade 3 data, baseline frequency may extend to 15 Hz or more.

An FIR or IIR bandpass digital filter may be used to pre-process the raw ECG before QRS detection. We prefer the use of FIR as an IIR filter of a high order, for example eighth order, sometimes rings when excited by the narrow width QRS complexes which may complicate the precise location of the R wave. The filter specifications used in the study are as follows:

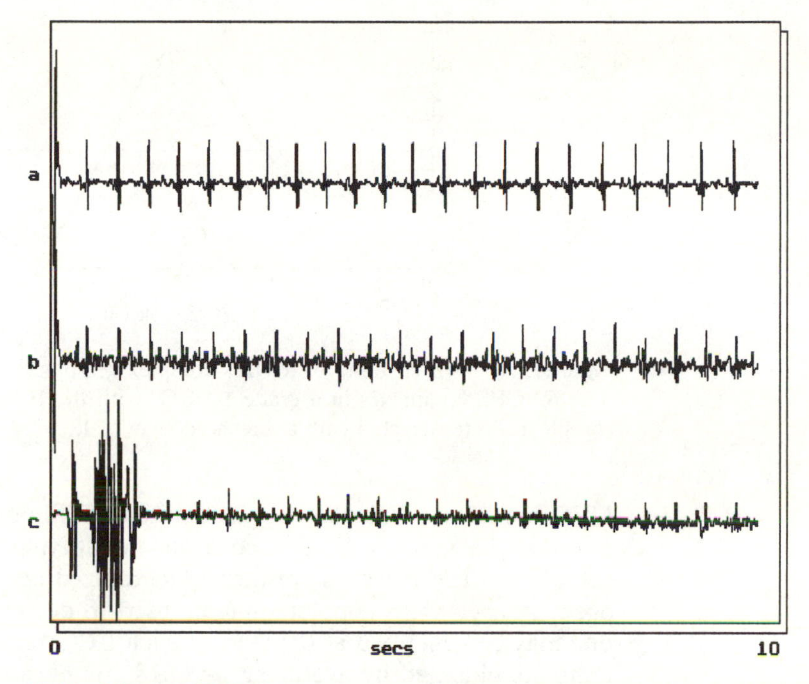

Figure 12.16 Filtered ECG data: (a) grade 1 (good); (b) grade 2 (average); (c) grade 3 (poor).

filter length	75
sampling frequency	500 Hz
stopbands	0–1 Hz, 47–250 Hz
passband	9–39 Hz
passband ripple	0.5 dB
stopband attenuation	30 dB

The filter coefficients were obtained using the optimal method described in Chapter 6. Figures 12.16(a) to 12.16(c) show the filtered ECG data. Compared with the corresponding unfiltered raw data, Figures 12.15(a) to 12.15(c), the baseline shifts as well as the high frequency noise have been reduced in the filtered data (ignoring the initial transients in the filtered data). In the grade 3 data, the ADC error appears as a burst which no doubt will confound most QRS detection algorithms; see Figure 12.16(c).

12.5.3 QRS template

Most QRS detection methods rely on the availability of a representative QRS template against which the incoming ECG signal is compared. The template may be generated from raw ECG data by detecting and averaging several good

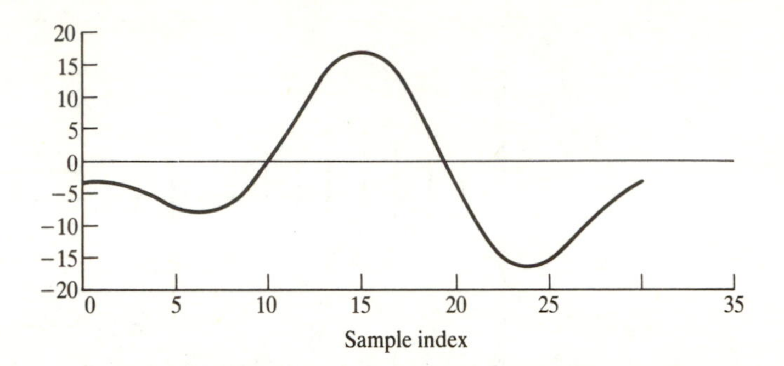

Figure 12.17 An example of a QRS complex template. This is obtained by averaging over 69 QRS complexes in a grade 1 ECG, with the R-waves synchronized. The QRS complexes were detected with a threshold level of 13.

QRS complexes. This may be done automatically or semimanually by visually examining a grade 1 ECG record and identifying good, unambiguous ECG complexes. The R waves are then synchronized and the QRS complexes averaged. A fixed QRS template may be used to detect QRS complexes or a new one may be generated at the start of each ECG record. An example of a QRS template obtained by averaging 69 QRS complexes in grade 1 data and then taking 31 samples (15 samples on either side of the R wave) of the averaged QRS complexes is shown in Figure 12.17.

In the study, templates of various lengths were tried. Typically, the length of the template, N, was between 11 and 31 samples, that is a width of between about 20 ms and 60 ms at a sampling rate of 500 samples s^{-1}. In this report, we will give results obtained with two templates of lengths 11 and 31 samples.

12.5.4 QRS detection methods

A general block diagram of the QRS detection process is given in Figure 12.18. The raw ECG data is first pre-processed to reduce the effects of noise. The pre-processed data samples are fed into a buffer one data point at a time. For each new data point fed into the buffer, the oldest data point is removed and the content of the buffer compared with a QRS template in the QRS detector. The output of the QRS detector is then thresholded. If this output exceeds the threshold value then a QRS is said to have occurred. Two conventional QRS detection methods are compared in this study, selected on the basis that they are either in practical use or potentially of practical use. Many other QRS detection methods exist.

The methods are

(1) average magnitude cross-difference (AMCD) (Lindecrantz *et al.*, 1988), currently used in a new foetal monitor described in Lindecrantz *et al.* (1988), and

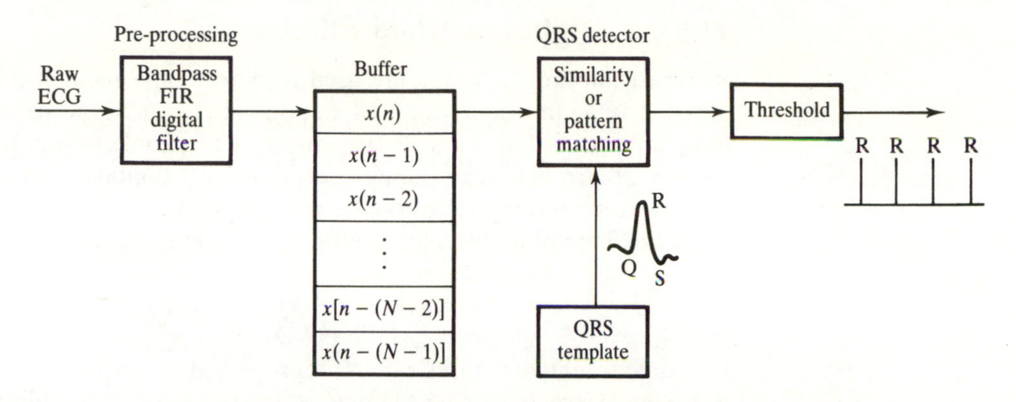

Figure 12.18 Concepts of QRS complex detection from raw ECG.

(2) matched filtering, which is a common QRS detection method and has been investigated by a number of workers (Azevedo and Longini, 1980; Favret, 1968): it is closely related to the correlation method.

12.5.4.1 Average magnitude cross difference

In this method, blocks of pre-processed foetal ECG data are compared against a template QRS complex as described above. The difference between corresponding samples in the ECG and the template are computed by waveform subtraction. The sum, $y(i)$, of the absolute values of the differences is then computed:

$$y(i) = \sum_{k=0}^{N-1} |x_t(k) - x_t - [x(k+i) - x_i]|, \quad i = 0, 1, \ldots \qquad \textbf{(12.8)}$$

where $x_t(k)$ are samples of the template QRS complex, $x(k+i)$ are samples of the ECG signal, N is the length of the template, and i is the time shift parameter. x_t is the mean value of the QRS template, and x_i the mean value of the ith data block for the ECG signal given by

$$x_t = \frac{1}{N} \sum_{k=0}^{N-1} x_t(i)$$

$$x_i = \frac{1}{N} \sum_{k=0}^{N-1} x(k+i)$$

When the ECG signal and QRS template are very similar in shape, that is in the neighbourhood of a QRS complex, the AMCD value, $y(i)$, becomes a minimum (theoretically zero).

12.5.4.2 Digital matched filtering

Matched filtering is commonly used to detect time recurring signals buried in noise. The main underlying assumptions in this method are that the signal is time limited and has a known waveshape. The problem then is to determine its time of occurrence. The impulse response of a digital matched filter, $h(k)$, is the time-reversed replica of the signal to be detected. Thus in our case, if $x_t(k)$ is the QRS template then the coefficients of the matched filter are given by

$$h(k) = x_t(N - k - 1), \quad k = 0, 1, \ldots, N - 1 \tag{12.9}$$

The digital matched filter can be represented as an FIR filter with the usual transversed structure, with the output and the input of the filter related as

$$\begin{aligned} y(i) &= \sum_{k=0}^{N-1} h(k)x(i - k) \\ &= \sum_{k=0}^{N-1} x_t(N - k - 1)x(i - k) \end{aligned}$$

where $x(i)$ are the samples of the input ECG signal, $x_t(k)$ are the samples of the QRS template, N is the filter length, $h(k)$ are matched filter coefficients, and i is the time shift index. It is evident that when the template and the QRS complex coincide, the output of the matched filter will be a maximum. Thus by searching the output of the matched filter for a value above a threshold the occurrence of the QRS can be tested.

12.5.5 Performance measure for QRS detection

To evaluate and compare the algorithms requires a measure of performance. Following Azevedo and Longini (1980) we define the performance measure as

$$\frac{(\text{total number of R waves} - \text{number of misses} - \text{number of false detections}) \times 100\%}{\text{total number of foetal R waves}} \tag{12.10}$$

For a given ECG record, the performance measure attains a value of 100% only if all the R waves in the record are correctly detected, that is no misses (undetected R waves) and no false detections (false alarms). For a given QRS detection method, the number of misses or false detections can be determined by comparing, visually, the output of the detector and the pre-processed ECG. Another alternative is the so-called 28-beat rule employed by clinicians to distinguish between a genuine change in the foetal heart rate and a false change due, for example, to an instrumentation error. According to this rule, an R wave causing a change in the frequency of the foetal heartbeat of more than

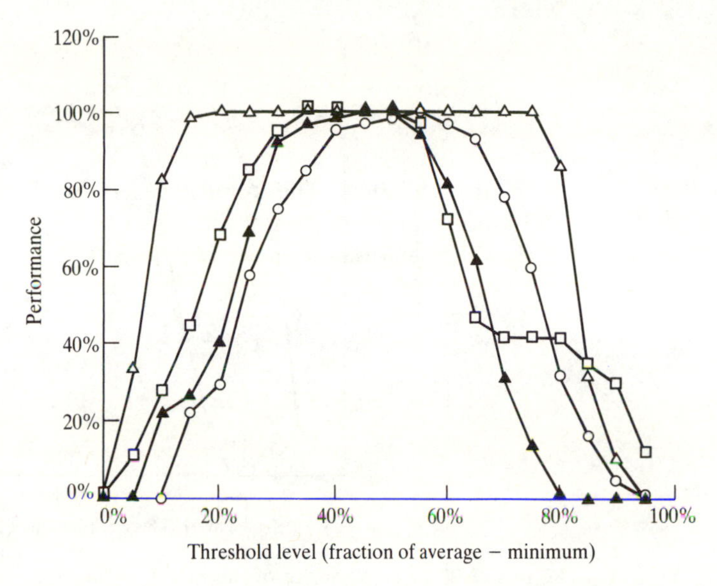

Figure 12.19 Performance of the average magnitude cross-difference method for grades 1 and 2 data, and templates of lengths 11 and 31: —×—, grade 1, T-11; —△—, grade 1, T-31; —◇—, grade 2, T-11; —⧖—, grade 2, T-31.

±28 beats per minute is indicative of either a missed or a false QRS complex. A baseline may be fitted to the FHR pattern to aid the application of the rule.

12.5.6 Results

Figures 12.19 and 12.20 show the performances of the AMCD and matched filtering methods, respectively, plotted against threshold levels for grades 1 and 2 data.

The performances of both methods are dependent on the threshold level used (each threshold was expressed as a fraction of the maximum input signal value) and the length of the QRS template. The best performance was achieved with a threshold level of about 50% for both methods. The wider template tends to perform better than the narrow template when the quality of data is good, but the main difference between them seems to be in their sensitivity to threshold levels.

Overall, there was little to choose between the AMCD and matched filtering methods in terms of their performance. With suitable threshold levels, both methods attained the following performances:

- 100% detection for all grade 1 ECG;
- >90% detection for grade 2 data;
- >60% detection for grade 3 data.

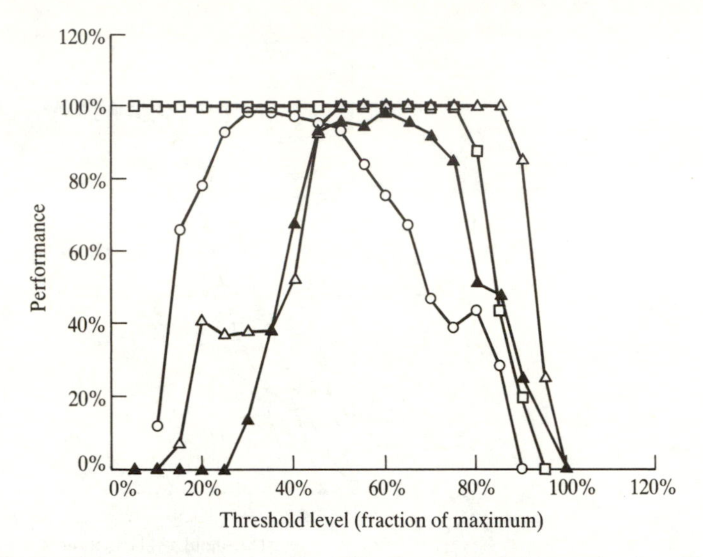

Figure 12.20 Performance of the matched filtering method for grades 1 and 2 data, and for templates of lengths 11 and 31: —×—, grade 1, T-11; —△—, grade 1, T-31; —◇—, grade 2, T-11; —⧖—, grade 2, T-31.

12.6 Real-time adaptive removal of ocular artefacts from human EEGs†

12.6.1 Introduction

12.6.1.1 Ocular artefacts and the need to remove them

The work described in this section is concerned with the on-line removal of ocular artefacts from the human electroencephalogram (EEG). The EEG is widely used in clinical and psychological situations, but it is often seriously contaminated by ocular artefacts (OAs) resulting from movements in the ocular systems (eyeball, eyelids, and so on) (Hillyard, 1974; Jervis *et al.*, 1985; Gratton *et al.*, 1983; Woestenburg *et al.*, 1983). In some cases, for example brain-damaged babies and patients with frontal tumours, it is difficult to distinguish between associated pathological slow waves in the EEG and ocular artefacts. The similarity between the OAs and signals of interest also makes it difficult to automate the analysis of the EEG by computer (Gotman *et al.*, 1975; Quilter *et al.*, 1977). A stimulus-related response, known as the contingent negative variation (CNV) (Hillyard, 1974), which has diagnostic usefulness for patients with Huntingdon's chorea (Jervis *et al.*, 1984), is very vulnerable to

†Adapted from Ifeachor *et al.*, 1986.

ocular artefacts (Hillyard, 1974). It is therefore necessary to remove the OA from the EEG so that the true EEG record can be analysed.

Although satisfactory OA removal is now possible off line, on-line OA removal has hitherto been unsatisfactory. The previously reported on-line methods required the cooperation of subjects which cannot always be guaranteed, involved time-consuming manual calibration, and at best can only deal with one type of OA as they assume a constant correction factor. In this section a new on-line system for removing OAs from the EEG signals is described which overcomes these disadvantages and offers additional advantages, such as flexibility. The system is based on the Motorola 68000 microprocessor and uses the numerically stable UD factorization algorithm which allows continuous adaptive OA removal. A description of the on-line algorithm, and of the hardware and software of the OAR system, is presented.

12.6.1.2 Methods for the removal and control of ocular artefacts

The problem of removing the OAs from the EEG is complicated by the similarity between them and some cerebral waves of interest, and by the spectral overlap between them. Of the various methods that have been proposed for removing or controlling the OA in the EEG signals, the electro-oculogram (EOG) subtraction methods are probably the best. In this chapter, the term EOG refers to the electric potential due to ocular movements measured between two skin electrodes placed close to the eyes. However, the various EOG subtraction techniques reported to date do not completely solve the problem, and new approaches are continually being developed (Hillyard, 1974; Jervis *et al*., 1980, 1985; Gratton *et al*., 1983; Woestenburg *et al*., 1983; Quilter *et al*., 1977; Barlow and Rémond, 1981; Girton and Kamiya, 1973; McCallum and Walter, 1968). All the techniques are based on the principle that the OA is additive to the background EEG. Thus, in discrete form

$$y(i) = \sum_{j}^{n} \theta_j x_j(i) + e(i) = \mathbf{x}^{\mathrm{T}}(i)\boldsymbol{\theta} + e(i) \qquad \textbf{(12.11)}$$

where

$$\mathbf{x}^{\mathrm{T}}(i) = [x_1(i) \quad x_2(i) \quad \ldots \quad x_n(i)]$$
$$\boldsymbol{\theta} = [\theta_1 \quad \theta_2 \quad \ldots \quad \theta_n]^{\mathrm{T}}$$

$y(i)$ and $x_j(i)$ are the samples of the measured EEG and the EOGs respectively, $e(i)$ is the 'true' EEG which may be regarded as an error term, and i is the sample number. θ_j are constants of proportionality which will be called the ocular artefact parameters, and n is the number of parameters in the model. The θ_j have also been called the transmission coefficients. $\mathbf{x}^{\mathrm{T}}(i)$ and $\boldsymbol{\theta}$ are the vectors of the EOGs and ocular artefact parameters, respectively, and ^{T} indicates transposition. If θ_j can be estimated then an estimate of $e(i)$ can be

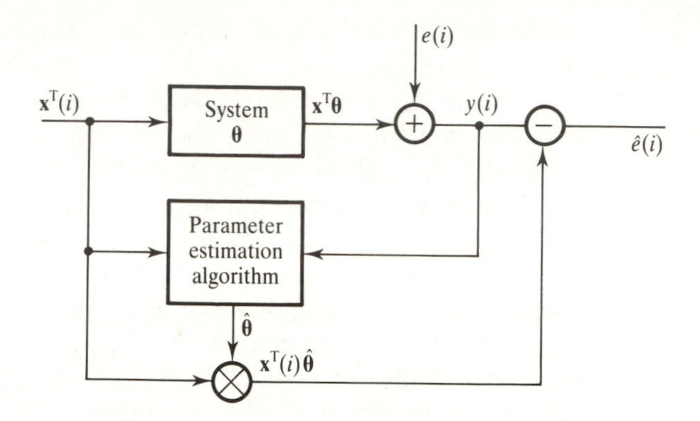

Figure 12.21 Block diagram representation of ocular artefact removal.

obtained as

$$\hat{e}(i) = y(i) - \sum_{j}^{n} \hat{\theta}_j x_j(i), \quad i = 1, 2, \ldots, m \tag{12.12}$$

where $\hat{\theta}_j$ are the estimates of θ_j and $\hat{e}(i)$ is the estimate of $e(i)$, and m is the number of samples used in the estimation. The problem then is one of estimating θ_j. This problem is illustrated in Figure 12.21. For a given type of ocular movement, the $\hat{\theta}_j$ are fairly constant but differ significantly between the different types of OAs (Gratton *et al.*, 1983; Corby and Kopell, 1972), although there is evidence to show that the $\hat{\theta}_j$ vary slowly, at least, even for a given type of OA (Gratton *et al.*, 1983). In general, there is no way of knowing the type of OA that will occur at a given time, and, because in many cases more than one type of OA occur simultaneously (Hillyard, 1974; Barry and Jones, 1965), $\hat{\theta}_j$ cannot be assumed constant so that it is best to estimate θ_j adaptively. The term 'adaptive' is used here to signify that the OA parameter estimates, and hence the OA removal should be automatically adjusted to changes in the OA. The various EOG subtraction techniques differ primarily in the way the θ_j are estimated, in the number of EOG signals that are used and the way these are measured (see Hillyard, 1974; Jervis *et al.*, 1985; Gratton *et al.*, 1983; Woestenburg *et al.*, 1983; Barlow and Rémond, 1981; Girton and Kamiya, 1973).

The EOG subtraction method can be carried out either on line, that is as the data is being acquired (Barlow and Rémond, 1981; Girton and Kamiya, 1973; McCallum and Walter, 1968), or off line (Jervis *et al.*, 1985; Gratton *et al.*, 1983; Woestenburg *et al.*, 1983). The main advantage of the off-line methods over previously reported on-line methods is that more sophisticated removal techniques can be employed. However, in applications requiring real-time processing and analysis the delay involved when off-line methods are employed is unacceptable. The trend in EEG signal processing is clearly towards real-time processing (see for example Harris, 1983; Barlow, 1979;

Takeda and Hata, 1985; Tomé *et al.*, 1985), and it is then necessary to remove the artefacts on line.

12.6.1.3 Off-line removal of ocular artefacts

In the off-line methods, estimates of θ are obtained by minimizing J, the sum of squares of the error term, that is $J = \sum_{i=1}^{m} e^2(i)$. This minimum leads to

$$\hat{\boldsymbol{\theta}}_m = [\mathbf{X}_m^{\mathrm{T}}\mathbf{X}_m]^{-1}\mathbf{X}_m^{\mathrm{T}}\mathbf{Y}_m \qquad \textbf{(12.13)}$$

where

$$\mathbf{Y}_m = [y(1) \quad y(2) \quad \dots \quad y(m)]^{\mathrm{T}}, \mathbf{X}_m = [\mathbf{x}^{\mathrm{T}}(1) \quad \mathbf{x}^{\mathrm{T}}(2) \quad \dots \quad \mathbf{x}^{\mathrm{T}}(m)]^{\mathrm{T}}$$
$$\hat{\boldsymbol{\theta}}_m = [\hat{\theta}_1 \quad \hat{\theta}_2 \quad \dots \quad \hat{\theta}_m]^{\mathrm{T}}, \mathbf{E}_m = [e(1) \quad e(2) \quad \dots \quad e(m)]^{\mathrm{T}}$$

This equation gives the ordinary least-squares (OLS) estimate of $\boldsymbol{\theta}$ which can be obtained using any suitable matrix inversion technique and forms the basis of the off-line OA removal methods. Having obtained $\hat{\boldsymbol{\theta}}_m$, estimates of the OA and hence the background EEG, $\hat{e}(i)$, can be obtained from Equation 12.12.

The OLS method described here can be extended to the multichannel case, where there is more than one EEG signal being corrected. Thus a system, with n EOG inputs and q measured EEG outputs, can be treated as q individual single-output subsystems, and the overall system identified in q separate ways.

12.6.1.4 On-line removal of ocular artefacts

A typical example of the previously reported on-line methods is depicted in Figure 12.22. In this method, due to Girton and Kamiya (1973), an initial calibration was made by adjusting the potentiometers while the subject moved his eyes repetitively in the vertical or horizontal planes until there was minimal amount of OA in the EEG trace. The device was then left at this setting during recording. A number of workers have used the method and found it unwieldy and inefficient in removing OA; see, for example, Gotman *et al.* (1975).

The previously reported on-line methods of removing ocular artefacts employed analogue methods (Barlow and Rémond, 1981; Girton and Kamiya, 1973; McCallum and Walter, 1968), requiring the cooperation of subjects which cannot always be guaranteed. They were, in general, inferior to the off-line methods in removing OAs (Jervis *et al.*, 1985; Gotman *et al.*, 1975), required the user to be familiar with the technique, and involved a time-consuming manual calibration of each EEG channel. In addition, these methods could be adjusted for only one type of OA.

In this section, an on-line system for removing ocular artefacts which overcomes the disadvantages mentioned will be described. It is a microcomputer-based system using a numerically stable on-line algorithm based on the efficient recursive least-squares technique.

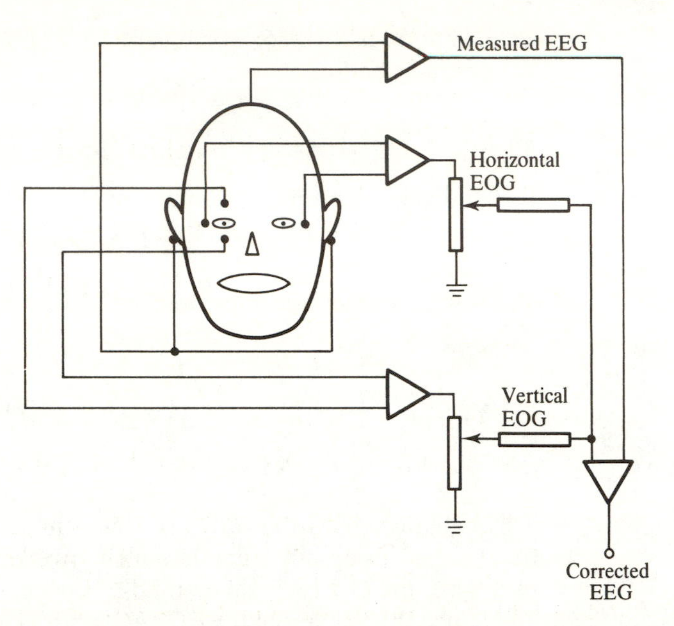

Figure 12.22 Typical example of previously reported on-line methods of removing artefacts (Girton and Kamiya, 1973).

12.6.2 On-line removal algorithms used in the OAR system

The computation of $\hat{\boldsymbol{\theta}}_m$ in Equation 12.13 requires the time-consuming computation of the inverse matrix. Clearly, the OLS approach is not suitable for real-time or on-line estimation for which on-line algorithms involving a fixed and restricted number of arithmetic operations and no direct matrix inversions are preferred. With the on-line algorithms, $\hat{\boldsymbol{\theta}}$ is updated at each sample point so that changes in ocular movements can be reflected in $\hat{\boldsymbol{\theta}}$ and the OA removal in this case can be viewed as adaptive filtering of OAs from the EEGs (Goodwin and Sin, 1984; Widrow *et al.*, 1975).

The two parameter estimation methods that are suitable for estimating $\boldsymbol{\theta}$ on line are the least-mean-square (LMS) (Goodwin and Sin, 1984; Widrow *et al.*, 1975) and recursive least-squares (RLS) algorithms (Peterka, 1975; Morris and Abaza, 1976; Young, 1974; Clarke, 1981). In terms of computation and storage, the LMS algorithm is more efficient than the RLS algorithm. In addition, it does not suffer from the numerical instability problem inherent in the RLS algorithm (see later). However, the RLS algorithm has superior convergence properties to those of the LMS algorithm and, for this reason, it is to be preferred (Goodwin and Sin, 1984; Cowan and Grant, 1984).

12.6.2.1 The LMS algorithm

The form of the LMS algorithm commonly used is (Goodwin and Sin, 1984; Widrow *et al.*, 1975; Cowan and Grant, 1984)

$$\hat{\boldsymbol{\theta}}(m+1) = \hat{\boldsymbol{\theta}}(m) + 2\mu\mathbf{x}(m+1)[y(m+1) - \mathbf{x}^{\mathrm{T}}(m+1)\hat{\boldsymbol{\theta}}(m)] \quad \textbf{(12.14)}$$

where $\hat{\boldsymbol{\theta}}(m)$ and $\hat{\boldsymbol{\theta}}(m+1)$ are the estimates of $\boldsymbol{\theta}$ at the mth and $(m+1)$th sample points, respectively, and μ is a constant that controls the rate of convergence and stability of the algorithm. For convergence μ should be within the limits

$$0 < \mu < 1/\lambda_{\max}$$

where $\lambda_{\max}$ is the maximum eigenvalue of the matrix $(\mathbf{X}_m^{\mathrm{T}}\mathbf{X}_m)$ of Equation 12.13. However, the convergence time of the algorithm is directly proportional to the ratio of the maximum to minimum eigenvalues of $(\mathbf{X}_m^{\mathrm{T}}\mathbf{X}_m)$ (Goodwin and Sin, 1984; Cowan and Grant, 1984), which can be very large when there is strong collinearity among the input variables, as is the case with the EOGs. However, the LMS algorithm has been widely used in biomedical applications to reduce noise or artefacts mainly because of its simplicity (Widrow *et al.*, 1975; Cowan and Grant, 1984).

12.6.2.2 The recursive least-squares algorithm

A suitable recursive least-squares (RLS) algorithm is obtained by exponentially weighting the data to remove gradually the effects of old data on the estimates (Peterka, 1975; Morris and Abaza, 1976; Young, 1974; Clarke, 1981). Thus

$$J = \sum_{i=1}^{m} \gamma^{m-i} e^2(i), \quad 0 < \gamma < 1 \quad \textbf{(12.15)}$$

Minimization of J with respect to the values of $\boldsymbol{\theta}$ leads to the following recursive least-squares algorithm (Peterka, 1975; Young, 1974; Clarke, 1981)

$$\hat{\boldsymbol{\theta}}(m+1) = \hat{\boldsymbol{\theta}}(m) + \mathbf{G}[y(m+1) - \mathbf{x}^{\mathrm{T}}(m+1)\hat{\boldsymbol{\theta}}(m)] \quad \textbf{(12.16a)}$$

$$\mathbf{P}(m+1) = \frac{1}{\gamma}\left[\mathbf{P}(m) - \frac{1}{\alpha}\mathbf{P}(m)\mathbf{x}(m+1)\mathbf{x}^{\mathrm{T}}(m+1)\mathbf{P}(m)\right] \quad \textbf{(12.16b)}$$

where

$$\alpha = \gamma + \mathbf{x}^{\mathrm{T}}(m+1)\mathbf{P}(m)\mathbf{x}(m+1)$$

$$\mathbf{x}^{\mathrm{T}} = [x_1(m+1) \quad x_2(m+1) \quad \ldots \quad x_n(m+1)]$$

$$\mathbf{G} = \mathbf{P}(m+1)\mathbf{x}(m+1) = \mathbf{P}(m)\mathbf{x}(m+1)/\alpha$$

and the argument m is used to emphasize the fact that the quantities are obtained at each sample point. γ is referred to as the forgetting factor and prevents the matrix $\mathbf{P}(m+1)$ from tending to zero (and $\hat{\boldsymbol{\theta}}(m+1)$ to a constant) with increased m, thus allowing the tracking of a slowly varying parameter. Typically, γ is between 0.98 and 1. Smaller values assign too much weight to the more recent data which leads to wildly fluctuating estimates.

There are, however, two main problems that may be encountered when the RLS algorithm is implemented directly. The first, referred to as 'blow-up', results if the signal is not 'persistently exciting' as, for example, when there is no ocular movement, leading to an exponential increase in the elements of **P** in Equation 12.16b. Thus

$$\lim_{m\to\infty} [P_{ij}(m+1)] = \lim_{m\to\infty}\left[\frac{P_{ij}(m)}{\gamma^m}\right] \to \infty \tag{12.17}$$

However, because of miniature ocular movements (which are always present), and other activities usually picked up in the EOG channels, this problem is not so serious in OA removal.

The second problem with the RLS is its sensitivity to computer roundoff errors, which results in a negative definite **P** matrix and eventually instability. For successful estimation, it is necessary that the matrix **P** be positive semi-definite which is equivalent to requiring in the off-line case that the matrix $(\mathbf{X}_m^{\mathrm{T}}\mathbf{X}_m)$ be invertible, but, because of differencing of terms in Equation 12.16b, positive definiteness of **P** cannot be guaranteed (Peterka, 1975; Clarke, 1981; Bierman, 1976). This problem is worse in multiparameter models, especially if the variables (EOGs in this case) are linearly dependent (Peterka, 1975) and when the algorithm is implemented on a small system with finite wordlength (Clarke, 1981).

The problem of numerical instability may be solved by suitably factorizing the matrix **P** such that the differencing of terms in Equation 12.16b is avoided. Such factorization algorithms are numerically better conditioned and have accuracies that are comparable with the RLS algorithms that use double precision (Bierman, 1976, 1977). Two such algorithms are the square root and the UD factorization algorithms. However, in terms of storage and computation the UD algorithm is more efficient, and is thus preferred. In fact, the UD algorithm is a square-root-free arrangement of the conventional square-root algorithm and thus shares the same properties as the latter.

12.6.2.3 UD factorization algorithm

In the UD method, $\mathbf{P}(m+1)$ is factored as

$$\mathbf{P}(m+1) = \mathbf{U}(m+1)\mathbf{D}(m+1)\mathbf{U}^{\mathrm{T}}(m+1) \tag{12.18}$$

where $\mathbf{U}(m+1)$ is a unit upper triangular matrix, $\mathbf{U}^{\mathrm{T}}(m+1)$ is its transpose and $\mathbf{D}(m+1)$ is a diagonal matrix. Thus, instead of updating **P**, its factors **U** and **D** are updated.

Using Equation 12.18, Equation 12.16b may be written as

$$\mathbf{P}(m+1) = \frac{1}{\gamma}\mathbf{U}(m)\left[\mathbf{D}(m) - \frac{1}{\alpha}\boldsymbol{v}\boldsymbol{v}^{\mathrm{T}}\right]\mathbf{U}^{\mathrm{T}}(m) \tag{12.19}$$

where

$$\boldsymbol{v} = \mathbf{D}(m)\mathbf{U}^{\mathrm{T}}(m)\mathbf{x}(m+1)$$

If the term in the square brackets is further factored into an upper triangular and diagonal matrices, such that

$$\bar{\mathbf{U}}(m)\bar{\mathbf{D}}(m)\bar{\mathbf{U}}^{\mathrm{T}}(m) = \mathbf{D}(m) - \frac{1}{\alpha}\boldsymbol{v}\boldsymbol{v}^{\mathrm{T}} \qquad \textbf{(12.20)}$$

where the bar is used to distinguish the **U** and **D** factors of $\mathbf{D}(m) - (1/\alpha)\boldsymbol{v}\boldsymbol{v}^{\mathrm{T}}$ from those of **P**, then

$$\mathbf{P}(m+1) = \frac{1}{\gamma}\mathbf{U}(m)\bar{\mathbf{U}}(m)\bar{\mathbf{D}}(m)\bar{\mathbf{U}}^{\mathrm{T}}(m)\mathbf{U}^{\mathrm{T}}(m) \qquad \textbf{(12.21)}$$

Comparing Equations 12.18 and 12.21, and noting that the product of upper triangular matrices is itself upper triangular and the symmetry in Equation 12.21, then

$$\mathbf{U}(m+1) = \mathbf{U}(m)\bar{\mathbf{U}}(m) \qquad \textbf{(12.22a)}$$

$$\mathbf{D}(m+1) = \frac{1}{\gamma}\bar{\mathbf{D}}(m) \qquad \textbf{(12.22b)}$$

Thus, the problem of updating $\mathbf{U}(m+1)$ and $\mathbf{D}(m+1)$ depends on finding appropriate recursive formulas for $\mathbf{U}(m)$ and $\mathbf{D}(m)$. Bierman (1976) has given an algorithm for updating $\mathbf{U}(m+1)$ and $\mathbf{D}(m+1)$ recursively for the Kalman filter, which uses the variance of the error term $e(i)$ but not γ. This algorithm has been trivially modified for the OA problem to incorporate γ instead, as has the presentation given above. The modified algorithm is given in the appendix.

The gain vector **G** obtained at step 10 of the appendix is used to update the parameter estimates, as indicated in Equation 12.16a. Thus, although $\mathbf{P}(m+1)$ can be obtained from the updated UD elements as in Equation 12.18, it is unnecessary to compute $\mathbf{P}(m+1)$ explicitly.

Some properties of the UD algorithm

To gain some insight into the UD algorithm, it is useful to write out the algorithm explicitly. Thus, for a two-parameter model, the algorithm of the appendix becomes

- step 1: $v_1 = x_1(m+1);\ v_2 = x_2(m+1) + U_{12}(m)x_1(m+1)$
- step 2: $b_1 = d_1(m)x_1(m+1);\ b_2 = d_2(m)v_2$
- step 3: $\alpha_1 = \gamma + b_1v_1 = \gamma + d_1(m)x_1^2(m+1)$
- step 4: $d_1(m+1) = d_1(m)/\alpha_1$
- step 5: $\alpha_2 = \alpha_1 + b_2v_2 = \alpha_1 + d_2(m)v_2^2$
- step 6: $U_{12}(m+1) = U_{12}(m+1) - b_1v_2/\alpha_1$

- step 7: $b_1 = b_1 + b_2 U_{12}(m)$

$$= d_1(m)x_1(m+1) + d_2(m)v_2 U_{12}(m)$$

- step 8: $d_2(m+1) = d_2\alpha_1/\gamma\alpha_2$

It is seen from step 3 that, provided that the starting values for the diagonal elements (that is $d_1(0)$ and $d_2(0)$) are positive, α_1 and hence $d_1(m+1)$ will always be positive. The same is true of α_2 (which is the same as α in Equation 12.16) and $d_2(m+1)$.

(1) *Positive definiteness of* ***P***. A matrix **P** is positive definite if and only if $\mathbf{x}^T\mathbf{P}\mathbf{x} > 0$, except when $x_1 = x_2 = \ldots = x_n = 0$ (Bajpai *et al*., 1973).

From Equations 12.16 and 12.18, $\alpha = \gamma + \mathbf{x}^T\mathbf{P}\mathbf{x} = \gamma + \mathbf{x}^T\mathbf{U}\mathbf{D}\mathbf{U}^T\mathbf{x}$, which for a two-parameter model becomes (see steps 3 and 5 above)

$$\alpha_2 = \alpha_1 + d_2 v_2^2 = \gamma + d_1(m)x_1^2(m+1) + d_2(m)v_2^2$$

so that, in this case,

$$\mathbf{x}^T\mathbf{P}\mathbf{x} = d_1(m)x_1^2(m+1) + d_2(m)v_2^2$$

Thus, it is seen that the sign of $\mathbf{x}^T\mathbf{P}\mathbf{x}$ depends on the signs of the diagonal elements (that is d_1 and d_2) which are always positive, as stated earlier, so that the matrix **P** satisfies the positive definiteness property. Thus, the UD factorization algorithm guarantees the positive definiteness property of **P**.

(2) *Blow-up problem*. If there is no data, that is $x_1 = x_2 = \ldots = x_n = 0$, then $\alpha_1 = \alpha_2 = \gamma$ (steps 3 and 5), so that both d_1 and d_2 will be continuously scaled by γ, which is less than unity, leading to an exponential increase in the diagonal elements. Thus it appears that the problem of blow-up is not eliminated by the matrix factorization as sometimes suggested. Other RLS schemes may be used to reduce the effects of blow-up (Goodwin and Sin, 1984), but, in the OA work, miniature ocular movements and other inherent systems noise ensure that the values of x never become zero indefinitely.

On-line OA removal using the UD and square-root algorithms has been simulated on a mainframe computer and these were found to give similar results to their off-line equivalents.

12.6.3 Hardware for the on-line ocular artefact removal system

In this section and the following, the on-line ocular artefact removal (OAR) system which uses the UD algorithm is described. First, target specifications for the system are set out. Next, a suitable system is described at system and block diagram level.

12.6.3.1 Instrument specification

The OAR system was designed with the following requirements in mind:

(1) compatibility with standard EEG machines;

(2) ability to provide continuous real-time OA removal in multichannel EEG signals (the OA removal should now be based on subjective criteria and should be adaptive);

(3) ability to output the corrected EEGs and/or the uncorrected EEGs and EOGs to the EEG machine, to allow instant comparison of the corrected and uncorrected EEGs;

(4) ability to avoid saturation, which would reduce the corrector's effectiveness, and the system should have some autoranging facility;

(5) the instrument should be suitable for use by unskilled persons.

These requirements can be met by a suitable microprocessor-based system that implements the UD algorithm. The use of a microprocessor-based instrument also offers the following advantages.

(1) Software-controlled design yields a very flexible system. Several OAR algorithms and models can be implemented on one system, and the models used in any application specified by the user.

(2) New models or ideas can be investigated by mere software modifications, without having to build a new instrument. Thus a software-controlled OAR system could be an excellent research aid.

(3) A programmed instrument allows the provision of housekeeping routines for self-checking, automatic calibration, reduction in overload problems, and so on. Data processing on the removal system may include digital filtering of the EOGs to reduce the effects of secondary artefacts (Hamer *et al.*, 1985).

12.6.3.2 System description

At the present, the OAR system can only process six channels of EEG and EOG signals, but is expandable to 20 channels (see later). The block diagram of the system is given in Figure 12.23. Each EEG/EOG signal from the auxiliary output of the EEG machine is first amplified and then bandlimited to 30 Hz by a lowpass filter which feeds a sample-and-hold circuit. The EEG/EOG signals are then simultaneously sampled at the positive transition of the sampling signal FS. Simultaneous sampling is employed to avoid the introduction of delays between corresponding time points. The negative transition of the signal FS then interrupts the processor, and signals the beginning of a cycle during which the 20 samples are sequentially selected by the multiplexer (MUX) and digitized by the analogue converter (ADC) under the control of the microprocessor (μP).

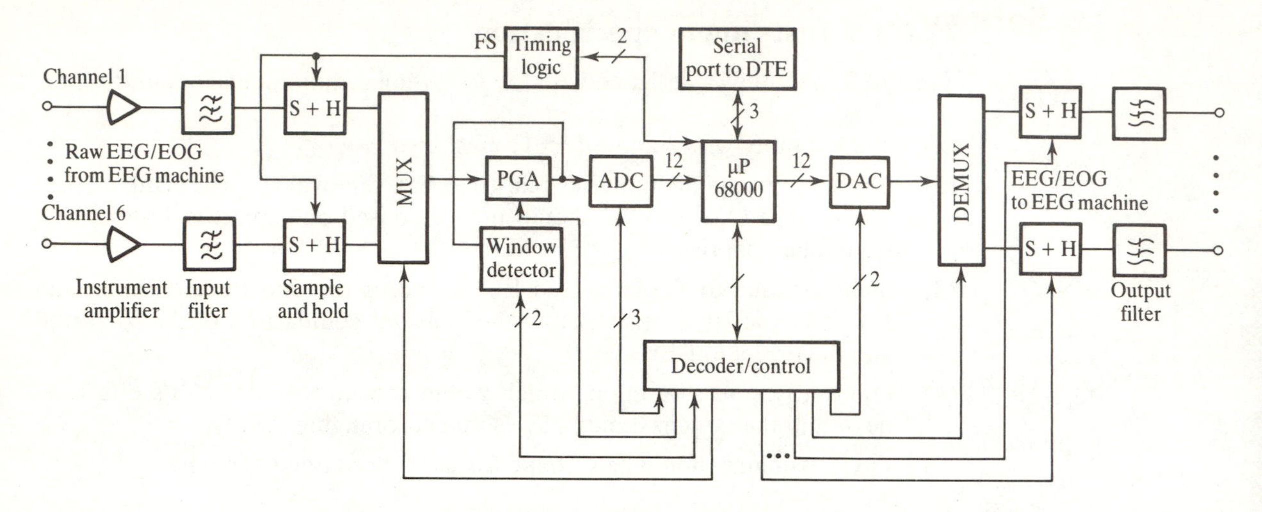

Figure 12.23 Block diagram of the microprocessor-based ocular artefact removal system.

The programmable gain amplifier (PGA) and the window detector are used to extend the dynamic range of the ADC to avoid overloading. ADC overload is a problem in the OA work and may lead to false parameter estimates. To avoid this, it is common practice to utilize only a fraction of the dynamic range of the ADC so that overloading is infrequent, and to discard all data in the region where overload occurred, but this can lead to wastage of data. The use of the PGA and window detector allows the gain to be changed dynamically. Thus, when the output of the PGA exceeds a predefined window limit, the gain of the PGA is set to a lower value which automatically halves the sample value and brings it to within the dynamic range of the ADC and thus avoids saturation. Account is taken of this before the digitized sample is saved in the memory.

The digitized samples are then processed by the OAR algorithm to obtain the corrected EEGs. The corrected EEG samples, together with the raw EEGs/EOGs if desired, are output to the auxiliary inputs of the EEG machine via the digital/analogue converter (DAC) and the associated network.

The multiplexer and the demultipliers are required to allow resource sharing by the input and output channels. An alternative approach would be to provide separate ADC and DAC for each channel. This approach reduces the system noise due to cross-talk between channels to a minimum, but was considered rather too expensive.

Each channel has a separate output sample-and-hold with a separate sampling signal line. The sample-and-hold is used to hold the analogue samples until the next analogue sample is obtained. This stretches the sample pulses and increases the signal power, but introduces aperture distortion, which is considered small in this case.

12.6.4 Software for the on-line ocular artefact removal system

The OAR system software consists of data acquisition and distribution routines, an on-line OA removal routine, software floating point arithmetic routines, and a supervising main program. The whole software occupies 3 kbytes of memory. The heart of the OAR system software, which is written in the 68000 microprocessor assembly language, is the UD algorithm described in Section 12.6.2. An overview of the software will be given here.

12.6.4.1 General description of software

The OAR system is interrupt driven. The interrupt signal is derived from the programmable timer module (PTM) on board the system controller, and has a frequency of 128 Hz (a frequency of 95 Hz was sometimes used to allow more time for computation, and the change in frequency was easily carried out in software). On interrupt, the OAR system software is used to acquire the EOG/EEG data, remove OAs from the EEG samples using the UD algorithm and output the corrected EEGs, and/or the raw data, to the EEG machine so that a paper chart record can be produced. The system operation is summarized in the flow-chart of Figure 12.24.

During the initialization phase, the user is invited by means of a visual display unit (VDU) to specify the various system constants, namely the number of EEG channels to be corrected for artefacts, the number of model parameters and hence the model that should be used in the removal algorithm, and the number of corrected EEG and/or raw EEG signals to be output to the EEG machine. Some EOG signals and parameter estimates may also be output to the EEG machine. These constants are checked and, if valid, are used to initialize the system. A default value is used for any constant that is invalid. After initialization, the program loops around endlessly until valid data is available (Figure 12.24(a)). The procedure described here applies to the prototype OAR system only. In future OAR systems, the user will not require a VDU, any selections would be made by push buttons, and the 'background' program would be replaced by a more useful housekeeping routine (see Section 12.6.3.1). A flag (data flag) is set in the interrupt service routine (see Figure 12.24(b)) each time the interrupt occurs, and this is the indication to the main program that valid data is now available. After the data has been acquired, the elements of the UD algorithm are updated, the OAs are removed from the EEGs, and the corrected EEGs and/or raw data are output to the EEG machine. Finally, the data flag is cleared to indicate that the current data samples have been successfully processed.

An error message is output to the VDU and the program halted if an interrupt occurs before the previous one has been serviced. This will normally occur if more parameters and/or EEG channels are specified than the software can process within the sampling interval of 8 ms, and prevents the accumulation of unserviceable interrupts and the eventual system failure.

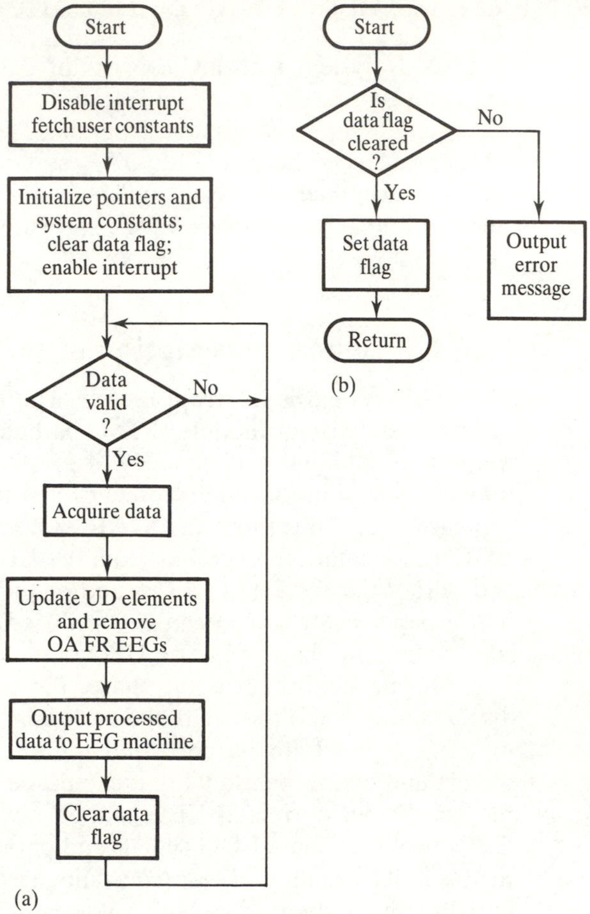

Figure 12.24 Ocular artefact removal system software: (a) main program; (b) interrupt service routine.

12.6.4.2 Arithmetic operations in the OAR system

Arithmetic operations in the OAR system are performed using the floating point (FP) format to take advantage of the increase in the dynamic range of the numbers that it affords, and to avoid the scaling problem associated with the fixed-point approach (see Section 12.7).

As speed is vital in this application, hardware floating point was considered the best approach, but it was found that hardware floating point devices available at the time were both expensive and too slow. Therefore, the number of EEG channels to correct was scaled down so that software FP could be used until fast FP devices became available.

A detailed description of the assembly language routines, which are of interest in their own right, in given in Section 12.7.

12.6.5 System testing and experimental results

A preclinical test on the OAR system was carried out at Freedom Fields Hospital, Plymouth. In the first phase, extensive tests of the reliability of the OAR system were made using six normal subjects. This phase was also a 'learning' phase, in which the behaviour of the system was understood and minor faults were identified. In the second phase, one uncooperative subject, whose EEGs contained spike and wave discharges, and two subjects, one of whom was uncooperative, whose EEGs contained slow waves, were used to assess how the OA removal process affected the spikes and slow waves. In all the tests carried out, the subjects were asked to perform ocular movement exercises which included repetitive and random blinking, vertical eye movements (VEMs) and horizontal eye movements (HEMs).

12.6.5.1 Experimental apparatus

Nine EEG signals were derived from several electrodes placed as shown in Figure 12.25. These were FP2–F4, F4–C4, C4–P4, FP1–F7, F3–C3, C3–P3, Fz–Cz, Cz referred to the right ear lobe or A2, and Cz–Pz. The EOG signals were derived from electrodes placed near the eyes, as shown in Figure 12.25(b).

The EEG and EOG signals were fed into an eight-channel EEG machine via the head box, and after amplification they were fed into the OAR system via a 37-way D-type connector. After removing the OA from the EEG signals, both the corrected and raw EEGs (with the means removed) and/or the EOGs were fed into the final amplifiers of the EEG machine, and thence to the paper chart for examination (see Figure 12.25(a)).

12.6.5.2 Models used in the tests

Several models were used in the tests, but only the three that gave the best results will be described here. Two of these utilized the EOGs derived from the electrode placement of Figure 12.25(b) and were also found to give the best OA removal in a previous study. These two models are as follows:

$$\begin{aligned} &\text{3D} \quad y(i) = \theta_1 \mathrm{VR}(i) + \theta_2 \mathrm{HR}(i) + \theta_3 \mathrm{HL}(i) + e(i) \\ &\text{4D} \quad y(i) = \theta_1 \mathrm{VR}(i) + \theta_2 \mathrm{HR}(i) + \theta_3 \mathrm{HL}(i) + \theta_4 \mathrm{HL}(i)\mathrm{HR}(i) + e(i) \end{aligned} \tag{12.23}$$

The third model, which will be called model 2H in line with earlier nomenclature, used the EOGs derived from the electrode placement pairs FP1–F7 and FP2–F8:

$$\text{2H} \quad y(i) = \theta_1 \mathrm{EOGR}(i) + \theta_2 \mathrm{EOGL}(i) + e(i) \tag{12.24}$$

It should also be mentioned that the selection switch on the EEG machine can be used to 'force' the OAR system to implement a variety of models by selecting the appropriate pairs of EOG electrodes to feed channels 1 to 4 (which are reserved for the EOGs).

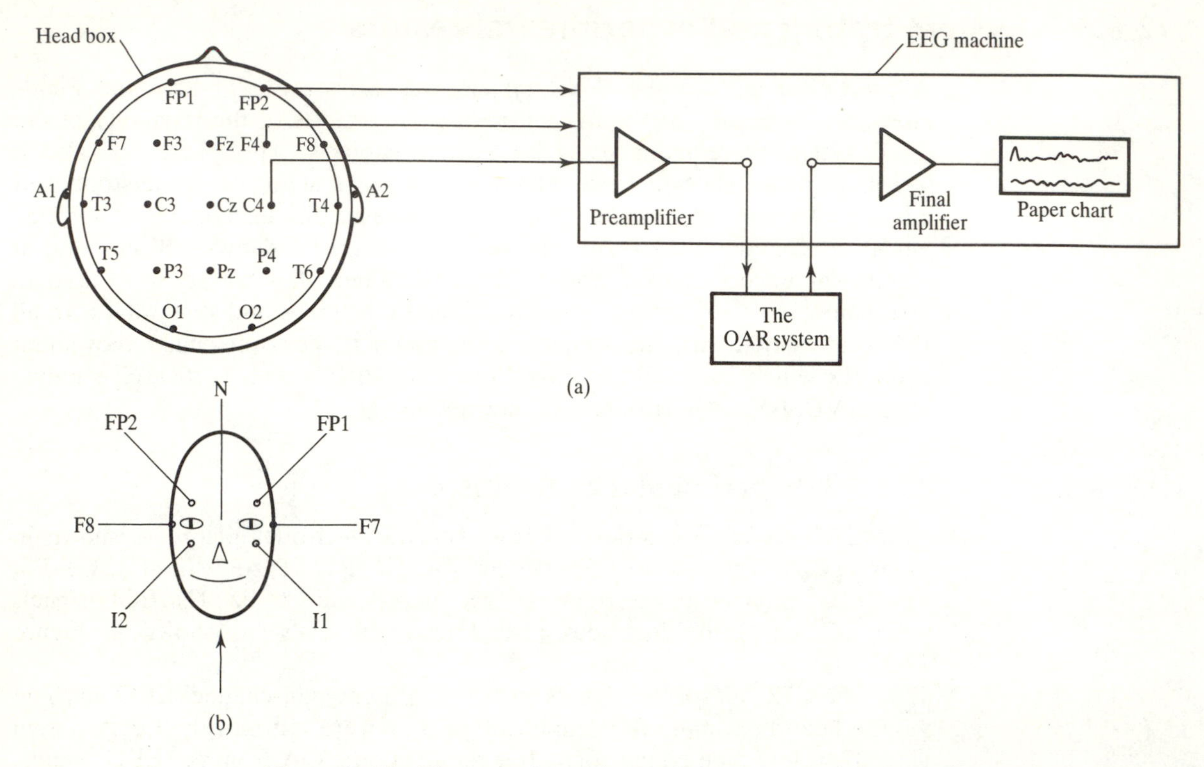

Figure 12.25 The experimental arrangement used. (a) Connection to the EEG machine; (b) EOG placements.

12.6.5.3 Experimental results

The OAR system was used to remove OAs from a number of other EEG electrodes (see Figure 12.25(a)) using model 2H (and occasionally models 3D and 4D). It was found that in all the cases studied the OA was satisfactorily removed, and this included all the frontal EEG channels where the OA was largest. Figure 12.26 gives results for four different electrodes (Fz–Cz, Cz–A2, F4–C4 and F3–C3) for blink experiments. In both Figures 12.26(a) and 12.26(b) two different EEG signals were simultaneously corrected for OA. Comparison of the corrected and uncorrected EEGs in both sets of figures showed that the system had satisfactorily removed the OAs (compare traces (v) and (vi) with traces (iii) and (iv) in each set of figures).

There was little OA contamination at the more posteriorly placed EEG electrodes, for example Cz–Pz and C4–P4. In these cases all the models performed equally well.

Figure 12.27 shows an EEG record containing epileptic spike and wave discharges as well as OAs obtained from an uncooperative subject. (In this and all other cases where the EEG contained abnormal waves, the raw EEG was fed direct to the final amplifier of the EEG machine as well as to the OAR system as described earlier, to allow an unbiased analysis of the records.)

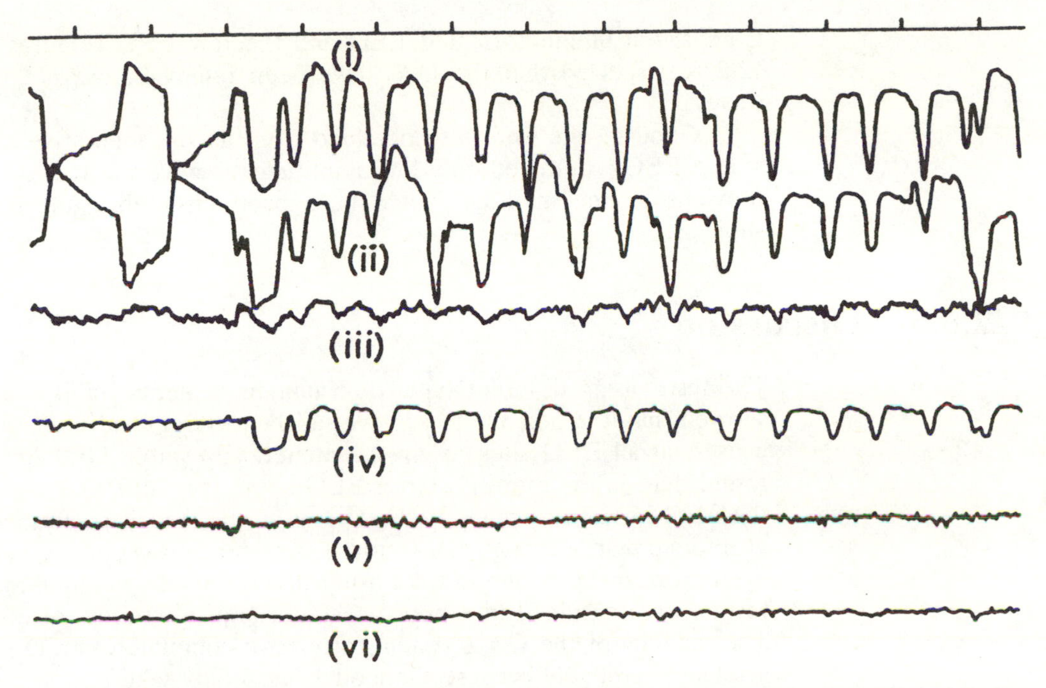

Figure 12.26 Simultaneous adaptive ocular artefact removal at a pair of EEG electrodes in a blink experiment. (i) and (ii) measured EOG signal for the right and left eyes; (iii) and (iv) measured EEGs at C_z–A_2 and F_z–C_z electrodes, (v) and (vi) corresponding EEGs with artefacts removed.

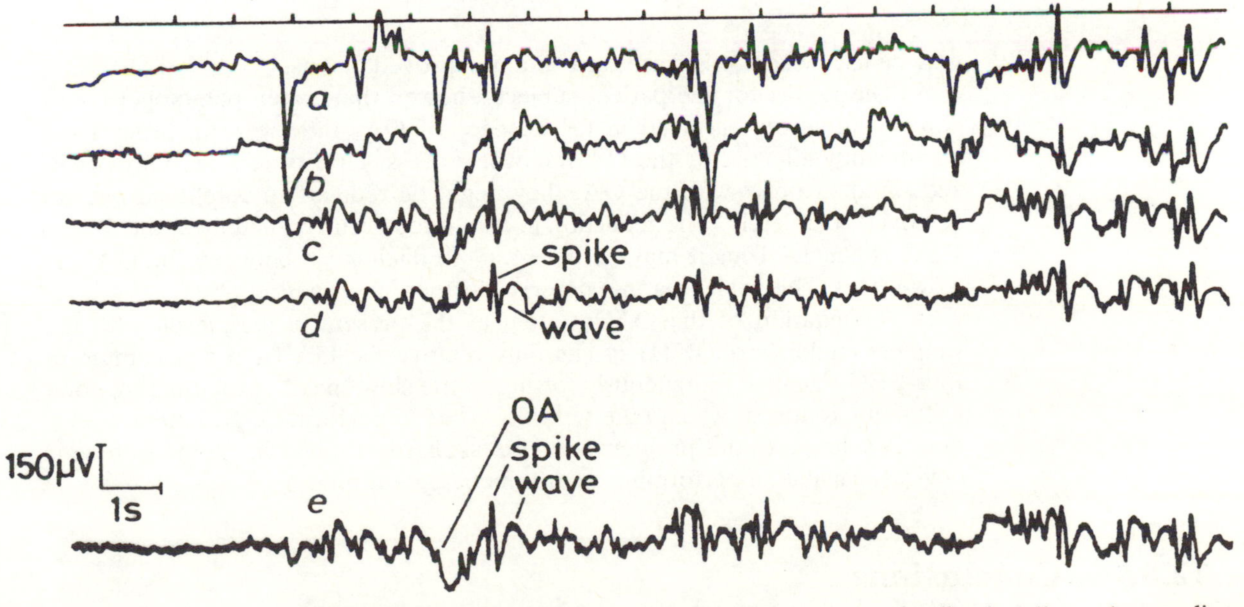

Figure 12.27 Ocular artefact removal in the presence of epileptic spike and wave discharges. (a) and (b) measured EOGs for the right and left eyes; (c) measured EEG; (d) EEG with artefacts removed; (e) raw EEG.

Comparison of the corrected EEG and the raw EEG (Figures 12.27(d) and 12.27(e)) showed that the OAs have been removed, but not the spike and waves.

Good results were also obtained from an uncooperative mental patient whose EEGs contained only low amplitude slow waves. It is noteworthy that previous on-line methods would have been unsuitable for the cooperative subjects.

12.6.6 Discussion

The tests, using different types of ocular movements (blinks, VEM, HEM), showed that it is possible to achieve satisfactory removal of OAs due to these causes, at all EEG sites, using the numerically stable UD algorithm. It was found that, in the more posterior EEG sites, very little OA contamination of the EEG was observed and in these cases all the models performed well. It was also found that, although satisfactory OA removal was obtained during vertical eye movement, OA due to rider artefact was not completely removed. Similar results were obtained in a previous study, and this result corroborates those. The inability of the OAR system to remove completely the OA due to rider artefact is probably because the models used only take into account simultaneous changes in the EOG and EEG. A dynamic model of the form (for a single input or EOG)

$$y(m) = \sum_{n=0}^{N-1} h(n)x(m-n) + e(n)$$

may be used where it is desired to obtain improved results.

The results for the patient subjects showed that, when pathological slow waves and spikes occurred in the absence of OA, they were in general not significantly affected by the OA removal process. However, when they occurred simultaneously with the OA, they might be reduced in amplitude but not removed completely. The reduction in amplitude occurred mainly at the frontal EEG channels. Thus it may be necessary to distinguish between the OA and slow waves. This is an area that deserves further investigation.

A limitation of the OAR system at the present is that, even with the simplest model (model 2H), it can only remove the OA from a maximum of four EEG signals simultaneously, owing to the slow speed of the floating point arithmetic routines, which take typically 70 μs to perform an arithmetic operation. A solution to this problem is to use fast hardware floating point arithmetic devices, capable of performing an arithmetic operation in 1 or 2 μs.

12.6.7 Conclusions

Preliminary results obtained with both normal and patient subjects showed that the OAR system gives satisfactory OA removal for blinks, vertical and horizontal eye movements and a bipolar EEG electrode montage. The use of the UD factorization algorithm and a software-controlled system enabled us to overcome the disadvantages of the previous on-line OA removal methods. Thus the OAR system is able to deal with multiple artefacts, does not need the cooperation of the subjects in a preliminary calibration and bases the removal criterion on a purely objective method. This system, which is the first of its kind, is compatible with standard EEG machines, so that it could be manufactured and sold as an accessory. However, the usefulness of the instrument can only be fully assessed after extensive clinical tests.

Although the OAR system was designed specifically for removing the OA from the EEG, it could be used as a general-purpose artefact (or noise) removal system in most physiological situations where both the contaminating and the contaminated signals can be separately measured. An example is the problem of measuring the foetal electrocardiogram (ECG) in the presence of large contaminating maternal ECG (Widrow *et al.*, 1975). Another example is the case where it is necessary to remove both the OA and the ECG artefacts from the EEG (Fortgens and De Bruin, 1983). In both applications the OAR system could be used to remove the artefacts, after possible minor modifications to the software and hardware. The OAR system, suitably programmed, could also be used in other signal processing applications; for example, digital filtering (Gotman *et al.*, 1977).

Since the design of the OAR system about 10 years ago considerable changes have taken place in DSP, especially in the area of DSP hardware. It is likely that, if the system was implemented now, a good DSP chip would be used. Today, a floating point digital signal processor such as the TMS320C30 or TMS320C40 would be appropriate for a time-critical system such as this. Despite the changes in DSP, the design principles and issues remain pertinent. We have emphasized these and hope the reader will benefit from our experience.

12.7 Fixed- and floating point implementation of DSP systems

12.7.1 Introduction

We have seen in previous chapters that the basic operations in DSP are multiplications, additions and delays (or shifts). For example, in digital FIR filtering the coefficients $h(k)$, $(k = 0, 1, \ldots, N-1)$, and the input data samples, $x(n)$, $(n = 0, 1, \ldots)$ are multiplied and the products added as follows:

$$y(n) = \sum_{k=0}^{N-1} h(k)x(n-k) = h(0)x(n) + h(1)x(n-1) + \ldots + h(N-1)x[n-(N-1)] \quad \textbf{(12.25)}$$

In practice, the arithmetic operations involved in DSP (such as those indicated above) are often carried out using fixed-point or floating point arithmetic (Rabiner and Gold, 1975). Other types of arithmetic are sometimes used, for example block floating point arithmetic, which seeks to combine the benefits of the two above. Fixed-point arithmetic is the most prevalent in DSP work because it leads to a fast and inexpensive implementation, but it is limited in the range of numbers that can be represented, and is susceptible to problems of overflow which may occur when the result of an addition exceeds the permissible number range (for example large-scale limit cycles in IIR filters and overload in high order FFTs). To prevent the results of arithmetic operations going outside the permissible number range, the operands are scaled. Such scaling degrades the performance of DSP systems, that is it reduces the signal-to-noise ratio achievable.

Floating point arithmetic is preferred where the magnitudes of variables or system coefficients vary widely (Flores, 1963). It allows a much wider dynamic range, and virtually eliminates overflow problems. Further, floating point processing simplifies programming. DSP algorithms developed on large machines, for example on personal or mainframe computers, in high level language can be implemented directly in DSP hardware with little change to the core algorithms. However, floating point arithmetic is more expensive and often slower, although high speed digital signal processors with built-in floating point processor (such as the Texas Instruments TMS320C30) are becoming widely available. Efficient floating point software routines are also available in the open literature (Texas Instruments, 1986). Thus the disparity in price and speed between fixed and floating point approaches has diminished significantly.

Increasingly, DSP techniques are being used in applications where both wide dynamic range and high precision are required. Fixed-point digital signal processors with large wordlengths (24 bits or more) may satisfy such requirements, but floating point processing provides a simpler and a more natural way of catering for these cases.

Application areas where wide dynamic range and high precision are required and the use of floating point arithmetic is necessary or desirable are many. An example is in real-time parametric equalization of digital audio signals using digital filters, where the values of filter coefficients vary widely as the user sweeps through the audio band and adjusts the equalizer parameters. In certain signal processing tasks (for example spectrum analysis in radar and sonar, seismology or biomedicine) a need often arises to resolve very low level components in a signal of wide dynamic range. In such cases both wide dynamic range and high precision are required. Other application examples include high resolution graphics stations and general engineering computations. Desirable maximum dynamic range and accuracy requirements in a number of applications are summarized in Table 12.2 (Weitek, 1984).

In the next few sections, we will discuss the basic concepts of the two types of arithmetic (that is fixed- and floating point) and present floating point implementations of two practical DSP systems. Several examples of fixed-point implementations have already been given in previous chapters.

Table 12.2 Dynamic range and accuracy requirements.

	Dynamic range (bits)	*Accuracy* (bits)
Noise cancelling	32	20
Radar processing	32	20
Broadcast quality picture processing	20	20
Image processing	30	20
Medical spectrum analysis	20	20
Seismic data processing	70	20

Table 12.3 A comparison of two's complement and offset binary number systems for a 4-bit wordlength.

Number	*Decimal fractions*	*Two's complement*	*Offset binary*
7	7/8	0111	1111
6	7/8	0110	1110
5	5/8	0101	1101
4	4/8	0100	1100
3	3/8	0011	1011
2	2/8	0010	1010
1	1/8	0001	1001
0	0	0000	1000
−1	−1/8	1111	0111
−2	−2/8	1110	0110
−3	−3/8	1101	0101
−4	−4/8	1100	0100
−5	−5/8	1011	0011
−6	−6/8	1010	0010
−7	−7/8	1001	0001
−8	−1	1000	0000

12.7.2 Fixed-point number system

12.7.2.1 Fixed-point representation

In DSP, variables are often represented as fixed-point, two's complement fractions; see for example Table 12.3. In this representation, the binary point is to the right of the MSB (most significant bit) which is also the sign bit. Each number lies in the range from -1 to $1-2^{-B}$, where B is the number of bits used to represent the number. A common representation in DSP is the so-called Q15 format which uses 16 bits (1 sign bit and 15 fractional bits):

0 110 0000 0000 0000
| binary point

Two's complement positive numbers are in natural binary form; see Table 12.3. A negative number is formed from the corresponding positive number by complementing all the bits of the positive number and then adding 1 LSB. For example, the two's complement representation of −3/8 is obtained from 3/8 (that is, 0011) as 1100 + 0001 = 1101.

When the input to a DSP system is from an ADC (analogue-to-digital converter) the data fed to the digital processor may be in offset binary form. Similarly, the output of the DSP system may need to be converted to offset binary if it feeds a DAC (digital-to-analogue converter). Conversion from offset binary to two's complement representation is trivially achieved by complementing the MSB of the offset binary code. For example, in Table 12.3, the offset binary code 1111, that is 7/8, is easily converted to two's complement code (0111) by complementing the MSB. In practice, the DSP chip bus is often wider than the ADC resolution. In this case, after conversion to two's complement, the sign bit is extended to fill the remaining spaces to the left. For example, the code (1111 1101) which has been two's complemented becomes (1111 1111 1111 1101) after sign extension.

12.7.2.2 Accuracy of fixed-point representation

In fixed-point, 2's complement representation, if each number is represented by B bits then a maximum of 2^B different numbers can be represented, with adjacent numbers separated by approximately 2^{-B}. It is useful to know the accuracy to which we can represent each number compared with decimal representation.

Given a decimal fraction, X, consisting of d digits, its accuracy is $\pm 0.5 \times 10^{-d}$. If we represent the same number in binary with B bits, its accuracy now becomes $\pm 0.5 \times 2^{-B}$. To retain the same accuracy for the two representations then:

$$0.5 \times 10^{-d} = 0.5 \times 2^{-B}, \text{ that is } \quad B = d \log_2 10 \simeq 3.3d \text{ bits} \qquad \textbf{(12.26)}$$

For example, suppose the decimal number 0.234 56 is to be represented in binary; then we require 3.3 × 5 = 17 bits to represent it as accurately as before. Table 12.4 summarizes the relationship between the number of bits in a binary system and their accuracies in decimal digits or places.

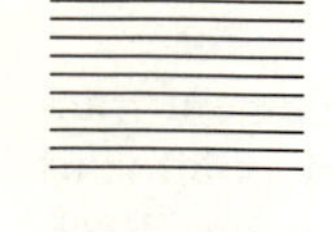

Example 12.1

Represent the decimal number, 0.956 24, as

(1) a Q3 number, and

(2) a Q4 number

Compare the errors in the two cases.

(3) Estimate the number of bits required to represent the decimal number above to retain the same accuracy.

Table 12.4 Relationship between the number of bits and accuracies in decimal digits.

Number of bits	*Accuracy (number of decimal digits)*
7	2.1
8	2.4
10	3
12	3.6
14	4.2
15	4.5
16	4.8
18	5.4
20	6.1
23	7.0
24	7.3
64	19.4

Solution

(1) A Q3 number is a 2's complement number with 1 sign bit and 3 fractional bits. To convert the decimal number to a Q3 format we simply multiply it by 2^3 and then round the product to the nearest permissible integer: $0.956\,24 \times 2^3 = 7.649\,92$, which is rounded to $7 = 0111$ (the highest permissible number).

(2) In this case, we represent the number using 1 sign bit and 4 fractional bits: $0.956\,24 \times 2^4 = 15.299\,84$, which is rounded to $15 = 01111$.

The errors in representing the number in parts (1) and (2) are, respectively, $0.649\,92/8 = 0.081\,24$ and $0.299\,84/16 = 0.018\,74$. The error in representing the number is often referred to as the coefficient quantization error.

(3) From Equation 12.26, we require $3.3 \times 5 = 16.5$ bits $\simeq 17$ bits.

12.7.2.3 Fixed-point arithmetic: multiplication

In fixed-point multiplication, use is made of the fact that the product of two fractions is also a fraction and that of an integer is also an integer. We will illustrate with an example.

Example 12.2

Find the square of 0.5625 using fixed-point two's complement arithmetic. Assume a Q4 format.

Solution

```
           0 1001 = 0.5625
           0 1001 = 0.5625
      0 0000
       0100 1
        000 00
         00 000
           0 1001
  ________________
  00.0101 0001  <-- this bit is lost when product is quantized
    ^
  binary point
```

After shifting left, to remove the extra sign bit, and rounding, we obtain the final answer: $0\,0101 = 0.25 + 2^{-4} = 0.3125$, instead of 0.316 406 25.

Example 12.2 shows that an additional sign bit is created after multiplication and that the product of two 5-bit numbers is 10 bits long, so that the result should be truncated or rounded to 5 bits before it can be saved in memory. In general, the product of two B-bit numbers is $2B$ bits long. For example, the 10-bit result is shifted left once to remove the extra sign bit and then rounded to 0.0101. The ability to round (or truncate) results as we have just done is one of the major attractions of two's complement fractional arithmetic as it means that overflow never occurs in multiplications. Such rounding (or truncation), however, introduces errors into the signal which can lead to instability or some undesirable side-effects in DSP systems with feedback.

12.7.2.4 Fixed-point arithmetic: addition

Addition of two fixed-point fractions is more difficult than multiplication. This is because the operands to be added must be in the same Q format and attention must be paid to the possibility of an overflow.

Example 12.3

Find the sum of the following two's complement numbers 0001 1001 and 0110 1101 0111 1101.

Solution

The operands are first expressed in the same Q format and then added:

```
0110 1101 0111 1101
0001 1001 0000 0000

1000 0110 0111 1101
|  overflow
```

One way to correct for the overflow is to shift the result one place to the right and then set an exponent flag. Thus, the answer becomes

```
1    0100 0011 0011 1110
↑
exponent
flag
```

Another alternative is to represent the result using double precision or to provide sufficient headroom to allow for growth due to overflow.

Example 12.4

Assuming a register length of 4 bits (1 sign bit and 3 data bits), find the sum of

(1) −0.25 and 0.75, and
(2) 0.5, 0.75, −0.5

Solution

From Table 12.3, $0.25_{10} = 0.010_2$, $0.5_{10} = 0.100_2$ and $0.75_{10} = 0.110_2$.

(1)

```
   0.75       0.110
  -0.25       1.110
   0.50     1 0.100
```

Answer: 0.100

(2)

```
   0.5       0.100
  +0.75      0.110
   1.25      1.010 ← partial sum

  -0.50      1.100
   0.75 1    0.110 ← final sum
```

Answer: 0.110

12.7.3 Floating point number system

12.7.3.1 Floating point representation

A binary floating point number X is represented as the product of two signed numbers, the mantissa M and the exponent, E:

$$X = M \cdot 2^E \tag{12.27}$$

where 2 is the base of the binary system.

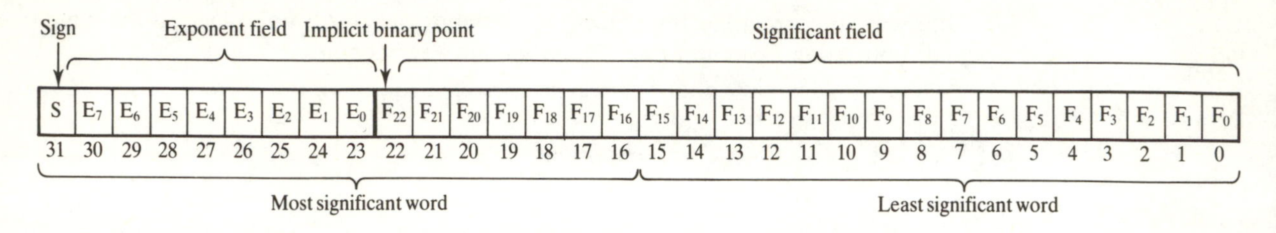

Figure 12.28 Floating point representation (IEEE single precision).

The exponent determines the range of numbers that can be represented, the mantissa the accuracy of the numbers. For example, if the exponent and the mantissa are represented by 8 and 16 bits, respectively, the range of the floating point numbers that can be represented in this simple case is from 0.5×2^{-128} to $1 - (2^{-15}) \times 2^{128}$. Of the 16 bits used to represent the mantissa, one bit is the sign bit and the least significant bit may be of a doubtful accuracy owing to rounding effects. Thus the accuracy of the floating point numbers is 1 in 2^{14} (0.61×10^4), that is about 4 decimal digits.

12.7.3.2 IEEE floating point format

One of the most widely used binary floating point system is the IEEE 754 Standard (Patterson and Hennessy, 1990; IEEE, 1985). The format for single precision is shown in Figure 12.28. In this case, floating point (FP) numbers with exponents in the range $0 < E < 255$ are said to be normalized.

The decimal equivalent, X, of a normalized IEEE FP number is given by

$$X = (-1)^s(1 \cdot F)2^{E-127}$$

where

- F is the mantissa in 2's complement binary fraction represented by bits 0 to 22,
- E is the exponent in excess 127 form, and
- $s = 0$ for positive numbers, $s = 1$ for negative numbers.

Two important features of the IEEE floating point format are the assumed 1 preceding the mantissa and the biased exponent.

Example 12.5

(1) For the FP number

```
sign
↘0 1000 0011.1100 . . . 00000
 |←exponent→|←——mantissa——→|
```

the exponent is $1000\,0011 = 131$, the mantissa is $0.1100\ldots = 0.75$, and $s = 0$. Therefore $X = (1.75)2^{131-127} = 1 \times 1.75 \times 24$.

(2) For the FP number

```
sign
↘1 0000 1111.0110 . . . 00000
 |←exponent→|←——mantissa——→|
```

The exponent is $0000\,1111 = 15$, the mantissa is $0.0110\ldots = 0.375$, and $s = 1$. Therefore $X = (1.375)2^{15-127} = -1 \times 1.375 \times 2^{-112}$.

12.7.3.3 Floating point arithmetic: additions and multiplications

If X_1 and X_2 are two floating point numbers to be added, where $X_1 = M_1 \times 2^{E_1}$ and $X_2 = M_2 \times 2^{F_2}$, then their sum X is given by

$$X = M \times 2^E$$

where

$$M = M_1 + M_2 \times 2^{E_1 - E_2};\ E = E_1, \text{ assuming that } X_1 > X_2 \qquad \textbf{(12.28)}$$

Before two floating point numbers are added, their exponents must be made equal. This is called alignment, and involves shifting right the mantissa of the smaller operand and incrementing its exponent until it equals that of the larger operand.

If X_1 and X_2 are two properly normalized FP numbers to be multiplied, where

$$X_1 = M_1 \times 2^{E_1},\ X_2 = M_2 \times 2^{E_2} \qquad \textbf{(12.29)}$$

then their product X is given by

$$X = M \times 2^E$$

where

$$M = M_1 \times M_2,\ E = E_1 + E_2$$

Thus the mantissas are multiplied and their exponents added. Since M_1 and M_2 are both normalized then their product, M, will be in the range $0.25 < M < 1$. Thus the product, M, cannot overflow but may not be properly normalized (mantissa underflow).

Example 12.6

(1) Find the sum of the two numbers A and B, where $A = 9.985 \times 10^4$ and $B = 5.6756 \times 10^2$.

(2) Find the product of the two numbers A and B, where $A = 2.75 \times 10^{-16}$ and $B = 4.5 \times 10^{10}$.

Solution

(1) First, the exponents of the two numbers are compared and if they are not equal, the mantissa of the operand with the smaller exponent is then shifted so that the two exponents are equal:

$$5.6756 \times 10^2 = 0.056\,756 \times 10^4$$

The mantissas are then added to give

$$M = 9.985 + 0.056\,756 = 10.041\,756; \; E = 10^4$$

and the sum is $10.041\,756 \times 10^4$. The sum is next normalized by moving the decimal point of the mantissa and then adjusting the exponent (if necessary) to give the final answer:

$$1.004\,175\,6 \times 10^5$$

(2) The mantissas are multiplied and the exponents added:

$$M = 2.75 \times 4.5 = 12.375; \; E = -16 + 10 = -6$$

giving the product $A \times B = 12.375 \times 10^{-6}$. The product is then normalized, where we have assumed that a properly normalized floating point number is less than 10, by adjusting the exponent and moving the position of the decimal point in the mantissa to obtain

$$A \times B = 1.2375 \times 10^{-5}$$

12.7.3.4 Floating point arithmetic routines and devices

Several floating point arithmetic routines and processors are available which can be used for DSP work. A particularly useful set of routines is available in Texas Instruments (1986). A summary of some features of current FP devices is given in Table 12.6. Both the C30 and the Weitek devices support the IEEE format.

Table 12.6 Timing for some floating devices.

	Multiply (μs)	*Add* (μs)	*FP word* (bits)
TMS32010 software	8.4	17.2	32
TMS32020 software	7.8	15.4	32
TMS320C30 (hardware)	0.06	0.06	32
Weitek 1032/1033	1	1	32

12.7.3.5 Errors in floating point arithmetic

In floating point arithmetic, roundoff errors occur in both addition and multiplications whereas in fixed point roundoff errors are only possible in multiplication. However, unlike fixed-point arithmetic overflow is very unlikely in floating addition because of the very wide dynamic range that is available: the more bits in the exponent the wider the dynamic range.

12.8 Equalization of digital audio signals†

Equalization of audio signals is an important functional requirement of mixing consoles used in many professional and semiprofessional audio applications, for example in studio recording, sound reinforcement in public address systems, and broadcasting. An audio equalizer is basically a set of filters with adjustable frequency responses which is used to shape the spectrum of the audio signals in a desired manner. In a traditional mixing console, equalization of audio signals is achieved by analogue filtering, but the trend is towards an all-digital mixing console because this offers improved sound quality and potential reduction in production costs in the future. In an all-digital mixer, the analogue filters would be replaced by equivalent real-time digital filters. We describe here a real-time semiparametric equalizer for audio signals implemented using a high speed floating point digital signal processor, the TMS320C30.

A standard parametric equalizer allows the user to sweep to a specific frequency in the audio band and to adjust the level of the audio signal at a single frequency or a range of frequencies. Three basic filter types are used.

- *Bell filter*. This allows the user to boost or attenuate a particular frequency in the audio band. The bell filter is basically a bandpass filter with adjustable gain, Q factor, and centre frequency. The centre frequency may vary in the range 20 Hz–16 kHz, the Q value between 0.5 and 3, and the gain in the interval ±15 dB. Figure 12.29(a) illustrates the bell characteristic.

†This section is based on the final year project of one of our former students, Mr Robin Clark.

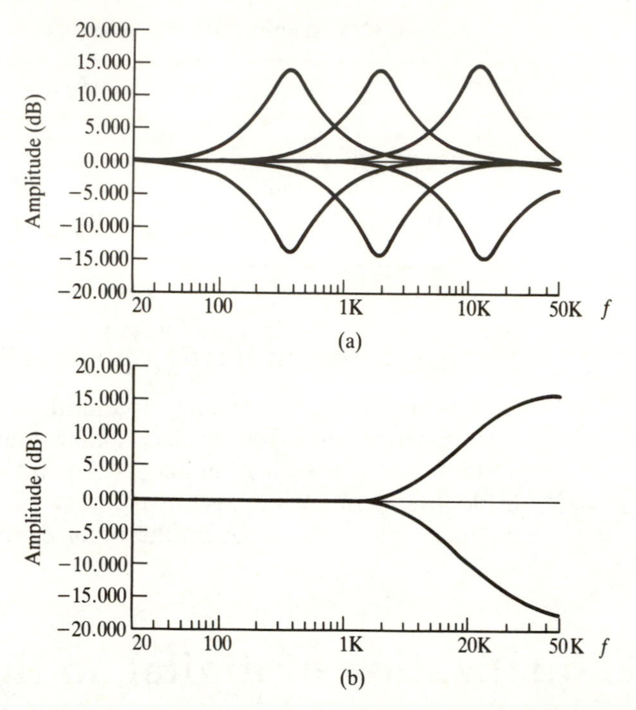

Figure 12.29 (a) Amplitude response of the bell filter for various centre frequencies and bandwidths. (b) Amplitude for high frequency shelving.

- *Shelf filter*. This allows adjustment of the gain and cutoff frequency of the equalizer over a range of frequencies at either the low or the high frequency end of the audio band. Low frequency shelf filters are used to boost or attenuate a band of low frequencies, for example between 20 and 500 Hz, whereas high frequency shelf filters are used to boost or attenuate a band of high frequencies, for example between 1.6 kHz and 16 kHz. The familiar treble or bass controls in the home hi-fi systems are essentially shelf filters with fixed responses. Figure 12.29(b) illustrates typical responses for a shelf filter.
- *Pass filters*. These are essentially lowpass and highpass filters with fixed cutoff frequencies and are used to remove low and/or high frequency noise from the audio signal.

Full equalization is achieved by the combined effects of the basic filters. In analogue equalizers, the characteristics of the filters (gain, centre frequencies, Q factor, and so on) are adjusted by the user via interactive controls (variable resistors). In a digital implementation, the adjustment is achieved by altering the digital filter coefficients in real time in response to changes in equalizer parameters.

Analysis of a typical analogue parametric equalizer showed that each filter type described above can be viewed as a Butterworth filter, with s-plane transfer functions of the following forms: for the bell filter,

$$H(s) = \frac{s^2 + As + \omega_n^2}{s^2 + Bs + \omega_n^2} \tag{12.30a}$$

for the low frequency shelf filter,

$$H(s) = \frac{s^2 + 2As + A^2}{s^2 + 2Bs + B^2} \tag{12.30b}$$

and for the high frequency shelf filter,

$$H(s) = \frac{A^2s^2 + 2A\omega_n^2 s + \omega_n^4}{B^2s^2 + 2B\omega_n^2 s + \omega_n^4} \tag{12.30c}$$

where

$$A = \frac{1}{(R + r)C}\left(3 - \frac{1}{K}\right)$$

$$B = \frac{1}{(R + r)C}\left(3 + \frac{1}{K}\right)$$

K is the gain/attenuation parameter, r is the frequency shifting parameter, and R and C are constants.

To achieve a performance similar to the traditional analogue equalizer, the analogue filters above are replaced by equivalent digital filters, by transforming each of the s-plane transfer functions given above using the bilinear z-transform technique (see Chapter 7). Each of the resulting z-transfer functions has the form

$$H(z) = \frac{az^2 + bz + c}{dz^2 + ez + f} \tag{12.31}$$

where

$$a = P^2 + AP + \omega_n^2,\ b = 2\omega_n^2 - 2P^2,\ c = P^2 - AP + \omega_n^2$$

$$d = P^2 + BP + \omega_n^2,\ e = 2\omega_n^2 - 2P^2,\ f = P^2 - BP + \omega_n^2$$

$$P = \frac{\omega_n}{\omega_p}, \quad \omega_p = \frac{2}{T}\tan\left(\frac{\omega_n T}{2}\right)$$

Simulation studies showed that floating point arithmetic should be used to cater for the large variations in the range of values of the filter coefficients as the equalizer parameters are adjusted over the audio range. Floating point arithmetic is also attractive in this application to make it easier to recompute the filter coefficients, via the BZT, in 'real time' to permit on-line adjustment of the equalizer characteristics.

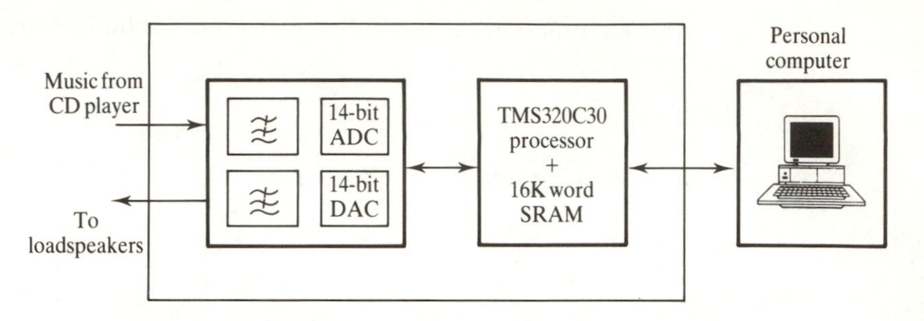

Figure 12.30 A simplified block diagram of the TMS320C30 EVM.

The equalizer was implemented in the PC (personal computer)-based TMS320C30 evaluation module (EVM), which contains a TMS320C30 processor, a 14-bit ADC/DAC module and a software development package. A simplified block diagram of the TMS320C30-based parametric equalizer is depicted in Figure 12.30. The analogue audio signal (for example, from a compact disc player) is digitized to 14 bits at a rate of about 18.9 kHz and passed to the C30 where it is digitally filtered by a bell, shelf and/or pass filter. The keyboard is used to adjust the parameters of the equalizer (gain, frequency, type of filter used). The VDU (visual display unit) displays dynamically the frequency response of the equalizer.

Several processes take place within the EVM and the PC. The C30 processor executes the filtering required to achieve equalization, and recomputes the filter coefficients for the equalizer when the user adjusts the equalizer parameters. The coefficient recalculation occurs in the background whereas the filtering is interrupt driven. The programs are written in ANSI C language, and the arithmetic operations are carried out using floating point (24-bit mantissa and 8-bit exponent). Fragments of the C codes for the filter and coefficient calculation are given in the appendix.

The performance of the equalizer was evaluated for various parameter settings using music from a CD player and was found to be quite effective. A restriction at present is that the EVM allows only a maximum of about 18.9 kHz sampling rate, making it impossible at this stage to use the equalizer at the professional audio rate of 44.1 kHz.

12.9 Adaptive ocular artefact filter

In Section 12.6, we discussed the design of a DSP system, the OAR, for artefact removal in neurophysiological signals using a real-time adaptive filter. The arithmetic operations in the adaptive filter (see Equations 12.32 and 12.33) were carried out in the OAR system in software floating point scheme. In this section, we describe the floating point arithmetic routines used. Although the routines were implemented for the Motorola 68000 microprocessor, details of the software floating point routines and the concepts on which they are based are of interest in their own right.

The two key equations in the adaptive artefact removal filter are

$$e(m+1) = y(m+1) - \sum_{k=1}^{n} \theta_m(k) x_{m+1}(k) \qquad \textbf{(12.32)}$$

$$\theta_{m+1}(k) = \theta_m(k) + Ge(m+1), \quad k = 1, 2, \ldots, n \qquad \textbf{(12.33)}$$

where $y(m+1)$ are the measured EEG sample at the discrete time $m+1$, $e(m+1)$ are the data sample of an estimate of the corrected EEG, $x_{m+1}(k)$, $k=1, 2, \ldots$, are the EOG data samples at the discrete time $m+1$, and $\theta_{m+1}(i)$, $i=1, 2, \ldots$, are the ocular artefact parameter estimates. The factor, G, is obtained from the factorization algorithm given the appendix.

The input and output data to and from the system were 12-bit integers (from 12-bit ADC and DAC), and so integer–floating point conversions are necessary.

12.9.1 Software floating point arithmetic routines

Floating point arithmetic was necessary in the OAR to exploit the wide dynamic range it offers and to avoid problems of scaling inherent in fixed point arithmetic. Further, the development of the OA removal algorithms was carried out on a mainframe using floating point arithmetic, and so its implementation in floating point follows naturally.

In the software floating point routines, the exponent and the mantissa are represented, respectively, by 8 and 16 bits. This is a suitable compromise between speed and accuracy. A similar choice of exponent and mantissa lengths has been used in a control application and was found to be adequate.

The range of the floating point numbers in this case is from 0.5×2^{-64} to $(1 - 2^{-15}) \times 2^{63}$. Of the 16 bits used to represent the mantissa, 1 bit is the sign bit and the least significant bit may be of a doubtful accuracy due to rounding effects. Thus, the accuracy of the floating point numbers is 1 in 2^{14} (0.61×10^4), that is about 4 decimal digits.

12.9.2 Floating point data format

The format for the floating point number (or word) is given in Figure 12.31. The mantissa is expressed as a 16-bit fractional two's complement value with the binary point assumed to the right of the sign bit. The exponent is 8 bits long in excess 64 form. The exponent normally has a range of $-64 \leqslant E \leqslant 63$, but by adding a fixed constant to the exponent (in this case 64) such that it is always positive the range becomes $0 \leqslant E \leqslant 127$. This ensures that when the mantissa is

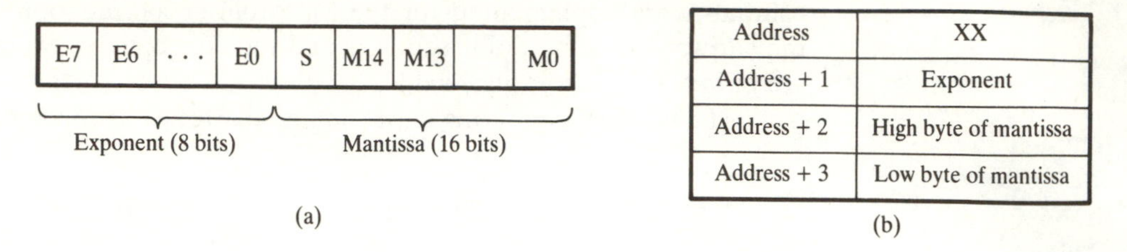

Figure 12.31 (a) Floating point data format and (b) floating point memory allocation.

zero the exponent is also zero, giving an all-zero floating point number. In addition, this form of representation has the advantage of making the detection of over- and underflow simple (Cope, 1975). However, when performing arithmetic operations the exponent has to be restored to its unbiased form.

All numbers going into or out of the floating point routines must be normalized. A normalized floating point number is one in which the sign bit and the bit to the right of the binary point (bits S and M14) of the mantissa are different. In our system, a normalized mantissa satisfies the condition $0.5 \leqslant |M| < 1$. If the result of an arithmetic operation produces a mantissa that is outside this range, it is said to have overflowed (mantissa overflow) if the result is equal to or greater than unity, and underflowed (mantissa underflow) if it is less than 0.5.

The floating point numbers in memory are allocated four consecutive bytes of memory (2 words), although only the first three are used as shown in Figure 12.31. This is to simplify reading and writing to the memory of the 68000 microprocessor. It also allows the number of bits in a floating point word (mantissa + exponent) to be increased easily, if desired.

12.9.3 Floating point arithmetic routines

The floating point arithmetic routines consist of three routines for the basic arithmetic operations: add/subtract, multiply and divide. Exit from these routines is made via a common routine, EXIT.

On entry into any of the arithmetic routines, registers D3 and D2 contain, respectively, the first and second operands. In the case of division, D3 and D2 contain respectively, the dividend and the divisor. On exit, D3 contains the result. Apart from D3, none of the registers is modified on exit by the floating point routines.

The 68000 microprocessor has an abundance of registers and these have been used freely to improve the execution speed. For the same reason, the use of some otherwise efficient instructions have been avoided. Arithmetic operations can be chained with the result of the previous operation left in D3 to be used as the first operand of the next operation.

12.9.3.1 EXIT routine

This routine is used to handle 'exceptions' such as overflow, to recover the sign of the result, to put the result in the correct floating point format and to return control to the calling routine. Three exceptions are handled.

(1) *Result is zero or too small*. If the result of an operation is zero or the number is too small to be represented (underflow) this routine sets the result to zero.

(2) *Result overflowed*. If there is overflow in the arithmetic operation, the result is set to the maximum floating point number possible, that is $(1 - 2^{-15}) \times 2^{63}$. In practice it is set to $007F7FF, and then the mantissa is given the appropriate sign.

(3) *Over-/underflow detection*. In excess 64 representation, the exponent of a valid floating point number is positive. Over-/underflow causes it to be negative. Thus the sign of the exponent is used to detect over-/underflow. In addition, underflow generates a carry bit whereas overflow does not. The EXIT routine determines which of the two has occurred and sets the result to either FPMAX or to zero, whichever is appropriate.

A listing of the EXIT routine, in 68000 microprocessor assembly language, is given in the appendix.

12.9.3.2 Floating point addition routine (FADD)

This routine is used to perform addition or subtraction. In the case of subtraction the second operand is negated in the calling routine with a single instruction (for example NEG.W D2) before entering the addition routine. Before adding two floating point numbers, their exponents must be made equal. This is called alignment, and involves shifting right the mantissa of the smaller operand and incrementing its exponent until it equals that of the larger operand. The aligned numbers are then added. Thus if X_1 and X_2 are two floating point numbers to be added, where $X_1 = M_1 \times 2^{E_1}$ and $X_2 = M_2 \times 2^{E_2}$, then their sum X is given by

$$X = (M_1 + M_2 \times 2^{E_1 - E_2}) \times 2^{E_1} = M \times 2^E$$

where $M = M_1 + M_2 \times 2^{E_1 - E_2}$, $E = E_1$, and $X_1 > X_2$. That is, the exponent of the result is the exponent of the larger operand and its mantissa is the sum of the mantissas of the two operands properly aligned.

The sum of the aligned mantissas, M, may not be properly normalized. It may be too large (mantissa overflow) or it may be too small (mantissa underflow). The former results when the mantissas of the original operands to be added (M_1, M_2) have the same signs, and the latter when they have opposite signs. Mantissa overflow and underflow are corrected by renormalizing the sum and adjusting the result exponent, E, but this may lead to exponent overflow

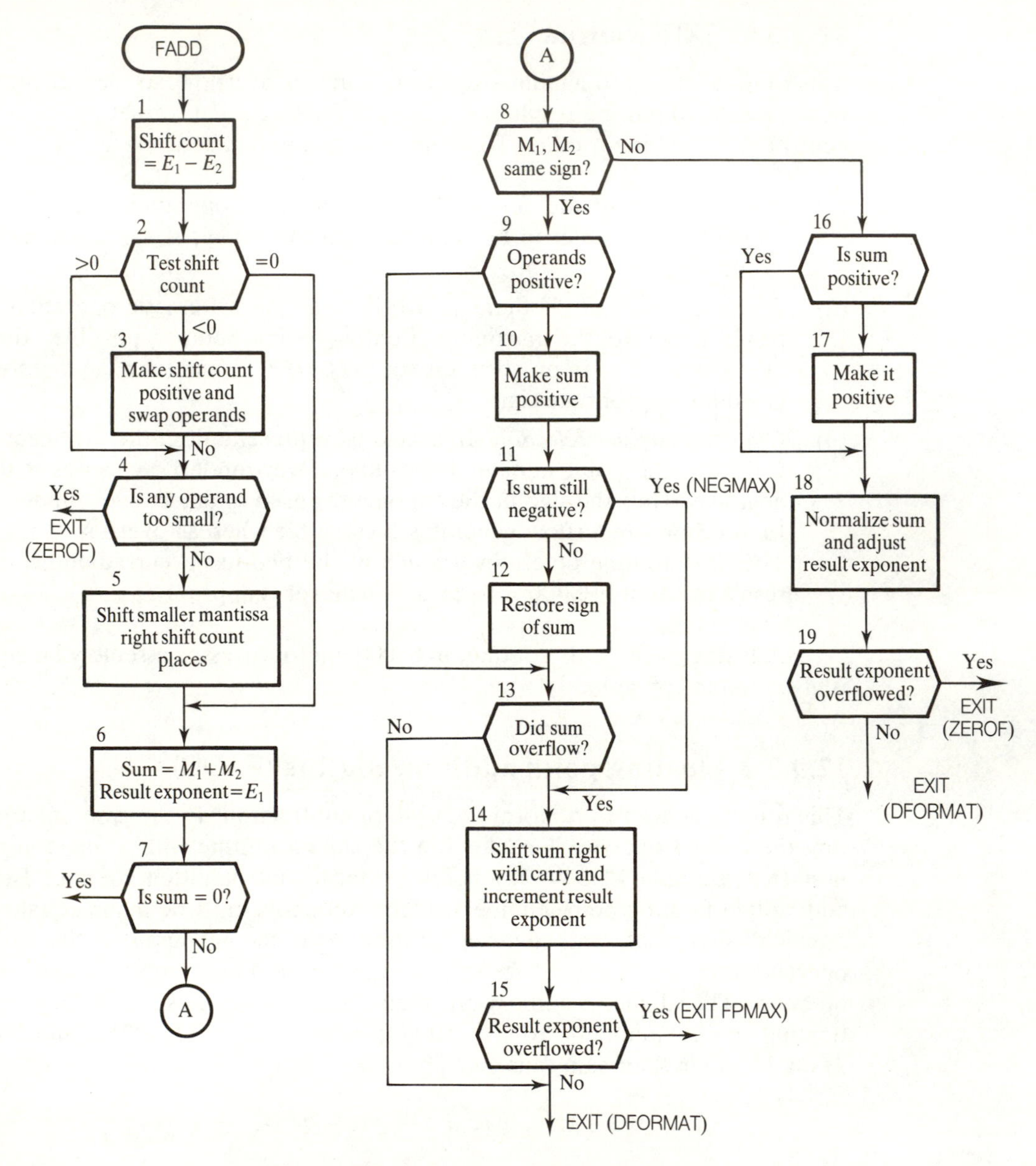

Figure 12.32 Floating point addition of two operands with exponents E_1 and E_2, and mantissas M_1 and M_2, respectively.

or underflow. Exponent overflow means that the result is too large to be represented and underflow that it is too small.

Figure 12.32 gives the flow chart for the FADD routine and the corresponding assembly program is given in the appendix. The first operation is to align the operands so that their exponents are equal (boxes 1 to 5). This is done by shifting the mantissa of the operands with the smaller exponent right an amount equal to the difference in the exponents (the shift count). If the two exponents were equal or their difference exceeds 15 (operand too small), no alignment is carried out. In the latter case the operation is terminated as the smaller operand is effectively zero and the result is simply the larger operand. Next, the aligned numbers are added (box 6) and the exponent of the result is set equal to the exponent of the larger operand.

If the sum is not zero, it is then tested for mantissa overflow or underflow. The two cases are treated separately and as mentioned earlier depend on the signs of the operands. Boxes 10 to 15 deal with mantissa overflow, while boxes 16 to 19 deal with mantissa underflow.

Mantissa overflow is detected by successively performing an exclusive-OR operation on the sign bits of the two original operands, the sum, M, and the carry bit. The result of this operation is a 1 if there was overflow and 0 otherwise. Overflow is corrected by shifting the sum, M, right with carry and incrementing the result exponent, E, by one (box 14). If the exponent overflows the result is set to FPMAX with the appropriate sign, otherwise the resulting number is now properly normalized. A special form of overflow may occur when the sum, M, is exactly -1 ($8000 in hexadecimal) or NEGMAX, and this is separately detected. NEGMAX is the maximum negative mantissa and has no positive equivalent. Thus when negated it still remains negative (boxes 10 and 11). When this condition is detected the sum, M, is normalized as in the ordinary mantissa overflow.

Mantissa underflow is corrected by successively shifting left the sum, M, and decrementing the exponent, E, until the sum is properly normalized (box 18). The exponent is tested for underflow at each shift. If exponent underflow occurs at any stage the operation is terminated and the result set to zero. The sum is made positive (box 17) before normalizing to simplify the operation. The sign is recovered in the EXIT routine where the result exponent, E, and the normalized sum, M, are put into the proper floating point format.

12.9.3.3 Floating point multiply routine (FMUL)

If X_1 and X_2 are two FP numbers to be multiplied, where

$$X_1 = M_1 \times 2^{E_1}$$

$$X_2 = M_2 \times 2^{E_2}$$

then their product, X, is given by $X = (M_1 \times M_2)2^{E_1+E_2} = M \times 2^E$ where $M = M_1 \times M_2$ and $E = E_1 + E_2$. Thus the mantissas are multiplied and their exponents added. Since M_1 and M_2 are both normalized then their product, M, will be in the range $0.25 \leq M < 1$. Thus the product, M, cannot overflow but may not be properly normalized (mantissa underflow).

The flowchart for the floating point multiply routine is given in Figure 12.33 and the assembly language program is listed in the appendix. The mantissas of the operands are multiplied and the product tested to see whether it is

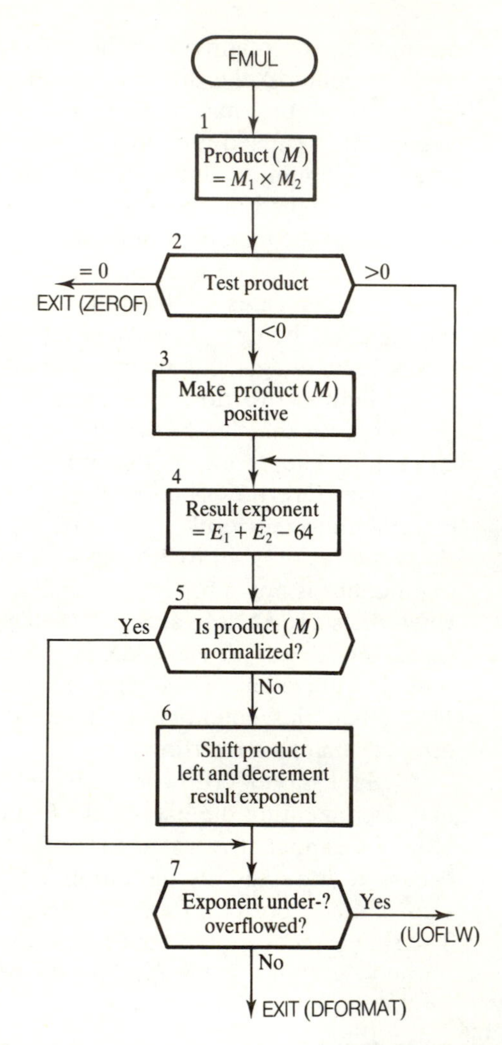

Figure 12.33 Floating point multiplication of two operands with exponents E_1 and E_2, and mantissas M_1 and M_2, respectively.

zero (boxes 1 and 2). If it is zero the operation is terminated and the result set to zero. The product, M, is double length (multiplication of two B-bit numbers gives a $2B$-bit result) and is thus rounded and reduced to single length. The exponents are then added and corrected for excess 64 (box 4). If the product, M, is not normalized it is then normalized by a single left shift and the exponent decremented by 1. The sign of the exponent is tested. If it is negative, then there was exponent overflow or underflow when the exponents were added. The EXIT routine then determines whether it was underflow or overflow that occurred and takes appropriate action. It is to be noted that exponent overflow or underflow was not tested for when the exponents were added as subsequent exponent adjustment could correct it. The normalized product, M, and the adjusted exponent give the result for the floating point multiplication, except when the exponent overflows or underflows.

12.9.3.4 Floating point divide routine (FDIV)

In floating point division the mantissas are divided and the exponents subtracted. Thus, division of X_1 by X_2 where

$$X_1 = M_1 \times 2^{E_1}$$

$$X_2 = M_2 \times 2^{E_2}$$

gives

$$X = \frac{M_1}{M_2} \times 2^{E_1 - E_2} = M \times 2^E$$

where $M = M_1/M_2$ and $E = E_1 - E_2$. To ensure that the quotient is within the permissible range M_1 must be less than M_2. This is readily achieved by shifting M_1 one place to the right. After division, the quotient, M, may not be properly normalized. If it is not, it is shifted left one place and the exponent, E, decremented by 1 and checked for underflow as before.

Figure 12.34 gives the flowchart for the floating point divide routine. The exponent, E, is obtained from the exponents of the operands and the bias is added (box 1). The two mantissas are then tested. If either is a zero, the operation is terminated and the result set to zero (boxes 2 to 4). The exponent of the result, E, is then tested for under- and overflow and treated in the same way as in the multiply. For easy comparison of the mantissas and normalization of the product the mantissas of the operands are always made positive (boxes 3 and 5). The sign is restored in the EXIT routine. Division is performed with the inherent divide instruction of the 68000 microprocessor (box 10). A listing of the assembly language program for the floating point divide routine is given in the appendix.

12.9.3.5 Execution times of the floating point routines

Estimates of the execution times of the floating point routines are given in Table 12.7. These times are only approximate as the execution times depend on the data.

12.10 Summary

In this chapter, we have presented in the form of case studies the design and development of the hardware used to implement some of the DSP algorithms

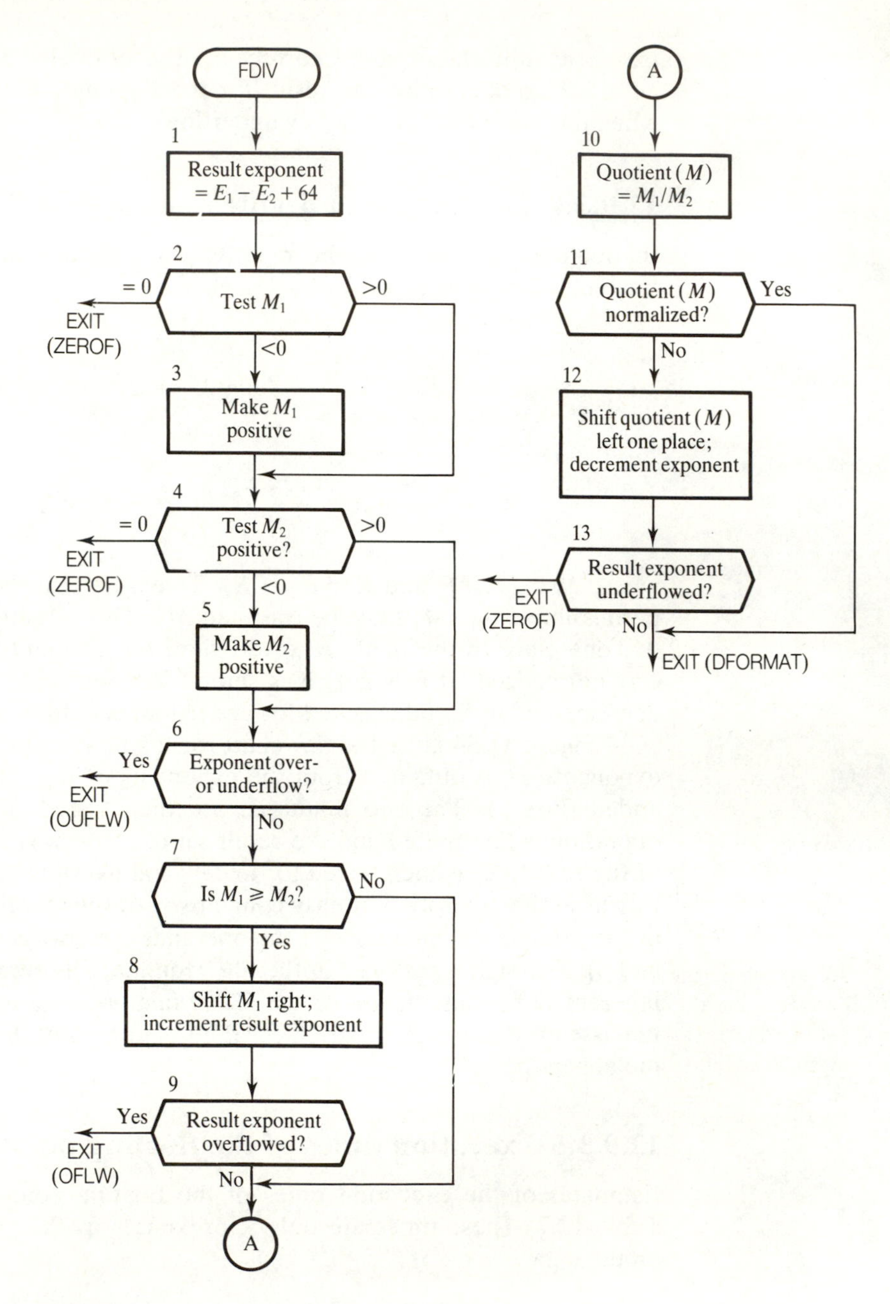

Figure 12.34 Floating point division of two operands with exponents E_1 and E_2, and mantissas M_1 and M_2, respectively.

Table 12.7 Execution times of floating point routines.

Routine	*Execution time* (μs)
Addition	70
Multiplication	70
Division	85

described in the book. In particular, we have described the design and applications of two home-made target boards: one based on the TMS320C10 and the other on the TMS320C25. A number of real-world applications of DSP have been described in the form of case studies.

Problems

12.1 Show, stating any assumptions made, that the autocorrelation function (ACF) of a signal contaminated by a random noise is the same as the ACF of the signal alone. Explain how this result may be used to detect hidden periodicities.

12.2 Prove, stating any reasonable assumptions, that the maximum signal-to-noise ratio at the output of a digital matched filter is independent of the waveshape of the input signal.

12.3 A recurring signal, buried in noise, is to be detected by digital matched filtering. Given below are the successive sample values of the noise-free signal and the noisy signal:

noise-free signal {−0.51, −0.35, −0.29, −0.25, −0.29, −0.39, −0.47}

noisy signal {−0.18, −0.06, 0.27, 0.69, −0.50, −0.44, −0.20, −1.46, −0.93, −1.46, −0.91, −0.39, −1.70}

Determine

(a) the coefficients of the digital matched filter,

(b) the output of the digital matched filter, and

(c) the improvement in signal-to-noise ratio, expressed in decibels, achievable by matched filtering.

Note: the variance, σ_0^2, of the noise at the filter output is given by

$$\sigma_0^2 = \sigma^2 \sum_{m=0}^{\infty} h^2(m)$$

where σ^2 is the variance of the noise at filter input, and the $\{h(m)\}$ are the filter coefficients.

References

Azevedo S. and Longini R.L. (1980). Abdominal-lead fetal electrocardiographic R-wave enhancement for heart rate determination. *IEEE Trans. Biomedical Engineering*, **27**(5), 255–60

Bajpai, A.C., Calus I.M. and Fairley J.A. (1973). *Mathematics for Engineers and Scientists*, Volume 2. New York NY: Wiley

Barlow J.S. (1979). Computerized clinical electroencephalography in perspective. *IEEE Trans.*, **26**, 377–91

Barlow J.S. and Rémond A. (1981). Eye movement artifact nulling in EEGs by multichannel on-line EOG subtraction. *Electroencephalography and Clinical Neurophysiology*, **52**, 418–23

Barry W. and Jones G.M. (1965). Influence of eyelid movement upon electro-oculographic recording of vertical eye movements. *Aerospace Medicine*, **36**, 855–8

Bierman G.J. (1976). Measurement updating using the *U-D* factorization. *Automatica*, **12**, 375–82

Bierman G.J. (1977). *Factorization Methods for Discrete Sequential Estimation*. New York NY: Academic Press

Clarke D.W. (1981). Implementation of self-tuning controllers. In *Self-Tuning and Adaptive Control* (Harris C. J. and Billings S. A. eds.), pp. 144–65. Stevenage, UK: Peter Peregrinus

Cope S.N. (1975). *Floating Point Arithmetic Routines and Macros for an Intel 8080 Microprocessor*. OUEL Report 1123/75

Corby J.C. and Kopell B.S. (1972). Differential contributions of blinks and vertical eye movements as artifacts in EEG recording. *Psychophysiology*, **9**, 640–44

Cowan C.F.N. and Grant P.M. (1984). Adaptive processing – an overview. In *IEE Colloq. Adaptive Processing and Biomedical Applications*, October 1984, Paper 1

Favret A.G. (1968). Computer matched filter location of fetal R-waves. *Medical and Biological Engineering*, **6**, 467–75

Flores I. (1963). *The Logic of Computer Arithmetic*. Englewood Cliffs NJ: Prentice-Hall

Fortgens C. and De Bruin M.P. (1983). Removal of eye movement and ECG artifacts from the non-cephalic reference EEG. *Electroencephalography and Clinical Neurophysiology*, **56**, 90–6

Girton D.G. and Kamiya A.J. (1973). A simple on-line technique for removing eye movement artifacts from the EEG. *Electroencephalography and Clinical Neurophysiology*, **34**, 212–8

Goodman G.C. and Sin K.S. (1984). *Adaptive Filtering, Prediction and Control*. Englewood Cliffs NJ: Prentice-Hall

Gotman J., Gloor P. and Ray W.F. (1975). A quantitative comparison of traditional reading of the EEG and interpretation of computer-extracted features in patients with supratentorial brain lesions. *Electroencephalography and Clinical Neurophysiology*, **38**, 623–39

Gotman J., Ives J.R. and Gloor P.A. (1977). A digital filter for eliminating EMG artifact from recordings of epileptic seizures. *Electroencephalography and Clinical Neurophysiology*, **43**, 475–6

Gratton G., Coles M.B.H. and Donchin E. (1983). A new method for off-line removal of ocular artifact. *Electroencephalography and Clinical Neurophysiology*, **55**, 468–84

Greene K.R. (1987). The ECG waveform. In *Balliere's Clinical Obstetrics and Gynaecology* (M. Whittle, ed.), Volume 1, pp. 131–55

Hamer C.F., Ifeachor E.C. and Jervis B.W. (1985). Digital filtering of physiological signals with minimal distortion. *Medical and Biological Engineering and Computation*, **23**, 274–8

Harris C.J. (1983). Brainwaves appear on T.V. in real-time. *Electronics*, (February 1983), 47–8

Hillyard S.A. (1974). Methodological issues in CNV research. In *Bioelectric Recording Techniques*, Part B (Thompson R. F. and Patterson M. M., eds.). New York NY: Academic Press

IEEE (1985). IEEE Standard for Binary Floating Point Arithmetic. *SIGPLAN Notices*, **22**(2), 9–25

Ifeachor E.C., Keith R.D.F., Westgate J. and Greene K.R. (1991). An expert system to assist in the management of labour. In *Proc. World Congress on Expert Systems* (Liebowitz J., ed.), Volume 4, pp. 2615–22. New York NY: Pergamon

Ifeachor E.C., Jervis B.W., Morris E.L., Allen E.M. and Hudson N.R. (1986). A new microcomputer-based online ocular artefact removal (OAR) system. *Proc. IEE*, **133**(5), 291–300

Jervis B.W., Allen E., Johnson T.E., Nichols M.J. and Hudson N.R. (1984). The application of pattern recognition techniques to the contingent negative variation for the differentiation of subject categories. *IEEE Trans. Biomedical Engineering*, **31**, 342–9

Jervis B.W., Nichols M.J., Allen E., Hudson N.R. and Johnson T.E. (1985). The assessment of two methods for removing eye movement artefact from the EEG. *Electroencephalography and Clinical Neurophysiology*, **61**, 444–52

Lindecrantz K.G., Lilja H. and Rosen K.G. (1988). New software QRS detector algorithm suitable for realtime applications with low signal to noise ratios. *J. Biomedical Engineering*, **10**, 280–3

McCallum W.C. and Walter W.G. (1968). The effects of attention and distraction on the contingent negative variation in normal and neurotic subjects. *Electroencephalography and Clinical Neurophysiology*, **25**, 319–29

Patterson D.A. and Hennessy J.L. (1990). *Computer Architecture: A Quantitative Approach*. San Mateo CA: Morgan Kaufmann

Peterka V. (1975). A square root filter for real-time multivariate regression. *Kybernetika*, **11**, 53–67

Quilter P.M., Macgillivray B.B. and Wadbrook D.G. (1977). The removal of eye movement artefact from EEG signals using correlation techniques. *IEE Conf. Publ.*, **159**, 93–100

Rabiner L.R. and Gold B. (1975). *Theory and Application of Digital Signal Processing*. Englewood Cliffs NJ: Prentice-Hall

Takeda H. and Hata S. (1985). Development of micro-computerized topographic EEG analyzer and its application to real time display. *Electroencephalography and Clinical Neurophysiology*, **61**, 98

Texas Instruments (1986). *Digital Signal Processing Applications with the TMS320 Family: Theory, Algorithms and Implementations*. Texas Instruments

Tomé A.M., Principe J.C. and Da Silva A.M. (1985). Micro analysis of spike and wave bursts in children's EEG. *Electroencephalography and Clinical Neurophysiology*, **61**, 113

Weitek (1984). *High Speed Digital Arithmetic VLS Application Seminar Notes*. Sunnyvale CA: Weitek

Welch P.D. (1967). The use of the FFT for estimation of power spectra: a method based on averaging over short, modified periodograms. *IEEE Trans. Audio and Acoustics*, **15**(2), 70–3

Widrow B., Glover J.R., McCool J.M., Kaunitz J., Williams C.S., Hearn R.H., Zeidler J.R., Dong E. and Goodlin R.C. (1975). Adaptive noise cancelling: principles and applications. *Proc. IEEE*, **63**, 1692–716

Woestenburg J.C., Verbaten M.N. and Slangen J.L. (1983). The removal of the eye-movement artifact from the EEG by regression analysis in the frequency domain. *Biological Psychology*, **16**, 127–47

Young P. (1974). Recursive approaches to time series analysis. *Bull. IMA*, **10**, 209–24

Bibliography

Clarke D.W. (1980). Some implementation considerations of self-tuning controllers. In *Numerical Techniques for Stochastic Systems* (Archetti F. and Cugiani M., eds.) pp. 81–101. Amsterdam: North-Holland

Clarke D.W., Cope S.N. and Gawthrop P.J. (1975). *Feasibility Study of the Application of Microprocessors to Self-tuning Controllers*. OUEL Report 1137/75

Kay S.M. (1987). *Modern Spectrum Estimation*. Englewood Cliffs NJ: Prentice-Hall

Marple S.L. Jr. (1987). *Digital Spectral Analysis With Applications*. Englewood Cliffs, NJ: Prentice-Hall

Motorola (1980). *16-bit Microprocessor User's Manual*. Austin TX: Motorola Semiconductor

Otnes R.K. and Enochson L. (1978). *Applied Time Series Analysis*, Volume 1. New York NY: Wiley

Rosen K.G. and Lindecrantz K.G. (1989). STAN, the Gothenburg model for fetal surveillance during labour by ST analysis of the fetal electrocardiogram. *Clinical Physiology and Physiological Measurement, Suppl. B*, **10**, 51–6

Stanley W.D., Dougherty G.R. and Dougherty R. (1984). *Digital Signal Processing* 2nd ed. Reston VA: Reston Publications

Verleger R., Gasser T. and Möcks J. (1982). Correction of EOG artifacts in event-related potentials of the EEG: aspects of reliability and validity. *Psychophysiology*, **19**, 472–80

Appendices

12A The modified UD factorization algorithm

- Step 1: $\mathbf{v} = \mathbf{U}^T(m)\mathbf{x}$.
- Step 2: $b_i = d_i(m)v_i, \quad i = 1, \ldots, n$.
- Step 3: $\alpha_1 = \gamma + b_1 v_1$.
- Step 4: $d_1(m + 1) = d_1(m)/\alpha_1$.

For $j = 2, \ldots, n$, recursively evaluate Equations 12.16–12.20.

- Step 5: $\alpha_j = \alpha_{j-1} + v_j b_j$.
- Step 6: $\rho_j = -v_j/\alpha_{j-1}$.

For $k = 1, 2, \ldots, j - 1$, recursively evaluate Equations 12.7 and 12.8.

- Step 7: $U_{kj}(m + 1) = U_{kj}(m) + b_k\rho_j$.
- Step 8: $b_k = b_k + b_j U_{kj}(m)$.
- Step 9: $d_j(m + 1) = d_j(m)\alpha_{j-1}/\alpha_j\gamma$.
- Step 10: $G = b/\alpha_n$ ($g_i = b_i/\alpha_n$, $i = 1, \ldots, n$).

The following points should be noted.

(1) α_n in step 10 is the value of α_j (step 5) after the nth iteration. This is also equal to α in Equation 12.6 of the RLS algorithm.

(2) The elements of $\mathbf{D}$ (that is the d_i in steps 4 and 9) can be stored along the diagonals of $\mathbf{U}$, since $U_{jj} = 1$. In addition, to save storage and for ease of programming, the elements of $\mathbf{U}$ (including those of $\mathbf{D}$) can be stored as a vector, even though the subscripts (k, j) indicate that $\mathbf{U}$ is a two-dimensional array.

12B Programs for the semiparametric equalizer and floating point arithmetic routines

Partial program for the equalizer is given in Program 12B.1 and assembly language floating point routines are given in Programs 12B.2–12B.5.

Program 12B.1 Partial C program for the semi parametric equalizer.

```
/* EQ procedure – 3 cascaded stages LP, MP, HP */
void para(void)
{
        if (filt_flag==OFF) output=input;          /* EQ bypass */
        else if (filt_flag==ON)                     /* EQ procedure */

        {
        X=0.05*((float)input);                      /* input scaling, convert to float X */
        HP=(HC*X)+(HB*HX1)+(HA*HX2)–(HE*HY1)–(HD*HY2);
        HY2=HY1;
        HY1=HP;                                     /* HF shelf algorithm */
        HX2=HX1;
        HX1=X;
        LP=(LC*X)+(LB*LX1)+(LA*LX2)–(LE*LY1)–(LD*LY2);
        LY2=LY1;
        LY1=LP;
        LX2=LX1;                                    /* LF shelf algorithm */
        LX1=X;
        MP=(MC*X)+(MB*MX1)+(MA*MX2)–(ME*MY1)–(MD*MY2);
        MY2=MY1;
        MY1=MP;
        MX2=MX1;                                    /* MF Bell algorithm */
        MX1=X;
        output=10*((int)MP);                        /* final output MF Bell, scaled by 10 */
        }
}
void recal_mid(void)
{
/* update MA,MB,MC,MD,ME */

        float Sn1,Sn2,Sn0;                          /* S plane numerator */
        float Sd1,Sd2,Sd0;                          /* S plane denominator */
        float Zn1,Zn2,Zn0;                          /* Z plane numerator */
        float Zd1,Zd2,Zd0;                          /* Z plane denominator */
        float p,t,A;

        A=1/((MF+2.7e+03)*4.7e–09);                 /* calculate freq required */

        Sn0=A*A;
        Sn2=1;
        Sn1=(3+MG)*A;                               /* apply freq and gain required */
        Sd0=Sn0;
        Sd1=(3–MG)*A;                               /* calc s-plane coeff's */
        Sd2=1;

        p=sqrt (Sd0);
        t=p/(tan(p/(37878)));                       /* calculate prewarp, t */

        Sn2=Sn2*t*t;                                /* prewarp: S=S/t */
        Sn1=Sn1*t;
        Sd2=Sd2*t*t;
        Sd1 = Sd1*t;
```

```
                                                /* BZT */
        Zn2=Sn2+Sn1+Sn0;                        /* calculate Z coeff's */
        Zn1=2*(Sn0-Sn2);                        /* subs S=(Z-1/Z+1) */
        Zn0=Sn2-Sn1+Sn0;                        /* times thro by (Z+1)^2 */
        Zd2=Sd2+Sd1+Sd0;                        /* equate coeff's of Z */
        Zd1=2*(Sd0-Sd2);
        Zd0=Sd2-Sd1+Sd0;

        MA=Zn0/Zd2;
        MB=Zn1/Zd2;                             /* update final coefficients */
        MC=Zn2/Zd2;                             /* for the MF Bell */
        MD=Zd0/Zd2;                             /* normalized thro Zd2 */
        ME=Zd1/Zd2;

}
void recal_low(void)
{
/* update LA,LB,LC,LD,LE */
        float Sn1,Sn2,Sn0;                      /* S plane numerator */
        float Sd1,Sd2,Sd0;                      /* S plane denominator */
        float Zn1,Zn2,Zn0;                      /* Z plane numerator */
        float Zd1,Zd2,Zd0;                      /* Z plane denominator */
        float p,t,A,B;

        A=(3+LG)/((LF+2.0e+03)*3.12e-08);
        B=(3-LG)/((LF+2.0e+03)*3.12e-08);
         Sn0=A*A;
         Sn2=1;
         Sn1=2*A;
         Sd0=B*B;
         Sd1=2*B;
         Sd2=1;

        p=sqrt (Sd0);
        t=p/(tan(p/(37878)));                   /* calculate prewarp, t */
                                                /* BZT */
        Sn2=Sn2*t*t;                            /* prewarp: S=S/t */
        Sn1=Sn1*t;
        Sd2=Sd2*t*t;
        Sd1=Sd1*t;

        Zn2=Sn2+Sn1+Sn0;                        /* calculate Z coeff's */
        Zn1=2*(Sn0-Sn2);                        /* subs S=(Z-1/Z+1) */
        Zn0=Sn2-Sn1+Sn0;                        /* times thro by (Z+1)^2 */
        Zd2=Sd2+Sd1+Sd0;                        /* equate coeff's of Z */
        Zd1=2*(Sd0-Sd2);
        Zd0=Sd2-Sd1+Sd0;

        LA=Zn0/Zd2;
        LB=Zn1/Zd2;                             /* update LF shelf coefficients */
        LC=Zn2/Zd2;                             /* normalized thro Zd2 */
        LD=Zd0/Zd2;
        LE=Zd1/Zd2;
}
void recal_high(void)                           /* HF shelf coefficients */
```

```
/* update HA,HB,HC,HD,HE */

        float Sn1,Sn2,Sn0;                    /* S plane numerator */
        float Sd1,Sd2,Sd0;                    /* S plane denominator */
        float Zn1,Zn2,Zn0;                    /* Z plane numerator */
        float Zd1,Zd2,Zd0;                    /* Z plane denominator */
        float p,t,A,B,C;

        C=1/((HF+2.7e+03)*4.7e-09);
        A=(0.2+HG)*C;                         /* calc splane coeff's */
        B=(0.2-HG)*C;                         /* due to EQ change */
        Sn0=C*C*C*C;
        Sn2=A*A;
        Sn1=2*A*C*C;
        Sd0=Sn0;                              /* splane coeff's */
        Sd1=2*B*C*C;
        Sd2=B*B;

        p=sqrt (Sd0);
        t=p/(tan(p/(37878)));                 /* calculate prewarp, t */

        Sn2=Sn2*t*t;                          /* prewarp: S = s/t */
        Sn1=Sn1*t;                            /* times thro by t^2 */
        Sd2=Sd2*t*t;
        Sd1=Sd1*t;

        Zn2=Sn2+Sn1+Sn0;                      /* calculate Z coeff's */
        Zn1=2*(Sn0-Sn2);                      /* subs S=(Z-1/Z+1) */
        Zn0=Sn2-Sn1+Sn0;                      /* times thro by (Z+1)^2 */
        Zd2=Sd2+Sd1+Sd0;                      /* equate coeff's of Z */
        Zd1=2*(Sd0-Sd2);
        Zd0=Sd2-Sd1+Sd0;

        HA=Zn0/Zd2;
        HB=Zn1/Zd2;                           /* NORMALIZE COEFFICIENTS */
        HC=Zn2/Zd2;                           /* update coeff's for HF shelf */
        HD=Zd0/Zd2;
        HE=Zd1/Zd2;

}
/* ================================================= */
```

Program 12B.2 The EXIT routine. Control is returned to the calling program via this routine. The routine also handles under-/overflow, recovers the sign and puts the result in the correct format.

```
*
*
* EXIT ROUTINE
* THIS ROUTINE HANDLES EXCEPTIONS FROM FLOATING POINT
* ARITHMETIC OPERATIONS AND RETURNS CONTROL TO THE CALLING ROUTINE.
*
DFORMAT     SWAP.W     D1                  ;PUT RESULT IN CORRECT FP
                                           ;FORMAT
```

```
          MOVE.W    D3,D1
          MOVE.L    D1,D3
SIGNS:    SUB.B     #1,D5                ;RECOVER SIGN OF MANTISSA
          BNE       EXITF
          NEG.W     D3
EXITF:    MOVEM.L   (A7)+,D0-D2/D4-D7
          RTS
ZEROF:    CLR.L     D3                   ;RESULT IS ZERO
          BRA       EXITF
UOFLW:    BTST      #15,D6               ;RESULT UNDER OR
                                         ;OVERFLOW?
          BNE       ZEROF                ;UNDERFLOW.SET RESULT TO
                                         ;ZERO
OFLW:     MOVE.L    FPMAX,D3             ;OVERFLOW.SET RESULT TO
                                         ;FPMAX
          BRA       SIGNS
```

Program 12B.3 The assembly language for the floating point addition.

```
* ---------------------------------------------------------------------
* SUBROUTINE FADD
* THIS SUBROUTINE ADDS TWO NORMALIZED FLOATING POINT NUMBERS
* IN D2 AND D3. THE RESULT IS RETURNED IN D3
*
FADD:     MOVEM.L   D0-D2/D4-D7,-(A7)
          MOVE.L    D3,D1                ;MAKE COPIES OF OPERANDS
          MOVE.L    D2,D0                ;
          SWAP.W    D1                   ;RETRIEVE EXPONENTS (E) OF
                                         ;OPERANDS
          SWAP.W    D0
          MOVE.B    D1,D4
          SUB.B     D0,D4                ;E2-E1=SHIFT COUNT(SC)
          BEQ       SCZRO                ;SC IS ZERO
          BPL       SCPOS                ;SC IS POSITIVE
          NEG.B     D4                   ;SC IS NEG;MAKE POSITIVE
          EXG       D0,D1                ;SWAP OPERANDS
          EXG       D2,D3
SCPOS:    CMP.B     #15,D4               ;ANY OPERAND TOO SMALL?
          BPL       EXITF                ;YES. RESULT=LARGER
                                         ;OPERAND
          ASR.W     D4,D2                ;NO. SHIFT SMALLER M RIGHT
                                         ;SC PLACES
SCZRO:    MOVE.W    D3,D6                ;SAVE SIGN OF 2ND OP
          ADD.W     D2,D3                ;OBTAIN SUM=M2+M1
          ROXR.W    #1,D5                ;SAVE CARRY FLAG
          TST.W     D3                   ;IS SUM=0?
          BEQ       ZEROF                ;IF YES, THEN EXIT
          EOR.W     D2,D6                ;NO. ARE THE TWO OPERNDS
                                         ;SAME SIGN?
          BMI       DFSIGN               ;OPERNDS ARE OPPOSITE
                                         ;SIGN
          MOVE.W    D5,D4                ;OPERNDS ARE SAME SIGN
          SUB.W     D5,D5
```

```
            TST.W       D2              ;IS SUM +VE OR -VE?
            BPL         OPRPOS          ;SUM IS E +VE.
            NEG.W       D3              ;SUM IS -VE. MAKE SUM +VE
            TST.W       D3              ;IS SUM STILL -VE?
            BMI         NEGMX           ;YES. SUM MUST BE NEGMAX
            NEG.W       D3              ;NO. RECOVER SIGN OF SUM
OPRPOS:     EOR.W       D4,D6           ;DID OVERFLOW OCCUR
                                        ;DURING ADDITION?
            EOR.W       D3,D6           ;
            BPL         DFORMAT         ;NO. THEN FORMAT RESULT
                                        ;AND EXIT
            ROXL.W      #1,D4
NEGMX:      ROXR.W      #1,D3           ;SHIFT SUM RIGHT WITH
                                        ;CARRY AND
            ADDQ.B      #1,D1           ;INCR EXP: E=E+1
            BMI         OFLW            ;CHECK THAT E DID NOT
                                        ;OVERFLOW
            BRA         DFORMAT         ;E IS OK. FORMAT RESULT
                                        ;AND EXIT.
DFSIGN:     SUB.B       D5,D5
            TST.W       D3              ;IS SUM POSITIVE?
            BPL         FNORMA          ;YES. NORMALIZE RESULT
            MOVEQ.L     #1,D5           ;NO. SET FLAG AND MAKE
                                        ;POSITIVE
            NEG.W       D3
FNORMA:     MOVEQ.L     #15,D0          ;SET UP REGS FOR NORMALIZTN
            MOVE.L      D0,D2
SHFTLA      ADD.W       D3,D3           ;NORMALIZE SUM AND
            DBMI        D0,SHFTLA
            SUB.B       D0,D2
            LSR.W       #1,D3           ;
            SUB.B       D2,D1           ;ADJUST EXPONENT
            BMI         ZEROF           ;E UNDERFLOW?
            BRA         DFORMAT         ;NO. FORMAT RESULT AND EXIT
```

Program 12B.4 The assembly language program for the floating point multiply.

```
* ---------------------------------------------------------------------------------------
* SUBROUTINE FMUL
* THIS SUBROUTINE MULTIPLIES TWO NORMALIZED FLOATING POINT
* NUMBERS IN D2 AND D3 AND RETURNS THE RESULT IN D3.
*
FMUL:       MOVEM.L     D0-D2/D4-D7,-(A7)
            MOVE.L      D3,D1           ;MAKE COPIES OF OPERANDS
            MOVE.L      D2,D0
            SWAP.W      D1              ;RETRIEVE THE EXPS.
            SWAP.W      D0
            SUB.B       D5,D5           ;CLEAR FLAG
            MULS        D2,D3           ;OBTAIN PRODUCT M1*M2
            ADD.L       D3,D3           ;SHIFT DLENGTH PRODUCT LEFT
            ADD.L       #32768,D3       ;ROUND RESULT
            SWAP.W      D3              ;CONVERT RESULT TO SINGLE
                                        ;LENGTH
```

```
              TST.W      D3          ;RESULT=0,+VE OR -VE?
              BEQ        ZEROF       ;RESULT IS =0. EXIT
              BPL        POSMUL      ;RESULT IS +VE.COMPUTE EXP.
              ADDQ.B     #1,D5       ;RESULT IS -VE.SET FLAG
              NEG.W      D3          ;MAKE POSITIVE
POSMUL:       ADD.B      D0,D1       ;E2+E1=E
              SUB.B      #64,D1      ;REMOVE XS64
              ROXR.W     #1,D6       ;SAVE CARRY FLAG
              BTST       #14,D3      ;IS PRODUCT NORMALIZED?
              BNE        TESTE       ;YES. TEST EXP.
              ADD.W      D3,D3       ;NO. NORMALIZE PRODUCT AND
              SUBQ.B     #1,D1       ;ADJUST EXPONENT
TESTE:        TST.B      D1          ;EXP UNDER/OVER FLOWED?
              BMI        UOFLW       ;YES. BRANCH TO UNDER/
                                     ;OVERFLOW TEST
              BRA        DFORMAT     ;NO. FORMAT DATA AND EXIT.
```

Program 12B.5 The assembly language program for the floating point division.

```
*
* SUBROUTINE FDIV
* THIS SUBROUTINE DIVIDES TWO NORMALIZED FLOATING POINT
* NUMBERS.
* ENTER WITH THE DIVIDEND IN D3 AND THE DIVISOR IN D2. THE RESULT
* IN THE CORRECT FLOATING POINT FORMAT IS RETURNED IN D3
*
FDIV:         MOVEM.L    D0-D2/D4-D7,-(A7)

              MOVE.L     D3,D1       ;MAKE COPIES OF OPERANDS
              MOVE.L     D2,D0
              MOVEQ.L    #0,D3       ;RETAIN ONLY M1 IN D3
              MOVE.W     D1,D3
              SWAP.W     D1          ;RETRIEVE THE EXPS.
              SWAP.W     D0
              ADD.B      #64,D1      ;ADD BIAS TO E1 (E1=E1+64)
              SUB.B      D0,D1       ;OBTAIN EXP. E=E1-E2
              ROXR.W     #1,D6       ;SAVE CARRY FLAG
              SUB.B      D5,D5       ;CLEAR SIGN FLAG
              MOVEQ.L    #1,D7
              TST.W      D2          ;IS M2=0,+VE OR -VE?
              BEQ        ZEROF       ;M2=0,THEN EXIT
              BPL        TESTM1      ;M2 IS +VE
              EOR.B      D7,D5       ;M2 IS -VE
              NEG.W      D2          ;MAKE M2 POSITIVE
TESTM1:       TST.W      D3          ;IS M1=0,+VE OR -VE?
              BEQ        ZEROF       ;M1=0,THEN EXIT
              BPL        EUFLOW      ;M1 IS +VE
              EOR.B      D7,D5       ;M1 IS -VE
              NEG.W      D3          ;MAKE M1 POSITIVE
EUFLOW:       TST.B      D1          ;DID EXP. OVER/UNDERFLOW?
              BMI        UOFLW       ;IF YES EXIT
              CMP.W      D2,D3       ;IS M2 .GE. M1?
              BLT        M2GT        ;YES. THEN IT IS OK
```

```
        ASR.W     #1,D3      ;NO. THEN DIVIDE M1 BY 2 AND
        ADDQ.B    #1,D1      ;INCR E1 BY 1
        BMI       OFLW
M2GT:   SWAP.W    D3         ;CONVERT M1 TO DOUBLE
                             ;LENGTH
        ASR.L     #1,D3
        DIVS      D2,D3      ;OBTAIN QUOTIENT OF M1/M2
        BTST      #14,D3     ;IS QUOTIENT NORMALIZED?
        BNE       DFORMAT    ;YES. THEN GO FORMAT RESULT
        ADD.W     D3,D3      ;NO. NORMALIZE AND
        SUBQ.B    #1,D1      ;ADJUST EXPONENT
        BMI       ZEROF      ;
        BRA       DFORMAT    ;GO FORMAT RESULT AND
                             ;EXIT
* ------------------------------------------------------------------------- *
```

Index

acoustic background noise 568
adaptive control 544
adaptive filter 541–76, 678
  adaptive algorithm, *see* LMS or RLS algorithms
  adaptive filtering 660–2
  applications of 561–68, 706–23
  C language implementation 571–75, 660–1
  concepts of 543–4
  configurations of 544–5
  error performance surface 548
  hardware implementation 552, 661–2
  time-sequenced 554
  when to use 542–3
adaptive filtering of ocular artefacts 561–3
adaptive jammer suppression 566–7
adaptive signal processing 541
aliasing 16–23, 388, 578, 583, 680–1
alternation theorem 307
amplitude spectral density 53–4
Analog Devices
  ADSP2100 629, 634–6
analogue-to-digital conversion 14–29, 253
analogue-to-digital converter (ADC) 3, 13–14, 18–29, 253, 492–3, 715–6
  noise 23, 26–8, 349, 425–7
  single bit 34, 520–1
analogue filter 41, 252, 383, 388–9, 394–47
anti-aliasing filter 16, 18–23, 494, 500–4
anti-image filter 31–2, 496
architecture for DSP
  hardware architecture 614–28
  Harvard architecture 617–8, 620
arithmetic
  block floating point 724
  fixed point 723–9
  floating point 723–4, 729–33, 737–44
artificial intelligence 37
assembly language programs
  TMS320C10-based FIR digital bandpass filter 672–5
  TMS320C25 fourth order digital IIR filter 675–8
  68000 floating point arithmetic routines 751–4
augmenting zeros 581, 605
auto-correlation, *see* correlation
autocovariance 579, 598, 601
autoregressive model 577–8
average magnitude cross difference 702–5
average signal power 26

band transformations 399–400
Bartlett 600
bell filter 733
Bessel function 292, 597
bilinear z-transform (BZT) method 388–416, 261–2
  calculating IIR filter coefficients: method 1 399–407
  calculating IIR filter coefficients: method 2 407–15
  designing highpass, bandpass, and bandstop filters 398–415
  illustration of 391–2
  use of classical analogue filters 394–398
biomedical signals, *see* signals
biquadratic transformation 400
bit-reversal 73, 655–7 (*see also* re-ordering of the data)
brain disease 577
butterfly 70–5
Butterworth filter 394–5, 407–8, 410–11, 740–1

C language programs for
  adaptive filtering 571–5
  auto- and cross-correlation 250
  cascade-to-parallel conversion 159, 179–81
  direct DFT 93–99
  FIR filter design 370–3
  frequency response 180–2
  impulse invariant method 482–87
  impulse response 144
  inverse z-transform 159–80
  multirate processing design 536–40
  PC disk for the book, *see* the Preface
  radix-2 decimation-in-time FFT 99–102
  semi parametric equalizer 748–54
cardiotocogram (CTG) 697–8
cascade realization 148–52, 419–24 (*see also* realization structure)
causal system 105, 106
Chebyshev filter 395–6, 407–8, 410
clock recovery 472–6
CNV 604–7, 706
coefficient quantization, *see* quantization
communication
  digital 542
  equalization of channel 542
  system 48
compact disc (CD) 39, 40
  comparison with LP records 39
  player 2, 40, 468, 492–3
  system 29, 521–2
component tolerances 2
computer-aided design (CAD) 253
contingent negative variation, *see* CNV
contour integral 121–3
control engineering 184
convolution 4–7, 16, 48, 63, 79, 128, 183–4, 213–37
  analytical 218–20
  application of 242–6
  circular 63, 183, 222
  frequency domain 220–3
  linear 63, 183, 222–4, 228, 231–2
  properties 221–2
  by sectioning 224–35
  theorem 64, 222–3, 225
convolution coding 245
Cooley–Tukey algorithm 70
correlation 4–9, 48, 183–213, 215, 226–28
  applications 199–206, 237–42
  auto- 6, 8–9, 183, 190–1, 196–199, 201, 203–6
  circular 63, 210–11
  cross- 6, 8, 10, 183, 184–213
  fast 186, 207–9
  recursive algorithm 211–12

by sectioning 225–35
theorem 207–9
correlation detector 184, 204
cross-correlation, *see* correlation

DAC 3, 13, 29–32, 41, 42, 45, 252, 357–8, 616
single bit 34
zero-order hold 30–2
data compression 3, 48, 79, 88–90, 245, 252
data windows, *see* window
decimation, *see* multirate process
deconvolution 48
delta sigma modulation 493
DFT 207–11, 581–2
C language program for 93–9
computational complexity 64–5
properties 62–4
(*see also* FFT)
difference equations 143–4, 148, 254, 264, 284, 381–2
differentiators 284
digital-to-analogue converter, *see* DAC
digital audio 468–9
digital control 469–70
digital filter design 251
digital filters
advantages and disadvantages 252–3
analysis of finite wordlength effects, *see* finite wordlength effects
coefficient calculation 261–2, 288–343
design example 269–70
design stages 258–69, 285–6, 376–7
filter specification 259–61, 285–7
implementation 268–70, 358–61, 461–2
realization 262–6, 344–8, 416–24
types of 254–8
(*see also* FIR and IIR filters)
digital frequency oscillators 470–1
digital touch-tone 471–2
discrete cosine transform 47–8, 79–80, 88–9
discrete Fourier transform, *see* DFT
discrete-time signals 103–5
discriminant analysis 606
DMA 616
dual-port RAMs 653
dynamic range 28, 724

ECG 529–31, 697, 699, 700–5
echo cancellation 563–5
echo cancellers 38, 252
EEG 1
electrocardiogram, *see* ECG
electroencephalogram, *see* EEG
electrooculogram, *see* EOG
elliptic filter 261, 395–7, 407–8, 410
energy spectral density 14, 53, 579
EOG 707
equalization 31, 733–6
equiripple filters 305–7
error correction 41
error rate 42
error spectral shaping 452–4, 456–7
errors
amplitude 30, 31, 34
aperture 31
aperture effect 24, 28
aperture uncertainty 34
droop 34
phase 20, 34
sample and hold 34
quantization, *see* quantization and finite wordlength effects
even functions 62
EVM 736
expected value 579, 600
exponent 729–32, 737–42

factorization algorithms 560–1
fast convolution 346
fast Fourier transform, *see* FFT
FFT
applications 139–83, 207–11, 531–2, 578, 580, 584–5, 615, 650–57, 689–96
butterfly 70–75, 650–2
C program for radix-2 decimation-in-time 99–102
comparison of DIT and DIF algorithms 78
constant geometry 652–5
data scrambling and bit reversal 655–7
decimation-in-frequency 77–8
decimation-in-time 65–77
implementation 77
in-place 78, 652–5
processing 650–7
TMS320C25 pseudocode 651
finite duration sequence 107
finite impulse response filters, *see* FIR filters
finite wordlength effects in 3, 253, 402, 425–61
adaptive filters 555–6
digital IIR filters 425–61
FIR digital filters 348–63
FIR digital filtering 636–42
FIR filters 49, 278–373
application 363, 370–3, 494–532, 700–1
design of 254–8, 261–2, 265–8, 285–359, 303–38
design example 359–61
features 279–83
fixed point, *see* arithmetic
floating point, *see* arithmetic
foetal monitoring 529–31
foldover frequency 18
forgetting factor 559
FORTRAN program 79, 307
Fourier integral 53
Fourier series 49–52
Fourier transform 48, 52–4, 56–7, 134
frequency domain 47
frequency resolution 581, 584–5, 600, 602
frequency response 48, 134–41, 154
frequency sampling filter
nonrecursive 317–26
recursive 326–35
with simple coefficients 328–35
frequency scaling 385, 390
frequency synthesis 13
frequency transformation 340–1
frequency units 139–42

general purpose hardware, *see* hardware
genetic algorithms 343
geometric series 108
Gibb's phenomenon 289
graphic equalization 468–9
group delay 280, 284–5, 512, 518, 520

Hadamard transform 79, 84–6
half band filters 339
hardware for DSP
algorithm-specific 615
application-specific 615
FFT processors 665–7
FIR digital filters 663–4
general purpose 614–25, 618, 620, 628–9
IIR digital filters 664–5
special purpose 614, 662–8
SWDS (software development system) 687–9
(*see also* Texas Instruments, Analog Devices, Motorola)
harmonic component 49, 57–60
Harr transform 79
Harvard architecture, *see* hardware
Hilbert transformer 284, 307
Huntington's disease 604

IIR digital filtering 642–50
IIR filters
analysis of finite wordlength effects, *see* finite wordlength effects
applications 468–76
basic features of 375–6
C programs for 483–7
design of 376–462
design program 415–19
detailed design example 462–67

image frequencies 16–18, 21, 23, 30–2, 34
image processing 48
imaging filter 680–1, 692
impulse invariant method 261–2, 383–8, 483–7
impulse response 39, 105, 109, 142–7, 155–6
infinite duration sequence 109
infinite impulse response filters, *see* IIR filters
INMOS
  A100 359, 627
  transputer 628
integer coefficient filters 328, 333, 343
integer programming 355
interpolation, *see* multirate processing
inverse discrete Fourier transform 60–1
inverse fast Fourier transform 76–7
inverse z-transform 109–27, 144
  long division methods 111–14
  partial fraction expansion (PFE) method 111, 114–20, 150–1, 176–9
  power series method 111–14, 144–5, 174–6
  recursive algorithm 112–14, 157–9
  residue method 111, 121–7

jump phenomenon 461

Kalman filter 544
Karhunen–Loeve transform 80
knowledge-based system 563

Laplace transform 48–9, 128, 139
lattice structure, *see* structure
least mean square, *see* LMS
least square 709
L'Hôpital's rule 288
limit cycles 458–61
linear phase 2, 42, 279–85, 331 345–6
linear prediction 544
linear predictive coding 35
linear system 104
linear time invariant (LTI) system 104, 242
LMS
  adaptive algorithm 550–57, 700–11
  complex LMS algorithm 556
  frequency domain 557
  hardware implementation 552
  implementation 550–2, 571–5
  practical limitations 553–6
  steepest descent algorithm 550, 552–3
loudspeaker 29, 36
loudspeaking telephones 565
magnitude versus frequency 47–8
mantissa 729–32, 737–42 (*see also* arithmetic)
matched filter 703–6, 202
microprocessor 13
mixing console 733
mean square error (MSE) 548–9
mean value 579
modulation 4, 10–11, 13, 40
Motorola
  DSP56000 269, 615, 618, 633–4
  DSP56200 359, 615, 664–5
  MC68000 13, 269, 359
moving average 578
multichannel system 20
multipath compensation 565–6
multiplexing 37
multiply-accumulate 33
multirate processing
  application examples 520–32
  concepts of 493–502
  current uses in industry 492–3
  decimation 492, 594–5, 501–4
  design example 506–8
  hardware implementation 657–60
  interpolation 495–7
  practical sampling rate converters 502–8
  software implementation 508–20
multirate system, *see* multirate processing
multirate techniques 31
music 39
muting 40

noise 28, 542
noise canceller 546, 554
normalized ac power 593
normalized frequencies 313
nonrecursive filter 243
nonstationary 549
Nyquist frequency 18, 137
Nyquist rate 56

ocular artefacts 706, 561–3
odd functions 62
on-chip data RAMs 628
on-chip memory/cache 628
oscilloscope 48
overflow error 425–41
overlap-add 235, 227–32, 236
overlap-save 235, 232–6
oversampling 20, 34, 492–3

parallel processing 627–8
parallel realization 148–52, 422–4 (*see also* realization structure)
Parkinson's disease 607
Parseval's theorem 62
partial fractions, *see* inverse z-transform
pass filter 734
pattern recognition 3, 48
phase delay 280, 284
phase spectra 52
picket fencing, *see* scalloping loss
pitfalls in spectral estimation 583–6
pipelining 33, 617–24
pits 39
pole 109, 114, 122–3, 125, 130–9, 142–4, 150, 384–5
  Butterworth filters of 394–5
  Chebychev filters of 396
  first order 116–17
  multiple-order poles 115
  pole–zero diagram 131–7, 154, 376, 379–82, 402–3, 406
  second order 126
pole–zero placement method 379–82
polyphase filtering 515
positive definite 559, 714
power series 292
power spectral density (PSD) 579–80, 597, 598, 694–6
probability density distribution 579
programs, *see* C language programs; assembly language programs
pseudonoise (PN) sequence 204–6
pulse code modulation (PCM) 13, 202, 203–4, 471, 521

QRS detection 700–5
quantization 26
  coefficient 349–57, 425, 427–32, 639–40
  noise level 19, 23, 26
  noise power 26
  step size 26
  uniform 25
  (*see also* ADC and finite wordlength effects)

radar signal processing 567
radix-2 FFT 78–9
radix-4 FFT 78
Raleigh test of circular variance 605
random waveform 193
real time DSP 13, 24, 34, 44, 614–5
realization diagrams, *see* realization structure
realization structure
  building blocks 418–9
  cascade 262–4, 271–2, 346–7, 419–22
  choice of 347–8
  fast convolution 346
  frequency sampling 265–6, 327–8, 346
  lattice 266–7, 273–4, 419
  linear phase 252, 345–8
  parallel 271–2, 422–4
  transversal 265–6, 273, 344–8

reconstruction filter 29
recording studio 42
recursive least squares, *see* RLS
Reed–Solomon coding 40
region of convergence (ROC) 106–9
Remez exchange algorithm 306–7
reordering the data 72–3 (*see also* bit-reversal)
residues 115, 121–2, 127
RLS 541, 544, 557–60, 563, 700–12, 714
ROM 36
Roundoff errors 349, 356–7, 425, 440–59

sample and hold 25, 28, 34, 55
sample rate 20, 41, 56
sampling function 51
sampling theorem 15–25, 137
Sande–Tukey algorithm 78
satellite 38
scaling 357, 433–40
  in cascade realization 436–8
  in parallel realization 438–40
scalloping loss or picket-fence effect 48, 77, 578, 583–4, 605
schizophrenia 604
shelve filter 734
shift invariant 105
signal
  analogue 2, 13, 14
  biomedical 1, 4, 542, 561, 563, 568
  buried in noise 572
  digital 14
  discrete-time 14, 103–5
  recovery 31
  sampled signal 14, 16
  speech 48
  types of 1–2, 14
  video 48
signal flow diagrams 147
$(\sin x)/x$ 289
  distortion 30–2, 680
  effect 31, 34
smearing, *see* spectral leakage
smoothing filter 29
sonar 49, 187, 240–2
space craft 237
special instructions 626–7
  application-oriented instructions 626–7
  zero-overhead looping 629, 634
special purpose hardware for DSP, *see* hardware
spectral analysis 48, 79
spectral leakage and smearing 48, 585–6
spectral resolution, *see* frequency resolution
spectroscopy 81
spectrum analyser 689–96
spectrum estimation
  applications 601–5
  Bartlett 580
  Blackman–Tukey method 577
  comparison 603–4
  modified periodogram 577
  non parametrics 577
  parametric methods 577, 603
  periodogram method 596–600
  pitfalls 583–6
  quality factor 579–80, 602
  Welch 580, 600–1
speech 1
recognition 4, 36
synthesis 13, 35
synthesizer 35–6
synthetic 35
s-plane 129, 338, 402, 405–7
spread spectrum 244, 542, 566–7
square root
  algorithm 714
  of complex number 404, 408, 487–8
stability 132, 142–3, 349–57, 427–9
stereo channels 39
stream processing 614
stochastic process 579
superposition 104
SWDS (software development system) 647, 687–9
symmetry 62, 281–4

target board, *see* hardware
telecommunications 4
television (TV) 37
Texas Instruments TMS320 family 628–32
  TMS320C10 630, 640–2, 674–6, 680–5
  TMS320C25 269, 630–1, 675–8, 685–690
  TMS320C30 630–2, 732–3, 736
  TMS320C40 723
time series analysis 579
tolerance scheme
  for FIR filters 260–1, 286–7, 504
  for IIR filters 377–9
transmission lines 80
transputer, *see* INMOS
truncation 441
T-test 606
twiddle factor 78, 650–1, (*see also* weighting factor)
two's complement 725–8 (*see also* arithmetic)

UD algorithm 712–15, 717, 748
unit circle 129, 143, 329, 376, 379–80, 389

variance 579, 593
VLSI 359
Von Neumann 615–16

Walsh–Hadamard transform 85–6
Walsh transform 47, 79–84, 88–90
weighting factor 65–74
Wiener filter 547–8
Wiener–Hopf 548
window
  Blackman 294, 299, 325–6
  Blackman–Harris 594, 596
  Blackman–Tukey 601–2
  choice of 594–97
  common functions 291–5
  Dolph–Chebyshev 594, 596
  Hamming 294, 594–5, 601
  Hanning 294, 594, 601
  Kaiser 292, 294–5, 299–300
  Kaiser–Bessel 591–2, 594–5, 605
  Poison window 594
  rectangular window 594
  Tukey (cosine tapered) 594–5
windowing 586–97
Winograd Fourier transform 79

zero 109, 130–9, 144, 376, 379–82
z-plane 129, 376, 379–80, 389, 405–6
z-transform 49, 103–82
  applications in DSP 130
  properties of the *z*-transform 127–30
  z-transform table 110, 119